Lecture Notes in Computer Science 16167

The series Lecture Notes in Computer Science (LNCS), including its subseries Lecture Notes in Artificial Intelligence (LNAI) and Lecture Notes in Bioinformatics (LNBI), has established itself as a medium for the publication of new developments in computer science and information technology research, teaching, and education.

LNCS enjoys close cooperation with the computer science R & D community, the series counts many renowned academics among its volume editors and paper authors, and collaborates with prestigious societies. Its mission is to serve this international community by providing an invaluable service, mainly focused on the publication of conference and workshop proceedings and postproceedings. LNCS commenced publication in 1973.

Emanuele Rodolà · Fabio Galasso · Iacopo Masi
Editors

Image Analysis and Processing – ICIAP 2025

23rd International Conference, Rome, Italy, September 15–19, 2025
Proceedings, Part I

Springer

Editors
Emanuele Rodolà
Sapienza University of Rome
Rome, Italy

Fabio Galasso
Sapienza University of Rome
Rome, Italy

Iacopo Masi
Sapienza University of Rome
Rome, Italy

ISSN 0302-9743 ISSN 1611-3349 (electronic)
Lecture Notes in Computer Science
ISBN 978-3-032-10184-6 ISBN 978-3-032-10185-3 (eBook)
https://doi.org/10.1007/978-3-032-10185-3

This Springer imprint is published by the registered company Springer Nature Switzerland AG
The registered company address is: Gewerbestrasse 11, 6330 Cham, Switzerland

Preface

The International Conference on Image Analysis and Processing (ICIAP) is a biennial scientific event promoted by the Italian Association for Computer Vision, Pattern Recognition and Machine Learning (CVPL, formerly GIRPR), the Italian Member Society of the IAPR. The 23rd edition of ICIAP (ICIAP 2025) was held in Rome, Italy, from September 15 to 19, 2025, and was jointly organized by the Department of Computer Science of Sapienza University of Rome. The event received strong support from the central administration of the university and was sponsored by numerous institutional, academic, and industrial partners. ICIAP 2025 continued its mission of offering a premier international forum for presenting and discussing the latest advances in Computer Vision, Image Analysis, Pattern Recognition, and Machine Learning.

The scientific program covered a broad and evolving spectrum of topics, spanning both foundational theories and impactful applications. In addition to the core areas, namely pattern recognition; machine learning and deep learning; 3D computer vision and geometry; image analysis, detection, and recognition; and video analysis and understanding, the conference featured contributions on biomedical and assistive technologies; digital forensics and biometrics; multimedia and cultural heritage; robot and embedded vision; EEG analysis; augmented, virtual, and mixed reality; geospatial and remote sensing analysis; and computer vision for UAVs. Several emerging themes were also prominent, including self-supervised and few-shot learning; domain adaptation and transfer learning; multimodal and cross-modal representation learning; explainability, trustworthiness, and fairness in AI; efficient and sustainable/edge AI; diffusion and generative models; anomaly and out-of-distribution detection; large-scale datasets and benchmarking; and reproducible research.

All submissions underwent a rigorous double-blind peer-review process. Each paper was evaluated by at least three independent reviewers and managed by one or more of the 29 Area Chairs. The review proceeded in two rounds with active discussion to ensure the quality and relevance of the accepted contributions. To guarantee fairness and transparency, submissions co-authored by Program Committee members or other chairs were handled with strict care: they were assigned to Area Chairs without conflict of interest and reviewed independently. The process included formal conflict-of-interest (COI) management, reviewer bidding and assignment, Area Chair management, at least three reviews per submission, followed by a meta-review and a discussion phase. Overall, the 2025 edition comprised 104 papers in the main conference and 105 papers across the workshops, reflecting both the vitality of the community and the selectivity of the program in a year with many high-quality submissions.

We were honored to host three distinguished invited speakers who enriched the scientific discourse with their keynote addresses: Francesco Locatello, Vittorio Ferrari, and Stefan Roth. Their lectures highlighted pressing challenges and innovative directions in contemporary computer vision and machine learning. ICIAP 2025 also hosted a wide array of workshops and tutorials, which further expanded opportunities for focused exchanges, hands-on training, and community building around specialized topics.

We would like to express our deepest gratitude to the area chairs, program committee members, external reviewers, and the entire organizing team for their dedication and professionalism. Their efforts were instrumental in maintaining the high scientific standard of the conference. We also thank all authors for submitting their valuable work and contributing to the success of this event. We hope that the proceedings collected in this volume will inspire further advancements and collaborations in the fields of image analysis, pattern recognition, computer vision, and machine learning.

September 2025

Emanuele Rodolà
Fabio Galasso
Iacopo Masi

Organization

General Chairs

Emanuele Rodolà	Sapienza University of Rome, Italy
Fabio Galasso	Sapienza University of Rome, Italy
Iacopo Masi	Sapienza University of Rome, Italy

Technical Program Chairs

Marco Cristani	University of Verona, Italy
Philippos Mordohai	Stevens Institute of Technology, USA
Paolo Soda	Università Campus Bio-Medico di Roma, Italy
Luisa Verdoliva	Università degli Studi di Napoli Federico II, Italy

Workshop Chairs

Giovanni Maria Farinella	University of Catania, Italy
Silvia Zuffi	IMATI-CNR, Italy

Tutorial Chairs

Alessio Del Bue	Italian Institute of Technology, Italy
Zorah Lähner	University of Siegen, Germany

Special Session Chair

Cosimo Distante	Institute of Applied Sciences and Intelligent Systems, CNR, Italy

Industrial Chairs

Ignas Budvytis	University of Cambridge, UK
Simone Scardapane	Sapienza University of Rome, Italy

Publicity and Social Media Chair

Antonino Furnari	University of Catania, Italy

Publication Chair

Danilo Avola	Sapienza University of Rome, Italy

Web Chairs

Antonio D'Orazio	Sapienza University of Rome, Italy
Daniele Pannone	Sapienza University of Rome, Italy

Local Arrangement Chairs

Danilo Avola	Sapienza University of Rome, Italy
Antonio D'Orazio	Sapienza University of Rome, Italy
Paolo Mandica	Sapienza University of Rome, Italy
Daniele Pannone	Sapienza University of Rome, Italy
Luca Scofano	Sapienza University of Rome, Italy
Daniele Trappolini	Sapienza University of Rome, Italy

CVPL Relation Chair

Luigi Cinque	Sapienza University of Rome, Italy

Steering Committee

Alberto Del Bimbo	University of Florence, Italy
Cosimo Distante	University of Salento, Italy
Nicu Sebe	University of Trento, Italy

Area Chairs

Video Analysis and Understanding

Mariella Di Miccoli	Polytechnic University of Catalonia, Spain
Niki Martinel	University of Udine, Italy

Pattern Recognition and Machine Learning

Simone Calderara	University of Modena and Reggio Emilia, Italy
Stéphane Lathuilière	Inria, Grenoble Alpes University, France
Riccardo Marin	TU Munich, Germany
Tatiana Tommasi	Politecnico di Torino, Italy

Deep Learning

Vishal Patel	Johns Hopkins University, USA
Indro Spinelli	Sapienza University of Rome, Italy

Multiview Geometry and 3D Computer Vision

Luca Cosmo	Ca' Foscari University of Venice, Italy
Federica Bogo	Meta AI, USA

Image Analysis, Detection, and Recognition

Tal Hassner	Meta AI, USA
Giuseppe Lisanti	University of Bologna, Italy

Multimedia

Lamberto Ballan	University of Padua, Italy

Biomedical and Assistive Technology

Simona Giunta	Sapienza University of Rome, Italy
Marco Leo	Institute of Applied Sciences and Intelligent Systems, CNR, Italy
Andrea Ravignani	Sapienza University of Rome, Italy

Digital Forensics and Biometrics

Irene Amerini	Sapienza University of Rome, Italy
Marilena De Marsico	Sapienza University of Rome, Italy

Image Processing for Cultural Heritage

Francesco Ragusa	University of Catania, Italy
Lorenzo Seidenari	University of Florence, Italy

Robot and Vision

Xavier Alameda-Pineda	Inria, Grenoble, France
Asako Kanezaki	TokyoTech, Japan

Brave New Ideas

Nicu Sebe	University of Trento, Italy
Fabrizio Silvestri	Sapienza University of Rome, Italy

AI and Science

Elena Agliari	Sapienza University of Rome, Italy
Elisa Tinti	Sapienza University of Rome, Italy

X-Realities

Andrea Giachetti	University of Verona, Italy
Marco Raoul Marini	Sapienza University of Rome, Italy

Embedded Vision

Luigi Capogrosso	University of Verona, Italy

Liasons

Latin America Liason

Olga Regina Pereira Bellon	Universidade Federal do Paraná, Brazil
Alexandre Xavier Falcão	University of Campinas, Brazil

Asia Liason

Luca Rossi	Hong Kong Polytechnic University, China

Endorsing Institutions

International Association for Pattern Recognition (IAPR)
Italian Association for Computer Vision, Pattern Recognition and Machine Learning (CVPL)

Contents

Incremental Object-Based Novelty Detection with Feedback Loop

Simone Caldarella[1(✉)], Elisa Ricci[1,3], and Rahaf Aljundi[2]

[1] University of Trento, Trento, Italy
simone.caldarella@unitn.it
[2] Toyota Motor Europe, Brussels, Belgium
[3] Fondazione Bruno Kessler, Trento, Italy

Abstract. Novelty Detection tackles the problem of identifying samples that do not belong to classes (*id*) observed by a given model during its training phase. In particular, *Object-based Novelty Detection* (*OND*) introduces an additional granularity compared to image-level novelty detection, allowing to process multiple objects at the same time. The majority of novelty detection approaches either assume full control over the initial (pre)training phase, or solely focus on post-processing of the pretrained model outputs at inference time. As a consequence, these solutions inevitably discard the precious knowledge within out-of-distribution (*ood*) data encountered after deployment. In this work, we assume that feedback in form of cheap annotations, confirming or rejecting the model's novelty decisions on detected objects, can be incrementally obtained. We then investigate how incorporating this feedback can strengthen models robustness. We propose both ① a novel setting and ② benchmark for *OND* with incremental feedback loop, focusing on the continuous incorporation of *ood* information over the model's life-cycle. Furthermore, we propose ③ a new method to effectively incorporate such feedback. Our approach, named iConP, consists of a lightweight novelty detection module, optimized continuously with the received feedback. Our method reduces the False Positive Rate (FPR) of the object detection model by ∼64%, when leveraging the continuously received feedback, while maintaining the object detection performance on known classes unaltered.

1 Introduction

Novelty Detection (ND), or Out-of-Distribution (OOD) detection, addresses the challenge of distinguishing novel (*ood*) samples from in-distribution (*id*) classes known during training. This task is critical for real-world deployment, as predefining all possible objects in dynamic environments is impossible. Unlike humans, neural networks inherently lack inference-time *id*/*ood* discrimination and exhibit *overconfidence* in incorrect predictions [42]. Prior work often normalizes output probabilities or pre-*softmax* logits to mitigate this [14,15,27]. While

S. Caldarella—Work mostly done during an internship at Toyota Motor Europe.

E. Rodolà et al. (Eds.): ICIAP 2025, LNCS 16167, pp. 1–16, 2026.
https://doi.org/10.1007/978-3-032-10185-3_1

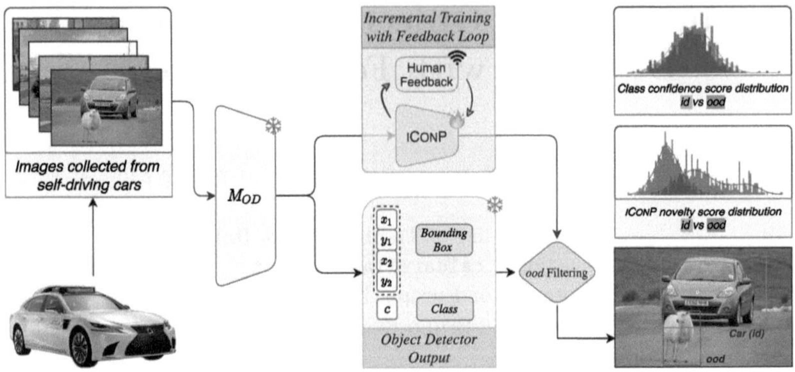

Fig. 1. Our proposed setting for *incremental* object-based novelty detection *with feedback loop*. During deployment, ICONP detects whether features extracted by an object detector M_{OD} represent *id* or *ood* objects. Predictions and images are sent to the cloud, where human annotators provide weak labels (accept/reject) for model decisions. This feedback improves *ood* robustness. On the right: Max Logit (baseline, no feedback) vs. ICONP after multiple feedback sessions

efficient, these methods fail to resolve inherent overconfidence, limiting real-world efficacy. Recent approaches leverage auxiliary *ood* data to enhance latent representations of *id*/*ood* distributions [3,8,16,17,27–31,40]. Limited exposure to representative *ood* samples improves performance over *id*-only methods. For example, Outlier Exposure [16] incorporates *ood* detection via exposed data as a model regularizer. However, all existing methods share a critical flaw: assuming single-step training while ignoring abundant post-deployment *ood* observations. We argue these assumptions are overly restrictive in real world, necessitating new solutions.

Consider autonomous cars with detection models trained to recognize classes like vehicles, pedestrians, and traffic signs. Given the vast array of potential road objects, identifying *id* samples is critical to avoid unsafe mispredictions. Predictions flagged as novel can be relayed to the cloud for human verification via weak annotations (requiring only binary novelty labels, *no class/bounding box details*). This feedback enhances robustness to unseen *ood* data. **Our core research question is optimizing the integration of these annotations.**

We tackle the underexplored problem of object-level novelty detection (ND) for real-world applications. Our model-agnostic framework enhances *ood* robustness in pretrained *id* detectors via test-time feedback. Feedback arrives incrementally in disjoint sessions, iteratively refining ND performance and forming a deploy-train loop (Fig. 1). We propose a lightweight novelty detection (ND) module atop a pretrained object detector, operating independently via latent representations of detected objects to classify *id*/*ood*; to ensure stable feedback and improved generalization, we replace naive Binary Cross Entropy training with Supervised Contrastive Loss [18] for representation separation, followed by a BCE-optimized projection for scoring. Furthermore, we find out that existing benchmarks often mix *id*/*ood* datasets, conflating domain shift with novelty detection. Worse, *id* evaluation data typically matches the detector's training set, violating real-world conditions where *id*/*ood* derives from the same unseen domain. To overcome this weakness, we build a benchmark where both *id* and *ood* are sampled from a test domain distinct from pretraining data.

Contributions. Our contributions are as follows: 1) we formulate the novel problem of incremental object-based novelty detection (*OND*) with feedback loop and 2) we introduce a novel approach based on a lightweight novelty detection module which shows robust performance and steady improvements from the received human feedback. To rigorously evaluate our approach we propose a new benchmark which overcomes the limitations of previous works [12,15,23]. 3) We show that existing novelty detection approaches suffer from severe performance drop when *id* and *ood* data come from the same testing distribution and from a different distribution than the one used for training the backbone model, and that our approach can effectively increase the robustness of the deployed object detection model as feedback is received.

2 Related Works

Novelty Detection. Existing approaches fall into two categories. 1) Post-processing: Methods like MSP (softmax probability) [15], energy scores [27], and Max Logit [14] operate on model outputs/features without modifying the base model. 2) Regularization: Techniques like autoencoder reconstruction errors [22] or contrastive learning with distributionally-shifted augmentations [39] modify training to enforce *id*/*ood* separability. Our contrastive approach uniquely optimizes embeddings to cluster *id* samples apart from *ood*. Another work, similar to ours within the scope of object-based novelty detection is VOS [12], which synthesizes outliers for energy-based training, we directly separate *id*/*ood* representations via an MLP. Other recent methods leverage auxiliary *ood* data: Outlier Exposure [16] combines cross-entropy losses, [27] optimizes energy margins, and Atom [8] mines *ood* data offline. Building on [4], we treat ND as binary classification but enhance BCE with supervised contrastive learning.

Continual and Active Learning. Continual learning avoids catastrophic forgetting via rehearsal [5,37] or regularization [1,20]. Inspired by GDumb [32], we rehearse features (storage: ~4KB/vector) for efficient updates (Sect. A.2). Unlike Continual ND [2] (evolving *id* classes) or IncDFM [36] (modeling *ood* as new *id*), we assume a fixed *id* set with incremental *ood* feedback. While related to active learning [34], our focus is robustness to unseen classes rather than sample selection.

3 Proposed Setting

In the proposed incremental object-based novelty detection (*OND*) with feedback loop we consider a pretrained object detection model M_{OD}. Several

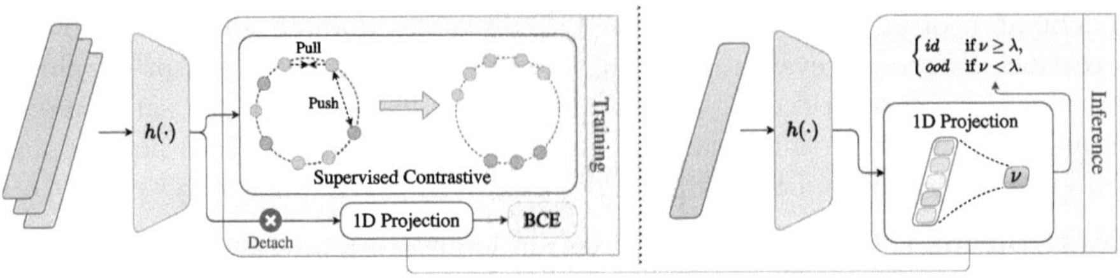

Fig. 2. Our proposed approach. During training (left) we optimize a light-weight ND module to project features from a pretrained object detection model into a hyper sphere, where *id* instances are grouped together and far from *ood* samples; jointly a scoring function is optimized as a 1D projection of learned latent features with BCE. At test time (right) *id* samples would be assigned high score indicating an instance of a known object as opposed to *ood* samples

existing object detectors are currently available, such as Faster-RCNN [35], YOLO [33], DETR [6] or even more recent open world object detection models such as Grounding DINO [26] and Yolo-World [9]. The object detection model is pretrained on large scale data and is deployed to detect instances of a predefined set of objects. This set is referred to as In-Distribution: $id = \{(img,\ b,\ y) :\ y\ \in\ \{c_1,\ ..,\ c_K\}\}$ where img is a given image, b is a bounding box, y is the corresponding class label and K is the number of the classes the model has to detect. The object detection pipeline is equipped with a novelty detection model M_{ND} used to reject detections of object instances that belong to classes not contained in the *id* set: $ood = \{(img,\ b,\ y) :\ y\ \notin\ \{c_1,\ ..,\ c_K\}\}$. Typically, M_{ND} predicts an *id* score that is high for *id* samples and low for novel unknown objects (*ood* samples).

Our setting (Fig. 1) aims to improve M_{ND} ability to detect *ood* instances, *e.g.*, unknown vehicles or objects. Initially, we train M_{ND} on a set of *id*/*ood* samples from the deployment environment (session S_0), refining it to local characteristics (*e.g.*, a specific city or factory). During deployment, M_{ND} filters unknown objects while recognizing *id* classes. Predictions are sent to the cloud for human annotation, labeling objects as known or unknown. These annotations retrain M_{ND} in subsequent sessions ($S_1, S_2, \ldots$), refining its detection capabilities iteratively. This process continues as long as errors occur and annotation budgets allow. Newly annotated *ood* data may belong to previously unseen classes, enabling M_{ND} to generalize to *unseen-ood* instances. Crucially, M_{OD}'s performance on *id* detection remains unaffected.

4 Method

Given an input image, the pretrained object detector M_{OD} extracts object proposals $\{\mathbf{x}_i\}_{i=1}^{P}$ with bounding boxes b_i. These representations $\{\mathbf{x}_i\}$ are passed to M_{ND}, a lightweight network parameterized by ϕ, which predicts the likelihood of a proposal belonging to *id* classes. M_{ND} maps each $\mathbf{x}_i$ to a detection score $\nu_i = f_\phi(\mathbf{x}_i)$. Training uses two sets: *id* and *ood* object proposals (Fig. 2).

4.1 The Proposed Approach

The work [4] theoretically analyzed *ood* training and scoring mechanisms, showing that a Bayes-optimal function for novelty detection is a binary discriminator between *id* and *ood* data. Inspired by this, we frame novelty detection as binary classification. However, unlike typical classification (generalizing to two specific classes), our *ood* set represents all possible non-*id* classes, requiring generalization to unseen *ood* classes.

We propose training M_{ND} to project *id* data into a space linearly separable from all *ood* data. While Binary Cross Entropy (BCE) is a straightforward solution, we hypothesize it lacks flexibility for refining decision boundaries with new feedback, as BCE ignores internal *id* similarities. Instead, we use Supervised Contrastive Loss (SupCon) [18], which learns embeddings where *id* samples cluster together, far from *ood* samples. This approach provides smoother *id*-*ood* separation and greater adaptability to new *ood* classes during retraining.

To this end, let f represents the function approximated by M_{ND}, we express f as a composition of two functions $f(\mathbf{x}_i) = g(h(x_i))$ were h maps $\mathbf{x}_i$ to a latent variable $\mathbf{z}_i$ and g maps $\mathbf{z}_i$ to its corresponding novelty score ν_i. We optimize the hidden representation $\mathbf{z}$ of M_{ND} using SupCon loss, as follows:

$$\ell_{SupCon} = \frac{1}{N}\sum_{i}^{N} \ell_{SupCon}(\mathbf{z}_i, P_i), \tag{1}$$

where:

$$\ell_{SupCon}(\mathbf{z}_i, P_i) = -\frac{1}{|P_i|}\sum_{\mathbf{z}_p \in P_i} \log \frac{\exp(\text{sim}(\mathbf{z}_i, \mathbf{z}_p)/\tau)}{\sum_{j \neq i} \exp(\text{sim}(\mathbf{z}_i, \mathbf{z}_j)/\tau)}, \tag{2}$$

and $\text{sim}(\mathbf{z}_i, \mathbf{z}_j) = \frac{\mathbf{z}_i^\top \mathbf{z}_j}{||\mathbf{z}_i|| \cdot ||\mathbf{z}_j||}$, P_i is the set of representations $\mathbf{z}_p$ forming positive pairs for the i-th sample (*i.e.*, the representations of proposals of the same set of objects, *id* or *ood*), the index j iterates over all samples, N is the total number of *id* and *ood* samples, and τ is a temperature parameter. We do not consider any data augmentation.

To estimate a novelty detection score, we propose to optimize a one dimensional projection g that associates *id* data with high scores as opposed to *ood* data. We approach this by minimizing BCE loss on the hidden representation ($\mathbf{z}_i$).

$$\ell_{BCE}(\mathbf{z}, y) = -\frac{1}{N}\sum_{i}^{N} \left[y_i \log(g(\mathbf{z}_i)) + (1 - y_i)\log(1 - g(\mathbf{z}_i))\right], \tag{3}$$

where $g(\mathbf{z}_i) = (1 + e^{-(\mathbf{w}^\top st(\mathbf{z}_i))})^{-1}$ is the novelty score ν_i, st denotes the stop gradient operation and $\mathbf{w}$ is the weight vector optimized for the 1D projection. Here, the binary cross entropy is only optimizing the final 1D projection to find the best separating direction, and it has no effect on the representations being trained. Our full training objective is:

$$\ell_{ND} = \ell_{SupCon} + \ell_{BCE}. \tag{4}$$

4.2 Incremental Updates

Our goal is to obtain a strong novelty detector that can be incrementally improved whenever feedback is received, without acting on the pretrained object detector. We attach our module M_{ND} to the pretrained object detector in order to receive the encoded object proposal features. The first step of training M_{ND} is based on offline gathered *id* and *ood* data from the expected deployment distribution as we explained earlier. We optimize M_{ND} using (4). Then, when a new annotated *id-ood* set is received, we finetune M_{ND} initialized with previous session's parameters.

Directly finetuning the model on a new set of *ood* classes will drift it from what it has learned before, causing forgetting of the previously captured decision boundaries [11]. Here to prevent forgetting, we store the representations $\{\mathbf{x}_i\}$ extracted by the OD model of the previously observed sets. We deploy Experience Replay [7] where representations of previous sets are replayed jointly with the new set samples. We do not use any explicit regularizer here to further control forgetting, as it was shown in [10] that representation trained by Supervised Contrastive loss is much less prone to forgetting than alternatives based on cross entropy. We leave the investigation of how more effective continual learning methods can be combined with our framework to future work.

Table 1. *OND* performance on VOC-COCO Offline (Classical and Test Domain) benchmarks *before the feedback loop*. Both iBCE and iConP improve significantly over compared methods. VOS [12] performs poorly on Test Domain benchmark

Method	VOC (ID) vs COCO (OOD)		COCO (ID) vs COCO (OOD)	
	FPR@95(↓)	AUROC (↑)	FPR@95 (↓)	AUROC(↑)
MSP [15]	74.32	78.47	81.62	74.12
Energy OOD [27]	64.30	77.60	70.14	75.97
Max Logit [14]	63.88	77.66	69.62	75.96
VOS [12]	49.11	87.54	81.70	74.44
iBCE	**22.96**	**95.22**	**44.78**	88.96
iConP	27.45	94.16	45.37	**89.90**

5 Experiments

5.1 Test-Domain Incremental Benchmark

Novelty detection evaluation benchmarks are constructed such that *id* classes are drawn from the same distribution as the training data for the pretrained model, while the *ood* classes are drawn from a new distribution, *i.e.*, a different dataset [12,15,23]. This setup is relatively easy as the novelty detection problem boils down to distinguishing samples from different datasets and that might result in a good but misleading performance. Some recent benchmarks for classification divide a given dataset into sets of known (*id*) and novel (*ood*) classes overcoming the dataset shift effect, however, the test distribution of a given model is assumed to be the same as the training distribution.

To emulate a real-world scenario, we introduce a benchmark called the Test-Domain Incremental Benchmark. The process of creating this benchmark relies on the initial pretraining set solely as a reference to select *id* classes and the key is to identify another dataset that contains large set of objects classes with *id* classes as a subset. We refer to the pretraining (or tuning) dataset of the main object detection model as D_{train}, containing a set of *id* classes $C_{id} = \{c_1, c_2, ..., c_K\}$. For evaluating the novelty detection performance, we consider a test dataset D_{test} that contains object classes $C_{test} = \{c_1, c_2, ..., c_K, ..., c_{K+J}\}$ where the *id* classes C_{id} form a subset of C_{test}.

Incremental Steps. The target is to simulate the scenario where M_{ND} predictions are sent to the cloud and their *id*-*ood* annotations are received on which a new training session is conducted. We split the test dataset D_{test} into disjoint groups of samples. Each group G_i contains two subgroups, an *id* subgroup G_i^{id} containing a unique set of images with instances of *id* object classes $C_{id} = \{c_1, c_2, ..., c_K\}$; and an *ood* subgroup G_i^{ood} containing a unique set of images and objects instances of distinct *ood* classes $C_i^{ood} \in \{c_{K+1}, ..., c_{K+J}\}$. We leave one group $G^{holdout}$ as a holdout set to evaluate the novelty detection performance on never seen set of images containing objects instances of *id* classes and never seen *ood* classes. Note that the constructed *ood* subsets $\{G_i^{ood}\}$ contain distinct set of classes. The first group G_0 is used in the first offline training session (S_0) of M_{ND}. At each later training session S_i, M_{ND} is provided with samples from G_i. Figure 3 illustrates our training and evaluation protocol to strictly assessing the possible improvements in the generalization performance of M_{ND} on a set of never seen *ood* classes and never seen *id* instances.

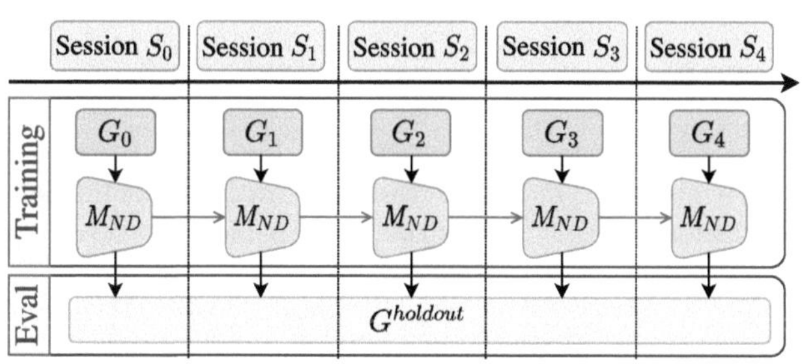

Fig. 3. Our incremental training and evaluation protocol. At each training session S_i a new group of *id* and *ood* samples G_i is received. Evaluation is performed on a distinct set, namely $G^{holdout}$

5.2 Datasets

We rely on three datasets. **VOC**, with over 10k images and 25k annotated objects across 20 classes [13]; **COCO**, with 123k images and nearly 900k instances across 80 classes, including 20 VOC classes [25]; and **OpenImages**, the largest collection with 16M bounding boxes for 600 classes on 1.9M images, including all VOC and 60 COCO classes [21]. We construct two test benchmarks, one with D_{train} as VOC and D_{test} being COCO, where we train the object detection model on VOC. The second is with D_{train} as COCO and D_{test} being OpenImages imitating a large scale scenario. The D_{test} in both cases is split into 6 independent groups $\{G_i\}$, corresponding to 5 training sessions $\{S_i\}$ plus the holdout group $G^{holdout}$ used for unseen *ood* novelty detection evaluation. We refer to the Appendix A.1 for more details on the sets construction.

5.3 Metrics

As we train a separate novelty detection module and we do not alter in any way the original object detection model, but only exploit the extracted features, we focused on the metrics that are directly related to the task on novelty detection, ignoring those related to the detection task (mAP and IoU).

FPR@95 ($\downarrow$), False Positive Rate when the True Positive Rate is set to 95%.
AUROC ($\uparrow$), area under the Receiver Operating Characteristic (ROC).

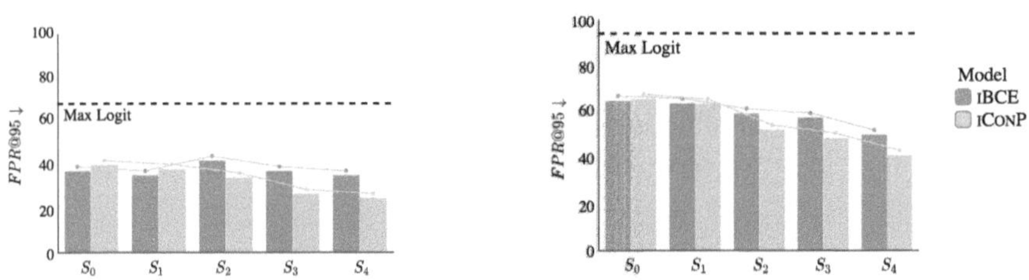

(a) VOC-COCO Benchmark (b) COCO-OpenImages Benchmark

Fig. 4. Novelty detection performance (FPR@95, **lower is better**) of Max Logit (dashed line), IBCE and ICONP on the Test-Domain Incremental benchmark, evaluated after each session on $G^{holdout}$. ICONP is capable of steady improvements as feedback is received contrary to IBCE

5.4 Baselines and Implementation Details

MSP [15]: Uses maximum *softmax* probability as the *id* score.
Energy OOD [27]: Computes a novelty score from the full logit distribution.
Max Logit [14]: Uses the maximum pre-*softmax* logit value.
VOS [12] (SOTA): Trains an MLP on energy values from *id* and synthetic *ood* features.

iBCE: incremental baseline trained with Binary Cross Entropy.
Details. For M_{OD}, on VOC we employed a Faster R-CNN based on ResNet-50 trained on VOC 2007, and VOC 2012 and for COCO trained backbone we used Faster R-CNN based on ResNet-50 with a Feature Pyramid Network [24] trained on COCO 2017. For M_{ND} we deploy an MLP of 4 hidden layers plus 1 layer for the final projection to one output logit, optimized with Adam [19], for iBCE, and with SGD, for iConP. Further details regarding the implementation are reported in the Appendix A.1.

5.5 Results

VOC-COCO Offline Benchmark. We first evaluate the different methods on a typical setting where *id* object instances of both train and test data are from VOC dataset and *ood* objects are sampled from COCO dataset. ND results in terms of FPR and AUROC are shown in Table 1. Among the methods that do not leverage auxiliary *ood* data, VOS [12] seems to perform the best. iConP achieves almost 50% reduction rate on the FPR metric compared to VOS [12]. Although we train on real *ood* data, it is evidence on our approach effectiveness.

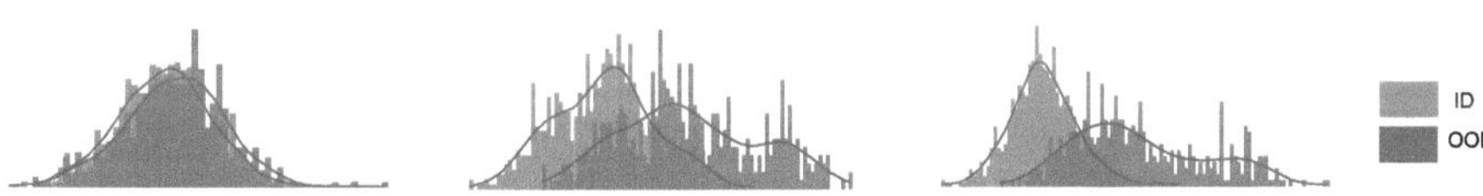

(a) Max Logit (b) iConP after S_0 (c) iConP after S_4

Fig. 5. The *id-ood* scores distributions on $G^{holdout}$ from COCO-OpenImages benchmark predicted by (a) Max Logit, (b) iConP after the initial training session S_0, and (c) iConP after S_4. Starting with overlapping distribution with Max Logit, iConP separates the *id-ood* and largely benefits from the received feedback

VOC-COCO Test-Domain Benchmark. Here we evaluate the different methodologies on a more realistic yet offline setting. We consider our test-domain benchmark (Sect. 5.1) where both *id* and *ood* images and their objects instances are drawn from the same test dataset, COCO, whereas the object detection model is trained on VOC. Results in terms of AUROC and FPR are reported in Table 1. First, it is clear that this is more challenging as the performance of all methods have dropped significantly. Surprisingly, VOS [12] is not the best performing among the offline methods. It is in fact on par with MSP and worse than Max Logit. This shows that VOS training on synthetic *ood* data does not generalize to the case of semantic shift only, *i.e.*, only different classes of objects among *id* and *ood*, and that VOS suffers from the shift between the training dataset and the test environment. Second, our method iConP still significantly improves the performance, we can see that the FPR drops from 77.6% by Max Logit to 45.37% by iConP with an AUROC increasing by 12%. Third, while both IBCE and ICONP perform closely, our target is to improve the ND performance as more feedback is received, and to show that it is possible to further enhance the generalization capabilities as more *ood* and *id* examples become available.

Test-Domain Incremental Benchmark. We evaluate our incremental pipeline using the test-domain incremental benchmark (Sect. 5.1). Max Logit, the best-performing offline method, serves as our baseline. Figure 4 shows the FPR of IBCE and ICONP on $G^{holdout}$ after each training session (full results in Appendix A.4). On VOC-COCO (Fig. 4a), ICONP initially matches IBCE but outperforms it significantly with more feedback. After 5 sessions, ICONP reduces FPR from 40.38% to 25.0%, compared to IBCE's 2.5% drop. For a larger-scale scenario, we train on COCO and test on OpenImages (60 *id* and 96 *ood* classes). Figure 4b shows ICONP improving notably by the third session, reducing FPR from 65.62% to 41.47%, versus IBCE's 64.88% to 50.18%. ICONP achieves 9% lower FPR and 2.5% higher AUROC (not shown). These results highlight the importance of optimizing *id* embeddings for separability, rather than directly optimizing decision boundaries as in IBCE. Figure 5 illustrates score distributions for COCO-OpenImages. Max Logit (a) shows overlapping *id*-*ood* distributions (FPR: 93.84%). ICONP achieves partial separation after S_0 (b) and near-optimal separation after feedback (c), demonstrating its effectiveness.

Fig. 6. Qualitative results on VOC-COCO after S_4: (Top) MaxLogit wrongly detects *ood* objects as *id*, causing overconfident misclassification by M_{OD}; (Bottom) ICONP correctly flags these *ood* cases, avoiding misleading predictions

Qualitative Results. To inspect how *ood* objects are treated by ICONP compared to Max Logit, Fig. 6 shows some qualitative examples. Relying on an offline *ood* detection method as Max Logit leads to serious mistakes and as a result confident yet wrong *id* object detections. On the contrary ICONP is much robust and can identify those objects as *ood* avoiding any mishandling of these cases.

6 Discussion

Reliance on Human-In-The-Loop. While our binary feedback loop incurs minimal annotation cost, efficiency can be improved via active learning to prioritize uncertain detections. iConP's lightweight design keeps compute overhead low, but extremely high-throughput settings (*e.g.*, millions of frames/day) may benefit from asynchronous feedback or on-device quantization. To address scalability, a two-tier labeling system could first query a larger expert model (*e.g.*, MLLM) for easy or uncertain samples, reserving human review for the hardest *oods*.
Dealing with Similar ID/OOD. In rare cases where *id* and *ood* classes are nearly indistinguishable (e.g., Toyota Corolla vs. Corolla Cross), naïve contrastive training may over-separate subtle differences. iConP mitigates this by (1) applying supervised contrastive loss with all *id* as positives and all *ood* as negatives, and (2) leveraging human-in-the-loop feedback to flag hard *ood* cases. Additionally, since both *id* and *ood* contain diverse classes, SupCon loss naturally averages across varied pairs, smoothing extreme gradients and preserving a stable feature space even for visually similar samples.

7 Conclusion

We showed in Table 1 that existing ND methods struggle when both *id* and *ood* objects come from the same testing distribution. Training our ND module on an initial subset of this distribution significantly improves detection. With feedback and new training sessions, iConP reduces FPR notably, from 40.38% to 25.00% on VOC-COCO and from 65.62% to 41.47% on COCO-OpenImages, demonstrating its ability to leverage new annotations and improve robustness. Ensuring ML safety in real-world settings is vital, especially for detecting unknown objects and avoiding errors. We emphasize the value of using abundant test data and low-cost human feedback. Our results show that incorporating feedback on *ood* decisions boosts pretrained model robustness, while our lightweight ND module updates maintain *id* performance and increase reliability on *ood* data.

Acknowledgments. This work was sponsored by the MUR PNRR project FAIR - Future AI Research (PE00000013), funded by NextGeneration, and the EU project ELIAS (No.01120237).

Disclosure of Interests. The authors have no competing interests to declare that are relevant to the content of this article

A Appendix

A.1 Implementation Details

For the object detection, we relied on the Detectron2 [43] library, which provides a variety of pretrained deep learning architectures for the most relevant computer vision tasks at the moment. Specifically, for our experiments on VOC based benchmarks we use `faster_rcnn_R_50_C4`, a Faster R-CNN model with ResNet-50 as backbone trained on Pascal VOC 2007 + VOC 2012, while for those on the COCO based benchmarks we use `faster_rcnn_R_50_FPN_3x`, a Faster R-CNN model with ResNet-50 and Feature Pyramid Network as a backbone trained on COCO 2017. As input to M_{ND} we use the hidden features predicted in the *roi* heads for each detected object after non-max suppression and objectiveness threshold.

iBCE is trained using ADAM optimizer with a $lr = 0.0005$ and a Cosine Annealing LR scheduler while employing early stopping (with a window size of 5). As for variants (presented in Table 2) with different regularization functions we tested M_{ND} with Dropout with $p = 0.5$, Mixup ($x_{mix} = x_i + \lambda \cdot x_j$) at input feature level with $\lambda \sim \mathcal{N}(\mu = 0.5, \sigma^2 = 0.2)$ as linear interpolation parameter, Manifold Mixup [41], where the Mixup happens at a random hidden layer L at each Mixup step.

For iConP we optimized the latent representation with SupCon loss [18], applied at the penultimate layer of the model, and BCE applied, in a disjoint fashion, to the last layer of the model, intended as a 1D projection to the *score space*. To alleviate any possible instability of SupCon, we initialized the weights of the model using Xavier initialization and a warmup for the first 10 epochs, varying the lr from 0.01 to 0.005. We found out the SGD optimizer to be a better match than Adam along with a Cosine Annealing LR scheduler, where the lr is updated before each epoch. Differently from the model trained with BCE, we keep a fixed number of epochs $e = 25$, since SupCon generally requires more number of epochs before obtaining a good features representation. Once a model has been trained on the first split G_0, it is incrementally updated employing the subsequent splits. During this process we reduce the learning rate by a factor of 10 for all our variants and we set a fixed number of epochs for both the iBCE and iConP based ones, respectively 5 epochs and 15 epochs, after which each model seems to converge.

In our benchmarks, VOC-COCO test domain and COCO-OpenImages incremental, we used respectively splits generated from COCO and OpenImages datasets. Starting from 80 classes in total in COCO dataset we first filter the 20 classes contained also in VOC as our *id* data and we keep the remaining 60 classes as *ood* data. We then generate G_0 by filtering 30 out of 60 *ood* classes and the remaining 30 have been evenly distributed to generate the latter 5 splits ($\{G_1, G_2, G_3, G_4, G^{holdout}\}$). Differently from COCO, given the large amount classes in OpenImages, after having extracted 60 (out of the 80) COCO *id* classes; we filter out 96 of the remaining 500+ *ood* classes to form our *ood* dataset. We chose to keep only *highly* represented classes with at least 5000 occurrences out of 15M bbox annotations, still maintaining the classical long tail shape distribution of *ood* classes. Similar as before, we keep half of the *ood* classes as our S_0 (48 classes), and the remaining 48 classes form our incremental sessions i.e., 10, 10, 10, 9, 9 classes respectively.

A.2 Computational Overhead

Since we rely on the features extracted from the frozen object detection model, the computational overhead of our novelty detection methodology appears negligible. Specifically, at inference time of our Novelty Detection module (150k parameters) represents the 1% of the average time required for the object detection inference. Similarly, the resources for one training session is minimal in time and compute, as it takes 10 s of training time for a few thousands of samples (on a budget gpu, i.e. Nvidia gtx 2080, with 5 training epochs, 8 GB of vram and a batch size of 512 samples). This illustrates the efficiency of our approach and that increasing a model robustness to *ood* samples can be quite cheap in terms of annotations and compute as opposed to updating the main object detection model performance.

A.3 Can iBCE Generalize Better When Combined with Different Regularization Functions?

In the experiments section, we have shown that iBCE exhibits poor performance improvements as more feedback is received and we have motivated our choice of loss with the fact that directly optimizing decision boundaries by binary cross entropy doesn't generalize well to unseen *ood* classes and doesn't allow enough flexibility when new samples and *ood* classes are received.

To further confirm the superiority of our design choice and motivation, we combine iBCE with different regularization techniques and inspect whether any improvements in generalization can be observed compared to iConP. We consider different regularization techniques namely, dropout [38], Mixup [44] and manifold mixup [41]. Table 2 reports the FPR and AUROC of iBCE with each regularizer and the combination of multiple regularizers in our test-domain incremental benchmark (VOC-COCO). It can be seen that some regularizers provide improvements at different training sessions but these improvements are not consistent and the overall performance is significantly inferior to that of our method iConP. Moreover, it is interesting to notice that, while iBCE trained models consistently show better initial performance, iConP outperforms all iBCE variants as the training progresses, suggesting that regularization is not sufficient to improve *ood* detection performance as training session are carried. This proves that iConP ability of steady performance improvements is not trivial.

Table 2. Novelty detection performance of iBCE with different regularization mechanisms and iConP on the test-domain incremental benchmark (VOC-COCO)

Method	S_0		S_1		S_2		S_3		S_4	
	FPR@95(↓)	AUROC(↑)	FPR@95(↓)	AUROC(↑)	FPR@95(↓)	AUROC(↑)	FPR@95(↓)	AUROC(↑)	FPR@95(↓)	AUROC(↑)
iBCE	37.50	91.37	35.57	91.65	42.30	90.98	37.50	92.22	35.57	92.65
iBCE + Dropout [38]	35.9	91.88	34.51	92.19	38.25	91.67	33.97	92.96	32.26	93.46
iBCE + Mixup [44]	43.91	89.35	40.17	91.59	41.03	91.07	36.43	92.43	34.83	92.89
iBCE + Manifold Mixup [41]	38.78	89.62	38.46	89.58	38.46	89.02	38.03	89.84	37.39	90.33
iBCE + Dropout + Mixup	**35.58**	92.58	**33.65**	92.57	34.72	91.99	31.52	93.17	31.2	93.67
iBCE + Dropout + Manifold Mixup	38.78	90.29	39.53	90.17	39.96	89.56	39.74	90.28	38.03	90.75
iConP	40.38	**93.11**	38.46	**93.89**	**34.61**	**93.75**	**26.92**	**94.33**	**25.00**	**94.19**

A.4 Detailed Results

Due to the space limits, in the main experiments we only show FPR metric in a plot for our test-domain incremental benchmark. Here we report the detailed numbers of both FPR and AUROC after each training session. Additionally we expand our evaluation protocol to evaluate on two holdout sets: 1) a holdout set of unseen ood classes $G^{holdout}$ (as in the main paper) and 2) a holdout set composed of both unseen and encountered ood classes of $G_{0,1,2,3,4}$ as it is natural to re-encounter some *ood* classes in practice, and it is important to see the ND performance on *unseen* samples of those previously encountered *ood* classes. Tables 3 and 4 report the FPR and AUROC for iBCE and iConP after each training session on VOC-COCO and COCO-OpenImages respectively.

On both holdout sets and and both incremental benchmarks, iConP shows steady improvements and superior performance to that of iBCE.

Table 3. *ND* performance on VOC-COCO incremental test-domain benchmark. As training sessions progress iConP benefits greatly from the feedback and reduce significantly the FPR while improving the AUROC

Method	Eval	S_0		S_1		S_2		S_3		S_4	
		FPR@95(↓)	AUROC(↑)	FPR@95(↓)	AUROC(↑)	FPR@95(↓)	AUROC(↑)	FPR@95(↓)	AUROC(↑)	FPR@95(↓)	AUROC(↑)
iBCE	$G_{0,1,2,3,4}$	**47.37**	**91.24**	**35.72**	92.42	29.12	94.01	26.79	94.92	24.85	95.28
	$G^{holdout}$	**37.50**	91.37	**35.57**	91.65	42.30	90.98	37.50	92.22	35.57	92.65
iConP	$G_{0,1,2,3,4}$	47.76	90.61	39.41	**92.57**	**21.16**	**95.44**	**19.22**	**96.41**	**17.86**	**96.54**
	$G^{holdout}$	40.38	**93.11**	38.46	**93.89**	**34.61**	**93.75**	**26.92**	**94.33**	**25.00**	**94.19**

Table 4. ND performance on COCO-OpenImages incremental test-domain benchmark. As training sessions progress iConP benefits greatly from the feedback and reduce significantly the FPR while improving the AUROC

Method	Eval	S_0		S_1		S_2		S_3		S_4	
		FPR@95(↓)	AUROC(↑)	FPR@95(↓)	AUROC(↑)	FPR@95(↓)	AUROC(↑)	FPR@95(↓)	AUROC(↑)	FPR@95(↓)	AUROC(↑)
iBCE	$G_{0,1,2,3,4}$	**75.77**	81.14	74.40	**82.68**	66.52	85.93	62.37	87.00	53.53	88.83
	$G^{holdout}$	**64.88**	82.45	63.76	**83.98**	59.40	86.74	57.53	87.48	50.18	89.10
iConP	$G_{0,1,2,3,4}$	77.00	**81.38**	**74.29**	82.56	**61.46**	**87.73**	**57.17**	**88.83**	**47.88**	**90.75**
	$G^{holdout}$	65.62	**83.23**	**63.51**	83.93	**52.42**	**88.68**	**48.81**	**89.49**	**41.47**	**91.58**

References

1. Aljundi, R., Babiloni, F., Elhoseiny, M., Rohrbach, M., Tuytelaars, T.: Memory aware synapses: learning what (not) to forget. In: Ferrari, V., Hebert, M., Sminchisescu, C., Weiss, Y. (eds.) ECCV 2018. LNCS, vol. 11207, pp. 144–161. Springer, Cham (2018). https://doi.org/10.1007/978-3-030-01219-9_9
2. Aljundi, R., Reino, D.O., Chumerin, N., Turner, R.E.: Continual novelty detection. In: CoLLas (2022)

3. Azzalini, D., Bonali, L., Amigoni, F.: A minimally supervised approach based on variational autoencoders for anomaly detection in autonomous robots. IEEE RAL **6**(2), 2985–2992 (2021)
4. Bitterwolf, J., Meinke, A., Augustin, M., Hein, M.: Breaking down out-of-distribution detection: many methods based on OOD training data estimate a combination of the same core quantities. In: ICML, pp. 2041–2074 (2022)
5. Buzzega, P., Boschini, M., Porrello, A., Abati, D., Calderara, S.: Dark experience for general continual learning: a strong, simple baseline. In: NeurIPS, vol. 33, pp. 15920–15930 (2020)
6. Carion, N., Massa, F., Synnaeve, G., Usunier, N., Kirillov, A., Zagoruyko, S.: End-to-end object detection with transformers. In: Vedaldi, A., Bischof, H., Brox, T., Frahm, J.-M. (eds.) ECCV 2020. LNCS, vol. 12346, pp. 213–229. Springer, Cham (2020). https://doi.org/10.1007/978-3-030-58452-8_13
7. Chaudhry, A., et al.: Continual learning with tiny episodic memories (2019)
8. Chen, J., Li, Y., Wu, X., Liang, Y., Jha, S.: ATOM: robustifying out-of-distribution detection using outlier mining. In: Oliver, N., Pérez-Cruz, F., Kramer, S., Read, J., Lozano, J.A. (eds.) ECML PKDD 2021. LNCS (LNAI), vol. 12977, pp. 430–445. Springer, Cham (2021). https://doi.org/10.1007/978-3-030-86523-8_26
9. Cheng, T., Song, L., Ge, Y., Liu, W., Wang, X., Shan, Y.: Yolo-world: real-time open-vocabulary object detection. In: CVPR (2024)
10. Davari, M., Asadi, N., Mudur, S., Aljundi, R., Belilovsky, E.: Probing representation forgetting in supervised and unsupervised continual learning. In: CVPR (2022)
11. De Lange, M., et al.: A continual learning survey: defying forgetting in classification tasks. IEEE TPAMI **44**(7), 3366–3385 (2021)
12. Du, X., Wang, Z., Cai, M., Li, Y.: Vos: learning what you don't know by virtual outlier synthesis. In: ICML (2022)
13. Everingham, M., Van Gool, L., Williams, C.K.I., Winn, J., Zisserman, A.: The PASCAL Visual Object Classes Challenge 2012 (VOC2012) Results
14. Hendrycks, D., et al.: Scaling out-of-distribution detection for real-world settings. arXiv preprint arXiv:1911.11132 (2019)
15. Hendrycks, D., Gimpel, K.: A baseline for detecting misclassified and out-of-distribution examples in neural networks. In: ICLR (2017)
16. Hendrycks, D., Mazeika, M., Dietterich, T.: Deep anomaly detection with outlier exposure. In: ICML (2019)
17. Ji, T., Sivakumar, A.N., Chowdhary, G., Driggs-Campbell, K.: Proactive anomaly detection for robot navigation with multi-sensor fusion. IEEE RAL **7**(2), 4975–4982 (2022)
18. Khosla, P., et al.: Supervised contrastive learning. arXiv preprint arXiv:2004.11362 (2020)
19. Kingma, D., Ba, J.: Adam: a method for stochastic optimization. In: ICLR (2015)
20. Kirkpatrick, J., et al.: Overcoming catastrophic forgetting in neural networks. Proc. Natl. Acad. Sci. **114**(13), 3521–3526 (2017)
21. Kuznetsova, A., et al.: The open images dataset v4: unified image classification, object detection, and visual relationship detection at scale. IJCV (2020)
22. Li, T., Wang, Z., Liu, S., Lin, W.Y.: Deep unsupervised anomaly detection. In: WACV (2021)
23. Liang, S., Li, Y., Srikant, R.: Enhancing the reliability of out-of-distribution image detection in neural networks. arXiv preprint arXiv:1706.02690 (2017)
24. Lin, T.Y., Dollár, P., Girshick, R., He, K., Hariharan, B., Belongie, S.: Feature pyramid networks for object detection. In: CVPR (2017)

25. Lin, T.-Y., et al.: Microsoft COCO: common objects in context. In: Fleet, D., Pajdla, T., Schiele, B., Tuytelaars, T. (eds.) ECCV 2014. LNCS, vol. 8693, pp. 740–755. Springer, Cham (2014). https://doi.org/10.1007/978-3-319-10602-1_48
26. Liu, S., et al.: Grounding Dino: marrying Dino with grounded pre-training for open-set object detection. arXiv preprint arXiv:2303.05499 (2023)
27. Liu, W., Wang, X., Owens, J., Li, Y.: Energy-based out-of-distribution detection. In: NeurIPS (2020)
28. Meinke, A., Hein, M.: Towards neural networks that provably know when they don't know. arXiv preprint arXiv:1909.12180 (2019)
29. Ming, Y., Fan, Y., Li, Y.: Poem: Out-of-distribution detection with posterior sampling. In: ICML (2022)
30. Papadopoulos, A.A., Rajati, M.R., Shaikh, N., Wang, J.: Outlier exposure with confidence control for out-of-distribution detection. Neurocomputing **441**, 138–150 (2021)
31. Pei, S., Zhang, X., Fan, B., Meng, G.: Out-of-distribution detection with boundary aware learning. In: Avidan, S., Brostow, G., Cissé, M., Farinella, G.M., Hassner, T. (eds.) ECCV 2022. LNCS, vol. 13684, pp. 235–251. Springer, Cham (2022). https://doi.org/10.1007/978-3-031-20053-3_14
32. Prabhu, A., Torr, P.H.S., Dokania, P.K.: GDumb: a simple approach that questions our progress in continual learning. In: Vedaldi, A., Bischof, H., Brox, T., Frahm, J.-M. (eds.) ECCV 2020. LNCS, vol. 12347, pp. 524–540. Springer, Cham (2020). https://doi.org/10.1007/978-3-030-58536-5_31
33. Redmon, J., Divvala, S., Girshick, R., Farhadi, A.: You only look once: unified, real-time object detection (2016)
34. Ren, P., et al.: A survey of deep active learning. ACM Comput. Surv. (CSUR) **54**(9), 1–40 (2021)
35. Ren, S., He, K., Girshick, R.B., Sun, J.: Faster R-CNN: towards real-time object detection with region proposal networks. In: NIPS (2015)
36. Rios, A., Ahuja, N., Ndiour, I., Genc, U., Itti, L., Tickoo, O.: incDFM: incremental deep feature modeling for continual novelty detection. In: Avidan, S., Brostow, G., Cissé, M., Farinella, G.M., Hassner, T. (eds.) ECCV 2022. LNCS, vol. 13685, pp. 588–604. Springer, Cham (2022). https://doi.org/10.1007/978-3-031-19806-9_34
37. Rolnick, D., Ahuja, A., Schwarz, J., Lillicrap, T., Wayne, G.: Experience replay for continual learning. In: NeurIPS, vol. 32 (2019)
38. Srivastava, N., Hinton, G., Krizhevsky, A., Sutskever, I., Salakhutdinov, R.: Dropout: a simple way to prevent neural networks from overfitting. JMLR **15**(56), 1929–1958 (2014)
39. Tack, J., Mo, S., Jeong, J., Shin, J.: CSI: novelty detection via contrastive learning on distributionally shifted instances. In: NeurIPS (2020)
40. Thulasidasan, S., Thapa, S., Dhaubhadel, S., Chennupati, G., Bhattacharya, T., Bilmes, J.: An effective baseline for robustness to distributional shift. In: ICMLA, pp. 278–285 (2021)
41. Verma, V., et al.: Manifold mixup: better representations by interpolating hidden states. In: ICML (2019)
42. Wei, H., Xie, R., Cheng, H., Feng, L., An, B., Li, Y.: Mitigating neural network overconfidence with logit normalization. In: ICLR, pp. 23631–23644. PMLR (2022)
43. Wu, Y., Kirillov, A., Massa, F., Lo, W.Y., Girshick, R.: Detectron2 (2019)
44. Zhang, H., Cisse, M., Dauphin, Y.N., Lopez-Paz, D.: mixup: beyond empirical risk minimization. arXiv preprint arXiv:1710.09412 (2017)

Zero-Shot Neural Architecture Search for Efficient Deep Stereo Matching

Alessio Mingozzi[1(✉)], Stefano Mattoccia[1], Matteo Poggi[1], and Fatma Güney[2]

[1] Alma Mater Studiorum University of Bologna, Bologna, Italy
alessio.mingozzi2@unibo.it
[2] KUIS AI Center, Koç University, Istanbul, Turkey

Abstract. This paper introduces a novel and efficient architecture for deep stereo matching obtained through Zero-Shot Neural Architecture Search (NAS). Although accurate and capable of good generalization across datasets, state-of-the-art iterative stereo models are often too computationally expensive for low-power devices. In order to address this limitation, this work employs NAS to efficiently explore a search space of different layer types and hyperparameters, including efficient residual layers from Ghost Modules. Instead of relying on extensive training, the method evaluates candidate architectures using a combined zero-cost proxy score based on the AZ-NAS score and the number of parameters, thus promoting the selection of smaller, efficient models. Applied to RAFT-Stereo, this process yields a significantly smaller – 1.14M parameters, compared to the original 11M – and substantially faster network. The resulting architecture maintains competitive performance on various stereo benchmarks while running 3× faster, demonstrating the effectiveness of Zero-Shot NAS in optimizing deep stereo networks for resource-constrained environments.

Keywords: Neural Architecture Search · Zero-Shot · Stereo Matching

1 Introduction

Stereo matching, a foundational problem in computer vision, aims to determine dense correspondences between a pair of rectified stereo images, thereby generating a dense disparity map. This cue is crucial for many applications, such as robotics and augmented reality, and purposely, many strategies have been proposed to obtain robust disparity maps. In contrast to the conventional use of 3D convolutional networks in stereo matching, RAFT-Stereo [18] utilizes 2D convolutions and a lightweight cost volume. The network employs multi-level GRUs to effectively propagate information across the image, enabling both real-time inference and accurate disparity estimation even when dealing with high-resolution images. Notably, RAFT-Stereo exhibits remarkable cross-dataset generalization capabilities, achieving impressive performance on real-world datasets even after

E. Rodolà et al. (Eds.): ICIAP 2025, LNCS 16167, pp. 17–28, 2026.
https://doi.org/10.1007/978-3-032-10185-3_2

training solely on synthetic data. Nevertheless, given its characteristics, the standard version struggles with low-performance or memory-constrained systems, such as edge devices.

Although the advent of deep neural networks has brought substantial advancements across different disciplines, their design often necessitates substantial human effort and expertise. To address this issue, Neural Architecture Search (NAS) aims to automate the design process of neural networks by identifying optimal network architectures through an iterative search process. However, since each candidate needs to be evaluated, this process can be computationally expensive and time-consuming. Purposely, One-Shot and Zero-Shot NAS paradigms have been proposed. The former strategy employs a hyper-network to share candidate operations, reducing training time significantly compared to traditional multi-shot approaches. The latter further enhances efficiency by eliminating training during the search phase, specifically, by employing various proxy scores to predict the accuracy of candidate network architectures during the initialization stage. These proxies, often grounded in theoretical analyses of deep neural networks, offer valuable insights into the characteristics of high-performing architectures. Employing Zero-Shot proxies that accurately reflect the nature of network performance, particularly in constrained scenarios, is essential to fully harness the potential of this strategy in the presence of a low computational budget.

The main contribution of this work is leveraging the Zero-Shot NAS paradigm to obtain a fast and lightweight network with a minimal loss in accuracy compared to RAFT-Stereo. The source code is available at https://github.com/amingozz/RAFT-StereoZero.

2 Related Work

We briefly review the literature relevant to our work.

Deep Stereo Matching. After years of developments in hand-crafted algorithms [31] already, end-to-end architectures starting with DispNetC [22] and developed with 2D [45] and 3D [12] convolutions became the dominant solutions in stereo matching. This led to the exploration of new paradigms, stereo video processing [47], fusion with active sensors [3,24], self-adapting models [25,26,36] or strategies to deal with challenging conditions such as non-Lambertian surfaces [7,27,46] or depth discontinuities [2,37,39]. The latest advances, thoroughly reviewed in [38], introduced recurrent architectures [18] inspired by RAFT [35]. This category represents the preferred choice nowadays, as it proved great generalization capabilities without requiring any explicit design choice [5]: among them, RAFT-Stereo [35] introduces an all-pair cost volume used to guide an iterative disparity estimation process. Despite the advances brought by deep learning, state-of-the-art architectures seldom fit low-power devices and when feasible, they are extremely slow.

Neural Architecture Search. Recent advancements in Automated Machine Learning (AutoML) [11] have notably accelerated network development. Neural

Architecture Search (NAS), a subset of AutoML, has proven successful in tasks such as classification [44], object detection [33], and semantic segmentation [20]. However, NAS demands substantial computational resources. To mitigate these computational demands, researchers have developed more efficient methods like Differentiable Architecture Search (DARTS) [20], which introduces a hierarchical NAS approach. DARTS has been applied to stereo, resulting in architectures like AutoDispNet [29] obtained in 42 GPU days. LEAStereo [6] further extended this approach to volumetric stereo matching, achieving state-of-the-art results. Recently, Zero-Shot NAS methods have emerged to reduce computational burdens by using low-cost proxies to evaluate candidate architectures without extensive training. Various proxies [1,21,34,42] have been developed to estimate architecture quality. ZenNAS [17] demonstrated the capability to measure architectural expressivity with minimal resources, identifying a competitive ImageNet architecture in 0.5 GPU days. Similarly, ZiCo [15] achieves better performances evaluating gradients over input data. Despite findings that many zero-shot proxies correlate similarly with test set accuracy compared to simpler metrics like parameter count and FLOPs [14], the recent AZ-NAS [13] proposes higher accuracy correlation by providing four distinct scores computed over a single forward pass, addressing the limitations of previous proxies and improving discrimination among top-performing architectures.

Given the absence of prior applications of Zero-Shot NAS methods to deep stereo problems and the high computational cost of traditional approaches, this work applies these methodologies for such tasks for the first time.

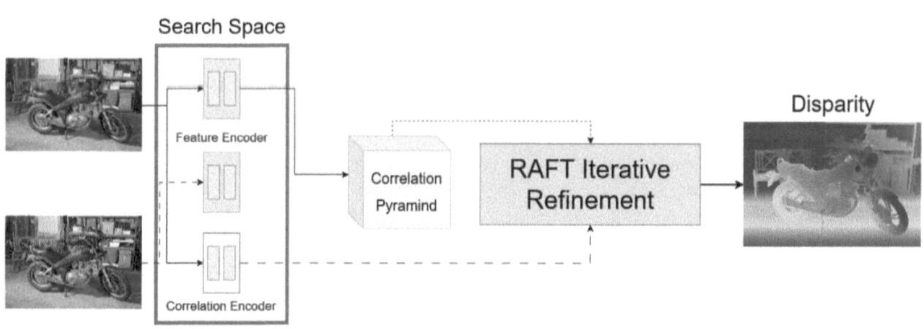

Fig. 1. RAFT-Stereo architecture. The search process is conducted across the two encoders.

3 Method

This section describes the methodology for obtaining a faster and lightweight RAFT-Stereo-like network by leveraging a NAS framework and a Zero-Shot strategy, in the form of an adapted AZ-NAS score.

3.1 Search Space

As shown in Fig. 1, RAFT-Stereo architecture consists of two primary components: the feature encoder and the context encoder, usually sharing the same structure, followed by the iterative refinement process carried out by ConvGRUs. Hence, the search space is constructed starting from these two modules. To achieve a smaller and more efficient architecture, we defined the global search space over two properties: the hyperparameters of the various layers, including the convolution feature dimensions or kernel size, and the layer type. Concerning the latter, the original RAFT-Stereo implementation uses 2D convolutions and residual blocks within the encoders. In contrast, we add depthwise convolutions and a more efficient residual layer derived from GhostModule [9,10] to the search space, as it can reduce computational cost and model size without sacrificing performance. In summary, the search space comprises the original RAFT-Stereo layers, depthwise convolutions, and GhostModules, with all their hyperparameters, as summarized in Fig. 2.

3.2 Search Strategy

NAS methodologies rely on a search strategy to define how to mutate the architecture to generate new candidates. Various strategies have been proposed in the literature, ranging from genetic algorithms to reinforcement learning. One such strategy is regularized evolution [28], which relies on evolutionary algorithms, a class of optimization techniques inspired by natural selection. This framework employs a randomized population of architectures as its starting point. During each search cycle, the architecture with the highest fitness score is selected as a parent for the new generation.

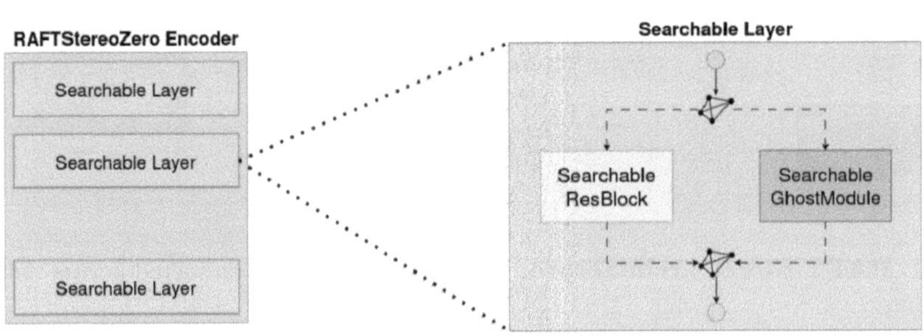

Fig. 2. Encoder search space. Layer selection is made between the original Resnet-like block and a GhostBlock.

3.3 Evaluation Strategy

To determine the best-performing architecture, evaluating and selecting candidates based on desired performance characteristics is crucial. An exhaustive evaluation of all candidates is computationally impractical, often requiring thousands of GPU days. To address this, techniques like One-Shot and Zero-Shot methodologies have been proposed. Specifically, we utilize the latter for efficiently ranking architectures, as it is the fastest and least resource-intensive. Among zero-shot methods, AZ-NAS scores [13] show a good correlation with test set accuracy, using different scores to capture various network characteristics and provide better ratings among candidates. However, zero-cost scores tend to optimize the expressive capabilities of networks, driving the search toward larger networks, which hinders the identification of smaller, more efficient combinations. To address this, we define a metric by summing AZ-Score and the number of parameters, aiming to find a network with the lowest parameters while retaining a high AZ-Score value. The AZ-Score comprises four metrics:

Expressivity($\mathcal{E}$)**.** It evaluates how features are distributed across orientations in the feature space, given randomly initialized network weights and Gaussian random inputs. Given the score of the l-th primary block, $s_l^{\mathcal{E}}$, as an entropy score, posing L1-normalized coefficients of PCs as probabilities:

$$s_l^{\mathcal{E}} = \sum_{i=1}^{c} -\hat{\lambda}_l(i) log \hat{\lambda}_l(i) \tag{1}$$

where $\hat{\lambda}_l(i)$ is a set of L1-normalized coefficient. The overall network score consists of the sum of all block-level scores:

$$s^{\mathcal{E}} = \sum_{l=1}^{L} s_l^{\mathcal{E}} \tag{2}$$

Progressivity($\mathcal{P}$). The difference of the block-level $\mathcal{E}$ scores from Eq.(1):

$$s^{\mathcal{P}} = \min_{l in 2,\ldots,L} s_l^{\mathcal{E}} - s_{l-1}^{\mathcal{E}} \tag{3}$$

Trainability($\mathcal{T}$)**.** Given the approximation of the Jacobian matrix A_l:

$$A_l^{\top} = \frac{1}{n} \sum_{p=1}^{n} g_{l-1}(p) g_l^{\top}(p), \tag{4}$$

The trainability score $s^{\mathcal{T}}$ is defined as:

$$s^{\mathcal{T}} = \frac{1}{L-1} \sum_{l=2}^{L} -\sigma_l - \frac{1}{\sigma_l} + 2, \tag{5}$$

which reaches its maximum value when the spectral norm of A_l, denoted as σ_l, is equal to 1 for all l.

Complexity($\mathcal{C}$)**.** Given the correlation between network dimension and performance, [13] proposes using the FLOPs count as a complexity proxy.

Algorithm 1. Evolutionary search using the combined score

```
Input: search space Z; number of NAS iteration T.
Output: selected architecture A*.
Randomly initialize first architecture A_1 for evolutionary search
while i = 1 ¡ T do
    Compute proxy scores s^E(i), s^T(i), s^P(i) for the architecture A_i.
    Append the architecture A_i to the population history 𝔸.
    Compute #PARAMS of A_i.
    Compute the score s(A_i).
    Generate a new candidate architecture A_{i+1} which belongs to the search space
    Z by mutating the top architecture based on s score.
end while
Compute the ranking over the history set 𝔸 and select the best candidate archi-
tecture A*
```

$$s^{\mathcal{C}}(i) = log_{10}(FLOPs(i)) \tag{6}$$

We use the log_{10} to reduce the summed score dimension.

Score Aggregation. Since the NNI Framework allows only a single floating point value as the score, we cannot rely on the ranking approach used by AZ-NAS, which instead will be used to select the final architecture. For this reason, we aggregate all scores into a single value. Given an architecture candidate, we compute the set of scores $\mathcal{M}$ and define the ith architecture score $s^{AZ}(i)$ as:

$$s^{AZ}(i) = \sum_{\mathcal{M} \in \mathcal{E},\mathcal{T},\mathcal{P},\mathcal{C}} s^{\mathcal{M}}(i) \tag{7}$$

In order to account for the architectural dimensions in the scoring process, the combined final score is formulated as:

$$s^{\Omega}(i) = ln(s^{AZ}(i)) - ln(\#PARAMS(i)) \tag{8}$$

We compute all proxy scores for the possible candidate architectures during the evolutionary search process. Subsequently, to generate new candidates, we mutate the top-scoring architecture based on the s^{Ω} value. Finally we select the best-performing architecture using the AZ ranking formula:

$$S^{\Omega}(i) = \sum_{\mathcal{M} \in \{\mathcal{E},\mathcal{P},\mathcal{T},\mathcal{C}\}} \log\left(\frac{\text{Rank}\left(s^{\Omega}(i)\right)}{m}\right) \tag{9}$$

where m is the number of candidate architecture. The chosen architecture A^{Ω} will be the first ranked candidate. The process is shown in Algorithm 1.

4 Experiments

In this section, we describe the application of our combined score within the neural architecture search process to identify an optimal stereo architecture. Then, we evaluate its performance against the original RAFT-Stereo, as well as with its more efficient variant, RAFT-Stereo *realtime* (RT), using a shared backbone for features context extraction and bi-level GRUs at $\frac{1}{8}$ and $\frac{1}{16}$ resolution [19].

4.1 Implementation Details

The architecture search was implemented in PyTorch using the NNI framework, using the NeRF stereo dataset [40] to compute scores over a single batch sample of the dataset during evaluation.

Search Process. Upon concluding the search process, which took 0.1 GPU days, we applied the ranking process over the architecture history to identify the network with the best overall score. The final architecture configuration, dubbed RAFT-StereoZero and shown in Fig. 3, counts 1.14M trainable parameters, in contrast to the original 11M, translating into lower memory occupancy. While RAFT-Stereo consists of ResBlocks of a variable number of channels, RAFT-StereoZero modules are mostly GhostModules counting 32 or 64 channels. Given the iterative nature of the prediction process, we further optimize RAFT-StereoZero complexity and speed by conducting a Hyperparameter Optimization (HPO) search to identify the best combination of training and validation iterations. Using the Tree-structured Parzen Estimator (TPE) algorithm [43], each combination was trained over a small subset of the training dataset, i.e., 400 samples for 5000 steps, with the error on Middlebury [30] and KITTI 2015 [23] used as evaluation scores, leading to 8 training and 10 validation iterations as the best trade-off between accuracy and speed.

Evaluation Metrics. To assess the accuracy of each model, we measure the average end-point error (EPE) between predicted and ground-truth disparity maps, as well as the percentage of pixels with error larger than a threshold (D1), fixed to 2, 1, and 3 for Middlebury, ETH3D, and KITTI datasets, respectively.

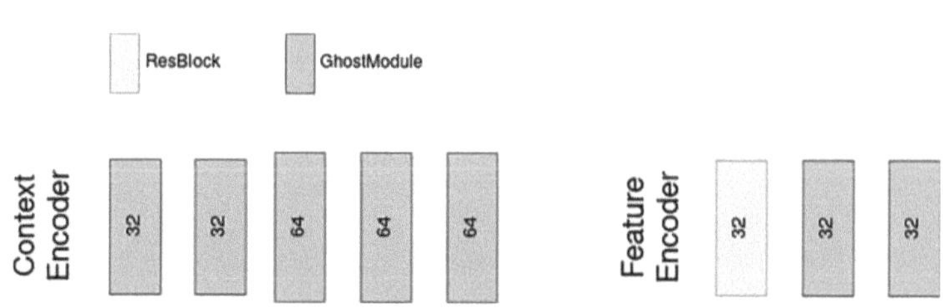

Fig. 3. RAFT-StereoZero architecture. The encoders consist predominantly of GhostModules, except for the initial layer in the feature encoder. Each layer employs distinct hyperparameter configurations.

Table 1. Ablation study – impact of the training data. We evaluate the impact of different training data configurations.

Model	Middlebury 2014		ETH3D		KITTI 2012		KITTI 2015	
	EPE	D1	EPE	D1	EPE	D1	EPE	D1
Setup A	3.21	18.87	0.67	16.92	0.90	**4.26**	1.51	**6.34**
Setup B	3.13	**18.19**	0.58	12.29	**0.89**	4.37	1.29	6.72
Setup C	**2.91**	18.89	**0.40**	**6.13**	0.90	4.51	**1.22**	6.44

4.2 Training

Once the search process is concluded, we trained the architecture using the same protocol as the original RAFT-Stereo, while exploring the use of more recent datasets – such as NERF-Stereo [40] and CREStereo [16] – alongside SceneFlow [22], Sintel Stereo [4], and Falling Things [41]. After balancing, our training set counts 760,836 samples during training. We trained RAFT-StereoZero on four A100 GPUs for over 800,000 steps, with a total batch size 32.

4.3 Evaluation

In order to assess the performance of RAFT-StereoZero compared to the original RAFT-Stereo, we run the evaluation over the zero-shot benchmark [40] composed of four standard different stereo datasets: Middlebury 2014 [30], ETH3D [32], KITTI 2012 [8] and KITTI 2015 [23].

Table 2. Quantitative results. When compared with the original architecture, RAFT-StereoZero achieves competitive results across different datasets.

Model	Middlebury 2014		ETH3D		KITTI 2012		KITTI 2015	
	EPE	D1	EPE	D1	EPE	D1	EPE	D1
RAFT-Stereo	1.57	11.50	0.27	2.61	0.83	3.86	1.13	5.68
RAFT-Stereo (Setup C)	1.79	11.78	0.88	24.17	0.76	2.88	1.08	4.86
RAFT-StereoZero	1.87	13.06	0.31	4.32	0.78	3.57	1.12	5.69
RAFT-Stereo (RT)	2.53	15.57	0.58	5.76	0.92	4.31	1.15	5.82
RAFT-StereoZero (RT)	2.93	18.59	0.49	9.28	0.89	4.03	1.13	5.62

Impact of Training Datasets. We study the effect of each additional dataset we introduce for training. Purposely, we train different instances RAFT-StereoZero following the same protocol, yet using three training data setup: first, we use just Sceneflow, Falling Things and Sintel Stereo (Setup A), then add NERF-Stereo (Setup B), and finally, we add CREStereo (Setup C). Table 1 reports the outcome of this evaluation. Specifically, we can see an average 8%

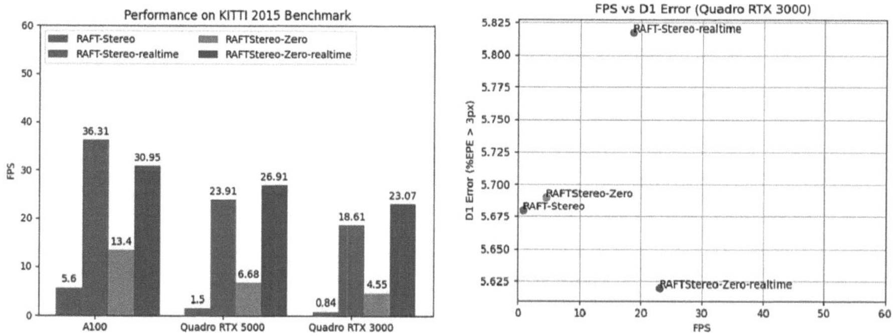

Fig. 4. Runtime analysis. Left: framerate measured on three different GPUS; right: accuracy-framerate plot on KITTI 2015 using Quadro RTX 3000 GPU.

relative improvement when adding NeRF Stereo, and a further 10% when adding CREStereo. In particular, the accuracy over the ETH3D benchmark benefits the most, with an improvement of 31% EPE and a 50% D1.

Accuracy Comparison with RAFT-Stereo. Table 2 reports a comparison between the original RAFT-Stereo and our model at the top. For a fair comparison, we also train a further instance of RAFT-Stereo on the same training data used for our model (Setup C): this yields some improvements on KITTI datasets, although being surprisingly ineffective on ETH3D. We can appreciate how RAFT-StereoZero achieves almost equivalent accuracy on any dataset with marginal drops only – lower than 0.1 EPE and about 1% D1 on Middlebury and KITTI datasets against RAFT-Stereo Setup C, about 0.03 EPE and 1.3% D1 on ETH3D against the original RAFT-Stereo.

At the bottom, we report a comparison between RAFT-Stereo (RT) variant and a *realtime* counterpart searched through our approach, referred to as RAFT-StereoZero (RT). Again, our model achieves marginal drops in accuracy on Middlebury and ETH3D, while gaining some accuracy on KITTI datasets.

The small drops exposed by both our models against their counterparts allow for training accuracy with efficiency, as we are going to discuss in the remainder.

Inference Speed. Figure 4 reports the accuracy-speed trade-off achieved by the original RAFT-Stereo models and our searched variants. On the left, the runtime required to process KITTI 2015 images when measured on three GPUs: nVidia A100, Quadro RTX 5000, and Quadro RTX 3000. RAFT-StereoZero achieves a speed-up ranging from 2.3× to 5.4× against RAFT-Stereo and an increase of 5FPS when considering (RT) variants. On the right, we report an accuracy-framerate plot concerning performance on the Quadro RTX 3000. It highlights that our models achieve faster and more accurate results on KITTI 2015 than their original counterparts.

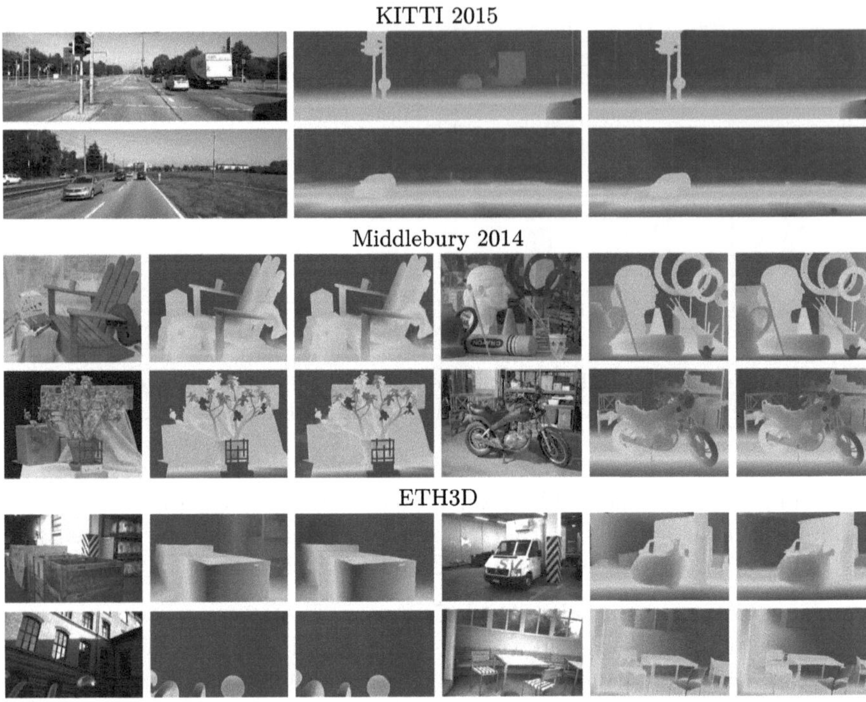

Fig. 5. Qualitative results. We report examples from KITTI 2015, Middlebury 2014 and ETH3D, showing the reference color image (left) and the disparity maps predicted by RAFT-Stereo (middle) and RAFT-StereoZero (right).

4.4 Qualitative Results

We conclude by reporting a qualitative comparison between the predictions by RAFT-Stereo and RAFT-StereoZero. Figure 5 shows examples from KITTI 2015, Middlebury 2014 and ETH3D datasets. We can appreciate how our model retains the generalization capability of the original RAFT-Stereo, as well as fine-grained details in the predicted disparity maps.

5 Conclusions

In this paper, we propose RAFT-StereoZero, a novel architectural configuration based on RAFT-Stereo [18] for two-view stereo, identified using the Zero-Shot Neural Architecture Search methodology. By leveraging the capabilities of the AZ-Score to evaluate the quality of candidate architectures, combined with a performance-focused approach, we have discovered a model that not only achieves competitive results but also exhibits faster performance than the original architecture.

References

1. Abdelfattah, M.S., Mehrotra, A., Dudziak, Ł., Lane, N.D.: Zero-cost proxies for lightweight NAS (2021). https://arxiv.org/abs/2101.08134
2. Aleotti, F., et al.: Neural disparity refinement for arbitrary resolution stereo. In: 3DV (2021)
3. Bartolomei, L., Poggi, M., Tosi, F., Conti, A., Mattoccia, S.: Active stereo without pattern projector. In: ICCV (2023)
4. Butler, D.J., Wulff, J., Stanley, G.B., Black, M.J.: A naturalistic open source movie for optical flow evaluation. In: Fitzgibbon, A., Lazebnik, S., Perona, P., Sato, Y., Schmid, C. (eds.) ECCV 2012. LNCS, vol. 7577, pp. 611–625. Springer, Heidelberg (2012). https://doi.org/10.1007/978-3-642-33783-3_44
5. Cai, C., Poggi, M., Mattoccia, S., Mordohai, P.: Matching-space stereo networks for cross-domain generalization. In: 3DV (2020)
6. Cheng, X., et al.: Hierarchical neural architecture search for deep stereo matching. In: NeurIPS (2020)
7. Costanzino, A., Ramirez, P.Z., Poggi, M., Tosi, F., Mattoccia, S., Di Stefano, L.: Learning depth estimation for transparent and mirror surfaces. In: CVPR (2023)
8. Geiger, A., Lenz, P., Urtasun, R.: Are we ready for autonomous driving? The Kitti vision benchmark suite. In: CVPR (2012)
9. Han, K., Wang, Y., Tian, Q., Guo, J., Xu, C., Xu, C.: GhostNet: more features from cheap operations (2020). https://doi.org/10.48550/arXiv.1911.11907
10. Han, K., et al.: GhostNets on heterogeneous devices via cheap operations (2022). https://doi.org/10.48550/arXiv.2201.03297
11. Hutter, F., Kotthoff, L., Vanschoren, J. (eds.): Automated Machine Learning. TSS-CML, Springer, Cham (2019). https://doi.org/10.1007/978-3-030-05318-5
12. Kendall, A., et al.: End-to-end learning of geometry and context for deep stereo regression. In: ICCV (2017)
13. Lee, J., Ham, B.: AZ-NAS: assembling zero-cost proxies for network architecture search. In: CVPR (2024)
14. Li, G., Hoang, D., Bhardwaj, K., Lin, M., Wang, Z., Marculescu, R.: Zero-shot neural architecture search: challenges, solutions, and opportunities. TPAMI 1–19 (2024)
15. Li, G., Yang, Y., Bhardwaj, K., Mărculescu, R.: ZiCo: zero-shot NAS via inverse coefficient of variation on gradients. ICLR (2023)
16. Li, J., et al.: Practical stereo matching via cascaded recurrent network with adaptive correlation. In: CVPR (2022)
17. Lin, M., et al.: Zen-NAS: a zero-shot NAS for high-performance image recognition. In: ICCV (2021)
18. Lipson, L., Teed, Z., Deng, J.: Raft-stereo: multilevel recurrent field transforms for stereo matching. In: 3DV (2021)
19. Lipson, L., Teed, Z., Deng, J.: RAFT-stereo: multilevel recurrent field transforms for stereo matching (2021). https://doi.org/10.48550/arXiv.2109.07547
20. Liu, H., Simonyan, K., Yang, Y.: DARTS: Differentiable Architecture Search (2019). https://doi.org/10.48550/arXiv.1806.09055
21. Lopes, V., Alirezazadeh, S., Alexandre, L.A.: EPE-NAS: efficient performance estimation without training for neural architecture search. In: ICANN (2021)
22. Mayer, N., et al.: A large dataset to train convolutional networks for disparity, optical flow, and scene flow estimation. In: CVPR (2016)
23. Menze, M., Geiger, A.: Object scene flow for autonomous vehicles. In: CVPR (2015)

24. Poggi, M., Pallotti, D., Tosi, F., Mattoccia, S.: Guided stereo matching. In: CVPR (2019)
25. Poggi, M., Tonioni, A., Tosi, F., Mattoccia, S., Di Stefano, L.: Continual adaptation for deep stereo. TPAMI **44**(9), 4713–4729 (2021)
26. Poggi, M., Tosi, F.: Federated online adaptation for deep stereo. In: CVPR (2024)
27. Ramirez, P.Z., et al.: Booster: a benchmark for depth from images of specular and transparent surfaces. TPAMI **46**(1), 85–102 (2024)
28. Real, E., Aggarwal, A., Huang, Y., Le, Q.V.: Regularized evolution for image classifier architecture search (2019). https://doi.org/10.48550/arXiv.1802.01548
29. Saikia, T., Marrakchi, Y., Zela, A., Hutter, F., Brox, T.: Autodispnet: improving disparity estimation with autoML. In: ICCV (2019)
30. Scharstein, D., et al.: High-resolution stereo datasets with subpixel-accurate ground truth. In: GCPR (2014)
31. Scharstein, D., Szeliski, R.: A taxonomy and evaluation of dense two-frame stereo correspondence algorithms. IJCV **47**, 7–42 (2002)
32. Schops, T., et al.: A multi-view stereo benchmark with high-resolution images and multi-camera videos. In: CVPR (2017)
33. Tan, M., Pang, R., Le, Q.V.: EfficientDet: scalable and efficient object detection (2020). https://doi.org/10.48550/arXiv.1911.09070
34. Tanaka, H., Kunin, D., Yamins, D.L., Ganguli, S.: Pruning neural networks without any data by iteratively conserving synaptic flow. In: NeurIPS (2020)
35. Teed, Z., Deng, J.: RAFT: recurrent all-pairs field transforms for optical flow. In: Vedaldi, A., Bischof, H., Brox, T., Frahm, J.-M. (eds.) ECCV 2020. LNCS, vol. 12347, pp. 402–419. Springer, Cham (2020). https://doi.org/10.1007/978-3-030-58536-5_24
36. Tonioni, A., Tosi, F., Poggi, M., Mattoccia, S., Stefano, L.D.: Real-time self-adaptive deep stereo. In: CVPR (2019)
37. Tosi, F., et al.: Neural disparity refinement. TPAMI (2024)
38. Tosi, F., Bartolomei, L., Poggi, M.: A survey on deep stereo matching in the twenties. arXiv preprint arXiv:2407.07816 (2024). https://arxiv.org/abs/2407.07816, extended version of CVPR 2024 Tutorial "Deep Stereo Matching in the Twenties" (https://sites.google.com/view/stereo-twenties)
39. Tosi, F., Liao, Y., Schmitt, C., Geiger, A.: SMD-nets: stereo mixture density networks. In: CVPR (2021)
40. Tosi, F., Tonioni, A., De Gregorio, D., Poggi, M.: NeRF-supervised deep stereo. In: CVPR (2023)
41. Tremblay, J., To, T., Birchfield, S.: Falling things: a synthetic dataset for 3D object detection and pose estimation. In: CVPR Workshops (2018)
42. Wang, C., Zhang, G., Grosse, R.: Picking winning tickets before training by preserving gradient flow (2020). https://arxiv.org/abs/2002.07376
43. Watanabe, S.: Tree-structured parzen estimator: understanding its algorithm components and their roles for better empirical performance 2023). https://doi.org/10.48550/arXiv.2304.11127
44. Wu, B., et al.: FBNet: hardware-aware efficient ConvNet design via differentiable neural architecture search (2019). https://doi.org/10.48550/arXiv.1812.03443
45. Yin, Z., Darrell, T., Yu, F.: Hierarchical discrete distribution decomposition for match density estimation. In: CVPR (2019)
46. Zama Ramirez, P., Tosi, F., Poggi, M., Salti, S., Di Stefano, L., Mattoccia, S.: Open challenges in deep stereo: the booster dataset. In: CVPR (2022)
47. Zhang, Y., Poggi, M., Mattoccia, S.: TemporalStereo: efficient spatial-temporal stereo matching network. In: IROS (2023)

DeepCS-TRD, a Deep Learning-Based Cross-Section Tree Ring Detector

Henry Marichal[1(✉)], Verónica Casaravilla[2,3], Candice Power[4], Karolain Mello[2,3], Joaquín Mazarino[2,3], Christine Lucas[2,3], Ludmila Profumo[2,3], Diego Passarella[2,3], and Gregory Randall[1]

[1] Instituto de Ingeniería Eléctrica, Facultad de Ingeniería, Universidad de la República, Montevideo, Uruguay
hmarichal93@gmail.com
[2] CENUR Noreste, Universidad de la República, Tacuarembo/Rivera, Uruguay
[3] CENUR Litoral Norte, Universidad de la República, Paysandú, Uruguay
[4] Ecoinformatics and Biodiversity, Department of Biology, Aarhus University, Aarhus, Denmark

Abstract. Here, we propose Deep CS-TRD, a new automatic algorithm for detecting tree rings in whole cross-sections. It substitutes the edge detection step of CS-TRD by a deep-learning-based approach (U-Net), which allows the application of the method to different image domains: microscopy, scanner or smartphone acquired, and species (*Pinus taeda*, *Gleditsia triachantos* and *Salix glauca*). Additionally, we introduce two publicly available datasets of annotated images to the community. The proposed method outperforms state-of-the-art approaches in macro images (*Pinus taeda* and *Gleditsia triacanthos*) while showing slightly lower performance in microscopy images of *Salix glauca*. To our knowledge, this is the first paper that studies automatic tree ring detection for such different species and acquisition conditions. The dataset and source code are available in https://hmarichal93.github.io/deepcstrd/.

Keywords: Tree rings detection · Dendrochronology · Deep learning · U-Net

1 Introduction

Most existing automatic tree ring measurement methods rely on images from cores (small cylindrical samples that capture a transect of the tree's growth rings) rather than entire cross-sections. However, extracting a core is not feasible for very small trees or shrubs, and analyzing entire cross-sections is essential for individuals with suppressed or irregular growth, where there may be wedging rings. Furthermore, modeling tree growth in forestry often requires 2D information, which can only be obtained from analyzing the entire cross-section.

Automatically delineating rings in images of tree cross-sections presents the challenge of generating a pattern of closed curves that accurately represent the

E. Rodolà et al. (Eds.): ICIAP 2025, LNCS 16167, pp. 29–41, 2026.
https://doi.org/10.1007/978-3-032-10185-3_3

ring boundaries, as seen in Fig. 1. These rings may exhibit irregular and asymmetric growth patterns, especially in trees and shrubs growing in extreme environments. Moreover, ring boundaries make up only a small portion of the disc, which may also contain perturbations such as cracks, knots, and fungal growth, further increasing the complexity of the task.

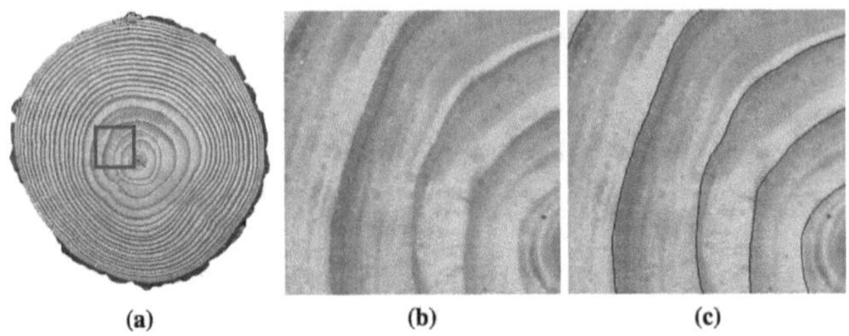

Fig. 1. Tree Ring Detection. Accurate detection of tree ring boundaries in images is critical. (a) Wood cross-section image. (b) Zoomed-in view of the red square in (a). (c) The same view with ring boundaries in blue. The ring thickness is set to 1 pixel. (Color figure online)

Figure 2 outlines the pipeline proposed in this work: a disc image is provided as input, and tree ring traces are generated as output. Once the tree ring curves are delineated, areas and perimeter can be computed, which is essential for forestry studies [7]. Marichal et al. [11] proposed an automatic method for tree ring delineation in wood cross-section images called CS-TRD. This method is based on classical edge detection techniques and was applied to two coniferous tree species, *Pinus taeda* and *Abies alba.* Furthermore, they introduced a model for wood cross-section images named *spider web*, see Fig. 2c. This model enforces constraints to ensure that tree rings do not intersect and remain roughly concentric.

In this work, the edge detection module from the CS-TRD method is replaced with a U-Net [18] model, a deep learning architecture designed explicitly for semantic segmentation, as illustrated in Fig. 2b. U-Net architecture has demonstrated remarkable results in tree ring delineation of cores [2] and in microscopy images of shrubs [4].

By replacing the edge detection module in the CS-TRD method with this deep learning-based model, we successfully extended its application to a diverse range of species and growth forms. Notably, this enhanced approach works with angiosperm trees (i.e. *Gleditsia triacanthos*) and shrubs prepared using microscopy (i.e. *Salix glauca*), samples illustrated in Fig. 3. We refer to this new method as DeepCS-TRD.

In addition, we introduce a new image dataset of the *Pinus taeda* species (UruDendro2 [9]), containing 53 samples, and another dataset comprising nine samples of *Gleditsia triacanthos* (UruDendro3a [8]), all with their corresponding

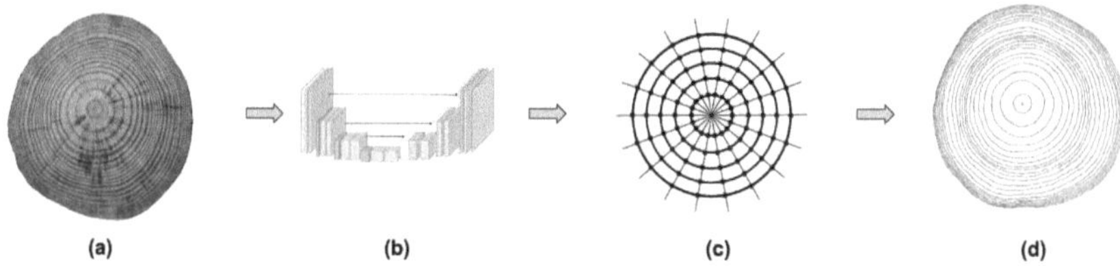

Fig. 2. DeepCS-TRD Tree-Ring detection pipeline. (a) Input disc image. (b) Deep Contour Detector. (c) *Spider web* model. (d) Tree-ring continuous curves.

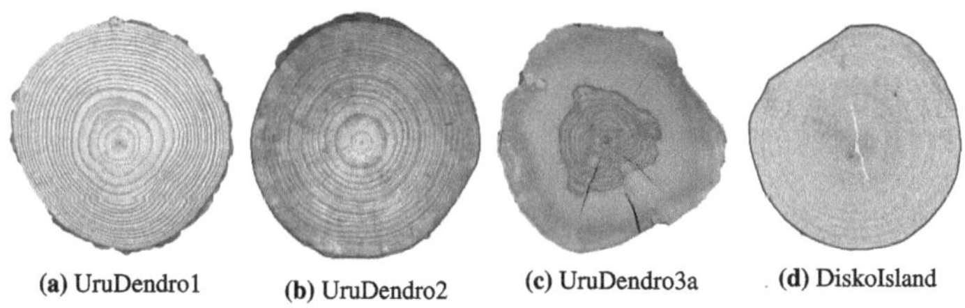

Fig. 3. Examples of the used datasets. Species are (a-b) *Pinus taeda*, (c) *Gleditsia triacanthos*, (d) *Salix glauca*. Acquisition conditions: (a-c) in a laboratory with a smartphone camera, sanded and polished. (d) microsection stained with 1 % safranin and 0.5 % astrablue, permanently fixed to a microscope slide with Eukitt, and scanned at 100 x magnification.

expert ring delineations. Generating these types of datasets is a highly time-consuming task that requires the expertise of dendrochronology specialists and is critical for algorithm training.

2 Previous Work

Though not yet widely adopted in the tree-ring community, deep learning approaches have become more prevalent in recent years and have naturally been applied to this problem, framing ring detection tasks as a segmentation problem. Two architectures are commonly used: U-Net and Mask R-CNN [17]. Gillert et al. [4] proposed a method for cross-section tree-ring detection called Iterative Next Boundary Detection Network (INBD), based on the U-NET architecture. This method was applied to high-resolution microscopy images of shrub cross-sections, detecting the annual rings individually at each iteration step, from the medulla to the bark. INBD was trained and tested on shrub microscopy images, using standard cross-entropy loss as the main loss for the ring detection network.

Besides the INBD method, most deep-learning-based approaches have been applied to core images. Polek et al. [14] used a Mask R-CNN deep learning architecture to detect rings in cores of coniferous species. During network training, they used mean average precision (mAP) as a loss function. Fabijańska et al. proposed a classic image-processing approach [3] (based on the linking of maximum image gradient pixels) and a convolutional neural network based on a U-Net architecture [2] for detecting tree rings in core images. When comparing both methods, they reported a significant improvement in precision and recall for the deep learning approach over the classic one. The U-Net network was trained to minimize the categorical cross-entropy loss function.

The scarcity of annotated tree ring image datasets remains a significant challenge, as the methods must be tailored to the specific characteristics of each species. Gillert et al. [4] made a dataset available of 213 high-resolution annotated images of cross-sections from three shrub species (*Dryas octopetala*, *Empetrum hermaphroditum*, and *Vaccinium myrtillus*), that were prepared using similar methods to those applied to the *Salix glauca* included in this study (see Fig. 3d). Furthermore, the CS-TRD publication included a publicly available dataset of 64 annotated images of *Pinus taeda* (see Fig. 3a). Kennel et al. [5] included a dataset of 7 cross-section images of *Abies alba*, although the ring annotations are not fully accessible. Regarding cores, Polek et al. [14] made available a dataset of 2601 image patches (similar to the red square in Fig. 1a) with ring annotations for *Picea abies*, a coniferous tree species.

Among the available methods, only INBD and CS-TRD are adapted for processing full cross-section images and provide accessible code, allowing for comparison with our proposed approach.

3 Approach

The proposed approach aims to extract a set of continuous, connected pixel curves from an RGB image of a wood cross-section, a disc. These curves serve as input to the *spider web* model presented in [11], enabling the reconstruction of digital tree-ring curves for the sample. An overview of the whole Tree-Ring detection pipeline is in Fig. 2.

The CS-TRD method [11] works as follows: (i) Several rays are traced departing from the center (the pith) outwards. (ii) Canny edge filtering is applied to the image. (iii) Filter out edges not belonging to the latewood to earlywood transitions, which correspond to the annual rings, using the angle between the edge normals and the rays. (iv) All edge chains that belong to the same ring are grouped by imposing the *spider web* structure through an iterative procedure that connects chains that do not cross each other, have strong edge information, and comply with a smoothness condition. The authors claim an F-Score of 89% in *Pinus taeda* and 97% in *Abies alba* species.

3.1 Algorithm

The DeepCS-TRD method maintains the principal steps of the CS-TRD but with a substantial modification: replacing the Canny-based edge detection step with a module based on the U-Net architecture [18] trained to segment ring boundaries (Fig. 2b). Algorithm 1 shows the pseudocode of this step. The method's inputs are an RGB image of a disc (Im_{in}) and a parameter to indicate the image is

Algorithm 1: Deep Contour Detector

Input: Im_{in}: RGB disc image;
$tile\_size$: Tile size (0 if no patching is applied);
$total\_rotations$: Number of rotations applied to the image;
cy, cx: x and y coordinates of the pith position;
Output: l_ch_s: list of continuous curve edges;
l_nodes_s: list of nodes of the curves;
1 $\Omega \leftarrow$ arange$(0, 360, 360/total_rotations)$; // Define rotation angles
2 $P \leftarrow$ Zero matrix of the same size as Im_{in}; // Initialize the probability map
`// Process each rotation angle`
3 **for** $\theta \in \Omega$ **do**
4     $I_{rotated} \leftarrow$ rotateImage$(Im_{in}, cy, cx, \theta)$; // Rotate the image around the pith position
5     $P_{rotated} \leftarrow$ ringSegmentationModel$(I_{rotated}, tile_size)$; // Apply the segmentation model. See Figure 4.
6     $P_{aux} \leftarrow$ rotateImage$(P_{rotated}, cy, cx, -\theta)$; // Undo the rotation of the probability map
7     $P \leftarrow P + P_{aux}$; // Accumulate the rotated probability map
8 $P \leftarrow P/total\_rotations$; // Average the probability map
9 $M \leftarrow P \geq 0.2$; // Threshold the accumulated probability map
10 $S \leftarrow$ skeletonize(M); // Skeletonize the binary mask
11 $m\_ch_e \leftarrow$ findContoursCurves(S); // Extract ring curves from the skeleton
12 $Nx, Ny \leftarrow$ getNormalComponent(m_ch_e); // Compute normal directions for each edge
13 $m\_ch_f \leftarrow$ filterEdges(m_ch_e, Nx, Ny); // Filter edge curves. See equation 1
14 $l\_ch_s, l\_nodes_s \leftarrow$ samplingCurves(m_ch_f); // Convert curves to the *spider web* format
15 **return** l_ch_s, l_nodes_s; // Return the results

split into tiles of ($tile\_size$) size, following the overlap-tile strategy described in [18]. Additionally, the method allows to apply a Test-Time Augmentations (TTA) [6] strategy, which consists of $total\_rotation$ rotations around the pith location (c_y, c_x). The method produces a list of continuous ring edge curves (l_ch_s) and a list of pixels nodes (l_nodes_s) which are the input to the *spyder web* model. In lines 1 and 2, the method initializes the TTA rotation domain angles Ω and the probability map P. Then, between lines 3 and 7, the inference loop iterates between the elements $\theta \in \Omega$. In line 4, Im_{in} is rotated an angle θ and stored in $I_{rotated}$. In line 5, the ring boundaries are generated with the *ring segmentation model* as illustrated in Fig. 4. First, if needed, the image is split into tiles with an overlapping of 10% (to avoid the tile border effects). The tiles are zero-padded when necessary to ensure compatibility with the stride size ($tile\_size$). Then, a probability map is computed for each tile with a U-Net network. Finally, the probability map for the whole disc ($P_{rotated}$) is built (with the original image size). The average between the tile's probability map is assigned to the overlapping area.

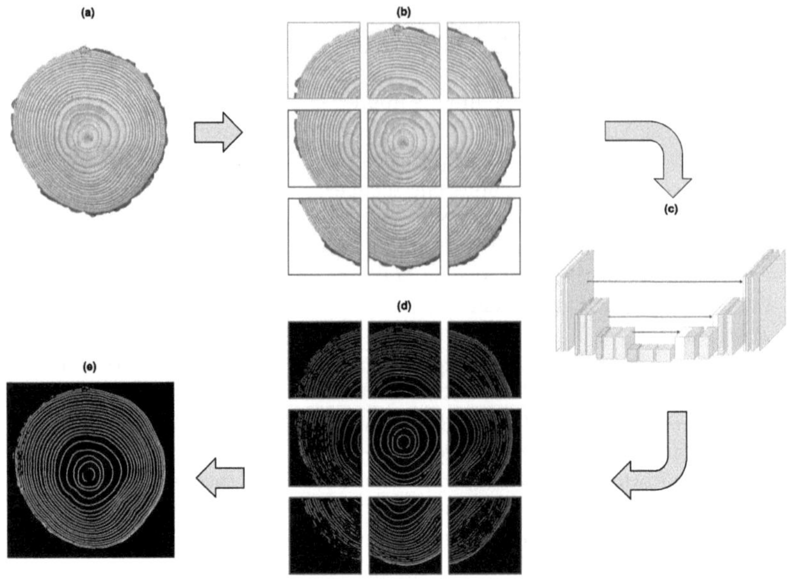

Fig. 4. Ring Segmentation Model Inference. The input to the model is a complete cross-section disc (a). (b) The disc is divided into square tiles. (c) and (d) Each tile is processed individually using the U-Net network to generate its corresponding probability map. Finally, the tile predictions are combined to reconstruct the probability map for the entire sample (e). This procedure corresponds to line 5 of Algorithm 1.

Once the probability map of the whole image ($P_{rotated}$) has been obtained, the rotation of angle θ needs to be inverted (line 6, angle -θ). Then, in line 7, the current probability map (P_{aux}) is accumulated into P. In line 8, the average probability map is obtained. In line 9, the probability map P is binarized with a 0.2 threshold to generate the mask M. This low threshold is deliberately chosen, considering an additional filtering step in line 13 that further refines the results. In line 10, this mask is skeletonized to improve the tree-ring curve precision. Finally, curves are extracted from the skeleton S using the Suzuki method [19] implemented in the function *findCountour* of the OpenCV library [1]. Figure 5 illustrates the lines 8 to 11 steps. Figure 5b shows the probability map P for a disc region. Brighter pixels indicate a higher probability of belonging to a ring. Figure 5c shows the skeleton S overlaid on the mask M, with colors representing curve labels. The skeleton corresponds to the center of the mask. Finally, Fig. 5d presents the extracted curves overlaid on the original image.

The normal component of each pixel belonging to a curve in m_ch_e is computed using a straightforward procedure (line 12). The pixels of each curve are ordered sequentially from one end to the other. The tangent direction of pixel p_0 is $\boldsymbol{T_{p_0}} = p_1 - p_{-1}$, where p_{-1} and p_1 are the preceding and succeeding pixels along the curve, respectively. The normal direction $\boldsymbol{N_{p_0}}$ is then computed as a vector perpendicular to $\boldsymbol{T_{p_0}}$, given by $\boldsymbol{N_{p_0}} = (-T_{p_0,y}, T_{p_0,x})$, where

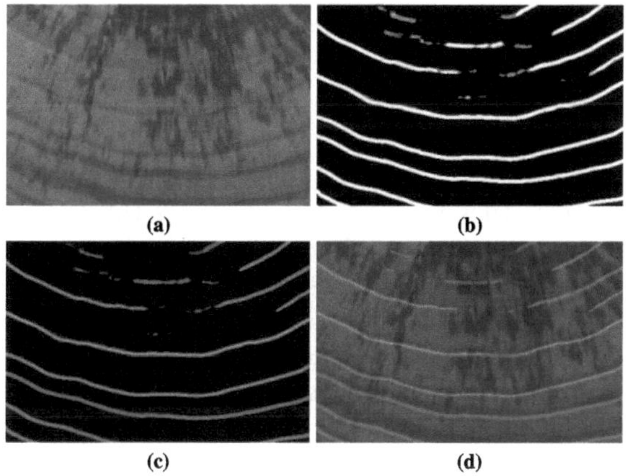

Fig. 5. Stages for extracting the ring edge curves. (a) Zoomed-in view of the RGB image. (b) The probability map of the image section is shown in (a). (c) Probability map with overlaid edge curves. (d) Extracted edge curves overlaid on the image.

$\boldsymbol{T_{p_0}} = (T_{p_0,x}, T_{p_0,y})$. This ensures that $\boldsymbol{N_{p_0}}$ is orthogonal to the curve at p_0. The point c is the pith location (c_x, c_y). In line 13, the angle δ between $\boldsymbol{cp_0}$ and $\boldsymbol{N_{p_0}}$ is computed as $\delta\left(\boldsymbol{cp_0}, \boldsymbol{N_{p_0}}\right) = \arccos\left(\frac{\boldsymbol{cp_0} \cdot \boldsymbol{N_{p_0}}}{\|\boldsymbol{cp_0}\| \|\boldsymbol{N_{p_0}}\|}\right)$.

We filter out all *pixels* p_0 with normals not collinear and outbound oriented concerning the ray's direction at that point:

$$\alpha \leq \delta(\boldsymbol{cp_0}, \boldsymbol{N_{p_0}}) \leq 180 - \alpha \quad (1)$$

We set $\alpha = 45$ degrees. When a pixel p_0 is removed, the curve to which p_0 belongs is split at that position. This step is straightforward because m_ch_e is a matrix of dimensions $N \times 2$. Each curve is separated from the others by a row with values $(-1, -1)$. Therefore, if a pixel does not satisfy the condition in Eq. 1, we insert the vector $(-1, -1)$ at its position.

Finally, in line 14, the curves in m_ch_f are converted into the format required by the *spider web* model, structured around a central *pith*, the origin of the *rays*. Each *curve* is represented as a set of points, with *nodes* in the intersections of the curve with the *rays*. The resulting sampled *curve* is called a *chain*. These *chains* have the notable property of being non-intersecting, inspired by the biological characteristics of tree rings observed in conifer species. More details can be found in [11].

Preprocessing. The background was removed from all images using the U^2-Net network [16] and manually corrected when necessary. Furthermore, portions of the image background were cropped to focus on the cross-section so that the minimum distance between the disc and the image edges is 50 pixels. Images from all datasets were resized to 1504×1504 pixels using Lanczos interpolation. To complete the image size, if the smallest dimension of the resized image does not reach the new dimension, a 255-value padding was applied, maintaining the aspect ratio.

3.2 Network Training

Non-overlapping square tiles were used for training the U-Net network. Only the patches with ring presence of each image sample were used. The tile size is a hyperparameter of the training step (see Sect. 5.2) where 60% of the samples were used for training, 20% for validation and 20% for assessing the full tree-ring detection pipeline (Fig. 2). Training and validation subsets were split into patches to train and select the network's hyperparameters. We also trained the network to predict the ring boundaries on the complete disc sample.

The network was trained over 100 epochs on an NVIDIA P100 GPU with 12 GB of VRAM and 40 GB of system RAM. The Adam optimizer was used with an initial learning rate of 1e−3, while a cosine annealing schedule was applied to adjust the learning rate dynamically during training. The Dice Loss function was employed to optimize the model's performance. The Resnet18 backbone, pre-trained on ImageNet, was also used to enhance feature extraction. The network's weights that result in the lowest validation loss during training are selected as the optimal model.

Regarding data augmentation, in 50% of the dataset images, randomly selected, we generate three augmented images: one with a random rotation around the pith, another with occlusions simulating cracks or fungi, and a third with elastic deformations.

4 Datasets

The datasets used in this study encompass a variety of species as shown in Fig. 3. Some species, like the conifer *Pinus taeda*, have clear ring contrasts, while *Gleditsia triacanthos* and *Salix glauca* are more challenging. Due to the differences in growth forms, the sample preparation methods also varied. While the cross sections from the trees in datasets UruDendro 1, 2, and 3, were sanded and polished as typical for tree samples, the cross-sections in the DiskoIsland dataset were prepared using standard protocols for shrubs. Specifically, 12–15 μm thick cross-sections were cut, stained with 1% safranin and 0.5% astrablue, and permanently fixed to microscope slides using Eukitt (BiOptica, Milan). The image acquisition methods also varied. The images from the UroDendro 1, 2, and 3 datasets were captured with smartphone cameras (iPhone 6 s and Huawei P20 Pro), while the images of the microsections from the DiskoIsland dataset were captured at 100x magnification using a Zeiss slide scanner (Axio Scan Z1, Zeiss, Germany (2.26 pixel/µm)).

Ring boundaries in the selected acquired images were annotated using the Labelme Tool [20], which allows users to mark polylines over the image and export the annotations in JSON format. Ring annotations were performed by users with different levels of expertise, from undergraduate students to expert professors, and in all cases, they were reviewed by at least one other expert. Finally, a binary image mask was generated from the ring boundary annotations, with a thickness of 3 pixels. Table 1 depicts the datasets, the number of samples and rings in each one, the species, and the acquisition method.

Table 1. Dataset description.

Collection	Samples	Tree-rings	Species	Acquisition
UruDendro1 [10]	64	1221	*Pinus taeda*	Smartphone
UruDendro2 [9]	53	1151	*Pinus taeda*	Smartphone
UruDendro3a [8]	9	216	*Gleditsia triacanthos*	Smartphone
DiskoIsland [15]	50	654	*Salix glauca*	Scanner

5 Results and Discussion

5.1 Metrics

We adopt the same evaluation metrics used by the authors of the INBD method: the mean Average Recall (mAR) and the Adapted Rand error (ARAND). False positives are preferred to false negatives because an incorrect ring can be deleted with just one click, while tracing a new one is more time-consuming. In this sense, the mAR metric is particularly suitable for this application.

5.2 Experiments

The DeepCS-TRD method involved two key parameters: *tile_size* and *total_rotations* (see Algorithm 1). We tested tile sizes of 64, 128, 256, 512, and 1504 (full resolution, no tiles) pixels for all the datasets. Figure 1a illustrates a tile size of 256px. For the *total_rotations* parameter, combinations of 0, 3, and 5 rotations were evaluated. The selection of this parameter was made by optimizing the validation subset. After experiments, parameter values were fixed in $tile\_size = 256$ and $total\_rotations = 5$ for the four datasets.

We noted that the thickness of the annotated ring boundary masks during training impacts the method's performance. The best value was 3-pixel in all datasets. In consequence, DeepCS-TRD and INBD methods are trained using ring boundary thicknesses of three pixels. The two networks in the INBD method were trained following the author's procedure during 100 epochs, setting the downsampling parameter to 1, which gave the best results.

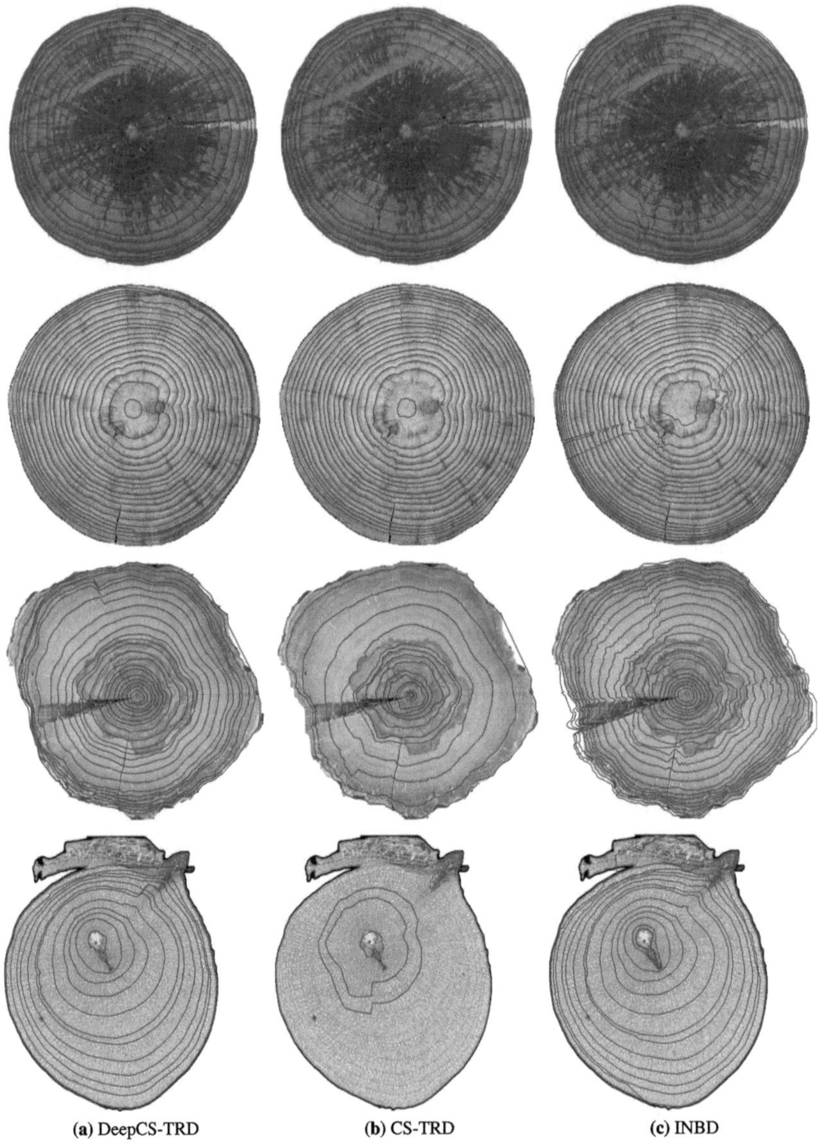

Fig. 6. Tree-Ring Delineation Results. Each column displays the ring boundaries produced by each method, shown in blue. Each row corresponds to a different disc sample. Column (a) DeepCS-TRD; Column (b) CS-TRD; Column (c) INBD. Note the presence of knots, cracks, fungus, and the differences between species. Note that the darker region in the third row does not correspond to the tree rings of that species.

Table 2. Results on the test set for the four used datasets and the available SOTA methods: INBD [4] and CS-TRD [11] as well as the proposed DeepCS-TRD. The best performance is in boldface. Uru1 stands for UruDendro1, and the same is done for UruDendro2 and UruDendro3a datasets. Disko stands for DiskIsland dataset.

	mAR ↑				ARAND ↓			
Method	Uru1	Uru2	Uru3a	Disko	Uru1	Uru2	Uru3a	Disko
CS-TRD	.787	.710	.007	.026	.093	.144	.466	.634
INBD	.846	.742	.200	**.735**	.081	.132	.494	**.099**
DeepCS-TRD	**.884**	**.809**	**.620**	.628	**.053**	**.105**	**.207**	.107

5.3 Results

This section evaluates the performance of INBD, CS-TRD, and DeepCS-TRD on the test set across all datasets. The INBD method was modified to accept the pith boundary ground truth as input to ensure a fair comparison, as both DeepCS-TRD and CS-TRD take the pith locations as input but modeled as a pixel [12].

Table 2 presents each method's mAR and ARAND values across all datasets. The proposed DeepCS-TRD method has the best results on all the UruDendro datasets. It achieves outstanding results with near-perfect ring detection from the pith to the bark in UruDendro1 and UruDendro2, as illustrated in Fig. 6. Despite the INBD method giving very good performance in most of the samples, in some situations, it generates notorious propagation errors in the presence of significant fungal growth, cracks, or knots because it begins delineating ring boundaries from the pith center (see rows 1 and 2). In row 1's sample, DeepCS-TRD detects some false rings near the center, but its remaining detections are highly accurate. In contrast, CS-TRD gives good results, but rings are inaccurate with this strong fungus presence. In the sample in row 2, both DeepCS-TRD and CS-TRD performed well.

In the UruDendro3a dataset, DeepCS-TRD performs well; in the sample shown in Fig. 6, row 3, it correctly delineates 15 out of 19 rings. In contrast, the INBD method detects only six rings. CS-TRD detects five rings but interestingly identifies the three innermost rings missed by the other methods (see the region around the pith in row 3 for all methods).

In the DiskoIsland dataset, DeepCS-TRD performs slightly worse than the INBD method on average. In the sample in row 4, the INBD method correctly detects all the rings, while DeepCS-TRD missed the second ring. CS-TRD performs poorly in this dataset.

There are differences regarding the training time for the DeepCS-TRD and INBD methods. Both methods' maximum required training time was in the UruDendro1 dataset, with 25 and 4 h for the INBD and DeepCS-TRD, respectively, using the same HW.

6 Conclusions

A new method for automatic Tree-Ring delineation has been proposed. It adapts the CS-TRD method by replacing the edge detector step with a U-Net deep convolutional network. This modification allows us to apply the *spider web* model to other species, such as *Gleditsia triachantos* or *Salix glauca*, with different, and often more difficult, annual ring patterns than for the coniferous species. Additionally, the proposed method also increases the performance of CS-TRD for *Pinus taeda* as can be seen in Table 2; in this case, the accuracy (mAR) has improved from 0.787 to 0.884 in UruDendro1 and from 0.710 to 0.809 in UruDendro2.

By applying data augmentation techniques and the overlap-tile strategy, a high ring detection rate has been achieved in UruDendro3a despite using a limited number of images. Additionally, satisfactory results have been obtained on the *Salix glauca* dataset. One advantage of this method over INBD is that INBD requires two training steps, whereas DeepCS-TRD requires one. Moreover, DeepCS-TRD significantly reduces training time, requiring just 4 h compared to 25 h for INBD on the UruDendro1 dataset.

Acknowledgments. The experiments presented in this paper used ClusterUY [13] (site: https://cluster.uy). This work was supported by project ANIIFMV176061. CCP was supported by IRFD (702700133B) and H2020 CHARTER (869471).

References

1. Bradski, G.: The OpenCV library. Dr. Dobb's J. Software Tools (2000)
2. Fabijańska, A., Danek, M.: DeepDendro - a tree rings detector based on a deep convolutional neural network. Comput. Electron. Agric. **150**, 353–363 (2018)
3. Fabijańska, A., Danek, M., Barniak, J., Piórkowski, A.: Towards automatic tree rings detection in images of scanned wood samples. Comput. Electron. Agric. **140**, 279–289 (2017)
4. Gillert, A., Resente, G., Anadon-Rosell, A., Wilmking, M., Von Lukas, U.F.: Iterative next boundary detection for instance segmentation of tree rings in microscopy images of shrub cross sections. In: IEEE/CVF Conference on Computer Vision and Pattern Recognition (CVPR), pp. 14540–14548 (2023)
5. Kennel, P., Borianne, P., Subsol, G.: An automated method for tree-ring delineation based on active contours guided by DT-CWT complex coefficients in photographic images: Application to abies alba wood slice images. Comput. Electron. Agric. **118**, 204–214 (2015)
6. Kimura, M.: Understanding test-time augmentation. In: Mantoro, T., Lee, M., Ayu, M.A., Wong, K.W., Hidayanto, A.N. (eds.) ICONIP 2021. LNCS, vol. 13108, pp. 558–569. Springer, Cham (2021). https://doi.org/10.1007/978-3-030-92185-9_46
7. Labrecque-Foy, J.-P., Angers-Blondin, S., Ropars, P., Simard, M., Boudreau, S.: The Use of basal area increment to preserve the multi-decadal climatic signal in shrub growth ring chronologies: a case study of Betula glandulosa in a rapidly warming environment. Atmosphere **14** (2013)

8. Marichal, H., et al.: UruDendro3a, a public dataset of 9 cross-section images and manual annual ring delineations of Gleditsia triachantos L.., 2025. V1. [Dataset]. https://doi.org/10.5281/zenodo.15652667
9. Marichal, H., Passarella, D., Lucas, C., Casaravilla, V., Profumo, L., Randall, G.: UruDendro2, a public dataset of 53 cross-section images and manual annual ring delineations of Pinus taeda L.. V1. [Dataset] (2025). https://doi.org/10.5281/zenodo.15652451
10. Marichal, H., et al.: Urudendro, a public dataset of 64 cross-section images and manual annual ring delineations of Pinus taeda L. Ann. Forest Sci. **82**, 25 (2025)
11. Marichal, H., Passarella, D., Randall, G.: CS-TRD: a Cross Sections Tree Ring Detection method. Preprint on On Line Image Processing (IPOL Journal): https://arxiv.org/abs/2305.10809 (2023)
12. Marichal, H., Passarella, D., Randall, G.: Automatic wood pith detector: local orientation estimation and robust accumulation. In: International Conference on Pattern Recognition (ICPR), pp. 1–15 (2025)
13. Nesmachnow, S., Iturriaga, S.: Cluster-UY: collaborative scientific high performance computing in Uruguay. In: Torres, M., Klapp, J. (eds.) ISUM 2019. CCIS, vol. 1151, pp. 188–202. Springer, Cham (2019). https://doi.org/10.1007/978-3-030-38043-4_16
14. PolÀček, M., Arizpe, A., Hüther, P., Weidlich, L., Steindl, S., Swarts, K.: Automation of tree-ring detection and measurements using deep learning. Methods Ecol. Evol. **14**, 2233–2242 (2023)
15. Power, C.C., Marichal, H., Treier, U.A., Resente, G., Normand, S., Randall, G.: Dataset of 50 cross-section images and manual annual ring delineations of Salix Glauca from western Greenland, 2025. V1. [Dataset] (2025). https://doi.org/10.5281/zenodo.15311335
16. Qin, X., Zhang, Z., Huang, C., Dehghan, M., Zaiane, O.R., Jagersand, M.: U2-net: going deeper with nested U-structure for salient object detection. Pattern Recogn. **106**, 107404 (2020)
17. Rettenberger, L., Münke, F.R., Bruch, R., Reischl, M.: Mask R-CNN outperforms U-net in instance segmentation for overlapping cells. Current Direct. Biomed. Eng. **9**, 335–338 (2023)
18. Ronneberger, O., Fischer, P., Brox, T.: U-net: convolutional networks for biomedical image segmentation. In: Navab, N., Hornegger, J., Wells, W.M., Frangi, A.F. (eds.) MICCAI 2015. LNCS, vol. 9351, pp. 234–241. Springer, Cham (2015). https://doi.org/10.1007/978-3-319-24574-4_28
19. Suzuki, S., Be, K.: Topological structural analysis of digitized binary images by border following. Comput. Vis. Graphics Image Process. **30**, 32–46 (1985)
20. Wada, K.: Labelme: image polygonal annotation with python. Zenodo (2024)

KairosAD: A SAM-Based Model for Industrial Anomaly Detection on Embedded Devices

Uzair Khan(✉), Franco Fummi, and Luigi Capogrosso

Department of Engineering for Innovation Medicine, University of Verona, Verona, Italy
{uzair.khan,franco.fummi,luigi.capogrosso}@univr.it

Abstract. In the era of intelligent manufacturing, anomaly detection has become essential for maintaining quality control on modern production lines. However, while many existing models show promising performance, they are often too large, computationally demanding, and impractical to deploy on resource-constrained embedded devices that can be easily installed on the production lines of Small and Medium Enterprises (SMEs). To bridge this gap, we present **KairosAD**, a novel supervised approach that uses the power of the Mobile Segment Anything Model (MobileSAM) for image-based anomaly detection. **KairosAD** has been evaluated on the two well-known industrial anomaly detection datasets, *i.e.*, MVTec-AD and ViSA. The results show that **KairosAD** requires 78% fewer parameters and boasts a 4× faster inference time compared to the leading state-of-the-art model, while maintaining comparable AUROC performance. We deployed **KairosAD** on two embedded devices, the NVIDIA Jetson NX, and the NVIDIA Jetson AGX. Finally, **KairosAD** was successfully installed and tested on the real production line of the Industrial Computer Engineering Laboratory (ICE Lab) at the University of Verona. The code is available at https://github.com/intelligolabs/KairosAD.

Keywords: Industrial Anomaly Detection · Efficient Deep Learning · Visual Foundation Models · Embedded Systems

1 Introduction

Image-based anomaly detection of industrial components plays a crucial role in enabling fully automated production and ensuring optimal quality control in manufacturing. In particular, image anomaly detection can be broadly classified into *semantic anomaly detection* and sensory anomaly detection [33]. Semantic anomaly detection emphasizes the overall semantics of the image, while sensory anomaly detection focuses on fine-grained details, such as minor scratches on objects. This article concentrates on semantic anomaly detection, a task commonly performed in Small and Medium Enterprises (SMEs), where it plays a crucial role in maintaining production efficiency and ensuring product quality [6,11].

E. Rodolà et al. (Eds.): ICIAP 2025, LNCS 16167, pp. 42–53, 2026.
https://doi.org/10.1007/978-3-032-10185-3_4

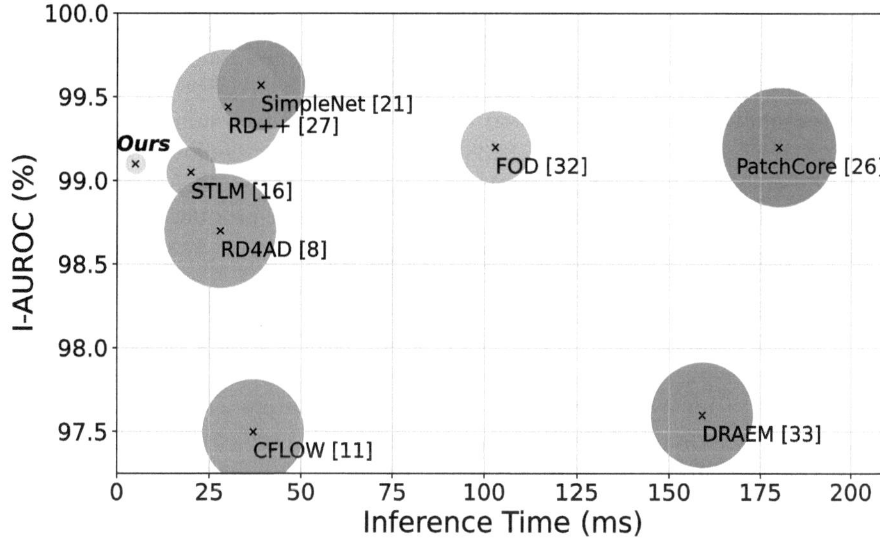

Fig. 1. Comparisons of different anomaly detection methods in terms of I-AUROC (vertical axis), inference time (horizontal axis), and the ratios of parameter numbers (circle radius), on the MVTec-AD dataset [2]. Our **KairosAD** achieves competitive I-AUROC results, while having 78% fewer parameters with respect to SimpleNet [24], *i.e.*, the model that achieves the highest AUROC. Instead, with respect to STLM [19], which represents the state-of-the-art of efficient deep learning method for anomaly detection, **KairosAD** has 35% of fewer parameters, achieves 4× speedup performance in terms of inference time, with an increase in I-AUROC performance as well.

Recently, several approaches, based on teacher-student networks [9], reconstructions [21], or memory banks [29], have been developed. Most of these anomaly detection methods are highly effective in classifying data into normal (fault-free) or anomalous categories. However, many of these architectures are large and computationally intensive. They demand significant processing power and memory, making them unsuitable for deployment on embedded devices [19].

However, SMEs, which often have limited infrastructure and budgets, often rely on low-cost edge computing devices for real-time quality control on their production lines [16]. This reliance on less powerful hardware creates a gap between the high performance of advanced anomaly detection models and their practical deployability in real-world manufacturing settings. As a result, many SMEs struggle to adopt state-of-the-art techniques due to constraints in computational resources and integration complexity.

To address this challenge, we introduce **KairosAD**, a novel supervised image-level anomaly detection model. **KairosAD** leverages the power of *Mobile Segment Anything Model (MobileSAM)* [37] to achieve accurate anomaly detection performance with a significantly reduced computational cost. Unlike existing methods that often require complex architectures and large parameter counts,

KairosAD is designed for simplicity and efficiency. In fact, as we can see from Fig. 1, **KairosAD** effectively balances precision, speed, and number of parameters, making it ideal for real-time deployment on embedded devices.

We evaluated **KairosAD** on two widely used industrial anomaly detection datasets: *MVTec-AD* [2] and *ViSA* [38]. Furthermore, we deployed **KairosAD** on several embedded devices, such as *NVIDIA Jetson NX* and *NVIDIA Jetson AGX*, and successfully installed and tested on the real production line of the Industrial Computer Engineering Laboratory (ICE Lab) at the University of Verona.

Finally, **KairosAD** demonstrates remarkable *eco-friendliness*, consuming only 0.0465 kWh of energy and producing only 0.0154 kg of CO_2 eq emissions in both the training and testing phases[1]. Even its power consumption remains optimized, with an average GPU usage of approximately 270W during training, highlighting the model's ability to achieve high performance while significantly reducing its environmental footprint compared to recent deep learning approaches.

2 Related Work

Over the years, anomaly detection in the industry domain has been extensively explored [22] (Sect. 2.1). Since **KairosAD** is based on MobileSAM and employs efficiency-driven methods to enable deployment on embedded systems, this section also delves into the literature on foundation models for vision (Sect. 2.2) and efficient deep learning techniques (Sect. 2.3).

2.1 Industrial Anomaly Detection

Although extensive research has been carried out on anomaly detection, industrial image data presents unique challenges [2].

Many industrial anomaly detection methods focus on image reconstruction and detect anomalies based on the reconstruction error [22]. Generative models such as autoencoders [3], variational autoencoders [23], and generative adversarial networks [1] are the architectures most widely used to reconstruct normal images from anomalous ones. Nonetheless, these methods face certain limitations, especially when reconstructing complex industrial textures and patterns.

Recent approaches leverage memory banks, where a core set of stored features from a pre-trained backbone is used to compute patch-level distances for anomaly detection [8]. PatchCore [29] is a foundational approach that extracts patch-level feature embeddings of normal images into a memory bank, detecting anomalous patches during inference through a patch query process. SoftPatch [14], based on PatchCore, introduced a patch-level filtering strategy in which patch features are filtered and weighted before being stored in the memory bank to reduce contamination from anomalous patches, thus enhancing model robustness. However,

[1] Results calculated by using: https://mlco2.github.io/codecarbon/.

most of these methods come at the cost of increased computational complexity and a large memory space, making them unsuitable for deployment on resource-constrained devices.

In terms of efficient industrial anomaly detection models, our main competitor is STLM [19]. Unlike STLM, which employs a two-branch architecture, we adopt just a single-branch design, reducing complexity and improving inference speed. Furthermore, while STLM focuses on both anomaly detection and localization, our method is specifically optimized for just the image-level task, prioritizing efficiency for real-time deployment. From a methodological perspective, STLM is based on feature distillation and contrastive learning, while **KairosAD** uses MobileSAM for a lightweight yet effective feature extraction, resulting in a simpler and more computationally efficient model.

2.2 Foundation Models for Vision

Multimodal foundation models have emerged as powerful tools across various tasks [20]. For visual anomaly detection, particularly relevant are multimodal approaches based on CLIP [27] and Segment Anything Model (SAM) [18], as well as vision-only models like DINO [7].

Specifically, CLIP learns visual concepts from natural language descriptions by training on image-text pairs using a contrastive learning objective that aligns embeddings from both modalities. This shared feature space enables downstream applications, such as zero-shot image classification, by comparing image embeddings to class-specific textual prompts.

Instead, DINO adopts a self-supervised student-teacher framework based on Vision Transformers (ViT), leveraging a multiview strategy to predict softened teacher output, resulting in robust and high-quality feature representations. DINOv2 [25] extends these ideas by incorporating patch-level reconstruction techniques and scaling to larger architectures and datasets.

Beyond these, SAM and its efficient variant, such as MobileSAM [37], have introduced a new paradigm in vision models, particularly for segmentation tasks. SAM is designed as a powerful promptable segmentation model that generalizes well across diverse images, enabling accurate object delineation with minimal supervision. MobileSAM builds upon SAM, optimizing it for edge devices by reducing computational demands while maintaining competitive performance. These models, though primarily designed for segmentation, have shown potential for various downstream tasks, including anomaly detection. As a result, we take advantage of their efficiency and adaptability to develop **KairosAD**, which is based on the MobileSAM image encoder.

2.3 Efficient Deep Learning

Over the past decades, a large amount of research has been invested in improving embedded technologies to enable real-time solutions for many complex applications. However, deploying learning models on tiny devices is substantially difficult due to severe architectural, energetic, and latency constraints [5].

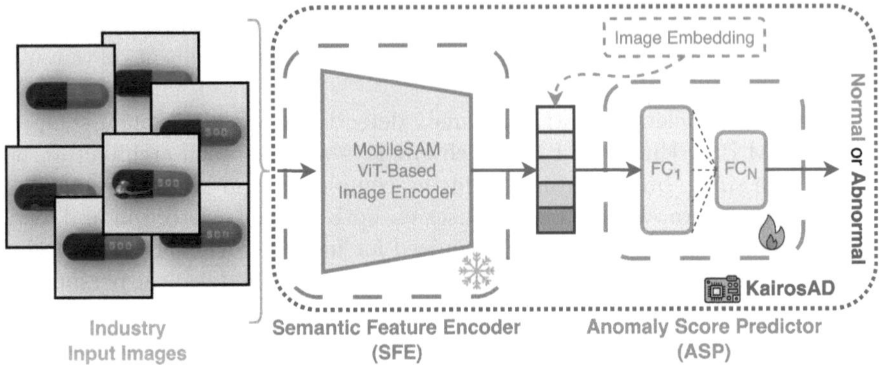

Fig. 2. The **KairosAD** architecture. Specifically, **KairosAD** is composed of a Semantic Feature Encoder (SFE) and an Anomaly Score Prediction (ASP). This simple but effective model allows us to perform efficient image-level anomaly detection on embedded devices.

Several techniques aim to reduce the size and computational cost of the model without sacrificing performance. These include pruning [31], quantization [10], and knowledge distillation [12]. Furthermore, in order to provide lightweight models capable of delivering acceptable performances for their intended applications, specialized techniques have been proposed in model architecture exploration, model simplification, and architectural modifications, such as Neural Architecture Search (NAS) [28], and the attention mechanism [4].

Although commonly associated with notable contributions to machine translation tasks, the attention mechanism has been adapted for a wide range of applications, allowing the model to selectively look for different features, facilitating the extraction of relevant information from complex and high-dimensional data [5]. One of the most significant advances that utilize this principle is the Transformer architecture [32].

Recent efforts to develop computationally efficient Transformers in vision have broadened their potential use in resource-constrained settings [15]. MobileSAM represents one of the latest advances, and for this reason, we decided to use the MobileSAM image encoder in **KairosAD**.

3 Methodology

In this section, we introduce **KairosAD**, a model designed for image-level industrial anomaly detection on embedded devices. The overall architecture is shown in Fig. 2.

3.1 The Semantic Feature Encoder (SFE) Sub-Network

The SAM model is based on a Vision Transformer (ViT-H) with 632 million parameters to generate high-quality image embeddings. However, this high computational cost makes the SAM encoder unsuitable for deployment on resource-constrained devices.

To address this limitation, MobileSAM [37] was introduced as a lightweight alternative, incorporating a more efficient image encoder with only 5.78 million parameters while maintaining comparable performance. The transition from the high-parameter ViT-H encoder to the significantly more compact MobileSAM encoder is achieved through a well-structured knowledge distillation process. In particular, instead of jointly optimizing both the encoder and the decoder, the authors separate the training of the image encoder from that of the mask decoder. This approach allows the lightweight encoder to inherit essential feature representations from the original SAM encoder. The optimization process also incorporates progressive compression techniques, where training is performed on a subset of the SA-1B dataset [18], using only a small fraction of the original data. This approach significantly reduces computational requirements while preserving the effectiveness of the model. Finally, instead of relying solely on transformer layers, the new lightweight encoder integrates convolutional blocks with inverted residuals in its early stages. This hybrid architecture enables efficient local feature extraction while maintaining the global context modeling capabilities of the Transformers. **KairosAD** incorporates all these optimizations and utilizes the MobileSAM image encoder as it is.

Formally, the **KairosAD** encoder can be represented as a function f_θ, parameterized by θ, that takes an input image $I \in \mathcal{R}^{H \times W \times C}$, and maps it to an embedding $E \in R^d$. This process can be formulated as:

$$E = f_\theta(I) \ , \tag{1}$$

where H, W, and C, denote the height, width, and number of channels of the input image, respectively, and $d = 3$ is the dimensionality of the output embedding.

3.2 The Anomaly Score Prediction (ASP) Sub-Network

To perform anomaly detection, the image embedding E is processed through a series of N_l fully connected linear layers. Let g_ϕ represent this transformation, parameterized by ϕ, mapping the embedding $E \in \mathbb{R}^d$ to an anomaly score $s \in [0, 1]$. This can be expressed as:

$$s - g_\phi(E) \ , \tag{2}$$

where g_ϕ consists of multiple linear layers followed by the Rectified Linear Unit (ReLU) activation function.

The model is trained using a Weighted Binary Cross-Entropy (WBCE) loss function, defined as:

$$\mathcal{L}_{\text{WBCE}} = -\frac{1}{N}\sum_{i=1}^{N} w_i \left[y_i \log \sigma(s_i) + (1 - y_i) \log(1 - \sigma(s_i))\right] \ , \tag{3}$$

where N is the number of samples, $y_i \in 0, 1$ represents the ground truth label (0 for normal, 1 for anomalous), and s_i is the predicted anomaly score for the i-th sample. The Sigmoid activation function is denoted by σ. The weight w_i is applied to the positive class and is calculated as the ratio of negative to positive samples, to ensure that the model adequately compensates for the class imbalance. Parameters ϕ are optimized to minimize this loss, ensuring that normal samples are assigned low scores while anomalous samples receive high scores.

4 Experiments

4.1 Experimental Details

Datasets. **KairosAD** has been evaluated on the two well-known industrial anomaly detection datasets, *i.e.*, MVTec-AD [2], and ViSA [38]. The MVTec AD contains 5,354 images distributed in 15 object categories, while the ViSA is comprised of 10,821 images distributed in 12 object categories. The anomalies present in both datasets are of various types, shapes, and scales.

Evaluation Metrics. In accordance with previous work [13,35,36], the anomaly detection performance is measured via the *Image-level Area Under the Receiver-Operator Curve (I-AUROC).* Furthermore, according to [19], we also report the results in terms of *inference time* and *number of parameters.*

Implementation Details. All the code is implemented in PyTorch [26]. On the MVTec-AD dataset, we train our models for 35 epochs, with a learning rate of 1×10^{-2}, using Adam [17] as an optimizer on an NVIDIA RTX 4090, using $N_l = 2$ fully connected layers to process image embedding E. Instead, on the ViSA dataset, we train our model for 50 epochs and use $N_l = 3$ fully connected layers; the other hyperparameters remain the same as in the MVTec-AD dataset.

4.2 Anomaly Detection on the MVTec-AD Dataset

The results for image-level anomaly detection on the MVTec-AD dataset are shown in Table 1. **KairosAD** significantly outperforms several recent approaches while maintaining a low computational footprint.

Specifically, compared to STLM [19] (our main competitor), which achieves an accuracy of 99.05%, **KairosAD** improves the I-AUROC performance while reducing the number of parameters from 16.56 M to 11.53 M, and the inference time from 20 ms to 5 ms, setting the new state-of-the-art for these two metrics. Furthermore, **KairosAD** also achieves competitive I-AUROC compared to SimpleNet [24] (the architecture with the highest I-AUROC), while having 78% fewer parameters and a speedup of more than 7×. RD++ [30] achieves the second highest I-AUROC (99.44%) but requires 154.87 M parameters, while **KairosAD** achieves a remarkably close accuracy with a 6× reduction in the parameter count. Similarly, PatchCore [29], achieves 99.20% of I-AUROC but requires 186.55 M parameters. With respect to this, **KairosAD** achieves a slightly slower average I-AUROC (99.10%) but is fully justified by significantly fewer parameters (11.53 M vs. 186.55 M) and a lower latency (5 ms vs. 180 ms).

Table 1. Anomaly detection in term of I-AUROC (%) on the MVTec-AD dataset [2]. The inference time is measured in ms, while the parameter count is expressed in millions. In gold, the best results. In 1Silver is the second best. Competitors' results are from [19]

Class	**DRAEM** [36] ICCV 2021	**CFLOW** [13] WACV 2022	**PatchCore** [29] CVPR 2022	**RD4AD** [9] CVPR 2022	**KairosAD** (ours)
Carpet	96.90	97.60	99.10	98.70	100
Grid	99.90	98.10	97.30	100	99.87
Leather	100	99.90	100	100	100
Tile	100	97.10	99.30	99.70	100
Wood	99.50	98.70	99.60	99.50	96.49
Bottle	98.00	99.90	100	100	98.82
Cable	90.90	97.60	99.90	96.10	100
Capsule	90.90	97.00	98.00	96.10	99.47
Hazelnut	100	100	100	100	99.17
Metal nut	100	98.50	99.90	100	92.43
Pill	97.10	96.20	97.50	98.70	100
Screw	98.70	93.10	98.20	97.80	99.63
Toothbrush	100	98.80	100	100	100
Transistor	100	92.90	99.90	95.50	99.74
Zipper	100	97.10	99.50	97.90	100
Average	97.60	97.50	99.20	98.70	99.10
Latency	159	37	180	28	**5**
Parameters	97	94.70	186.55	150.64	**11.53**
Class	**SimpleNet** [24] CVPR 2023	**RD++** [30] CVPR 2023	**FOD** [35] ICCV 2023	**STLM** [19] ACM MCCA 2025	**KairosAD** (ours)
Carpet	99.70	100	100	99.48	100
Grid	99.70	100	100	95.57	99.87
Leather	100	100	100	100	100
Tile	99.80	99.70	100	100	100
Wood	100	99.30	99.10	100	96.49
Bottle	100	100	100	100	98.82
Cable	99.90	99.20	99.50	98.98	100
Capsule	97.70	99.00	100	98.69	99.47
Hazelnut	100	100	100	100	100
Metal nut	100	100	100	100	92.43
Pill	99.00	98.40	98.40	98.20	100
Screw	99.00	98.90	96.70	99.47	99.63
Toothbrush	99.70	100	94.40	99.06	100
Transistor	100	98.50	100	97.58	99.74
Zipper	99.90	98.60	99.70	99.74	100
Average	**99.57**	99.44	99.20	99.05	99.10
Latency	39	30	103	20	**5**
Parameters	52.88	154.87	28.83	16.56	**11.53**

4.3 Anomaly Detection on the ViSA Dataset

Table 2. Anomaly detection in term of I-AUROC (%) on the ViSA dataset [38]. In gold, the best results. Silver the second best. Competitors' results are from [34].

Class	**DRAEM** [36] ICCV 2021	**PatchCore** [29] CVPR 2022	**RD4AD** [9] CVPR 2022	**SimpleNet** [24] CVPR 2023	**KairosAD** (ours)
Candle	89.6	98.7	96.2	96.9	98.1
Capsules	89.2	68.8	91.8	89.5	69.7
Cashew	88.3	97.7	98.7	94.8	90.7
Chewinggum	96.4	99.1	99.3	100	88.3
Fryum	94.7	91.6	96.9	96.6	99.1
Macaroni1	93.9	90.1	98.7	97.6	95.5
Macaroni2	88.3	63.4	91.4	83.4	89.7
Pcb1	84.7	96.0	96.7	99.2	96.5
Pcb2	96.2	95.1	97.2	99.2	87.2
Pcb3	97.4	93.0	96.5	98.6	95.9
Pcb4	98.9	99.5	99.4	98.9	99.1
Pipe fryum	94.7	99.0	99.6	99.2	97.8
Average	92.4	91.0	**96.9**	96.2	92.3
Latency	159	37	180	39	5
Parameters	97	94.70	186.55	52.88	**10.55**

The results for image-level anomaly detection on the ViSA dataset are shown in Table 2. Due to space limitations, not all competitors listed in Table 1 are included; we have selected the top performers while maintaining the time frame mentioned above. STLM [19] is not reported as it only reports overall ViSA results (93.83%), without providing class-specific performance.

Even in this case, **KairosAD** achieves the best inference speed and lowest parameter count among all methods compared. Although **KairosAD** ranks fourth in overall I-AUROC performance (92.3%), it outperforms the top two models in several individual classes. However, its performance is notably affected by the "Capsules" class, which remains a key limitation and will be the focus of future research.

4.4 Deployment of KairosAD on the ICE Lab at the University of Verona

We deployed **KairosAD** on various embedded devices, including NVIDIA Jetson NX, and NVIDIA Jetson AGX, achieving inference times of 211 ms and 218 ms, respectively. These devices were also the ones we installed and tested on the real production line of ICE Lab at the University of Verona.

In Fig. 3, we present the qualitative results of **KairosAD** deployed on the production line, together with the results of other datasets, further validating the robustness and versatility of the model in different industrial settings.

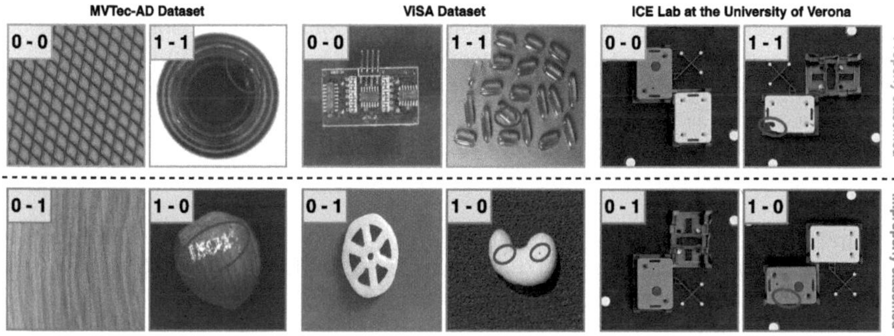

Fig. 3. Qualitative results of **KairosAD** on the MVTec-AD, ViSA, and the real production line of the ICE Lab at the University of Verona. For each image, we report the [True Label] - [Prediction] on the top left (where 0 stands for normal and 1 stands for abnormal). The anomalies are circled in blue. (Color figure online)

5 Conclusions

In this article, we present **KairosAD**, an efficient image-based anomaly detection model optimized for real-time deployment on embedded devices. Using MobileSAM, **KairosAD** achieves a balance of high accuracy and low computational cost, outperforming the state-of-the-art methods in inference speed and parameter efficiency. We demonstrated its effectiveness on multiple industrial datasets and successfully deployed it on different embedded devices on a real production line. This work provides a practical solution for anomaly detection in resource-constrained environments, with potential for broader applications in industrial settings.

Acknowledgments.. This study was carried out within the "SUPREME: Sistemi Umano centrici per Prodotti e pRocEssi Manufatturieri Evoluti" Funded by PR Veneto FESR 2021-2027, Obiettivo Specifico 1.1, Azione 1.1.1 "Rafforzare la ricerca e l'innovazione (in collaborazione) tra imprese e organismi di ricerca" Sub A "Rafforzare la ricerca e l'innovazione tra imprese e organismi di ricerca".

References

1. Akcay, S., Atapour-Abarghouei, A., Breckon, T.P.: GANomaly: semi-supervised anomaly detection via adversarial training. In: Jawahar, C.V., Li, H., Mori, G., Schindler, K. (eds.) ACCV 2018. LNCS, vol. 11363, pp. 622–637. Springer, Cham (2019). https://doi.org/10.1007/978-3-030-20893-6_39
2. Bergmann, P., Fauser, M., Sattlegger, D., Steger, C.: MVTec AD – a comprehensive real-world dataset for unsupervised anomaly detection. In: Conference on Computer Vision and Pattern Recognition (CVPR) (2019)
3. Bergmann, P., Löwe, S., Fauser, M., Sattlegger, D., Steger, C.: Improving unsupervised defect segmentation by applying structural similarity to autoencoders.

In: Proceedings of the 14th International Joint Conference on Computer Vision, Imaging and Computer Graphics Theory and Applications (2019)
4. Brauwers, G., Frasincar, F.: A general survey on attention mechanisms in deep learning. IEEE Trans. Knowl. Data Eng. **35**(4), 3279–3298 (2023)
5. Capogrosso, L., Cunico, F., Cheng, D.S., Fummi, F., Cristani, M.: A machine learning-oriented survey on tiny machine learning. IEEE Access **12**, 23406–23426 (2024)
6. Capogrosso, L., et al.: Diffusion-based image generation for in-distribution data augmentation in surface defect detection. In: 19th International Conference on Computer Vision Theory and Applications (VISAPP) (2024)
7. Caron, M., Touvron, H., Misra, I., Jégou, H., Mairal, J., Bojanowski, P., Joulin, A.: Emerging properties in self-supervised vision transformers. In: International Conference on Computer Vision (ICCV) (2021)
8. Damm, S., Laszkiewicz, M., Lederer, J., Fischer, A.: AnomalyDINO: boosting patch-based few-shot anomaly detection with DINOv2. In: Winter Conference on Applications of Computer Vision (WACV) (2024)
9. Deng, H., Li, X.: Anomaly detection via reverse distillation from one-class embedding. In: Conference on Computer Vision and Pattern Recognition (CVPR) (2022)
10. Gholami, A., Kim, S., Dong, Z., Yao, Z., Mahoney, M.W., Keutzer, K.: A Survey of Quantization Methods for Efficient Neural Network Inference, pp. 291–326. Chapman and Hall/CRC (2022)
11. Girella, F., Liu, Z., Fummi, F., Setti, F., Cristani, M., Capogrosso, L.: Leveraging latent diffusion models for training-free in-distribution data augmentation for surface defect detection. In: International Conference on Content-Based Multimedia Indexing (CBMI) (2024)
12. Gou, J., Yu, B., Maybank, S.J., Tao, D.: Knowledge distillation: a survey. Int. J. Comput. Vision **129**(6), 1789–1819 (2021). https://doi.org/10.1007/s11263-021-01453-z
13. Gudovskiy, D., Ishizaka, S., Kozuka, K.: CFLOW-AD: real-time unsupervised anomaly detection with localization via conditional normalizing flows. In: Winter Conference on Applications of Computer Vision (WACV) (2022)
14. Jiang, X., et al.: SoftPatch: unsupervised anomaly detection with noisy data. In: Advances in Neural Information Processing Systems (NeurIPS) (2022)
15. Khan, S., Naseer, M., Hayat, M., Zamir, S.W., Khan, F.S., Shah, M.: Transformers in vision: a survey. ACM Comput. Surv. **54**(10), 1–41 (2022)
16. Khan, U., Cheng, D., Setti, F., Fummi, F., Cristani, M., Capogrosso, L.: A comprehensive survey on deep learning-based predictive maintenance. ACM Trans. Embedded Comput. Syst. (2025)
17. Kingma, D.P., Ba, J.: Adam: a method for stochastic optimization. In: International Conference for Learning Representations (ICLR) (2014)
18. Kirillov, A., et al.: Segment Anything. In: International Conference on Computer Vision (ICCV) (2023)
19. Li, C., Qi, L., Geng, X.: A SAM-guided two-stream lightweight model for anomaly detection. ACM Trans. Multimed. Comput. Commun. Appl. **21**(2), 1–23 (2025)
20. Li, C., Gan, Z., Yang, Z., Yang, J., Li, L., Wang, L., Gao, J.: Multimodal foundation models: from specialists to general-purpose assistants. Found. Trends Comput. Graph. Vision **16**(1-2), 1–214 (2024)
21. Liang, Y., Zhang, J., Zhao, S., Wu, R., Liu, Y., Pan, S.: Omni-frequency channel-selection representations for unsupervised anomaly detection. IEEE Trans. Image Process. **32**, 4327–4340 (2023)

22. Liu, J., Xie, G., Wang, J., Li, S., Wang, C., Zheng, F., Jin, Y.: Deep industrial image anomaly detection: a survey. Mach. Intell. Res. **21**(1), 104–135 (2024)
23. Liu, W., Li, R., Zheng, M., Karanam, S., Wu, Z., Bhanu, B., Radke, R.J., Camps, O.: Towards visually explaining variational autoencoders. In: Conference on Computer Vision and Pattern Recognition (CVPR) (2020)
24. Liu, Z., Zhou, Y., Xu, Y., Wang, Z.: SimpleNet: a simple network for image anomaly detection and localization. In: Conference on Computer Vision and Pattern Recognition (CVPR) (2023)
25. Oquab, M., et al.: DINOv2: learning robust visual features without supervision. Trans. Mach. Learn. Res. (2024)
26. Paszke, A., et al.: Pytorch: an imperative style, high-performance deep learning library. In: Advances in Neural Information Processing Systems (NeurIPS) (2019)
27. Radford, A., et al.: Learning transferable visual models from natural language supervision. In: 38th International Conference on Machine Learning (ICML) (2021)
28. Ren, P., Xiao, Y., Chang, X., Huang, P.y., Li, Z., Chen, X., Wang, X.: A comprehensive survey of neural architecture search: challenges and solutions. ACM Comput. Surv. **54**(4), 1–34 (2021)
29. Roth, K., Pemula, L., Zepeda, J., Scholkopf, B., Brox, T., Gehler, P.: Towards total recall in industrial anomaly detection. In: Conference on Computer Vision and Pattern Recognition (CVPR) (2022)
30. Tien, T.D., Nguyen, A.T., Tran, N.H., Huy, T.D., Duong, S.T., Nguyen, C.D.T., Truong, S.Q.H.: Revisiting reverse distillation for anomaly detection. In: Conference on Computer Vision and Pattern Recognition (CVPR) (2023)
31. Vadera, S., Ameen, S.: Methods for pruning deep neural networks. IEEE Access **10**, 63280–63300 (2022)
32. Vaswani, A., et al.: Attention is all you need. In: Advances in Neural Information Processing Systems (NeurIPS) (2017)
33. Yang, J., Zhou, K., Li, Y., Liu, Z.: Generalized out-of-distribution detection: a survey. Int. J. Comput. Vision **132**(12), 5635–5662 (2024)
34. Yao, H., Liu, M., Yin, Z., Yan, Z., Hong, X., Zuo, W.: GLAD: towards better reconstruction with global and local adaptive diffusion models for unsupervised anomaly detection. In: European Conference on Computer Vision (ECCV) (2024)
35. Yao, X., Li, R., Qian, Z., Luo, Y., Zhang, C.: Focus the discrepancy: intra- and inter-correlation learning for image anomaly detection. In: International Conference on Computer Vision (ICCV) (2023)
36. Zavrtanik, V., Kristan, M., Skocaj, D.: DRAEM - a discriminatively trained reconstruction embedding for surface anomaly detection. In: International Conference on Computer Vision (ICCV) (2021)
37. Zhang, C., Han, D., Qiao, Y., Kim, J.U., Bae, S.H., Lee, S., Hong, C.S.: Faster Segment Anything: Towards Lightweight SAM for Mobile Applications. arXiv preprint arXiv:2306.14289 (2023)
38. Zou, Y., Jeong, J., Pemula, L., Zhang, D., Dabeer, O.: SPot-the-difference self-supervised pre-training for anomaly detection and segmentation. In: European Conference on Computer Vision (ECCV) (2022)

SegDT: A Diffusion Transformer-Based Segmentation Model for Medical Imaging

Salah Eddine Bekhouche[1(✉)], Gaby Maroun[1], Fadi Dornaika[1,2], and Abdenour Hadid[3]

[1] University of the Basque Country UPV/EHU, San Sebastian, Spain
{sbekhouche001,gmaroun001}@ikasle.ehu.eus, fadi.dornaika@ehu.eus
[2] IKERBASQUE, Basque Foundation for Science, Bilbao, Spain
[3] Sorbonne University Abu Dhabi, Abu Dhabi, UAE
abdenour.hadid@sorbonne.ae

Abstract. Medical image segmentation is crucial for many healthcare tasks, including disease diagnosis and treatment planning. One key area is the segmentation of skin lesions, which is vital for diagnosing skin cancer and monitoring patients. In this context, this paper introduces SegDT, a new segmentation model based on diffusion transformer (DiT). SegDT is designed to work on low-cost hardware and incorporates Rectified Flow, which improves the generation quality at reduced inference steps and maintains the flexibility of standard diffusion models. Our method is evaluated on three benchmarking datasets and compared against several existing works, achieving state-of-the-art results while maintaining fast inference speeds. This makes the proposed model appealing for real-world medical applications. This work advances the performance and capabilities of deep learning models in medical image analysis, enabling faster, more accurate diagnostic tools for healthcare professionals. The code is made publicly available at https://github.com/Bekhouche/SegDT

Keywords: Medical image segmentation · Diffusion Transformer · Skin lesion segmentation

1 Introduction

Skin cancer is a major global health concern, and its early detection is crucial for improving survival rates. Medical image segmentation plays a vital role in this process by enabling precise lesion boundary delineation. Among the successful approaches to automatic image segmentation is deep learning which has revolutionized the field, with Convolutional Neural Networks (CNNs) like U-Net [18] and DeepLabV3+ [3] demonstrating interesting performance. However,

The supports of TotalEnergies and Sorbonne University Abu Dhabi are fully acknowledged (Abdenour Hadid).

E. Rodolà et al. (Eds.): ICIAP 2025, LNCS 16167, pp. 54–66, 2026.
https://doi.org/10.1007/978-3-032-10185-3_5

CNNs can struggle with long-range dependencies, limiting their ability to segment complex or irregularly shaped lesions. As an alternative, Transformers [20], inspired by their success in Natural Language Processing (NLP) and vision, offer an appealing solution by capturing global context through self-attention, with models like TransUNet [2] and Swin-UNet [1] showing improved accuracy (although their high computational cost can hinder real-world applications). Diffusion models, on the other hand, have recently emerged as a powerful technique, achieving state-of-the-art results in various medical imaging tasks [21] by iteratively refining segmentation through a denoising process. While highly accurate, their computational cost and long inference times pose a challenge for real-world deployment.

To address some of these limitations, we propose SegDT, an extra small Diffusion Transformer (DiT) model designed for efficient segmentation on low-cost GPUs. SegDT integrates rectified flow to accelerate inference while maintaining high segmentation accuracy.

The main contributions of this work are: **(i)** A compact DiT architecture optimized for resource-constrained GPUs is proposed; **(ii)** A methodology for the incorporation of rectified flow for efficient inference with reduced sampling steps is described; and **(iii)** Extensive experiments are conducted on three benchmarking datasets, achieving state-of-the-art performances, compared to exiting methods.

2 Related Work

This section discusses relevant prior work in medical image segmentation, focusing on CNNs, Transformers, hybrid architectures, and diffusion models, with an emphasis on approaches related to skin lesion segmentation and efficient architectures.

2.1 CNN-Based Segmentation

CNNs have been the dominant approach in medical image segmentation for many years. U-Net [18], with its encoder-decoder structure and skip connections, revolutionized the field by enabling effective learning with limited data. Its success stems from the ability to capture both local and global context, crucial for accurate segmentation. Many variants and extensions of U-Net have been proposed, such as DeepLabV3+ [3], which incorporates convolutions to capture multi-scale information, and ResUNet++ [11], which leverages residual connections for improved training stability. However, CNNs, due to their limited receptive field, often struggle with capturing long-range dependencies, which can be critical for segmenting complex or irregularly shaped lesions, like those encountered in skin cancer imaging. While these methods can achieve impressive results, their limitations motivated researchers to introduce Transformers based methods.

2.2 Transformer-Based Segmentation

Transformers, initially developed for NLP [20], have recently shown remarkable success in computer vision [7]. Their key strength lies in the self-attention mechanism, which allows them to capture long-range dependencies and global context effectively. In medical image segmentation, TransUNet [2] was a pioneering work, combining the strengths of transformers and U-Nets. It utilizes a transformer encoder to extract global features and a U-Net decoder to preserve local details. While TransUNet demonstrated the potential of transformers, its computational complexity can be a limiting factor. Swin-UNet [1] addressed this issue by introducing a hierarchical Swin Transformer architecture with shifted windows, enabling efficient computation of self-attention. This approach balances local and global feature extraction, achieving competitive performance with reduced computational cost. However, transformer-based models can still be resource-intensive, motivating the development of more resource- efficient architectures.

2.3 Hybrid Architectures: CNN-Transformers

Recognizing the complementary strengths of CNNs and transformers, researchers have explored hybrid architectures. DS-TransUNet [14] incorporates deformable self-attention into the TransUNet framework, allowing the model to focus on relevant regions in the image. BRAU-Net++ [13] combines convolutional feature extractors with transformer-based global reasoning, achieving state-of-the-art results on several benchmarks. These hybrid approaches aim to leverage the local feature extraction capabilities of CNNs and the global context modeling of transformers. While effective, these methods often come with increased complexity and resource requirements. MobileUNETR [17] takes a different approach by focusing on efficiency. It proposes a lightweight transformer-based model optimized for mobile and edge devices. While sacrificing some accuracy compared to larger models, it achieves faster inference, making it suitable for resource-constrained environments.

2.4 Diffusion Models for Segmentation

Diffusion models have recently emerged as a powerful tool for image generation and segmentation. They operate by progressively adding noise to an image until it becomes pure noise (forward diffusion) and then learning to reverse this process to generate the original image (reverse diffusion). In medical image segmentation, MedSegDiff [21] integrates transformers within a diffusion framework, enabling the model to capture fine-grained anatomical details. MedSegDiff-V2 [22] further improves upon this by incorporating multi-resolution features. While these diffusion-based models have demonstrated impressive performance, their computational cost and long inference times can be a limiting factor for real-time deployment.

Our work addresses this limitation by introducing SegDT, a Diffusion Transformer (DiT) model that incorporates rectified flow to accelerate inference, making it more suitable for real-world applications. Unlike previous diffusion models that often rely on U-Net architectures, we explore a more efficient transformer-based design. Furthermore, we consider a rectified flow approach to achieve high-quality segmentations with fewer sampling steps, significantly reducing the inference time. This is a key difference between our work and previous diffusion-based segmentation methods.

3 Proposed Method

This section details SegDT, our proposed architecture for segmenting skin lesions in medical images. Figure 1 illustrates the SegDT architecture, which comprises a Variational Autoencoder (VAE) encoder, a DiT, and a VAE decoder. The VAE components are based on the pretrained Tiny AutoEncoder for Stable Diffusion (TAESD)[1], chosen for its compact size and computational efficiency. TAESD's small footprint and fast encoding/decoding speeds contribute to SegDT's overall efficiency by generating a compact latent representation, compressing the width and height 8 times. This smaller latent space allows for a more efficient DiT, reducing the computational burden and further accelerating the segmentation process. The DiT component is based on the DiT-XS (extra-small) variant, which employs a patch size of 2, as detailed in [16]. DiTs have demonstrated strong performance in image generation, making them a suitable foundation for our segmentation task. SegDT is designed to address the challenges of precise lesion boundary delineation in noisy medical images - a critical requirement for accurate dermatological diagnosis and treatment planning. Furthermore, the model's compact architecture and the integration of rectified flow during inference enable rapid and efficient analysis by reducing the required number of inference steps.

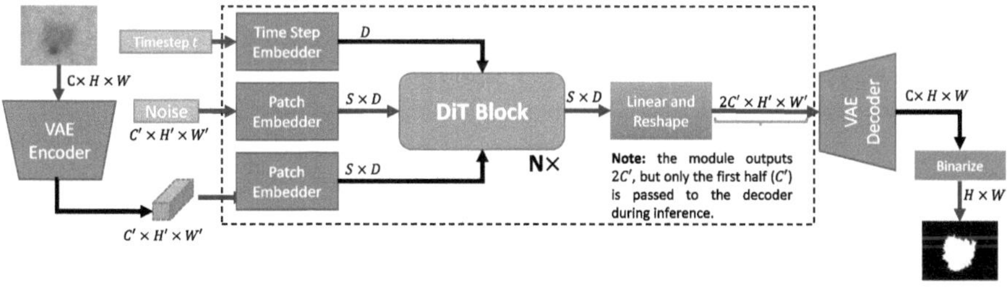

Fig. 1. Overview of the SegDT inference architecture for medical image segmentation.

[1] https://github.com/madebyollin/taesd.

3.1 Rectified Flow for Efficient Inference

In the forward diffusion process, a segmentation label $x_0 \in \mathcal{X}$ (represented by its encoded latent representation $z_0 \in \mathcal{Z}$) is gradually perturbed by adding Gaussian noise over T timesteps. This results in a sequence of noisy segmentations $x_1, x_2, \ldots, x_T \in \mathcal{X}$ (with corresponding noisy latent representations $z_1, z_2, \ldots, z_T \in \mathcal{Z}$). The reverse diffusion process aims to reconstruct the original segmentation mask x_0 (or z_0) from the noisy x_T (or z_T), conditioned on a latent representation $y \in \mathcal{Y}$ of the corresponding medical image. Inspired by efficient diffusion sampling techniques like those used in Denoising Diffusion Implicit Models (DDIMs) [19], we employ a rectified flow approach to learn a modified, more efficient reverse process.

Instead of directly predicting the noise added at each timestep, our model learns a velocity field $v_\theta(z, t, y) : \mathcal{Z} \times [0, T] \times \mathcal{Y} \rightarrow \mathcal{Z}$ in the latent space $\mathcal{Z}$. This velocity field represents the direction and magnitude of the change required to move from z_t to z_{t-1}, conditioned on z_t, the time step $t \in [0, T]$, and the latent representation of the image y. The reverse diffusion process then follows this learned velocity field. We approximate the reverse trajectory using a numerical integration method, such as Euler's method.

$$z_{t-1} = z_t + v_\theta(z_t, t, y)\Delta t, \tag{1}$$

where Δt is the size of the integration step. In our normalized formulation, we assume that the continuous time interval is $[0, 1]$. When this interval is divided into discrete timesteps of T, each step corresponds to an increment of $\Delta t = \frac{1}{T}$.

The velocity field v_θ is learned by minimizing a loss function. Given a noisy latent representation z_t (obtained by adding noise to the encoded ground truth mask z_0), the model predicts the velocity $v_\theta(z_t, t, y)$. This prediction is compared to a target velocity derived from the known noise added to z_0 to obtain z_t. A common metric for this comparison is the Mean Squared Error (MSE). This training process enables the model to accurately predict the velocity needed to reverse the diffusion process.

Learning a velocity field, rather than directly predicting noise, can be seen as a form of rectification, promoting a smoother and more predictable reverse diffusion process. It also provides increased flexibility for incorporating conditioning information, such as the anatomical context provided by y. Although the concept of learning a velocity field for diffusion has been explored (e.g., [15]), the specific term "rectified flow" and its precise definition are still evolving within the diffusion modeling community. Our use of this term emphasizes the goal of a more direct, efficient path from noisy latent to segmentation.

3.2 Transformer-Based Diffusion Model (SegDT)

This section describes the architecture of SegDT, a transformer-based diffusion model for image segmentation. SegDT employs a pre-trained VAE to map images into a latent space. This latent representation is then processed by a DiT. SegDT employs 12 DiT blocks.

During training, the input to the DiT consists of latent representations of ground truth segmentation masks. In contrast, during inference, the model generates segmentations by processing randomly sampled, noisy latent vectors initialized with a fixed seed. Critically, the VAE decoder component is not utilized during training, as the loss function is computed directly within the latent space. This approach significantly reduces training time.

The SegDT architecture comprises the following components:

VAE Encoder. A pre-trained VAE encoder is employed to map input images of size $C \times H \times W$ into a lower-dimensional latent space. During training, this encoder also processes corresponding ground-truth masks in the same manner. The resulting latent representation has dimensions $C' \times H' \times W'$, where $C' = 4$, and the spatial dimensions are reduced by a factor of 8, such that $H' = H/8$ and $W' = W/8$. This latent representation serves as the input to the DiT model.

Patch Embedder. A patch embedding module, utilizing a shared architecture for both input and conditional tensors, is applied. For the input, it processes the latent representation from the VAE encoder. During inference, when ground-truth masks are not available, it processes a randomized tensor for the conditional input. The input tensor is flattened and then linearly projected to a $S \times D$ dimensional embedding space, where $D = 192$ is the embedding dimension derived from the DiT's token length. The number of patches, S, is determined by $S = \frac{H' \times W'}{P^2}$, where $P = 2$ is the patch size used in the DiT model, and $S = 256$ represents the total number of patches.

Time Step Embedder. A time step embedding module is used to generate a D-dimensional vector (where $D = 192$, consistent with the patch embedding dimension) that encodes the diffusion time step t. This embedding vector is then used to condition the DiT blocks, providing information about the current stage of the diffusion process.

DiT Blocks. A series of DiT blocks form the core of SegDT. Each block receives the embedded latent representation $\mathtt{z}$, the timestep embedding $\mathtt{t}$, and the condition embedding $\mathtt{y}$ as input. These blocks progressively refine the latent representation $\mathtt{z}$. The detailed architecture of each DiT block is shown in Fig. 2. Each DiT block comprises:

- **Adaptive Layer Normalization (adaLN):** Normalizes activations and then modulates them using the parameters α_1, α_2 (for scaling), β_1, β_2 (for shifting), and γ_1, γ_2 (for scaling) to shift and scale the normalized activations.
- **Self-Attention:** Models global spatial relationships within the image embedding.
- **Cross-Attention:** Integrates contextual guidance from the condition embedding y.

- **Feed-Forward Networks (FFN):** Enhances feature representations. The FFN hidden layer size is determined by `mlp_ratio` × `hidden_size`. GELU with an approximate tanh calculation is used as the activation function.
- **DropPath Regularization:** Applies DropPath regularization to prevent overfitting.

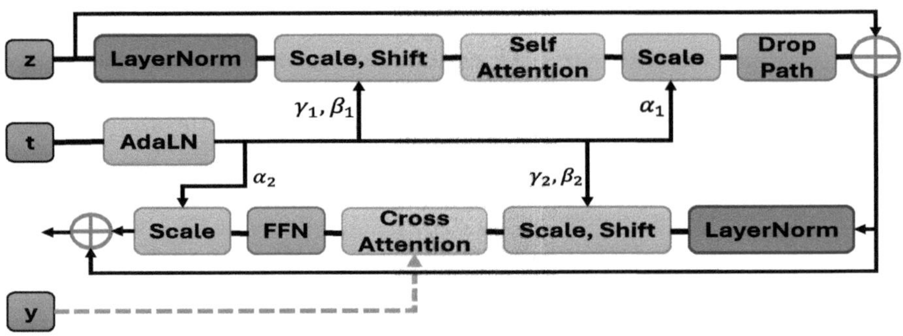

Fig. 2. Detailed architecture of a single DiT block.

Linear and Reshape. After the DiT blocks, a linear layer is applied to the $S \times D$ output, transforming it to a tensor of shape $S \times 8C'$. This linear transformation projects each patch embedding from dimension D to a new dimension of $8C'$ (equivalent to $2C' \times P^2$). Subsequently, this tensor is reshaped to $2C' \times H' \times W'$, aligning with the VAE's latent spatial dimensions ($H' \times W'$) but with twice the channel depth ($2C'$ instead of C'). Within this $2C' \times H' \times W'$ tensor, the first C' channels represent the predicted noise, and the subsequent C' channels represent the predicted variance, primarily used during training.

VAE Decoder. At inference time, after this denoising process, the model's output $2C'$ (only the first C' channels) is fed into a pre-trained VAE decoder. This decoder then reconstructs an image from this representation, producing a 3-channel image of size $C \times H \times W$. Finally, this reconstructed image is converted into the final segmentation mask through binarization.

4 Experimental Analysis

4.1 Datasets

The ISIC 2016 [10], 2017 [6], and 2018 [5] challenges are widely recognized benchmarks in medical image segmentation, specifically for skin lesion analysis. These challenges provided large, labeled datasets for detecting and segmenting melanoma from dermoscopic images. The ISIC 2016 dataset included 900 training images and 335 testing images, each with corresponding segmentation masks, allowing evaluation of lesion detection and boundary accuracy. The ISIC 2017 dataset expanded on this with 2000 training images, 150 validation images, and 600 testing images, all with segmentation masks. The ISIC 2018 dataset further increased in size, containing 2594 training images, 100 validation images, and 1000 testing images, ensuring a more robust assessment of segmentation models.

4.2 Metrics

The metrics widely used for evaluating skin lesion segmentation on these datasets include: Dice Similarity Coefficient (Dice), Intersection over Union (IoU), Pixel Accuracy (ACC), Sensitivity (SE), and Specificity (SP). These metrics assess the overlap and classification accuracy between predicted segmentation masks (SM) and ground truth masks (GM).

$$\text{Dice} = \frac{2|SM \cap GM|}{|SM| + |GM|} \tag{2}$$

$$\text{IoU} = \frac{|SM \cap GM|}{|SM \cup GM|} \tag{3}$$

$$\text{ACC} = \frac{TP + TN}{TP + FP + TN + FN} \tag{4}$$

$$\text{SE} = \frac{TP}{TP + FN} = \frac{|SM \cap GM|}{|GM|} \tag{5}$$

$$\text{SP} = \frac{TN}{TN + FP} \tag{6}$$

where $|SM \cap GM|$ is the number of shared pixels, $|SM|$ and $|GM|$ are the total pixels in each mask, and TP, TN, FP, and FN are true positives, true negatives, false positives, and false negatives, respectively.

4.3 Implementation Details

SegDT was trained with Adam Optimizer with a learning rate of 1e-4 and a batch size of 32 for 100 epochs. The learning rate was reduced by a factor of 10 after 50 epochs. During training, no explicit data enhancement techniques were used. The experiments were carried out on two NVIDIA RTX 3090 GPUs. The input images were bilinearly resized to achieve the target dimensions of the 256×256 pixels and then normalized to the [0, 1] range. In the model, these dimensions correspond to $C = 3$, $H = 256$, and $W = 256$. The final segmentation mask was obtained by binarizing the reconstructed image using a fixed threshold of 0.2. This threshold was selected based on performance on a held-out validation set. We explored thresholds ranging from 0.1 to 0.5 in increments of 0.05 and selected 0.2 because it yielded the best overall performance in terms of the Dice score on the validation set. The `mlp_ratio` and `hidden_size` parameters for the DiT blocks were set to 4 and 192, respectively.

4.4 Results and Discussion

To validate SegDT's accuracy and generalization, we conducted extensive experiments on the ISIC 2016, 2017, and 2018 skin lesion image datasets. We compared SegDT's performance with several state-of-the-art methods and analyzed the performance of different methods under different scenarios and challenging settings.

Segmentation Performance. Table 1 presents the results of the segmentation in the three ISIC datasets. SegDT achieves state-of-the-art or highly competitive performance across all datasets and most metrics. In ISIC 2016, SegDT achieves the highest Dice score (94. 76%), IoU (91. 40%) and accuracy (97. 08%). Although DU-Net + achieves slightly higher sensitivity, SegDT demonstrates significantly higher specificity (99. 44%), indicating its ability to accurately identify healthy tissue, which is crucial in clinical settings.

In ISIC 2017, SegDT again achieved competitive results, with the highest Dice score (91. 70%) and accuracy (95. 49%). Although DU-Net+ shows slightly better IoU and Sensitivity, SegDT maintains significantly higher Specificity (98.74%).

In the largest dataset, ISIC 2018, SegDT achieves the highest Dice score (94. 51%) and IoU (90. 43%), and competitive accuracy and sensitivity. In particular, SegDT achieves the highest Specificity (97.43%), further highlighting its strength in identifying healthy tissue.

The inclusion of GFLOPs and parameter counts in Table 1 highlights the efficiency of SegDT. Compared to DU-Net+, SegDT achieves significantly better performance with drastically fewer GFLOPs (3.68 vs. 54.00) and parameters (9.95M vs. 39.00M). This demonstrates SegDT's ability to achieve superior results with a much lighter computational footprint, making it more suitable for resource-constrained environments. Although MobileUNETR and SLP-Net have lower GFLOPs and parameter counts, SegDT offers a better balance between performance and efficiency. SegDT's performance is achieved with a relatively small increase in computational cost compared to the most efficient models, but with a significant improvement in segmentation accuracy, especially in terms of specificity. This trade-off is often desirable in medical image analysis, where accuracy is paramount.

Table 1. Segmentation results on the ISIC 2016, 2017, and 2018 datasets.

Method	Dataset	Dice↑	IoU↑	ACC↑	SE↑	SP↑	Flops (G)↓	Params (M)↓
MobileUNETR [17]	2016	92.80	87.47	96.59	93.03	96.87	**1.30**	**3.00**
GU-Net [4]	2016	89.75	82.41	-	-	-		
AM-Net [24]	2016	93.29	88.44	95.57	88.73	97.02	-	-
DU-Net+ [12]	2016	92.46	85.25	96.89	**95.75**	95.64	54.00	39.00
SegDT	2016	**94.76**	**91.40**	**97.08**	93.35	**99.44**	3.68	9.95
MobileUNETR [17]	2017	86.84	79.00	94.46	85.18	96.93	**1.30**	**3.00**
GU-Net [4]	2017	93.94	88.98	-	-	-	-	-
PCCTrans [8]	2017	84.65	-	93.28	-	-	38.50	50.80
AM-Net [24]	2017	87.30	80.02	95.13	81.70	97.99	-	-
DU-Net+ [12]	2017	90.46	**85.20**	95.54	**94.07**	93.85	54.00	39.00
SegDT	2017	**91.70**	84.70	**95.49**	87.39	**98.74**	3.68	9.95
SynergyNet-8s2h [9]	2018	89.31	80.68	94.91	87.28	97.37	-	-
PCCTrans [8]	2018	90.03	-	96.49	-	-	38.50	50.80
BRAU-Net++ [13]	2018	90.10	84.01	95.61	92.24	91.18	22.45	50.76
MobileUNETR [17]	2018	90.74	84.56	94.40	92.55	95.03	1.30	3.00
GU-Net [4]	2018	92.42	86.46	-	-	-	-	-
DU-Net+ [12]	2018	92.93	88.13	**97.41**	95.12	96.76	54.00	39.00
SLP-Net [23]	2018	88.21	80.61	93.87	89.30	95.36	**2.30**	**0.20**
SegDT	2018	**94.51**	**90.43**	96.81	**95.21**	**97.43**	3.68	9.95

SegDT offers a compelling advantage in inference speed compared to IDDPM. SegDT achieves segmentation quality comparable to IDDPM with only 15 inference steps, while IDDPM requires 35 steps. This substantial reduction in steps, facilitated by SegDT's use of rectified flow, reduces the inference time by almost 2 times, making SegDT significantly faster and crucial for practical applications.

Qualitative Analysis. Figure 3 shows example segmentation results in the ISIC datasets, including both well-segmented cases and challenging cases. The left examples demonstrate SegDT's ability to accurately delineate lesion boundaries even with variations in shape, size, and texture. The challenging cases on the right highlight some limitations. For instance, in some cases, SegDT struggles with lesions that have irregular borders or are very small. This could be due to the limited receptive field of the transformer blocks or the difficulty in capturing fine-grained details in extremely small lesions. Further investigation is needed to address these limitations.

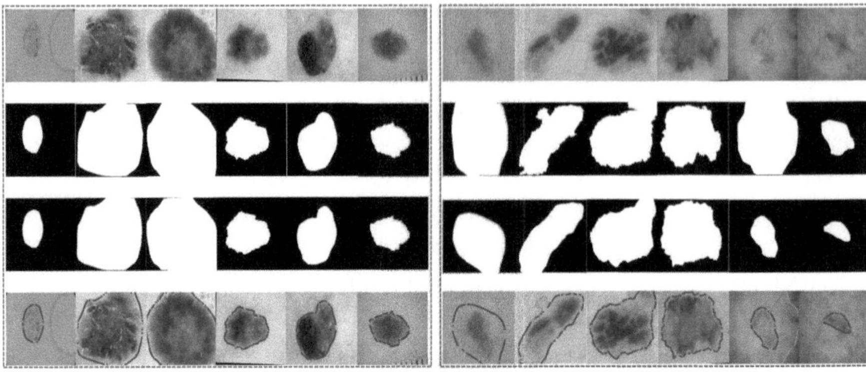

Fig. 3. Segmentation results of SegDT on the ISIC datasets. The figure shows examples of well-segmented cases (left) and challenging cases (right). Each example consists of six columns: the first two are from ISIC 2016, the middle two from ISIC 2017, and the last two from ISIC 2018. Within each column, the first row shows the original image, the second row the ground truth segmentation mask, and the third row the predicted segmentation mask. The fourth row visualizes the segmentation on the original image, with the ground truth contour in green and the predicted mask contour in blue. (Color figure online)

5 Conclusion

This paper presented SegDT, a novel DiT model for efficient medical image segmentation, specifically targeting skin lesion segmentation. SegDT leverages a compact DiT architecture and incorporates rectified flow for accelerated inference on resource-constrained GPUs. Our approach addresses the limitations of existing diffusion models, which often suffer from high computational costs and long inference times, hindering their practical application in real-world clinical settings.

SegDT demonstrates its effectiveness on ISIC datasets (2016, 2017, 2018) by achieving competitive, state-of-the-art segmentation performance with significantly faster inference speeds. This efficiency, enhanced by the use of rectified flow, is crucial for rapid clinical analysis in real-world applications. Furthermore, SegDT's compact architecture enables deployment on low-cost GPUs, broadening the accessibility of advanced medical image segmentation and contributing to the democratization of cutting-edge diagnostic tools.

Future work will focus on further optimizing SegDT's architecture and exploring additional techniques to enhance its performance and efficiency. We also plan to investigate the model's generalizability to other medical image segmentation tasks and datasets. Furthermore, we will explore incorporating additional clinical information, such as patient metadata, to further improve segmentation accuracy and clinical utility. We believe that SegDT represents a promising step towards the development of efficient and accurate medical image segmentation models for real-world clinical applications.

References

1. Cao, H., et al.: Swin-unet: Unet-like pure transformer for medical image segmentation. In: European Conference on Computer Vision, pp. 205–218. Springer (2022)
2. Chen, J., et al.: Transunet: Transformers make strong encoders for medical image segmentation. arXiv preprint arXiv:2102.04306 (2021)
3. Chen, L.C., Papandreou, G., Kokkinos, I., Murphy, K., Yuille, A.L.: Deeplab: semantic image segmentation with deep convolutional nets, atrous convolution, and fully connected crfs. IEEE Trans. Pattern Anal. Mach. Intell. **40**(4), 834–848 (2018)
4. Cheng, D., et al.: Gu-net: causal relationship-based generative medical image segmentation model. Heliyon **10**(18) (2024)
5. Codella, N., et al.: Skin lesion analysis toward melanoma detection 2018: A challenge hosted by the international skin imaging collaboration (isic). arXiv preprint arXiv:1902.03368 (2019)
6. Codella, N.C., et al.: Skin lesion analysis toward melanoma detection: A challenge at the 2017 international symposium on biomedical imaging (isbi), hosted by the international skin imaging collaboration (isic). In: 2018 IEEE 15th International Symposium on Biomedical Imaging (ISBI 2018), pp. 168–172. IEEE (2018)
7. Dosovitskiy, A., et al.: An image is worth 16x16 words: Transformers for image recognition at scale. arXiv preprint arXiv:2010.11929 (2020)
8. Feng, Y., Su, J., Zheng, J., Zheng, Y., Zhang, X.: A parallelly contextual convolutional transformer for medical image segmentation. Biomed. Signal Process. Control **98**, 106674 (2024)
9. Gorade, V., Mittal, S., Jha, D., Bagci, U.: Synergynet: Bridging the gap between discrete and continuous representations for precise medical image segmentation. In: Proceedings of the IEEE/CVF Winter Conference on Applications of Computer Vision, pp. 7768–7777 (2024)
10. Gutman, D., Codella, N.C., Celebi, E., Helba, B., Marchetti, M., Mishra, N., Halpern, A.: Skin lesion analysis toward melanoma detection: A challenge at the international symposium on biomedical imaging (isbi) 2016, hosted by the international skin imaging collaboration (isic). arXiv preprint arXiv:1605.01397 (2016)
11. Jha, D., Riegler, M.A., Johansen, D., Halvorsen, P., Johansen, H.D.: Resunet++: an advanced architecture for medical image segmentation. IEEE Access **7**, 21455–21467 (2019)
12. Kaur, R., Ranade, S.K.: Du-net+: a fully convolutional neural network architecture for semantic segmentation of skin lesions. SIViP **19**(1), 152 (2025)
13. Lan, L., Cai, P., Jiang, L., Liu, X., Li, Y., Zhang, Y.: Brau-net++: U-shaped hybrid cnn-transformer network for medical image segmentation. arXiv preprint arXiv:2401.00722 (2024)
14. Lin, A., Chen, B., Xu, J., Zhang, Z., Lu, G., Zhang, D.: Ds-transunet: dual swin transformer u-net for medical image segmentation. IEEE Trans. Instrum. Meas. **71**, 1–15 (2022)
15. Liu, X., Gong, C., Liu, Q.: Flow straight and fast: Learning to generate and transfer data with rectified flow. arXiv preprint arXiv:2209.03003 (2022)
16. Peebles, W., Xie, S.: Scalable diffusion models with transformers. In: Proceedings of the IEEE/CVF Inter. Conf. on Computer Vision, pp. 4195–4205 (2023)
17. Perera, S., Erzurumlu, Y., Gulati, D., Yilmaz, A.: Mobileunetr: a lightweight end-to-end hybrid vision transformer for efficient medical image segmentation. arXiv preprint arXiv:2409.03062 (2024)

18. Ronneberger, O., Fischer, P., Brox, T.: U-net: Convolutional networks for biomedical image segmentation. In: International Conference on Medical Image Computing and Computer-Assisted Intervention (MICCAI), pp. 234–241. Springer (2015)
19. Song, J., Meng, C., Ermon, S.: Denoising diffusion implicit models. arXiv preprint arXiv:2010.02502 (2020)
20. Vaswani, A., et al.: Attention is all you need. Advances in neural information processing systems **30** (2017)
21. Wu, J., et al.: Medsegdiff: Medical image segmentation with diffusion probabilistic model. In: Medical Imaging with Deep Learning, pp. 1623–1639. PMLR (2024)
22. Wu, J., Ji, W., Fu, H., Xu, M., Jin, Y., Xu, Y.: Medsegdiff-v2: Diffusion-based medical image segmentation with transformer. In: Proceedings of the AAAI Conference on Artificial Intelligence, vol. 38, pp. 6030–6038 (2024)
23. Yang, B., Zhang, R., Peng, H., Guo, C., Luo, X., Wang, J., Long, X.: Slp-net: an efficient lightweight network for segmentation of skin lesions. Biomed. Signal Process. Control **101**, 107242 (2025)
24. Yang, Z., Chen, R., Lin, C.: Am-net: a network with attention and multi-scale feature fusion for skin lesion segmentation. IEEE Sensors J. (2025)

FS-SAM2: Adapting Segment Anything Model 2 for Few-Shot Semantic Segmentation via Low-Rank Adaptation

Bernardo Forni[1(✉)], Gabriele Lombardi[2], Federico Pozzi[2], and Mirco Planamente[2]

[1] University of Pavia, Pavia, Italy
bernardo.forni01@universitadipavia.it
[2] ARGO Vision, Milano, Italy
{gabriele.lombardi,federico.pozzi,mirco.planamente}@argo.vision

Abstract. Few-shot semantic segmentation has recently attracted great attention. The goal is to develop a model capable of segmenting unseen classes using only a few annotated samples. Most existing approaches adapt a pre-trained model by training from scratch an additional module. Achieving optimal performance with these approaches requires extensive training on large-scale datasets. The Segment Anything Model 2 (SAM2) is a foundational model for zero-shot image and video segmentation with a modular design. In this paper, we propose a **F**ew-**S**hot segmentation method based on **SAM2** (FS-SAM2), where SAM2's video capabilities are directly repurposed for the few-shot task. Moreover, we apply a Low-Rank Adaptation (LoRA) to the original modules in order to handle the diverse images typically found in standard datasets, unlike the temporally connected frames used in SAM2's pre-training. With this approach, only a small number of parameters is meta-trained, which effectively adapts SAM2 while benefiting from its impressive segmentation performance. Our method supports any K-shot configuration. We evaluate FS-SAM2 on the PASCAL-5^i, COCO-20^i and FSS-1000 datasets, achieving remarkable results and demonstrating excellent computational efficiency during inference. Code is available at https://github.com/fornib/FS-SAM2

Keywords: Few-Shot Semantic Segmentation · Segment Anything Model 2 · Low-Rank Adaptation

1 Introduction

Semantic Segmentation is one of the most important tasks in Computer Vision. The goal is to assign a predefined class label to each pixel in an image. Recent developments in Deep Neural Networks have made tremendous progress in the semantic segmentation task [3]. However, most advancements can be attributed to the availability of large amounts of training samples for each target class. In

E. Rodolà et al. (Eds.): ICIAP 2025, LNCS 16167, pp. 67–78, 2026.
https://doi.org/10.1007/978-3-032-10185-3_6

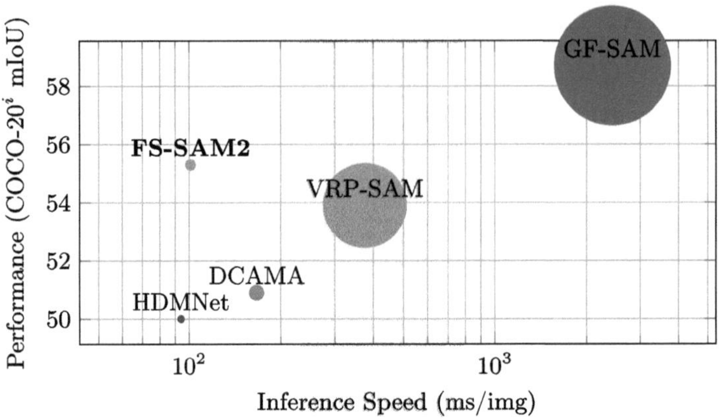

Fig. 1. Performance-efficiency comparison of FSS models. Size is proportional to the number of parameters.

real-world applications, providing labeled data for novel classes can be expensive or even impractical, especially when pixel-wise labeling is required. In low-data scenarios, most methods exhibit poor generalization performance due to overfitting. To deal with this challenging setting, *Few-shot* Semantic Segmentation (FSS) aims to train a semantic segmentation model that adapts to novel classes using a limited amount of samples [22]. Specifically, the model is evaluated on segmentation performances for a novel class within a query image, given a support set that contains only a few annotated images of that class. Recent approaches are based on a pre-trained image encoder to extract query and support features, which are then combined to decode the target segmentation [16,19,23,24,30]. These methods either train a complex and ad-hoc module with a relatively small number of training samples, or employ huge encoders to extract more general features. The latter approach typically demonstrates better performances, but these performances are obtained at the cost of a significantly lower inference speed. In practical applications, real-time latency is often a critical constraint, and it forces a trade-off between efficiency and performance.

The Segment Anything Model 2 (SAM2) [20] is a *Vision Foundation Model* for class-agnostic image and video segmentation, that can utilize different kinds of user-provided prompts, such as points, boxes or masks. Thanks to a powerful model architecture, based on a Hiera image encoder [21], and exploiting a large-scale training process for video segmentation, SAM2 boasts strong performance with fast inference speed. An internal module, called memory attention, enables SAM2 to track a prompted object across video frames by matching their features. In this paper, we propose to leverage this matching capability for the few-shot task by treating support images as annotated video frames, and using the query image as the subsequent frame to be segmented. We adapt SAM2's memory and encoder modules by employing Low-Rank Adaptation (LoRA), a fine-tuning strategy that introduces a small number of trainable parameters in each module. This approach enables effective model adaptation and enhanced feature extraction without the need to train from scratch any ad-hoc module.

Our contributions are as follows:

1. We propose FS-SAM2: a simple yet effective repurposing of the SAM2 architecture, demonstrating how its video-based design can be seamlessly adapted to the few-shot semantic segmentation task.
2. We apply Low-Rank Adaptation (LoRA) to adapt SAM2 for efficient and robust learning in the few-shot setting with a simple training strategy and minimal parameter requirements.
3. We conduct extensive empirical studies on PASCAL-5^i, COCO-20^i and FSS-1000 datasets to validate our approach. FS-SAM2 achieves performance comparable to state-of-the-art methods while being computationally efficient.

2 Background and Related Works

2.1 Few-Shot Semantic Segmentation

FSS is the application of the few-shot learning paradigm [7] to the semantic segmentation task, with the aim of improving generalization capabilities in low-data scenarios. A *support* set, consisting of few images with segmentation masks of a novel target class, is provided as input for model adaptation, and a *query* image containing that class is used for evaluation. Recent methods condition the query image on the novel class by matching query features with support features during inference. In *prototypical matching*, the target class is represented by one feature vector (or a few) extracted from the support set, which is used to classify each query feature [26,27,32]. This heavy compression of the support set features helps with regularization, but may contain an incomplete or inaccurate representation of the target class. Instead, *pixel-wise matching* involves computing element-wise correlations between query features and support set features, thus providing more detailed context [9,16,19,23,28,31]. In DCAMA [23], cross-attention is used to compute pixel-wise matching and aggregate support masks into multi-scale query features, which are fed into a convolutional decoder to extract the segmentation mask.

2.2 SAM-Based Few-Shot Segmentation

SAM [11] is a segmentation model composed of three modules: an image encoder that extracts embeddings from the input image, a prompt encoder that computes embeddings based on user-provided prompts, and a mask decoder that combines these embeddings to generate the segmentation mask. SAM2 [20] extends the segmentation task to videos by adding a memory encoder module. It encodes image embeddings of previous frames with their predicted masks. A memory attention module conditions each new frame on these encoded embeddings using pixel-wise matching. Recent methods [14,24,30] successfully adapt SAM to the FSS setting by leveraging the prompt capabilities, although they often require an additional encoder to achieve satisfactory segmentation performance. For example, VRP-SAM [24] directly generates prompt embeddings that are applied to the query image by training a separate network based on PFENet [26] and the cross-attention of learnable vectors. Similarly, GF-SAM [30] matches image features to generate a prior query mask, which is then refined by clustering SAM's output masks derived using points from the prior mask. GF-SAM displays outstanding performance, but is overall very resource-intensive as it employs a separate DINOv2 encoder [18] and repeatedly applies SAM's mask decoder. Furthermore, GF-SAM does not perform any in-domain training but relies solely on the generalization capabilities of DINOv2, which has been shown [2] to outperform most semantic models. Other methods [1,33] use SAM2 for medical images by repurposing the memory attention module with a selection criterion to identify a small subset of samples to serve as annotated frames. These methods work without additional training, as medical datasets generally consist of similar images, analogous to video frames from SAM2's pre-training.

2.3 Fine-Tuning Strategies

Vision foundation models are pre-trained on large-scale datasets and excel at extracting high-quality image representations. Most FSS methods use these models as the *backbone* of the network, typically kept frozen to reduce overfitting and followed by an ad-hoc module, meta-trained for the few-shot task. However, as shown in [25], fine-tuning just a small amount of parameters in the backbone may improve model generalization on learning novel classes. One of the most effective and widely adopted parameter-efficient fine-tuning methods is Low-Rank Adaptation (LoRA) [10,29]. This method consists of adapting selected linear layers by projecting initial features into a low-rank representation and back to output size for point-wise summation with the original output features.

$$\hat{y} = \hat{W}x = Wx + BAx \tag{1}$$

where $\hat{y} \in \mathbb{R}^{d_{\text{out}}}$ is the updated layer output, $\mathbf{x} \in \mathbb{R}^{d_{\text{in}}}$ is the layer input, $W \in \mathbb{R}^{d_{\text{out}} \times d_{\text{in}}}$ is the weight matrix associated with the linear layer, $A \in \mathbb{R}^{r \times d_{\text{in}}}$ is the down-projection, $\mathbf{B} \in \mathbb{R}^{d_{\text{out}} \times r}$ is the up-projection, and $r \ll \min\{d_{\text{in}}, d_{\text{out}}\}$ is the rank. LoRA minimizes the number of trainable parameters by fine-tuning only the low-rank projections while keeping the original parameters frozen. After fine-tuning, the low-rank updates can be incorporated into existing layers by replacing the original weights W with the updated ones $\hat{W}$. This integration allows the model to utilize the benefits of fine-tuning without retaining separate LoRA components during inference, thereby maintaining the original model's computational efficiency and simplifying deployment.

3 Methodology

3.1 Problem Setting

FSS in the K-shot regime is the task of learning a segmentation model $\mathcal{F}_\Theta$ trained on a meta-training dataset $\mathcal{D}_{train}$ and evaluated on a separate meta-test dataset

$\mathcal{D}_{test}$. These datasets consist of images with associated segmentation masks of a single class, where the segmented classes are disjoint between the two datasets. Following previous works [7,26,27], the episodic learning paradigm is applied to both datasets $\mathcal{D}_{train}$ and $\mathcal{D}_{test}$, ensuring that the training process mimics the evaluation process.

For each dataset, let l be a single class that is not the background. The support set is defined as $S = \{(I_i, M_i)\}_{i=1}^{K}$, where K is the number of shots indexed by i, $I_i \in \mathbb{R}^{H_i \times W_i \times 3}$ are different RGB images in the dataset containing class l and $M_i \in \{0,1\}^{H_i \times W_i}$ are their corresponding binary segmentation mask for class l. The query image $I_q \in \mathbb{R}^{H_q \times W_q \times 3}$ is an RGB image from the same dataset but distinct from those in S, and the query mask $M_q \in \{0,1\}^{H_q \times W_q}$ is the corresponding mask for class l. An episode consists of a support set together with a query image. For each episode, the task is to generalize from the support set and accurately segment the class l in the query image.

$$\hat{M}_q = \mathcal{F}_\Theta(I_q, S) \in \{0,1\}^{H_q \times W_q} \qquad \mathrm{IoU}_l = \frac{\sum_q |M_q \cap \hat{M}_q|}{\sum_q |M_q \cup \hat{M}_q|} \tag{2}$$

where $\hat{M}_q$ is the predicted binary segmentation of the query image for class l. During evaluation on the meta-test dataset, the episodes are independent of each other, ensuring no information is shared between them. The standard evaluation metric for FSS is mean Intersection-over-Union (mIoU): the IoU is calculated as a pixel sum over all query images q, the mIoU is computed as the average IoU across all foreground classes: $\mathrm{mIoU} = \frac{1}{C}\sum_{l=1}^{C} \mathrm{IoU}_l$ where C is the total number of classes in the evaluation dataset.

3.2 FS-SAM2

In this work, we present a simple yet effective strategy to adapt the SAM2 for the FSS task, taking advantage of its extensive pre-training and computational efficiency. We repurpose SAM2's video segmentation capabilities by treating the support set S as annotated video frames and the query image I_q as the subsequent frame to be segmented. We notice that SAM2's pre-training, based on temporally connected frames that are very similar to each other, heavily relies on these frame similarities. This results in suboptimal FSS performance on standard datasets, where the support and query images can significantly differ. To address this issue, we employ a meta-training strategy to learn a small amount of parameters and effectively adapt the model.

The overall architecture of FS-SAM2 is illustrated in Fig. 2, the modules are all inherited from SAM2. Given a support image I_s with its corresponding mask M_s and a query image I_q, high-quality features are independently extracted from both images using the *image encoder*. The support mask and features are downsampled and combined within the *memory encoder*, a convolutional module that incorporates the segmentation mask into the image features. The *memory attention* conditions the query features on the encoded support features through

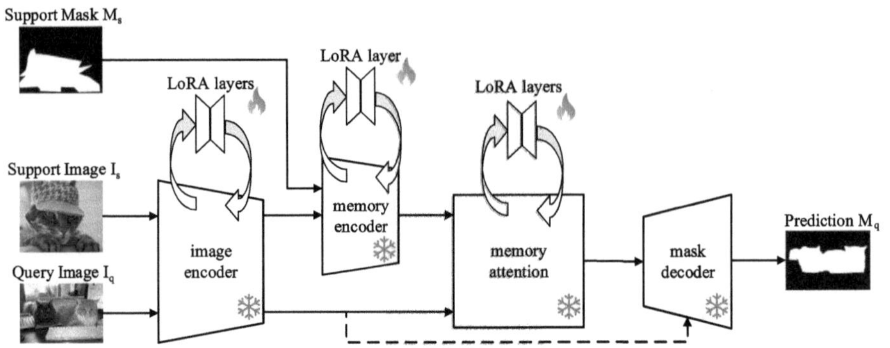

Fig. 2. FS-SAM2 framework for 1-shot semantic segmentation. The SAM2 video segmentation pipeline is repurposed by treating the support and query images as consecutive input frames, taking advantage of pixel-wise matching in the *memory attention* module. The model is efficiently specialized by meta-training only LoRA layers.

pixel-wise matching, utilizing repeated cross-attention and self-attention operations on the image embedding. The output embedding with high-level query features is used to generate the prediction mask M_q in the *mask decoder*, a module based on a lightweight transformer architecture [4] with convolutional upsampling. The prompt encoder is ignored, as only default token embeddings are used for the mask decoder and geometric prompts are not required.

We apply LoRA to image encoder, memory encoder and memory attention. The mask decoder is kept as-is, since its extensive class-agnostic pre-training makes it already suitable for the task. During meta-training, the unlabeled query image is matched with labeled support images using the memory attention module. The predicted labels for the query image are then compared to the ground-truth labels to compute the loss, which is used to fine-tune the parameters. This approach allows fine-tuning by meta-training only a small fraction of parameters, while effectively specializing the model for the few-shot setting.

3.3 Extension to K-Shots

Our approach accommodates any K-shot configuration without the need to train a different model for each K. We treat all the K support images as previously annotated frames and apply the same framework. Specifically, each support image and mask is encoded by means of the *memory encoder*, these encodings are then concatenated along the spatial dimension before applying the *memory attention*.

4 Experimental Results

4.1 Datasets

We evaluate FS-SAM2 performance and generalization capabilities by conducting extensive experiments on three widely used public datasets: PASCAL-5^i [22], with 20 classes from the PASCAL VOC 2012 [5] and Extended SDS [8] datasets; COCO-20^i [17,27], which is based on the MS COCO [13] dataset with 80 classes; and FSS-1000 [12], which contains 1000 classes. PASCAL-5^i and COCO-20^i classes are evenly split in 4 folds evaluated using cross-validation, while FSS-1000 is split into 520 classes for training, 240 for validation, and 240 for testing. Following recent works [16,26], we also evaluate the domain-shift scenario, where each model trained on a COCO-20^i fold is tested on PASCAL-VOC images that do not overlap in classes for that fold. To ensure consistency with other benchmarks, we conducted tests using 1000 randomly sampled episodes for each fold.

To produce an episode, the query image is first selected, then a class in the image is randomly sampled, and finally a support set corresponding to that class is randomly sampled, where other classes are marked as background in the segmentation masks. For COCO-20^i, following previous works [16,23,24,30], the class is first randomly sampled, and then a query image and a support set are randomly sampled with that class.

4.2 Implementation Details

We mostly follow the implementation of VRP-SAM [24]. We use Binary Cross-Entropy (BCE) loss and Dice loss to supervise the learning during meta-training. We employ the AdamW optimizer [15] ($\beta_1 = 0.9, \beta_2 = 0.999$) with a cosine learning rate decay schedule. The model is trained for 50 epochs with an initial learning rate of $1e^{-4}$ and batch size of 16. Each epoch corresponds to the size of the meta-train dataset, with early stopping validation on the meta-test dataset. We adopt the same training augmentations as some previous works [19,26]. Images are first resized to 1024×1024 before being processed by SAM2, the predicted masks are resized to their original resolution for evaluation. The performance of the model is evaluated using mean Intersection-over-Union (mIoU) for each fold.

We conducted all the experiments on the Hiera-B+ version of SAM2 initialized with SAM2.1 pre-trained weights. We apply LoRA to the Q, V, K, O projection layers of the transformer blocks. We set the LoRA rank to 4 in the image encoder, while in the memory attention and memory encoder we use a rank of 32 to introduce a comparable amount of trainable parameters given the small amount of layers. The architecture totals 81 million parameters, with only 750 thousand trainable. The training was done with two NVIDIA A100-64GB GPUs and casting to bfloat16.

4.3 Achieved Results

We present performance comparisons of FS-SAM2 against other state-of-the-art (SOTA) methods for a comprehensive evaluation. To further illustrate the improvements of our method, we also compare with VRP-SAM2, which is our implementation of VRP-SAM [24] using SAM2 as the backbone instead of SAM.

Table 1 presents performance comparisons on the PASCAL-5^i dataset. In the 1-shot scenario, our model achieves a new SOTA mIoU of 73.4%, surpassing VRP-SAM by 1.5%, the most comparable method to ours. The VRP-SAM

Table 1. Performance results for FSS on PASCAL-5^i dataset with mIoU metric. The best results are highlighted in **bold** and the second-best results are underlined.

Method	Backbone	PASCAL-5^i 1-shot					PASCAL-5^i 5-shot				
		fold-0	fold-1	fold-2	fold-3	mean	fold-0	fold-1	fold-2	fold-3	mean
PFENet [26]	RN50	61.7	69.5	55.4	56.3	60.8	63.1	70.7	55.8	57.9	61.9
HSNet [16]	RN50	64.3	70.7	60.3	60.5	64.0	70.3	73.2	67.4	67.1	69.5
DCAMA [23]	Swin	72.2	73.8	64.3	67.1	69.3	75.7	77.1	72.0	74.8	74.9
FPTrans [32]	DeiT	72.3	70.6	68.3	64.1	68.8	76.7	79.0	**81.0**	75.1	78.0
Matcher [14]	SAM,DINOv2	67.7	70.7	66.9	67.0	68.1	71.4	77.5	74.1	72.8	74.0
HDMNet [19]	RN50	71.0	75.4	68.9	62.1	69.4	71.3	76.2	71.3	68.5	71.8
VRP-SAM [24]	SAM,RN50	73.9	78.3	70.6	65.0	71.9	-	-	-	-	72.9
GF-SAM [30]	SAM,DINOv2	71.1	75.7	69.2	**73.3**	72.1	**81.5**	**86.3**	79.7	**82.9**	**82.6**
VRP-SAM2	SAM2,RN50	74.0	77.4	69.3	63.5	71.0	75.6	74.6	69.7	61.8	70.4
FS-SAM2	SAM2	**74.0**	**79.6**	**73.5**	66.7	**73.4**	75.7	79.6	75.3	68.8	74.8

Table 2. Performance results for FSS on the COCO-20^i dataset with mIoU metric. The best results are highlighted in **bold** and the second-best results are underlined.

Method	Backbone	COCO-20^i 1-shot					COCO-20^i 5-shot				
		fold-0	fold-1	fold-2	fold-3	mean	fold-0	fold-1	fold-2	fold-3	mean
PFENet [26]	RN50	36.5	38.6	34.5	33.8	35.8	36.5	43.3	37.8	38.4	39.0
HSNet [16]	RN50	36.3	43.1	38.7	38.7	39.2	43.3	51.3	48.2	45.0	46.9
DCAMA [23]	Swin	49.5	52.7	52.8	48.7	50.9	55.4	60.3	59.9	57.5	58.3
FPTrans [32]	DeiT	44.4	48.9	50.6	44.0	47.0	54.2	62.5	61.3	57.6	58.9
Matcher [14]	SAM,DINOv2	52.7	53.5	52.6	52.1	52.7	60.1	62.7	60.9	59.2	60.7
HDMNet [19]	RN50	43.8	55.3	51.6	49.4	50.0	50.6	61.6	55.7	56.0	56.0
VRP-SAM [24]	SAM,RN50	48.1	55.8	**60.0**	51.6	53.9	-	-	-	-	-
GF-SAM [30]	SAM,DINOv2	**56.6**	**61.4**	59.6	**57.1**	**58.7**	**67.1**	**69.4**	**66.0**	**64.8**	**66.8**
VRP-SAM2	SAM2,RN50	47.1	55.5	55.2	52.8	52.7	50.8	60.0	57.7	57.7	56.6
FS-SAM2	SAM2	49.3	57.6	58.3	56.1	55.3	55.1	60.6	58.7	59.3	58.4

method does not benefit from the newer SAM2 backbone, as it likely requires some fine-tuning of the backbone for the few-shot task. It is worth noting that GF-SAM [30], which utilizes a DINOv2-L [18] backbone pre-trained on a dataset containing PASCAL-VOC, still underperforms compared to our method in the 1-shot setting. We observe a slight improvement in segmentation performance in the 5-shot setting, indicating that our method can leverage contextual information from multiple support images. However, further gains could be achieved by explicitly training the model in a 5-shot regime.

Table 2 shows results on the COCO-20^i dataset and demonstrates the effectiveness and generalization capabilities of FS-SAM2. Our model outperforms VRP-SAM [24] by 1.4% in the 1-shot benchmark and shows a notable increase

Table 3. Performance results for FSS on FSS-1000 dataset and domain-shift from COCO-20^i to PASCAL-VOC dataset with mIoU metric. The best results are highlighted in **bold** and the second-best results are underlined.

Method	Backbone	FSS-1000		domain-shift	
		1-shot	5-shot	1-shot	5-shot
PFENet [26]	RN50	80.8	81.4	61.1	63.4
HSNet [16]	RN50	85.5	87.8	61.6	68.7
VAT [9]	RN50	<u>90.1</u>	<u>90.7</u>	64.5	69.7
DCAMA [23]	Swin	88.2	88.8	-	-
FPTrans [32]	DeiT	-	-	69.7	79.3
MSDNet [6]	RN50	-	-	<u>72.1</u>	<u>74.2</u>
FS-SAM2	SAM2	**90.6**	**91.7**	**77.2**	**79.7**

Fig. 3. Qualitative comparison results on the COCO-20^i dataset. Our meta-training strategy successfully adapts SAM2, outperforming baseline methods.

in the 5-shot benchmark. While GF-SAM [30] is the current SOTA on this benchmark, we note that it does not undergo meta-training for in-domain knowledge but instead relies on DINOv2 pre-training and is significantly more resource-intensive than FS-SAM2, as illustrated in Fig. 1.

We also evaluate our method on the FSS-1000 dataset, that contains a large number of different classes, and on the domain-shift scenario, where the base training data and the testing data have a significant domain gap. The results presented in Table 3 clearly show the robustness of FS-SAM2 and demonstrate that LoRA successfully adapted the model to the few-shot setting.

Figure 3 presents a qualitative comparison of our approach with VRP-SAM and the SAM2 without meta-training. The examples highlight how the SAM2 struggles with dissimilar support-query pairs. In contrast, FS-SAM2 consistently generates accurate and complete masks, capturing fine details and entire object regions that VRP-SAM misses. Our method often generates cleaner boundaries, even refining minor inaccuracies present in the ground truth (GT) annotations.

Table 4. Performance comparison of different fine-tuning strategies on COCO-20^i 1-shot task averaged over 4 folds. The LoRA rank applied to the module is indicated in parentheses. The column #params indicates the number of parameters updated during meta-training.

Strategy	mIoU	#params
No training	26.2	0
Fine-tuning mem.	51.8	7,306,912
LoRA mem.(32)	49.7	475,136
LoRA enc.(4)	54.6	265,356
LoRA enc.(32)	54.8	1,536,560
LoRA enc.(4), mem.(4)	<u>55.2</u>	326,028
LoRA enc.(4), mem.(32)	**55.3**	750,732
LoRA enc.(4), mem.(32), dec.(32)	55.1	1,127,564
LoRA enc.(4), fine-tuning mem.	54.2	7,572,268

Table 5. Same as Table 4, but the support image is set to be identical to the query image during the meta-test.

Strategy	mIoU
No training	**95.4**
Fine-tuning mem.	64.0
LoRA mem.(32)	<u>85.2</u>
LoRA enc.(4)	82.5
LoRA enc.(4), mem.(32)	80.0

4.4 Ablation Study

We conducted ablation studies on the COCO-20^i dataset in 1-shot regime to analyze the contributions of various adaptation strategies. This analysis aims to provide a deeper understanding of the role and importance of each module.

Table 4 compares different fine-tuning strategies applied to various modules of the model. We observe that the model performs inadequately without meta-training. Fine-tuning only the memory modules results in a significant improvement, demonstrating that the model can be effectively adapted for the FSS task. Utilizing LoRA only in the image encoder yields improved performance, indicating that this strategy reduces the importance of feature position information for subsequent pixel-wise matching while improving generalization for novel classes. Applying LoRA to both the image encoder and the memory modules results in the most effective strategy, achieving 55.3% mIoU. Using a smaller rank of 4 for LoRA in the memory modules, or extending LoRA to the mask decoder, does not significantly impact performance.

Table 5 compares fine-tuning strategies in the scenario where the support set is not randomly sampled but is the same as the query image and mask during the evaluation on the meta-test dataset. This setup more closely mimics SAM2's original pre-training of similar frames and is beneficial for assessing the impact of meta-training. Without any meta-training the performance is nearly perfect. Moreover, we note that output masks are qualitatively better than the input masks. Fine-tuning the memory modules severely deteriorates performance, indicating that the model has reduced the reliance on the similarity of the images when matching novel classes. Applying LoRA better retains the learned capabilities while maintaining high performance on the FSS task.

5 Conclusions

In this paper, we introduced FS-SAM2, a new framework for SAM2 adaptation to the few-shot semantic segmentation task. SAM2's video segmentation capabilities can be effectively repurposed to condition the query image on the support set by training a small number of parameters and without requiring additional backbones or ad-hoc modules. Adaptation with LoRA further improves generalization performance on novel classes. Overall, our method demonstrates very competitive results while being computationally efficient.

For future research directions, it would be worthwhile to investigate alternative 5-shot training methods and explore more parameter-efficient fine-tuning strategies. Furthermore, we aim to extend our approach to the multi-class setting, a relatively under-explored task that offers significant practical advantages, as SAM2 lacks inter-class communication capabilities.

Acknowledgements. We acknowledge ISCRA for awarding this project access to the LEONARDO supercomputer, owned by the EuroHPC Joint Undertaking, hosted by CINECA (Italy).

References

1. Bai, Y., Yu, Q., Yun, B., Jin, D., Xia, Y., Wang, Y.: Revsam2: prompt sam2 for medical image segmentation via reverse-propagation without fine-tuning. arXiv:2409.04298 (2024)
2. Bensaid, R., Gripon, V., Leduc-Primeau, F., Mauch, L., Hacene, G.B., Cardinaux, F.: A novel benchmark for few-shot semantic segmentation in the era of foundation models. arXiv:2401.11311 (2024)
3. Csurka, G., Volpi, R., Chidlovskii, B.: Semantic image segmentation: two decades of research. Found. Trends Comput. Graph. Vis. **14**(1–2), 1–162 (2022)
4. Dosovitskiy, A., et al.: An image is worth 16x16 words: Transformers for image recognition at scale. ICLR (2021)
5. Everingham, M., Van Gool, L., Williams, C.K., Winn, J., Zisserman, A.: The pascal visual object classes (voc) challenge. IJCV **88**, 303–338 (2010)
6. Fateh, A., Mohammadi, M.R., Motlagh, M.R.J.: Msdnet: Multi-scale decoder for few-shot semantic segmentation via transformer-guided prototyping. arXiv:2409.11316 (2024)
7. Fei-Fei, L., Fergus, R., Perona, P.: One-shot learning of object categories. IEEE TPAMI **28**(4), 594–611 (2006)
8. Hariharan, B., Arbeláez, P., Bourdev, L., Maji, S., Malik, J.: Semantic contours from inverse detectors. In: ICCV, pp. 991–998. IEEE (2011)

9. Hong, S., Cho, S., Nam, J., Lin, S., Kim, S.: Cost aggregation with 4d convolutional swin transformer for few-shot segmentation. In: ECCV. Springer (2022)
10. Hu, E.J., et al.: Lora: Low-rank adaptation of large language models. ICLR (2022)
11. Kirillov, A., et al.: Segment anything. In: ICCV, pp. 4015–4026 (2023)
12. Li, X., Wei, T., Chen, Y.P., Tai, Y.W., Tang, C.K.: Fss-1000: a 1000-class dataset for few-shot segmentation. In: CVPR, pp. 2869–2878 (2020)
13. Lin, T.Y., Maire, M., Belongie, S., Hays, J., Perona, P., Ramanan, D., Dollár, P., Zitnick, C.L.: Microsoft coco: Common objects in context. In: ECCV, pp. 740–755. Springer (2014)
14. Liu, Y., Zhu, M., Li, H., Chen, H., Wang, X., Shen, C.: Matcher: Segment anything with one shot using all-purpose feature matching. arXiv:2305.13310 (2023)
15. Loshchilov, I., Hutter, F.: Decoupled weight decay regularization. ICLR (2018)
16. Min, J., Kang, D., Cho, M.: Hypercorrelation squeeze for few-shot segmentation. In: ICCV, pp. 6941–6952 (2021)
17. Nguyen, K., Todorovic, S.: Feature weighting and boosting for few-shot segmentation. In: ICCV. pp. 622–631 (2019)
18. Oquab, M., et al.: Dinov2: Learning robust visual features without supervision. arXiv:2304.07193 (2023)
19. Peng, B., Tian, Z., Wu, X., Wang, C., Liu, S., Su, J., Jia, J.: Hierarchical dense correlation distillation for few-shot segmentation. In: CVPR. pp. 23641–23651 (2023)
20. Ravi, N., Gabeur, V., et al.: Sam 2: Segment anything in images and videos. arXiv:2408.00714 (2024)
21. Ryali, C., et al.: Hiera: a hierarchical vision transformer without the bells-and-whistles. In: ICML, pp. 29441–29454 (2023)
22. Shaban, A., Bansal, S., Liu, Z., Essa, I., Boots, B.: One-shot learning for semantic segmentation. arXiv:1709.03410 (2017)
23. Shi, X., Wei, D., Zhang, Y., Lu, D., Ning, M., Chen, J., Ma, K., Zheng, Y.: Dense cross-query-and-support attention weighted mask aggregation for few-shot segmentation. In: ECCV, pp. 151–168. Springer (2022)
24. Sun, Y., Chen, J., Zhang, S., Zhang, X., Chen, Q., Zhang, G., Ding, E., Wang, J., Li, Z.: Vrp-sam: Sam with visual reference prompt. In: CVPR, pp. 23565–23574 (2024)
25. Sun, Y., Chen, Q., He, X., Wang, J., Feng, H., Han, J., Ding, E., Cheng, J., Li, Z., Wang, J.: Singular value fine-tuning: few-shot segmentation requires few-parameters fine-tuning. NeurIPS **35**, 37484–37496 (2022)
26. Tian, Z., Zhao, H., Shu, M., Yang, Z., Li, R., Jia, J.: Prior guided feature enrichment network for few-shot segmentation. IEEE TPAMI **44**(2), 1050–1065 (2020)
27. Wang, K., Liew, J.H., Zou, Y., Zhou, D., Feng, J.: Panet: Few-shot image semantic segmentation with prototype alignment. In: ICCV, pp. 9197–9206 (2019)
28. Xu, Q., Zhao, W., Lin, G., Long, C.: Self-calibrated cross attention network for few-shot segmentation. In: ICCV, pp. 655–665 (2023)
29. Zanella, M., Ben Ayed, I.: Low-rank few-shot adaptation of vision-language models. In: CVPR, pp. 1593–1603 (2024)
30. Zhang, A., Gao, G., Jiao, J., Liu, C.H., Wei, Y.: Bridge the points: Graph-based few-shot segment anything semantically. arXiv:2410.06964 (2024)
31. Zhang, G., Kang, G., Yang, Y., Wei, Y.: Few-shot segmentation via cycle-consistent transformer. NeurIPS **34**, 21984–21996 (2021)
32. Zhang, J.W., Sun, Y., Yang, Y., Chen, W.: Feature-proxy transformer for few-shot segmentation. NeurIPS **35**, 6575–6588 (2022)
33. Zhu, J., Qi, Y., Wu, J.: Medical sam 2: Segment medical images as video via segment anything model 2. arXiv:2408.00874 (2024)

Confidently Biased (ConB): A Per-sample Confidence Approach for Unsupervised Model Debiasing

Massimiliano Ciranni[1], Aliaksandr Shcharbitski[1], Vittorio Murino[2,3], and Vito Paolo Pastore[1,2](✉)

[1] MaLGa, DIBRIS, Università degli Studi di Genova, Genova, Italy
[2] Istituto Italiano di Tecnologia, Genova, Italy
vito.paolo.pastore@unige.it
[3] Università di Verona, Verona, Italy

Abstract. The presence of bias in data may affect the performance of deep neural networks, impacting their generality and accuracy on unseen data not presenting bias spurious correlations. In such scenarios, models are likely to learn shortcuts corresponding to bias rather than semantic attributes. To address this issue, assuming no prior information on bias is available, we propose a two-step method where an auxiliary model is trained with a selective gradient back-propagation strategy to provide bias pseudo-labels, later employed within a modified group robustness framework, optimizing worst-case subpopulation performance. Exploiting popular real-world datasets, we show how our method is effective in mitigating model bias dependency and improving generalization while outperforming available state-of-the-art methods.

Keywords: Model Debiasing · Per-Sample Confidence · Image Classification

1 Introduction

The generalization of deep neural networks is affected when dealing with biased datasets [22]. In this case, a model can learn unwanted spurious correlations between data and targets, performing poorly with unseen samples that do not present such correlations [10]. To provide an example, let us consider an image classification problem and, specifically, the *hiking* class of the Biased Action Recognition (BAR) dataset [21]. Here, most training samples present images of people hiking on rock walls. As a consequence, a deep neural network trained on such a dataset ends up learning that a rock wall implies you are climbing, failing to correctly predict the same action for images with a subject climbing on other objects, such as an ice wall. Samples presenting bias attributes (rock walls in our example) are denoted as *bias-aligned*, and the ones not showing bias spurious correlations are named *bias-conflicting*.

E. Rodolà et al. (Eds.): ICIAP 2025, LNCS 16167, pp. 79–90, 2026.
https://doi.org/10.1007/978-3-032-10185-3_7

Mitigating a model dependency on bias when making predictions is the objective of model debiasing [22]. A common approach (supervised debiasing) consists in re-weighting training samples according to bias labels, specifically, weighing bias-conflicting samples more than the bias-aligned ones, thus discouraging the model from learning solely bias shortcuts [3,14,24]. In real-world scenarios, bias is generally unknown, thus, model debiasing can only be performed in an unsupervised framework with respect to bias labels [11].

Within unsupervised debiasing, a popular approach consists in two-step methods, where a bias-identification phase provides bias pseudo-labels to be employed for reweighting samples, for instance upweighting or upsampling those predicted as bias-conflicting [14,20,25]. The bias-identification step accuracy is fundamental for the debiased model generalization [22]. Many different solutions have been proposed for precise bias identification, including clustering in the feature space of the biased model [25], anomaly detection [22], bias-amplification [6,14] or model predictions [20]. Furthermore, bias features are typically *easier* to learn than semantic ones, and thus, are learned *earlier* by a model trained on biased data (as empirically shown in [21]).

Based on this assumption, we propose Confidently Biased (ConB), a two-step method for unsupervised model debiasing. Our first bias-identification step leverages a model intentionally biased exploiting a gradient selection strategy. Specifically, we only backpropagate the gradients corresponding to samples with the lowest loss, which are supposed to predominantly be bias-aligned. The resulting model is employed to obtain bias binary pseudo-labels (i.e., bias-aligned or bias-conflicting), further used within an adapted version of GroupDRO [24], a popular group robustness approach to maximize accuracy on the worst group (i.e., bias-conflicting), typically employed in supervised model debiasing [5]. When evaluated on three popular biased benchmark datasets, ConB appears effective and versatile, providing state-of-the-art competitive performance with different types of bias and percentages of bias-aligned training samples. To summarize our main contributions are the following: (i) we propose a gradient-selective strategy for training an intentionally biased auxiliary model with amplified dependency on bias when making predictions. The proposed approach consists in backpropagating only the gradients for samples corresponding to the lowest losses; (ii) We design a two-step method that we refer to as Confidently Biased (ConB) incorporating the bias-amplified auxiliary model for bias identification and a debiasing scheme adapting GroupDRO to the unsupervised framework, based on the identified bias pseudolabels. ConB obtains state-of-the-art performances on three popular real-world biased benchmark datasets.

The rest of the paper is organized as follows: Sect. 2 outlines the context of unsupervised model debiasing and current state-of-the-art solutions, while we detail the core part of our work, *per-sample confidence* and our debiasing pipeline (ConB), in Sect. 3. A detailed description of our experimental setting, including the employed benchmark datasets and our obtained results, can be found in Sect. 4. We conclude our study with a dedicated ablation study, summing up our contributions and future research directions in Sect. 5.

2 Related Work

Common approaches to Model Debiasing can be roughly divided into *Supervised* and *Unsupervised* Debiasing. In this context, supervision refers to *bias-supervision*, i.e. the employment of ground-truth annotations on biased attributes while training a target model. Such additional information can be leveraged for employing sample reweighing [17], specific regularization or ensembling techniques [18,26], or robust/group-invariant optimization approaches [2,24]. However, having prior per-sample information on the presence of biased attributes is highly uncommon in realistic scenarios. As such, unsupervised debiasing tries to tackle the problem of bias by only leveraging class information. The core challenge in unsupervised debiasing is to provide during training a distinction between samples containing biased attributes (bias-aligned) and those that do not (bias-conflicting samples) that is as precise as possible. To do so, an auxiliary model is trained specifically to attach onto bias attributes, either optimized in tandem with the target model training in an end-to-end fashion ([19,21]) or trained in a separate step and then employed to provide a reference signal for a subsequent debiasing phase (two-step methods, [4,7,10,20,22]). Indirect information from the intentionally biased model (e.g., its predictions [20], per-sample losses [12,21], per-sample gradients [1], or mined from feature embeddings [4,22,25]) can be used to drive the training of the target model, so that it learns to rely less on the shortcut enabled by the presence of bias, and more on the true semantic attributes of interest.

In this context, a commonly shared assumption is that bias-related features are easier to learn than semantic ones, with the effect that they are learned earlier during the training process of a naively trained model, preventing it from learning generalizable features. This has a relevant side-effect for training an optimal intentionally biased auxiliary mode, as it will risk memorizing the training set, including many bias-conflicting samples, if not properly regularized. As such, its loss/mispredictions/gradients will not be systematically different between bias-aligned and bias-conflicting training samples, and the model will not be useful at all for mining bias information from the training set.

To this end, a widespread solution is to employ Generalized Cross-Entropy (GCE) Loss [29] for training the auxiliary model [10,12,21], as it promotes the learning of training samples for which the model is more confident on, highly reducing the contribution from model's predictions with low-confidence. For instance, in Learning from Failure (LfF) [21], the auxiliary model is trained with GCE in parallel with the target one, allowing to compute a loss-weighing factor for each training sample that is much higher for bias-conflicting samples than for bias-aligned ones. As the intentionally biased model can degrade its performance as training goes on, in [15] the authors propose to use an ensemble of models trained on disjointed subsets of the training set with GCE for identifying bias-conflicting samples to be upsampled. In [22], for each target class, a dedicated anomaly detector is fitted to recognize out-of-distribution feature embeddings, to isolate a pure selection of bias-conflicting samples. In [12], a *bootstrap ensemble* strategy is employed to learn a committee of biased auxiliary classifiers, trained on top of a frozen backbone trained with self-supervision. While these approaches are able to partially mitigate the memorization side-effect, in the literature it is not yet clear how to avoid challenging hyper-parameter tuning or computationally heavy ensembling techniques for providing the best bias-mining performance [12,19,21,22,27]. Recently, model debiasing methods leveraging vision and language information have been emerging, such as [5,8,13]. Domino optimizes cross-modality embeddings to highlight dataset subsets that share common attributes and result in consistent failures from the model. In Bias2Text [13], BLIP2 [16] is used to caption the entire validation set of the target dataset to extract *bias keywords*, later used for debiasing using CLIP [23] in zero-shot classification and GroupDRO. Requiring a validation set containing conflicting samples is not always applicable (e.g. BAR has no validation set, and BFFHQ's validation split contains only aligned images).

In this paper, we propose a simpler yet more effective strategy to obtain an effective intentionally biased auxiliary model. Differently from the previously cited methods, we base our method on the concept of *per-sample confidence*, directly addressing bias-conflicting training samples memorization and allowing for improved debiasing performance over three real-world popular benchmark datasets.

3 Approach

We propose *ConB*, a novel machine learning approach designed to accurately identify bias-conflicting samples and ensure robust model performance across diverse and underrepresented data groups. Our approach can be framed as a two-step debiasing method, where the training of a biased model for inferring *group* pseudo-labels is followed by the actual debiasing step [4,20,27]. See Fig. 1 for a schematic overview.

3.1 Bias Identification Based on Per-Sample Confidence

Consider an image classification problem, and its related training dataset $\mathcal{D}_{\text{tr}} = \{(\boldsymbol{x}_i, y_i)\}_{i=1}^{N}$. Whereas the standard approach of training a deep neural network f_θ ERM through Cross-Entropy (CE) Loss would result in good average performance, using the same model in inference on an unbiased test set results in dramatic failures on less-represented training subpopulations, with measurable generalization gaps. To tackle this, we customize standard ERM training for learning an effective bias-amplified model, proposing a selective gradient back-propagation strategy.

At each training iteration of the bias-identification model, we can compute the per-sample loss of each minibatch $\mathcal{B} = \{(\boldsymbol{x}_i, y_i)\}_{i=1}^{k}$ as

$$\mathcal{L}_{\mathcal{B}} = \{\ell_i = \mathcal{L}(f(\boldsymbol{x}_i, \boldsymbol{\theta}), y_i)\}_{i=1}^{k} \tag{1}$$

Our goal is to obtain a model that has learned only bias information, hence we should ideally propagate back only the learning signal coming from aligned samples. To do so, we rank the set of per-sample losses $\mathcal{L}_{\mathcal{B}}$ such that $\ell_{(1)} \leq \ell_{(2)} \leq \cdots \leq \ell_{(k)}$, so that we can select from it a subset of very low-loss contributions, coming from the easier-to-learn biased-aligned population of the training set. We define such subset $\mathcal{L}'_{\mathcal{B}} \subset \mathcal{L}_{\mathcal{B}}$ as the bottom α fraction of samples in $\mathcal{L}_{\mathcal{B}}$, with $\alpha \in (0, 1)$. This selection process runs separately for each class in the batch, to avoid accidentally focusing on biases from specific target classes. The set $\mathcal{B}'$ of mini-batch samples corresponding to the selected losses in $\mathcal{L}'_{\mathcal{B}}$ will be the only ones propagating gradient to the auxiliary model, resulting in the final optimization objective in Eq. 2:

$$\boldsymbol{\theta}^* = \arg\min_{\boldsymbol{\theta}} \sum_{(\boldsymbol{x}_i, y_i) \in \mathcal{B}'} \mathcal{L}(f(\boldsymbol{x}_i, \boldsymbol{\theta}), y_i). \quad (2)$$

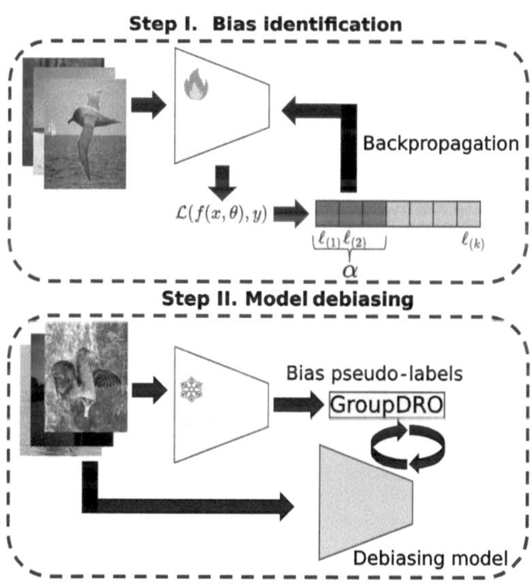

Fig. 1. Schematic representation of our proposed method, ConB. First, the intentionally biased auxiliary model is trained with a per-sample confidence selection approach. Then, the same model is frozen and used in inference on the training set to produce bias pseudo-labels, further exploited for debiasing a final bias-mitigated model with the GroupDRO objective.

3.2 Bias Mitigation

Once the auxiliary model is trained, we run an evaluation process on the training set to categorize samples as bias-aligned or bias-conflicting, depending on their classification outcomes. Samples that are correctly classified by the biased model are considered *bias-aligned*, whereas misclassified samples are identified as *bias-conflicting*. As a debiasing scheme capable of leveraging the obtained bias pseudo-labels, we adapt GroupDRO [24], a group robustness optimization method generally employed in supervised debiasing, and thus expecting bias annotations by design. Alongside the target class, we encode each class subgroup as either aligned or conflicting depending on the previously obtained pseudo-labels, forming a total of $2 \cdot \mathcal{C}$ groups, where $\mathcal{C}$ denotes the number of classes in the dataset. The GroupDRO objective is formulated as

$$\min_{\boldsymbol{\theta}} \max_{g \in \mathcal{G}} \mathbb{E}_{(\boldsymbol{x},y)\sim g} \left[\mathcal{L}(f(\boldsymbol{x}, \boldsymbol{\theta}), y) \right]. \tag{3}$$

where $\mathcal{G}$ represents the set of all groups, aiming to learn a model that generalizes well across all subpopulations.

4 Experiments

4.1 Datasets

We evaluate our approach with three popular biased datasets, namely Waterbirds [24], BAR [21], and BFFHQ [10]. This selection allows us to test our method across diverse types and severity of biases, including gender, background, and context bias. In this context, bias severity can be measured as the proportion of bias-aligned samples in the training set, usually indicated with $\rho \geq 0.95$.

Fig. 2. Sample images from Waterbirds. The top row shows bias-aligned samples while bias-conflicting samples are at the bottom.

Waterbirds. The Waterbirds dataset [24] is a benchmark designed to study spurious correlations and model bias in image classification. It contains two primary classes: waterbirds (e.g., ducks) and landbirds (e.g., sparrows). A deliberate bias is introduced by pairing 95% of waterbirds with water backgrounds and 95% of landbirds with land backgrounds ($\rho = 0.95$). This enables a shortcut where models trained on the dataset often rely on background features rather than bird morphology for classification. For instance, a waterbird with a land background is frequently misclassified as a landbird during evaluation. It contains 4,795 training images and an unbiased test set of 5,794 samples. Samples from this dataset are shown in Fig. 2.

Fig. 3. Sample images from BAR. The top row shows bias-aligned samples while bias-conflicting samples are at the bottom.

BAR. The Biased Action Recognition (BAR) dataset [21] comprises 2,595 real-world images across six action classes (climbing, diving, fishing, racing, throwing, vaulting), each strongly correlated with specific environmental contexts (e.g., "diving" with underwater scenes). BAR intentionally embeds background-place biases to mimic real-world data imbalances. The training set is made of 1,941 images, while the test set has 654 images. Neither a validation set nor explicit bias annotations are available for this dataset. Figure 3 depicts a selection of images from the six classes.

Fig. 4. Sample images from BFFHQ. The top row shows bias-aligned samples while bias-conflicting samples are at the bottom.

BFFHQ. Derived from the FFHQ dataset, the Biased FFHQ (BFFHQ) [10] introduces intentional age-gender correlations to study algorithmic fairness in facial attribute prediction (see Fig. 4 for some sample images). It overrepresents young women and older men, introducing a gender bias when predicting if the subject is young or old. It consists of 21,200 images, with 19,200 used for training, 1,000 for validation, and 1,000 for testing. This dataset presents the most severe bias scenario in our analysis, with $\rho = 0.995$, meaning that only 0.5% of training images are not aligned with the gender bias.

4.2 Experiment Details

Following standard comparison in the relevant literature [13,19,21] we choose ResNets [9] as the backbone architecture for our experiments. ResNet-18 is used for BFFHQ and BAR, while ResNet-50 is used for the Waterbirds dataset. All models are initialized with ImageNet-pretrained weights.

During the bias-identification step, we train the models using Cross Entropy (CE) loss and optimize with the AdamW optimizer. The learning rate is set to 10^{-5}, and the batch size is fixed at 128. The training proceeds for 50 epochs on Waterbirds and 30 epochs on BAR and BFFHQ datasets, with a gradient selection parameter α equal to 0.75 for BAR and BFFHQ and 0.7 for Waterbirds.

In the debiasing step with the Group DRO technique, the learning rate for BAR, Waterbirds, and BFFHQ are $5 \cdot 10^{-4}$,$1 \cdot 10^{-5}$, $1 \cdot 10^{-4}$, respectively. The batch sizes are set to 32, 128, and 128 for BAR, Waterbirds, and BFFHQ datasets, respectively. Training is conducted for 50 epochs, with weight decay values set to 0.01 for BAR and 1 for both Waterbirds and BFFHQ.

4.3 Results

We evaluate ConB on three popular benchmark datasets, affected by different types of bias and proportions of bias-aligned-bias-conflicting training samples. As evaluation metrics, we adopt Worst-Group Accuracy for Waterbirds, following [13,24]. For BFFHQ, we adopt Conflicting Accuracy, which measures the average accuracy of only bias-conflicting test samples. In BAR no ground truth bias information is available, thus we measure just the average test accuracy. Table 1 summarizes our results.

On BAR, we reach an average accuracy of $70.93 \pm 0.95\%$, surpassing recent state-of-the-art works including MoDAD ($69.83 \pm 0.72\%$) and DebiAN ($69.88 \pm 2.92\%$). In waterbirds, ConB achieves a worst-group accuracy of $90.31 \pm 0.46\%$, which is highly competitive with DRO-B2T ($90.7 \pm 0.3\%$), which leverages vision language models. On the challenging BFFHQ dataset, where only 0.5% of training samples are bias-conflicting, ConB delivers a compelling performance, with $78.13 \pm 1.92\%$ of conflicting accuracy, far exceeding other methods and outperforming the other two closest results by 9.80% (MoDAD) and by 10.57% (BiasEnsemble). These results empirically show ConB's capabilities to maintain robust performance in datasets where training bias-conflicting samples are as low as 0.5%.

4.4 Ablation Studies

We investigate the impact of varying the parameter α on the worst-group accuracy of the Waterbirds dataset. The parameter α, which decides what part of

Table 1. Comparison of debiasing methods on the three examined benchmark datasets. Best results are highlighted in bold, second best are underlined.

Method	Waterbirds Worst	BFFHQ Conflicting	BAR Average
ERM+CE	62.60 ± 0.30	60.13 ± 0.46	58.00 ± 1.39
JTT [20]	83.80 ± 1.20	62.20 ± 1.34	68.53 ± 3.29
LfF [21]	78.00	62.97 ± 3.22	62.98 ± 2.76
LWBC [12]	–	–	62.03 ± 0.74
CNC [28]	88.50 ± 0.30	–	–
GEORGE [25]	76.20 ± 2.00	–	–
BiasEnsemble [15]	–	67.56 ± 2.11	–
DebiAN [19]	–	62.80 ± 0.60	<u>69.88 ± 2.92</u>
BiaSwap [10]	–	58.87	52.44
MoDAD [22]	–	<u>68.33 ± 2.89</u>	69.83 ± 0.72
DRO-B2T [13]	**90.70 ± 0.30**	–	–
ConB (Ours)	<u>90.31 ± 0.46</u>	**78.13 ± 1.92**	**70.93 ± 0.95**

data in each batch to use in backpropagation, is systematically adjusted across a range of values from 0.1 to 0.95 to evaluate its influence in bias-identification and thus the final performance of the debiased model. Worst-group accuracy is the employed metric for this analysis.

Table 2. Ablation study on α for worst-group accuracy.

α	Worst Group Acc. (%)	α	Worst Group Acc. (%)
0.10	78.23	0.70	90.78
0.25	85.46	0.75	90.33
0.50	89.10	0.80	90.11
0.60	89.72	0.85	89.41
0.65	90.29	0.90	87.45
		0.95	83.33

The results reported in Table 2 indicate a clear trend of improvement in worst-group accuracy as α increases, up to a certain threshold, beyond which performance begins to plateau and then decline. For small values of α, such as 0.1 and 0.25, the worst-group accuracy is lower, with values of 78.23% and 85.46%, respectively. This suggests that the debiased model struggles to perform well for the most challenging group, which indicates a less precise bias pseudo-labeling. However, as α increases to 0.6, worst-group accuracy rises significantly to 89.72%, indicating that the model becomes more robust and better balances its predictions across groups.

The highest worst-group accuracy is observed at α in the interval for 0.65 to 0.80, with accuracies between 90.11% and 90.78%. However, further increases in α beyond this range lead to a slight decline in worst-group accuracy. This decline indicates that excessively large values of α may allow memorization of some bias-conflicting samples during training of the biased model, introducing additional errors into the bias identification process.

5 Conclusion

The generalization of deep neural networks may be affected by the presence of bias in data. In this paper, to mitigate this issue, we present ConB, a novel unsupervised debiasing pipeline based on a per-sample confidence selection for selective gradient backpropagation strategy during training. Designed as a two-step debiasing method, we rely on an intentionally biased auxiliary model to obtain bias pseudo-labels in a training set without bias annotations. For optimizing the auxiliary model we devise a selective backpropagation strategy, where only the gradients coming from low-loss samples are utilized for training the auxiliary model, thus promoting the learning of just the biased concepts. From the auxiliary model misclassifications, we estimate bias-aligned and bias-conflicting information, encoding it as subgroup annotations to minimize a GroupDRO objective. Our bias mitigation strategy proves effective, reaching state-of-the-art results on three popular debiasing image classification tasks. Our method does not require cumbersome hyperparameter tuning or a bias-annotated validation set, providing a practical solution for real-world debiasing applications. In future work, we plan to further investigate how bias features are propagated through the network, aiming at increasing model robustness towards spurious correlations.

References

1. Ahn, S., Kim, S., Yun, S.Y.: Mitigating dataset bias by using per-sample gradient. In: The Eleventh International Conference on Learning Representations (2023). https://openreview.net/forum?id=7mgUec-7GMv
2. Arjovsky, M., Bottou, L., Gulrajani, I., Lopez-Paz, D.: Invariant risk minimization (2020). https://arxiv.org/abs/1907.02893
3. Bahng, H., Chun, S., Yun, S., Choo, J., Oh, S.J.: Learning de-biased representations with biased representations. In: International Conference on Machine Learning (ICML) (2020)
4. Capitani, G., Bolelli, F., Porrello, A., Calderara, S., Ficarra, E.: Clusterfix: a cluster-based debiasing approach without protected-group supervision. In: Proceedings of the IEEE/CVF Winter Conference on Applications of Computer Vision, pp. 4870–4879 (2024)
5. Ciranni, M., et al.: Say my name: a model's bias discovery framework. arXiv preprint arXiv:2408.09570 (2024)

6. Ciranni, M., Pastore, V.P., Di Via, R., Tartaglione, E., Odone, F., Murino, V.: Diffusing debias: synthetic bias amplification for model debiasing. arXiv preprint arXiv:2502.09564 (2025)
7. Creager, E., Jacobsen, J.H., Zemel, R.: Environment inference for invariant learning. In: International Conference on Machine Learning, pp. 2189–2200. PMLR (2021)
8. Eyuboglu, S., et al.: Domino: discovering systematic errors with cross-modal embeddings. arXiv preprint arXiv:2203.14960 (2022)
9. He, K., Zhang, X., Ren, S., Sun, J.: Deep residual learning for image recognition. In: Proceedings of the IEEE Conference on Computer Vision and Pattern Recognition (CVPR) (2016)
10. Kim, E., Lee, J., Choo, J.: Biaswap: removing dataset bias with bias-tailored swapping augmentation. In: Proceedings of the IEEE/CVF International Conference on Computer Vision, pp. 14992–15001 (2021)
11. Kim, N., Hwang, S., Ahn, S., Park, J., Kwak, S.: Learning debiased classifier with biased committee. Adv. Neural. Inf. Process. Syst. **35**, 18403–18415 (2022)
12. Kim, N., Hwang, S., Ahn, S., Park, J., Kwak, S.: Learning debiased classifier with biased committee. In: Koyejo, S., Mohamed, S., Agarwal, A., Belgrave, D., Cho, K., Oh, A. (eds.) Advances in Neural Information Processing Systems, vol. 35, pp. 18403–18415. Curran Associates, Inc. (2022), https://proceedings.neurips.cc/paper_files/paper/2022/file/750046157471c56235a781f2eff6e226-Paper-Conference.pdf
13. Kim, Y., Mo, S., Kim, M., Lee, K., Lee, J., Shin, J.: Discovering and mitigating visual biases through keyword explanation. In: Proceedings of the IEEE/CVF Conference on Computer Vision and Pattern Recognition, pp. 11082–11092 (2024)
14. Lee, J., Park, J., Kim, D., Lee, J., Choi, E., Choo, J.: Revisiting the importance of amplifying bias for debiasing. In: Proceedings of the AAAI Conference on Artificial Intelligence, vol. 37, no. 12, pp. 14974–14981 (2023). https://doi.org/10.1609/aaai.v37i12.26748. https://ojs.aaai.org/index.php/AAAI/article/view/26748
15. Lee, J., Park, J., Kim, D., Lee, J., Choi, E., Choo, J.: Revisiting the importance of amplifying bias for debiasing. In: Proceedings of the AAAI Conference on Artificial Intelligence, vol. 37, pp. 14974–14981 (2023)
16. Li, J., Li, D., Savarese, S., Hoi, S.: Blip-2: bootstrapping language-image pre-training with frozen image encoders and large language models. In: International conference on machine learning, pp. 19730–19742. PMLR (2023)
17. Li, Y., Vasconcelos, N.: Repair: removing representation bias by dataset resampling. In: Proceedings of the IEEE/CVF Conference on Computer Vision and Pattern Recognition, pp. 9572–9581 (2019)
18. Li, Z., et al.: A whac-a-mole dilemma: shortcuts come in multiples where mitigating one amplifies others. In: Proceedings of the IEEE/CVF Conference on Computer Vision and Pattern Recognition, pp. 20071–20082 (2023)
19. Li, Z., Hoogs, A., Xu, C.: Discover and mitigate unknown biases with debiasing alternate networks. In: European Conference on Computer Vision, pp. 270–288. Springer, Heidelberg (2022). https://doi.org/10.1007/978-3-031-19778-9_16
20. Liu, E.Z., et al.: Just train twice: improving group robustness without training group information. In: Meila, M., Zhang, T. (eds.) Proceedings of the 38th International Conference on Machine Learning. Proceedings of Machine Learning Research, vol. 139, pp. 6781–6792. PMLR (2021). https://proceedings.mlr.press/v139/liu21f.html

21. Nam, J., Cha, H., Ahn, S., Lee, J., Shin, J.: Learning from failure: de-biasing classifier from biased classifier. Adv. Neural. Inf. Process. Syst. **33**, 20673–20684 (2020)
22. Pastore, V.P., Ciranni, M., Marinelli, D., Odone, F., Murino, V.: Looking at model debiasing through the lens of anomaly detection. In: 2025 IEEE/CVF Winter Conference on Applications of Computer Vision (WACV), pp. 2548–2557 (2025). https://doi.org/10.1109/WACV61041.2025.00253
23. Radford, A., et al.: Learning transferable visual models from natural language supervision. In: International Conference on Machine Learning, pp. 8748–8763. PMLR (2021)
24. Sagawa*, S., Koh*, P.W., Hashimoto, T.B., Liang, P.: Distributionally robust neural networks. In: International Conference on Learning Representations (2020). https://openreview.net/forum?id=ryxGuJrFvS
25. Sohoni, N., Dunnmon, J., Angus, G., Gu, A., Ré, C.: No subclass left behind: fine-grained robustness in coarse-grained classification problems. Adv. Neural. Inf. Process. Syst. **33**, 19339–19352 (2020)
26. Tartaglione, E., Barbano, C.A., Grangetto, M.: End: entangling and disentangling deep representations for bias correction. In: Proceedings of the IEEE/CVF Conference on Computer Vision and Pattern Recognition, pp. 13508–13517 (2021)
27. Zarlenga, M.E., Sankaranarayanan, S., Andrews, J.T., Shams, Z., Jamnik, M., Xiang, A.: Efficient bias mitigation without privileged information. arXiv preprint arXiv:2409.17691 (2024)
28. Zhang, M., Sohoni, N.S., Zhang, H.R., Finn, C., Re, C.: Correct-n-contrast: a contrastive approach for improving robustness to spurious correlations. In: Chaudhuri, K., Jegelka, S., Song, L., Szepesvari, C., Niu, G., Sabato, S. (eds.) Proceedings of the 39th International Conference on Machine Learning. Proceedings of Machine Learning Research, vol. 162, pp. 26484–26516. PMLR (2022). https://proceedings.mlr.press/v162/zhang22z.html
29. Zhang, Z., Sabuncu, M.: Generalized cross entropy loss for training deep neural networks with noisy labels. Adv. Neural Inf. Process. Syst. **31** (2018)

A Reinforcement Learning Framework for Optimizing Predictive Maintenance of Photovoltaic Systems

Andrés F. Pérez[1] and Giulia Lombardi[2(✉)]

[1] University of Manitoba, Winnipeg, MB, Canada
[2] University of Trento, Trento, Italy
giulia.lombardi@unitn.it

Abstract. Photovoltaic (PV) systems are essential to sustainable energy, but maintaining long-term efficiency requires data-efficient tools for predictive maintenance—especially when historical data is scarce. We propose a reinforcement learning (RL) framework combining digital twin simulations with a power loss regression model to estimate degradation from environmental inputs. The agent learns maintenance policies by maximizing an interpretable, economically grounded reward based on the net value of the PV system. We train our agent using three different RL algorithms (PPO, Maskable PPO, and Recurrent PPO) on observed and clear-sky irradiance data from real PV systems. All strategies outperform the no-maintenance baseline, with RPPO improving the net value by up to 37%, showing that effective, generalizable policies can be learned even under data constraints.

Keywords: Renewable energy · Climate change mitigation · Reinforcement learning · Time-series analysis · Photovoltaic systems

1 Introduction

Photovoltaic (PV) systems are central to the global transition towards low-carbon energy [2]. However, their long-term efficiency and reliability critically depend on timely and cost-effective maintenance. Traditional approaches relying on reactive or scheduled maintenance often result in suboptimal interventions or prolonged downtimes. Predictive maintenance offers a more efficient alternative by leveraging data analytics, machine learning, and forecasting to anticipate performance degradation and proactively schedule interventions, minimizing losses and maximizing energy yield [5,7].

This task naturally aligns with pattern recognition, as it requires detecting and modeling temporal trends in system degradation and power loss. In this work, we specifically target gradual degradation patterns caused by environmental and operational stressors such as soiling, wiring deterioration, or loose connections rather than sudden component failures. These slow-evolving effects demand robust time-series modeling and sequential decision-making under uncertainty.

E. Rodolà et al. (Eds.): ICIAP 2025, LNCS 16167, pp. 91–103, 2026.
https://doi.org/10.1007/978-3-032-10185-3_8

To address this problem, we adopt a reinforcement learning (RL) approach tailored to model gradual degradation in PV systems. Our framework combines digital twin simulations with with power loss estimates derived from a regression network, allowing the agent to detect and respond to evolving patterns of performance loss. A key innovation lies in the reward formulation, which moves beyond traditional energy-centric objectives to optimize the overall Net Value of the system by accounting for residual asset value, energy output, maintenance, and downtime costs. We implement our approach within a custom simulation-based environment and evaluate it across both site-specific and cross-site settings demonstrating that data unavailability is not a limitation. Given the model-free nature of the environment and the need for robustness and training stability in tasks involving discrete actions and dense, delayed reward signals, we adopt Proximal Policy Optimization (PPO) as our baseline [3,15,18]. We further evaluate two enhanced variants: Maskable PPO (MPPO), which incorporates action masking to respect budget constraints, and Recurrent PPO (RPPO), which leverages an LSTM-based policy to capture temporal dependencies. Under the assumption of prolonged effects of maintenance actions on the system's Net Value, our simulations show that it is possible to achieve both energy and Net Value surplus through a monthly planned maintenance policy under budget constraints. In particular, our best-performing model (RPPO) yields a 37% average Net Value improvement compared to the no-maintenance baseline, demonstrating the effectiveness of intelligent, temporally-aware maintenance planning in maximizing long-term system value.

2 Related Work

Detecting a failing solar panel is essential for maintaining the overall system efficiency. Several methods are available for this purpose, including performance monitoring, visual inspections, external measurement tools, and specialized devices that enable real-time monitoring of individual panels [5]. According to [10], defect detection techniques can be broadly categorized into Imaging-Based Techniques (IBTs) and Electrical Testing Techniques (ETTs) which is the focus of this work. Approaches based on AI/ML are applicable across both categories, with a considerable amount of research within IBTs devoted to computer vision techniques [1,13,16]. In the context of ETTs, few studies explore fault diagnosis using performance metrics. Most research has focused on estimating solar energy output using regression models and microclimate parameters [5], which typically do not account for losses arising from defects in PV modules or other system components.

RL offers a natural framework for predictive maintenance, as it is specifically designed to optimize sequential decision-making under uncertainty by maximizing a cumulative return. In recent years, RL has been increasingly used to optimize maintenance strategies for degrading systems [21], with applications spanning the industrial and energy sectors [14]. In the context of PV systems, it has proven its potential to provide dynamic and cost-effective solutions, outperforming traditional approaches [4,17,19]. However, the lack of failure data for performing age-based maintenance optimization remains a significant challenge [21]. Current RL-based techniques are often designed as stand-alone components for predictive maintenance, rather than being integrated within broader ensemble frameworks. These approaches tend to be site-specific, relying on predefined stochastic models for grid transition probabilities [17], failure probabilities [21], or inferred losses from environmental and internal variables [19].

To address these limitations, we propose a data-driven framework that fully integrates digital twin modeling, power loss estimation, and model-free RL within a unified analytical pipeline for predictive maintenance. Unlike prior approaches that rely on predefined failure or transition models, our method operates directly on real performance data, enabling maintenance policies to adapt dynamically to different sites without retraining on site-specific modeling assumptions.

3 Method

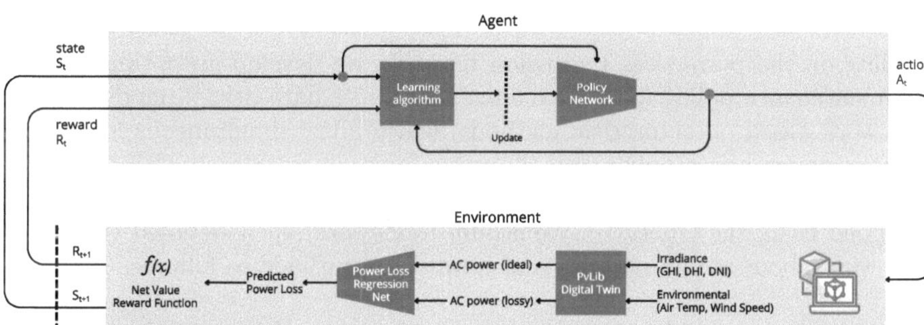

Fig. 1. Proposed framework architecture for predictive maintenance of PV systems. The environment integrates a digital twin model with a power loss regression network to simulate power loss from environmental inputs. A model-free RL agent interacts with this environment to learn a maintenance policy, guided by a Net Value reward function that accounts for intervention costs, downtime, and power loss.

With the goal of detecting defective PV systems, we design a modular architecture illustrated in Fig. 1 combining sequence regression and reinforcement learning. A sequence model is trained to regress the power loss from multivariate time series data. The estimated power loss is then used by a reinforcement learning agent to inform policy learning for preventive maintenance, enabling decisions that account for degradation trends inferred from temporal patterns.

3.1 Power Loss Regression Network

We address power loss estimation as a regression problem using a Temporal Convolutional Network (TCN) [12]. Given a multivariate time series

$$X = [x_1, x_2, \ldots, x_T]^\top \in \mathbb{R}^{T\times 2} \tag{1}$$

where T is the length of the time series and each x_t contains the ideal and actual AC power values at time step t. The TCN encodes the sequence along the time axis and the resulting embedding is then fed into a Multi-Layer Perceptron (MLP) regression network, producing the final power loss prediction.

The TCN consists of 4 convolutional layers, each with a kernel size of 3 and a hidden size of 64 channels. The layers use causal, dilated convolutions, where the dilation factor doubles at each layer (i.e., 1, 2, 4, 8) to capture increasingly long-range temporal dependencies. All convolutional layers are followed by ReLU activations and dropout with a rate of 0.2 for regularization. The final output is produced by the MLP layer to produce a scalar prediction.

This architecture is particularly effective for sequences with strong autocorrelation as it is the case of power residuals, as it models temporal structures more efficiently than LSTMs. Additionally, it outperforms Transformers by avoiding the computational overhead of attention mechanisms, making it more sample-efficient for small datasets or sequences with strong local dependencies.

3.2 Policy Learning for Preventive Maintenance

Building on the power loss regression network, we develop an optimal predictive maintenance policy for PV systems adopting a data-driven, model-free RL framework based on actorcritic methods, which allow simultaneous learning of both the maintenance policy and the value function associated with preventive interventions. We implement this framework within a custom RL environment developed using the OpenAI Gymnasium framework [9]. A detailed theoretical framework about states, actions, and rewards, is outlined as follows.

States. A state s_t encodes the system's condition at time t through four key components: predicted power loss from the Power Loss Regression Network, system lifetime, residual value, and cumulative maintenance cost. These components jointly describe the system's degradation status and the financial impact of past interventions, with further details on the economic terms provided in Sect. 3.2.

Actions. The RL agent chooses among four discrete maintenance actions—none (0), minor (1), intermediate (2), and major (3)—with increasing cost and duration of effect. Actions impact the system state via simulated degradation and recovery dynamics, with effects decaying over time at rates μ_i, where $0 < \mu_3 < \mu_2 < \mu_1$ and $\mu_0 = 0$.

Rewards. The reward R_{t+1} incorporates three core components: the system's residual value, the maintenance cost, and the downtime cost. These components jointly reflect both the system's operational performance and its financial sustainability, and are detailed in the following. The residual value $C_{res,t+1}$ captures the combined effect of natural degradation and past maintenance actions, and is defined as:

$$C_{res,t+1} = \min\left(C_0 \cdot \left(e^{-\lambda(t+1)} + \sum_{t_i \le t+1} I_{A_{t_i}} \cdot e^{-\mu_{A_{t_i}}(t+1-t_i)}\right), C_0\right) \tag{2}$$

where C_0 is the initial system cost, λ is the natural degradation rate, and each maintenance action A_{t_i} contributes a temporary recovery $I_{A_{t_i}}$ that decays over time with rate $\mu_{A_{t_i}}$. The residual value is upper-bounded by C_0 and follows an exponential decay structure commonly used in PV reliability modeling [20]. Figure 2 illustrates both short-term maintenance recovery and long-term natural degradation.

The downtime cost $C_{\text{down},t+1}(S_t, A_t)$ is computed as a function of three elements: the expected downtime duration associated with the selected action, the unit energy cost, and the average AC power loss predicted by the digital twin model. Formally, it is defined as:

$$C_{\text{down},t+1}(S_t, A_t) = d(A_t) \cdot e \cdot \mathbb{E}(L_{t+1}) \tag{3}$$

where $d(A_t + 1) \in [0, 1]$ is a downtime coefficient associated with the action A_t, representing the fraction of the time unit during which the system is non-operational due to maintenance; e is the average unit energy cost, and $\mathbb{E}(L_{t+1})$ denotes the average predicted AC power loss during time step $t + 1$.

The maintenance cost $C_{\text{maint},t+1}(A_t)$ captures the dynamic nature of intervention expenses and is defined as

$$C_{\text{maint},t+1}(A_t) = \mathrm{M}(A_t) \cdot \mathbb{E}(l_{t+1}) \tag{4}$$

where $\mathrm{M}(A_t)$ represents the maximum unitary cost associated with the selected maintenance level A_t, and $\mathbb{E}(l_{t+1})$ is the average power loss observed at time t. This approach ensures that maintenance costs are sensitive to system severity, leading to greater costs for higher degradation levels.

By combining all components defined in Eqs. 2, 3, and 4, the Net Value over a decision horizon of length T is defined as:

$$NV = C_0 + \sum_{t=1}^{T} \frac{C_{res,t}(A_t) - C_{\text{maint},t}(A_t) - C_{\text{down},t}(S_t, A_t)}{(1+r)^t}, \tag{5}$$

where the discount rate r accounts for the time value of money.

According to Eq. 5, the reward is defined as:

$$R_{t+1}(S_t, A_t) = \frac{(C_{res,t+1}(A_t) - C_{\text{maint},t+1}(A_t) - C_{\text{down},t+1}(S_t, A_t))}{C_0}, \quad \text{with } R_0 = 1. \tag{6}$$

By setting the discount factor $\gamma = \frac{1}{1+r}$, the total discounted return $G_t = \sum_{k=0}^{T-t} \gamma^k R_{t+k}$ at time $t = 0$ directly corresponds to the Net Value of the system, thus embedding the long-term economic objective into the reward structure.

In fact, is it easy to see that G_0 can be rewritten as

$$G_0 = r_0 + \sum_{k=1}^{T} \gamma^k R_k = \frac{1}{C_0} \left(C_0 + \sum_{k=0}^{T} \frac{C_{res,k} - C_{\text{maint},k} - C_{\text{down},k}}{(1+r)^k} \right) = \frac{NV}{C_0}. \quad (7)$$

4 Experimental Results

Our goal is to learn a policy that generates maintenance schedules based on AC power yield and local environmental and meteorological data, with the objective of maximizing the long-term Net Value of a PV system under a constrained maintenance budget. To evaluate the effectiveness of our approach, we conduct a series of experiments targeting the two main components of our method. First, we validate the power loss regression network to determine its accuracy and its reliability as a component of the full system. Second, we assess the RL-based maintenance policy by evaluating: 1) the benefit of using an optimized maintenance schedule, 2) the performance of the policy when environmental data is unavailable, and 3) its robustness when deployed in sites lacking historical yield data.

4.1 The NIST Dataset

To evaluate the performance of the proposed framework, we use the NIST Campus PV Arrays and Weather Stations dataset [6], which features three grid-connected photovoltaic (PV) arrays located on the National Institute of Standards and Technology (NIST) campus in Gaithersburg, Maryland, USA. The dataset offers comprehensive, one-minute averaged PV performance and meteorological data collected between 2015 and 2018. It also includes hourly images of the sky, which enable the evaluation and validation of short-term irradiance forecasts, and hourly images of the arrays, which support the observation of shading, snow, and any obstructions or signs of damage.

All three arrays employ identical monocrystalline silicon modules, but differ in mounting configurations. The *Parking Lot Canopy Array* is mounted on east/west-facing canopies over a parking area, the *Ground Mount Array* is installed on tilted ground supports in an open field, and the *Roof Tilted Array* is placed on tilted, weighted racks atop a flat building roof. Each array is equipped with research-grade instrumentation to monitor electrical output, temperature, and a variety of meteorological variables including wind speed, ambient temperature, and solar irradiance components. Additionally, a comprehensive weather station on a nearby building provides detailed measurements of solar components, spectral distribution, ultraviolet and infrared radiation, along with standard meteorological conditions.

4.2 Power Loss Estimation

A major challenge in training the proposed power regression network is the lack of publicly available datasets that include both ideal and actual yield power with detailed loss annotations. To address this, we turn to digital twin technology, leveraging *pvlib*, a widely recognized open-source library for simulating the performance of PV systems. This library implements the PVWatts [8] multiplicative power losses model, where all losses are categorized into a single total system loss percentage $L_{\text{total}} = \left[1 - \prod_i \left(1 - \frac{L_i}{100}\right)\right]$, which aggregates various loss mechanisms into one simplified value used as a derating factor on the DC power output.

Aiming at generating site-specific energy production simulations, *pvlib* requires solar irradiance measurements including Direct Normal Irradiance (DNI), Global Horizontal Irradiance (GHI), and Diffuse Horizontal Irradiance (DHI), along with optional meteorological variables such as wind speed, air temperature, and surface albedo. To model varying degrees of power loss, we sample from predefined ranges for three key loss types chosen due to their high maintenance impact and potential for significant efficiency improvement: soiling, wiring, and connections. We simulate ten daily energy production runs for each of the three PV arrays in the NIST dataset, yielding 14,600 time series for the *Canopy* and *Ground* sites, and 11,740 for *Roof*, which has data only until March 19, 2018.

We train the power regression network on each site-specific dataset where data from 2015 is used for training and validation with an 80/20 split, and data from the following years is used as the test set. In all cases we trained for 100 epochs with batches of 32 elements using the Adam optimizer with learning rates of $1 \times 10_2$. As a baseline, we use a function that computes the step-wise mean proportion between the ideal and actual yield power, mimicking the multiplicative losses model of *PVWatts*. The performance of these models for all three grid PV array sites is summarized in Table 1 where we report the coefficient of determination or goodness of fit (R^2) and the Mean Squared Error (MSE).

Table 1. Performance of the power loss estimation model highlighting accuracy (R^2) and error (MSE) for all three NIST PV arrays.

	Canopy		Ground		Roof	
Metric	Baseline	Ours	Baseline	Ours	Baseline	Ours
MSE	0.2249	**0.0020**	0.2217	**0.0013**	23671.01171875	**0.0022**
R^2 score	–3.3586	**0.9604**	–3.2506	**0.9748**	–476549.0625	**0.9568**

The proposed regression network consistently outperforms the baseline across all sites, with the negative R^2 scores meaning that our model outperforms the baseline, and the baseline performs worse than simply predicting the mean (in the traditional R^2 sense). The largest error is reported at the *Roof* site, with an

MSE and R^2 score several orders of magnitude larger than those achieved by the model at the other sites. This means that the step-wise mean proportion baseline is making large errors, indicating that the simple proportional assumption does not hold well, either because the ratio is unstable over time or because the baseline fails to model important temporal patterns (such as sudden drops, ramp ups, or seasonal effects).

In comparison with the other sites, the PV system at the *Roof* site packs a distinct inverter type, which may increase sensitivity to disturbances. However this should be further investigated. For the remainder of this section we use the best model per site. Compared to the other sites, the PV system at the *Roof* site uses a distinct inverter type, which may increase sensitivity to disturbances. This could explaining the poor performance of our simplistic baseline model, however, this should be further investigated. For the remainder of this section, we use the best model per site.

4.3 Predictive Maintenance Policy Evaluation

Using the site-specific power regression networks trained in the previous section we train the RL predictive maintenance model earlier described. We evaluate three PPO-based algorithms: Vanilla PPO (PPO), Maskable PPO (MPPO) with action masking for budget constraints, and Recurrent PPO (RPPO) with an LSTM-based policy to model temporal dependencies.

Agents are trained in a custom RL environment built with the OpenAI Gymnasium framework [9], simulating a 30-day horizon with four discrete maintenance actions. Natural degradation follows an exponential decay with rate $\lambda = 0.0005$, while temporary recoveries from maintenance are modeled using action-specific decay rates $\mu = [0.0, 0.658, 0.329, 0.219]$. Actions incur downtime of 0–1 day and costs of \$0–\$300. Energy is valued at \$0.13/kWh, with a monthly maintenance budget of \$700 and a discount rate of $r = 0.05$.

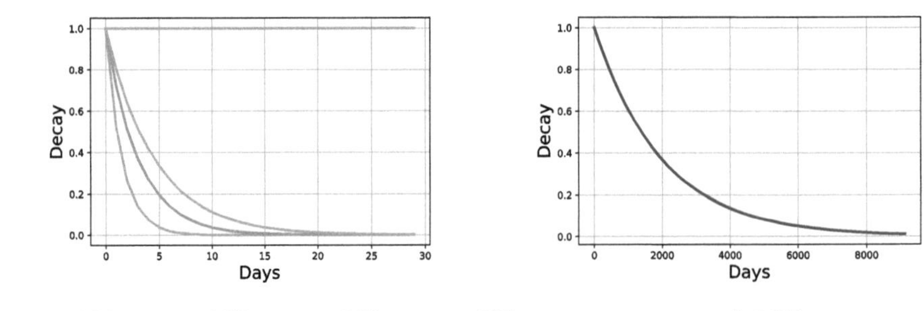

Fig. 2. Exponential decay functions used in the environment. Left: Decay of maintenance effects over a one-month horizon for different decay rates μ. Right: Natural PV system degradation over a 25-year horizon, modeled with $\lambda = 0.0005$.

Each model is trained for a maximum of 50,176 steps with early stopping based on reward stability. Training uses a rollout horizon of 256, batch size of 1024, learning rate $3 \cdot 10^{-4}$, and discount factor $\gamma = (1+r)^{-1} \approx 0.952$. The policy network consists of two fully connected layers with 64 units; RPPO includes an additional LSTM module to model temporal dependencies. All algorithms are trained across three PV sites (*Canopy, Ground, Roof*).

To validate the robustness of our approach in scenarios where real environmental measurements are unavailable, we use the Ineichen and Perez clear-sky model [11] to estimate irradiance data. Models are trained on both estimated and observed inputs, resulting in a total of 18 experimental configurations.

Trained models are evaluated across all calendar months of 2017 to assess maintenance policy performance. We report two primary metrics, the average monthly Net Value (US$) and the average monthly energy production (kW). Table 2 summarizes the results across all experimental configurations, including the no-maintenance baseline.

Table 2. Achieved episodic average energy production and Net PV system Value across all grid arrays sites using different learned maintenance policies. IRR: Real irradiance values from weather station data. CS: Estimated irradiance values according to the clear-Sky model [11].

Site	Metric	No Maint.		PPO		MPPO		RPPO	
		IRR	CS	IRR	CS	IRR	CS	IRR	CS
Canopy	AC Power (kW)	29.79	43.35	29.80	43.36	29.79	43.38	30.22	44.80
	Net Value (US$)	32.1K	32.1K	32.1K	32.1K	32.1K	32.3K	**44.0K**	49.3K
Ground	AC Power (kW)	44.86	73.65	44.88	73.66	45.16	73.69	44.98	73.83
	Net Value (US$)	32.1K	32.1K	32.1K	32.1K	**39.7K**	32.4K	35.1K	35.0K
Roof	AC Power (kW)	10.61	16.63	10.61	16.64	10.61	16.64	10.61	17.09
	Net Value (US$)	32.1K	32.1K	32.1K	32.1K	32.1K	32.1K	**32.1K**	47.8K

All strategies outperform or match the no-maintenance baseline, confirming the effectiveness of RL for predictive maintenance. RPPO achieves the highest Net Value on real observed data at both the *Canopy* ($44,045.90) and *Roof* ($32,192.76) sites, leveraging its LSTM-based policy to identify degradation trends and schedule well-timed, medium-level interventions. MPPO, which incorporates action masking, follows a more reactive approach with more frequent high-level actions, reaching the best performance at the *Ground* site ($39,707.41). PPO, in contrast, is the weakest performer, as the limited number of training timesteps—designed to reflect real-world data scarcity—proves insufficient for learning an effective maintenance strategy. In conclusion, RPPO's advantage becomes even more evident with clear-sky inputs, where reduced environmental noise further amplifies the benefits of temporal modeling.

Due to space constraints, we present qualitative results from a representative case. Figure 3 shows the behavior of RPPO on the *Canopy* installation, offering a year-long overview (2017) of the agent's maintenance actions, reward trajectory, and energy production. The first plot indicates that the agent follows a proactive strategy, with frequent low- and medium-level interventions and only a few major actions across the year. The second plot shows a consistent improvement in reward compared to the no-maintenance baseline, highlighting the cumulative benefits of well-timed interventions. Finally, the energy production curve reveals that maintenance actions often coincide with periods of high irradiance, suggesting that the agent effectively learns to leverage favorable conditions to enhance long-term system performance.

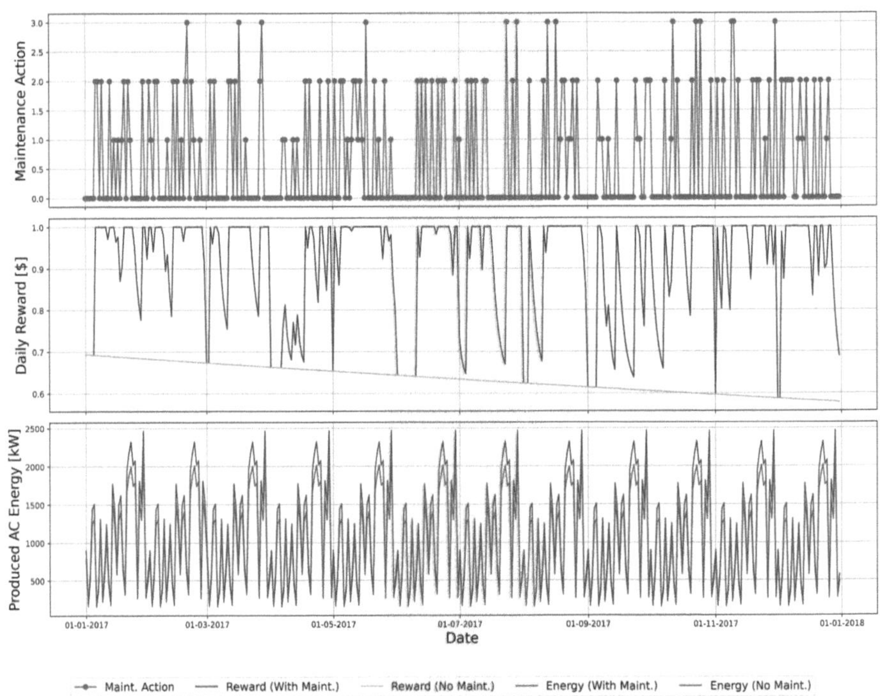

Fig. 3. Maintenance policy learned by the RPPO agent for the *Canopy* installation. Top: daily maintenance actions selected by the agent throughout 2017. Middle: comparison of daily scaled reward trajectories with and without maintenance. Bottom: produced AC energy under both strategies.

To assess generalization, we perform a cross-site evaluation by testing RPPO models trained on one site against unseen target sites, using real irradiance data. As shown in Table 3, models trained on the *Canopy* site consistently yield the best transfer performance in terms of both Net Value and energy output, outperforming single-site baselines. This suggests that maintenance policies can be effectively transferred between nearby installations—such as the NIST arrays, which are located within a 2 km radius—providing a practical solution in scenarios where historical data is limited. However, differences in local degradation dynamics or operational regimes may still affect policy effectiveness, especially when training is performed over short time horizons.

Table 3. Performance of the Predictive Maintenance Policy Network in a cross-site setting.

Train Set	Canopy		Ground		Roof	
Test Set	Ground	Roof	Canopy	Roof	Canopy	Ground
AC Power (kW)	45.52	10.77	29.83	10.63	29.81	44.88
Net Value (US$)	**44.8K**	**44.9K**	33.8K	34.0K	32.8K	32.2K

5 Conclusions

We proposed an RL framework for predictive maintenance of PV systems, integrating digital twin simulations and a time series analysis model for power loss estimation. The key innovation lies in a reward formulation that optimizes the system's Net Value, incorporating energy production, asset depreciation, maintenance costs, and downtime. Experimental results on three real-world PV arrays demonstrate that our best-performing agent (RPPO) achieves a monthly NV increase of over 37% compared to the no-maintenance baseline, confirming the effectiveness of proactive interventions under constrained budgets.

Nonetheless, several limitations remain. The current agent operates day by day without foresight beyond the immediate timestep, limiting long-term planning. Moreover, the power loss model in pvlib relies on a multiplicative formulation that lacks expressiveness in capturing overlapping degradation phenomena. Another important constraint is the dependency on pvlib's internal component models: the simulation can only be replicated for PV systems whose module and inverter specifications are included in the pvlib database.

Future work will focus both on reinforcing the theoretical backbone of the framework and on overcoming the identified limitations, while also enabling practical deployment. On the theoretical side, we plan to incorporate environmental forecasts—e.g., via learned models or reanalysis datasets like ERA5—not only to support multi-day decision-making, but also to anticipate environmentally induced losses such as soiling or shading. In the absence of annotated datasets or dedicated on-site sensors required for fine-grained diagnostics, correlating observed losses with environmental variables offers a promising first step toward distinguishing different degradation types. We also aim to validate the framework across a broader range of PV installations to assess its generalization capabilities under heterogeneous operational conditions. On the deployment side, the learned policy could be embedded into standalone applications or integrated into existing monitoring platforms developed by PV manufacturers, utilities, or service providers. These tools could offer actionable recommendations based on local conditions and historical performance, supporting both small-scale users and utility-scale operators in optimizing maintenance schedules. By enabling interpretable and adaptive decision-making, such tools would bridge the gap between data-driven strategies and real-world operations, promoting wider adoption in both residential and industrial PV systems.

Reproducibility

To enable reproducibility of our results and to further motivate research on predictive maintenance of PV systems, we release publicly our code and models `https://github.com/afperezm/pv-predictive-mainten ance` .

Acknowledgments. The authors would like to thank Dr. Mostafa Farrokhabadi from the University of Calgary for his support in the development of earlier versions of this article through the Climate Change AI (CCAI) mentorship program at ICLR 2025. We are also grateful to Alessandra Capelli, engineer at the Municipality of Bolzano, for the insightful discussions that contributed to shaping this work.

References

1. Al-Mashhadani, R., et al.: Deep learning methods for solar fault detection and classification: a review. Sol. Cells **11**, 12 (2021)
2. Ali, B.M., et al.: Sustainable strategies for preventive maintenance and replacement in photovoltaic power systems: enhancing reliability, efficiency, and system economy. Unconvent. Res. **6**, 100170 (2025). https://doi.org/10.1016/j.uncres.2025.100170. https://www.sciencedirect.com/science/article/pii/S2666519025000366
3. Andrychowicz, M., et al.: What matters for on-policy deep actor-critic methods? A large-scale study. In: International Conference on Learning Representations (ICLR) (2021)
4. Bacalhau, E.T., Casacio, L., Barbosa, F., Yamada, F., Guimarães, L.: Preventive maintenance policy in photovoltaic systems using reinforcement learning. In: Proceedings of the 12th IMA International Conference on Modelling in Industrial Maintenance and Reliability (2023). https://api.semanticscholar.org/CorpusID:264940945
5. Bosman, L.B., Leon-Salas, W.D., Hutzel, W., Soto, E.A.: Pv system predictive maintenance: challenges, current approaches, and opportunities. Energies **13**(6) (2020). https://doi.org/10.3390/en13061398. https://www.mdpi.com/1996-1073/13/6/1398
6. Boyd, M., Chen, T., Dougherty, B.: NIST Campus Photovoltaic (PV) Arrays and Weather Station Data Sets (2017). https://doi.org/10.18434/M3S67G
7. Carvalho, T., Soares, F., Vita, R., Francisco, R., Basto, J.P., Alcalã, S.G.S.: A systematic literature review of machine learning methods applied to predictive maintenance. Comput. Ind. Eng. **137** (2019). https://doi.org/10.1016/j.cie.2019.106024
8. Dobos, A.P.: Pvwatts version 5 manual (2014). https://doi.org/10.2172/1158421. https://www.osti.gov/biblio/1158421

9. Foundation, F.: Gymnasium: The next generation of openai gym (2023). https://gymnasium.farama.org/. Accessed 29 Aug 2024
10. Hijjawi, U., Lakshminarayana, S., Xu, T., Piero Malfense Fierro, G., Rahman, M.: A review of automated solar photovoltaic defect detection systems: Approaches, challenges, and future orientations. Sol. Energy **266**, 112186 (2023). https://doi.org/10.1016/j.solener.2023.112186. https://www.sciencedirect.com/science/article/pii/S0038092X23008204
11. Ineichen, P., Perez, R.: A new airmass independent formulation for the linke turbidity coefficient. Sol. Energy **73**(3), 151–157 (2002). https://doi.org/10.1016/S0038-092X(02)00045-2. https://www.sciencedirect.com/science/article/pii/S0038092X02000452
12. Lea, C., Vidal, R., Reiter, A., Hager, G.D.: Temporal convolutional networks: a unified approach to action segmentation. In: Hua, G., Jégou, H. (eds.) ECCV 2016. LNCS, vol. 9915, pp. 47–54. Springer, Cham (2016). https://doi.org/10.1007/978-3-319-49409-8_7
13. de Oliveira, A.K.V., Aghaei, M., Rüther, R.: Automatic inspection of photovoltaic power plants using aerial infrared thermography: a review. Energies **15**(6) (2022). https://doi.org/10.3390/en15062055. https://www.mdpi.com/1996-1073/15/6/2055
14. Perera, A., Kamalaruban, P.: Applications of reinforcement learning in energy systems. Renew. Sustain. Energy Rev. **137**, 110618 (2021). https://doi.org/10.1016/j.rser.2020.110618. https://www.sciencedirect.com/science/article/pii/S1364032120309023
15. Raffin, A., Hill, A., Ernestus, A., Gleave, A., Kanervisto, A., Dormann, N.: Stable-baselines3: reliable reinforcement learning implementations (2021). https://github.com/DLR-RM/stable-baselines3. Accessed 14 Apr 2025
16. Rana, N., Arora, S.: A review on surface defect detection of solar cells using machine learning. In: Marriwala, N., Tripathi, C.C., Jain, S., Mathapathi, S. (eds.) Soft Computing for Intelligent Systems, pp. 385–395. Springer, Singapore (2021)
17. Rocchetta, R., Bellani, L., Compare, M., Zio, E., Patelli, E.: A reinforcement learning framework for optimal operation and maintenance of power grids. Appl. Energy **241**, 291–301 (2019). https://doi.org/10.1016/j.apenergy.2019.03.027. https://www.sciencedirect.com/science/article/pii/S0306261919304222
18. Schulman, J., Wolski, F., Dhariwal, P., Radford, A., Klimov, O.: Proximal policy optimization algorithms. arXiv preprint arXiv:1707.06347 (2017)
19. Uhanto, U., Yandri, E., Hilmi, E., Saiful, R., Hamja, N.: Predictive maintenance with machine learning: a comparative analysis of wind turbines and pv power plants. Heca J. Appl. Sci. **2**(2), 87-98 (2024). https://doi.org/10.60084/hjas.v2i2.219. https://heca-analitika.com/hjas/article/view/219
20. Yan, W., Liu, W., Kong, W.: Reliability evaluation of PV modules based on exponential dispersion process. Energy Rep. **7**, 3023–3032 (2021)
21. Zhang, P., Zhu, X., Xie, M.: A model-based reinforcement learning approach for maintenance optimization of degrading systems in a large state space. Comput. Ind. Eng. **161**, 107622 (2021). https://doi.org/10.1016/j.cie.2021.107622. https://www.sciencedirect.com/science/article/pii/S036083522100526X

Unsupervised Segmentation for Reliable Estimation of Stratum Corneum Thickness Using LC-OCT

Raoul Missodey[1,2(✉)], Rémy Leconge[1], Samuel Ralambondrainy[2], Sylvie Treuillet[1], Yves Lucas[1], Jean-Hubert Cauchard[2], and Franck Bonnier[2]

[1] Laboratoire PRISME, Polytech Orléans, 12 rue de Blois, 45067 Orléans, France
[2] LVMH Recherche, 185 avenue de Verdun, 45800 Saint Jean de Braye, France
raoul.missodey@etu.univ-orleans.fr, fbonnier@research.lvmh-pc.com

Abstract. The thickness of the stratum corneum (SC) is an important biomarker for assessing the quality of the skin barrier function. Thanks to LC-OCT imaging, it is now possible to determine SC thickness, in vivo, non-invasively and with a precision comparable to that of histology (about 1 μm). In this paper, we propose a reliable, unsupervised and fully automatic method for SC segmentation, incorporating hair removal. We compare it to the reference method for estimating SC thickness by LC-OCT based on deep learning model U-Net. Differences between the two methods are evaluated on a database of 900 images of healthy skin acquired from one hundred volunteers. The mean absolute error lies between 0.41 μm and 0.65 μm, i.e. around half the resolution of the imaging system. The strong similarity between the two methods is confirmed by a Student's test on the hypothesis that the difference is less than the resolution of the imaging system, with p-value of less than 0.001. We can therefore conclude that our method is as efficient and accurate as the reference one using deep learning. Being unsupervised, our method has the advantage of not requiring manual annotation of a large dataset for model training. Furthermore, our method integrates hair removal in a fully automatic way, whereas U-Net segmentation must be followed by manual hair removal. These methods surpass the state-of-the-art of other imagery and could therefore help reduce the number of skin biopsies.

Keywords: Segmentation · 3D LC-OCT · Skin

1 Introduction

The skin is a complex organ with a multilayered structure, which envelops the entire body with a surface area of around 2 m^2 for an adult. The stratum corneum (SC) is the first layer with a thickness between 10 and 30 μm depending on the location. SC is in direct contact with the external environment and plays a crucial barrier role again external threats and in controlling water flow. The SC thickness

E. Rodolà et al. (Eds.): ICIAP 2025, LNCS 16167, pp. 104–115, 2026.
https://doi.org/10.1007/978-3-032-10185-3_9

is a key biomarker of the integrity of the skin barrier function and its ability to regulate the penetration of active molecules from topical dermo-cosmetic products. Skin aging can lead to modifications and potential impairments (dryness, micro wrinkles,etc.), which, when measured, can serve as indicators for better treatment using suitable skincare products. Therefore, the measurement of SC thickness can be used as a biomarker for testing product efficacy.

Recent imaging techniques make it possible to measure SC thickness non-invasively. In particular, Line Field Confocal Optical Coherence Tomography (LC-OCT) is an in vivo 3D imaging technology that delivers optical biopsies of the skin with high isotropic resolution ($\sim$1 μm) and a depth of analysis up to 500 μm, through the superficial layers of the skin: stratum corneum (SC), epidermis, and papillary dermis. Calculating thickness of skin layers requires reliable segmentation in images. Published reference techniques for measuring the thickness of the SC are often based on deep learning (DL) and time-consuming manual annotation of large databases, which can be affected by inter-observer variability. Additionally, the presence of hairs or follicles in the skin interferes with the measurement, and requires a manual post-processing to remove hairs from images after DL segmentation. To adress these limitations, this paper proposes an alternative unsupervised segmentation method that incorporates hair detection and thus enables reliable and fully automatic estimation of SC thickness in LC-OCT. The proposed method is compared with the state-of-the-art DL-based method and achieves similar accuracy on a large dataset of healthy skin 3D images. The rest of the paper is organized as follows. Section 2 reviews the relative work. The image database and segmentation methodology developed are described in Sect. 3. Section 4 presents the experiments carried out and the results obtained. The paper concludes with a discussion and a conclusion.

2 Relative Work

Among the non-invasive in vivo imaging techniques well-established in dermatology are reflectance confocal microscopy (RCM), High-Frequency Ultrasound (HFUS) imaging and Optical Coherence Tomography (OCT), Confocal laser scanning microscopy (CLSM) or Multiphoton Laser Scanning Microscopy (MLSM). These technologies offer different balances between spatial resolution and imaging depth for skin structure investigation. Although HFUS imaging [4] has been widely used for the characterization of the collagen network in the dermis, the technique suffers from a limited resolution of a few tens of micrometers depending on the frequency used, making it irrelevant for studying finer microstructures such as the stratum corneum (SC) and epidermis. Recent advances in OCT revealed greater precision for quantifying intradermal structures between 3 and 7.5μm [9,19]. RCM [8] offers higher resolution, comparable to histology ($\sim$1μm), but it is limited to a depth of 200-250μm. Since 2020, Line Field Confocal Optical Coherence Tomography (LC-OCT, DAMAE Medical) has emerged as a powerful imaging tool by combining the depth of the OCT vertical section with the spatial resolution of RCM [6,17]. Despite a shallower penetration depth than OCT, LC-OCT offers the advantage of 3D exploration of the skin (1200 μm × 500 μm × 350 μm), with an isotropic spatial resolution about 1μm, and quasi-real-time acquisition of optical biopsies of the skin in vivo comparable to histology [18]. In LC-OCT images, skin structures with a high refractive index, such as keratin-rich corneocytes, shine brighter facilitating better observation and segmentation of the SC (see Fig. 1).

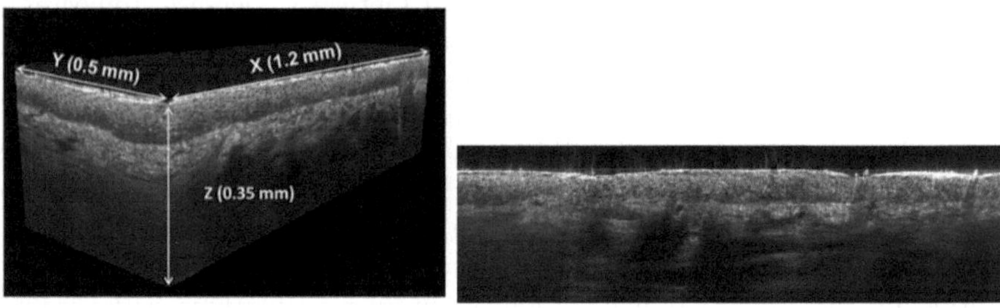

Fig. 1. (Left) 3D stack of LC-OCT images on healthy skin from [1]; (Right) Example of B-Scan.

Various automated or semi-automated approaches for skin layers segmentation have been proposed according to the different imaging technologies. A large part seeks to separate dermis and epidermis or detect dermal-epidermal junction (DEJ), far fewer studies deal with SC segmentation. A conventional algorithm based on speckle noise reduction, hysteresis-based edge detection and morphological filtering was unable to correctly handle skin layer segmentation in OCT images in the case of intersections with veins, sweat ducts or hair follicles [2]. In [21], a generic 3D graph-based segmentation was applied with an improved cost function across B-scan to deal with the problem of shadowing due to the presence of hair. Experiments on a dataset of 26 volumes acquired on anterior and posterior forearms and face from healthy subjects report mean absolute errors (MAE) of 2.48, 3.29 and 1.97 pixels for the methods [12,13,21] respectively. In the best case scenario, the error is therefore about 5-6μm.

More recently, a deep learning (DL) method was tested to segment epidermis in images, based on the U-Net architecture [5]. Post-processing consisting of a Savitzky-Golay and Fourier filtering is included to remove follicular structures. An error of 10.25 μm on epidermal thickness is reported on 270 images acquired with an ultra high resolution OCT system which achieves 6 μm and 2 μm for lateral and axial resolution. Other work [11,15] has tested U-Net models with different pre-trained encoders such as VGG, ResNet or Inception to segment different skin layers in OCT images. With a refined U-Net, a 4-part segmentation (air, stratum corneum, epidermis and dermis) is achieved with fairly good overall accuracy, but the authors point out the challenge in correctly delineating the SC layer as it has a smaller number of pixels than the other layers [15]. The authors have gone further than skin layer segmentation [16], proposing quantitative detection of keratinocyte cell nuclei [14]. 2D and 3D U-Net models were also compared for delineating skin layers boundaries in OCT images [20]. Although the 3D model shows greater accuracy in delineating papillary patterns, the study does not confirm that 3D information can improve the accuracy of layer segmentation.

To the best of our knowledge, the only work proposing an estimation of SC thickness by LC-OCT is based on U-Net model [3]. This DL segmentation method requires the annotation of a large labeled dataset for the machine learning. Manual segmentation is observer-dependent, and can present high inter- and intra-observer variability. Moreover, a manual post-processing step is required to remove hairs or follicles in the images so as not to interfere with the measurement of skin thickness (see Fig. 1). Therefore, the use of a fully automated unsupervised method integrating hair removal, as proposed in this paper, overcomes these drawbacks.

3 Data and Methods

3.1 Data

The experiments use the same dataset as [3], which was collected as part of a study on the efficacy of cosmetic products. Consequently, only healthy cases are taken into account. The LC-OCT images were acquired using a deepLiveTM system (DAMAE MEDICAL, France) on three different facial zones (temple, cheekbone and mandible), from one hundred healthy Caucasian female volunteers, aged from 20 to 70 years (with a distribution of 20 women for each decade). Three acquistions were performed for each areas. The data collection comprises 900 stacks of $1200 \times 500 \times 350$ voxels (Fig. 1). Both vertical images (B-scans) and horizontal ones (C-scans) have a spatial resolution of $\sim$1 μm per pixel.

3.2 Segmentation Method

A flowchart of the proposed approach is shown in Fig. 2. Starting with the input of 3D stack, the process is divided into two main flows: SC layer detection is realized in each vertical images B-scans (xz) and hair detection is realized in horizontal images C-scans (xy). The segmentations obtained in the B and C scans are merged. In this way, the thickness of the SC is calculated on the layer segmented in each B scan after hair removal, so as not to bias the thickness estimate. The image processing carried out in each flow is detailed below.

Stratum Corneum Detection
Traditional automatic contour detection algorithms such as Canny, Prewitt or Sobel often require image-specific adjustments to achieve optimum results. They have been tested, but failed to deliver satisfactory results in the LC-OCT skin images. Conversely, the SC detection in vertical images B scan (xz) based on eigenvalues analysis of the Hessian matrix has yielded good results. The Hessian matrix contains the second partial derivatives of the image intensity; its eigenvalues describe the local curvatures of the image intensity at a given point. They

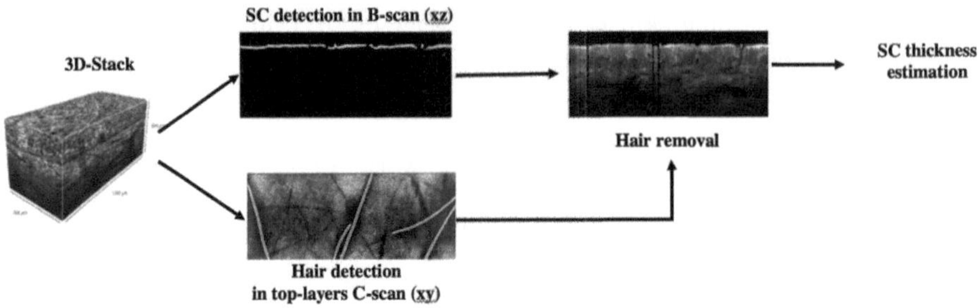

Fig. 2. Stratum corneum segmentation flow

can thus be used to detect points of high curvature in the image that correspond to contours or corners.

In linear scale spaces theory, the second derivative at a scale σ is defined as the convolution of raw image data $I(x, y)$ and derivative of a Gaussian filter (1).

$$\frac{\partial^2 I_\sigma(x,y)}{\partial} = I(x,y) * \frac{\partial^2 G_\sigma(x,y)}{\partial} \tag{1}$$

Figure 3 shows an example of Hessian features at different scales, using Gaussian kernels from the B scan of a 22 years old caucasian female (right cheek), illustrated in Fig. 1.

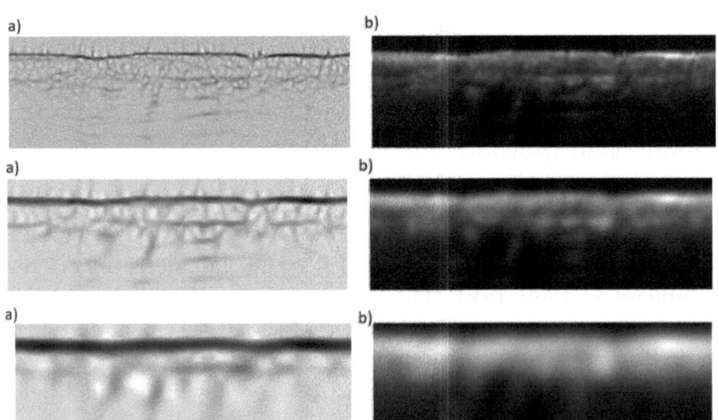

Fig. 3. a) Multiscale Hessian features in a B-scan image: a) First eigenvector e_1 from Hessian and b) Blurred image by Gaussian filtering f_σ, with scales increasing from top to bottom, i.e. σ values equal to 5.22, 11.43, and 25.

As shown in Fig. 4, the finest scale $\sigma_1 = 5.22$ enhances accurately the SC, which can be segmented by using 2-class Kmeans pixel clustering [7]. The value of 5.22 was set empirically, observing that below, the SC was undersegmented, and above, oversegmented. However, the clustering brings some artifacts in depth. To remove them, we filtered the first eigenvector of the Hessian e_1 by multiplying it by the blurred image calculated previously f_σ, before applying the Kmeans. As it can be observed in the next two rows of the Fig. 4, the best filtering is obtained for the Gaussian kernel at the scale of $\sigma_2 = 11.42$. A larger-scale filtering with $\sigma_3 = 25$ gives more artifacts. We therefore apply the 2-class Kmeans clustering to the filtered images $e_1 \times f_{\sigma_2}$ from each B-scan image.

Finally, morphological filtering operations were performed to remove small disconnected objects and to join some disconnected part of the SC layer in the image.

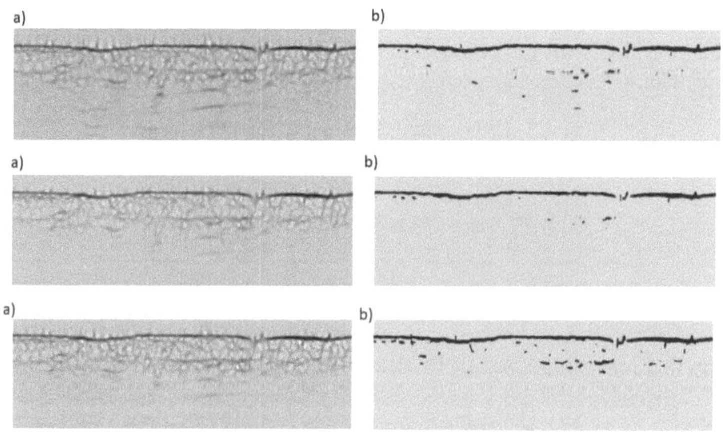

Fig. 4. SC segmentation by Kmeans: (1st line) a) the first eigenvector from Hessian e_1 and b) 2-class Kmeans pixel clustering from e_1; (2nde line) a) the filtered first eigenvector $e_1 \times f_{\sigma_2}$ with $\sigma_2 = 11.42$; b) 2-class Kmeans pixel clustering from $e_1 \times f_{\sigma_3}$ a) the filtered first eigenvector 2-class Kmeans pixel clustering with $\sigma_3 = 25$; b) 2-class Kmeans pixel clustering from $e_1 \times f_{\sigma_3}$.

Hair Removal

Hair removal is absolutely necessary to avoid bias in the estimation of SC thickness. Hairs are present in the top layers of the LC-OCT 3D Stack with various directions, as observed in Fig. 5, which shows some C scan (xy) at different depths. The hairs appear as very shiny filaments.

To recover all hair structures that may interfere with SC segmentation, a Principal Composant Analysis (PCA) [10] is applied to the 20 first C scans (xy). The image corresponding to the first component PC_1 represents the maximum variance and thus includes the whole hairs as shown in Fig. 6.

As hairs are tubular structures, they can thus be detected from a local curvature analysis using the eigenvalues of the Hessian matrix, applied at the appropriate scale. Figure 7 shows the multiscale Hessian features extracted from PC_1,

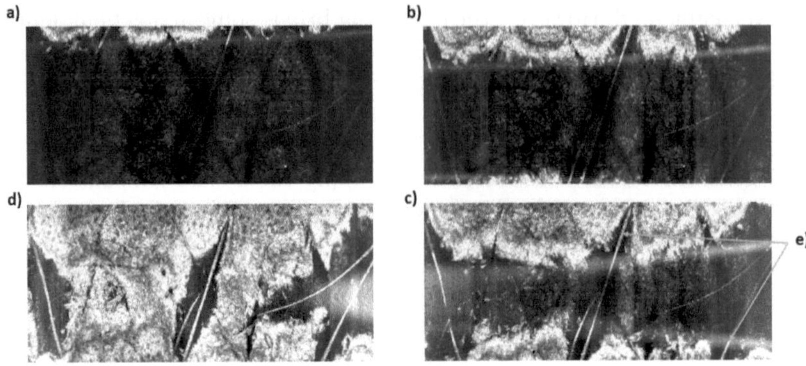

Fig. 5. Horizontal C scan (xy) images at differents depth. a) 1 μm, b) 5 μm, c) 9 μm, d) 13 μm e) Hairs.

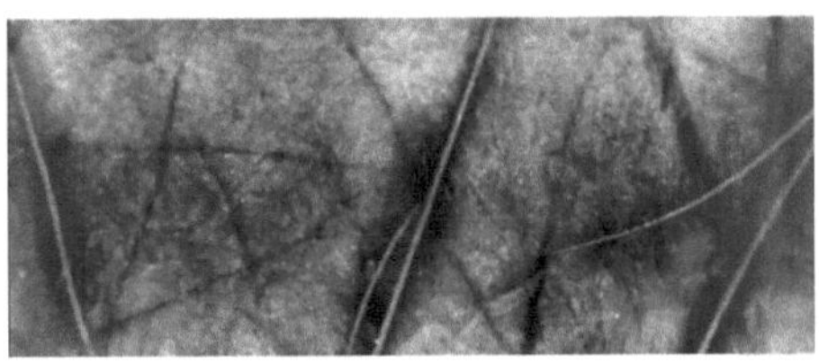

Fig. 6. First component of PCA applied to the 20 first C scans (PC_1).

the first component of PCA applied to the 20 first C scans (xy). The finest scale $\sigma = 5.22$ is suitable for detecting hairs, and Kmeans algorithm is applied directly to the PC_1 image. After some morphological filtering, it is possible to obtain a precise segmentation of the hairs (see Fig. 8).

The segmented hairs in PC_1, assimilated to the C scan (xy), enable the pixels of the corresponding x-column to be set to zero in the B scans (xz), where SC segmentation is applied (as shown in Fig. 9).

The Fig. 10 shows the final SC segmentation without and with hair removal. Note that without blanking of the columns corresponding to the hairs, the thickness estimate would have been biased.

4 Experiments and Results

The proposed segmentation method was tested on the dataset presented in Sect. 3.1, which comprises 900 LC-OCT stacks; the three areas of healthy skin (temple, cheekbone, and mandible) were scanned three times from one hundred volunteers. For comparison with the reference method based on the U-Net [3], the same protocol was applied for calculating SC thickness. After segmentation with hair removal, the SC thickness estimate is averaged from all B scans in a 3D LC-OCT stack.

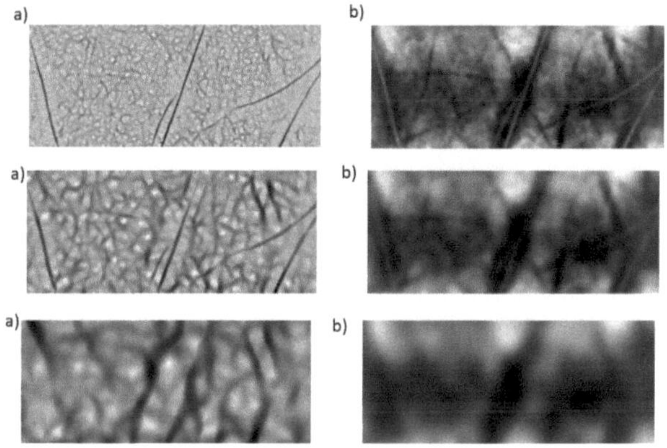

Fig. 7. Multiscale Hessian features in PC_1 (the first component of PCA applied to the 20 first C scans): a) First eigenvector from Hessian and b) Blurred image by Gaussian filtering with scales increasing from top to bottom, i.e. σ values equal to 5.22, 11.43, and 25.

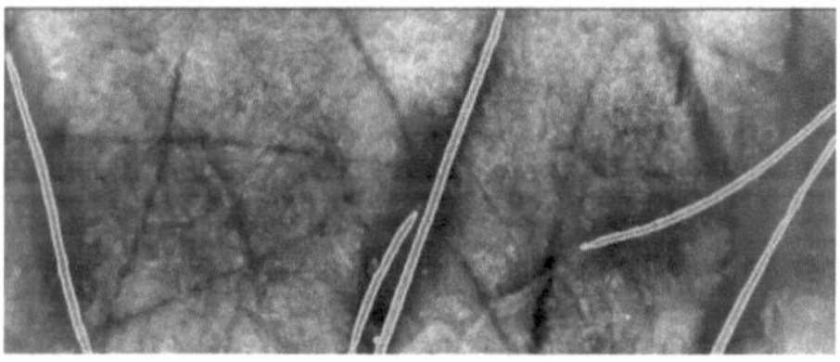

Fig. 8. Hairs segmentation in PC_1 image (the first component of PCA applied to the 20 first C scans), outlined in yellow. (Color figure online)

The scatter plot in Fig. 11 matches the thickness estimates (in μm) given by our proposed method (Y-axis) with the ones given by the DL method U-Net (X-axis) for the 100 individuals tested, with different markers for each skin zone. The strength of the relationship between the two series of measurements is good with a correlation coefficient r equals to 0.67 ($p < 0.001$) for cheekbone, 0.70 ($p < 0.001$) for temple, and 0.79 ($p < 0.001$) for mandibule.

Table 1 provides comparative statistics between the two methods. The mean and standard deviation of the estimates given by each method are computed in μm for each skin zone, i.e. averaged on 300 LC-OCT stacks. The means are similar with a standard deviation that is always lower for the proposed method. SC thickness on the cheekbone is slightly lower than that measured on the temple or mandible for both methods, which is consistent.

Differences between the two methods are evaluated by mean, min and max of absolute error over all B-scans in a volume and over all volumes for each skin area. The MAE are between 0.41μm and 0.65μm, around half the resolution of the LC-OCT imaging system. These results indicate a strong similarity between the two methods, confirmed by Student's t-test on the hypothesis that the difference

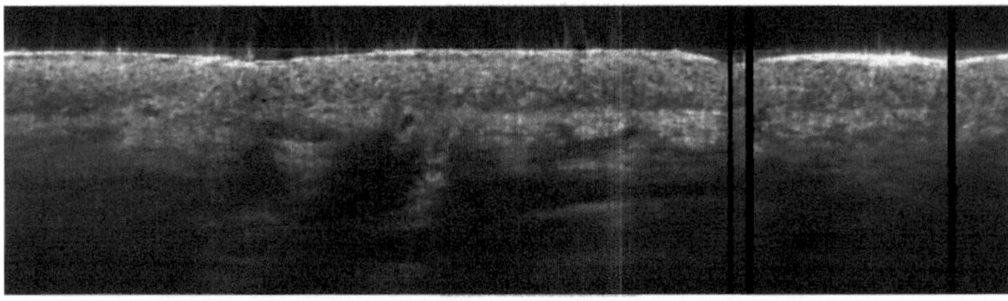

Fig. 9. Column blanking in B-scan from hair segmentation in PC_1 image

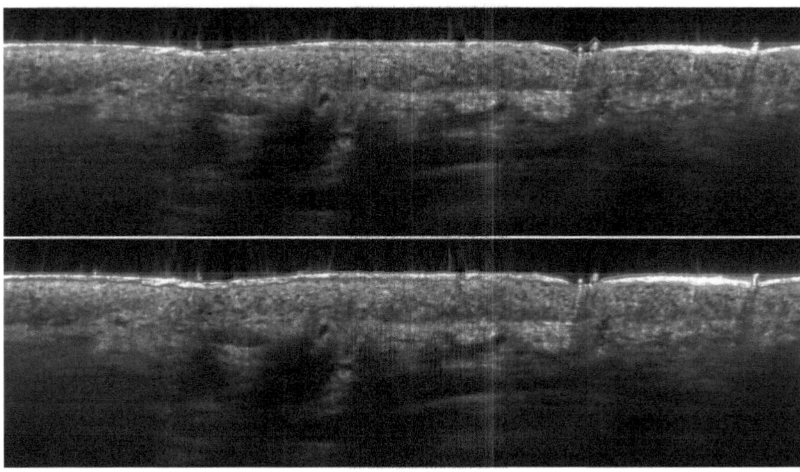

Fig. 10. Stratum corneum segmentation: without hair removal (Top), with hair removal (Bottom).

between the two sets of measurements is on average less than the resolution of the imaging system (one micrometer), with p-values of less than 0.001 for all 3 zones.

We can therefore conclude that our unsupervised segmentation method is as efficient and accurate as deep learning U-Net method.

5 Conclusion

The thickness of the stratum corneum (SC) is an important biomarker for assessing the quality of the skin barrier function. The study of its variations within a population can be used, for example, to analyze the penetration of active drug molecules or the effect of moisturizing or anti-aging skin care products. Thanks to LC-OCT imaging, it is now possible to determine SC thickness, in vivo, non-invasively, and with a precision comparable to that of histology. In this paper, a new reliable, unsupervised and fully automatic method is presented for SC segmentation in vertical images, incorporating hair removal in the scanned volume.

Table 1. Comparative statistics of SC thickness estimates for all dataset (3 skin areas scanned 3 times for the 100 individuals, i.e. 300 estimates from LC-OCT stacks for each skin zone).

Skin Area	Cheekbone	Temple	Mandibule
OURS: Mean ± SD	12.18 ± 0.54	13.82 ± 0.78	14.01 ± 0.81
U-Net: Mean ± SD	12.36 ± 0.67	13.56 ± 0.98	13.48 ± 0.93
Mean Abs. Error	0.4088	0.6050	0.6498
Min Abs. Error	0.0017	0.0202	0.0126
Max Abs. Error	1.6189	1.9986	1.7955

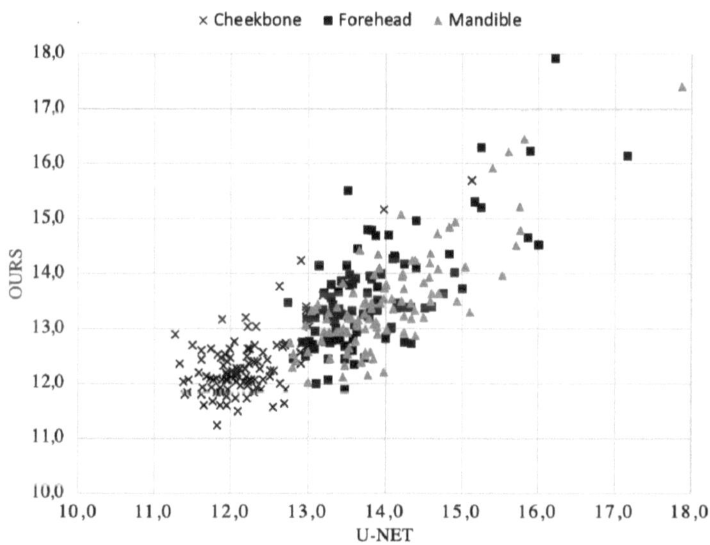

Fig. 11. Comparison of thickness estimates (in μm) given by the two methods for all dataset (3 skin zones x 100 individuals).

Experiments carried out on a database of 900 images of healthy skin acquired from one hundred volunteers show that the proposed method performs as well as the reference method based on U-Net and deep learning for estimating SC thickness. Being unsupervised, our method has the advantage of not requiring manual annotation of a large dataset for model training, with the risk of inter- and intra-observer variability. Furthermore, our method incorporates hair removal in a fully automatic way, whereas U-Net segmentation must be followed by manual hair removal to avoid distorting thickness calculations. These methods surpass the state-of-the-art of other imagery and could therefore help reduce the number of skin biopsies. Ongoing tests on other panels of Asian and Afro-American populations demonstrate the generalizability of the proposed method. Future work will extend segmentation to other skin layers, such as the epidermal junction, and explore the possibility of taking into account contextual information across B-scans during segmentation.

Ethics Statement. The study was approved by the French Ethics Committee "Comité de protection des personnes sud méditerranée I" (IDRCB : 2021-A00101-40). A written informed consent was obtained from all participants.

References

1. Assi, A., et al.: Comparison of facial skin ageing in healthy Asian and caucasian females quantified by in vivo line-field confocal optical coherence tomography 3D imaging. Skin Res. Technol. **30**(9), e13643 (2024)
2. Avanaki, M.R., Hojjatoleslami, A.: Skin layer detection of optical coherence tomography images. Optik **124**(22), 5665–5668 (2013)
3. Bonnier, F., et al.: Line-field confocal optical coherence tomography coupled with artificial intelligence algorithms to identify quantitative biomarkers of facial skin ageing. Sci. Rep. **13**(1), 13881 (2023)
4. Czajkowska, J., Badura, P., Korzekwa, S., Płatkowska-Szczerek, A.: Deep learning approach to skin layers segmentation in inflammatory dermatoses. Ultrasonics **114**, 106412 (2021). https://doi.org/10.1016/j.ultras.2021.106412
5. Del Amor, R., et al.: Automatic segmentation of epidermis and hair follicles in optical coherence tomography images of normal skin by convolutional neural networks. Front. Med. **7**, 220 (2020)
6. Dubois, A., et al.: Line-field confocal time-domain optical coherence tomography with dynamic focusing. Opt. Express **26**(26), 33534–33542 (2018)
7. Forgy, E.W.: Cluster analysis of multivariate data: efficiency versus interpretability of classifications. Biometrics **21**, 768–769 (1965)
8. Hames, S.C., Ardigò, M., Soyer, H.P., Bradley, A.P., Prow, T.W.: Automated segmentation of skin strata in reflectance confocal microscopy depth stacks. PLoS One **11**(4), e0153208 (2016)
9. Holmes, J., Welzel, J.: OCT in Dermatology, pp. 2189–2207. Springer International Publishing, Cham (2015). https://doi.org/10.1007/978-3-319-06419-2_75
10. Hotelling, H.: Analysis of a complex of statistical variables into principal components. J. Educ. Psychol. **24**(6), 417 (1933)
11. Kumar, P., Dhara, S., Gope, A., Chatterjee, J., Mandal, S.: Deep learning based skin-layer segmentation for characterizing cutaneous wounds from optical coherence tomography images. In: 45th Annual International Conference of the IEEE Engineering in Medicine & Biology Society (EMBC), pp. 1–4 (2023)
12. Li, A., et al.: Epidermal segmentation in high-definition optical coherence tomography. In: 37th Annual International Conference of the IEEE Engineering in Medicine and Biology Society (EMBC), pp. 3045–3048 (2015). https://doi.org/10.1109/EMBC.2015.7319034
13. Li, K., Wu, X., Chen, D., Sonka, M.: Optimal surface segmentation in volumetric images-a graph-theoretic approach. IEEE Trans. Pattern Anal. Mach. Intell. **28**(1), 119–134 (2006). https://doi.org/10.1109/TPAMI.2006.19
14. Liu, C.H., et al.: Quantitative assessment of in vivo nuclei and layers of human skin by deep learning-based oct image segmentation. Biomed. Opt. Express **16**(4), 1528–1545 (2025)

15. Liu, X., Chuchvara, N., Liu, Y., Rao, B.: Real-time deep learning assisted skin layer delineation in dermal optical coherence tomography. OSA Continuum **4**(7), 2008–2023 (2021)
16. Liu, X., Ouellette, S., Jamgochian, M., Liu, Y., Rao, B.: One-class machine learning classification of skin tissue based on manually scanned optical coherence tomography imaging. Sci. Rep. **13**(1), 867 (2023)
17. Ogien, J., Daures, A., Cazalas, M., Perrot, J.-L., Dubois, A.: Line-field confocal optical coherence tomography for three-dimensional skin imaging. Front. Optoelectron. **13**(4), 381–392 (2020). https://doi.org/10.1007/s12200-020-1096-x
18. Pedrazzani, M., et al.: Comparison of line-field confocal optical coherence tomography images with histological sections: validation of a new method for in vivo and non-invasive quantification of superficial dermis thickness. Skin Res. Technol. **26**(3), 398–404 (2020)
19. Schuetzenberger, K., et al.: Comparison of optical coherence tomography and high frequency ultrasound imaging in mice for the assessment of skin morphology and intradermal volumes. Sci. Rep. **9**(13643) (2019). https://doi.org/10.1038/s41598-019-50104-4
20. Shishkova, V., et al.: Segmentation of 3D oct images of human skin using neural networks with U-Net architecture. Architecture. Sovrem Tekhnologii Med. **17**(1), 6–16 (2025). https://doi.org/10.17691/stm2025.17.1.01
21. Srivastava, R., Yow, A.P., Cheng, J., Wong, D.W.K., Tey, H.L.: Three-dimensional graph-based skin layer segmentation in optical coherence tomography images for roughness estimation. Biomed. Opt. Express **9**(8), 3590–3606 (2018). https://doi.org/10.1364/BOE.9.003590

MATE: Multimodal Agent that Talks and Empathizes

Niyati Rawal[1(✉)], Matteo Xia[2], David Tessaro[2], Lorenzo Baraldi[1], and Rita Cucchiara[1]

[1] University of Modena and Reggio Emilia, Modena, Italy
niyati.rawal@unimore.it
[2] Omitech S.r.l., Padua, Italy

Abstract. Recently, Large Language Models (LLMs) like ChatGPT have become quite popular with the general public, enabling humans to chat with artificial agents. However, these LLMs do not understand human emotions as their responses often sound neutral and their goal is to support users in various capacities. In this work, we introduce Empathetic-HRI, a novel dataset comprising paired face images, contextual text, and emotion labels to facilitate multimodal empathetic response generation. In addition, we propose MATE, a multimodal agent that takes the human's facial expressions and text as an input and responds back to the human in an empathetic manner. MATE is based on a transformer architecture, where the inputs are fed into a transformer encoder. The embeddings that are the output from the transformer encoder are then fed into the transformer decoder (Llama3). To evaluate the performance of our model, we conduct a survey and ask the participants if the responses given by the proposed model are appropriate and empathetic. The results are compared with ChatGPT and validated using a paired-samples T test. As the results are statistically significant, we prove that the responses given by MATE are both appropriate and empathetic.

Keywords: Vision-and-Language · Emotions · Dialogue

1 Introduction

Large Language Models (LLMs) such as ChatGPT [25] and Llama [35] have gained popularity and allowed the general public to interact with artificial agents in terms of natural language. Despite this, these models do not understand human emotions, and it seems that the goal of the models is to assist humans in some way or another. For example, when ChatGPT was given the following prompt: "I am scared of talking in front of so many people", it responded that it was normal to feel scared when talking in front of a large group. But with a few strategies, one can build confidence and manage that fear. This is followed by a list of points on how one can overcome stage fright. Humans do not necessarily

E. Rodolà et al. (Eds.): ICIAP 2025, LNCS 16167, pp. 116–127, 2026.
https://doi.org/10.1007/978-3-032-10185-3_10

Fig. 1. We propose MATE, a multimodal agent that talks and empathizes. MATE takes the human's facial expressions and text as an input and responds back to the human in an empathetic manner. The empathetic output given by the robot consists of the textual response as well as a facial expression.

always give each other tips to improve something. Just a small encouragement saying "You can do this. I believe in you." might be sufficient in this case.

As social robots are getting increasingly popular in a plethora of domains [23, 31], it is crucial for these robots to understand human emotions and respond to humans in an empathetic manner. Deep learning algorithms are already quite good at recognizing human facial expressions and even being able to classify human emotions coming from a multimodal input signal combining modalities such as facial image, body gestures, text, audio and physiological signals [19,29]. When these deep learning algorithms are plugged in to the robots, the robots are capable of recognizing human emotions. Some of these robots can even express emotions through facial expressions or gestures [26,37].

In Human-Human Interaction (HHI), there is parallel empathy and reactive empathy [7]. Parallel empathy is when a human responds to another human with the same facial expression, such as a human smiles when being smiled at [5,13]. Reactive empathy is when a human generates an emotion in response to the emotion of another human being. Robots should also be capable of responding to the human with both parallel and reactive empathy depending on the situation. For example, a robot could be happy when the human shares a happy news or it could be sad when the human is angry at it.

In this work, we introduce Empathetic-HRI, a novel dataset comprising paired face images, contextual text, and emotion labels to facilitate multimodal empathetic response generation. Additionally, we propose a computational model that responds to a human empathetically by considering a multimodal input of the person's facial expressions and the text content of what the person says (see Fig. 1). The computational model that we propose is inspired by HHI and is based on a transformer encoder-decoder structure. The encoder part of the transformer takes the person's facial expressions, the textual prompt, and the target emotion as input. The embeddings that are the output from the transformer encoder are then fed into the decoder part as input. The decoder part of the transformer is fineutuned on Llama3 [9] model that returns the textual response to a given input prompt. We conduct a survey where the responses given by the model are rated as appropriate and empathetic on a five-point likert scale. Finally, we conduct a statistical test to validate that the results are statistically significant.

The contributions of this work are as follows.

- We create a new Empathetic-HRI dataset between two people where the second person responds to the first person using parallel or reactive empathy.
- We propose a novel computational model MATE that is based on a transformer architecture for generating empathetic textual responses.
- We show that the responses generated by MATE are both appropriate and empathetic by conducting a survey where participants are asked to rate responses corresponding to various prompts.

2 Related Work

2.1 Multimodal Vision-and-Language Models (MVLM)

While early language models (e.g., RNNs [30], LSTMs [14]) struggled with long-range dependencies, transformer-based LLMs (e.g., Llama3 [9]) can generate coherent text over hundreds of tokens without encountering vanishing-gradient issues. LLMs are autoregressive models and their goal is to predict the next token.

After the recent success of LLMs, there has been a growing trend to propose models with a multimodal input consisting of vision and language. Some models deploy transformer architecture for performing diverse vision-language tasks, including Visual Question Answering (VQA) and image retrieval [4]. Recently, Vision and Language Understanding is not just limited to images but is also getting extended to videos [34,36]. GPT-4V [1], Llava [16], Flamingo [2] are some of the models that use multimodal input and generate text. Recent models such as Video-Llama, Video-ChatGPT etc. use videos instead of images and text as an input to generate an output text [18,38]. There are few works that generate multimodal output with multimodal input [32,33].

2.2 Empathy in Social Robotics

Robots are being used in a wide variety of domains where they have to interact with humans at hotel receptions, in hospitals etc. [8,23]. Social robots are expected to understand human emotions coming from multimodal signals such as facial expressions, body gestures and voice. The robots are then, expected to respond in an empathetic manner, showing their emotions through facial expressions, body gestures and voice, so as to not cause harm to the human being in any manner and appear as natural in a conversation as possible. In HHI, there is parallel empathy and a reactive empathy. Some researchers have tried to model parallel and reactive empathy in virtual agents and robots [3,21]. For normal emotion intensity, the type of empathetic reaction, parallel or reactive is determined by the type of emotion [3]. For example, positive emotions i.e. happiness and surprise are responded with parallel empathy towards the user and negative emotions i.e. anger, fear, sadness and disgust are responded with reactive empathy towards the user.

Previous studies have finetuned VLMs on empathetic datasets [27,28]. In this work, we take inspiration from HHI and propose a transformer encoder-decoder model to generate the dialogue output token-by-token.

3 Method

We propose a novel computational model, MATE, a multimodal agent that talks and empathizes. The model is based on a transformer encoder-decoder structure. The facial image, the input text containing what the person says and a target emotion is used as an input and converted to embeddings using a transformer encoder. These embeddings are used as prompts for the transformer decoder part that is a pretrained Llama3 [9] model. We finetune Llama3 to output the dialogue response for the given input prompt.

3.1 Creating the Empathetic-HRI Dataset

To create the Empathetic-HRI dataset, we prompt ChatGPT as follows: "Create a two line dialogue between two people `[SITUATION]` such that person 1 feels `[EMOTION 1]` and person 2 gives a `[EMOTION 2]` response". For the list of situations we considered none, at work, in school, while travelling, in a hospital, at a ski resort, in a coffee shop, at the gym, at a party, at the beach, at a concert, at a wedding, at a supermarket, at a museum, at a sports concert and at a movie theater. For the combinations of `[EMOTION 1]` and `[EMOTION 2]`, in general, the positive emotions are responded with parallel empathy and the negative emotions are responded with reactive empathy. Table 1 shows the possible combinations of `[EMOTION 1]` and `[EMOTION 2]`. We pair the language data with images from KDEF dataset [17], showing various facial expressions to make the dataset multimodal. In the end, the Empathetic-HRI dataset has 2800 dialogues, showing various emotions.

3.2 Transformer Model Architecture for Empathetic Response Generation

The proposed model comprises of a transformer encoder and decoder structure. As shown in Fig. 2, first, the image v is fed to ResNet-152 [12] pre-trained on VGG-Face Dataset [6] to extract the facial features. These facial features are then passed through a linear layer and converted to embeddings. The current emotion $\mathcal{E}_h$ of the person (e.g. "happy") is converted into a global vector using Glove42B tokens [24]. This global vector is also passed through a linear layer and converted into embeddings. Similarly, the emotion with which the agent should reply $\mathcal{E}_a$

Table 1. To build our own Empathetic-HRI dataset, we prompt ChatGPT to "Create a two line dialogue between two people `[SITUATION]` such that person 1 feels `[EMOTION 1]` and person 2 gives a `[EMOTION 2]` response". The combinations of `EMOTION 1` and `EMOTION 2` are listed below.

EMOTION 1	Empathy Type	EMOTION 2
happy	parallel	happy
surprise	parallel	surprise
sad	parallel	sad
sad	reactive	neutral
angry	reactive	sad
angry	reactive	neutral
neutral	parallel	neutral
neutral	reactive	happy
fearful	reactive	neutral
disgusted	reactive	neutral

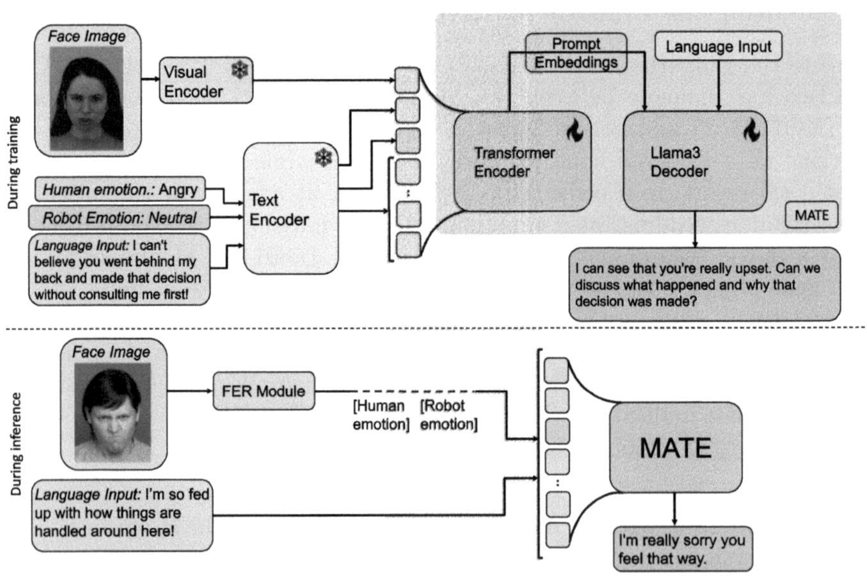

Fig. 2. The training process of the proposed computational model is as follows. The facial features, ground truth facial expression label of the human, the emotion of the agent and what the human says are fed as input to the transformer encoder model which gives the embeddings as an output. These prompt embeddings are fed into the Llama3 decoder and the response is generated token by token. During inference, the `Facial Expression` is given by the Facial Expression Recognition (FER) Module.

(the emotion corresponding to EMOTION 2 in the dataset) is also converted into a global vector using Glove42B tokens which is then passed through a linear layer and converted into embeddings. For tokenizing the words of the human $\mathcal{T}$, Llama3 tokenizer is used which are also converted into embeddings. The four kinds of embeddings for faces, emotion of the person, emotion of the agent, as well as the words of the person are concatenated and sent as input to the transformer encoder model. Formally,

$$z = \texttt{Encoder}\left(\left[\underbrace{v}_{\text{Image}}, \underbrace{\mathcal{E}_h}_{\text{Human Emotion}}, \underbrace{\mathcal{E}_r}_{\text{Agent Emotion}}, \underbrace{\mathcal{T}}_{\text{Human Text}}\right]\right) \tag{1}$$

where v denotes the set of visual features for the facial expression image, $\mathcal{E}_h$ indicates the ground truth emotion label of the human, $\mathcal{E}_a$ indicates the emotion of the agent, $\mathcal{T}$ denote the human text input tokens.

The embeddings that are an output of the transformer encoder are then fed into the transformer decoder model (i.e. Llama3) as prompt embeddings. The prompt embeddings that are fed into the Llama3 decoder are concatenated with the embeddings of the language input. Formally,

$$\hat{y} = \texttt{Llama3}\left(\left[\underbrace{P_v}_{\text{Image}}, \underbrace{P_h}_{\text{Human Emotion}}, \underbrace{P_a}_{\text{Agent Emotion}}, \underbrace{\mathcal{T}}_{\text{Human Text}}, \texttt{BOS}, \underbrace{i_1, .., i_m,}_{\text{Instruction}} \texttt{EOS}\right]\right) \tag{2}$$

where P_v denotes the prompt embeddings of visual features, P_h indicates the prompt embeddings of the human emotion label, P_a indicates the prompt embeddings of the agent emotion label, $\mathcal{T}$ denote the human text input tokens, BOS and EOS are begin of string and end of string tokens respectively. Consequently, $(i_1, ..., i_m)$ denotes the tokens that correspond to the response.

The Llama3 decoder is fine-tuned to respond to a given query and facial expression in a supervised manner. The goal is to reduce the cross-entropy loss between the predicted responses and the ground truth responses. Specifically,

$$\begin{aligned} l_n &= -\sum_{c=1}^{C} y_{n,c} \log \frac{\exp(x_{n,c})}{\Sigma_{i=1}^{C} \exp(x_{n,i})} \\ L &= \frac{\sum_{n=1}^{N} l_n}{N}, \end{aligned} \tag{3}$$

where x is the input, y is the target, C is the number of tokens in the dictionary, and N spans the minibatch dimension.

During the testing phase, first the facial expression of the person is recognized using FIXR [20] that is trained on three different datasets [11,15,22]. Depending upon the recognized facial expression, the emotion of the agent is decided using Table 1. If there is more than one possibility of reacting with an emotion for a given human emotion, e.g. when the human emotion is detected as sad, the agent's emotion is chosen between sad and neutral, where sad is a

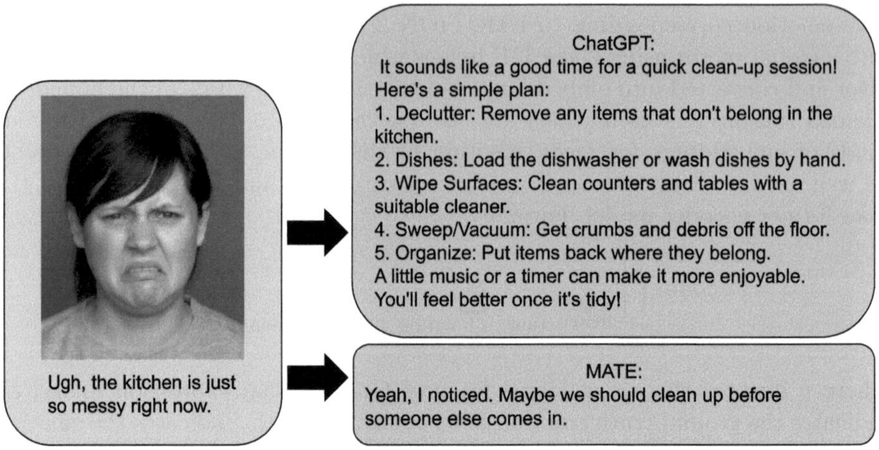

Fig. 3. A qualitative sample where the prompt consist of a disgusted emotion. The sample compares the response of MATE with that of ChatGPT-4o for a given human facial expression and input prompt.

parallel emotion type while neutral is a reactive emotion type. Similarly to the method during fine-tuning of the model, first the embeddings for the four inputs, i.e. facial features of the human, recognized facial expression, agent's emotion, and the words that the human says are extracted. These are then concatenated and sent as an input to the transformer encoder. After getting the embeddings from the transformer encoder model, these become an input for the transformer decoder model i.e. Llama3, followed by the embeddings of the language input tokens. Llama3 predicts the output response token-by-token until it reaches the `EoS` token.

4 Experimental Evaluation

AdamW optimizer with a learning rate $0.2e-5$ is used to fine-tune the model.

4.1 Qualitative Results

In Fig. 3, an example of a face image from the FACES dataset [10] is presented along with its corresponding input text that is fed into the proposed model and compared to the response given by ChatGPT-4o. The example is of disgusted emotion, where the human exclaims that the kitchen is so messy. The algorithm correctly recognizes the human emotion as disgust and responded with a neutral emotion that it noticed and that they should clean up before someone else came in. On the other hand, ChatGPT gave a long list of tips on how to clean up a space.

In addition to this qualitative sample, more qualitative samples of afraid, angry, happy, neutral, and sad expressions are presented in Table 2. All these

Table 2. Sample inputs of the proposed model and their corresponding outputs are provided below.

Human	Input	Agent	MATE
afraid	I heard some strange, unusual noise in the streets last night. It was a bit scary.	neutral	It's probably just the sound of the city. We live in a busy area. Nothing to worry about.
angry	I can't believe my boss ignored all my suggestions in the meeting today, like my ideas don't matter.	neutral	I understand how that feels. I've been there before, and it hurts to feel that your hard work isn't valued.
happy	I finally booked my dream vacation to New Zealand for next month. I've been wanting to go there for years.	happy	That's amazing! I've heard it's absolutely stunning. You're going to have the time of your life.
neutral	I finished the report earlier than I thought. Can you help me check it?	neutral	Sure, let's take a look together.
sad	I didn't get the job I was hoping for even though I really thought I had a good chance this time	sad	The news hit me like a punch in the gut. I know how much you wanted this, and I'm so sorry you didn't get it.

Table 3. Comparison of performance of different models to perform quantitative evaluation of emotional accuracy

Model	Accuracy	Average Cosine Similarity
bert	88.89%	0.9444
all-mpnet-base-v2	77.78%	0.4663
Model	Accuracy	Confidence (> 75%)
emotion classifier	70.59%	41.18%

samples are also accompanied by their corresponding facial images, but in the table, the images are omitted to save space. As can be seen in the table, all the responses given by MATE are appropriate and empathetic.

4.2 Quantitative Evaluation

In this section, we provide a numerical metric quantifying how well the generated responses match the agent's emotion label. We generate the responses to 17 prompts and face images in the FACES dataset [10] using MATE. For each generated response, we calculate the closest emotion via cosine similarity using BERT embeddings and sentence-transformer model embeddings. In addition, we also use an emotion-specific model for the prediction of agent's emotion. This emotion classifier is emotion-english-distilroberta-base model. Table 3 provide the results for these evaluations. 88.89% of responses matched the intended emotional category with a cosine similarity threshold of 0.8 using BERT embeddings. However, the accuracy was lower when sentence-transformer model and emotion classifier were used.

4.3 Human Evaluation

For this study, the responses to certain prompts and face images in the FACES dataset [10] are obtained using MATE. To compare the responses of the model with those by ChatGPT-4o, the agent's responses are recorded for 17 different prompts with varying emotions, and an anonymized survey is conducted asking human participants to rate whether the responses given by the agent are appropriate and empathetic. If the responses are not empathetic, human participants could write in free text how they would respond to the given query.

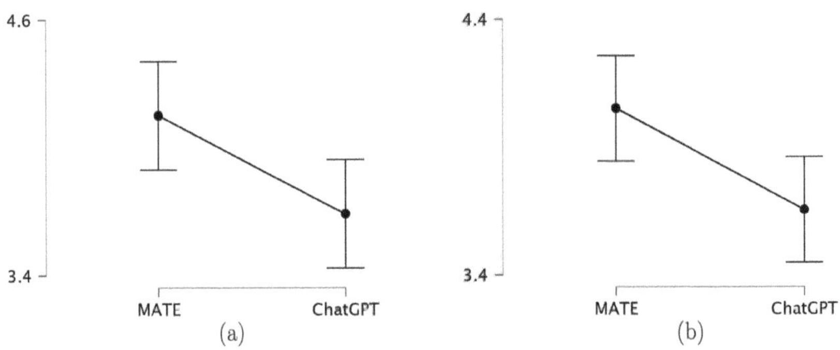

Fig. 4. Comparison of responses generated by MATE and those generated by ChatGPT using the paired-samples T-test. The appropriateness of the responses is measured in (a), and the empathetic nature of the responses is measured in (b).

The survey that was created consisted of the 17 qualitative examples. Each qualitative example consisted of a human image, the language text that the human said and responses given my MATE as well as by ChatGPT-4o. The survey was answered by 20 human participants. Of these, 14 were men and six were women. The survey data was collected anonymously, and each participant had to answer mainly two questions for every qualitative example. The first question was whether they agreed that the response given by the robot was appropriate. This question had to be answered on five-point Likert scale where one meant strongly disagree and five meant strongly agree that the response given was appropriate. The second question if the response given by the robot was empathetic. This question also had to be answered on five-point Likert scale where one meant strongly disagree and five meant strongly agree that the response given was empathetic. However, if the human participant said that the responses given by both the models were not empathetic, they had free space to write in text how they would have responded to the given input prompt.

To analyze the survey results of the participants, the paired-samples T-test is conducted. For the first questions where the participants had to select to what extent they agreed the responses given by ChatGPT and MATE were appropriate, the t-value of MATE − ChatGPT was 2.683 and the p-value was 0.015. As the p-value was less than 0.05, the results are statistically significant. Figure 4(a) show the comparison plot of the two models.

For the second question of the survey where the participants had to select to what degree they agreed that the responses given by ChatGPT and MATE were empathetic, the t-value of MATE − ChatGPT was 2.847 and the p-value was 0.010. As the p-value was less than 0.05, the results are statistically significant. Figure 4(b) show the comparison plot of the two models.

5 Conclusion and Future Work

We create our own Empathetic-HRI dataset by prompting the ChatGPT to create a two line dialogue between two people such that person 1 feels `[EMOTION 1]` and person 2 gives a `[EMOTION 2]` response. The dialogue was generated for various kinds of situations such as at work, at school, while travelling etc. After creating the dataset, we finetune our proposed multimodal model that is based on a transformer encoder-decoder structure to generate responses for given prompts by considering both the facial expression and what the human says. We generated responses for 17 prompts with different emotions using our proposed method and those generated by ChatGPT. A survey was conducted, asking human participants their opinions on whether the responses given by the proposed model were appropriate and empathetic in comparison to the responses given by ChatGPT. We performed paired-samples T test and found that the results of the survey were significant indicating that the responses given by our model are both appropriate and empathetic. In future, we would like to conduct HRI experiment in real time.

Acknowledgement. We acknowledge the CINECA award under the ISCRA initiative, for the availability of high-performance computing resources. This work has been conducted with the support of the PNRR project "Fit for Medical Robotics (Fit4MedRob)" funded by the Italian Ministry of University and Research and the European project MINERVA, funded by European High-Performance Computing Joint Undertaking (JU) under grant agreement No 101182737.

References

1. Achiam, J., et al.: GPT-4 technical report. arXiv preprint arXiv:2303.08774 (2023)
2. Alayrac, J.B., et al.: Flamingo: a visual language model for few-shot learning. NeurIPS **35**, 23716–23736 (2022)
3. Bagheri, E.: Improving human-robot interaction through empathy and transparency (2022)
4. Bao, H., et al.: VLMo: unified vision-language pre-training with mixture-of-modality-experts. In: Advances in Neural Information Processing Systems, vol. 35, pp. 32897–32912 (2022)

5. Barger, P.B., Grandey, A.A.: Service with a smile and encounter satisfaction: emotional contagion and appraisal mechanisms. Acad. Manag. J. **49**(6), 1229–1238 (2006)
6. Cao, Q., Shen, L., Xie, W., Parkhi, O.M., Zisserman, A.: VGGFace2: a dataset for recognising faces across pose and age. In: 2018 13th IEEE International Conference on Automatic Face & Gesture Recognition (FG 2018), pp. 67–74. IEEE (2018)
7. Davis, M.: Empathy: A Social Psychological Approach (2018)
8. Donnermann, M., Schaper, P., Lugrin, B.: Social robots in applied settings: a long-term study on adaptive robotic tutors in higher education. Front. Robot. AI **9**, 831633 (2022)
9. Dubey, A., et al.: The Llama 3 herd of models. arXiv preprint arXiv:2407.21783 (2024)
10. Ebner, N.C., Riediger, M., Lindenberger, U.: Faces–a database of facial expressions in young, middle-aged, and older women and men: development and validation. Behav. Res. Methods **42**, 351–362 (2010)
11. Goodfellow, I.J., et al.: Challenges in representation learning: a report on three machine learning contests. In: Lee, M., Hirose, A., Hou, Z.-G., Kil, R.M. (eds.) ICONIP 2013. LNCS, vol. 8228, pp. 117–124. Springer, Heidelberg (2013). https://doi.org/10.1007/978-3-642-42051-1_16
12. He, K., Zhang, X., Ren, S., Sun, J.: Deep residual learning for image recognition. In: CVPR, pp. 770–778 (2016)
13. Hennig-Thurau, T., Groth, M., Paul, M., Gremler, D.D.: Are all smiles created equal? How emotional contagion and emotional labor affect service relationships. J. Mark. **70**(3), 58–73 (2006)
14. Hochreiter, S., Schmidhuber, J.: Long short-term memory. Neural Comput. **9**(8), 1735–1780 (1997)
15. Li, S., Deng, W., Du, J.: Reliable crowdsourcing and deep locality-preserving learning for expression recognition in the wild. In: CVPR, pp. 2852–2861 (2017)
16. Liu, H., Li, C., Wu, Q., Lee, Y.J.: Visual instruction tuning. In: NeurIPS, vol. 36 (2024)
17. Lundqvist, D., Flykt, A., Öhman, A.: Karolinska directed emotional faces. PsycTESTS Dataset **91**, 630 (1998)
18. Maaz, M., Rasheed, H., Khan, S., Khan, F.S.: Video-ChatGPT: towards detailed video understanding via large vision and language models. arXiv preprint arXiv:2306.05424 (2023)
19. Maharjan, R.S., Rawal, N., Romeo, M., Baraldi, L., Cucchiara, R., Cangelosi, A.: Multimodal emotion recognition in conversation via possible speaker's audio and visual sequence selection. In: ICASSP 2025 - 2025 IEEE International Conference on Acoustics, Speech and Signal Processing (ICASSP), pp. 1–5 (2025). https://doi.org/10.1109/ICASSP49660.2025.10888172
20. Maharjan, R.S., Romeo, M., Cangelosi, A.: Faces are domains: domain incremental learning for expression recognition. In: 2023 International Joint Conference on Neural Networks (IJCNN), pp. 1–8. IEEE (2023)
21. McQuiggan, S.W., Robison, J.L., Phillips, R., Lester, J.C.: Modeling parallel and reactive empathy in virtual agents: an inductive approach. In: Proceedings of the 7th International Joint Conference on Autonomous Agents and Multiagent Systems, vol. 1, pp. 167–174. Citeseer (2008)
22. Mollahosseini, A., Hasani, B., Mahoor, M.H.: AffectNet: a database for facial expression, valence, and arousal computing in the wild. IEEE Trans. Affect. Comput. **10**(1), 18–31 (2017)

23. Pai, R.Y., Shetty, A., Dinesh, T.K., Shetty, A.D., Pillai, N.: Effectiveness of social robots as a tutoring and learning companion: a bibliometric analysis. Cogent Bus. Manage. **11**(1), 2299075 (2024)
24. Pennington, J., Socher, R., Manning, C.D.: Glove: global vectors for word representation, pp. 1532–1543 (2014)
25. Radford, A., Narasimhan, K., Salimans, T., Sutskever, I., et al.: Improving language understanding by generative pre-training. OpenAI (2018)
26. Rawal, N., Koert, D., Turan, C., Kersting, K., Peters, J., Stock-Homburg, R.: ExGenNet: learning to generate robotic facial expression using facial expression recognition. Front. Robot. AI **8**, 730317 (2022)
27. Rawal, N., et al.: Intelligent multimodal artificial agents that talk and express emotions. In: International Workshop on Human-Friendly Robotics, pp. 240–254. Springer (2024)
28. Rawal, N., et al.: Multimodal dialogue for empathetic human-robot interaction. In: International Conference on Social Robotics (2025)
29. Rawal, N., Stock-Homburg, R.M.: Facial emotion expressions in human-robot interaction: a survey. Int. J. Soc. Robot. **14**(7), 1583–1604 (2022)
30. Rumelhart, D.E., Hinton, G.E., Williams, R.J.: Learning representations by back-propagating errors. Nature **323**(6088), 533–536 (1986)
31. Sawik, B., et al.: Robots for elderly care: Review, multi-criteria optimization model and qualitative case study. In: Healthcare, vol. 11, p. 1286. MDPI (2023)
32. Tang, Z., Yang, Z., Khademi, M., Liu, Y., Zhu, C., Bansal, M.: CoDi-2: in-context interleaved and interactive any-to-any generation. In: CVPR, pp. 27425–27434 (2024)
33. Tang, Z., Yang, Z., Zhu, C., Zeng, M., Bansal, M.: Any-to-any generation via composable diffusion. In: NeurIPS, vol. 36 (2024)
34. Tong, Z., Song, Y., Wang, J., Wang, L.: VideoMAE: masked autoencoders are data-efficient learners for self-supervised video pre-training. In: Advances in Neural Information Processing Systems, vol. 35, pp. 10078–10093 (2022)
35. Touvron, H., et al.: Llama: open and efficient foundation language models. arXiv preprint arXiv:2302.13971 (2023)
36. Xu, H., et al.: Videoclip: contrastive pre-training for zero-shot video-text understanding. arXiv preprint arXiv:2109.14084 (2021)
37. Yang, D., Sato, W., Liu, Q., Minato, T., Namba, S., Nishida, S.: Optimizing facial expressions of an android robot effectively: a Bayesian optimization approach. In: 2022 IEEE-RAS 21st International Conference on Humanoid Robots (Humanoids), pp. 542–549. IEEE (2022)
38. Zhang, H., Li, X., Bing, L.: Video-llama: an instruction-tuned audio-visual language model for video understanding. arXiv preprint arXiv:2306.02858 (2023)

Intrinsic Training Signals for Federated Learning Aggregation

Cosimo Fiorini(✉), Matteo Mosconi, Pietro Buzzega, Riccardo Salami, and Simone Calderara

AImageLab - University of Modena and Reggio Emilia, Modena, Italy
{cosimo.fiorini,matteo.mosconi,pietro.buzzega,riccardo.salami, simone.calderara}@unimore.it

Abstract. Federated Learning (FL) enables collaborative model training across distributed clients while preserving data privacy. While existing approaches for aggregating client-specific classification heads and adapted backbone parameters require architectural modifications or loss function changes, our method uniquely leverages intrinsic training signals already available during standard optimization. We present LIVAR (Layer Importance and VARiance-based merging), which introduces: *i)* a variance-weighted classifier aggregation scheme using naturally emergent feature statistics, and *ii)* an explainability-driven LoRA merging technique based on SHAP analysis of existing update parameter patterns. Without any architectural overhead, LIVAR achieves state-of-the-art performance on multiple benchmarks while maintaining seamless integration with existing FL methods. This work demonstrates that effective model merging can be achieved solely through existing training signals, establishing a new paradigm for efficient federated model aggregation. The code will be made publicly available upon acceptance.

Keywords: Federated Learning · Model Merging · Explainability

1 Introduction

The limited ability of deep models to effectively learn from sparse and distributed real-world data has necessitated the development of Federated Learning (FL), a paradigm in which clients train on their local data distributions and share model updates with a central server. The server aggregates these updates without accessing raw local data, thereby preserving privacy. FL has found practical applications in diverse domains, such as manufacturing, where factories collaborate to build a unified model without sharing proprietary sensor data [20], and healthcare, where hospitals cannot exchange sensitive patient information [24]. In recent years, Parameter-Efficient Fine-Tuning (PEFT) [3] has emerged as

Supplementary Information The online version contains supplementary material available at https://doi.org/10.1007/978-3-032-10185-3_11.

E. Rodolà et al. (Eds.): ICIAP 2025, LNCS 16167, pp. 128–142, 2026.
https://doi.org/10.1007/978-3-032-10185-3_11

a training paradigm well-suited to FL. PEFT techniques leverage pre-trained architectures [21] to reduce the number of trainable parameters while maintaining performance comparable to full fine-tuning. Beyond computational efficiency, PEFT is particularly advantageous in FL, as it minimizes communication overhead by transmitting fewer parameters.

Recent studies employing pre-trained architectures highlight distinct contributions of fine-tuning the backbone versus the final classification layer, with evidence suggesting adapting the classifier plays a more critical role in model performance [32]. Consequently, some FL methods employ specialized strategies for tuning the classifier [17] while aggregating backbone parameters via simple averaging [19]. Others utilize PEFT techniques, like LoRA [7], to fine-tune the backbone [25] and subsequently merge updated vectors. Research explored various approaches for determining module aggregation coefficients [29,30], including learnable weighting mechanisms [28] and closed-form solutions [22].

Alternatively, other approaches employ separate training and aggregation solutions for the backbone and classification head [23]. While we consider this decoupled strategy superior, we also advocate for exploring more lightweight solutions for both classifier head merging and backbone module aggregation. We argue that leveraging naturally available training signals can establish both efficient and effective Federated Learning baselines.

Building upon these foundations, we present LIVAR, a novel Federated Learning framework that implements a decoupled approach for aggregating both backbone parameters and classification heads. For classifier merging, we propose variance-weighted heads, where each client's class-specific head is weighted by the variance of features from correctly predicted training samples of that class. Intuitively, higher variance indicates richer class-specific information, and this measure is obtained effortlessly during local training.

For backbone adaptation, LIVAR employs LoRA modules, integrating them with a novel explainability-driven aggregation scheme. We're the first to propose applying SHAP values [15] to determine optimal merging coefficients for federated LoRA modules. Aggregation weights are derived from layer-wise gradients and parameter update patterns during client training; larger cumulative updates signify more substantial adaptations of specific network components. These naturally emergent signals – available without architectural modifications – serve as proxies for each client's contribution to particular model parameters. To implement this, we leverage a proxy dataset to construct an explainability curve that directly maps these adaptation patterns to their optimal merging coefficients.

We extensively evaluate LIVAR on in-distribution and out-of-distribution datasets w.r.t to pre-training. Our method achieves state-of-the-art performance in all in-distribution settings and remains competitive otherwise. Furthermore, we demonstrate that our classifier weighting scheme remains compatible with existing head-specific strategies, enabling seamless integration with these approaches. Beyond FL, our work highlights the broader potential of explainability-driven module merging, opening new research directions in model aggregation.

In summary, our contributions are as follows:

- We propose a variance-weighted aggregation strategy for classification heads.
- We conceive explainability-driven merging coefficients for the federated aggregation of LoRA modules.
- We propose LIVAR, achieving state-of-the-art results across multiple benchmarks while maintaining seamless integration with existing approaches.

2 Related Works

Federated Learning addresses the challenges of training models in a distributed environment, where local data are private and cannot be exchanged. Federated Averaging (FedAvg) [19] established the foundation by averaging client model weights, yet its simplicity makes it vulnerable to highly heterogeneous data streams, which inherently introduce local client divergence. To overcome this challenge, techniques such as Scaffold [10] introduce control variates that compensate for local updates, thereby reducing client drift. Other approaches include FedProx [14], which incorporates a proximal term to constrain the deviation between local and global models, and MOON [13], which employs contrastive loss to align the representations learned by local and global models. Similarly, CCVR [17] enhances the classifier on the server side by rebalancing it with synthetic features, thereby stabilizing updates across heterogeneous clients and promoting convergence. FedProto [26] leverages prototypical representations to regularize the local training process, enabling the maintenance of class-specific information, while GradMA [16] uses a gradient-memory mechanism to guide local update processes by incorporating historical gradient information, thereby stabilizing update and accelerating model convergence.

Model Merging methodologies have emerged as powerful alternatives to conventional weight averaging in FL systems. Fisher Averaging [18] leverages the Fisher Information Matrix to weight client contributions, thus prioritizing more informative updates during aggregation. RegMean [9] presents a closed-form solution for merging model weights, enabling the identification of a global model that best approximates the responses of the clients, demonstrating strong potential for federated settings. Following the introduction of LoRA (Low-Rank Adaptation) [7], a growing body of research has emerged to address the challenge of optimal coefficient selection for module merging [8,28–30], including several approaches specifically designed for Federated Learning scenarios [22]. While these methods achieve promising results, they commonly rely on auxiliary architectural elements, computationally expensive procedures, or customized loss functions. In contrast, our methodology introduces a more streamlined solution that utilizes naturally emerging training signals to effectively aggregate both the LoRA-adapted backbone and the classification head without requiring additional components.

3 Preliminaries

LoRA. Low-Rank Adaptation [7] is an efficient parameter adaptation technique that significantly reduces the number of trainable parameters in deep neural networks. For a given layer with weight matrix $W \in \mathbb{R}^{d \times k}$, LoRA decomposes the weight update ΔW into two trainable low-rank matrices: $B \in \mathbb{R}^{d \times r}$ and $A \in \mathbb{R}^{r \times k}$, where the rank $r \ll min(d, k)$. The adapted weights are computed as:

$$W' = W + BA \tag{1}$$

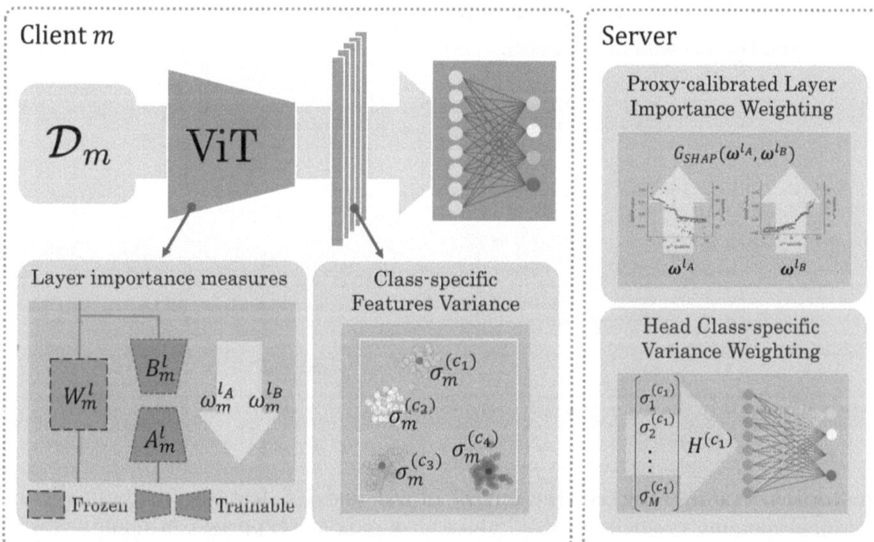

Fig. 1. LIVAR employs a dual weighting strategy architecture. The left diagram depicts client-side information collection during training, the right diagram shows the distinct weighting strategies used on the server side.

During adaptation, the original weights W remain frozen while only the low-rank matrices A and B are updated. This decomposition reduces the number of trainable parameters from $d \times k$ to $r \times (d + k)$, resulting in significant memory savings compared to full fine-tuning.

Problem Setting. We address a C-class classification problem in Federated Learning where data are distributed non-IID across M clients. Let $\mathcal{D}_m$ represent the data partition for client m. The training process consists of sequential communication rounds; at each round, clients perform local optimization for a specified number of epochs, and then the server aggregates the optimized parameters before redistributing them to clients. Let f_θ denote the global model parameterized by θ. The training procedure aims to find the optimal parameter set θ that minimizes the loss across the entire dataset:

$$\underset{\theta}{\text{minimize}} \quad \frac{1}{M} \sum_{m=1}^{M} \mathbb{E}_{(x,y)\sim\mathcal{D}_m} \mathcal{L}(f_\theta(x), y) \tag{2}$$

4 LIVAR

We present LIVAR, a Federated Learning framework with distinct aggregation strategies for the backbone and the classification head. Each client m maintains parameters $\theta_m = \{W_m, H_m\}$, where W_m represents the backbone weights and H_m denotes the classification head.

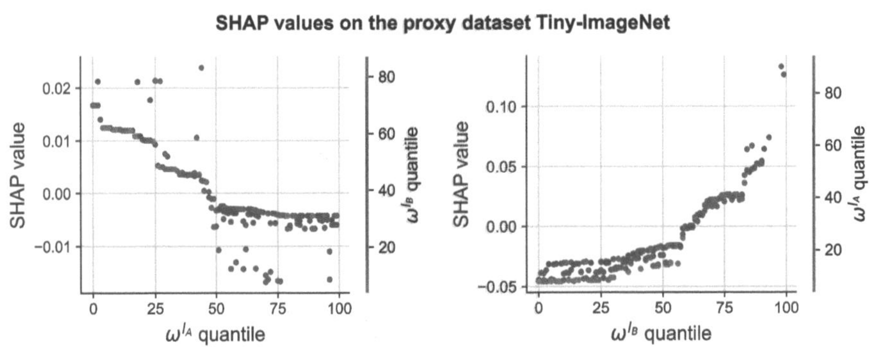

Fig. 2. SHAP values represent the impact of $\omega_m^{l_A}$ and $\omega_m^{l_B}$ features in the prediction of optimal merging coefficients λ_m^l. Values on x axis are expressed in quantiles with respect to the values of all clients and all layers.

For the backbone, we employ Low-Rank Adaptation with our novel *Proxy-Calibrated Layer Importance Weighting* mechanism, using explainability-driven coefficients to merge client updates. The classification head uses *data-variance-aware* merging that automatically adapts to client data heterogeneity, ensuring robustness under non-IID distributions. The dual merging strategy is depicted in Fig. 1.

4.1 Proxy-Calibrated Layer Importance Weighting

Each client m learns L low-rank adaptations $\Delta W_m = \{\Delta W_m^1, \ldots, \Delta W_m^L\}$, where each adaptation $\Delta W_m^l = B_m^l A_m^l$ follows the standard LoRA decomposition. The server aggregates these client updates through a layer-wise weighted combination:

$$\Delta W^l = \sum_{m=1}^{M} \alpha_m^l \cdot \Delta W_m^l \tag{3}$$

where α_m^l represents the layer-specific merging coefficients that constitute the target of our estimation procedure.

These α_m^l coefficients are derived through a three-stage process. First, we establish a ground truth reference using a proxy dataset (*i.e.*, Tiny-ImageNet [12]) by training five federated clients in parallel with a centralized model trained on the full dataset. This process yields the reference parameters $\Delta W_{\text{joint}}^l$ of the centralized model. The fundamental premise is that optimal aggregation should minimize the distance between federated combinations and centralized training outcomes. Formally, for each layer l, we solve the constrained regression problem:

$$\underset{\lambda_m^l}{\text{minimize}} \quad \|\Delta W_{joint}^l - \sum_{m=1}^{5} \lambda_m^l \Delta W_m^l\|_2^2 \quad \text{subject to} \quad \lambda_m^l \geq 0 \tag{4}$$

Second, we generalize this solution by establishing a mapping from trainable parameters to optimal coefficients. Our method builds on the key insight that a layer's contribution to loss variation effectively captures its importance during server-side aggregation, providing a reliable basis for estimating the optimal merging coefficients λ_m^l. Following [31], we compute layer importance measures $\omega_m^{l_A}$ and $\omega_m^{l_B}$ as the running sums of Hadamard products between gradients and parameter updates for matrices A_m^l and B_m^l respectively. To find the relationship between layer importance scores $\{\omega_m^{l_A}, \omega_m^{l_B}\}$ and actual coefficients λ_m^l, we train a Random Forest regressor [1] to predict λ_m^l from the importance signals. This enables a robust, explainable, and efficient estimation of coefficients on new datasets, without relying on a centralized (joint) reference model.

Third, we employ SHAP analysis [15] to interpret the trained model (see Fig. 2), revealing distinct roles of the A and B matrices the former initialized via Kaiming uniform distribution [4] to explore directions, the latter initialized to zeros to control direction utilization. The SHAP values are discretized through a quantile-based function G_{SHAP} to yield the final coefficients:

$$\alpha_m^l = G_{SHAP}(\omega_m^{l_A}, \omega_m^{l_B}) \tag{5}$$

Additional implementation details regarding the quantile-based discretization procedure are provided in Appendix A.

Beyond the Proxy Dataset. The function $G_{SHAP}(\cdot)$, which estimates merging coefficients from importance signals $\omega_m^{l_A}$ and $\omega_m^{l_B}$ observed during training, enables federated aggregation across arbitrary datasets. Our quantile-based discretization ensures robust generalization beyond the proxy dataset used for calibration. We validate this generalization capability through a posteriori analysis of SHAP value distributions across multiple benchmarks in Sect. 5.4.

For server-side aggregation, the coefficients α_m^l computed via Eq. (5) are applied to combine client updates ΔW_m^l, as in Eq. (3), with each client's importance measures $\omega_m^{l_A}$ and $\omega_m^{l_B}$ derived from their local training process.

Algorithm 1. LIVAR training for a single round

Input: M clients, L LoRA layers, $G_{SHAP}(\cdot)$ proxy-calibrated function, C classes.
Client side:
for client $m \in \{1, \dots, M\}$ **do**
  $\Delta W_m \leftarrow$ *(training)* ▷ Collect $(\omega_m^{l_A}, \omega_m^{l_B})$ during training
  Collect $\sigma_m^{(c)}$ over well classified examples
  Send $(\omega_m^{l_A}, \omega_m^{l_B})$ and $\sigma_m^{(c)}$ to the server
end for
Server side:
for layer $l \in \{1, \dots, L\}$ **do**
  $\alpha_m^l \leftarrow$ use Equation (5)
  $\Delta W^l \leftarrow$ use Equation (3) ▷ Proxy-Calibrated Layer Importance Weighting
end for
for class $c \in \{1, \dots, C\}$ **do**
  $\Sigma^{(c)} = \sum_m (\sigma_m^{(c)})$
  $H^{(c)} \leftarrow$ use Equation (6) ▷ Head Class-Specific Variance Weighting
end for

4.2 Head Class-Specific Variance Weighting

The classification head aggregation component of LIVAR operates on the principle that feature-space variance correlates with class representation quality. At the end of client-side training, we compute the mean variance $\sigma_m^{(c)}$ of pre-logits features for correctly classified examples of each class c, serving as a proxy for how comprehensively each client represents the class.

At the server, aggregation weights are derived from normalized variance contributions across clients. For each class c, the server first computes the total variance $\Sigma^{(c)} = \sum_m \sigma_m^{(c)}$ across all clients, then performs the weighted aggregation as:

$$H^{(c)} = \sum_m \frac{\sigma_m^{(c)}}{\Sigma^{(c)}} H_m^{(c)} \tag{6}$$

This variance-weighted scheme prioritizes classification heads whose training data exhibits broader coverage of each class's feature distribution. The complete training procedure is formally specified in Algorithm 1.

5 Experimental Studies

5.1 Evaluation Settings

We conducted a comprehensive evaluation of LIVAR on both in-distribution datasets (CIFAR-100 [11], ImageNet-R [6]) and out-of-distribution benchmarks (EuroSAT [5], CUB-200 [27]) relative to the pre-training domain. Data was distributed across 10 clients using a *distribution-based* label imbalance setting governed by a Dirichlet distribution with parameter β. We evaluated all methods under three distribution-based imbalance scenarios: $\beta \in \{0.2, 0.5, 1.0\}$, where

Table 1. Comparison of approaches on in-distribution datasets.

	CIFAR-100			ImageNet-R		
Distribution β	1.0	0.5	0.2	1.0	0.5	0.2
FedAvg	93.2	92.93	91.56	84.38	81.68	78.73
RegMean	93.36	93.09	92.11	84.12	83.35	81.18
FisherAvg	<u>93.4</u>	93.34	92.22	84	83.78	82.67
FedProto	92.99	92.9	89.88	84.27	82.2	76.97
CCVR	93.27	<u>93.38</u>	<u>92.35</u>	<u>84.87</u>	<u>84.17</u>	<u>82.8</u>
LIVAR	**93.57**	**93.58**	**92.92**	**85.45**	**84.35**	**83.05**

lower values indicate greater data heterogeneity. For all experiments, we evaluated model performance using a global test set.

We benchmark LIVAR against five state-of-the-art approaches spanning complementary research directions: FedAvg [19], CCVR [17], and FedProto [26] represent federated learning methods, while RegMean [9] and Fisher Merging [18] constitute established model merging techniques. Our implementation leverages a ViT-B/16 [2] backbone pre-trained on ImageNet-21K [21] across all methods. For LIVAR, we applied LoRA adaptation to all *qkv*, *mlp*, and *multi-head attention* projection layers. For all datasets, we standardized the training protocol with 5 communication rounds, each comprising 5 local epochs.

5.2 Results of LIVAR

We report comparative results in terms of average accuracy across all evaluated methods.

In Distribution. As reported in Table 1, for both in-distribution datasets, we observe consistent trends: LIVAR achieves state-of-the-art results across all β settings, while CCVR consistently ranks second. For both datasets, higher degrees of distribution imbalance yield more pronounced differences in performance – while LIVAR and CCVR maintain their leading performance, the model merging approaches follow closely, and other methods fail to remain robust under these harder conditions.

Out of Distribution. Table 2 shows that the EuroSAT dataset proves less informative for evaluating Federated Learning with ViT-based methods, as all models achieve similar accuracy levels – likely due to the dataset's limited 10-class structure. In contrast, CUB-200 presents a more challenging benchmark where LIVAR and CCVR significantly outperform competing approaches. LIVAR achieves state-of-the-art results for moderate data imbalance ($\beta \in \{1, 0.5\}$), with CCVR consistently ranking second. However, under extreme imbalance ($\beta = 0.2$), CCVR surpasses LIVAR, demonstrating the advantages of its classifier rebalancing strategy over our variance-based approach.

Table 2. Comparison of approaches on out-of-distribution datasets.

	EuroSAT			CUB-200		
Distribution β	1.0	0.5	0.2	1.0	0.5	0.2
FedAvg	98.35	**98.36**	97.43	73.51	78.01	61.98
RegMean	98.44	98.28	97.47	74.85	70.94	59.8
FisherAvg	**98.49**	97.98	96.86	73.11	75.16	75.92
FedProto	98.31	98.15	97.11	79.05	74.51	62.79
CCVR	98.44	**98.36**	**97.83**	85.42	84.79	**84.66**
LIVAR	98.47	98.17	97.53	**86.24**	**85.73**	82.86

Method Composability. We investigate LIVAR's compatibility with existing federated learning techniques through integration with CCVR on CUB-200 – a dataset representing a significant domain shift from ViT's pre-training distribution. While CCVR alone outperforms LIVAR in the extreme imbalance setting ($\beta = 0.2$), we demonstrate their complementary strengths when combined: applying LIVAR's layer-specific aggregation followed by CCVR's server-side calibration with synthetic data yields superior performance to either method in isolation. As shown in Table 3, this synergy highlights how LIVAR's feature variance head weighting provides foundational benefits that enhance subsequent refinement techniques like CCVR. This composability represents a significant advantage, as it allows practitioners to combine multiple techniques to address the unique challenges of different federated learning scenarios.

Table 3. Evaluation of the combined performance of LIVAR and CCVR [17] on CUB-200 across varying β-distribution parameter.

Distribution β	1.0	0.5	0.2
CCVR	85.42	84.79	84.66
LIVAR	86.24	85.73	82.86
LIVAR +CCVR	**86.52**	**86.42**	**86.37**

Table 4. Ablation study on LIVAR components.

α^l	$\sigma^{(c)}$	Avg (std)
✗	✗	79.16 (0.16)
✓	✗	79.81 (0.12)
✗	✓	90.63 (0.07)
✓	✓	90.90 (0.07)

5.3 Ablation Study on LIVAR Components

We systematically evaluate LIVAR's components by measuring Average Accuracy and standard deviation on CIFAR-100 with extreme label imbalance ($\beta = 0.05$). Starting with a baseline of FedAvg-aggregated ViT-B/16 parameters – including both LoRA residuals ΔW^l and the classification head H – Table 4 demonstrates progressive improvements through incremental enhancements. The

Proxy-Calibrated Layer Importance Weighting (α^l) achieves statistically significant accuracy gains beyond the standard deviation across five runs (reported in parentheses in Table 4). Subsequent incorporation of Head Class-Specific Variance Weighting ($\sigma^{(c)}$) delivers substantially greater improvement, corroborating the classification head's critical role. The full LIVAR framework combines these components synergistically to achieve peak performance.

5.4 A Posteriori Analysis of SHAP Results

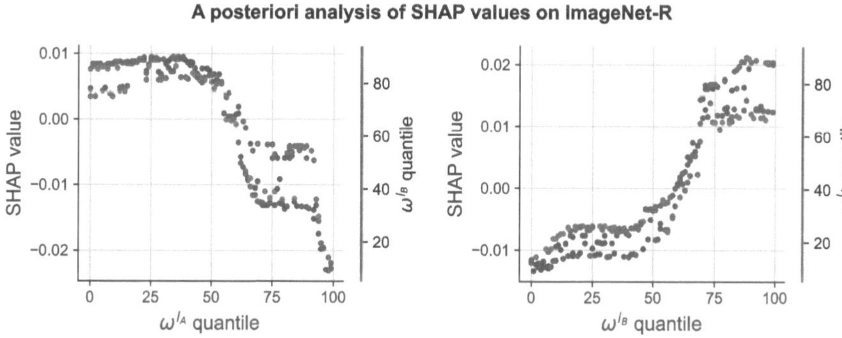

Fig. 3. A posteriori analysis on $\omega_m^{l_A}$ and $\omega_m^{l_B}$ impact in predicting λ_m^l coefficients on ImageNet-R. The trend of both features' impact is consistent with the one observed in the proxy dataset.

To verify the generalizability of our proxy-calibrated SHAP analysis explained in Sect. 4.1, we repeated the calibration procedure across all four datasets: CIFAR-100, ImageNet-R, EuroSAT, and CUB-200. Importantly, this analysis does not involve recalibrating the $G_{SHAP}(\cdot)$ function for each specific dataset – since such per-dataset calibration would be impossible in a federated setting where the joint data distribution is inaccessible – but instead demonstrates that the original proxy-based calibration effectively generalizes to different data domains.

Following the same methodology that produced the results in Fig. 2, we generated comparable analyses for each dataset. As shown in Fig. 3 for ImageNet-R, the SHAP value trends remain consistent: ω^{l_A} exhibits decreasing or stable influence with increasing feature values, while ω^{l_B} maintains its clear increasing trend across all datasets. These consistent patterns (fully documented in Appendix B) confirm the proxy dataset's representativeness, despite variations in specific impact magnitudes.

The observed consistency, coupled with the variations in absolute values, provides empirical justification for our ω^l discretization approach. This design choice ensures LIVAR's robustness when applied to diverse datasets while effectively preventing overfitting to the specific characteristics of the proxy dataset used for calibration.

6 Conclusions

We introduce LIVAR, a Federated Learning framework that advances model aggregation through two novel strategies. For backbone adaptation, we develop an explainability-driven merging procedure for LoRA modules based on training-emerging signals, while our classifier aggregation employs feature variance to weight client contributions. The method demonstrates state-of-the-art performance across multiple benchmarks while maintaining seamless compatibility with existing approaches.

Looking ahead, we identify promising opportunities to extend this work through local explainability techniques for coefficient estimation, potentially removing the need for proxy datasets and enabling more dataset-specific merging. We believe these directions will not only benefit Federated Learning but could also influence broader model fusion research, particularly for explainable model adaptation.

A Details on $G_{SHAP}(\omega^{l_A}, \omega^{l_B})$

Table 5. Calibrated values for $G_{SHAP}(\omega^{l_A}, \omega^{l_B})$ based on ω^{l_A} and ω^{l_B} quantiles.

	$Q_B < 60$	$60 \leq Q_B \leq 80$	$Q_B > 80$
$Q_A < 25$	0.038	0.100	0.193
$25 \leq Q_A \leq 50$	0.038	0.089	0.118
$Q_A > 50$	0.038	0.078	0.093

The merging coefficients α_m^l, detailed in the main paper, are derived through a discretization procedure applied to the importance signals $\omega_m^{l_A}$ and $\omega_m^{l_B}$. To mitigate overfitting on the proxy dataset, a discretization of these values is performed instead of directly utilizing the SHAP-explained regression model's predictions.

Specifically, these values are obtained by summing a SHAP base value of 0.063 with the SHAP values of individual features, the latter being approximated through analysis of the SHAP proxy dataset figure presented in the main paper.

At the conclusion of each round, the server calculates the quantile values Q_A and Q_B for the complete distribution of $\omega_m^{l_A}$ and $\omega_m^{l_B}$, respectively, across all clients m and layers l. Subsequently, for each layer, the corresponding value is assigned based on the intervals specified in Table 5, and a normalization is applied across each layer to ensure that the sum of the resulting SHAP values after the discretization of $\omega_m^{l_A}$ and $\omega_m^{l_B}$ equals 1. This procedure defines the function $G_{SHAP}(\cdot)$, which is executed server-side at the end of each communication round, returning α^l coefficients.

Conceptually, this approach rewards clients exhibiting high ω^{l_B} and low ω^{l_A} values, assigns moderate weights to clients with either high levels of both ω^{l_B} and ω^{l_A} or intermediate ω^{l_B} levels, and assigns minimal weights to clients with low ω^{l_B} values.

B Robustness in the Choice of $\omega_m^{l_A}$ and $\omega_m^{l_B}$

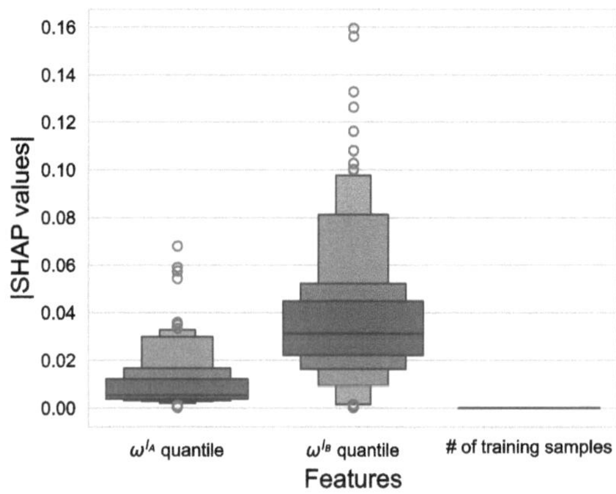

Fig. 4. Evaluation of efficacy of the number of training samples in λ_m^l prediction.

The selection of $\omega_m^{l_A}$ and $\omega_m^{l_B}$ as predictors of the optimal weighting coefficients λ_m^l stems from the intuition that a layer's contribution to loss variation can effectively capture its importance during server-side aggregation. We further investigate whether the number of examples seen during training could provide additional useful information for predicting the optimal coefficients.

Specifically, we initially trained the Random Forest using three features: $\omega_m^{l_A}$, $\omega_m^{l_B}$, and the number of training samples for each client. When evaluating the SHAP values of these three features, it became evident that the impact of the number of samples is negligible, as shown in Fig. 4. This result is directly attributable to the fact that λ_m^l coefficients are estimated layer-wise. More importantly, this finding supports our hypothesis that $\omega_m^{l_A}$ and $\omega_m^{l_B}$ are sufficient and appropriate choices for predicting optimal merging coefficients.

C Additional a posteriori Analysis of SHAP Results

As demonstrated in the main paper, the SHAP trend observed for the ω^{l_A} and ω^{l_B} quantiles is consistent between the Tiny-ImageNet proxy dataset and other distributions. Figures Fig. 5, Fig. 6, and Fig. 7 further illustrate that this trend is consistently maintained across the CIFAR-100, EuroSAT, and CUB-200 datasets. Specifically, for CIFAR-100, the negative impact of ω^{l_A} at high quantiles is more pronounced at the upper extreme of the distribution. The impact of ω^{l_B} is generally similar across the datasets examined, with the notable exception of CUB-200, where its relative impact is lower in magnitude. Consequently, the non-decreasing trend for ω^{l_B} is not fully maintained across the entire distribution for this dataset. This observation is likely attributable to the dissimilarity between the CUB-200 dataset and the pretraining data, suggesting that the change in impact layer direction, represented by the A matrices, exerts a more significant influence on the optimal merging strategy in this case.

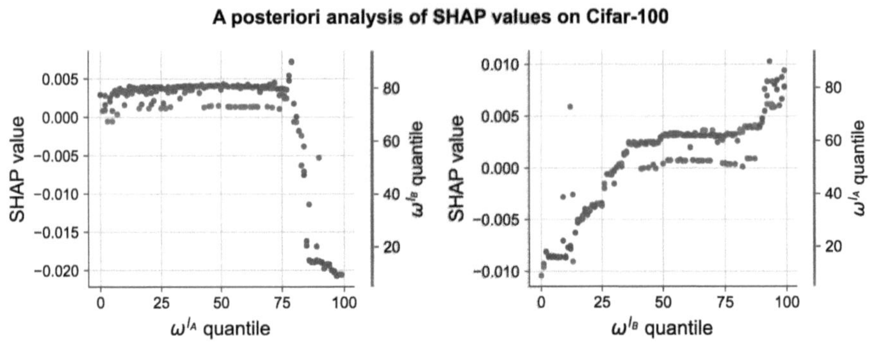

Fig. 5. A posteriori analysis on $\omega_m^{l_A}$ and $\omega_m^{l_B}$ impact in predicting λ_m^l coefficients on CIFAR-100.

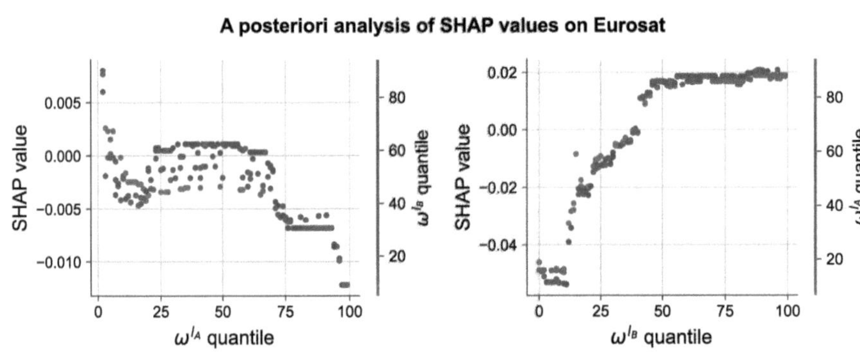

Fig. 6. A posteriori analysis on $\omega_m^{l_A}$ and $\omega_m^{l_B}$ impact in predicting λ_m^l coefficients on EuroSAT.

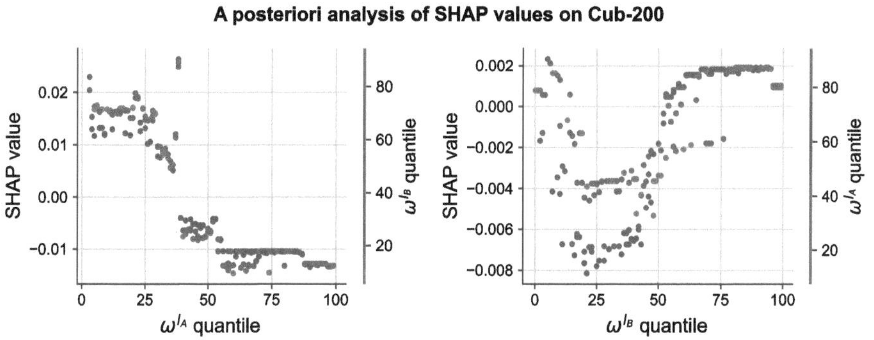

Fig. 7. A posteriori analysis on $\omega_m^{l_A}$ and $\omega_m^{l_B}$ impact in predicting λ_m^l coefficients on CUB-200.

References

1. Breiman, L.: Random forests. Mach. Learn. (2001)
2. Dosovitskiy, A., et al.: An image is worth 16x16 words: transformers for image recognition at scale. In: International Conference on Learning Representation (2021)
3. Fu, Z., Yang, H., So, A.M.C., Lam, W., Bing, L., Collier, N.: On the effectiveness of parameter-efficient fine-tuning. In: AAAI (2023)
4. He, K., Zhang, X., Ren, S., Sun, J.: Delving deep into rectifiers: surpassing human-level performance on imagenet classification. In: International Conference on Computer Vision (2015)
5. Helber, P., Bischke, B., Dengel, A., Borth, D.: Eurosat: a novel dataset and deep learning benchmark for land use and land cover classification. IEEE J. Sel. Topics Appl. Earth Observat. Remote Sens. (2019)
6. Hendrycks, D., et al.: The many faces of robustness: a critical analysis of out-of-distribution generalization. In: International Conference on Computer Vision (2021)
7. Hu, E.J., et al.: Lora: low-rank adaptation of large language models. In: International Conference on Learning Representation (2022)
8. Huang, C., Liu, Q., Lin, B.Y., Pang, T., Du, C., Lin, M.: Lorahub: efficient cross-task generalization via dynamic lora composition. arXiv preprint arXiv:2307.13269 (2023)
9. Jin, X., Ren, X., Preotiuc-Pietro, D., Cheng, P.: Dataless knowledge fusion by merging weights of language models. In: International Conference on Learning Representation (2023)
10. Karimireddy, S.P., Kale, S., Mohri, M., Reddi, S., Stich, S., Suresh, A.T.: Scaffold: stochastic controlled averaging for federated learning. In: International Conference on Machine Learning (2020)
11. Krizhevsky, A.: Learning multiple layers of features from tiny images. Master's thesis, University of Tront (2009)
12. Le, Y., Yang, X.: Tiny imagenet visual recognition challenge. CS 231N (2015)
13. Li, Q., He, B., Song, D.: Model-contrastive federated learning. In: IEEE Conference on Computer Vision and Pattern Recognition (2021)

14. Li, T., Sahu, A.K., Zaheer, M., Sanjabi, M., Talwalkar, A., Smith, V.: Federated optimization in heterogeneous networks. In: Proceedings of Machine Learning and Systems (2020)
15. Lundberg, S.M., Lee, S.I.: A unified approach to interpreting model predictions. Adv. Neural Inf. Process. Syst. (2017)
16. Luo, K., Li, X., Lan, Y., Gao, M.: Gradma: a gradient-memory-based accelerated federated learning with alleviated catastrophic forgetting. In: IEEE Conference on Computer Vision and Pattern Recognition (2023)
17. Luo, M., Chen, F., Hu, D., Zhang, Y., Liang, J., Feng, J.: No fear of heterogeneity: classifier calibration for federated learning with non-iid data. Adv. Neural Inf. Process. Syst. (2021)
18. Matena, M.S., Raffel, C.A.: Merging models with fisher-weighted averaging. Adv. Neural Inf. Process. Syst. (2022)
19. McMahan, B., Moore, E., Ramage, D., Hampson, S., Arcas, B.A.: Communication-efficient learning of deep networks from decentralized data. In: International Conference on Artificial Intelligence and Statistics (2017)
20. Pruckovskaja, V., et al.: Federated learning for predictive maintenance and quality inspection in industrial applications. In: 2023 Prognostics and Health Management Conference (PHM). IEEE (2023)
21. Ridnik, T., Ben-Baruch, E., Noy, A., Zelnik-Manor, L.: Imagenet-21k pretraining for the masses. In: Conference on Neural Information Processing Systems Datasets and Benchmarks Track (2021)
22. Salami, R., Buzzega, P., Mosconi, M., Bonato, J., Sabetta, L., Calderara, S.: Closed-form merging of parameter-efficient modules for federated continual learning. In: International Conference on Learning Representation (2025)
23. Salami, R., Buzzega, P., Mosconi, M., Verasani, M., Calderara, S.: Federated class-incremental learning with hierarchical generative prototypes. arXiv preprint arXiv:2406.02447 (2024)
24. Sheller, M.J., et al.: Federated learning in medicine: facilitating multi-institutional collaborations without sharing patient data. Sci. Rep. (2020)
25. Sun, Y., Li, Z., Li, Y., Ding, B.: Improving lora in privacy-preserving federated learning. arXiv preprint arXiv:2403.12313 (2024)
26. Tan, Y., et al.: Fedproto: federated prototype learning across heterogeneous clients. In: AAAI (2022)
27. Wah, C., Branson, S., Welinder, P., Perona, P., Belongie, S.: Caltech-ucsd birds-200-2011. Technical Report. CNS-TR-2011-001, California Institute of Technology (2011)
28. Wu, X., Huang, S., Wei, F.: Mixture of lora experts. In: International Conference on Learning Representation (2024)
29. Yadav, P., Tam, D., Choshen, L., Raffel, C.A., Bansal, M.: Ties-merging: resolving interference when merging models. Adv. Neural Inf. Process. Syst. (2024)
30. Yang, E., et al.: Adamerging: adaptive model merging for multi-task learning. In: International Conference on Learning Representation (2024)
31. Zenke, F., Poole, B., Ganguli, S.: Continual learning through synaptic intelligence. In: International Conference on Machine Learning. PMLR (2017)
32. Zhang, G., Wang, L., Kang, G., Chen, L., Wei, Y.: SLCA: slow learner with classifier alignment for continual learning on a pre-trained model. In: International Conference on Computer Vision (2023)

Disentangled Latent Spaces Facilitate Data-Driven Auxiliary Learning

Geri Skenderi[1(✉)], Luigi Capogrosso[2], Andrea Toaiari[2], Matteo Denitto[3], Franco Fummi[2], and Simone Melzi[4]

[1] Bocconi Institute for Data Science and Analytics, Bocconi University, Milan, Italy
geri.skenderi@unibocconi.it
[2] Department of Engineering for Innovation Medicine, University of Verona, Verona, Italy
[3] HUMATICS - SYS-DAT Group, Verona, Italy
[4] Department of Informatics, Systems and Communication, University of Milano-Bicocca, Milan, Italy

Abstract. Learning to solve auxiliary tasks concurrently with a principal task of interest can improve performance when data is scarce or the principal task is complex. This idea is inspired by the improved generalization capability induced by solving multiple tasks simultaneously, leading to a robust shared representation. However, selecting optimal auxiliary tasks typically requires manual design or costly meta-learning approaches. We propose **Detaux**, a framework that discovers an unrelated auxiliary classification task via weakly supervised disentanglement at the representation level. Isolating variations relevant to the principal task in one subspace while generating orthogonal subspaces with high separability allows us to discover auxiliary labels by clustering in these subspaces, allowing a transition from Single-Task Learning (STL) to Multi-Task Learning (MTL). In particular, the original labels associated with the principal task and the newly discovered ones can be fed into any MTL framework. Experiments and ablation studies highlight the effectiveness of **Detaux** and reveal an unexplored link between disentangled representations and MTL. The source code is available at https://github.com/intelligolabs/Detaux.

Keywords: Representation Learning · Auxiliary Learning · Multi-Task Learning

1 Introduction

Human learning is often thought to arise from a combination of evolutionary and acquired processes that are used together and can be transferred from one problem to another. Inspired by this, MTL [5] represents a machine learning paradigm where multiple tasks are learned together to improve the downstream capability of a model using shared knowledge that derives from considering different aspects of the input. Specifically, this is achieved by jointly optimizing the

E. Rodolà et al. (Eds.): ICIAP 2025, LNCS 16167, pp. 143–155, 2026.
https://doi.org/10.1007/978-3-032-10185-3_12

model parameters across different tasks on the same data, allowing the model to learn task-specific and shared representations simultaneously. As a result, MTL can lead to better generalization, improved efficiency at inference time, and improved performance on individual tasks by exploiting their underlying relationships. However, there are still many open challenges, such as understanding which tasks work well together [10,35,40] or understanding the impact of different tasks on loss functions [6].

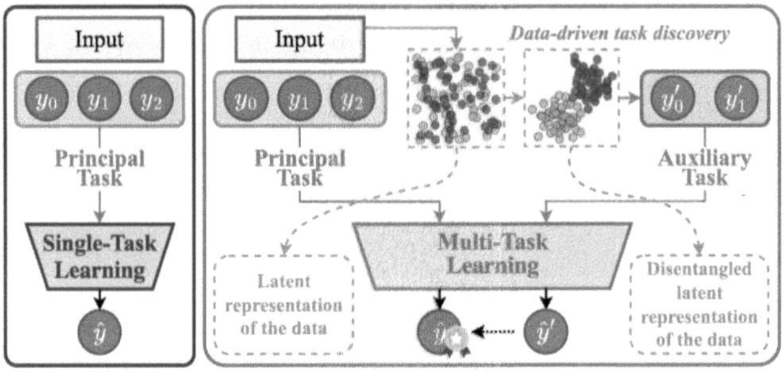

Fig. 1. Overview of our data-driven auxiliary task discovery framework. The figure shows the difference between STL (left) and MTL (right) with **Detaux**, which uses an auxiliary task generated through disentanglement on the data to improve performance on the principal task.

A particular form of MTL, dubbed *auxiliary learning*, has garnered considerable interest in recent years [22]. Auxiliary learning consists of using an additional set of tasks that lead to a shared representation with the goal of enhancing performance on the principal task. Therefore, these auxiliary tasks are tasks of minor interest, or even irrelevant compared to the principal task we want to solve, and can thus be seen as implicit regularizers [38]. At the state-of-the-art, auxiliary tasks are generated by leveraging meta-learning [24,32], which requires an a priori definition of the hierarchy of the desired auxiliary tasks and is often computationally inefficient. In practice, both [24] and [21] have to train two neural networks simultaneously: a label generation model to produce the auxiliary labels and a MTL model to solve the principal task alongside the auxiliary task. The question we try to answer in this work is *can we discover, with no prior knowledge, an additional auxiliary task directly from the data structure in order to improve the performance of the principal task?*

In this paper, we explore this problem by proposing **Detaux**, a weakly supervised strategy that discovers auxiliary classification tasks and allows solving a STL classification problem in a MTL fashion, as depicted in Fig. 1. Specifically, **Detaux** is capable of individuating unrelated auxiliary tasks: unrelatedness in MTL means to have two or more tasks whose features have *no* semantic intersection, as proven to be effective in the literature [17,34,41,42]. Our method takes root in the intuition that the two groups of tasks, the principal task and the auxiliary tasks, are known to be unrelated and assumes the claim that joint learning of unrelated tasks can improve the performance on the principal task.

The method of [34] exploits a linear classifier and requires the knowledge of the labels for both the principal task and the auxiliary tasks to force their respective representations to be orthogonal. Our method aims to follow a similar process while not requiring full supervision and using non-linear parametrizations through neural networks. **Detaux** generates auxiliary tasks so that their labels implicitly drive an MTL network to understand the unrelatedness between the tasks. Our idea is to work in a specific representation space, *i.e.*, a product manifold, to reveal the auxiliary tasks for a given principal task. To do this, we get inspiration from [11], who discovered the product manifold as a convenient representation basis for disentanglement.

As depicted in Fig. 2, we first extract task-specific features using a weakly supervised disentanglement procedure that implements projections in orthogonal subspaces of the latent representation. Then, we identify a subspace where the respective projections are maximally separated. Finally, we generate new labels via a clustering procedure on this subspace, leading to theoretically provable linear independence between task features (in the optimal scenario). In particular, this makes our proposed framework agnostic to the choice of the MTL model, given that the latter acts directly on the primary and generated auxiliary labels, as shown in the bottom right of Fig. 2.

In the experimental section, we tested three different MTL models with the **Detaux** framework on four different image datasets, revealing its flexibility.

2 Related Work

Representing data in a space where different components are independent is a long-standing research topic in machine learning. The rise of deep learning, which is based on learning representations, has made this concept even more relevant and useful in understanding the latent space [2]. Recent literature has proposed several characterizations of disentanglement, whether that is in terms of group theory [15], metric and product spaces [11], or permutations of element-wise, nonlinear functions [16]. [26] showed that completely unsupervised disentanglement is not possible due to the inability of the models to identify factors of variation. Soon after, the same authors proposed weak supervision and access to few labels to bypass this limitation [27,28]. In **Detaux**, we place ourselves in the same setting as in [11] but control and force the disentanglement by supervision only on the known principal task.

[31] reports a connection between disentangled representations and MTL, showing that disentangled features can improve the performance of MTL networks, especially on data with previously unseen properties. Disentanglement is obtained by adversarial learning, forcing the encoded features to be minimally informative about irrelevant tasks. In this case, the tasks to be disentangled are known a priori, while in our case, only the principal task task is known. [39] proposes a novel concept called "Knowledge Factorization". Exploiting the

knowledge contained in a pre-trained multi-task network (called teacher), the idea is to train disentangled single-task networks (called students) to reduce the computational effort required by the final single-task network. The factorization of the teacher's knowledge is dual: they provide structural factorization and representation factorization. In structural factorization, they split the net into a common knowledge network and a task-specific network based on mutual information. Finally, [30] explores the degree of disentanglement of MTL models in a controlled, semi-synthetic setting. Initially, a set of task labels is created using a randomly initialized Multi-Layer Perceptron (MLP) starting from the latent factors of parametric disentanglement datasets [3]. The authors successively train a separate neural network to solve these artificially created tasks and understand how disentangled the representations are w.r.t. the original latent factors.

In this work, we show that disentanglement in a representation space can be used as a general prior to MTL. After using disentanglement to mine for auxiliary tasks, an MTL model extracts a model-specific embedding that takes advantage of the combination of the principal and newly discovered labels, improving downstream performance on the principal task.

3 Mathematical Background

3.1 Disentanglement Framework

At a high level, disentangled representation learning aims to learn a representation of the data where any latent factor is independent of the others; that is, we have a factorized representation [2]. There are different ways to formalize this concept [15]. In this work, we rely on the approach proposed by [11]. The primary assumption behind this approach is the manifold hypothesis, *i.e.*, that high-dimensional data lies near a lower-dimensional manifold. Assuming that independent factors generate the data, it is reasonable to see this manifold as a product manifold $\mathcal{M} = \mathcal{M}_1 \times \mathcal{M}_2 \times \ldots \times \mathcal{M}_k$, with each $\mathcal{M}_i, i \in \{1 \ldots k\}$ representing at most one latent factor of the data. This latter concept is formalized adequately by relying on metric space topology to learn a weak isometry between the data and the learned product manifolds, as follows.

Definition 1 *(Product Manifold Disentanglement [11]). Let $\mathcal{M} = \mathcal{M}_1 \times \mathcal{M}_2 \times \ldots \times \mathcal{M}_k$ be the data product manifold, embedded in $\mathcal{X}$. Furthermore, assume that we have access to a metric that endows these two spaces with the properties of a metric space. A representation z in some product space $\mathcal{Z} = S_1 \times \ldots \times S_k$, such that $dim(\mathcal{Z}) \ll dim(\mathcal{X})$, is* disentangled *with respect to $\mathcal{M}$ if there exists a diffeomorphism (a bijection with a smooth inverse) $\tilde{g} : \mathcal{Z} \to \mathcal{M}$ such that $\forall x_1, x_2 \in \mathcal{M}; \forall i \in 1, \ldots, k$:*

$$d_{\mathcal{M}_i}(x_1^i, x_2^i) > 0 \implies d_{\mathcal{S}_i}(s_1^i, s_2^i) > 0 \,,$$

$$d_{\mathcal{M}_i}(x_1^i, x_2^i) = 0 \implies d_{\mathcal{S}_i}(s_1^i, s_2^i) = 0 \implies s_1^i = s_2^i \,,$$

where x_j^i is the projection of x_j on $\mathcal{M}_i$ and $s_j^i = \Pi_i \tilde{g}^{-1}(x_j)$ with Π_i being the projection onto the subspace $\mathcal{S}_i \subset \mathcal{Z}$.

According to Definition 1, given a pair (x_1, x_2) known to differ only in the h-th latent factor, their learned representations are considered disentangled if they have fixed projections in all the submanifolds $\{\mathcal{M}_i\}_{i=1}^k$, except the h-th. A simple pictorial understanding can be obtained by imagining any product manifold composed of k 1D connected submanifolds (without boundary). This manifold could be represented in a latent space as $\mathbb{R}^k$ with respect to Definition 1. In practice, this approach has a significant advantage when it comes to its application in generating auxiliary tasks, namely that each submanifold can have a different dimensionality as long as the sum of the dimensions of the submanifolds equals that of the product manifold. Therefore, we can generalize the intuitive idea of an "axis of variation" and look for auxiliary tasks in a higher-dimensional space instead of being limited to 1D representation axes as in Variational Auto-Encoders (VAE)-based methods [14,18].

3.2 Implementation

In practice, we consider a finite-dimensional normed vector space $\mathbf{Z} \subseteq \mathbb{R}^d$, containing the disentangled latent representation ($\mathbf{Z}$ is a particular case of a smooth manifold). The disentangled latent representation takes the form of a Cartesian product space $\mathbf{Z} = \mathcal{S}_1 \times \mathcal{S}_2 \times \ldots \times \mathcal{S}_k$, such that $\forall i, j \in \{1 \ldots k\}$, with $i \neq j$, $\mathcal{S}_i \cap \mathcal{S}_j = \{0\}$. As mentioned above, each subspace generalizes the notion of an "axis of variation". An encoder network $f : \mathcal{M} \rightarrow \mathcal{Z}$ produces the product space, and a set of non-linear operators $\{p_i\}_{i=1}^k$ projects the representation to the respective subspace. The representations in each subspace are then aggregated, and a decoder g maps the resulting vectors back to the input data space. Each $\mathbf{S}_i \subseteq \mathcal{R}^d$ is defined so that it has the same ambient dimensionality as the product space $\mathbf{Z}$. Using a specific $L1$ regularization term encourages orthogonal and sparse representations for each $\mathbf{S}_i$, which can then be summed to produce a latent code (each subspace will have few non-zero entries, and the non-zero entries in one subspace will be zero in the others). Subsequently, this latent code is fed into a decoder g that approximates the inverse of f. Thus, the decoder approximates $\tilde{g}$ in Definition 1. To wrap up, the representation framework operates in the following way:

$$x \xrightarrow{f} z \xrightarrow[i=1\ldots k]{\{p_i\}} \{s_i\} \xrightarrow[i=1\ldots k]{\sum_i} \tilde{z} \xrightarrow{g} \tilde{x} \ , \tag{1}$$

where $\tilde{z}$ and $\tilde{x}$ are the aggregated latent representation and the reconstructed input, respectively. This process is depicted in Fig. 2, inside the red rectangle.

In the following, we describe how this parametrization can be trained via differentiable constraints that lead to a minimization problem resulting in Definition 1. Due to space constraints, we will provide only a synthesized textual description of the loss functions. Their formalizations and additional details can be found in [11]. The encoder f receives non-i.i.d. data pairs $(x^{(1)}, x^{(2)})$ and produces the latent representations $(z^{(1)}, z^{(2)})$. The pair is sampled through weak supervision, requiring a pair of images known to vary in at least one latent factor (this is crucial to later isolate the change from $x^{(1)}$ to $x^{(2)}$ in one subspace). Furthermore, a set of k neural networks $p_i, i \in \{1 \dots k\}$ are trained simultaneously to map the latent codes in the subspaces $\{\mathcal{S}_i\}_{i=1}^k$, each of which contains the corresponding submanifold $\{\mathcal{M}_i\}_{i=1}^k$. An initial warm-up phase trains f and g only to minimize the data reconstruction error, needed to learn the global data manifold $\mathcal{M}$. Subsequently, four differentiable constraints that refer to different aspects of Definition 1 are added, leading to the following optimization problem:

$$\mathcal{L} = \mathcal{L}_{rec} + \beta_1(\mathcal{L}_{dist} + \mathcal{L}_{spar}) + \beta_2\mathcal{L}_{cons} + \beta_3\mathcal{L}_{reg} \ , \tag{2}$$

where β_1, β_2, and β_3 are Lagrange multipliers. $\mathcal{L}_{rec}$ corresponds to a common *reconstruction loss*, implemented in practice as the squared error between the input and the reconstructed images. The *distance loss*, $\mathcal{L}_{dist}$, is a contrastive term. It is built based on an oracle function $\mathcal{O} : \mathbf{Z} \times \mathbf{Z} \to \{1, \dots, k\}$, which calculates the subspace $\mathcal{S}_i$ where the projections of the images in the pair (x_1, x_2) are the furthest apart. Then, it encourages the projection representation of the two input images in the subspaces not selected by $\mathcal{O}$ to be as close as possible while retaining the separation in $\mathcal{S}_i$. The oracle function is a crucial part of our method, and the custom version we implemented in **Detaux** will be defined in Eq. 3. $\mathcal{L}_{spar}$ is a $L1$ constraint that promotes sparsity and orthogonality between subspaces. In our finite-dimensional setting, this constraint is equivalent to imposing that the product space is a direct sum of the subspaces. $\mathcal{L}_{cons}$, namely the *consistency loss*, encourages each projector p_i to be invariant to changes in subspaces $\mathcal{S}_j, j \neq i$. Together with $\mathcal{L}_{dist}$, this constraint encourages the metric definition of disentanglement. Finally, the *regularization* loss $\mathcal{L}_{reg}$ introduces a penalty which ensures that the choices of the oracle $\mathcal{O}$ are uniformly distributed among the subspaces to avoid representation collapse (due to the initial warm-up period).

4 Methodology

4.1 The Principal Task-Based Oracle

We assume the existence of a labeled image dataset $D = \{(x^{(i)}, y^{(i)}) \,|\, \forall i \in \{1 \dots N\}, x^{(i)} \in \mathcal{R}^{w \times h \times c}, y^{(i)} \in \mathbb{N}\}$, where w is the width, h the height, c the number of channels, and N the number of tuples (image, label). We consider a classification task whose objective is to learn a mapping $x^{(i)} \to y^{(i)}$ that maps each image to its corresponding label. To discover auxiliary tasks from new visual concepts, we must have a way to accommodate the (known) variation related to the principal task in one subspace and fix it there. To achieve this, we define a principal task-based oracle $\hat{\mathcal{O}} : \mathbf{Z} \times \mathbf{Z} \to \{1, \dots, k\}$, which ensures that the α-th subspace will contain all the variations in the data corresponding to pairs $(x^{(1)}, x^{(2)})$ that differ in their principal task label. Note that we do not inject direct knowledge of these labels, but only whether or not they differ, as follows:

$$\hat{\mathcal{O}}(z^{(1)}, z^{(2)}) = \begin{cases} \alpha & \text{if } y^{(1)} \neq y^{(2)} \\ \underset{i \in \{1,\dots,k\} \setminus \alpha}{\operatorname{argmax}} \ d(s_i^{(1)}, s_i^{(2)}) & \text{otherwise} \end{cases} , \tag{3}$$

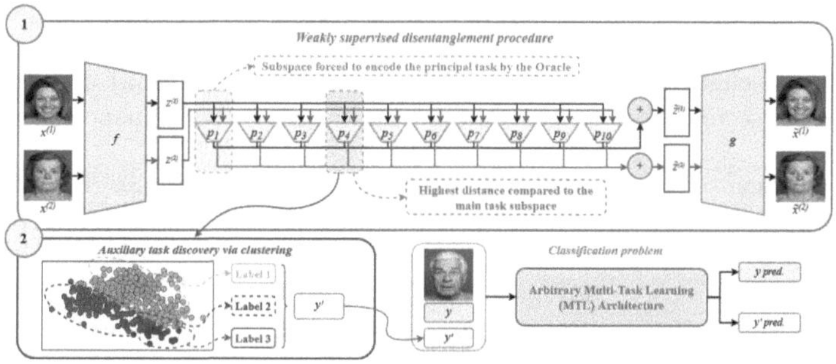

Fig. 2. Detaux steps: 1) We use weakly supervised disentanglement to isolate the structural features specific to the principal task in one subspace (red rectangle). 2) We identify the subspace with the most disentangled factor of variation w.r.t. the principal task, and through a clustering module, we obtain new labels (blue rectangle). These can be used to create a new classification task that can be combined with the principal task in any MTL model (bottom right). (Color figure online)

where $d(s_i^{(1)}, s_i^{(2)})$ is the distance between the projections of $(z^{(1)}, z^{(2)})$ in the i-th subspace $\mathcal{S}_i$. Our new oracle implies that distance and regularization losses will always force the variation in the data to be encoded in $\mathcal{S}_\alpha$ if $y^{(1)} \neq y^{(2)}$, and otherwise in a different subspace. The choice of the subspace for the case $y^{(1)} = y^{(2)}$ is made by looking at where the distance between the projections is maximal, as this is where the difference between the pair in that latent factor will be encoded. Due to the consistency loss, the remaining subspaces can encode other variations while remaining invariant to the ones related to the principal task and contained in $\mathcal{S}_\alpha$. The combination of these constraints will lead us to discover a proper representation in which unknown tasks correspond to (possibly) multiple subspaces, orthogonal to those of the principal task. We set $\alpha = 1$ in practice, as this is a simple and intuitive choice, but any other value $\in \{1...k\}$ is perfectly suitable. In practice, we approximate the *argmax* operator in Eq. 3 by applying a *softmax* at low temperature to the normalized pairwise Euclidean distances. This produces a discrete probability distribution over the distances, which can then be used to weigh the contributions of the projections in each subspace.

4.2 Auxiliary Task Discovery

Having forced the known principal task variation into a subspace of choice, we look to find new auxiliary tasks in the remaining subspaces. Intuitively, we wish to have a disentangled subspace that exhibits a clustering tendency over the projected data. This notion is implicitly built into the disentanglement loss function in Eq. 2, mainly due to the distance loss $\mathcal{L}_{dist}$. Let $\mathcal{S}_j$ (obtained from the projector p_j) be the subspace where $\mathcal{L}_{dist}$ is minimal after training. We apply a clustering algorithm to the latent representation found in $\mathcal{S}_j$, as shown inside the blue rectangle of Fig. 2. After clustering, we obtain a set of discrete pseudo-labels, determining the new auxiliary task. Given that the disentanglement procedure

Table 1. Classification accuracy on the FACES [9], CIFAR-10 [20], SVHN [33], and Cars [19] datasets. (*) The results are the ones reported in the original paper since we encountered challenges in replicating the performance. In **bold**, the best results. Underlined the second best. In parentheses, the change in performance over STL.

Learning Paradigm	**FACES** [9] ↑	**CIFAR-10** [20] ↑	**SVHN** [33] ↑	**Cars** [19] ↑
STL	0.915	0.844	0.956	0.711
MAXL [24]	0.933 (+0.018)	0.868 (+0.024)	0.953 (-0.003)	0.638 (-0.073)
AuxiLearn [32]	0.915 (+0.000)	0.811 (-0.033)	0.943 (-0.013)	0.644* (-0.067)
MTL-HPS [5] + **Detaux (ours)**	0.951 (+0.036)	0.848 (+0.004)	0.954 (-0.002)	0.789 (+0.078)
NDDR [12] + **Detaux (ours)**	0.932 (+0.017)	0.872 (+0.028)	0.952 (-0.004)	0.712 (+0.001)
MTI [36] + **Detaux (ours)**	**0.978 (+0.063)**	**0.910 (+0.066)**	**0.961 (+0.005)**	**0.807 (+0.096)**

already indicates how much each subspace might contain different clusters (via $\mathcal{L}_{dist}$), choosing the subspace $\mathcal{S}_j$ makes it so that the image embeddings are already well separated, providing a big advantage for the clustering procedure.

Although it is possible to use an arbitrary clustering algorithm, we would like it to support clusters of arbitrary shapes and not directly specify the number of clusters. Therefore, we utilize HDBSCAN [4] because it allows us to cluster data points based on their density (essentially pairwise proximity) without explicitly specifying the number of clusters. This algorithm can associate points that cannot be assigned to any cluster to a "noise" cluster, which we can retain as an additional label of the auxiliary task, making it very robust. If HDBSCAN finds only one cluster, we denote the run as unsuccessful and stop the procedure, as training it would lead to trivial MTL results. Otherwise, we have discovered a novel task and its corresponding labels $y' \in \mathbb{N}$, which can be used with any MTL model, as shown in the bottom right part of Fig. 2. In this work, we limit ourselves to finding only one auxiliary task, as this is common practice [21] and because the use of multiple auxiliary tasks increases the computational load of MTL. Scaling to more tasks is the subject of future work. At this stage, we have enriched our dataset with an additional set of labels, obtaining $D' = \{\, (x_i, y_i, y_i') \,|\, \forall i \in \{1 \dots N\},\, x_i \in \mathcal{R}^{w \times h \times c},\, y_i, y_i' \in \mathbb{N}\}$.

5 Experiments

Implementation Details. We fix the batch size at 32, the learning rate at 0.0005 and use the AdamW [29] optimizer. The disentanglement model is trained for 400 epochs, with the first quarter being used as a warm-up period where only $\mathcal{L}_{rec}$ is active. The multipliers $\beta_1, \beta_2, \beta_3$ follow an exponential warm-up routine after the reconstruction-only phase so that the constraints they control are gradually introduced into the optimization procedure. The projectors $p_i, i \in \{1 \dots k\}$ are implemented as two layers MLPs. Finally, all the MTL models are trained for 150 epochs. All experiments were performed on NVIDIA RTX 3090 GPUs.

5.1 Comparison with STL and Meta-Learning Approaches

As mentioned previously, image pairs are sampled only on the basis of available principal task labels. In the multi-class FACES dataset [9], this corresponds to the person's facial expression. In CIFAR-10 [20], SVHN [33], and Cars [19], it corresponds to the only annotated class. We used a ResNet-18 [13] encoder-decoder architecture to obtain high-quality reconstructions. The number of subspaces k is set to 10. During the auxiliary task discovery phase, we set the minimum cluster size hyperparameter of HDBSCAN to $0.01 * N$ for all the datasets. We compare our approach with two different auxiliary learning methods, *i.e.*, MAXL [24] and AuxiLearn [32]. Unlike these auxiliary learning architectures that exploit a meta-learning procedure, our discovered auxiliary task can be exploited interchangeably with any MTL model.

To have as much control over the experiments as possible and focus on the benefits of our discovered auxiliary task, we choose parameter-sharing MTL networks to ensure that the gains are due to the new task and not advanced learning dynamics. Modern MTL research has shifted towards more complex ideas, which is why these well-known approaches are dated at or before 2021. This dichotomy is also discussed in [23]. Thus, we select three different models: the standard Hard Parameter Sharing for MTL (MTL-HPS) [5], weighing the losses to give more importance to the main task, as explained in [6], NDDR [12], and MTI [36]. All these models have a loss term composed as a summation of each task's classification loss. During backpropagation, the gradient alters any shared parameters between the two tasks while looking to maximize performance on both, which is what drives the improved generalization capability. In particular, MTI was proposed to operate with an HRNet backbone [37]. This type of network performs multi-resolution fusion, starting from a high-resolution convolution stream and gradually adding high-to-low-resolution convolution streams one by one. Since the datasets on which we operate contain mainly images of size $\leq 224 \times 224$, learning HRNet from scratch results in low-quality representations. Therefore, we use the official MTI code, which uses an HRNet pre-trained on ImageNet [8].

Table 1 summarizes the results. MTI, with our generated auxiliary labels, displays the best performance. Furthermore, even simple MTL models, such as MTL-HPS and NDDR, achieve superior results compared to MAXL and AuxiLearn. Most notably, we outperform STL with at least one of the MTL+**Detaux** models in all the datasets, whereas MAXL and AuxiLearn have significant performance discrepancies between the datasets. For the sake of fairness, we report that we exploit pre-trained backbones for the MTL models on the Cars dataset, which contains very complex images, being a fine-grained classification dataset. Only for this data set, we change the encoder f so that it does not produce a dense representation in the bottleneck layer but a compressed feature map through 1×1 convolution and subsequent flattening.

5.2 Research Questions

How Important is the Disentangled Representation for Auxiliary Task Discovery? On the FACES dataset, we compare the auxiliary task generated

Table 2. *Left:* Classification accuracy when using different clustering algorithms to generate the auxiliary task labels. All the MTL results are obtained using the MTL-HPS [5] model. In parentheses, the change in performance over STL. *Right:* Normalized and Adjusted Mutual Information between the principal and auxiliary task labels generated using **Detaux**.

	Clustering	**FACES** [9] ↑	**Cars** [19] ↑
STL	–	0.915	0.711
MTL	HDBSCAN [4]	0.951	0.789
MTL	KMeans [25]	0.953 (+0.038)	0.789 (+0.078)
MTL	KMeans++ [1]	0.934 (+0.019)	0.790 (+0.079)
MTL	MeanShift [7]	0.963 (+0.048)	0.783 (+0.072)

Dataset	**Normalized MI** ↓	**Adjusted MI** ↓
FACES [9]	0.1405	0.1390
CIFAR-10 [20]	0.0033	0.0031
SVHN [33]	0.0033	0.0030
Cars [19]	0.0311	0.0085

by **Detaux** with the auxiliary task resulting from the clustering in the latent space of an autoencoder that only learns to reconstruct. Without disentanglement, MTL-HPS can only reach 0.9 accuracy, worse than the 0.915 obtained by STL. This reveals that performing auxiliary task mining on the entangled autoencoder space provides a less informative auxiliary task to the multi-task network compared to our approach.

Is It Possible to Use Other Clustering Algorithms? We use two versions of the KMeans algorithm [1,25], which assume a flat geometry, and MeanShift [7], which works well even in non-flat geometries. We set k equal to the number of clusters found by HDBSCAN and estimate the bandwidth parameter of MeanShift for each dataset [7]. The results are presented in Table 2 (left) and clearly show that our method is flexible to the choice of the clustering algorithm, improving the performance in all cases compared to STL.

Are the Generated Labels Correlated? We calculate the Normalized Mutual Information (NMI) and Adjusted Mutual Information (AMI) between the principal task and the auxiliary task labels generated by **Detaux**. The results in Table 2 (right) indicate that the two sets of labels are not correlated, with the labels from FACES showing minimal correlation.

6 Conclusion

In this paper, we propose a novel perspective on the utility of disentangled representations, utilizing them as a proxy for auxiliary learning in order to improve the accuracy of a principal task, originally solvable only in a STL fashion. Our proposed framework facilitates the weakly supervised discovery of a new task from a factorized representation. This newly discovered task can be incorporated into any MTL framework, and we empirically show that this offers improved performance on the principal task.

Acknowledgments. Geri Skenderi is funded by the European Union through the Next Generation EU - MIUR PRIN PNRR 2022 Grant P20229PBZR. Furthermore, this study was also carried out within the "COMMIS5.0: Chatbot e interfacce umanocentriche per l'OttiMizzazione della Manifattura e l'Innovazione nei Servizi 5.0" Funded by PR Veneto FESR 2021-2027, Obiettivo Specifico 1.1, Azione 1.1.1 "Rafforzare la ricerca e l'innovazione (in collaborazione) tra imprese e organismi di ricerca" Sub A "Rafforzare la ricerca e l'innovazione tra imprese e organismi di ricerca".

References

1. Arthur, D., Vassilvitskii, S.: k-means++: The Advantages of Careful Seeding. Technical Report, Stanford InfoLab (2006)
2. Bengio, Y., et al.: Representation learning: a review and new perspectives. IEEE Trans. Pattern Anal. Mach. Intell. **35**(8), 1798–1828 (2013)
3. Burgess, C., Kim, H.: 3D Shapes Dataset (2018)
4. Campello, R.J.G.B., Moulavi, D., Sander, J.: Density-based clustering based on hierarchical density estimates. In: Pei, J., Tseng, V.S., Cao, L., Motoda, H., Xu, G. (eds.) PAKDD 2013. LNCS (LNAI), vol. 7819, pp. 160–172. Springer, Heidelberg (2013). https://doi.org/10.1007/978-3-642-37456-2_14
5. Caruana, R.: Multitask learning. Mach. Learn. **28**, 95–133 (1998)
6. Cipolla, R., et al.: Multi-task learning using uncertainty to weigh losses for scene geometry and semantics. In: Conference on Computer Vision and Pattern Recognition (CVPR) (2018)
7. Comaniciu, D., Meer, P.: Mean shift: a robust approach toward feature space analysis. IEEE Trans. Pattern Anal. Mach. Intell. **24**(5), 603–619 (2002)
8. Deng, J., et al.: ImageNet: a large-scale hierarchical image database. In: Conference on Computer Vision and Pattern Recognition (CVPR) (2009)
9. Ebner, N.C., et al.: FACES-A database of facial expressions in young, middle-aged, and older women and men: development and validation. Behav. Res. Methods **42**(1), 351–362 (2010)
10. Fifty, C., et al.: Efficiently identifying task groupings for multi-task learning. In: Advances in Neural Information Processing Systems (NeurIPS) (2021)
11. Fumero, M., et al.: Learning disentangled representations via product manifold projection. In: International Conference on Machine Learning (ICML) (2021)
12. Gao, Y., et al.: NDDR-CNN: layerwise feature fusing in multi-task CNNs by neural discriminative dimensionality reduction. In: Conference on Computer Vision and Pattern Recognition (CVPR) (2019)
13. He, K., et al.: Deep residual learning for image recognition. In: Conference on Computer Vision and Pattern Recognition (CVPR) (2016)
14. Higgins, I., et al.: Beta-VAE: learning basic visual concepts with a constrained variational framework. In: International Conference on Learning Representations (ICLR) (2017)
15. Higgins, I., et al.: Towards a definition of disentangled representations. arXiv preprint arXiv:1812.02230 (2018)
16. Horan, D., et al.: When is unsupervised disentanglement possible? In: Advances in Neural Information Processing Systems (NeurIPS) (2021)
17. Jayaraman, D., et al.: Decorrelating semantic visual attributes by resisting the urge to share. In: Conference on Computer Vision and Pattern Recognition (CVPR) (2014)
18. Kim, H., Mnih, A.: Disentangling by factorising. In: International Conference on Machine Learning (ICML) (2018)

19. Krause, J., et al.: 3D object representations for fine-grained categorization. In: International Conference on Computer Vision Workshops (ICCVW) (2013)
20. Krizhevsky, A., Hinton, G.: Learning Multiple Layers of Features from Tiny Images. Technical Report, University of Toronto (2009)
21. Li, Y., Shan, S.: Meta auxiliary learning for facial action unit detection. IEEE Trans. Affect. Comput. **14**(3), 2526–2538 (2023)
22. Liebel, L., Körner, M.: Auxiliary tasks in multi-task learning. arXiv preprint arXiv:1805.06334 (2018)
23. Lin, B., Zhang, Y.: LibMTL: a Python library for deep multi-task learning. J. Mach. Learn. Res. **24**(1), 9999–10005 (2023)
24. Liu, S., et al.: Self-supervised generalisation with meta auxiliary learning. In: Advances in Neural Information Processing Systems (NeurIPS) (2019)
25. Lloyd, S.: Least squares quantization in PCM. IEEE Trans. Inf. Theory **28**(2), 129–137 (1982)
26. Locatello, F., et al.: Challenging common assumptions in the unsupervised learning of disentangled representations. In: International Conference on Machine Learning (ICML) (2019)
27. Locatello, F., et al.: Disentangling factors of variation using few labels. In: International Conference on Learning Representations (ICLR) (2020)
28. Locatello, F., et al.: Weakly-supervised disentanglement without compromises. In: International Conference on Machine Learning (ICML) (2020)
29. Loshchilov, I., Hutter, F.: Decoupled weight decay regularization. In: International Conference on Learning Representations (ICLR) (2018)
30. Maziarka, L., et al.: On the relationship between disentanglement and multi-task learning. In: European Conference on Machine Learning and Principles and Practice of Knowledge Discovery in Databases (ECML PKDD) (2023)
31. Meng, Q., Pawlowski, N., Rueckert, D., Kainz, B.: Representation disentanglement for multi-task learning with application to fetal ultrasound. In: Wang, Q., et al. (eds.) PIPPI/SUSI -2019. LNCS, vol. 11798, pp. 47–55. Springer, Cham (2019). https://doi.org/10.1007/978-3-030-32875-7_6
32. Navon, A., et al.: Auxiliary learning by implicit differentiation. In: International Conference on Learning Representations (ICLR) (2021)
33. Netzer, Y., et al.: Reading digits in natural images with unsupervised feature learning. In: Advances in Neural Information Processing Systems Workshop (NeurIPSW) (2011)
34. Paredes, B.R., et al.: Exploiting unrelated tasks in multi-task learning. In: 15th International Conference on Artificial Intelligence and Statistics (AISTATS) (2012)
35. Standley, T., et al.: Which tasks should be learned together in multi-task learning? In: International Conference on Machine Learning (ICML) (2020)
36. Vandenhende, S., Georgoulis, S., Van Gool, L.: MTI-Net: multi-scale task interaction networks for multi-task learning. In: Vedaldi, A., Bischof, H., Brox, T., Frahm, J.-M. (eds.) ECCV 2020. LNCS, vol. 12349, pp. 527–543. Springer, Cham (2020). https://doi.org/10.1007/978-3-030-58548-8_31
37. Wang, J., et al.: Deep high-resolution representation learning for visual recognition. IEEE Trans. Pattern Anal. Mach. Intell. **43**(10), 3349–3364 (2020)
38. Yan, H., et al.: Auxiliary learning and its statistical understanding. arXiv preprint arXiv:2501.03463 (2025)
39. Yang, X., et al.: Factorizing knowledge in neural networks. In: European Conference on Computer Vision (ECCV) (2022). https://doi.org/10.1007/978-3-031-19830-4_5

40. Zamir, A., et al.: Taskonomy: disentangling task transfer learning (IJCAI). In: 28th International Joint Conference on Artificial Intelligence (2019)
41. Zheng, Y., et al.: Exploiting related and unrelated tasks for hierarchical metric learning and image classification. IEEE Trans. Image Process. **29**, 883–896 (2020)
42. Zhou, D., et al.: Hierarchical classification via orthogonal transfer. In: International Conference on Machine Learning (ICML) (2011)

Deep Multimodal Fusion for 2D-3D Vineyard Scene Understanding

Mirko Usuelli, Cristian Sbrolli(✉), Alessio Braccini, Matteo Frosi, and Matteo Matteucci

Dipartimento di Elettronica, Informazione e Bioingegneria, Politecnico di Milano, Milano, Italy
{mirko.usuelli,cristian.sbrolli,alessio.braccini,matteo.frosi, matteo.matteucci}@polimi.it

Abstract. This study addresses 3D scene understanding in vineyard agriculture through multimodal data fusion for robotic applications. We present an automated annotation pipeline designed to overcome dataset limitations in agricultural environments, enabling robust semantic perception. After benchmarking state-of-the-art architectures, we propose a multimodal framework that integrates zero-shot 2D segmentation, with a fine-tuned 3D scene understanding model (CLIP2Scene) to improve semantic segmentation of sparse 3D point clouds. Experimental results highlight how combining 2D-3D semantic data with geospatial (GPS) information generates detailed, semantically enriched vineyard maps. Our work underscores the effectiveness of multimodal fusion in enhancing agricultural robotics, offering scalable solutions for precision farming. This approach not only improves operational efficiency but also promotes sustainable practices by enabling data-driven insights in complex, unstructured environments.

Keywords: 3D Scene Understanding · Agricultural Robotics · Multimodal Learning

1 Introduction

Recent advances in computer vision and deep learning have boosted robotic perception in areas like autonomous driving and virtual assistants. Yet, agriculture, with challenges like seasonal variability and labor shortages, remains underexplored. Effective 3D scene understanding is crucial for robots to navigate and perform tasks such as pruning, harvesting, and crop monitoring in complex environments.

Using only one sensor is often insufficient. LiDAR excels at depth estimation but struggles with fine details and adverse weather, while RGB cameras provide high resolution but are sensitive to lighting and geometric inconsistencies due to spatial perspective.

This study introduces a deep learning multimodal fusion approach that integrates 2D camera and 3D LiDAR data to enhance semantic perception for autonomous operations in agriculture. The key contributions are:

E. Rodolà et al. (Eds.): ICIAP 2025, LNCS 16167, pp. 156–167, 2026.
https://doi.org/10.1007/978-3-032-10185-3_13

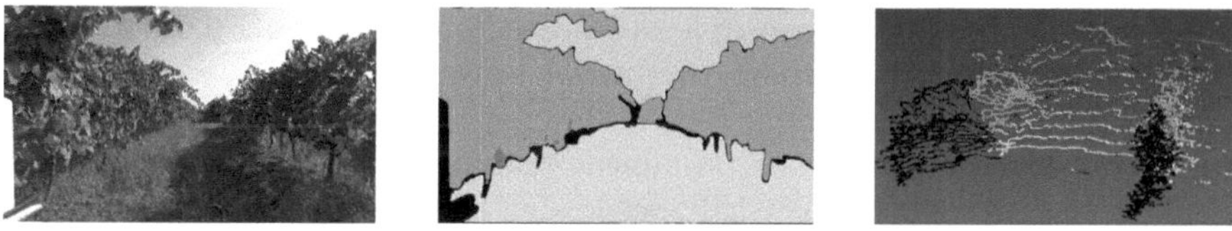

(a) RGB image from the vineyard in the BLT dataset.

(b) Segmented 2D labels generated by the LangSAM pipeline.

(c) Projected 3D labels onto the input point cloud.

Fig. 1. Results of the annotation pipeline on the BLT dataset. (a) Original RGB image, (b) segmented 2D labels, and (c) projected 3D labels onto the corresponding point cloud.

- **Automatic Semantic Annotation Pipeline:** A zero-shot pipeline that leverages pretrained open-vocabulary image segmentation models to annotate point clouds from datasets with aligned images and LiDAR scans. This process generates the necessary annotations for fine-tuning models on scene understanding tasks.
- **Multimodal Fusion for Scene Understanding:** A novel deep learning framework that fuses features from both RGB images and LiDAR data. It combines a zero-shot image backbone with a fine-tuned 3D backbone, exploiting the complementary strengths of each modality to enhance prediction accuracy.
- **Real-World Deployment and Evaluation:** The framework is deployed in two vineyard scenarios–one using a low-cost 16-plane LiDAR and another with a higher-resolution 64-plane LiDAR–demonstrating its robustness and adaptability across different sensor configurations.

We conduct our experiments on two unannotated datasets, which we annotate using our proposed scene understanding annotation pipeline. The first dataset is the Bacchus Long-Term (BLT) dataset [14], which provides a comprehensive 3D representation of the same vineyard across multiple seasons. The second dataset was collected in Piacenza, Italy, using an autonomous robot operating in a local vineyard.

2 Related Works

Recent progress in perception has moved beyond single-sensor approaches, increasingly embracing multimodal methods that combine LiDAR, RGB cameras, and even text inputs. This trend stems from the observation that each modality carries unique advantages: LiDAR excels at capturing 3D geometry, while RGB images provide high-resolution visual details, and text offers semantic richness for class-level understanding. By weaving together these modalities, researchers aim to construct more robust and generalizable models capable of performing tasks under varied conditions.

In particular, a surge of research has focused on aligning 3D data with textual and visual modalities through contrastive learning [16,18,19], enabling zero-shot and few-shot 3D understanding by leveraging large-scale vision-language models trained on 2D images.A prominent example of such multimodal fusion including the 3D domain is CLIP2Scene [2], which takes advantage of the pre-trained knowledge of the Contrastive Language-Image Pre-Training (CLIP) model. By aligning 2D and 3D representations through consistency-regularized training, CLIP2Scene performs zero-shot and few-shot 3D semantic segmentation with minimal supervision. In parallel, the Multimodal Interlaced Transformer (MIT) [20] employs a weakly supervised approach, relying solely on scene-level labels to iteratively cross-attend features from both 2D images and 3D point clouds. While CLIP2Scene leverages textual grounding via a vision-language model, MIT's design emphasizes attention mechanisms for fusing visual and geometric cues, highlighting the complementary roles that different modalities and architectural choices can play.

Underpinning these multimodal strategies is a broader landscape of research in zero-shot learning (ZSL) for 3D scenes. Although zero-shot learning has been extensively explored in the 2D image domain [1,17,21,23], its application to 3D data has only recently gained momentum [3,5,13]. Early efforts primarily targeted tasks such as point cloud classification, with more recent works like [22,24] and CLIP2Scene [2] moving toward semantic segmentation.These studies demonstrate how language-image models can be adapted to expand the label space of 3D data, enabling recognition of previously unseen categories and offering a valuable solution in situations where manual annotation is limited or impractical.

Complementing these advancements in zero-shot learning, another active area of research emphasizes self-supervised representation learning and cross-modal distillation. In this domain, models aim to learn robust 3D features without relying on extensive human-labeled data. Building on contrastive learning paradigms [1,4,6], recent methods transfer knowledge from pre-trained 2D backbones to 3D encoders [12,15]. Such techniques leverage the granularity and semantic richness of image features to guide point-cloud representation learning, often bypassing the need for large-scale 3D annotations.

Despite these advancements, applications in agricultural robotics remain relatively unexplored. While multimodal and zero-shot methods show promise across automotive, indoor, and general 3D benchmarks, the agricultural setting presents unique challenges like seasonal variation, complex plant structures, and the scarcity of robust, labeled datasets. This gap offers fertile ground for new approaches that integrate multimodal, zero-shot, and self-supervised paradigms to improve agricultural tasks such as yield estimation, automated harvesting, and real-time navigation in challenging farm environments.

3 Proposed Approach

The availability of agricultural datasets for training and evaluating deep learning models is severely limited, particularly in comparison to the abundance of

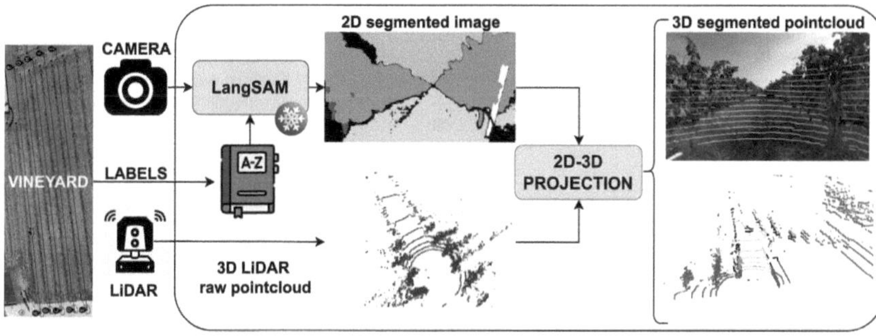

Fig. 2. LiDAR data annotation pipeline: synchronized RGB frontal images and 360° LiDAR point clouds are captured while navigating vineyard rows. Open-world agricultural labels are used to guide LangSAM in generating 2D masks. These masks are then projected onto corresponding 3D points using a projection module, completing the labeling process sequentially for each frame.

datasets available in domains such as autonomous driving or indoor scene understanding. Moreover, while existing datasets provide aligned 2D and 3D data captured from cameras and LiDAR sensors, they lack pointwise 3D semantic labels. This limitation hinders the fine-tuning and evaluation of models in agricultural scenarios. To address this gap, our approach introduces a general framework for generating fine-grained pointwise semantic labels for point cloud data by leveraging zero-shot 2D semantic segmentation models. We propose a multimodal fusion model for 3D scene segmentation that integrates information from both 2D and 3D modalities.

3.1 Zero-Shot 3D Semantic Scene Labeling Pipeline

We propose a two-stage zero-shotpipeline for generating fine-grained 3D semantic labels to enable 3D scene understanding. We show the pipeline in Fig. 2. It processes point clouds and images registered from sensors, along with the desired textual class labels, through the following stages:

2D Segmentation. The first stage applies zero-shot semantic segmentation to 2D images using the Language Segment-Anything Model (LangSAM) architecture, which generates segmentation masks from input RGB images and textual class prompts. LangSAM integrates two key components: GroundingDINO (GrDINO) [11], which produces bounding boxes for specified object classes in a zero-shot manner, and the Segment Anything Model (SAM) [9], which refines these bounding boxes into precise segmentation masks. The resulting masks are then consolidated into a pixel-level map, as illustrated in Fig. 1b.

3D Point Cloud Annotation. In the second stage, we propagate the 2D annotations to the 3D point cloud by implementing the method from [7]. First,

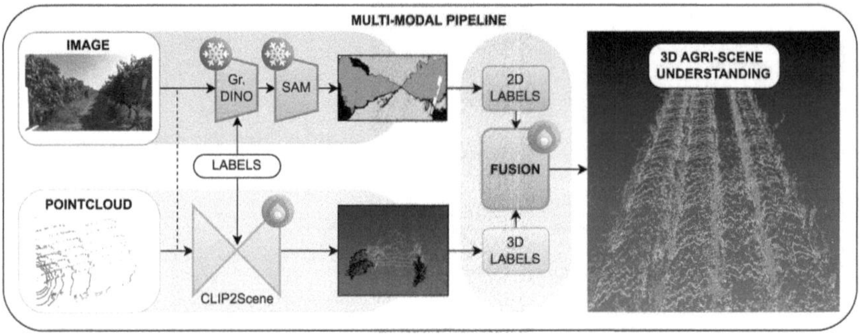

Fig. 3. Multimodal pipeline overview. Synchronized 2D images and 3D point clouds are processed using agricultural label prompts. Grounding DINO and SAM segment objects in 2D, while CLIP2Scene generates 3D annotations. The fused 2D3D labels, combined with GPS data, provide the robot with a rich semantic understanding of its environment.

points behind the RGB camera are discarded to reduce computational overhead. We then transform the point cloud from the LiDAR reference frame to the RGB camera reference frame. We then project the transformed 3D points onto the 2D segmentation masks, and assign semantic labels to each 3D point based on the corresponding mask regions. This automated process facilitates efficient annotation of large-scale 3D point cloud datasets using zero-shot 2D segmentation. An example annotation from the BLT dataset is visualized in Fig. 1c.

After developing the data-labeling pipeline, we turned our attention to selecting a suitable model for 3D scene understanding. Due to the limited availability of large-scale agricultural datasets, we focused on pretrained multimodal 2D-3D models for initialization. In particular, we evaluated two state-of-the-art architectures: CLIP2Scene and MIT. Both models were reimplemented and tested on the ScanNet dataset, a standard benchmark in the field.

Our experiments revealed that the fine-tuned CLIP2Scene model achieved the highest mean Intersection over Union (mIoU) with a score of 0.6547, significantly outperforming the MIT model, which reached 0.3671, and the annotation-free version of CLIP2Scene, which only achieved 0.2099. This evaluation exposed a key limitation of the MIT model: its training strategy tends to penalize underrepresented semantic classes, leading to reduced performance in those categories.

Given these results, we selected CLIP2Scene as the backbone for our 3D scene understanding pipeline. We annotated the training split of the BLT dataset using our zero-shot labeling approach and fine-tuned the CLIP2Scene 3D encoder with the Adam optimizer, a learning rate of 3×10^{-4}, and a batch size of 32.

3.2 Selecting an Optimal 2D Backbone for Segmentation

To enhance 3D segmentation with 2D features from the camera, we aim to select an appropriate 2D backbone. A natural starting point is the 2D CLIP image encoder used in CLIP2Scene. While this encoder is not utilized during inference by the original authors, it has demonstrated effectiveness in extracting 2D segmentation maps. However, its direct application to segmentation poses challenges due to uni-modality information loss. Specifically, the CLIP encoder is optimized to maximize shared information between the 2D and 3D modalities, often at the expense of modality-specific features.

This phenomenon, observed in prior studies on contrastive learning [8,10], arises from the InfoNCE loss function, which encourages cross-modal alignment but often disregards intra-modal details. Consequently, while the learned features excel at aligning modalities, they may overlook the distinctive characteristics intrinsic to individual modalities.

In contrast, our approach seeks to preserve and integrate the unique, complementary features of both 2D and 3D modalities to improve scene understanding. To achieve this, we leverage CLIP2Scene for its robust 3D feature extraction capabilities while incorporating a separate model specifically tailored for 2D pixel-level segmentation.

Given the demonstrated zero-shot effectiveness of the LangSAM framework, we adopt its two-stage architecture as described in Sect. 3.1. This pipeline processes a user-defined list of object classes alongside a 2D RGB image through two sequential steps: Bounding Box Localization, powered by GroundingDINO [11], and Segmentation Mask Generation, facilitated by SAM [9].

Although this approach yields high-quality segmentation results, it incurs significant computational cost-processing a single image with seven target classes takes approximately 20 s on an Nvidia GTX 1080Ti. To mitigate this overhead, we introduce two key optimizations:

1. **Parallel Bounding Box Predictions**: We modify the GroundingDINO processing to generate bounding boxes for all target classes simultaneously, rather than sequentially.
2. **Efficient SAM Variant**: We employ a more compact version of SAM (SAM ViT-b), as we find that it maintains the same level of precision given the relatively small image regions being processed.

These optimizations reduce the average processing time from approximately 20 s to under 1 s, representing a 95% decrease in computational cost. This improvement significantly enhances the pipeline's efficiency without compromising segmentation accuracy. We will refer to this network as the F-LangSAM, which stands for Faster-LangSAM.

3.3 Deep Multimodal 2D-3D Fusion

Once defined the predictors for both 2D and 3D modalities, we define an ensembled model combining the two predictions. This multimodal approach ensures that all available data modalities are exploited for improved semantic interpretation. We report an overview of the model pipeline in Fig. 3.

The core of our ensemble architecture lies in the fusion module, which integrates outputs from CLIP2Scene and F-LangSAM into a unified prediction.

CLIP2Scene generates a labeled point cloud, while F-LangSAM provides 2D segmentation masks. To align F-LangSAM's 2D output with the 3D domain, we employ a 2D-to-3D projection technique detailed in Sect. 3.1. The process for annotating 3D points from a 2D mask is defined as:

$$\text{annotated}_{3D} = \Pi(\text{mask}_{2D}), \tag{1}$$

where $\Pi(\cdot)$ maps pixels from the 2D mask to their corresponding 3D points. This mapping involves a series of transformations. First, a point in the LiDAR coordinate system is transformed into the camera coordinate system using a matrix that combines rotation and translation. Next, the point in the camera coordinate system is projected onto the image plane by applying the camera's intrinsic parameters, which include the focal lengths and the coordinates of the principal point. Finally, the resulting image coordinates are normalized by dividing them by the depth value from the camera. The transformation from LiDAR to camera coordinates is performed by a roto-translation matrix that incorporates both a rotation component and a translation component, while the intrinsic camera parameters are organized into a matrix defined by the focal lengths and the principal point coordinates.

This sequence transforms LiDAR points into the camera frame, projects them onto the 2D image plane, and normalizes them to pixel coordinates (u, v), enabling the retrieval of color or semantic information from the 2D mask. The resulting annotated point cloud aligns seamlessly with CLIP2Scene's output, allowing the fusion module to handle predictions consistently across both models.

Once both CLIP2Scene and F-LangSAM produce their respective sets of logits, the fusion module combines them according to either an arithmetic or a neural strategy. In the arithmetic approach, the raw logits from each model, denoted $\mathbf{l_{img}}$ and $\mathbf{l_{3D}}$, are normalized by a sigmoid function. The normalized outputs are then scaled by weighting coefficients α and $(1 - \alpha)$, resulting in

$$\hat{\mathbf{p}}(\mathbf{l_{img}}, \mathbf{l_{3D}}) = \alpha \frac{1}{1 + e^{-\mathbf{l_{img}}}} + (1 - \alpha) \frac{1}{1 + e^{-\mathbf{l_{3D}}}}, \tag{2}$$

where $0 \leq \alpha \leq 1$. For each dataset, we systematically evaluate discrete weight combinations within the range $[0, 1]$, incremented by 0.1, to determine the value of α that achieves the highest predictive accuracy. The results for each dataset are presented and analyzed in Sect. 5.

In the neural fusion method, a Multi-Layer Perceptron (MLP) is employed to optimize the combination of logits generated by CLIP2Scene and F-LangSAM. However, while CLIP2Scene produces logits for all classes, F-LangSAM outputs logits exclusively for the classes it detects. To address this, a tailored vector representation is constructed to handle overlapping classes (e.g., grapes and foliage), with logits for classes not identified by F-LangSAM explicitly set to zero.

$$\mathbf{v}_{\text{F-LangSAM}} = [z_1, z_2, \ldots, z_n], \quad \mathbf{v}_{\text{C2S}} = [c_1, c_2, \ldots, c_n], \tag{3}$$

where z_i is the logit from F-LangSAM for the i-th class (or zero if absent) and c_i is the logit from CLIP2Scene for the i-th class. These vectors are then

Table 1. Performance of the Ensemble Model on the Bacchus Long-Term (BLT) and Piacenza Vineyard (PV) datasets. Results are reported for different weight configurations (3D–2D) and prediction methods. Metrics include Intersection over Union (IoU) for individual classes, mean IoU (mIoU), and corrected mIoU, which excludes absent classes. The best result for each metric is highlighted with a **green** background, the second-best with a yellow background, and the third-best with an orange background, except for zero values, which are not highlighted.

Dataset	BLT Dataset				PV Dataset			
Weights (3D - 2D)	0.1–0.9	Only 3D(CLIP2Scene)	Only 2D (LangSAM)	Learnable	0.8–0.2	Only 3D (CLIP2Scene)	Only 2D (LangSAM)	Learnable
Foliage	0.812	0.277	**0.813**	0.004	**0.637**	0.628	0.227	0.288
Ground	**0.865**	0.439	0.849	0.219	0.099	0.099	0.078	**0.329**
Row Terminal	**0.025**	0.017	0.000	0.000	0.000	0.000	0.000	0.000
Grape	**0.581**	0.024	0.534	0.000	0.000	0.000	0.000	0.000
Unannotated	**0.918**	0.703	0.881	0.749	**0.926**	0.911	0.653	0.605
mIoU	**0.7430**	0.4943	0.7253	0.4246	**0.5231**	0.5204	0.4227	0.4604
mIoU Corrected	**0.6315**	0.2920	0.6154	0.1944	**0.5540**	0.5528	0.3193	0.4073

concatenated into

$$\mathbf{v}_{\text{fusion}} = [\mathbf{v}_{\text{F-LangSAM}}, \mathbf{v}_{\text{C2S}}]. \tag{4}$$

An MLP with learnable parameters θ maps this fused vector to the final prediction space:

$$f_\theta : \mathbb{R}^{2n} \to \mathbb{R}^n, \quad \hat{\mathbf{p}} = f_\theta(\mathbf{v}_{\text{fusion}}). \tag{5}$$

Due to the modest size of the dataset, a more compact MLP tends to generalize effectively. A notable advantage of this data-driven fusion is that it does not require manual hyperparameter tuning; the optimal weighting is inferred automatically during training, thereby offering a more flexible and streamlined approach to combining network predictions.

3.4 Complete Vineyard Map Generation

Once obtained our ensemble network which produces single scan point cloud segmentations, we define a method to construct a complete vineyard map. To this end, we utilize point cloud data and Global Positioning System (GPS) traces. GPS coordinates are converted to Universal Transverse Mercator (UTM) format for Cartesian representation. The robot's heading is estimated from the yaw angles between consecutive GPS entries, allowing precise placement of each point cloud segment within a global coordinate system.

A subset of point clouds is used to prevent overlaps and manage sensor noise due to the robot's slow movement and high-frequency sensor acquisition. After applying 3D semantic labeling, extraneous elements such as "Ground" can be removed to focus on specific agricultural features, such as grape locations along the vine. This process yields a dense and semantically meaningful 3D reconstruction of the vineyard, supporting tasks such as vine monitoring and yield estimation.

4 Datasets

In this study, we utilize two distinct datasets that differ significantly both in their agricultural context and the properties of sensors used for data acquisition:

- **BLT Dataset**: The Bacchus-Long-Term (BLT) [14] dataset provides multimodal information, including LiDAR, images, and GPS data, captured across the vineyard growing season in Greece using a UGV robot. It features a 16-plane LiDAR and consists of monthly recordings from March to September. This dataset is employed to fine-tune our proposed approach using data from the literature and to evaluate its performance on additional scenes within the BLT dataset.
- **PV Dataset**: The Piacenza Vineyard (PV) dataset, collected in Piacenza, Italy, adopts a similar structure to the BLT dataset but features a higher-resolution 64-plane LiDAR. It includes repeated recordings of two vineyard rows, each 50 m long, conducted five times during September, prior to harvest. This dataset is used to test and validate our approach in varied conditions and scenarios.

5 Results

The performance of the proposed approach is assessed using the Intersection over Union (IoU) metric along with its average value, the mean IoU (mIoU). For each class, IoU is defined as the ratio of correctly predicted pixels (true positives) to the total count of pixels that are either correctly predicted or incorrectly assigned (including both false positives and false negatives). The overall performance is then determined by averaging the IoU scores across all classes. Additionally, when some classes are missing in the predictions, a corrected version of this average, termed `mIoU_corr`, is used. This adjusted measure only includes the classes that appear in both the ground truth and the predictions.

Table 1 reports the per-class results of experiments conducted on both the test set of BLT and on the PV dataset. Figure 4 shows instead the $\mathrm{mIoU}_{\mathrm{corr}}$ for different combinations of weights on each dataset, justifying the best settings reported on the table. For the BLT dataset, the model is heavily reliant on the 2D features from F-LangSAM, which alone reaches much higher scores than using CLIP2Scene alone. However, reliance on F-LangSAM exclusively fails to surpass the ensemble performance, which gaps the baseline with 1.6% higher accuracy when using the 0.9–0.1 weighting. This indicates that a contribution from CLIP2Scene remains crucial for capturing elements otherwise occluded in 2D data. We also note that frequent classes such as Ground and Foliage achieve higher IoU values, while underrepresented categories like Grape and Row Terminal perform less consistently. Despite the domain gap between the pre-training corpus and real-world vineyard data, the results demonstrate the promise of zero-shot methods for agricultural applications with limited annotated data.

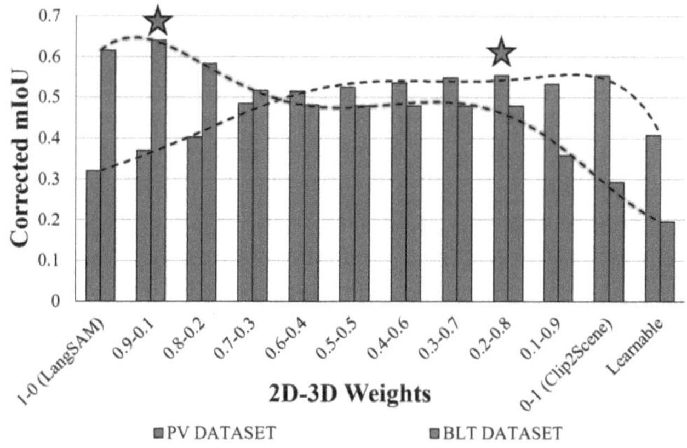

Fig. 4. mIoU results on the BLT and PV dataset for different weight combinations.

For the Piacenza Vineyard (PV) dataset, we perform experiments in a zero-shot manner using the model trained on the BLT dataset. Among the baselines, the highest accuracy is achieved with a combination of 80% CLIP2Scene and 20% F-LangSAM. Unlike the BLT dataset, the PV dataset is captured with a high-resolution LiDAR sensor, generating dense point clouds. This increased point cloud density significantly improves training quality, leading to higher prediction accuracy. As shown in Fig. 4, the trends for the two datasets reveal markedly contrasting patterns.

Interestingly, while the 3D backbone is fine-tuned on the BLT dataset and the 2D backbone remains zero-shot, the 3D backbone demonstrates a more prominent role in the PV dataset (on which it was not trained) compared to the BLT dataset. In contrast, the BLT dataset benefits more significantly from the zero-shot 2D backbone, which provides complementary features. These results suggest that the ensemble architecture is particularly advantageous for sparse or low-resolution point clouds, while detailed LiDAR scans enable robust 3D scene understanding with reduced reliance on 2D cues.

Overall, these findings underscore the flexibility of the proposed ensemble method, which adapts effectively to diverse sensor configurations and ensures reliable semantic parsing of vineyard environments under varying conditions.

6 Conclusion

This study presents a multimodal deep learning framework for 2D3D agricultural scene analysis, addressing vineyard data scarcity via an automated annotation pipeline using zero-shot 2D segmentation and 3D label propagation. Experiments validate its effectiveness in generating large-scale labeled datasets for supervised learning. We evaluate CLIP2Scene and Multimodal Interlaced Transformer (MIT), revealing challenges in adapting these models to unstructured, seasonal agricultural environments due to domain mismatch.

To address these limitations, we propose an Ensemble Model integrating CLIP2Scene (geometric LiDAR analysis) with 2D F-LangSAM (RGB/text-based fine-grained detection). The ensemble outperforms individual models on sparse LiDAR data (BLT dataset), particularly for foliage and grapes, but shows marginal gains on high-resolution LiDAR (Piacenza Vineyard), emphasizing its value in low-data scenarios.

Finally, we combine annotated 3D data with GPS to create semantically rich vineyard maps, enabling precision agriculture applications like yield estimation, targeted monitoring, and optimized pesticide use. This approach advances sustainable farming through enhanced field visibility and resource efficiency.

Acknowledgements. Mirko Usuelli's work was conducted within the Agritech National Research Center and received funding from the European Union Next-GenerationEU (PIANO NAZIONALE DI RIPRESA E RESILIENZA (PNRR) MISSIONE 4 COMPONENTE 2, INVESTIMENTO 1.4 – D.D. 1032 17/06/2022, CN00000022). This manuscript represents the authors' views only; the EU and European Commission are not responsible for them.

Matteo Matteucci's work has received co-funding from the European Union's Digital Europe Programme under grant agreement N° 101100622, as part of the Agrifood TEF initiative.

References

1. Caron, M., Misra, I., Mairal, J., Goyal, P., Bojanowski, P., Joulin, A.: Unsupervised learning of visual features by contrasting cluster assignments. In: Advances in Neural Information Processing Systems (NeurIPS), vol. 33, pp. 9912–9924 (2020)
2. Chen, R., et al.: Clip2scene: towards label-efficient 3d scene understanding by clip (2023)
3. Chen, R., et al.: Zero-shot point cloud segmentation by transferring geometric primitives. arXiv preprint arXiv:2210.09923 (2022)
4. Chen, R., et al.: Towards 3d scene understanding by referring synthetic models. arXiv preprint arXiv:2203.10546 (2022)
5. Cheraghian, A., Rahman, S., Chowdhury, T.F., Campbell, D., Petersson, L.: Zero-shot learning on 3d point cloud objects and beyond. Int. J. Comput. Vision **130**(10), 2364–2384 (2022)
6. Grill, J.B., et al.: Bootstrap your own latent: a new approach to self-supervised learning. In: Advances in Neural Information Processing Systems (NeurIPS), vol. 33, pp. 21271–21284 (2020)
7. Hartley, R.I., Zisserman, A.: Multiple View Geometry in Computer Vision, 2nd edn. Cambridge University Press, Cambridge (2004). ISBN: 0521540518
8. Jiang, Q., et al.: Understanding and constructing latent modality structures in multi-modal representation learning. In: Proceedings of the IEEE/CVF Conference on Computer Vision and Pattern Recognition (CVPR), pp. 7661–7671 (2023)
9. Kirillov, A., et al.: Segment anything. arXiv preprint arXiv:2304.02643 (2023)
10. Liang, V.W., Zhang, Y., Kwon, Y., Yeung, S., Zou, J.Y.: Mind the gap: understanding the modality gap in multi-modal contrastive representation learning. In: Koyejo, S., Mohamed, S., Agarwal, A., Belgrave, D., Cho, K., Oh, A. (eds.) Advances in Neural Information Processing Systems, vol. 35, pp. 17612–17625. Curran Associates, Inc. (2022)

11. Liu, S., et al.: Grounding dino: marrying dino with grounded pre-training for open-set object detection. arXiv preprint arXiv:2303.05499 (2023)
12. Liu, Y.C., et al.: Learning from 2d: contrastive pixel-to-point knowledge transfer for 3d pretraining. arXiv preprint arXiv:2104.04687 (2021)
13. Michele, B., Boulch, A., Puy, G., Bucher, M., Marlet, R.: Generative zero-shot learning for semantic segmentation of 3d point clouds. In: Proceedings of the International Conference on 3D Vision (3DV), pp. 992–1002 (2021)
14. Polvara, R., et al.: BLT dataset: acquisition of the agricultural bacchus long-term dataset with automated robot deployment. J. Field Rob. Agric. Rob. Ag 4.0 (2024)
15. Sautier, C., Puy, G., Gidaris, S., Boulch, A., Bursuc, A., Marlet, R.: Image-to-lidar self-supervised distillation for autonomous driving data. In: Proceedings of the IEEE/CVF Conference on Computer Vision and Pattern Recognition (CVPR), pp. 9891–9901 (2022)
16. Sbrolli, C., Matteucci, M.: No captions, no problem: captionless 3d-clip alignment with hard negatives via clip knowledge and llms. In: 35th British Machine Vision Conference 2024, BMVC 2024, Glasgow, UK, 25–28 November 2024. BMVA (2024). https://papers.bmvc2024.org/0032.pdf
17. Xie, S., Gu, J., Guo, D., Qi, C.R., Guibas, L., Litany, O.: Pointcontrast: unsupervised pre-training for 3d point cloud understanding. In: Proceedings of the European Conference on Computer Vision (ECCV), pp. 574–591 (2020)
18. Xue, L., et al.: Ulip: learning unified representation of language, image and point cloud for 3d understanding. arXiv preprint arXiv:2212.05171 (2022)
19. Xue, L., et al.: Ulip-2: towards scalable multimodal pre-training for 3d understanding (2023)
20. Yang, C.K., Chen, M.H., Chuang, Y.Y., Lin, Y.Y.: 2d-3d interlaced transformer for point cloud segmentation with scene-level supervision (2024)
21. Éloi Zablocki, Bordes, P., Piwowarski, B., Soulier, L., Gallinari, P.: Context-aware zero-shot learning for object recognition. In: Proceedings of the International Conference on Machine Learning (ICML), pp. 7292–7303 (2019)
22. Zhang, R., et al.: Pointclip: point cloud understanding by clip. In: Proceedings of the IEEE/CVF Conference on Computer Vision and Pattern Recognition (CVPR), pp. 8552–8562 (2022)
23. Zhang, Z., Girdhar, R., Joulin, A., Misra, I.: Self-supervised pretraining of 3d features on any point-cloud. In: Proceedings of the IEEE/CVF International Conference on Computer Vision (ICCV), pp. 10252–10263 (2021)
24. Zhou, C., Loy, C.C., Dai, B.: Extract free dense labels from clip. In: Proceedings of the European Conference on Computer Vision (ECCV), pp. 696–712 (2022)

There is no Model to Beat Them All: Recommendations for Deep Learning Model Selection When Training on Synthetic Images

Joachim Rüter(✉), Johann C. Dauer, and Umut Durak

German Aerospace Center (DLR), Institute of Flight Systems, Brunswick, Germany
{joachim.rueter,johann.dauer,umut.durak}@dlr.de

Abstract. Synthetic training images generated with game-engines are a promising approach to enable the use of deep learning perception models in domains that lack diverse datasets. However, previous works have shown significant performance drops when these models are deployed to real-world scenarios and definite reasons and influences are not yet found. This paper builds on previous work investigating the influence of the model architecture on the sim-to-real generalizability and extends it by addressing key limitations. Based on an extensive study of 378 trained variations of 27 semantic segmentation models on an autonomous driving and an aerial dataset as well as the current literature, this work is the first to provide practical recommendations for selecting deep learning models when training on simulation images.

Keywords: Synthetic Data · Sim-to-Real Gap · Deep Learning

1 Introduction

Training state-of-the-art deep learning semantic segmentation models requires large and diverse datasets. However, for many use-cases, a lack of such datasets hinders their deployment, particularly for semantic segmentation tasks where manual annotation is extremely time-consuming. Furthermore, in safety-critical domains like aviation, diverse image datasets cannot even be recorded because of safety issues and high costs, especially for edge-cases. To mitigate both problems, some works investigate the use of synthetic images generated using modern game-engines. Game-engines allow to simulate diverse situations and to automatically extract ground truth labels. In this work, these synthetic images extracted form a game-engine are called *simulation images.* While using simulation images is a promising approach, deep learning perception models trained on them usually have problems generalizing to real-world images. This is shown by a sharp performance drop when the models are deployed to the real-world.

Recently, [29] was the first to investigate the influence of the deep learning model architecture on the generalization capability from simulation to real-world

E. Rodolà et al. (Eds.): ICIAP 2025, LNCS 16167, pp. 168–180, 2026.
https://doi.org/10.1007/978-3-032-10185-3_14

images. On the one hand, this question is reasonable as the model itself is an integral part of a deep learning system. On the other hand, many new deep learning model architectures have been developed in the past years accomplishing accuracy improvements. However, it has not been studied whether these improvements also translate to improved generalization capabilities.

While [29] provides first insights, it has three main limitations. First, only one dataset for the visually simple use-case of drogue detection in aerial refueling was considered which is not publicly available. Second, only object detection and no other tasks like semantic segmentation were studied. Third, the investigated models were relatively old and did not include any modern transformer architectures. While it is reasonable to first look at a simple use-case and older models when investigating general trends, these points hinder the generalizability of the results. Especially since transformers seem particularly promising for generalizing from simulation to real-world images as discussed in Sect. 2.

Against this background, this work builds upon the results of [29] and investigates the generalizability from simulation to real-world images of deep learning perception models further by mitigating the limitations named above. To do so, the task of semantic segmentation is considered, newer models including transformers are evaluated and two additional datasets are used. To facilitate research and improve the practical relevance of the results, publicly available datasets from the autonomous driving and Uncrewed aircraft system (UAS) sector are used. Overall, this work investigates the question whether the generalizability always aligns with the overall model performance or whether some semantic segmentation architectures or backbones have less problems generalizing from simulation training images to real-world images. Furthermore, influencing factors are identified and first recommendations for practical use are given.

The main contributions of this work are as follows: (i) investigate the sim-to-real generalizability of 378 trained variations of 27 semantic segmentation models on two datasets; (ii) give first practical recommendations for deep learning model selection when training on simulation images.

2 Related Work

Recent years have seen growing interest in using synthetic images from game engines to train deep learning models for various semantic perception tasks due to challenges associated with missing real-world data. Early research efforts primarily focused on autonomous driving, where synthetic images were used for semantic segmentation [24,25,35] and object detection [22,35]. Researchers have also applied game engine-based simulation images from the perspective of a UAS to semantic segmentation [11] and various detection tasks [13,15,30].

Numerous studies have shown that using simulated images in combination with real-world ones can enhance model performance in general [6,13,24,25,35] or on rare cases [1,28] and reduce the amount of real-world data required [11, 22,24]. Nevertheless, existing research highlights that synthetic images alone are generally insufficient to train deep learning models that generalize well to real-world data, as models often suffer significant performance drops when applied outside the synthetic domain. This issue, commonly referred to as *sim-to-real gap* has been observed across many domains like autonomous driving [6,25,35], as well as UAS perception [11,13,15]. While some works have reported minimal gaps for specific applications [30], the sim-to-real gap remains a widespread issue.

In response, research efforts have been dedicated to bridging this gap [13, 15,27,31,35]. As described in Sect. 1, on model side, [29] was the first work to systematically investigate the influence of the deep learning model architecture on the sim-to-real gap and found some influencing factors.

Recently, it has also been found that vision transformers seem to be more robust to domain shifts [21] and give more emphasis to shape information [21,23] while CNNs often exploit texture to make decisions [8,21,23]. Combined with the finding that the shape of certain objects in game-engines looks realistic but the texture does not [31], transformers seem promising for having better generalization capabilities from simulation to real-world images. This assumption is not yet extensively studied but seems to be supported in a first study [26].

3 Experimental Setup

3.1 Method and Metrics

To investigate the generalization from simulation to real-world images of deep learning perception models, the models are trained on datasets containing only simulation images. All trained models are evaluated on corresponding evaluation datasets containing only real-world images using mean Intersection over Union (mIoU). This measure is important for practical applications as it directly quantifies the final performance on a real-world dataset.

However, for a model with bad performance on real-world images, this metric does not take into account whether the model learned useful features but is just not able to generalize to real-world images or whether it also has bad performance on simulation images. Therefore, to quantify the generalization to real-world images in reference to the performance on simulation images, we furthermore follow [29] by also investigating the proxy *sim-to-real gap*. In this work, it is transferred to the use-case of semantic segmentation by defining it as the difference in mIou of the model m on the real-world and the simulation evaluation datasets R_{test} and S_{test} as

$$g := \text{mIoU}(m, S_{\text{test}}) - \text{mIoU}(m, R_{\text{test}}). \tag{1}$$

It gives a sim-to-real gap in percent points (pp) and has the advantage of not requiring any model training on real-world images and of allowing to calculate the sim-to-real gap for one model in isolation.

As this work aims to investigate the sim-to-real generalizability in correspondence with the best possible real-world performance of different semantic segmentation models trained only on simulation images, real-world validation datasets are used to make sure that the model that performs best on real-world images is selected. It is important to note that the use of a real-world validation dataset does not influence the training itself as the weights of the final model are adjusted *only* using simulation images.

3.2 Datasets

KITTI. As the real-world dataset for the autonomous driving use-case, the dataset from [14] is used. It provides semantic segmentation annotations for 445 images from the KITTI dataset [7] that contains images recorded from a camera mounted on top of a car. From the annotated images, 146 images are used as the real-world validation and 299 images as the real-world evaluation dataset. As the simulation dataset, the Virtual KITTI 2.0 (VKITTI2) dataset [2] is used. It rebuilds 5 scenes from the KITTI dataset in the Unity game-engine and extracts 2,115 simulation images. For evaluation, scene number 2 containing 223 images is used, as it covers a wide variety of potential situations. The remaining 1,892 synthetic images from the other scenes are used for training.

While the VKITTI2 dataset is rebuild after KITTI, there are some variations in the class annotations that have to be resolved. Overall, our final datasets contain the classes *Sky*, *Building*, *Vehicle*, *Vegetation*, *Sign/Pole* (containing poles, signs, and traffic lights), *Ground* (containing roads, sidewalks and terrain) and *Other* (containing fences, pedestrians, cyclists, guards rails, and the misc and ignore classes). Because of minimal variations in real-world image sizes, all images are center cropped to the largest common size of 1226×370 pixels.

Ruralscapes. As the real-world dataset for the UAS use-case, the *Ruralscapes* dataset [20] is used. It contains images taken from an UAS with a tilted camera at multiple altitudes in a rural area. The authors designated 816 images for training and 331 for evaluation. In this work, a subset of 20% of the training images, i.e. 164 images, is used as the real-world validation dataset and the 331 designated evaluation images are used as the real-world evaluation dataset. As the simulation dataset, the dataset from [11] is used. It was generated using the Unreal Engine and stylistically replicates the Ruralscapes dataset. In this work, 1,569 simulation images are available for training and 449 for evaluation.

While the dataset is rebuild after [20], variations in class annotations are matched per [11] to *Street*, *Building*, *Car*, *Human*, *Greenery*, and *Background*, where *Tree* and *Grass* are further merged to *Greenery* because of stated annotation issues. To reduce computational load during training and evaluation, all images are resized to 960×540 pixels using linear interpolation.

3.3 Model Configurations and Training Settings

In this work, following 27 semantic segmentation models are considered: UPerNet [36] with ResNet-50 and ResNet-101 [10], ConvNeXt-B, -S and -T [17], Swin-B, -S and -T [16], Twins-PCPVT-B, -L, -S and Twins-SVT-B, -S [5], DeiT-B Multi Level Neck (MLN), and -S MLN [33], as well as each of the architectures PSPNet [38], FCN [18], DeepLabV3 [3], and DeepLabV3+ [4] with each of the backbones ResNet-50, ResNeSt-101 [37], and MobileNetV2 [32], respectively.

All models are taken from the MMSegmentation[1] library and come with weights pre-trained on ADE20K [39]. Using pre-trained weights is common prac-

[1] https://github.com/open-mmlab/mmsegmentation.

tice across both academia and industry to improve training stability and reduce training time. In this context, ADE20K is a widely used, generic pre-training dataset that serves as a basic initialization but does not mirror the specific distributions targeted during evaluation. Importantly, the actual training of the final models is performed exclusively on synthetic images. This setup reflects practical scenarios in which pre-trained models but no task-specific real-world data are available and therefore enhances the practical relevance of our findings.

Each of these models is trained with a range of learning rates and optimizers to minimize potential bias of the results because of unsuitable training hyperparameters. Overall, each model is trained seven times with following parameters: Adam with learning rate (LR) of 0.001, AdamW with LR of 0.001 and weight decay (WD) of 0.01, Nadam with LR of 0.002, momentum of 0.9 and momentum decay of 0.004, Stochastic Gradient Descent (SGD) with LR of 0.01, momentum of 0.9 and WD of 0.0005, and SGD with LRs of 0.01, 0.001, and 0.0001.

Each model is trained for a maximum of 200 epochs with a training batch size of two and early stopping with a patience of 10. The model performing best on the validation dataset is used for further evaluation on the evaluation dataset.

4 Evaluation

4.1 All Experiments

Figures 1 and 2 show the performances of all trained model variations. It can be seen that the sim-to-real generalizability differs significantly between the trained models and variations. The general trend is similar to [29]: The better a model performs on the simulation images, the better it performs on the real-world images. However, the increase in performance on the real-world images is lower than the increase on the simulation images, i.e., the sim-to-real gap also tends to increase when performance on simulation images increases. In the figures, the sim-to-real gap, as described in Sect. 3.1, is symbolized by the distance in y-axis direction of a datapoint to the identity line.

Furthermore, it can be seen that the gap can also be negative, i.e., the performance on the real-world images is better than on the simulation images even though the final model was trained only on simulation images. However, this occurs less often for these two datasets than for the simple object detection use-case in [29]. This is probably rooted in the complexity of the used datasets.

It can be seen that the sim-to-real gap is dataset-dependent, with Ruralscapes showing a larger gap than KITTI. The larger gap is probably rooted in the fact that the Ruralscapes simulation training and evaluation datasets, although disjoint, are drawn from the same distribution, leading to high performance on the simulation evaluation dataset and, consequently, an increased sim-to-real gap. In contrast, the KITTI simulation evaluation dataset looks considerably different from training, leading to a lower performance on the simulation evaluation dataset and, consequently, a lower gap.

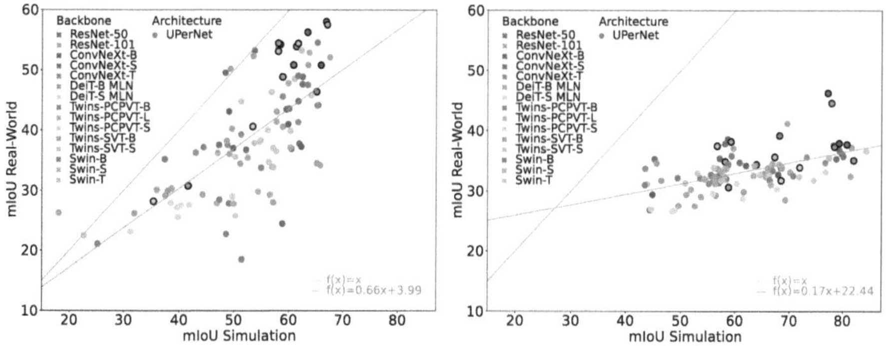

(a) KITTI dataset (b) Ruralscapes dataset

Fig. 1. Results of the trained semantic segmentation models using UPerNet on the respective evaluation datasets. X-axis gives mIou on the simulation, y-axis on the real-world evaluation dataset. Pink: Identity line. Turquoise: Linear regression through all datapoints. Bold edges: Best model w.r.t. real-world mIoU for each backbone. (Color figure online)

4.2 UPerNet with Different Backbones

To evaluate the influence of the backbone on the sim-to-real generalizability, an UPerNet architecture is trained with different backbones. Figure 1 shows all trained model variations using UPerNet. The numerical values for the variations with the best real-world mIou for each backbone are given in Table 1.

In general, when looking at the best performing models, it can be seen that the backbone has a significant influence on the sim-to-real gap. It varies between 3.6 and 18.7%-points (pp) on the KITTI dataset and between 19.2 and 47.0 pp on the Ruralscapes dataset.

When looking at the absolute performance, the experiments show that Swin and ResNet models generally achieve the best performance in terms of real-world mIou when trained on the simulation images. However, there are no clear indications for the hypothesis that transformers outperform CNN-based models in their generalization capabilities, despite their shape-bias combined with the realistic representation of shapes in game-engines as discussed in Sect. 2. Contrary, CNN-based ResNets regularly seem to have the best generalization capabilities on both datasets showing a smaller gap than the other models in Fig. 1. A potential explanation could be rooted in the relatively small size of the simulation datasets as transformers usually have a strong dependence on large datasets and struggle to outperform CNNs on smaller ones, possibly because of their inherent lack of inductive biases [9].

Model Size. Extending the findings from [29], which showed that deeper ResNets are not necessarily better at generalizing from simulation to real-world images for object detection, our experiments show that this also holds true for semantic segmentation. While on real-world KITTI, the best ResNet-50 and -101

Table 1. Results of the model variations with UPerNet that perform best on the real-world KITTI (left) and Ruralscapes (right) evaluation dataset with the respective backbone. Sim-to-real gap (mIoU Diff) in percent points, mIoU in percent. Best values for each column in bold, worst in italic. Tw abbreviates Twins.

Backbone	Optim	mIoU Sim	mIoU Real	mIoU Diff
Swin-B	SGD	66.8	**58.0**	8.8
Swin-S	SGD	**67.0**	57.5	9.5
ConvNeXt-S	SGD	63.4	56.3	7.1
ResNet-101	SGD	58.1	54.5	**3.6**
Tw-SVT-S	SGD	61.6	54.4	7.2
ResNet-50	SGD	58.4	54.3	4.1
Tw-SVT-B	SGD	61.3	53.9	7.4
Swin-T	SGD	58.0	53.1	4.9
Tw-PCPVT-B	SGD	60.8	50.8	10.0
ConvNeXt-B	Adam	65.9	50.8	15.1
Tw-PCPVT-L	SGD	58.9	48.8	10.1
ConvNeXt-T	Adam	65.1	46.3	*18.7*
Tw-PCPVT-S	SGD	53.5	40.6	12.9
DeiT-B MLN	AdamW	41.6	30.8	10.8
DeiT-S MLN	SGD	*35.4*	*28.2*	7.2

Backbone	Optim	mIoU Sim	mIoU Real	mIoU Diff
ResNet-50	SGD	77.2	**46.2**	31.0
ResNet-101	SGD	77.9	44.6	33.3
Swin-B	SGD	68.2	39.2	29.0
Swin-S	SGD	59.2	38.2	21.1
ConvNeXt-B	Adam	79.3	37.9	41.4
ConvNeXt-S	AdamW	80.8	37.7	43.1
Tw-SVT-S	SGD	*56.7*	37.5	**19.2**
ConvNeXt-T	AdamW	78.4	37.3	41.1
Swin-T	SGD	67.3	35.6	31.6
Tw-SVT-B	AdamW	**82.0**	35.1	*47.0*
Tw-PCPVT-L	SGD	58.2	34.9	23.3
Tw-PCPVT-B	SGD	63.9	34.4	29.5
Tw-PCPVT-S	Adam	71.9	33.9	38.0
DeiT-S MLN	Nadam	68.5	31.8	36.7
DeiT-B MLN	Adam	58.8	*30.6*	28.2

variants perform similar, with only a 0.2 pp difference, the best ResNet-50 variant outperforms the best ResNet-101 variant by 1.6 pp on real-world Ruralscapes.

However, the hypothesis that smaller models are better at generalizing cannot be generally confirmed. For example, the experiments show that for Swin transformers, bigger models seem to have better sim-to-real generalization capabilities. For the other models, the influence of the model size seems quite mixed but for many of them, the smallest model variants generalize worse than bigger counterparts. A possible explanation is that the given semantic segmentation tasks are much harder than drogue detection in front of a simple background. Therefore, models with a bigger capacity are needed to reach a good level of performance. It may be expected that there is a specific capacity above which the model will again tend to overfit leading to a bigger sim-to-real gap. However, this could not be shown in these experiments.

Optimizer. For sim-to-real generalization, SGD seems to be the best optimizer. The nine best performing models on the KITTI dataset and the four best performing models on the Ruralscapes dataset all use SGD as optimizer, as shown in Table 1. Similar advantages of SGD were found in [29]. In contrast, the models with the two largest sim-to-real gaps on KITTI as well as the four largest on Ruralscapes use Adam or AdamW as optimizer suggesting overfitting to the

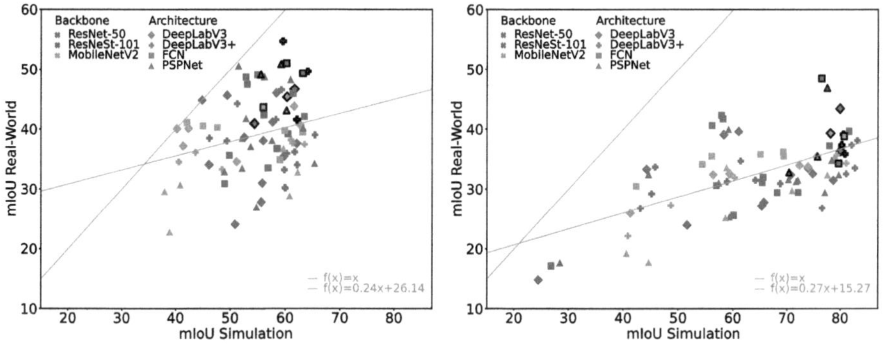

(a) KITTI dataset (b) Ruralscapes dataset

Fig. 2. Results of the trained semantic segmentation models with different semantic segmentation architectures and different backbones on the respective evaluation datasets. X-axis gives mIou on the simulation, y-axis on the real-world evaluation dataset. Pink: Identity line. Turquoise: Linear regression through all datapoints. Bold edges: Best model w.r.t. real-world mIoU for each architecture backbone combination. (Color figure online)

simulation data. This seems to be further supported by the literature, as some works indicate that Adam can lead to more overfitting compared to SGD [19,34].

4.3 Comparison of Different Architectures

To evaluate the influence of the semantic segmentation architecture on the sim-to-real generalizability, different architectures with the same backbone are trained and compared. To further enhance the generalizability of the results, the experiments are run with a ResNet-50, a ResNeSt-101, and a MobileNetV2 backbone. All 168 trained variations are shown in Fig. 2. The numerical values for the variations with the best real-world mIoU are given in Table 2.

For the best models, the sim-to-real gap varies by 6.6 pp for ResNet-50, by 8.1 pp for MobileNetV2 and by 8.6 pp for ResNeSt-101 on the KITTI dataset. On the Ruralscapes dataset, it varies by 13.4 pp for ResNet-50, by 4.1 pp for MobileNetV2 and by 7.8 pp for ResNeSt-101. Compared to the range of sim-to-real gaps for different backbones in Table 1, where the gap varied by 15.1 pp on the KITTI dataset and by 27.9 pp on the Ruralscapes dataset, the range of the gap is much smaller. The variation of the real-world mIoU is also much smaller for different architectures than for different backbones.

Overall, the findings from [29] that the architecture has a smaller influence on the sim-to-real gap than the backbone can be generalized to semantic segmentation.

Table 2. Results of the model variations with the respective architecture that perform best on the real-world KITTI (left) and Ruralscapes (right) evaluation dataset with the respective backbone. Sim-to-real gap (mIoU Diff) in percent points, mIoU in percent. Best values for each column in bold, worst in italic. DLabV3 abbreviates DeepLabV3.

Backbone	Architect	Optim	mIoU Sim	mIoU Real	mIoU Diff
ResNet-50	PSPNet	SGD	*59.2*	**50.9**	**8.4**
	DLabV3+	SGD	**64.1**	49.7	14.4
	FCN	SGD	63.2	49.3	13.9
	DLabV3	SGD	61.7	*46.7*	*15.0*
MobileNetV2	DLabV3	Nadam	60.3	**45.4**	15.0
	FCN	SGD	*56.0*	43.6	**12.4**
	PSPNet	Nadam	60.2	43.1	17.1
	DLabV3+	Adam	**62.1**	*41.6*	*20.5*
ResNeSt-101	DLabV3+	SGD	59.6	**54.7**	**4.9**
	FCN	SGD	**60.2**	51.0	9.2
	PSPNet	SGD	55.6	49.1	6.5
	DLabV3	AdamW	*54.4*	*40.9*	*13.5*

Backbone	Architect	Optim	mIoU Sim	mIoU Real	mIoU Diff
ResNet-50	FCN	SGD	*76.5*	**48.5**	**28.0**
	PSPNet	SGD	77.5	46.9	30.7
	DLabV3	SGD	79.9	43.4	36.5
	DLabV3+	AdamW	**80.6**	*39.1*	*41.4*
MobileNetV2	DLabV3	AdamW	78.2	**39.3**	**38.8**
	FCN	Adam	**80.7**	38.9	41.9
	DLabV3+	SGD	80.3	37.4	*42.9*
	PSPNet	Nadam	*75.7*	*35.4*	40.3
ResNeSt-101	DLabV3	Adam	80.0	**36.4**	43.6
	DLabV3+	AdamW	**80.9**	35.9	45.0
	FCN	AdamW	79.7	34.3	*45.4*
	PSPNet	AdamW	*70.4*	*32.8*	**37.6**

5 Recommendations

In this section, recommendations for model selection when training on synthetic images are derived. They are based on the results from this work as well as the results from [29].

Focus on the Backbone: From the experiments and the literature, it can be concluded that the backbone has a significant influence on the sim-to-real generalizability of deep learning object detection as well as semantic segmentation models. The influence of the object detection or semantic segmentation architecture is much smaller. Therefore, try a few different architectures but focus mainly on experimenting with different backbones.

Try a ResNet Backbone (First): Although many new deep learning models are developed every year, they do not necessarily improve in their sim-to-real generalizability. In the object detection use-case of [29] and in the UAS semantic segmentation use-case, a model with a ResNet backbone gave the best results on the real-world evaluation datasets. In the autonomous driving semantic segmentation use-case, it gave the second best results behind a Swin transformer. In general, ResNets seem to have strong generalization capabilities showing a smaller sim-to-real gap than the other models.

Try a Transformer: As described in Sect. 2, transformers seem promising for generalizing from simulation to real-world images. The experiments currently do not confirm this assumption unrestrictedly but on the autonomous driving dataset and in [26], transformers already show the best overall performance on the real-world evaluation datasets. As mentioned in Sect. 4.2, especially for large training datasets, transformers should be considered.

For Simple Tasks, Use Smaller Models: For visually simple use-cases like the drogue detection use-case in front of a simple background, larger models do not seem to be better at generalizing from simulation training images to real-world ones. For this task, the model with the fewest weights, a MobileNetV3 [12], often had the smallest sim-to-real gap [29]. This result probably translates to similar visually simple tasks like object detection on a conveyor.

Harder Tasks May Need Bigger Models: The recommendation above seems to generalize to harder tasks for ResNet models, as ResNet-101 did not outperform ResNet-50 on real-world images in this work. However, for some harder tasks, bigger models seem necessary. On the harder tasks of this work, the performance of the Swin transformer on the real-world images declined when smaller model variants were used.

Use SGD Optimizer: For all three considered use-cases, most of the models with the best sim-to-real generalizability used SGD optimizer compared to Adam, AdamW and Nadam. Furthermore, the models with the worst sim-to-real gaps in Table 1 and 2 used Adam or AdamW optimizer.

Experiment and Iterate: The experiments show that the generalization capability from simulation training images to real-world images differs significantly between different models and that there is no model to beat them all. While these recommendations provide first guidelines, some experimentation and iteration can help to improve the results on real-world images for your specific use-case further.

6 Summary and Future Work

This work investigated the sim-to-real generalization capabilities of deep learning perception models, i.e. the ability to generalize from simulation training images to real-world evaluation images. For this purpose, numerous semantic segmentation models were trained only on simulation images and evaluated on real-world images. Generalization problems were shown and influences analyzed. While definite reasons for the generalization problems are still not fully understood, this work offers first practical recommendations for model selection when training on simulation images. They are based on 378 trained variations of 27 semantic segmentation models trained on two publicly available dataset as well as 144 trained variations of 12 different object detection models from [29].

While it was discussed that transformer models have promising characteristics that may help to generalize from simulation to real-world images, the models still show a significant performance drop on real-world images. Currently, there are no clear indications that newer models or transformer architectures improve the overall sim-to-real generalization capability significantly compared to older CNN-based architectures like ResNet. As this hinders the usage of deep learning models in use-cases where diverse real-world datasets are missing, we hope that this work helps to facilitate the idea of explicitly developing models with better sim-to-real generalization capabilities. It was also shown in this work, that the generalization is dataset dependent. Therefore, future work should investigate the interaction of the used datasets on the sim-to-real generalization capabilities as well as influencing factors.

References

1. Beery, S., et al.: Synthetic examples improve generalization for rare classes. In: IEEE/CVF Winter Conference on Applications of Computer Vision (2020)
2. Cabon, Y., Murray, N., Humenberger, M.: Virtual KITTI 2. arXiv preprint arXiv:2001.10773 (2020)
3. Chen, L.C., Papandreou, G., Schroff, F., Adam, H.: Rethinking atrous convolution for semantic image segmentation. arXiv preprint arXiv:1706.05587 (2017)
4. Chen, L.C., Zhu, Y., Papandreou, G., Schroff, F., Adam, H.: Encoder-decoder with atrous separable convolution for semantic image segmentation. In: ECCV (2018)
5. Chu, X., et al.: Twins: revisiting the design of spatial attention in vision transformers. Adv. Neural Inf. Process. Syst. **34** (2021)
6. Gaidon, A., Wang, Q., Cabon, Y., Vig, E.: Virtual worlds as proxy for multi-object tracking analysis. In: IEEE/CVF Conference on Computer Vision and Pattern Recognition (2016)
7. Geiger, A., Lenz, P., Stiller, C., Urtasun, R.: Vision meets robotics: the kitti dataset. Int. J. Rob. Res. **32**(11) (2013)
8. Geirhos, R., Rubisch, P., Michaelis, C., Bethge, M., Wichmann, F.A., Brendel, W.: Imagenet-trained CNNs are biased towards texture; increasing shape bias improves accuracy and robustness. arXiv preprint arXiv:1811.12231 (2018)
9. Han, K., et al.: A survey on vision transformer. IEEE Trans. Pattern Anal. Mach. Intell. **45**(1) (2022)
10. He, K., Zhang, X., Ren, S., Sun, J.: Deep residual learning for image recognition. In: IEEE/CVF Conference on Computer Vision and Pattern Recognition (2016)
11. Hinniger, C., Rüter, J.: Synthetic training data for semantic segmentation of the environment from uav perspective. Aerospace **10**(7) (2023)
12. Howard, A., et al.: Searching for mobilenetv3. In: IEEE/CVF International Conference on Computer Vision (2019)
13. Kiefer, B., Ott, D., Zell, A.: Leveraging synthetic data in object detection on unmanned aerial vehicles. In: 26th International Conference on Pattern Recognition (2022)
14. Krešo, I., Čaušević, D., Krapac, J., Šegvić, S.: Convolutional scale invariance for semantic segmentation. In: German Conference on Pattern Recognition (2016)
15. Krump, M., Stütz, P.: UAV based vehicle detection on real and synthetic image pairs: performance differences and influence analysis of context and simulation parameters. In: MESAS (2022)
16. Liu, Z., et al.: Swin transformer: hierarchical vision transformer using shifted windows. In: IEEE/CVF International Conference on Computer Vision (2021)

17. Liu, Z., Mao, H., Wu, C.Y., Feichtenhofer, C., Darrell, T., Xie, S.: A convnet for the 2020s. In: IEEE/CVF Conference on Computer Vision and Pattern Recognition (2022)
18. Long, J., Shelhamer, E., Darrell, T.: Fully convolutional networks for semantic segmentation. In: IEEE Conference on Computer Vision and Pattern Recognition (2015)
19. Loshchilov, I., Hutter, F.: Decoupled weight decay regularization. arXiv preprint arXiv:1711.05101 (2019)
20. Marcu, A., Licaret, V., Costea, D., Leordeanu, M.: Semantics through time: semi-supervised segmentation of aerial videos with iterative label propagation. In: Asian Conference on Computer Vision (2020)
21. Naseer, M.M., Ranasinghe, K., Khan, S.H., Hayat, M., Shahbaz Khan, F., Yang, M.H.: Intriguing properties of vision transformers. Adv. Neural Inf. Process. Syst. **34** (2021)
22. Nowruzi, F.E., Kapoor, P., Kolhatkar, D., Hassanat, F.A., Laganiere, R., Rebut, J.: How much real data do we actually need: analyzing object detection performance using synthetic and real data. arXiv preprint arXiv:1907.07061v1 (2019)
23. Park, N., Kim, S.: How do vision transformers work? arXiv preprint arXiv:2202.06709 (2022)
24. Richter, S.R., Vineet, V., Roth, S., Koltun, V.: Playing for data: ground truth from computer games. In: ECCV (2016)
25. Ros, G., Sellart, L., Materzynska, J., Vazquez, D., Lopez, A.M.: The SYNTHIA dataset: a large collection of synthetic images for semantic segmentation of urban scenes. In: IEEE/CVF Conference on Computer Vision and Pattern Recognition (2016)
26. Ruis, F.A., et al.: Improving object detector training on synthetic data by starting with a strong baseline methodology. In: Synthetic Data for Artificial Intelligence and Machine Learning, Tools, Techniques, and Applications II (2024)
27. Rüter, J., Durak, U., Dauer, J.C.: Investigating the influence of image augmentations on the sim-to-real generalization of deep learning perception models. In: Synthetic Data for Computer Vision Workshop @CVPR (2025)
28. Rüter, J., Maienschein, T., Schirmer, S., Schopferer, S., Torens, C.: Filling the gaps: using synthetic low-altitude aerial images to increase operational design domain coverage. Sensors **24**(4) (2024)
29. Rüter, J., Durak, U., Dauer, J.C.: Investigating the sim-to-real generalizability of deep learning object detection models. J. Imaging **10**(10) (2024)
30. Rüter, J., Schmidt, R.: Using only synthetic images to train a drogue detector for aerial refueling. In: MESAS (2023)
31. Saleh, F.S., Aliakbarian, M.S., Salzmann, M., Petersson, L., Alvarez, J.M.: Effective use of synthetic data for urban scene semantic segmentation. In: ECCV (2018)
32. Sandler, M., Howard, A., Zhu, M., Zhmoginov, A., Chen, L.C.: Mobilenetv2: inverted residuals and linear bottlenecks. In: IEEE/CVF Conference on Computer Vision and Pattern Recognition (2018)
33. Touvron, H., Cord, M., Douze, M., Massa, F., Sablayrolles, A., Jégou, H.: Training data-efficient image transformers & distillation through attention. In: International Conference on Machine Learning (2021)
34. Wilson, A.C., Roelofs, R., Stern, M., Srebro, N., Recht, B.: The marginal value of adaptive gradient methods in machine learning. arXiv preprint arXiv:1705.08292 (2018)
35. Wrenninge, M., Unger, J.: Synscapes: a photorealistic synthetic dataset for street scene parsing. arXiv preprint arXiv:1810.08705 (2018)

36. Xiao, T., Liu, Y., Zhou, B., Jiang, Y., Sun, J.: Unified perceptual parsing for scene understanding. In: ECCV (2018)
37. Zhang, H., et al.: Resnest: split-attention networks. In: IEEE/CVF Conference on Computer Vision and Pattern Recognition (2022)
38. Zhao, H., Shi, J., Qi, X., Wang, X., Jia, J.: Pyramid scene parsing network. In: IEEE Conference on Computer Vision and Pattern Recognition (2017)
39. Zhou, B., et al.: Semantic understanding of scenes through the ADE20K dataset. Int. J. Comput. Vision **127** (2019)

SVGauge: Towards Human-Aligned Evaluation for SVG Generation

Leonardo Zini[1,2], Elia Frigieri[1], Sebastiano Aloscari[1], Marcello Generali[2], Lorenzo Dodi[2], Robert Dosen[2], and Lorenzo Baraldi[1,2(✉)]

[1] University of Modena and Reggio Emilia, Modena, Italy
{leonardo.zini,elia.frigieri,sebastiano.aloscari, lorenzo.baraldi}@unimore.it
[2] Doxee S.p.A, Modena, Italy
{mgenerali,ldodi,rdosen}@doxee.com

Abstract. Generated Scalable Vector Graphics (SVG) images demand evaluation criteria tuned to their symbolic and vectorial nature – criteria that existing metrics such as FID, LPIPS, or CLIPScore fail to satisfy. In this paper, we introduce SVGauge, the first human-aligned, reference-based metric for text-to-SVG generation. SVGauge jointly measures (*i*) visual fidelity, obtained by extracting SigLIP image embeddings and refining them with PCA and whitening for domain alignment, and (*ii*) semantic consistency, captured by comparing BLIP-2-generated captions of the SVGs against the original prompts in the combined space of SBERT and TF-IDF. Evaluation on the proposed SHE benchmark shows that SVGauge attains the highest correlation with human judgments and reproduces system-level rankings of eight zero-shot LLM-based generators more faithfully than existing metrics. Our results highlight the necessity of vector-specific evaluation and provide a practical tool for benchmarking future text-to-SVG generation models.

Keywords: Human Evaluation Metric · Text-to-SVG Generation · Scalable Vector Graphics

1 Introduction

The objective of text-to-SVG generation is to produce vectorial graphical representations conditioned on natural language prompts, accurately conveying the intended semantics while respecting the stylistic and structural properties of SVGs. Unlike raster images, SVGs offer a symbolic, abstract, and resolution-independent representation of visual concepts, often emphasizing geometric structures and minimalistic design choices. As such, the task demands not only understanding the textual description but also capturing fine-grained visual abstractions and compositional relationships between elements.

Recent advances in generative modeling have considerably improved the quality of SVG synthesis, with innovative strategies including rasterization-then-vectorization pipelines [12], differentiable vector renderers [11], and latent diffusion frameworks adapted to the vector space [7,22,24]. Additionally, there has

E. Rodolà et al. (Eds.): ICIAP 2025, LNCS 16167, pp. 181–193, 2026.
https://doi.org/10.1007/978-3-032-10185-3_15

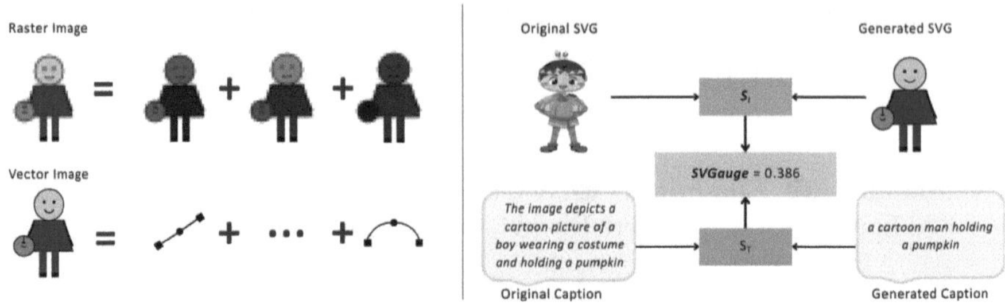

Fig. 1. Comparison between raster (*e.g.*, JPG) and vector images (*e.g.*, SVG) and an overview of SVGauge for SVG generation evaluation.

been a growing interest in leveraging Large Language Models (LLMs) for autoregressive SVG token generation [17,19,20,23], harnessing the human-readable nature of SVG markup for more interpretable synthesis.

As generation quality continues to progress, the need for effective evaluation becomes more critical. Current evaluation methods largely rely on metrics designed for raster images, such as Fréchet Inception Distance (FID) [6], Learned Perceptual Image Patch Similarity (LPIPS) [26], or DINOv2-based similarities [13], or on text-image alignment scores like CLIPScore [5]. However, these approaches often fall short when applied to SVGs: they fail to capture the symbolic, geometric, and stylistic nuances intrinsic to vector graphics, and can exhibit strong sensitivity to low-level changes despite semantic equivalence.

In light of these limitations, we argue that evaluating text-to-SVG generation requires dedicated metrics that specifically account for the distribution of SVG images and align closely with human perceptual and semantic judgment. In response, we propose SVGauge, a novel metric for assessing the quality of text-to-SVG generations, designed to robustly capture both visual similarity in a vector-aware embedding space and semantic preservation via multimodal text analysis (Fig. 1). Our approach begins by rasterizing both the reference and generated SVGs to enable feature extraction via pre-trained vision backbones, followed by domain adaptation techniques including PCA and whitening. Visual similarity is computed by comparing embeddings in this adapted space, while semantic consistency is evaluated through a captioning loop that uses a multimodal LLM to generate descriptions of the SVGs, which are then compared using a combination of Sentence-BERT embeddings and TF-IDF-weighted similarity.

Extensive experiments demonstrate that SVGauge achieves superior alignment with human judgment across a wide range of SVG generation scenarios, outperforming conventional raster-based metrics. We show that our dual-axis design provides robustness and increased alignment with human evaluations, better reflecting the qualities valued in SVG synthesis. As a complementary contribution, we also develop and release the first dataset with prompt-SVG pairs annotated with human evaluations.

2 Related Works

Here, we provide a concise overview of the most relevant works related to SVG generation using Latent Diffusion Models or LLMs, and quality measurement.

Latent Diffusion Models. A common approach to SVG generation synthesizes a raster image from a text prompt using latent diffusion models, followed by vectorization via traditional tools like LIVE [12], VTracer, or Potrace. DiffVG [11] introduced a differentiable renderer enabling SVG optimization through backpropagation with image-based losses such as Score Distillation Sampling (SDS). VectorFusion [7] initializes an SVG with fixed paths and optimizes it using latent SDS gradients, while SVGDreamer [24] improves initialization using activation maps but at the cost of a complex pipeline. Recently, SVGFusion [22] proposed a unified architecture based on a Vector-Space Diffusion Transformer and a Vector-Pixel Autoencoder to improve both efficiency and quality.

Large Language Models. The structured nature of SVGs has also motivated LLM-based generation approaches. IconShop [20] trained a Transformer decoder to produce simple icons from text, while LLM4SVG [23] built a large dataset of 250,000 auto-captioned SVGs and fine-tuned compact LLMs with instruction-tuned prompts. Chat2SVG [19] developed a prompting pipeline for SVG generation, and StarVector [17] proposed a multimodal model that translates raster images into SVGs by predicting SVG tokens with a large language model.

Evaluation Methods for SVG Generation. Current evaluation methods for text-to-SVG rely on metrics developed for raster images, such as DINOv2-based similarity [13], FID [6], and LPIPS [26]. Text-image alignment metrics like CLIPScore [5] have also been employed. However, these methods can be unreliable for SVGs: being built on backbones optimized for natural images, they often fail to capture the abstract, symbolic, or geometric semantics typical of SVGs. For instance, an icon with a unique stylistic interpretation might be penalized by CLIPScore even if it faithfully represents the input caption. These limitations result in reduced alignment with human perception, which ultimately hinders the development of generative models, as they are evaluated with metrics that poorly reflect true quality and fail to reward meaningful improvements.

3 Proposed Approach

We argue for moving beyond the application of metrics developed for images coming from the raster domain and instead propose a metric specifically designed for the SVG domain, carefully aligned with human judgment. Our approach takes a dual perspective, accounting for both visual similarity and semantic preservation, and demonstrates state-of-the-art alignment with human evaluations (Fig. 2).

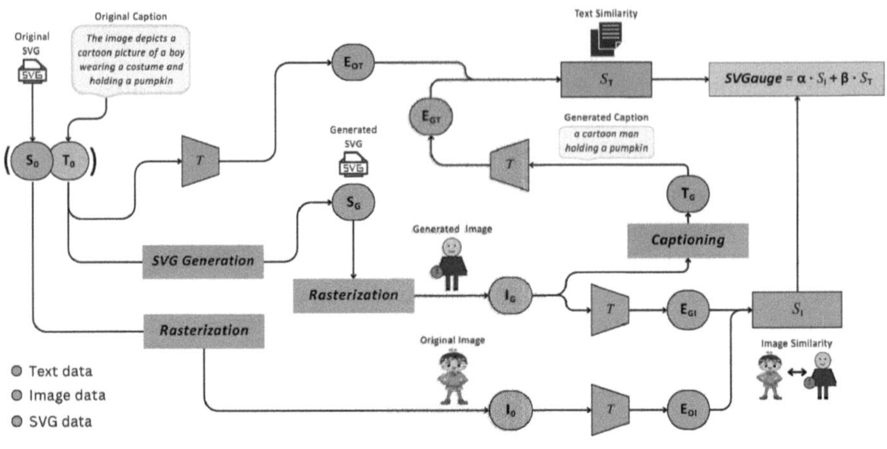

Fig. 2. Overview of the SVGauge metric for text-to-SVG generation evaluation.

Overview of our Approach. Our approach for text-to-SVG evaluation begins with a reference pair (T_O, S_O), where T_O represents a natural language prompt describing a visual concept, and S_O is an SVG image that accurately depicts this concept. Given a generator $\mathcal{G}$ which takes the prompt T_O as input, we evaluate the quality of its generation, $S_G = \mathcal{G}(T_O)$ through a dual-axis evaluation that captures both the visual resemblance of the generated vectorial image, and the semantic consistency between the reference and generated outputs.

Rasterization and Image Encoding. Since SVGs are vectorial representations and not directly compatible with standard image similarity techniques, we first rasterize both the reference S_O and generated S_G into pixel-based images, denoted as I_O and I_G, respectively. This rasterization step enables the use of pre-trained vision backbones operating on image tensors.

Each rasterized image is passed through a visual backbone, producing high-dimensional embeddings E_{OI} and E_{GI}. To enhance spatial localization, we extract a grid of feature vectors (*e.g.*, from the final self-attention layer of a ViT backbone) and compute their average. Formally, the embedding for the reference image is defined as

$$E_{OI} = \frac{1}{H \times W} \sum_{i,j} \text{backbone}(I_O)_{[i,j]}, \tag{1}$$

where (H, W) denotes the feature grid dimensions. The embedding for the generated image is computed analogously.

To further refine these embeddings for similarity computation and adapt them to the domain of vectorial images, we apply a two-stage post-processing pipeline consisting of PCA and whitening, which have been demonstrated to improve retrieval performance in high-dimensional spaces [8]. Specifically, PCA projects each embedding onto a lower-dimensional subspace by (*i*) centering the data via mean subtraction, μ, computed over the distribution of vectorial images, *i.e.*,

$$X_{\text{centered}} = X - \mu,$$

where μ is the mean vector over a distribution of vector images, and (*ii*) projecting onto the principal components P associated with the largest eigenvalues of the covariance matrix:

$$X_{\text{PCA}} = P^{\mathsf{T}} X_{\text{centered}}.$$

This projection not only reduces noise but also emphasizes the most discriminative directions, as it ensures that the projected components are decorrelated – *i.e.*, for any two different eigenvectors p_i and p_j, $p_i^{\mathsf{T}} p_j = 0$. Further, being the projection estimated on a distribution of SVG images, it also reprojects the original embeddings into a subspace which is tailored to the distribution of vectorial images.

Following PCA, we apply a whitening transformation to further normalize the embeddings. Operating in the reduced d'-dimensional subspace identified by PCA, whitening rescales each principal component to unit variance, effectively normalizing the covariance matrix to the identity. Let $\lambda_1, \ldots, \lambda_{d'}$ denote the eigenvalues associated with the selected eigenvectors. The whitening transformation is defined as

$$\hat{X} = \operatorname{diag}\left(\lambda_1^{-\frac{1}{2}}, \ldots, \lambda_{d'}^{-\frac{1}{2}}\right) X_{\text{PCA}},$$

where each component is individually scaled to achieve decorrelation and variance equalization. The resulting embeddings $\hat{X}$ exhibit a balanced distribution, ensuring that no direction dominates the similarity computation and all feature dimensions contribute equally.

Visual Similarity Evaluation. Overall, the combination of averaging, PCA, and whitening yields compact and robust image representations, aligning with the principles of negative evidence and decorrelation [6,8]. Finally, we compute the cosine similarity between the embedding of the reference and generated image, to obtain a visual similarity score S_I. The aforementioned similarity score quantifies how similar the generated image I_G is to the reference I_O in the learned embedding space [26].

Semantic Evaluation via Captions. While visual similarity provides a measure of surface-level resemblance, it may not fully capture whether the generated image semantically reflects the input prompt. An image can be structurally similar to the reference yet fail to convey the intended meaning.

To address this, we introduce a semantic evaluation loop. A multimodal LLM, such as BLIP-2 [10], is employed to generate a caption T_G for the generated image

I_G, capturing its semantic content. We then compare T_G to the original prompt T_O by encoding both with Sentence-BERT (SBERT) [16], obtaining dense sentence embeddings E_{GT} and E_{OT}. Semantic similarity is initially measured via cosine similarity between these embeddings.

However, SBERT may overestimate similarity for short or generic sentences. To mitigate this, we integrate a TF-IDF weighting mechanism that emphasizes rare, informative terms. By doing so, matches on distinctive words are rewarded, while overlaps on common or generic terms are downweighted. The final semantic similarity score is computed as

$$S_T = \text{CosineSim}(E_{OT}, E_{GT}) \cdot \Big(0.8 + 0.2\,\text{CosineSim}\big(V_{OT}, V_{GT}\big)\Big), \quad (2)$$

where E_{OT}, E_{GT} are the SBERT-space embeddings of the reference and generated texts, respectively, and V_{OT}, V_{GT} are their corresponding TF-IDF vectors. The TF-IDF factor is normalized within the range $[0.8, 1]$ to act as a gentle rescaling term that adjusts similarity based on the content informativeness of the caption match.

Combined Evaluation Metric. Our final evaluation metric combines both visual and semantic similarities to form a unified score, defined as

$$\text{SVGauge} = \alpha \cdot S_I + \beta \cdot S_T.$$

Here, α and β are scalar weights that allow us to tune the importance of each component based on the use case. For instance, applications that prioritize aesthetic consistency may favor $\alpha > \beta$, while semantic-critical applications (*i.e.*, educational content generation) may prefer $\beta > \alpha$.

Why Both Similarities Matter. Existing metrics, primarily designed for raster images, struggle to capture the intrinsic characteristics of vector graphics. Image similarity in the projected space effectively captures local visual cues specific to vectorized representations, while textual similarity provides a global measure of semantic alignment.

For example, a text-to-SVG model may generate an SVG that differs structurally from the reference (*e.g.*, a triangle with rounded edges instead of sharp corners) but preserves the intended concept. In this case, the semantic similarity S_T would remain high, while the image similarity S_I would drop. Conversely, a model might replicate the visual structure but misrepresent the intended meaning (*e.g.*, generating a star when a flower was described), resulting in high S_I but low S_T. By combining both scores, our metric remains robust to such discrepancies and achieves closer alignment with human judgment of SVG quality and semantic fidelity.

4 Experimental Evaluation

In this section, we assess the quality and human-alignment level of the proposed metric for text-to-SVG generation. Using a newly constructed dataset consisting of SVG-prompt pairs and human ratings, we explore various methodological combinations and benchmark our metric against established metrics (LPIPS [26], DINO similarity [13], FID [6], and CLIPScore [5]). Further, we also apply our metric in system-level correlation experiments, where we assess eight LLM-based generators under zero-shot conditions.

4.1 The SHE Dataset

To assess the alignment of the proposed metric with the human judgment, we create and release the SVG Human-Evaluation dataset (SHE)[1] – a collection consisting of 333 SVG-prompt pairs, each associated with around eight generations coming from different models, evaluated with human scores. The collection of the original SVG images is initially obtained through web scraping and then manually assessed to ensure diversity in image complexity – ranging from simple black icons to elaborate illustrations. The prompts accompanying the SVG images are instead automatically generated through a state-of-the-art Multimodal LLM, Idefics3 [9], and subsequently validated through human annotation.

Table 1. Dataset generation and user study statistics. Here, "% Generated" indicates the percentage of times that the generator returned SVG code, "% Correct syntax" the percentage of times the generated SVG code was syntactically correct, and "% Whites" the percentage of times the generated SVG code resulted in a fully white image. We also report the average human score obtained over images generated by each model.

	% Generated	% Correct Syntax	% Whites	Human Score
DeepSeek-R1-8B [3]	93.4	100.0	21.9	1.21
Llama 3.1 8B [2]	98.8	100.0	6.1	1.83
Gemma2-9B [18]	92.8	100.0	64.6	1.90
Ministral 8B	93.4	100.0	1.6	1.80
Mistral 24B	100.0	96.4	0.3	2.49
DeepSeek-R1-70B [3]	98.5	98.8	5.3	2.23
Llama 3.1 70B [2]	99.7	100.0	3.6	2.46
ChatGPT-4o [1]	100.0	100.0	0.0	3.37

To collect generated SVG images for the subsequent collection of human judgment, we select eight state-of-the-art LLMs, covering both small-scale models (Gemma2-9B [18], Llama 3.1 8B [2], DeepSeek-R1-8B [3], and Ministral 8B) and larger models (ChatGPT-4o [1], Llama 3.1 70B [2], Mistral 24B, and DeepSeek-R1-70B [3]). Generated outputs that failed to produce syntactically correct SVG code were excluded from the dataset, while syntactically valid yet visually blank outputs were retained to increase diversity and fairness of evaluation in system-level evaluations. In Table 1, we report statistics on the generated results across

[1] The dataset will be publicly released upon paper acceptance.

the selected models. As can be observed, smaller LLMs typically struggle even with basic SVG image generation, irrespective of output quality, while larger models tend to provide syntactically correct SVG generations.

We then conducted a human evaluation study to collect human judgment annotations. The study involved 40 participants with diverse backgrounds, ranging from AI researchers to non-experts. Participants were asked to evaluate the correlation between input prompts and the corresponding generated SVG images on a scale from 1 (completely unrelated) to 5 (completely related). The study resulted in a total of 2,461 annotations, with an average of 57 evaluations per user. Visually blank generations were always evaluated with the lowest score. Finally, the dataset is split into training and test subsets containing 2,000 and 461 samples, respectively. Table 1 also reports the average human rating obtained by each generator.

4.2 Quantitative Results

Implementation Details. To assess the role of the visual backbone for computing the visual similarity score, we consider different image encoders, namely *DINOv2-base* [13], *MAE-vit-base* [4], *SigLIP-base-patch16-224* [25], and *CLIP-vit-base-patch32* [15]. SVG images are rasterized with a fixed white background, using the input resolutions employed by each encoder. Unless otherwise stated, we always consider the grid of activations coming from the last self-attention layer of the architecture, ignoring the `[CLS]` token. For PCA computation, we retain the first 128 largest eigenvectors. Additionally, to assess the role of the captioner when encoding the generated image, we consider three captioning models, namely *Florence2-base* [21], *Idefics3* [9], and *Blip2* [10].

To measure the correlation between the ratings of SVGauge and human evaluations, we employ the Spearman, Kendall, and Pearson correlation coefficients.

Ablation Studies. We first evaluate the impact of the selected image encoder and the choice of captioning model. In Table 2, we present aggregated human correlation values, when employing feature vectors extracted from different visual encoders, and when verifying the captioner employed to generate the description of the generated image. In particular, we test by employing the `[CLS]` token of the last self-attention layer, or by taking the average of the grid of features at the last self attention layer (excluding the `[CLS]`) of CLIP, DINOv2 and SigLIP.

To represent the overall performance of each experimental combination, we aggregate the human correlation values across different values of α and β (at evenly spaced values between 0 and 1.0) and by taking the average of the Spearman, Kendall, and Pearson correlation values. This yields a single aggregated score, which is formally computed as

$$\frac{1}{C \times P} \sum_{\alpha} \sum_{\beta} S_{\alpha,\beta} + K_{\alpha,\beta} + P_{\alpha,\beta},$$

where C is 3 (the number of correlation coefficients) and P is 11 (number of α,β combinations).

Table 2. Average Human Correlation with different captioners and image encoders.

	CLS token		
	Florence2 [21]	Idefics3 [9]	Blip2 [10]
CLIP [15]	33.0	26.5	34.1
DINOv2 [13]	36.2	29.7	37.1
SigLIP [25]	**37.5**	**30.5**	**38.8**
	Mean Feature Grid		
MAE [4]	0.8	−5.2	−1.7
CLIP [15]	32.5	23.2	33.9
DINOv2 [13]	35.2	28.3	36.1
SigLIP [25]	**42.0**	**35.4**	**42.9**

Table 3. Effects of transformation in the features space of each image encoder, when using Blip2 [10] as captioner.

	GeM Pooling			PCA	
	$p = 1$	$p = 2$	$p = 4$	w/o whit.	w/ whit.
MAE [4]	−1.7	30.4	31.6	32.1	32.1
CLIP [15]	33.9	34.4	34.9	39.7	39.6
DINOv2 [13]	36.1	36.1	36.2	38.7	38.5
SigLIP [25]	**42.9**	**36.3**	**36.2**	**44.1**	**44.3**

As shown in Table 2, Blip2 [10] on average emerges as the most effective captioning model across all image encoders, while SigLIP [25] proves to be the most effective image encoder. Interestingly, employing the average of the feature grid always provides better results than the [CLS] token, underlining the effectiveness of considering spatial-aware features.

We then conduct further experiments to assess the impact of the different transformations applied to the visual feature vectors. Table 3 reports the results obtained when applying a Generalized Mean Pooling [14] (with $p = 1, 2, 4$) and PCA, with and without whitening. PCA is applied by transforming the average of the feature grid into a lower-dimensional subspace. When PCA is combined

Table 4. Correlation with human judgment for SVGauge with SigLIP [25] and Blip2 [10] for different coefficients α and β. Correlation used are Spearman ($S\rho$), Kendall ($K\tau$) and Pearson (Pr). The higher the better for all correlation scores.

Coefficients		Training set			Test set		
α	β	$S\rho$	$K\tau$	Pr	$S\rho$	$K\tau$	Pr
1.0	0.0	42.6	33.1	52.2	38.2	29.3	46.3
0.9	0.1	45.4	35.2	53.6	41.8	31.9	48.3
0.8	0.2	47.5	36.7	55.0	44.8	34.2	50.4
0.7	0.3	48.8	37.8	56.1	47.3	36.2	52.2
0.6	0.4	**49.2**	**38.0**	56.8	**48.3**	**37.0**	53.7
0.5	0.5	48.8	37.7	**57.0**	**48.4**	**37.1**	54.6
0.4	0.6	47.3	36.4	56.5	47.2	36.1	**54.7**
0.3	0.7	45.0	34.6	55.1	45.1	34.4	53.8
0.2	0.8	42.3	32.3	52.8	42.3	32.3	52.0
0.1	0.9	39.1	29.8	49.7	39.4	30.1	49.4
0.0	1.0	35.8	27.1	46.0	36.5	27.7	46.3
Overall Mean		44.7	34.4	53.7	43.6	33.3	51.0

with whitening, data is centered by subtracting the mean, and each component is scaled to achieve unit variance, resulting in decorrelated features and alignment with the distribution of SVG images.

As can be noticed from Table 3, the best performance is achieved when combining PCA and whitening with SigLIP [25] as visual feature extractor, and Blip2 [10] as captioner. Table 4 further illustrates the correlation with human evaluation across different values of α and β on both training and test sets, underscoring the robust generalization capability of our proposed approach. Based on the results obtained from the training set, we select $\alpha = 0.6$ and $\beta = 0.4$ as the default values of our final metric configuration.

Comparison with the State-of-the-Art. Finally, we compare our proposed metric against established alternatives for text-to-SVG generation – namely FID [6], LPIPS [26], DINO similarity [13], and CLIPScore [5], by performing experiments both at the system level and instance level. Table 5 shows that, as per human evaluation scores, ChatGPT-4o [1] is the best-performing model, followed by Llama 3.1 70B [2] and Mistral 24B. Notably, our metric produces a model ranking that closely mirrors human judgments, assigning the highest score to ChatGPT-4o [1], followed by Mistral 24B and Llama 3.1 70B [2]. In contrast, existing older metrics such as FID [6] and LPIPS [26] either poorly correlate with human ratings or produce rankings that are not aligned with observed semantic relevance. Moreover, those metrics struggle to correctly rank generator models, which may be attributed to the absence of SVG-like data in their training distributions. It is also worth noting that lower FID scores indicate better performance, yet the average FID values observed in our experiments are significantly high, suggesting a poor fit of this metric to the SVG domain (*e.g.*, icons, illustration, etc.).

Table 5. System-level comparison between eight different generator models and different metrics.

	FID	LPIPS	DINO Sim.	CLIP Score†	Human	SVGauge ↑
DeepSeek-R1-8B [3]	656.43	0.61	0.32	0.65	1.28	**0.11**
Ministral 8B	497.95	0.56	0.40	0.70	1.86	**0.17**
Llama 3.1 8B [2]	541.41	0.59	0.38	0.66	1.66	**0.18**
Gemma2-9B [18]	800.39	0.59	0.29	0.62	1.66	**0.15**
Mistral 24B	485.95	0.56	0.40	0.71	2.31	**0.23**
DeepSeek-R1-70B [3]	502.27	0.58	0.42	0.72	2.23	**0.18**
Llama 3.1 70B [2]	458.15	0.54	0.44	0.71	2.38	**0.22**
ChatGPT-4o [1]	878.37	0.52	0.78	0.50	3.42	**0.25**

† Reference free

This trend is also quantitatively confirmed in the system-level correlations reported in Table 6, where our metric achieves the highest correlation with

Table 6. Comparison of system-level correlations. Correlation between the mean of human judgements for each generator model with other metrics.

	FID	LPIPS	DINO Sim.	CLIP Score†	SVGauge
Sρ	-7.14	-76.2	76.2	80.9	**91.0**
Kτ	-7.14	-57.1	57.1	64.2	**83.6**
Pr	27.7	-71.6	77.9	86.3	**93.1**

Table 7. Instance-level correlation between CLIPScore [5] and SVGauge both in reference-free and reference based versions.

	Sρ	Kτ	Pr
LPIPS	-14.0	-10.7	-16.5
Dino Sim.	34.6	26.5	39.3
SVGauge Ref. based	**48.4**	**37.1**	**53.7**
CLIP Score†	**37.8**	**28.7**	40.6
SVGauge Ref. free	36.5	27.7	**46.3**

human judgments across Spearman, Kendall, and Pearson scores. These results highlight the limitations of state-of-the-art automatic metrics in capturing the semantic alignment between SVG generated images and text prompts, and highlight the effectiveness of our proposed approach in producing evaluations that are more consistent with human perception.

In Table 7, we report instance-level correlations in terms of Spearman, Kendall and Pearson correlation coefficients for four metrics (LPIPS [26], DINO similarity [13], CLIPScore [5], and SVGauge) in both reference-based and reference-free settings. For the purpose of this experiment and fairness of evaluation, we also show a reference-free version of SVGauge, where we remove the visual similarity portion of the metric (*i.e.*, setting $\alpha = 0$). Reference-based SVGauge achieves the strongest agreement with human judgments. The reference-free variant of SVGauge trails CLIPScore [5] in ranking metrics but surpasses it in Pearson correlation. Overall, our reference-free approach offers a competitive alternative to CLIPScore [5]. These results further underline that current spread-wise evaluation methods remain unreliable when applied to SVG based content.

4.3 Qualitative Results

Finally, in Fig. 3 we showcase a qualitative comparison between SVGauge, CLIPScore [5] and DINO similarity [13]. In the first row, although the generated SVG is visually and semantically unrelated to the reference image, CLIPScore [5] and DINO similarity [13] still give relatively high scores. In contrast, SVGauge correctly assigns a low score, better aligned with human perception. On the other hand, in the second row, when the generated SVG accurately preserves the core semantics required by the prompt, despite the lower quality image, SVGauge correctly assigns a moderate score. These examples highlight that SVGauge features an improved sensitivity to semantic consistency, and demonstrates a significant ability to distinguish between meaningful and misleading generations in comparison with other existing metrics.

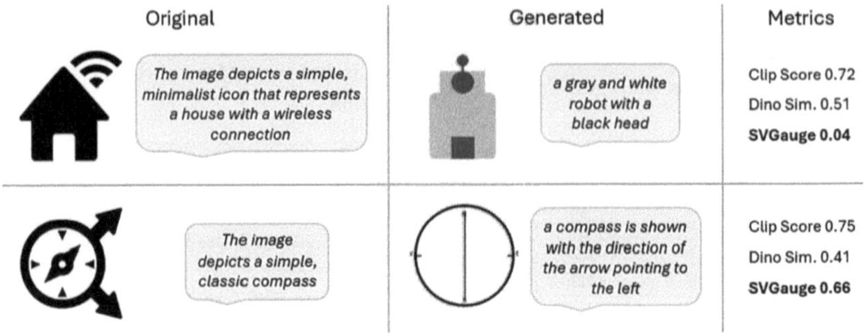

Fig. 3. Sample qualitative results.

5 Conclusion

We proposed the first evaluation framework specifically designed for text-to-SVG generation. By capturing both visual similarity and semantic alignment, our proposed metric SVGauge can provide quantitative evaluations with significant alignment with human perception. As a complementary contribution, we also developed the first dataset with prompt-SVG pairs annotated with human evaluations. We believe that our approach can serve as a foundation for future benchmarks and model evaluations in structured generative domains.

Acknowledgments. We acknowledge the CINECA award under the ISCRA initiative, for the availability of high-performance computing resources. This work has been conducted with the support of the PRIN 2022-PNRR project "MUCES" (CUP E53D23016 290001), the PRIN 2022 project "MUSMA" (CUP E53D23008310001) and the European project MINERVA, funded by European High-Performance Computing Joint Undertaking (JU) under grant agreement No 101182737.

References

1. Achiam, J., et al.: GPT-4 Technical Report. arXiv preprint (2023)
2. Grattafiori, A., et al.: The llama 3 herd of models. arXiv preprint (2024)
3. Guo, D., et al.: Deepseek-r1: incentivizing reasoning capability in llms via reinforcement learning. arXiv preprint (2025)
4. He, K., Chen, X., Xie, S., Li, Y., Dollár, P., Girshick, R.: Masked autoencoders are scalable vision learners. In: CVPR (2022)
5. Hessel, J., Holtzman, A., Forbes, M., Bras, R.L., Choi, Y.: CLIPScore: a reference-free evaluation metric for image captioning. In: EMNLP (2021)
6. Heusel, M., Ramsauer, H., Unterthiner, T., Nessler, B., Hochreiter, S.: Gans trained by a two time-scale update rule converge to a local nash equilibrium. In: NeurIPS (2017)
7. Jain, A., Xie, A., Abbeel, P.: Vectorfusion: text-to-svg by abstracting pixel-based diffusion models. In: CVPR (2023)

8. Jégou, H., Chum, O.: Negative evidences and co-occurrences in image retrieval: the benefit of PCA and whitening. In: ECCV (2012)
9. Laurençon, H., Marafioti, A., Sanh, V., Tronchon, L.: Building and better understanding vision-language models: insights and future directions. arXiv preprint (2024)
10. Li, J., Li, D., Savarese, S., Hoi, S.: Blip-2: bootstrapping language-image pre-training with frozen image encoders and large language models. In: ICML (2023)
11. Li, T.M., Lukáč, M., Gharbi, M., Ragan-Kelley, J.: Differentiable vector graphics rasterization for editing and learning. TOG (2020)
12. Ma, X., et al.: Towards layer-wise image vectorization. In: CVPR (2022)
13. Oquab, M., et al.: Dinov2: learning robust visual features without supervision. TMLR (2024)
14. Radenović, F., Tolias, G., Chum, O.: Fine-tuning cnn image retrieval with no human annotation. IEEE TPAMI (2019)
15. Radford, A., et al.: Learning transferable visual models from natural language supervision. In: ICML (2021)
16. Reimers, N., Gurevych, I.: Sentence-bert: sentence embeddings using siamese bert-networks. arXiv preprint (2019)
17. Rodriguez, J.A., et al.: Starvector: generating scalable vector graphics code from images and text. In: AAAI (2024)
18. Team, G.: Gemma. Kaggle (2024)
19. Wu, R., Su, W., Liao, J.: Chat2SVG: vector graphics generation with large language models and image diffusion models. arXiv preprint (2024)
20. Wu, R., Su, W., Ma, K., Liao, J.: IconShop: text-guided vector icon synthesis with autoregressive transformers. TOG (2023)
21. Xiao, B., et al.: Florence-2: advancing a unified representation for a variety of vision tasks. In: CVPR (2023)
22. Xing, X., Hu, J., Zhang, J., Xu, D., Yu, Q.: Svgfusion: scalable text-to-svg generation via vector space diffusion. arXiv preprint (2025)
23. Xing, X., Hu, J., Zhang, L., Guotao, J., Xu, D., Yu, Q.: Empowering llms to understand and generate complex vector graphics. arXiv preprint (2024)
24. Xing, X., Zhou, H., Wang, C., Zhang, J., Xu, D., Yu, Q.: Svgdreamer: text guided svg generation with diffusion model. In: CVPR (2024)
25. Zhai, X., Mustafa, B., Kolesnikov, A., Beyer, L.: Sigmoid loss for language image pre-training. In: ICCV (2023)
26. Zhang, R., Isola, P., Efros, A.A., Shechtman, E., Wang, O.: The unreasonable effectiveness of deep features as a perceptual metric. In: CVPR (2018)

A Second-Order Perspective on Pruning at Initialization and Knowledge Transfer

Leonardo Iurada(✉), Beatrice Occhiena, and Tatiana Tommasi

Politecnico di Torino, Torino, Italy
leonardo.iurada@polito.it

Abstract. The widespread availability of pre-trained vision models has enabled numerous deep learning applications through their transferable representations. However, their computational and storage costs often limit practical deployment. Pruning-at-Initialization has emerged as a promising approach to compress models before training, enabling efficient task-specific adaptation. While conventional wisdom suggests that effective pruning requires task-specific data, this creates a challenge when downstream tasks are unknown in advance. In this paper, we investigate how data influences the pruning of pre-trained vision models. Surprisingly, pruning on one task retains the model's zero-shot performance also on unseen tasks. Furthermore, fine-tuning these pruned models not only improves performance on original seen tasks but can recover held-out tasks' performance. We attribute this phenomenon to the favorable loss landscapes induced by extensive pre-training on large-scale datasets.

Keywords: Neural Network Pruning · Transfer Learning

1 Introduction

The surge in digital data availability has fueled the development of large-scale *foundation* models which are now indispensable tools across a broad spectrum of applications. Through extensive training, these models learn rich representations with impressive generalization capabilities, achieving strong performance on a variety of downstream tasks with minimal fine-tuning, and often even in zero-shot settings. Nonetheless, their sheer size and complexity continue to pose significant challenges regarding computational and storage resources, leading to a growing interest in model compression strategies to enhance efficiency.

The vast majority of literature on model *pruning* focuses on reducing the inference latency for real-time deployment, although there is also a growing interest in approaches to eliminate unnecessary weights before the network's training. *Pruning-at-initialization* (PaI) identifies sparse sub-networks based on network topology and theoretically estimated learning dynamics. It is effective both when the initialization is random (learning from scratch) and when starting from a pre-trained model. Indeed, compressing networks before fine-tuning has

E. Rodolà et al. (Eds.): ICIAP 2025, LNCS 16167, pp. 194–206, 2026.
https://doi.org/10.1007/978-3-032-10185-3_16

the potential to reduce the cost of model adaptation, thereby improving accessibility to foundation models especially for resource-constrained devices. However it comes with the unique challenge of preserving the pre-trained model's original broad generalization regardless of the task-specific data used to guide the pruning process.

We dedicate this work to an in-depth investigation of whether and to what extent it is possible to directly compress pre-trained vision models without compromising their transferability. The study is particularly relevant for scenarios in which the downstream application is not known in advance. In such cases, pruning is guided by data from a *source* task, while performance is evaluated on a different but related *target* task. This setting enables a rigorous assessment of knowledge retention and addresses a significant gap in the current PaI literature.

Our contributions can be summarized as follows:

1. We set up a controlled environment and run an extensive analysis on the transfer performance of large-scale pre-trained pruned models, considering both convolutional neural networks and vision transformers, encompassing all major PaI techniques for unstructured pruning.
2. We offer theoretical insights about the transfer learning potential of pruned models by analyzing the loss landscape and its curvature, studying the effects of weight perturbation induced during pruning and subsequent fine-tuning.
3. Notably, we observe distinct trends between architectures but minimal variation across their pre-training configurations with a general conclusion about the possibility of pruning on a task and retaining performance on other tasks with distinct labeling functions, without requiring any data from the latter.

2 Related Works

Neural Network Pruning. The challenge of reducing neural network parameters while preserving performance has been studied since the 1980 s [14,20]. The *Lottery Ticket Hypothesis* [11] conjectured that within large networks there exist smaller sub-networks ("winning tickets") that, if correctly identified and trained, can match the performance of the original full model. The Iterative Magnitude Pruning (IMP) algorithm discovers these sub-networks through alternating training and pruning phases, with high computational costs. This led to the development of more efficient paradigms such as *Pruning-at-Initialization* (PaI) which searches for "winning tickets" before any training, favoring a more efficient downstream task adaptation. PaI approaches fall into two categories. *Data-agnostic* methods evaluate the parameters' importance without considering training data. Notable examples include SynFlow [31], which assesses parameter relevance through synaptic saliency, NTK-SAP [34], which leverages Neural Tangent Kernel (NTK) theory for more precise training dynamics estimation and PHEW [26], which scores parameter importance using biased random walks in a single-pass. *Data-driven* methods incorporate training data and architecture information to guide pruning. SNIP [22] scores parameters based on their impact on the initial loss, while GraSP [32] uses gradient norm changes to identify important parameters. Furthermore, PX [17] improved on NTK-based data-agnostic methods by incorporating task data information.

Transfer Learning in the Context of Pruning. Pruning is often applied to obtain highly task-specific models optimized for a given task. However, in many real-world scenarios, only limited or even no task-specific data may be available during pruning. While this challenge could be addressed by *data-agnostic* PaI methods on randomly initialized networks, it has been shown that such approaches fail in the context of pre-trained vision models [17]. Recent literature has explored combining pruning with transfer learning to address this issue. In the NLP domain, [6] applied structured pruning and re-training to BERT [7], leveraging transfer learning under limited task-specific data. Although they operate in a low-data regime, their approach still assumes access to transfer task data. Similarly, [13] compresses BERT during pre-training and analyzes the effects on downstream transfer. They find that pruning once during pre-training is sufficient, obviating the need for task-specific pruning for each downstream task. In the vision domain, [29] demonstrates a correlation between the number of unpruned parameters in the network and the entropy of its learned representation. The authors of [23] propose an alternative strategy by jointly tuning the weights and structure of pre-trained vision models to better align with the transfer task. However, their method requires access to both the source and transfer task data during pruning and re-training.

In this work, we focus on vision models extensively pre-trained on large-scale datasets covering a wide range of tasks. We apply PaI methods, simulating in a controlled setting the scenario where downstream tasks are unknown in advance. Specifically, we use data from a single *source* task for pruning and re-training, then evaluate the resulting models on multiple held-out *target* tasks. By analyzing how feature quality is affected by pruning, we aim to understand the role of data and to which extent compressed representations allow for transfer learning.

3 Transfer Learning of Pruned Pre-trained Models

Background & Problem Formulation. Consider a neural network $f : \mathbb{R}^d \times \mathbb{R}^p \to \mathbb{R}^C$ taking inputs $\boldsymbol{x} \in \mathbb{R}^d$ and parametrized by weights $\boldsymbol{\theta} = \{\boldsymbol{W}, \boldsymbol{\xi}\} \subseteq \mathbb{R}^p$. We assume our $f(\boldsymbol{x}, \boldsymbol{\theta}) = \varphi(\boldsymbol{x}, \boldsymbol{W})^\top \boldsymbol{\xi}$ to be composed of a feature extractor $\varphi : \mathbb{R}^d \times \{\boldsymbol{W}\} \to \mathbb{R}^h$, mapping inputs to h-dimensional feature vectors and a one-versus-rest linear classifier with weight matrix $\boldsymbol{\xi} \in \mathbb{R}^{h \times C}$, on which we take the maximal value (over the C classes) as the final model's prediction.

For simplicity, we assume two multi-class *classification* tasks with their overall data support $\mathcal{D} = \mathcal{D}_s \cup \mathcal{D}_t$ composed by two disjoint sets ($i.e.\mathcal{D}_s \cap \mathcal{D}_t = \varnothing$). One defines the *source* task $\mathcal{D}_s = \{(\boldsymbol{x}_i^s, f_s^\star(\boldsymbol{x}_i^s) = y_i^s)\}_{i \leq n_s}$ and the other is the *transfer* task $\mathcal{D}_t = \{(\boldsymbol{x}_i^t, f_t^\star(\boldsymbol{x}_i^t) = y_i^t)\}_{i \leq n_t}$. Here n_s, n_t indicate the respective data cardinality for the two sets and $f_s^\star : \mathcal{D}_s \to \mathcal{Y}_s$ is the target function that maps inputs to the ground truth labels. By analogy we define $f_t^\star : \mathcal{D}_t \rightarrow \mathcal{Y}_t$ for the transfer task.

Let us denote with $\boldsymbol{\theta}_0$ the parameters obtained via extensive pre-training on a large corpus of data. The problem of *unstructured* neural network pruning [22] can be formalized as finding a binary mask $\boldsymbol{c} \in \{0,1\}^p$ that is optimal for

$$\min_{\boldsymbol{c}} \mathbb{E}_{(\boldsymbol{x},y)\sim\mathcal{D}}[L(f(\boldsymbol{x}, \mathcal{A}(\boldsymbol{\theta}_0, \boldsymbol{c}) \odot \boldsymbol{c}), y)] \quad \text{s.t. } \boldsymbol{c} \in \{0,1\}^p, \quad \frac{\|\boldsymbol{c}\|_0}{p} \leq k = 1 - q \quad (1)$$

where L is the loss function (*e.g.*MSE), q is the desired sparsity ratio, $\mathcal{A}$ is an optimization algorithm (*e.g.*SGD) that takes as input the mask and the initial weights and returns the pruned weights at convergence, and $\odot$ denotes the element-wise product. In the following, we will focus on the problem in Eq. (1) by assuming to only have access to the *source* task data $\mathcal{D}_s$.

3.1 A Second-Order Perspective On Pruning Pre-Trained Models

Solving the problem in Eq. (1) poses significant challenges. However, assuming $\boldsymbol{\theta} = \boldsymbol{\theta}_0$ is a point of local minimum for the empirical risk $L(\mathcal{D}, \boldsymbol{\theta})$, the second-order approximation of the loss around $\boldsymbol{\theta}_0$ allows to derive a tractable closed-form solution [14]. Specifically, its second-order Taylor expansion around $\boldsymbol{\theta}_0$ is

$$L(\mathcal{D}, \boldsymbol{\theta}) = L(\mathcal{D}, \boldsymbol{\theta}_0) + (\boldsymbol{\theta} - \boldsymbol{\theta}_0)^\top \mathbf{H}(\boldsymbol{\theta} - \boldsymbol{\theta}_0) + \mathcal{O}(\|\boldsymbol{\theta} - \boldsymbol{\theta}_0\|^3) , \quad (2)$$

with $\mathbf{H} \triangleq \nabla^2_{\boldsymbol{\theta}} L(\mathcal{D}, \boldsymbol{\theta}_0)$ identifying the Hessian around $\boldsymbol{\theta}_0$. Given that the Hessian is positive semi-definite for $\boldsymbol{\theta} = \boldsymbol{\theta}_0$ and considering the quadratic nature of Eq. (2), such expansion defines a locally convex basin for $L(\mathcal{D}, \boldsymbol{\theta})$. Thus, as done by [14], we can derive the optimal parameter to remove by deriving in closed-form the incurring increase in loss $\varepsilon_{[j]}$ of removing the j-th weight $\boldsymbol{\theta}_{[j]}$ and the optimal update $\boldsymbol{\delta}_{[j]}$ to adjust for its removal,

$$\varepsilon_{[j]} = \frac{\boldsymbol{\theta}^2_{[j]}}{[\mathbf{H}^{-1}]_{[jj]}}, \quad \boldsymbol{\delta}_{[j]} = -\frac{\boldsymbol{\theta}_{[j]}}{[\mathbf{H}^{-1}]_{[jj]}} \cdot \mathbf{H}^{-1}_{[:,j]} . \quad (3)$$

Although optimal, these quantities assume access to the whole $\mathcal{D}$, rather than only $\mathcal{D}_s$ (as our assumption). However, recent insights on the regularities induced by large-scale pre-training [18] reveal that, if it is true that $\mathcal{D}_s, \mathcal{D}_t$ are similar to data seen during pre-training, then, we can make the following predictions (which we will empirically test via controlled experiments in Sect. 4).

P1. We expect that *parameters not important for $\mathcal{D}_s$ are also not important for $\mathcal{D}_t$*. Thus, guiding pruning using only $\mathcal{D}_s$ is enough to compress models in an approximately optimal way.
P2. If **P1** is verified and re-training after pruning can recover performance on $\mathcal{D}_s$ (*e.g.*via SGD or using $\boldsymbol{\delta}_{[j]}$), then it will also recover performance on $\mathcal{D}_t$, as the updated pruned weights will also have a lower empirical risk on the *target* task, given they will be closer to $\boldsymbol{\theta}_0$ (which are optimal for all tasks) in the Riemann manifold [21] induced by $\mathbf{H}$.

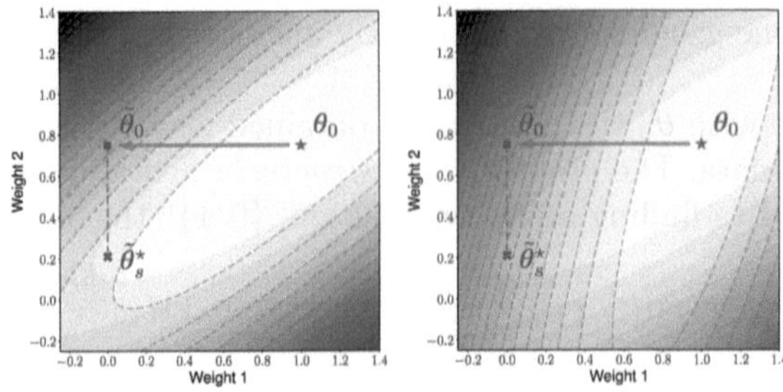

Fig. 1. Toy loss landscapes of pruned models. We visualize the loss landscapes of toy models with two parameters. Dashed blue lines indicate the *transfer* task's loss, not observable during pruning and re-training. Blue shades represent the *source* task's loss. Both losses are modeled as quadratic functions measuring squared distance from the shared minimum $\boldsymbol{\theta}_0$, with rotated principal directions to simulate varying task alignment. $\tilde{\boldsymbol{\theta}}_0$ indicates the pruned weights, while $\tilde{\boldsymbol{\theta}}^*_s$ the pruned and re-trained weights, both for the *source* task. The red arrow indicates the weight perturbation, resulting from pruning a weight, while the red dashed line the re-training on the *source* task. **(Left)** Aligned tasks. **(Right)** Misaligned tasks. Light colors indicate lower loss. (Color figure online)

We visualize our two conjectures **P1** and **P2** in Fig. 1. If the *source* and *transfer* tasks' loss landscapes are aligned, pruning on one or the other is the same. Furthermore, re-training, after pruning, only on *source* moves the pruned weights back again in a point of lower loss also for the *transfer* task. Conversely, if the two tasks have misaligned landscapes (in an extreme case, orthogonal) then pruning and re-training will not help with transferability as the pruned solution moves orthogonal to the contour lines of the *transfer* task.

In the following section we will cast the problem of Eq. (1) as a *Pruning-at-Initialization* problem [17,22,31,32,34] and introduce the main methodology we will adopt to assess our predictions **P1** and **P2**.

3.2 Adopted Pruning-at-Initialization Pipeline

Given a task $\mathcal{D}_i$ on which we want to compress the weights $\boldsymbol{\theta}_0$, the general pipeline of *Pruning-at-Initialization* (PaI) methods to tackle the problem of Eq. (1) starts by assigning each weight $j \in 1, \ldots, p$ a saliency score of the form

$$S(\boldsymbol{\theta}_{0[j]}) = \frac{\partial F}{\partial \boldsymbol{\theta}_{[j]}} \cdot \boldsymbol{\theta}_{0[j]} \tag{4}$$

where F is a function of the weights evaluated at $\boldsymbol{\theta} = \boldsymbol{\theta}_0$, capturing some property of the network. For instance, in SNIP [22] the saliency is the empirical risk at-initialization: $F = \mathbb{E}_{(\boldsymbol{x},y)\sim\mathcal{D}_i}[L(f(\boldsymbol{x}, \boldsymbol{\theta}_0 \odot \mathbf{c}), y)]$. Thus, it assigns to each parameter a score which reflects how the initial loss would change when removing that specific parameter from the network. Once the saliency scores are assigned to each weight, we keep the top-k weights by setting the masks with the highest saliency scores to one ($i.e. \forall j \in \text{top}_k(S),\ \boldsymbol{c}_{[j]} = 1$), while the other weights will be pruned ($i.e. \forall j \notin \text{top}_k(S),\ \boldsymbol{c}_{[j]} = 0$). Then, the pruned model $f(\cdot, \boldsymbol{\theta}_0 \odot \boldsymbol{c})$ is re-trained on $\mathcal{D}_i$ to recover its original performance.

Setting $S(\boldsymbol{\theta}_{0[j]}) = \hat{\varepsilon}_{[j]}$ as in Eq. (3) provides an optimal score. However, computing this quantity requires to calculate outer products between gradients to estimate the Hessian which incurs a complexity of approximately $\mathcal{O}(p^2)$ in both time and memory. Moreover, computing matrix inverses further raises the complexity to $\mathcal{O}(p^3)$, with the number of parameters p ranging from millions to billions. To bypass this, standard practice is to adopt Hessian approximations [12,14,20,30] that incur in a lower cost by sacrificing some precision. Specifically,

Isotropic. By approximating the Hessian as the identity matrix $\mathbf{H} = I$, the score is only based on the weight magnitude. Thus, $\hat{\varepsilon}^{mag}_{[j]} = \boldsymbol{\theta}^2_{[j]}$. This is the cheapest approximation possible for the Hessian, which assumes the curvature of the loss landscape to be isotropic around $\boldsymbol{\theta}_0$.

Diagonal [20,30]**.** This approximation assumes each parameter of the model as independent. Computing only the diagonal elements of the Hessian incurs in a cost of approximately $\mathcal{O}(p)$ in both time and memory and can be done efficiently by computing per-example gradients as $\mathbf{H}_{[jj]} = \mathbb{E}_{(\boldsymbol{x},y)\sim\mathcal{D}_i}[\big(\nabla_{\boldsymbol{\theta}_{[j]}} L((\boldsymbol{x}, y), \boldsymbol{\theta}_0)\big)^2]$. Additionally, the inverse of a diagonal matrix is the reciprocal of its diagonal elements, bypassing the $\mathcal{O}(p^3)$ cost of computing the inverse. Thus, under this approximation, $\hat{\varepsilon}^{diag}_{[j]} = \boldsymbol{\theta}^2_{[j]} \cdot \mathbf{H}_{[jj]}$.

Block-Diagonal [12,14]**.** By grouping parameters layer-wise into blocks, this approximation assumes independence between layers, discarding cross-layer curvature terms. For layer l with parameters $\boldsymbol{\theta}_{[l]} \in \mathbb{R}^{p_l}$, its Hessian block $\mathbf{H}_{[l]} \in \mathbb{R}^{p_l \times p_l}$ is approximated as $\mathbf{H}_{[l]} \approx \mathbf{A}_{[l]} \otimes \mathbf{B}_{[l]}$, where $\otimes$ denotes the Kronecker product, $\mathbf{A}_{[l]} \in \mathbb{R}^{m_l \times m_l}$ and $\mathbf{B}_{[l]} \in \mathbb{R}^{h_l \times h_l}$ are small symmetric matrices derived from the expectation of gradient outer products over the data distribution $\mathcal{D}_i$ [9,24]. Specifically, for weight gradients $\nabla_{\boldsymbol{\theta}_{[l]}} L$ reshaped into a $m_l \times h_l$ matrix (with $m_l \cdot h_l = p_l$), from the chain rule $\mathbf{A}_{[l]}$ and $\mathbf{B}_{[l]}$ correspond to the covariances of the layer's input activations and gradient signals, respectively:

$$\mathbf{A}_{[l]} = \mathbb{E}_{(\boldsymbol{x},y)\sim\mathcal{D}_i}\left[\boldsymbol{a}_{[l-1]} \cdot \boldsymbol{a}_{[l-1]}^\top\right], \quad \mathbf{B}_{[l]} = \mathbb{E}_{(\boldsymbol{x},y)\sim\mathcal{D}_i}\left[\nabla_{\boldsymbol{z}_{[l]}} L \cdot \nabla_{\boldsymbol{z}_{[l]}} L^\top\right], \tag{5}$$

where $\boldsymbol{a}_{[l-1]} \in \mathbb{R}^{m_l}$ is the input to layer l and $\nabla_{\boldsymbol{z}_{[l]}} L \in \mathbb{R}^{h_l}$ the gradient of the loss w.r.t.layer l's pre-activation output $\boldsymbol{z}_{[l]}$. The Kronecker structure reduces memory from $\mathcal{O}(p_l^2) = \mathcal{O}((m_l \cdot h_l)^2)$ to $\mathcal{O}(m_l^2 + h_l^2)$ for an $m_l \times h_l$ weight matrix, and enables efficient inversion via $\mathbf{H}^{-1}_{[l]} \approx \mathbf{A}^{-1}_{[l]} \otimes \mathbf{B}^{-1}_{[l]}$, replacing the $\mathcal{O}(p_l^3)$ cost of direct inversion with a lower cost of $\mathcal{O}(m_l^3 + h_l^3)$. This allows layer-wise computation of $\hat{\varepsilon}^{block}_{[j]}$ using Eq. (3), with per-parameter scores derived from local blocks rather than the full Hessian.

Finally, as we only have access to $\mathcal{D}_s$, we restrict the computation of the Hessian on the data of the *source* task. Given these approximations, the resulting closed-form update $\boldsymbol{\delta}_{[j]}$ may no longer be optimal. To address this, we align with the PaI framework and, instead, we update the remaining weights using SGD.

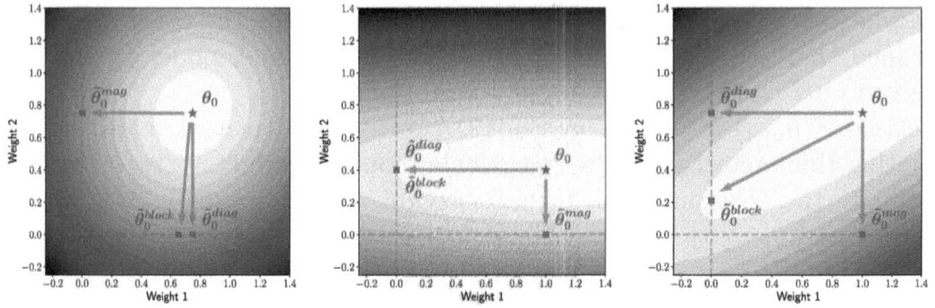

Fig. 2. Toy loss landscapes based on different Hessian eigenstructures. We visualize loss landscapes of toy models with two parameters, modeled as quadratic functions that compute the squared distance from the minimum $\boldsymbol{\theta}_0$. The curvature is defined by rescaled eigenvalues of the Hessian to simulate different task geometries. **(Left)** Isotropic Hessian. **(Center)** Diagonal Hessian. **(Right)** Hessian with non-zero cross-terms. Pruned weight configurations, denoted $\tilde{\boldsymbol{\theta}}_0$, are obtained using Hessian-based importance scores detailed in Sect. 3.2 ($\hat{\varepsilon}^{mag}_{[j]}$, $\hat{\varepsilon}^{diag}_{[j]}$, $\hat{\varepsilon}^{block}_{[j]}$). The red arrow indicates the weight perturbation, resulting from pruning. Light colors indicate lower loss. (Color figure online)

To intuitively understand when these three Hessian approximations might capture sufficiently well the true Hessian structure, we visualize in Fig. 2 three cases. In the left plot, the loss landscape is isotropic so different pruning criteria lead to similar outcomes, and re-training would have little effect. In the center plot, the landscape is anisotropic, but aligned with the canonical basis. Here, magnitude pruning may be suboptimal, but re-training would not significantly help. In the right plot, the Hessian has non-zero cross-terms, creating a tilted landscape. In this case, re-training would be essential to get back to a lower-loss region.

In the next section, we define our protocol for measuring how pruning affects the feature quality of pre-trained vision models, and how any resulting degradation influences transfer learning performance.

3.3 Feature Quality Evaluation Protocol

Inspired by [19], given a task $\mathcal{D}_k$ we measure *feature quality* as the ability of a classifier to linearly separate the features $\boldsymbol{\Phi}_k = [\varphi(\boldsymbol{x}_1, \boldsymbol{W}_0), \ldots, \varphi(\boldsymbol{x}_n, \boldsymbol{W}_0)]^\top \in \mathbb{R}^{n \times h}, \forall \boldsymbol{x} \in \mathcal{D}_k$ into their correct classes. As done in [1,10,28], to avoid introducing extra trainable parameters, we fit one linear classifier per task, obtaining $\boldsymbol{\xi}^\star_{k=s,t}$ by solving the following Ridge Regression [2] problem:

$$\boldsymbol{\xi}^\star_k = \arg\min_{\boldsymbol{\xi} \in \mathbb{R}^{h \times C}} \|Y_k - \boldsymbol{\Phi}_k \boldsymbol{\xi}\|_2^2 + \alpha \|\boldsymbol{\xi}\|_2^2 \ , \tag{6}$$

where $Y_k \in \mathbb{R}^{n \times C}$ is the matrix of the stacked one-hot-encoding vectors of the corresponding C classes and $\alpha > 0$ is an hyperparameter controlling the strength of the regularization term. This problem admits a closed-form solution:

$$\boldsymbol{\xi}^\star_k = (\boldsymbol{\Phi}_k^\top \boldsymbol{\Phi}_k + \alpha I)^{-1} \boldsymbol{\Phi}_k^\top Y_k \ , \tag{7}$$

where I is the $h \times h$ identity matrix. As our primary focus is not on the effect of regularization, we set $\alpha = 1$ to ensure the invertibility of $(\boldsymbol{\Phi}_k^\top \boldsymbol{\Phi}_k + \alpha I)$ [2].

We keep the obtained classifiers $\boldsymbol{\xi}_s^\star, \boldsymbol{\xi}_t^\star$ frozen for all the subsequent operations and, based on which data the network is processing, we load the respective classifier. Thus, to quantify the degradation in feature quality due to pruning,

1. Unpruned: we first obtain baseline predictions from the original unpruned model $\forall \boldsymbol{x} \in \mathcal{D}_k,\ f(\boldsymbol{x}, \{\boldsymbol{W}_0, \boldsymbol{\xi}_k^\star\}) = \varphi(\boldsymbol{x}, \boldsymbol{W}_0)^\top \boldsymbol{\xi}_k^\star$, for all tasks $k = s, t$;

2. Pruned: we assume having access only to the *source* task $\mathcal{D}_s$ and calibrate the mask $\hat{\boldsymbol{c}} \in \{0,1\}^{p-h\times C}$ for pruning. We obtain the predictions with the pruned model $\forall \boldsymbol{x} \in \mathcal{D}_k,\ f(\boldsymbol{x}, \{\boldsymbol{W}_0 \odot \hat{\boldsymbol{c}}, \boldsymbol{\xi}_k^\star\}) = \varphi(\boldsymbol{x}, \boldsymbol{W}_0 \odot \hat{\boldsymbol{c}})^\top \boldsymbol{\xi}_k^\star$, for all tasks $k = s, t$;

3. Pruned & Fine-tuned: we fit the remaining unpruned weights on the *source* task $\mathcal{D}_s$, obtaining $\tilde{\boldsymbol{W}}_s^\star$ and evaluate the performance of the fine-tuned compressed model $\forall \boldsymbol{x} \in \mathcal{D}_k,\ f(\boldsymbol{x}, \{\tilde{\boldsymbol{W}}_s^\star, \boldsymbol{\xi}_k^\star\}) = \varphi(\boldsymbol{x}, \tilde{\boldsymbol{W}}_s^\star)^\top \boldsymbol{\xi}_k^\star$, for all tasks $k = s, t$.

4 Experiments

In this section, we present the results to test our predictions **P1** and **P2**, conjectured in Sect. 3.1. We describe our adopted experimental details and the results of our analysis that provide a positive answer to our hypotheses.

Dataset. We adopt ImageNet-10 [16,33], consisting of 10 classification tasks with 10 classes each, derived as subsets of the ImageNet-1k dataset [5] (see Tab. 1 for full details). As we aim to understand the degradation of feature quality induced by pruning, all procedures were performed exclusively on the test split (50 examples per class) to minimize external factors such as distribution shifts or class unbalancing that would complicate the experimental evaluation (Table 1).

Table 1. We define 10 multi-class classification tasks from ImageNet-1k [5], each comprising 10 semantically related classes. To ensure semantic consistency within each task, classes are grouped based on the WordNet Hierarchy [25]. For each task, we report the task number, the class names, and their corresponding ImageNet class indices.

ImageNet-10 - Selected Multi-class Classification Tasks

	Class 1	Class 2	Class 3	Class 4	Class 5	Class 6	Class 7	Class 8	Class 9	Class 10
Task 1 (Food)	cheeseburger (933)	carbonara (959)	hot dog (934)	pizza (963)	ice cream (928)	burrito (965)	mashed potato (935)	cucumber (943)	pretzel (932)	spaghetti (940)
Task 2 (Geo Formation)	sandbar (977)	coral reef (973)	volcano (980)	promontory (976)	valley (979)	lakeside (975)	bubble (971)	geyser (974)	cliff (972)	alp (970)
Task 3 (Birds)	great gray owl (24)	robin (15)	rooster (7)	hen (8)	ostrich (9)	brambling (10)	goldfinch (11)	bulbul (16)	bald eagle (22)	jay (17)
Task 4 (Structures)	castle (483)	patio (706)	bridge (821)	obelisk (682)	totem (863)	dock (536)	lighthouse (437)	fountain (562)	sawmill (634)	dam (525)
Task 5 (Clothing)	trench coat (869)	lab coat (617)	poncho (735)	bow tie (457)	sweatshirt (841)	kimono (614)	jersey (610)	miniskirt (655)	bikini (445)	bulletproof vest (465)
Task 6 (Furniture)	toilet seat (861)	china cabinet (495)	table lamp (846)	desk (526)	studio couch (831)	stove (827)	folding chair (559)	pool table (736)	bookcase (453)	wardrobe (894)
Task 7 (Equipment)	cellular telephone (487)	cd player (485)	oscilloscope (688)	printer (742)	polaroid camera (732)	computer keyboard (508)	coffeepot (505)	mouse (673)	ipod (605)	paintbrush (696)
Task 8 (Vehicles)	bullet train (466)	school bus (779)	police van (734)	garbage truck (569)	fire engine (555)	pickup truck (717)	limousine (627)	ambulance (407)	go-kart (573)	tractor (866)
Task 9 (Land Animals)	king snake (56)	mud turtle (35)	zebra (340)	lion (291)	tiger (292)	red fox (277)	timber wolf (269)	cardigan (264)	ice bear (296)	brown bear (294)
Task 10 (Aquatic Animals)	tench (0)	goldfish (1)	great white shark (2)	tiger shark (3)	hammerhead (4)	electric ray (5)	stingray (6)	king crab (121)	crayfish (124)	jellyfish (107)

Architectures & Pre-trained Models. We focus on both residual networks (ResNet-50) [15] and vision transformers (ViT-B/16) [8] architectures, leveraging different pre-training paradigms: supervised (on ImageNet-1k [5]), self-supervised contrastive (MoCoV2 [4]), self-supervised student-teacher distillation (DINO [3]), and vision-language alignment (CLIP [27]).

Baselines. To study how pruning affects transferability, based on [17] we resort to data-driven PaI methods: SNIP [22], GraSP [32] and PX [17], although we skip the latter on ViT, as its score applies only for networks in the NTK regime, which is problematic due to their highly non-linear self-attention layers [35].

Implementation Details. We follow standard PaI procedures [17,22,31] as outlined in Sect. 3.2, while the evaluation follows [19] as outlined in Sect. 3.3. In detail, we first fit one classifier $\boldsymbol{\xi}_k^\star$ per task $k = 1, \ldots, 10$ on the unpruned model and keep them frozen at all times. Then, at each run one task is elected as *source* while the other nine are held-out as *transfer* tasks. We prune the pre-trained encoder $\boldsymbol{W}_0$ by calibrating the pruning mask $\boldsymbol{c}$, accessing only the *source* task $\mathcal{D}_s$ when needed. Then, we re-train the pruned encoder on $\mathcal{D}_s$, aligning with [17]. We run SGD on MSE loss for 90 epochs with 0.9 momentum and 10^{-4} weight decay. The step size is 5×10^{-5} (except for ViT DINO, where 10^{-4} was used) and it is decayed by a factor $\times 10$ at epochs 30, 60, 80. For ResNet models we avoid updating the normalization statistics. Finally, akin to [34], we evaluate each method at {36.00, 47.52, 59.04, 66.42, 73.80, 78.52, 83.22} sparsities (%).

4.1 Impact of Source-Only Pruning on *Source* and *Transfer* Tasks

In Fig. 3, each cluster along the horizontal axis corresponds to a target sparsity level. Within each cluster, the six colored bars represent different pruning scores. Solid bars show zero-shot accuracy of the pruned model; black-hatched bars indicate accuracy after fine-tuning on the *source* task.

Pruning on *source* preserves *transfer* (P1). Across all ResNet models (ImageNet-1k, MoCoV2, DINO), magnitude ($\hat{\varepsilon}_{[j]}^{mag}$) and block Hessian ($\hat{\varepsilon}_{[j]}^{block}$) pruning maintain both source and transfer performance up to high sparsities, with parallel trends between solid bars in the top and bottom rows, thus supporting prediction **P1**: these criteria prune largely redundant weights. In contrast, diagonal Hessian ($\hat{\varepsilon}_{[j]}^{diag}$) pruning, SNIP, GraSP, and PX show sharper declines, indicating the removal of critical parameters.

A similar yet distinct pattern holds for ViTs: DINO models retain transfer accuracy under diagonal and block Hessian pruning, whereas CLIP models degrade sooner, suggesting lower compressibility. For ViTs, diagonal Hessian is the only score besides block Hessian that reliably preserves source and transfer performance, again supporting **P1**.

Fine-tuning on *source* helps (and never hurts) *transfer* (P2). Examining the hatched bars in Fig. 3, fine-tuning on the *source* task never reduces transfer accuracy; often it improves it. This is especially notable for MoCoV2, which has greater headroom for transfer gains. In DINO-pretrained models, gains are smaller, reflecting their already strong zero-shot transfer performance, thus, supporting **P2**: brief source-task fine-tuning enhances or preserves transfer ability. Compared to Hessian-based scores, which approximate the optimal pruning conditions, SNIP, GraSP and PX exhibit significantly steeper drops in both source and transfer accuracy at moderate sparsity.

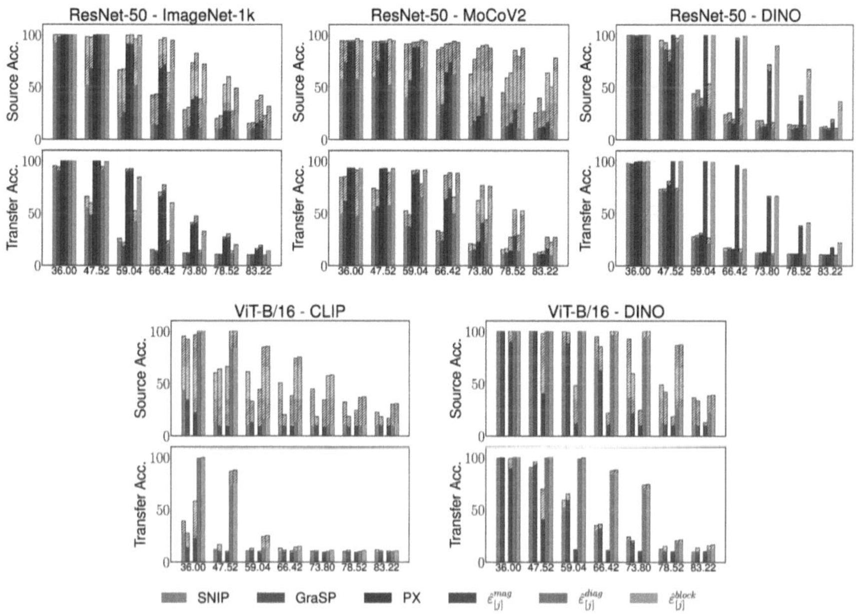

Fig. 3. Source and transfer performance of pruned models. We report the average accuracy of models pruned at increasing sparsity ratios using different scores, where pruning is performed solely based on the *source* task. Solid bars indicate zero-shot accuracy on the pruned model. Hatched bars represent the accuracy after re-training the pruned model on the *source* task; in all cases, re-training leads to improved performance. For each model, we show the average accuracy on the *source* task (top) and the average accuracy across the *transfer* tasks (bottom). Accuracy is averaged over 10 runs, where in each run a different task is designated as the *source*, and the remaining nine are treated as *transfer* tasks.

To further test our findings and provide a more interpretable perspective, we visualize the loss landscapes of two tasks, one designated as the *source* and the other as the *transfer* task, in Fig. 4. We consistently choose tasks 1 and 10 for this analysis, as we observed similar patterns across other task pairs. Focusing on ResNet-50 models, we find that magnitude pruning ($\hat{\varepsilon}_{[j]}^{mag}$) results in more favorable loss landscapes. This observation supports prediction **P1** for ResNet-50. Moreover, we see that fine-tuning tends to move the solution toward lower-loss regions not only for the *source* task but also for the *transfer* task, particularly in the case of ResNet-50 MoCoV2. In contrast, this trend does not hold as clearly for ResNet-50 DINO, aligning with prediction **P2**. On the other hand, diagonal Hessian-based pruning ($\hat{\varepsilon}_{[j]}^{diag}$) appears less effective, as it tends to prune important directions in weight space and results in less favorable loss landscapes. Still, when re-training improves performance on the source task, it generally helps move toward better solutions for the transfer task as well. Switching to ViT models, we observe the opposite: diagonal Hessian-based pruning performs bet-

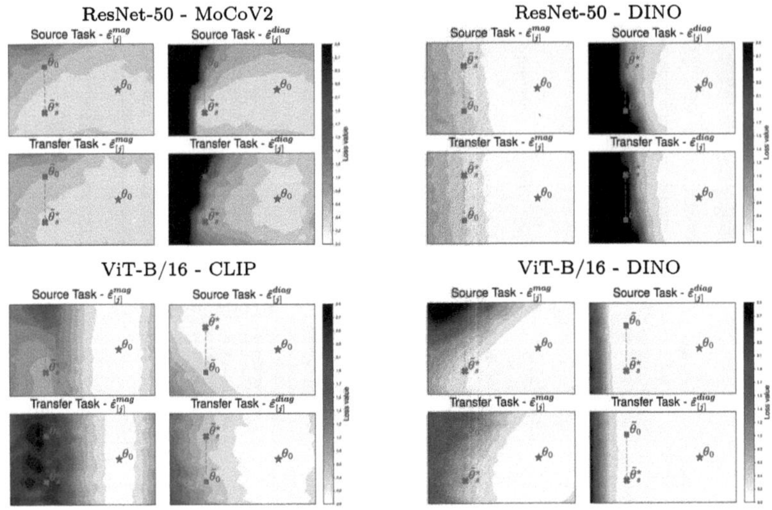

Fig. 4. Loss landscapes of real pre-trained models. We visualize the effect of pruning and re-training on Task 1 (*source*) and compare it to the corresponding behavior on Task 10 (*transfer*), using the pruning scores $\hat{\varepsilon}^{mag}_{[j]}$ and $\hat{\varepsilon}^{diag}_{[j]}$. Each plot is a projection onto the subspace spanned by the top two principal eigenvectors of the covariance matrix, computed from the concatenated weights $[\boldsymbol{\theta}_0, \tilde{\boldsymbol{\theta}}_0, \tilde{\boldsymbol{\theta}}^{\star}_s]$: where $\boldsymbol{\theta}_0$ are the original pre-trained weights, $\tilde{\boldsymbol{\theta}}_0$ are the weights after pruning (on the source task), and $\tilde{\boldsymbol{\theta}}^{\star}_s$ are the weights after both pruning and re-training (also on the source task). ResNet models are pruned to 66.42% sparsity, and ViT models to 47.52%.

ter than magnitude-based pruning, though the overall trends remain consistent. Notably, for ViT-B/16 DINO, the lack of dramatic differences seen in Fig. 3 can be attributed to already high zero-shot performance on both source and transfer tasks, limiting the observable gains.

5 Conclusion

Our results challenge the prevailing view that pruning without task-specific guidance necessarily harms transferability. We show that vision models retain zero-shot performance on *transfer* tasks even when compressed on *source* task data. Moreover, fine-tuning these pruned models often enhances *transfer* performance, suggesting a surprising alignment in the loss landscapes across tasks. This robustness holds across architectures, though ViTs require more sophisticated Hessian-based criteria, while ResNets perform best with simpler magnitude pruning.

Nonetheless, our controlled setup omits real-world complexities. Also, theoretical guarantees for safe pruning thresholds remain unexplored. Future research should address these gaps by linking pruning-induced perturbations to the geometry of the loss landscape and testing PaI under dynamic or heterogeneous conditions. In summary, we show that the rich representations learned during large-scale pre-training provide a robust foundation for efficient compression, enabling lighter, faster models that retain their capacity to adapt *almost* for free.

Acknowledgements. L.I. acknowledges the grant received from the European Union Next-GenerationEU (Piano Nazionale di Ripresa E Resilienza (PNRR)) DM 351 on Trustworthy AI. T.T. acknowledges the EU project ELSA - European Lighthouse on Secure and Safe AI. This study was carried out within the FAIR - Future Artificial Intelligence Research and received funding from the European Union Next-GenerationEU (PIANO NAZIONALE DI RIPRESA E RESILIENZA (PNRR) MISSIONE 4 COMPONENTE 2, INVESTIMENTO 1.3 D.D. 1555 11/10/2022, PE00000013). This manuscript reflects only the authors' views and opinions, neither the European Union nor the European Commission can be considered responsible for them. We acknowledge the CINECA award under the ISCRA initiative for the availability of high-performance computing resources and support.

References

1. Björck, Å.: Numerical methods for least squares problems. SIAM (2024)
2. Boyd, S.: Convex Optimization. Cambridge UP (2004)
3. Caron, M., Touvron, H., Misra, I., Jégou, H., Mairal, J., Bojanowski, P., Joulin, A.: Emerging properties in self-supervised vision transformers. In: ICCV (2021)
4. Chen, X., Fan, H., Girshick, R., He, K.: Improved baselines with momentum contrastive learning. arXiv preprint arXiv:2003.04297 (2020)
5. Deng, J., Dong, W., Socher, R., Li, L.J., Li, K., Fei-Fei, L.: ImageNet: a large-scale hierarchical image database. In: CVPR (2009)
6. Dery, L., Grangier, D., Hannun, A.: Transfer learning for structured pruning under limited task data. In: NeurIPS Workshop (2023)
7. Devlin, J., Chang, M.W., Lee, K., Toutanova, K.: BERT: pre-training of deep bidirectional transformers for language understanding. In: ACL (2019)
8. Dosovitskiy, A., et al.: An image is worth 16×16 words: transformers for image recognition at scale. In: ICLR (2021)
9. Eschenhagen, R., Immer, A., Turner, R., Schneider, F., Hennig, P.: Kronecker-factored approximate curvature for modern neural network architectures. In: NeurIPS (2023)
10. Fanì, E., Camoriano, R., Caputo, B., Ciccone, M.: Accelerating heterogeneous federated learning with closed-form classifiers. In: ICML (2024)
11. Frankle, J., Carbin, M.: The lottery ticket hypothesis: finding sparse, trainable neural networks. In: ICLR (2019)
12. Frantar, E., Alistarh, D.: SparseGPT: massive language models can be accurately pruned in one-shot. In: ICML (2023)
13. Gordon, M.A., Duh, K., Andrews, N.: Compressing BERT: studying the effects of weight pruning on transfer learning. In: ACL Workshop (2020)
14. Hassibi, B., Stork, D.: Second order derivatives for network pruning: optimal brain surgeon. In: NeurIPS (1992)
15. He, K., Zhang, X., Ren, S., Sun, J.: Deep residual learning for image recognition. In: CVPR (2016)
16. Huang, H., Ma, X., Erfani, S.M., Bailey, J., Wang, Y.: Unlearnable examples: making personal data unexploitable. In: ICLR (2021)

17. Iurada, L., Ciccone, M., Tommasi, T.: Finding lottery tickets in vision models via data-driven spectral foresight pruning. In: CVPR (2024)
18. Iurada, L., Ciccone, M., Tommasi, T.: Efficient model editing with task-localized sparse fine-tuning. In: ICLR (2025)
19. Kang, B., Xie, S., Rohrbach, M., Yan, Z., Gordo, A., Feng, J., Kalantidis, Y.: Decoupling representation and classifier for long-tailed recognition. In: ICLR (2020)
20. LeCun, Y., Denker, J., Solla, S.: Optimal brain damage. In: NeurIPS (1989)
21. Lee, J.M.: Riemannian manifolds: an introduction to curvature, vol. 176. Springer Science & Business Media (2006)
22. Lee, N., Ajanthan, T., Torr, P.H.: SNIP: single-shot network pruning based on connection sensitivity. In: ICLR (2019)
23. Liu, B., Cai, Y., Guo, Y., Chen, X.: TransTailor: pruning the pre-trained model for improved transfer learning. In: AAAI (2021)
24. Martens, J., Grosse, R.: Optimizing neural networks with Kronecker-factored approximate curvature. In: ICML (2015)
25. Miller, G.A.: WordNet: a lexical database for English. In: ACM (1995)
26. Patil, S.M., Dovrolis, C.: PHEW: constructing sparse networks that learn fast and generalize well without training data. In: ICML (2021)
27. Radford, A., et al.: Learning transferable visual models from natural language supervision. In: ICML (2021)
28. Rifkin, R., Yeo, G., Poggio, T.: Regularized least-squares. Adv. Learn. Theory: Methods, Models Appl. **190**, 131 (2003)
29. Spadaro, G., et al.: Shannon strikes again! entropy-based pruning in deep neural networks for transfer learning under extreme memory and computation budgets. In: ICCV Workshop (2023)
30. Sun, M., Liu, Z., Bair, A., Kolter, J.Z.: A simple and effective pruning approach for large language models. In: ICLR (2024)
31. Tanaka, H., Kunin, D., Yamins, D.L., Ganguli, S.: Pruning neural networks without any data by iteratively conserving synaptic flow. In: NeurIPS (2020)
32. Wang, C., Zhang, G., Grosse, R.: Picking winning tickets before training by preserving gradient flow. In: ICLR (2020)
33. Wang, S., Veldhuis, R., Brune, C., Strisciuglio, N.: What do neural networks learn in image classification? A frequency shortcut perspective. In: ICCV (2023)
34. Wang, Y., Li, D., Sun, R.: NTK-SAP: improving neural network pruning by aligning training dynamics. In: ICLR (2023)
35. Wu, Y., Liu, F., Chrysos, G., Cevher, V.: On the convergence of encoder-only shallow transformers. In: NeurIPS (2023)

Depth and Event-Based Approaches for Human Detection and Pose Estimation

Hao Quan[1(✉)], Milad Goudarzi[1], Chiara Plizzari[1], Simone Mentasti[1], Francesca Palermo[2], Diana Trojaniello[2], and Matteo Matteucci[1]

[1] Politecnico di Milano, Milan, Italy
hao.quan@polimi.it
[2] EssilorLuxottica, Milan, Italy

Abstract. RGB cameras, while widely used for human detection and pose estimation, face challenges in real-world applications due to sensitivity to illumination changes, low-light conditions, and motion blur. Moreover, deploying state-of-the-art models is hindered by privacy concerns and the high computational demands of RGB images, making them less suitable for real-time, resource-constrained environments. This paper investigates depth and event-based cameras as alternatives to RGB for human detection and pose estimation. These modalities offer advantages such as improved low-light performance and reduced privacy concerns. Event cameras, which capture pixel-level intensity changes at high temporal resolution, minimize motion blur and consume significantly less power and memory than RGB cameras, making them ideal for edge applications. We simulate depth and event-based data using the MS COCO dataset and retrain state-of-the-art models. Results show depth data performs similarly to RGB, while event-based data, though slightly less accurate, remains competitive. These findings highlight the potential of depth and event-based cameras for efficient, privacy-preserving human detection and pose estimation in real-world applications.

Keywords: Human detection · Human Pose estimation · Event-based Cameras · Depth

1 Introduction

Human detection and pose estimation are critical computer vision tasks that enable technologies to understand and interact with the physical world. These tasks have diverse applications, including autonomous vehicles [26,29] and remote sensing systems [19,31]. By observing human position, motion, and actions in the surroundings, these technologies enhance user awareness of their environment, improving overall user experience.

Over the past decade, deep learning-based models, particularly the YOLO (You Only Look Once) [22] and YOLO-Pose [14] families, have gained popularity for their speed and accuracy in real-time human detection and pose estimation.

E. Rodolà et al. (Eds.): ICIAP 2025, LNCS 16167, pp. 207–219, 2026.
https://doi.org/10.1007/978-3-032-10185-3_17

However, state-of-the-art person detection and pose estimation models face significant challenges when deployed in real-world scenarios, such as lighting variations, complex backgrounds, occlusions, particularly on edge devices like smart eyewear [18]. Wearable devices are often subject to strong variations in lighting conditions, including low-light environments [13], which can hinder the effectiveness of RGB-based models. Alternative data sources, including depth sensors and event-based cameras [6], present promising solutions to these challenges. Figure 1 provides a comparison of the RGB, depth, and event-based modalities for the same scene. These modalities offer distinct advantages, including improved performance in low-light and complex background conditions, lower power consumption compared to RGB cameras, and a reduced risk of revealing sensitive visual details, making them more suitable for privacy persevering applications. However, research in this domain remains limited, with notable gaps in publicly available datasets, optimized methodologies, and practical deployment strategies.

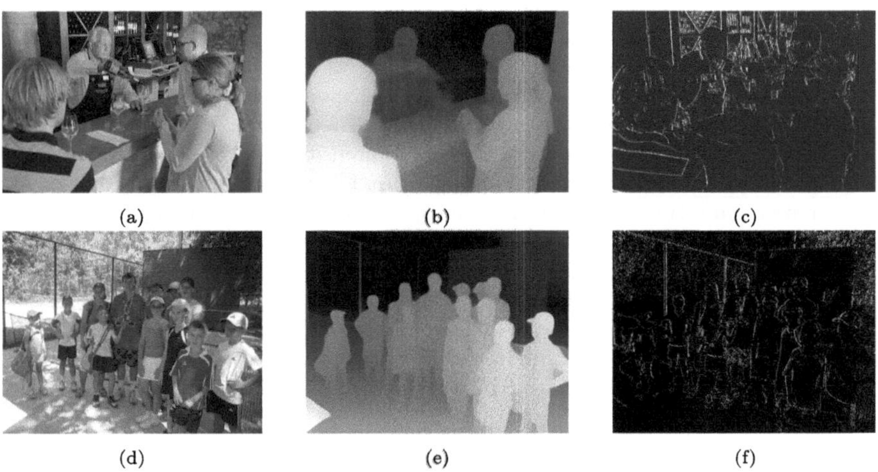

Fig. 1. Visualization of the same scene from the COCO dataset [11] captured in three modalities: (a)(d) RGB provides a standard visual representation, (b)(e) depth encodes object distances from the camera, and (c)(f) the event-based modality detects brightness changes over time, enhancing dynamic scene perception.

This paper investigates how these unconventional modalities can achieve performance comparable to RGB-based approaches for human detection and pose estimation, while offering additional benefits such as lower power consumption, robustness to complex visual conditions and enhanced privacy. We used the large-scale MS COCO object detection dataset [11] as our benchmark. To simulate the depth and event-based data modalities, we employed Depth Anything [30] and the Prophesee Metavision SDK [21], respectively. We then re-trained YOLO and YOLO-Pose 5 on various modalities of the COCO dataset, including RGB, depth, and event-based data, for the human detection and human pose estimation tasks respectively. Results show that the performance of the depth-based modality is on-par with that of the traditional RGB modality, while the performance of the event-based modality is slightly lower than RGB. However, event-based cameras still provided competitive results, all while offering the advantages, such as robustness to illumination changes and complex backgrounds and lower power consumption.

The key contributions of this work are:

- We simulated event-based and depth data from the RGB images in the MS COCO dataset and released it to the research community, facilitating further advancements in human detection and human pose estimation with multiple modalities.
- We benchmarked YOLO / YOLO-Pose 5 human detection and pose estimation algorithms across different image modalities, including RGB, depth, and event-based data using the MS COCO dataset.
- Our experiments demonstrated that re-trained YOLO and YOLO-Pose models perform well on depth and event-based data from the simulated MS COCO datasets, making them promising for future work in human detection and human pose estimation tasks.

2 Related Works

2.1 Human Detection and Pose Estimation

Human Detection. Human detection has seen considerable advancements with various deep learning-based methodologies. Generally, these techniques can be categorized into two main approaches: region proposal-based two-stage methods and one-stage methods. Two-stage methods, such as R-CNN [7] and Faster R-CNN [23], first generate region proposals and then classify them, getting high detection accuracy at the cost of increased computational complexity. On the other hand, YOLO (You Only Look Once) [22] algorithm is a popular one-stage detection framework that integrates region proposal and classification into a single step, prioritizing speed and simplicity. Other notable one-stage method include Single-Shot Refinement Neural Network (RefineDet) [34], which improves detection accuracy by refining predictions. Transformer-based approaches such as DETR [3] use self-attention for solving detection tasks in an end-to-end manner, directly predicting bounding boxes by modelling relatinships between image regions without need for hand-crafted anchors or region proposals.

Human Pose Estimation. Human pose estimation algorithms divide into 2D and 3D approaches. However, the limited computing resources of edge devices make 2D methods more suitable for edge devices. Among those, OpenPose [2] and HRNet [4] have established benchmarks in the field by using multistage architectures and high-resolution feature maps to predict human keypoints with high accuracy. YOLO-Pose [14] extends the capabilities of the YOLO object detection framework by integrating pose estimation into a unified architecture, enabling simultaneous detection and pose estimation.

In this paper, we use YOLO and YOLO-Pose due to their efficiency and adaptability, which make them well-suited for real-time applications and deployment on edge devices. Furthermore, we leverage these algorithms to explore the potential of human detection and pose estimation on alternative data modalities beyond RGB, such as depth and event-based data.

2.2 Event-Based Human Detection and Pose Estimation Tasks

While conventional human detection is well-established in computer vision, *event-based human detection* remains in its early stages. It holds great promise for real-time applications due to the low latency and power efficiency of event cameras, but also poses new challenges due to the shift from frame-based to event-based data. Morever, due to the challenges in data collection and annotation, the availability of annotated event-based datasets is significantly limited compared to other common modalities, limiting the research. PEDRo [1] offers a large-scale event-based human detection dataset from mobile robots, which was collected from public spaces with different lighting and meteorological conditions. eTraM [28] provides a novel event-based dataset for traffic monitoring that includes human detection annotations. Besides the event-based modality, MMPedestron [35] provides multi-modal data sources (RGB, IR, Depth, and LiDAR) for pedestrian detection. All the above approaches explore event-based human detection by converting event streams into frame-like representations followed by standard detection algorithms.

Recent works have adapted state-of-the-art *pose estimation models for event-based data.* EventHPE [36] is a two-stage deep learning framework for event-based 3D human pose and shape estimation. The first FlowNet stage infers optical flow from events, while the second ShapeNet stage estimates 3D shapes using both event data and flow. Yu et al. [32] proposed an adaptive vision transformer with adaptive patch sampling and adaptive token reduction, optimizing efficiency while maintaining accuracy. They also provided a large-scale event-based human pose estimation dataset EventMM HPE.

While event-based human detection and pose estimation have typically been treated as separate tasks, we propose a unified framework that jointly tackles both, leveraging not only RGB, but also depth and event-based data for robust performance in diverse real-world scenarios. To support this effort, we introduce a large-scale dataset specifically designed to enable the joint study of event-based human detection and pose estimation. Unlike prior works that focus mainly on pedestrian detection in constrained or driving environments, our dataset captures a broad range of contexts, covering varied activities and settings.

3 Event-Based Vision

Event-Based Data. Unlike traditional cameras that capture frames at fixed intervals, event cameras asynchronously record brightness changes at each pixel. Pixels in event cameras [6] operate independently and respond to variations in the continuous log-brightness function $L(\mathbf{u}, t)$, where $\mathbf{u} = (x_k, y_k)^T$ represents the spatial coordinates of a pixel, and t denotes the timestamp at which the brightness change occurs. An event is represented as a tuple $e_k = (x_k, y_k, t_k, p_k)$, where (x_k, y_k) denotes the pixel location, t_k is the timestamp, and $p_k \in \{-1, 1\}$ represents the polarity of the brightness change (indicating increase or decrease). An event is triggered whenever the change in log-brightness at pixel $\mathbf{u} = (x_k, y_k)^T$ exceeds a predefined threshold C, given that a sufficient time interval has elapsed since the last event at the same pixel. This is mathematically formulated as:

$$\Delta L(\mathbf{u}, t_k) = L(\mathbf{u}, t_k) - L(\mathbf{u}, t_k - \Delta t_k) \geq p_k C. \tag{1}$$

Thus, the output of an event-based camera consists of a continuous stream of events, represented as the sequence:

$$\mathcal{E} = \{(x_k, y_k, t_k, p_k) \mid t_k \in \tau\}, \tag{2}$$

where τ defines the time interval within which events occur.

Event-Based Representation. To generate a frame at a precise time t_k, events should be accumulated over a longer period of time (between the time $t_k - \Delta t_k$ and time t_k). Note that Δt_k is usually called accumulation time. To generate a frame with the accumulated events, for each event occurring between the time $t_k - \Delta t_k$ and time t_k, a white pixel is stored if the polarity of the event is positive, and a grey pixel if the polarity is negative. As it can be seen in Fig. 1 the background is set to black. This is the so-called *event counts* representation [15]. The resulting representation is an image of size $H \times W$.

4 Methodology

This work focuses on human detection and pose estimation from RGB, depth, and event-based images. In this section, we first define the tasks (Sect. 4.1) and describe the dataset used (Sect. 4.2), followed by an overview of the pipeline and models (Sect. 4.3). This section also includes a description of our approach to simulate depth and event-based data, as well as the evaluation metrics used.

4.1 Task Definition

Human Detection. Human detection aims to identify and localize individuals within a given input image. The model takes a two-dimensional image as input and predicts the regions containing humans by outputting bounding boxes defined by four coordinates $(x_{\min}, y_{\min}, x_{\max}, y_{\max})$. Additionally, each detection is associated with a confidence score ranging from 0 to 1, indicating the likelihood of the detected region containing a person.

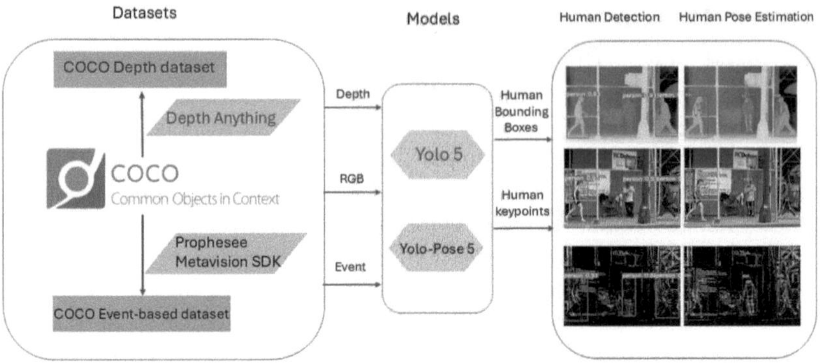

Fig. 2. Starting from the MS COCO RGB dataset, we simulate depth data generated using Depth Anything, and event data with the Prophesee Metavision SDK [21]. We then use YOLO 5 and YOLO-Pose 5 for producing human detection and pose estimation results across depth, RGB, and event-based modalities.

Human Pose Estimation. Human pose estimation involves predicting the 2D coordinates of 17 keypoints corresponding to human joints from images. Given an input image, the model estimates a predefined set of keypoints, such as shoulders, elbows, and knees, along with their respective confidence scores. The output consists of keypoint coordinates (x_i, y_i) for each detected individual, typically visualized as a skeletal structure.

4.2 Datasets

We use the MS COCO dataset [11] as the baseline for our experiments. It includes 118,287 training images, 5,000 validation images, and 40,670 test images, with a total of 164,957 images spanning 80 object categories. Each image contains multiple object instances, and for human-related tasks, the dataset provides 17 keypoint annotations per person, covering joints such as the head, shoulders, elbows, wrists, hips, knees, and ankles–enabling detailed pose estimation.

Building on this, we introduce depth-based and event-based versions of the dataset, empowering the community to explore human detection and pose estimation using multiple modalities.

4.3 Pipeline

Overview. The complete pipeline of the methodology is shown in Fig. 2. Specifically, starting from the original MS COCO RGB dataset, we introduced two simulated datasets: a MS COCO Depth dataset and a MS COCO Event-based dataset. These datasets enable a comprehensive evaluation of human detection and pose estimation under diverse visual inputs. We employ YOLO [22] and YOLO-Pose [14] in all experiments, as they offer a unified framework for both human detection and pose estimation–making them well-suited for our joint task.

We test these models across RGB, depth, and event-based modalities, demonstrating the adaptability of our approach to different data modalities.

Model Description. You Only Look Once (YOLO) [22] is a real-time object detection algorithm that performs object localization and classification in a single forward pass of a deep neural network. Unlike region-based approaches, YOLO formulates detection as a regression problem, dividing the input image into an $S \times S$ grid, where each cell predicts B bounding boxes, objectness scores, and class probabilities. The final detection is obtained by filtering predictions using non-maximum suppression [17]. This one-stage detection framework enables high-speed inference while maintaining competitive accuracy, making it suitable for applications. Given an input image I, YOLO predicts a set of bounding boxes $\mathcal{B}$, where each box b_i is defined as:

$$b_i = (x_i, y_i, w_i, h_i, c_i),$$

where (x_i, y_i) represents the center coordinates, (w_i, h_i) the width and height, and c_i the confidence score

YOLO-Pose [14] extends the YOLO detection framework to predict keypoints for human pose estimation. In addition to detecting objects, the model regresses multiple keypoint coordinates per detected entity, enabling real-time multi-person pose estimation. Given an input image I, YOLO Pose predicts a set of bounding boxes $\mathcal{B}$, where each detected entity b_i is defined as:

$$b_i = (x_i, y_i, w_i, h_i, c_i, K_i),$$

where (x_i, y_i) represents the bounding box center coordinates, (w_i, h_i) the width and height, c_i the confidence score, and $K_i = \{(k_{x1}, k_{y1}), \ldots, (k_{xN}, k_{yN})\}$ represents the set of N keypoints associated with the object. Each keypoint $k_j = (k_{xj}, k_{yj})$ corresponds to a specific body joint.

Model Selection. The selection of the specific model versions was based on specific criteria: (1) implementation in plain PyTorch without proprietary dependencies, excluding YOLO 9–11 [27]; (2) consistency between YOLO and YOLO-Pose to streamline modifications, ruling out YOLO 6 due to the lack of a publicly released YOLO-Pose 6 [10]; and (3) community validation, favoring YOLO/YOLO-Pose 5, which was released at CVPR 2022 [14] and includes optimized versions for edge devices. YOLO/YOLO-Pose 8 lacks similar validation. After evaluating available versions, YOLO/YOLO-Pose 5 was selected for its balance of precision, parameters, and model size, making it ideal for real-time human detection and pose estimation on edge devices.

Data Simulation. Since the original MS COCO dataset does not include depth or event-based modalities, we generated corresponding versions by utilizing Depth Anything [30] for depth simulation and the Prophesee library [21] for event-based modality. We make both datasets publicly available to the community, with the aim of supporting further research and development in human detection and pose estimation using non-RGB modalities.

Metrics. For human detection, we use mean Average Precision (mAP) 0.5–0.95 as the evaluation metric. For human pose estimation, we also employed mean Average Precision (mAP) as the evaluation metric, but with a key difference in its computation. Instead of using bounding boxes, mAP for pose estimation is based on the alignment of predicted keypoints with ground truth keypoints, evaluated using the Object Keypoint Similarity (OKS) metric [16].

5 Experiments

In this section, we describe the experiments conducted to evaluate the proposed methods. We begin with implementation details (Sect. 5.1), followed by results across depth, event-based, and RGB modalities on the MS COCO dataset [11] (Sect. 5.2). Finally, we present some qualitative results (Sect. 5.3).

5.1 Implementation Details

The hyperparameters used in the experiments are consistent with the original YOLO 5 and Yolo-Pose 5 settings on the COCO RGB modality. Specifically, the models were trained using a batch size of 16 and SGD optimizer [24] with a one-cycle learning rate policy [25], where the learning rate follows a cosine decay schedule [12]. The initial learning rate was set to 0.01, reaching a peak of 0.002 before gradually decreasing throughout the training process.

All experiments were conducted using an NVIDIA RTX A6000 GPU. We used the models with best mAP during training for validation.

5.2 Results

Depth. Table 1 presents the performance on the human detection task of the YOLO 5 s, YOLO 5 m, and YOLO 5l models on both RGB data and the simulated depth version of the MS COCO dataset, as described in Sect. 4.3. For the COCO RGB dataset, the models achieved mAP scores of 66.22%, 69.15%, and 70.16%, respectively. When evaluated on the depth modality, the mAP scores across IoU thresholds from 0.5 to 0.95 were 64.3%, 66.28%, and 66.58%, respectively. Similarly, experiments on pose estimation task with YOLO-Pose 5 s, YOLO-Pose 5 m, and YOLO-Pose 5l were conducted on the simulated depth COCO dataset. The mAP scores at IoU thresholds from 0.5 to 0.95 were 61.29%, 62.1%, and 65.15%, compared to 63.44%, 67.06%, and 69.47% on the MS COCO RGB dataset for YOLO-Pose 5 s, YOLO-Pose 5 m, and YOLO-Pose 5l, respectively. For the smaller model YOLO 5 s, the performance gap between RGB and depth is only around 2%. As the model size increases, the gap becomes slightly more pronounced. For example, in the case of YOLO 5l, the performance difference between RGB and depth increases to approximately 4%. These trends are consistent across YOLO-Pose models, with larger models also showing a slightly larger gap between modalities. These results indicate that while the performance of YOLO and YOLO-Pose models on the depth modality of the MS COCO dataset is slightly lower than their performance on the RGB modality, the differences are minimal, indicating comparable effectiveness across modalities.

Table 1. mAP results of re-training and evaluating human detection YOLO 5 and human pose estimation YOLO-Pose 5 on different modalities of the COCO dataset.

Model	RGB (%)	Depth (%)	Event (%)
Human Detection			
YOLO 5 s	66.22	64.30	44.45
YOLO 5 m	69.15	66.28	47.90
YOLO 5l	70.16	66.58	48.82
Human Pose Estimation			
YOLO-Pose 5 s	63.44	61.29	43.03
YOLO-Pose 5 m	67.06	62.10	47.30
YOLO-Pose 5l	69.47	65.15	49.14

Event. In Table 1 we also present results obtained on the simulated event-based data. We re-trained and evaluated YOLO 5 s, YOLO 5 m, and YOLO 5l on the simulated event-based MS COCO dataset, obtaining mAP scores 44.45%, 47.9%, and 48.82%, respectively. When compared to performance on the MS COCO RGB dataset, the models achieved mAP scores of 66.22%, 69.15%, and 70.16%, respectively. The mAP scores for YOLO-Pose 5 s, YOLO-Pose 5 m, and YOLO-Pose 5l were 43.03%, 47.3%, and 49.14%, respectively, which were slightly lower than the results on the RGB modality, where the scores were 63.44%, 67.06%, and 69.47%, respectively. Notably, the event modality has historically lagged significantly behind RGB in classification tasks. For instance, the recent N-ImageNet benchmark [9] reports an accuracy of 48.94% for the best-performing event-based architecture, markedly lower than the >90% accuracy achieved by RGB models [5,8,20,33]. Our results indicate that, although YOLO and YOLO-Pose perform worse on event-based data compared to RGB data, the performance gap is not substantial. As demonstrated in the qualitative results in the next section, event-based data remains a competitive modality for human detection and pose estimation tasks.

5.3 Qualitative Results

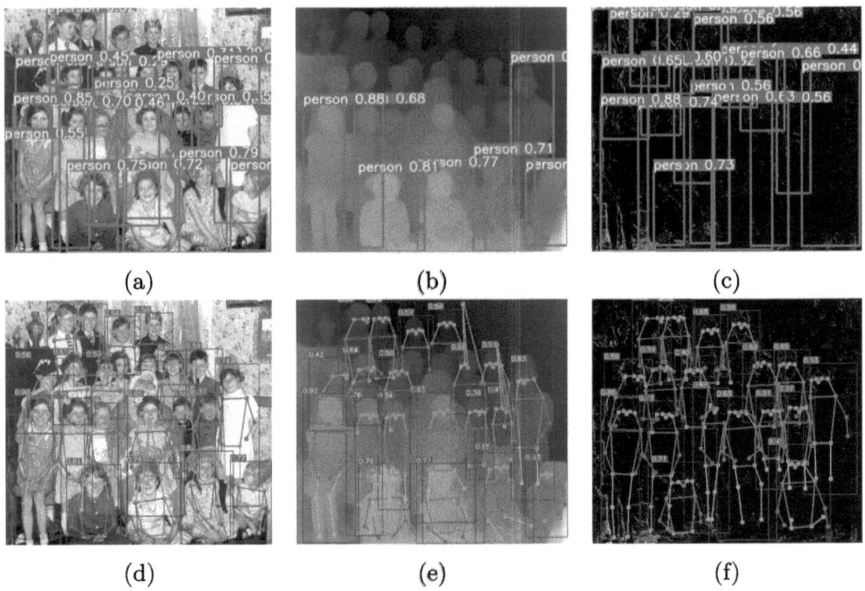

Fig. 3. Results from our re-trained YOLO 5 model for human detection (top row: a-c) and YOLO-Pose 5 for pose estimation (bottom row: d-f) on a MS COCO scene across RGB, depth, and event modalities. Red bounding boxes and scores show detected people. Blue boxes, green keypoints, and colored skeletons visualize estimated poses. (Color figure online)

For these analyses, we used our re-trained YOLO 5 and YOLO-Pose 5 models, which achieved the highest mAP scores on the RGB, depth, and event-based modalities of the MS COCO validation set (Sect. 5.2). These models were applied for human detection and pose estimation on the same MS COCO images represented in RGB (Fig. 3a, Fig. 3d), depth (Fig. 3b, Fig. 3e), and event-based (Fig. 3c, Fig. 3f) formats. The results show strong performance in RGB for both tasks. In the depth modality, accuracy remains high with few people, but in crowded scenes, pose estimation detects more individuals (Fig. 3e) than the detection model (Fig. 3b). However, YOLO-Pose may misassign head joints when people are close together, as seen in the upper right of Fig. 3e, highlighting challenges in resolving closely positioned individuals. While Table 1 indicates that event-based data yields lower mAP than RGB and depth, qualitative results remain competitive, especially as crowd size increases (Fig. 3c, Fig. 3f). These findings support depth and event-based modalities as effective alternatives for human detection and pose estimation.

6 Conclusion

In this work, we explored the potential of various data modalities for human detection and pose estimation. Traditional RGB data struggles in challenging lighting and complex backgrounds, so we investigated alternatives like depth and event-based data. Our experiments showed that re-trained YOLO and YOLO-Pose models perform well on non-RGB data modalities, including depth and event-based data, using the MS COCO dataset. We conclude that depth and event-based data can effectively handle lighting and background challenges, making them viable alternatives to RGB for these tasks.

Acknowledgements. This work was conducted at Smart Eyewear Lab, a joint research center between EssilorLuxottica and Politecnico di Milano.

References

1. Boretti, C., Bich, P., Pareschi, F., Prono, L., Rovatti, R., Setti, G.: PEDRo: an event-based dataset for person detection in robotics. In: Proceedings of the IEEE/CVF Conference on Computer Vision and Pattern Recognition, pp. 4065–4070 (2023)
2. Cao, Z., Simon, T., Wei, S.E., Sheikh, Y.: Realtime multi-person 2D pose estimation using part affinity fields. In: Proceedings of the IEEE Conference on Computer Vision and Pattern Recognition, pp. 7291–7299 (2017)
3. Carion, N., Massa, F., Synnaeve, G., Usunier, N., Kirillov, A., Zagoruyko, S.: End-to-end object detection with transformers. In: European Conference on Computer Vision, pp. 213–229. Springer (2020)
4. Cheng, B., Xiao, B., Wang, J., Shi, H., Huang, T.S., Zhang, L.: HigherHRNet: scale-aware representation learning for bottom-up human pose estimation. In: Proceedings of the IEEE/CVF Conference on Computer Vision and Pattern Recognition, pp. 5386–5395 (2020)
5. Dai, Z., Liu, H., Le, Q.V., Tan, M.: CoAtNet: marrying convolution and attention for all data sizes. Adv. Neural. Inf. Process. Syst. **34**, 3965–3977 (2021)
6. Gallego, G., et al.: Event-based vision: a survey. TPAMI **44**(1), 154–180 (2020)
7. Girshick, R., Donahue, J., Darrell, T., Malik, J.: Rich feature hierarchies for accurate object detection and semantic segmentation. In: Proceedings of the IEEE Conference on Computer Vision and Pattern Recognition, pp. 580–587 (2014)
8. He, K., Chen, X., Xie, S., Li, Y., Dollár, P., Girshick, R.: Masked autoencoders are scalable vision learners. In: CVPR, pp. 16000–16009 (2022)
9. Kim, J., Bae, J., Park, G., Zhang, D., Kim, Y.M.: N-ImageNet: towards robust, fine-grained object recognition with event cameras. In: Proceedings of the IEEE/CVF International Conference on Computer Vision, pp. 2146–2156 (2021)
10. Li, C., et al.: YOLOv6: a single-stage object detection framework for industrial applications. arXiv preprint arXiv:2209.02976 (2022)
11. Lin, T.Y., et al.: Microsoft COCO: common objects in context. In: ECCV, pp. 740–755. Springer (2014)
12. Loshchilov, I., Hutter, F.: SGDR: stochastic gradient descent with warm restarts. arXiv preprint arXiv:1608.03983 (2016)

13. Mai, C., et al.: DBCG-Net: dual branch calibration guided deep network for UAV images semantic segmentation. IEEE J. Sel. Top. Appl. Earth Observ. Remote Sens. (2024)
14. Maji, D., Nagori, S., Mathew, M., Poddar, D.: YOLO-Pose: enhancing yolo for multi person pose estimation using object keypoint similarity loss. In: CVPR, pp. 2637–2646 (2022)
15. Maqueda, A.I., Loquercio, A., Gallego, G., García, N., Scaramuzza, D.: Event-based vision meets deep learning on steering prediction for self-driving cars. In: Proceedings of the IEEE Conference on Computer Vision and Pattern Recognition, pp. 5419–5427 (2018)
16. Microsoft: COCO keypoint detection challenge. https://cocodataset.org/#keypoints-2020 (2020). Accessed 20 Jan 2025
17. Neubeck, A., Van Gool, L.: Efficient non-maximum suppression. In: 18th International Conference on Pattern Recognition (ICPR'06), vol. 3, pp. 850–855. IEEE (2006)
18. Palermo, F., et al.: Advancements in context recognition for edge devices and smart eyewear: sensors and applications. IEEE Access (2025)
19. Pan, J., et al.: MSFA-Net: multiple spatial-channel feature aggregation network for change detection and a UAV-CD dataset. In: IGARSS, pp. 10328–10332. IEEE (2024)
20. Pham, H., Dai, Z., Xie, Q., Le, Q.V.: Meta pseudo labels. In: CVPR, pp. 11557–11568 (2021)
21. Prophesee: https://www.prophesee.ai/ (2024)
22. Redmon, J.: You only look once: unified, real-time object detection. In: Proceedings of the IEEE Conference on Computer Vision and Pattern Recognition (2016)
23. Ren, S., He, K., Girshick, R., Sun, J.: Faster R-CNN: towards real-time object detection with region proposal networks. IEEE Trans. Pattern Anal. Mach. Intell. **39**(6), 1137–1149 (2016)
24. Robbins, H., Monro, S.: A stochastic approximation method. Ann. Math. Stat., 400–407 (1951)
25. Smith, L.N.: A disciplined approach to neural network hyper-parameters: Part 1–learning rate, batch size, momentum, and weight decay. arXiv preprint arXiv:1803.09820 (2018)
26. Song, Z., et al.: Robustness-aware 3D object detection in autonomous driving: a review and outlook. TPAMI (2024)
27. Ultralytics: YOLO. https://www.ultralytics.com/ (2025). Accessed 12 Jan 2025
28. Verma, A.A., Chakravarthi, B., Vaghela, A., Wei, H., Yang, Y.: eTraM: event-based traffic monitoring dataset. In: CVPR, pp. 22637–22646 (2024)
29. Xian, T., et al.: A scale-temporal interaction network for remote sensing image change detection and a UAV-CD dataset. In: IGARSS, pp. 8603–8607. IEEE (2024)
30. Yang, L., Kang, B., Huang, Z., Xu, X., Feng, J., Zhao, H.: Depth anything: unleashing the power of large-scale unlabeled data. In: CVPR, pp. 10371–10381 (2024)
31. Ying, Z., et al.: Large-scale high-altitude UAV-based vehicle detection via pyramid dual pooling attention path aggregation network. IEEE Trans. Intell. Transp. Syst. (2024)
32. Yu, N., et al.: Adaptive vision transformer for event-based human pose estimation. In: Proceedings of the 32nd ACM International Conference on Multimedia, pp. 2833–2841 (2024)
33. Zhai, X., Kolesnikov, A., Houlsby, N., Beyer, L.: Scaling vision transformers. arXiv preprint arXiv:2106.04560 (2021)

34. Zhang, S., Wen, L., Bian, X., Lei, Z., Li, S.Z.: Single-shot refinement neural network for object detection. In: CVPR, pp. 4203–4212 (2018)
35. Zhang, Y., Zeng, W., Jin, S., Qian, C., Luo, P., Liu, W.: When pedestrian detection meets multi-modal learning: Generalist model and benchmark dataset. In: ECCV, pp. 430–448. Springer (2025)
36. Zou, S., et al.: EventHPE: event-based 3D human pose and shape estimation. In: CVPR, pp. 10996–11005 (2021)

Comparative Analysis of Lightweight Vision Models on Embedded Accelerator Devices

Greta Corti[1(✉)], Christian Veronesi[1], Pietro Bartoli[1], Sara Pidò[1], Andrea Giudici[1], Francesca Palermo[2], Diana Trojaniello[2], Matteo Matteucci[1], and Simone Mentasti[1]

[1] Politecnico di Milano, 20133 Milan, Italy
{greta.corti,christian.veronesi,pietro.bartoli,sara.pido,andrea.giudici, matteo.matteucci,simone.mentasti}@polimi.it
[2] EssilorLuxottica, 20133 Milan, Italy
{francesca.palermo,diana.trojaniello}@luxottica.com

Abstract. The rapid growth of edge devices demands efficient neural network accelerators for low-latency, energy-efficient AI tasks like object detection and pose estimation. Despite benefits such as improved privacy and reduced cloud dependence, edge AI faces challenges including hardware heterogeneity, resource constraints, and the balance between real-time performance and accuracy. This paper benchmarks STMicroelectronics STM32N6, Luxonis OAK-FCC 4P, and SONY IMX500 on object detection and human pose estimation tasks, evaluating power consumption, inference time, accuracy, and memory usage. Using YOLOv5 and YOLOv8 models in Small and Nano configurations on COCO and COCO-Pose datasets, the study highlights the trade-offs between efficiency and accuracy. Results show IMX500 leads in inference time and power efficiency but is limited in memory compared to N6 and OAK-FCC 4P. This study provides a comprehensive evaluation of the suitability of different hardware architectures, guiding the optimization of lightweight neural network deployment on resource-constrained platforms.

Keywords: Lightweight Neural Network · Neural Network Accelerators · Hardware Benchmarking · Low-Power Edge Devices

1 Introduction

The increasing availability of edge devices, e.g., IoT sensors, mobile phones, drones, has driven the demand for low-latency, efficient neural network accelerators capable of enabling real-time AI tasks such as object detection and pose estimation [14,38], reducing reliance on cloud computing [37] and addressing privacy concerns [20]. However, deploying AI on the edge introduces challenges, notably hardware heterogeneity, limited memory, restricted power budgets, and the difficulty of achieving high accuracy under strict resource constraints [10,24]. Selecting an appropriate neural accelerator thus requires careful consideration of these

E. Rodolà et al. (Eds.): ICIAP 2025, LNCS 16167, pp. 220–233, 2026.
https://doi.org/10.1007/978-3-032-10185-3_18

(a) Luxonis OAK-FCC 4P (b) STMicroelectronics STM32N6 (c) SONY IMX500

Fig. 1. Target devices selected for benchmarking analysis.

trade-offs. For example, certain edge devices, such as the Luxonis OAK-FCC 4P board [22], shown in Fig. 1a, and STMicroelectronics STM32N6 [32], shown in Fig. 1b, are designed to facilitate efficient processing and minimize power consumption through the incorporation of dedicated neural compute engines. These devices are designed to handle real-time inference for applications such as object detection and pose estimation while operating within strict power budgets. In contrast, other devices prioritize on-chip inference, integrating AI capabilities directly into sensors or compact processors with highly constrained memory footprints, such as the SONY IMX500 camera [30], shown in Fig. 1c.

While the hardware landscape evolves rapidly, there is still limited research exploring their potential and performances of these platforms on state-of-the-art neural network architectures. This highlights a critical gap where further investigation is needed to advance the field and enable more effective and efficient AI deployment on edge devices. To address this, we benchmark three representative edge devices, STM32N6, OAK-FCC 4P, and IMX500, on object detection and human pose estimation using You Only Look Once (YOLO) models (5 and 8 versions, Nano and Small variants), trained on COCO and COCO-Pose datasets. Key contributions of this paper are:

- Benchmarking accuracy, inference speed, memory usage, and power consumption across the three devices using YOLO.
- Providing an in-depth analysis of quantization challenges in YOLO models and proposing possible solutions to address them.
- Providing insights for optimizing lightweight neural networks for deployment on resource-constrained platforms.

2 Related Works

Recent advancements in neural network accelerators have enabled real-time inference on resource-constrained devices, playing a key role in edge computing, IoT, and autonomous systems [9,24]. Efficient hardware-software co-design has become critical to balancing latency, energy consumption, and accuracy. AI accelerators can be implemented in several forms:

- **Dedicated ASICs**: Specialized chips such as Google's Edge TPU [11] and Intel's Movidius Myriad X [15] deliver efficient AI inference.
- **FPGAs**: Configurable circuits that can be repeatedly programmed after manufacturing.
- **Embedded GPUs**: General-purpose processors (e.g., NVIDIA Jetson [26]) able to perform mathematical calculations at high speeds in parallel.

These accelerators are optimized for the execution of parallel AI workloads, including matrix multiplication and convolution, which are prevalent operations in deep learning models. The market offers a variety of artificial intelligence accelerators, each with distinct features and performance characteristics. These accelerators differ in their processing capabilities, power consumption, and supported AI frameworks, underscoring the importance of selecting an appropriate accelerator to align with the specific requirements of the edge AI application.

System-on-Chip. Application-Specific Integrated Circuits (ASICs) are specialized chips designed to perform dedicated tasks, e.g. digital recording [7], video compression [25], neural network inference [18]. Modern ASICs integrate microprocessors, memory, and peripherals, earning the term System-on-Chip (SoC).

Nowadays, there has been a growing trend in energy-efficient SoC designs for real-time neural network models. Among these, Intel's Movidius Myriad X Vision Processing Unit (VPU) [10,29] remains a popular benchmark, although its performance for lightweight object detection and pose estimation remains underexplored relative to newer platforms like STM32N6 and IMX500.

Lightweight Deep Neural Networks. Deep neural networks, powered by GPU acceleration, have achieved state-of-the-art performance in object detection and pose estimation [6,8,28,33,34]. Models such as YOLOv4 [6], YOLOv5 [28], and EfficientDet [34] for object detection, as well as OpenPose [8] and HRNet [33] for human pose estimation, have exhibited remarkable accuracy and speed when executed on high-performance GPU platforms. However, their high computational and memory demands limit their deployment on edge devices. To address this, lightweight networks such as YOLO Nano and YOLO Small offer significant reductions in model size and complexity, enabling real-time inference at the edge. This decrease in parameters, however, is accompanied by a corresponding decline in accuracy. YOLO Nano models reduce parameters by 80–85% compared to Medium configurations, with an associated 5–15% accuracy drop; Small models achieve a 30–40% reduction with a smaller accuracy loss of 3–8%. These models are designed to balance the reduction of accuracy with the acceleration of inference and the reduction of memory usage.

3 Methodology

In this section, we first describe the neural network models utilized in our experiments for edge device comparison. We then compare different edge device architectures, summarized in Table 1.

Table 1. Specifications of the Target Edge Devices (-: values are not available)

Specification	OAK-FCC 4P	STM32N6	IMX500
Memory Available (MB)	323	64	8
TOPS	1.4	0.6	-
Max Power Consumption (W)	4.5	3	3
W/TOPS	3.2	5	-
Supported Frameworks	Caffe, ONNX, TensorFlow	ONNX, TensorFlow	PyTorch, TensorFlow

3.1 YOLO Detection and Pose Models

To evaluate the performance of various accelerators, we employed two key computer vision tasks: object detection and pose estimation, both essential for edge computing where efficiency and accuracy are critical.

YOLO was chosen as the reference model due to its state-of-the-art performance and broad adoption. We used the PyTorch Ultralytics implementation [16] for object detection (YOLOv5 and YOLOv8, Nano and Small versions) and for human pose estimation (YOLOv8 Nano and Small).

Although newer versions of YOLO (e.g., v11, v12) have been recently released, YOLOv5 and YOLOv8 were preferred due to their extensive validation, stable performance, and widespread support in real-world applications. Furthermore, they offer a balanced trade-off between model complexity and hardware compatibility, which is crucial for edge deployment.

The models were resized from 640×640 pixels to 224×224 to fit the memory constraints of the target devices. All models were retrained from the original weights using Adam optimizer [17] with default hyperparameters. Object detection models were trained using COCO (Common Objects in Context) dataset [19], while pose estimation models were trained on a specialized version, COCO-Pose dataset. All models were successfully deployed, although Non-Maximum Suppression was performed on a PC due to hardware limitations.

3.2 Edge AI Hardware Platforms Deployment

The following section provides a comprehensive analysis of the hardware components of the three devices considered in our study: OAK-FCC 4P, STM32N6, and IMX500. Additionally, we present an overview of the deployment pipelines used for the models on each edge device.

Luxonis' OAK-FCC 4P. The OAK-FFC 4P, designed by Luxonis, is a versatile prototyping platform that supports interconnection with various camera modules. It integrates the Movidius Myriad X SoC inside the Robotics Vision Core 2 [23], enabling neural network inference directly on the device without external computation. The board features approximately 4.5 W maximum power consumption, 1.4 TOPS AI capability, and around 323 MB available memory.

The Myriad X VPU includes Intel's Neural Compute Engine, supporting Floating Point (FP) 16 and 8-bit fixed-point operations, and enabling deployment of networks in Caffe [2], TensorFlow [35], and ONNX [27] formats.

Since PyTorch models are not natively supported, a dedicated deployment pipeline was required. Models were first converted to ONNX format using PyTorch tracing, then transformed into a `.blob` file using Luxonis' BlobConverter [1]. BlobConverter optimizes the model into OpenVINO IR format and compiles it into a final BLOB file. The `.blob` was executed on the device using Luxonis' DepthAI Python API [21]. The full pipeline is illustrated in Fig. 2a.

STMicroelectronics' STM32N6. The STM32N6 (N6) is a Microcontroller Unit (MCU) developed by ST-Microelectronics using 16 nm FFC technology. Designed for energy-efficient AI and ML workloads in battery-powered devices, it features an ARM Cortex-M55 CPU operating up to 800 MHz with vector processing, TrustZone, and Arm Helium extensions for DSP and ML efficiency [4]. In addition to its CPU, the N6 integrates ATON, a high-performance CNN accelerator comprising convolution, scalar, activation, pooling, and decompression units, along with a reconfigurable buffer and high-throughput AXI interfaces. An epoch controller further streamlines configuration and reduces CPU load.

As shown in Fig. 2b, the deployment pipeline is similar to that of the OAK-FCC 4P. Models are converted from PyTorch to ONNX, then statically quantized to 8-bit precision using the ONNX Runtime library [3] by inserting QuantizeLinear and DeQuantizeLinear operators. The quantized model is subsequently compiled using ST's STEdgeAI tool, producing a `.c` file for deployment and an AI runtime model to verify whether the model was compiled optimally.

SONY IMX500. The IMX500, developed by SONY, is the first Intelligent Vision Sensor with integrated edge processing. It features a stacked structure combining an image sensor, high-performance DSP, and on-chip static RAM, enabling high-speed AI inference without remote server transmission.

In this study, we used the IMX500 integrated with a Raspberry Pi, following SONY's documentation for specifications and deployment. While the processor's TOPS are unspecified, its power consumption remains below 3 W, with an available memory of 8 MB. The IMX500 supports models in both TensorFlow and PyTorch formats. For deployment, PyTorch models were quantized and compressed using Sony's Model Compression Toolkit (MCT) [31] into ONNX format, then converted into binary files via the IMX500 Converter [12]. The resulting `.zip` file includes a memory report and is packaged into an RPK file using the IMX500 Packager [13], then deployed to the Raspberry Pi camera through a Python application. The complete pipeline is shown in Fig. 2c.

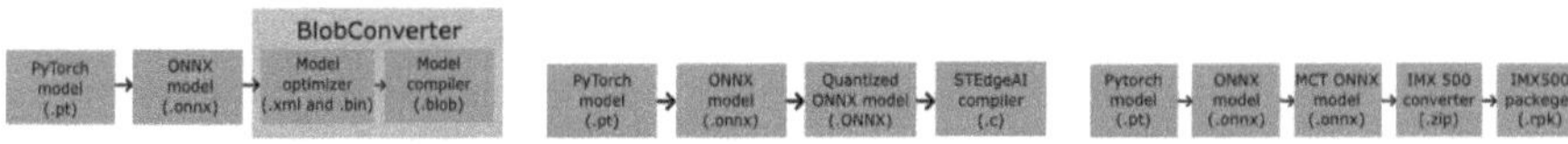

(a) OAK-FCC 4P (b) STM32N6 (c) IMX500

Fig. 2. Target devices deployment pipelines.

4 Experiments and Results

In this section, we compare the available memory, model utilization, and hardware limitations of the three edge devices introduced in Sect. 3.2, followed by an evaluation of inference speed for real-time performance. Next, we assess post-deployment accuracy and possible degradation. Finally, we examine power consumption to evaluate energy efficiency for low-power applications.

4.1 Memory Usage

As shown in Sect. 3.2 and Table 1, memory varies across devices: OAK-FCC 4P supports up to 323 MB, STM32N6 up to 64 MB, and IMX500 up to 8 MB.

Table 2 shows the memory footprint of YOLO models for object detection and human pose estimation across the devices. While the initial model sizes are identical, the deployed memory usage differs due to device-specific compiler optimizations and memory management strategies. Notably, Nano models occupy about 50% of the IMX500's memory, but only around 6% on STM32N6 and 3% on OAK-FCC 4P. The larger footprint on OAK-FCC 4P is due to its compiler using FP16 precision, whereas STM32N6 and IMX500 employs int8 quantization. Overall, the IMX500's limited memory significantly restricts its ability to deploy larger models, unlike STM32N6 and OAK-FCC 4P, which comfortably handle more complex networks. With a maximum of 8 MB of memory, the IMX500 has

Table 2. Memory usage (MB) of object detection and pose estimation models evaluated on target devices. ((E): the model exceeds the available memory of the device; -: the model is not currently capable of being deployed on the device.)

Model	OAK-FCC 4P ↓	STM32N6 ↓	IMX500 ↓
Object detection models			
YOLO v5 n	7.07	**3.86**	-
YOLO v5 s	22.54	**10.80**	-
YOLO v8 n	8.02	4.35	**4.05**
YOLO v8 s	26.38	**12.78**	11.87 (E)
Human pose estimation models			
YOLO v8 n	8.14	4.41	**4.15**
YOLO v8 s	27.15	**13.24**	12.27 (E)

already reached 50% of its capacity with YOLO Nano models and this limits its suitability for deployment of larger models on this hardware. In contrast, the other two devices can handle larger models with greater ease.

4.2 Inference Time

Given an input image, we compared the inference times by measuring how long each device takes to produce a prediction for the estimation models. This allows for an evaluation on their performance and an understanding of their suitability for running neural models in real-time, key objective for edge accelerators.

Each device is able to measure this value internally and reporting it to the user. The inference times, in Table 3, are the results of the arithmetic mean of inferences performed on images from the COCO dataset for object detection and the COCO-Pose dataset for human pose estimation.

The values calculated on the OAK-FCC 4P and the STM32N6 are highly similar, differing by only a few milliseconds for the Nano models of both object detection and human pose estimation. Both devices are capable of running the models in real-time, achieving an inference rate of approximately 45 Hz.

The YOLO v8 Nano model on IMX500 demonstrates superior performance in terms of processing speed and frequency when compared to the other two devices, operating at approximately 200 Hz.

Table 3. Inference times (ms) of object detection and human pose estimation models evaluated on target devices.

Model	OAK-FCC 4P ↓	STM32N6 ↓	IMX500 ↓
Object detection models			
YOLO v5 n	22.04	**21.14**	-
YOLO v5 s	**37.67**	47.84	-
YOLO v8 n	21.39	23.44	**5.55**
YOLO v8 s	**40.62**	54.38	-
Human pose estimation models			
YOLO v8 n	22.56	24.39	**5.84**
YOLO v8 s	**43.25**	57.64	-

4.3 Accuracy Evaluation

During training, model weights and activations are optimized in FP32, whereas edge accelerators typically operate with lower-precision formats such as integers. Accurate deployment required the pipelines described in Sect. 3.2. To assess model performance during the deployment process, we measured accuracy after each conversion step. As reference metric, we used mean Average Precision with

Table 4. Accuracy results considering mAP@[0.50:0.95] in percentage of the two models after conversions into ONNX and quantized ONNX.

Model	PyTorch ↑	ONNX ↑	Quantized ONNX ↑
Object detection models			
YOLO v5 n	22.0	21.8	**10.6**
YOLO v5 s	28.4	28.2	3.07
YOLO v8 n	23.6	23.4	3.5
YOLO v8 s	**29.3**	**29.1**	7.9
Human pose estimation models			
YOLO v8 n	26.6	26.6	3.1
YOLO v8 s	**35.9**	**35.8**	**7.2**

Table 5. Accuracy results considering mAP@[0.50:0.95] in percentage of the two models considering the three solutions for the quantization problem.

Model	ONNX ↑	ONNX norm bbox ↑	Mixed quant ONNX ↑	Quant ONNX norm bbox ↑	Quant ONNX multiple outputs ↑	MCT quant ONNX multiple outputs ↑
Object detection models						
YOLO v5 n	21.8	21.5	19.9	17.8	17.2	**20.0**
YOLO v5 s	28.2	27.9	**16.4**	9.9	10.2	-
YOLO v8 n	23.4	23.2	14.2	8.6	12.5	**22.0**
YOLO v8 s	29.1	28.7	**17.0**	13.4	13.1	-
Human pose estimation models						
YOLO v8 n	26.6	26.1	**15.8**	4.9	14.1	15.6
YOLO v8 s	35.8	35.3	**22.3**	9.2	20.0	-

Intersection over Union (IoU) thresholds from 0.50 to 0.95 (mAP@[0.50:0.95]) over the 80 COCO object detection classes and 17 COCO-Pose keypoints.

As discussed in Sect. 3.1, all models were retrained using resized inputs due to hardware memory limitations. Table 4 shows that conversion to ONNX format preserves accuracy, while quantization with ONNX Runtime causes a significant drop of at least 70%. Further investigation revealed that the degradation was caused by the structure of YOLO model heads. Bounding box coordinates (ranging from 0 to image size) and confidence scores (ranging from 0 to 1) are quantized with different scale factors, leading to instability and accuracy loss.

To address the quantization issue, three potential solutions were identified:

- Exclude the final head layers from quantization, keeping them in FP32;
- Remove the final concatenation, separately quantize bounding box and score outputs;
- Normalize bounding box coordinates relative to the input image size, bringing them into the [0,1] range to stabilize full quantization.

Table 5 summarizes the results of these approaches. The "ONNX norm bbox" column shows accuracy after normalization, which causes a slight decline due to internal model changes. The "MCT quant ONNX multiple outputs" column reports results using Sony's MCT tool, while the other columns correspond to ONNX Runtime quantization methods. Compared to Table 4, all strategies significantly improve quantized model accuracy, bringing it closer to the original ONNX results. Among them, MCT achieves the highest overall accuracy due to the internally implemented methods. Particularly, the MCT framework applies graph optimizations, like BatchNorm folding, during calibration collects activation statistics for estimating the quantization thresholds for each layers and finally it uses symmetric quantization with power-of-two thresholds. These steps help improve accuracy compared to ONNXRuntime, which performs standard PTQ without advanced graph transformations or threshold tuning. Mixed quantization via ONNX Runtime also performs well but is unsuitable for deployment, as the partially floating-point head cannot run on accelerators. Therefore,

Table 6. Accuracy results considering mAP@[0.50:0.95] in percentage of the two models on OAK-FCC 4P and STM32N6 on the COCO dataset.

Model	OAK-FCC 4P ↑	STM32N6 norm bbox ↑	STM32N6 multiple outputs ↑
Object detection models			
YOLO v5 n	**21.8**	17.8	17.2
YOLO v5 s	**28.2**	9.9	10.2
YOLO v8 n	**23.4**	8.6	12.5
YOLO v8 s	**29.1**	13.4	13.1
Human pose estimation models			
YOLO v8 n	**26.6**	4.9	14.1
YOLO v8 s	**35.7**	9.2	20.0

Table 7. Accuracy results considering mAP@[0.50:0.95] in percentage of the two models with multiple outputs on OAK-FCC 4P, STM32N6, and IMX500 on the custom datasets captured by IMX500.

Model	OAK-FCC 4P ↑	STM32N6 ↑	IMX500 ↑
Object detection model			
YOLO v8 n	**20.3**	10.8	19.5
Human pose estimation model			
YOLO v8 n	**32.8**	21.2	22.4

for STM32N6 deployment, full quantization strategies were preferred despite slightly lower accuracy. The OAK-FCC 4P did not suffer from this issue, as its compiler maintains high accuracy after quantization, as shown in Table 6. In contrast, STM32N6 models exhibited noticeable degradation. However, analysis shows the accuracy loss originates from the quantization process, not from ST's compiler. We observed a significant drop in accuracy for the YOLOv5s model when using ONNXRuntime PTQ, which might be related to numerical instabilities introduced during the quantization process. A deeper analysis of the quantization methods and their impact on robustness is planned as future work.

IMX500 results are excluded here, as COCO and COCO-Pose datasets could not be directly transmitted to the Raspberry Pi AI Camera during experiments. A dedicated validation setup was created for this device: dataset images were displayed on a screen while a script allowed users to navigate images, capture them through the IMX500 to account for camera distortions, run inference, and compute accuracy. Direct comparison with Table 6 is not feasible due to slight perspective changes. Therefore, models were re-evaluated on OAK-FCC 4P and STM32N6 using the IMX500-captured dataset. Table 7 presents the results.

For object detection, OAK-FCC 4P and IMX500 show similar performance, whereas STM32N6 exhibits about half the accuracy, mainly due to the quantization from FP32 to integer. As shown in Fig. 3, N6 predictions display many errors, while OAK-FCC 4P and IMX500 maintain reliable detections.

For human pose estimation, the gap narrows: STM32N6 and IMX500 achieve comparable accuracy, although OAK-FCC 4P remains superior with a mAP of 32.8%. As illustrated in Fig. 4, OAK-FCC 4P predictions closely match ground truth, while the other devices show greater keypoint localization errors.

Table 6 and Table 7 further demonstrate that screen characteristics do not impact the experiments. Indeed, when evaluating the same device on the same dataset but with different cameras, we observe that the accuracy stays consistent.

(a) Ground Truth (b) OAK-FCC 4P (c) STM32N6 (d) IMX500

Fig. 3. Qualitative comparison of object detection predictions on the same image of COCO validation dataset captured by IMX500 on target devices.

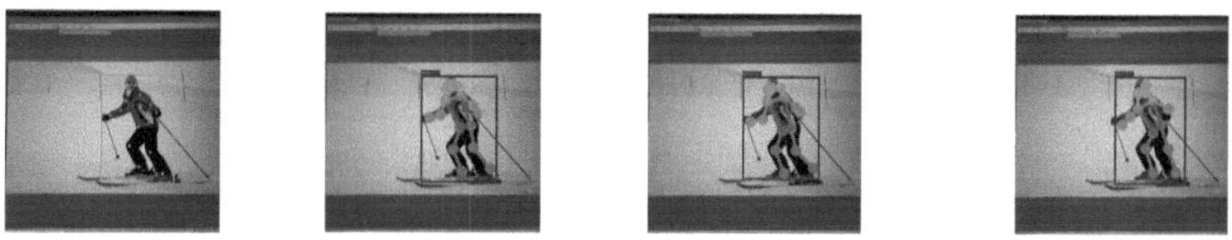

(a) Ground Truth (b) OAK-FCC 4P (c) STM32N6 (d) IMX500

Fig. 4. Qualitative comparison of pose estimation predictions on the same image of COCO-Pose validation dataset captured by IMX500 on target devices.

4.4 Power Consumption

Power consumption is crucial for battery-powered devices, as excessive energy usage can compromise integration in IoT applications. Edge AI devices often show significant power variability due to differences in data paths, preprocessing, peripherals, and firmware, complicating cross-device comparisons [5].

This section analyzes the power usage of the devices listed in Table 1. To accurately measure inference energy, a GPIO pin was raised at the start and lowered at the end of inference (when supported), defining a precise measurement window and minimizing signal transition effects [36]. While rise and fall times were negligible, each device required specific ad-hoc adaptations to enable reliable power measurements.

IMX500. The IMX500 camera core's energy consumption was measured, but firmware limitations prevented using a GPIO signal to isolate inference. Instead, a MIPI signal, which goes high when the camera is idle, monitored activity. A 220 $m\Omega$ shunt resistor was inserted into the core's supply path to enable precise oscilloscope measurements. Figure 5c shows the setup diagram.

OAK-FCC 4P. For the OAK-FCC 4P, the inference phase was isolated using a Raspberry Pi GPIO pin, which was set high before inference and reset low afterward, according to the block diagram reported in Fig. 5a. This was enabled by modifying the device management script, and the timing accuracy was verified against the device's logs. To measure energy consumption, a 33 $m\Omega$ resistor was

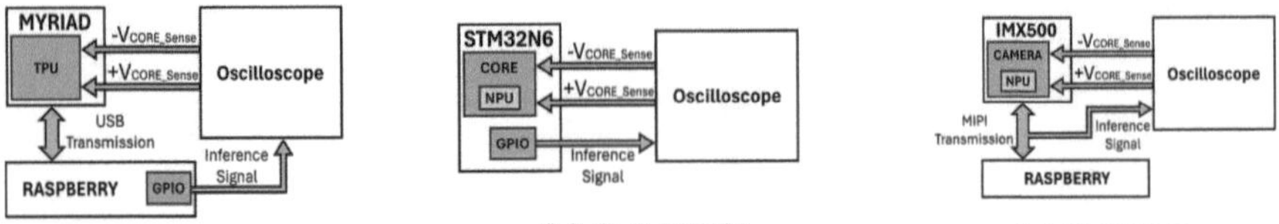

(a) OAK-FCC 4P (b) STM32N6 (c) IMX500

Fig. 5. Block diagrams of the measurement setup used for target devices.

Table 8. Energy consumption of the considered target devices for object detection and human pose estimation models (values in mJ).

Model	OAK-FCC 4P ↓	STM32N6 ↓	IMX500 ↓
Object detection models			
YOLO v5 n	30.5	**6.23**	-
YOLO v5 s	49.1	**16.62**	-
YOLO v8 n	30.22	7.33	**4.5**
YOLO v8 s	54.03	**18.06**	-
Human pose estimation models			
YOLO v8 n	30.14	7.13	**4.7**
YOLO v8 s	58.26	**19.78**	-

added in series with the core after cutting the power supply trace, following the same approach used for the IMX500.

STM32N6. For the STM32N6, the energy consumption of the core was measured using a GPIO to isolate the inference phase exclusively. The core voltage was fixed at 900 mV, with clock frequencies set to 1 GHz for the NPU, 800 MHz for the CPU, 400 MHz for the system RAMs, and 900 MHz for the NPU RAMs. To evaluate the performance of the N6, the STM32N6 Discovery kit was configured following the block diagram reported in Fig. 5b. Leveraging these configurations, the power consumption can be calculated as $P_{\text{core}}(t) = I_{\text{core}}(t) \cdot V_{\text{core}}$ where I_{core} and V_{core} are the core current and the core voltage, respectively. Since the inference execution time is marked by the trigger signal, the total energy consumption during inference is computed as $E_{core} = \int_{t_0}^{t_1} P_{\text{core}}(t) \cdot dt$ where t_0 and t_1 represent the start and end of the trigger signal, respectively, E_{core} represents the total energy consumed by the core over this period. Table 8 shows the measured mJ when different versions of YOLO models are executed.

5 Conclusion and Future Works

This paper presented a benchmarking analysis of lightweight neural network models, specifically YOLO Nano and Small, on three hardware accelerators:

OAK-FCC 4P, STM32N6, and IMX500. We evaluated key metrics including accuracy, inference speed, memory usage, and power consumption, highlighting the trade-offs of deploying models on edge devices. The IMX500's primary limitation is its constrained memory, restricting the use of YOLO Small models, while STM32N6 and OAK-FCC 4P can accommodate larger networks. In inference speed, IMX500 outperforms the others, while STM32N6 and OAK-FCC 4P deliver real-time performance on Nano models. The OAK-FCC 4P achieves the highest accuracy, although quantization introduced some degradation. In terms of energy consumption, IMX500 is the most efficient, followed closely by STM32N6. Our experiments highlight the need for more robust post-training quantization tools for the N6 platform, as current ONNXRuntime PTQ methods can significantly impact model accuracy, especially for complex architectures. Future work will focus on improving quantization techniques to exploit hardware capabilities and expanding the benchmarking to new emerging accelerators.

Acknowledgments. This work was carried out in the Smart Eyewear Lab, a Joint Research Center between EssilorLuxottica and Politecnico di Milano.

References

1. Blobconverter: Convert models for luxonis devices. https://blobconverter.luxonis.com/
2. Caffe. https://caffe.berkeleyvision.org/
3. GitHub - microsoft/onnxruntime: ONNX Runtime: cross-platform, high performance ML inferencing and training accelerator — github.com. https://github.com/microsoft/onnxruntime
4. arm: Arm® CoreSight™ ETM-M55 Technical Reference Manual
5. Banbury, C., Reddi, V.J., Torelli, P., et al.: Mlperf tiny benchmark (2021)
6. Bochkovskiy, A., Wang, C.Y., Liao, H.Y.M.: Yolov4: Optimal speed and accuracy of object detection. arXiv preprint arXiv:2004.10934 (2020)
7. Byun, K.J., Hahn, M., Kim, K.S.: Implementation of 13 kbps qcelp vocoder asic. In: AP-ASIC'99. First IEEE Asia Pacific Conference on ASICs (Cat. No. 99EX360), pp. 258–261. IEEE (1999)
8. Cao, Z., Simon, T., Wei, S.E., Sheikh, Y.: Realtime multi-person 2D pose estimation using part affinity fields. In: Proceedings of the IEEE Conference on Computer Vision and Pattern Recognition, pp. 7291–7299 (2017)
9. Carrión, D.S., Prohaska, V.: Exploration of tpus for ai applications. arXiv preprint arXiv:2309.08918 (2023)
10. Dunkel, E., et al.: Benchmarking deep learning inference of remote sensing imagery on the qualcomm snapdragon and intel movidius myriad x processors onboard the international space station. In: IGARSS 2022 - IEEE International Geoscience and Remote Sensing Symposium, pp. 5301–5304. Kuala Lumpur, Malaysia (2022)
11. Google: Introduction to cloud tpu | google cloud. https://cloud.google.com/tpu/docs/intro-to-tpu
12. Group, S.S.S.: Imx500 converter: Aitrios. https://developer.aitrios.sony-semicon.com/en/raspberrypi-ai-camera/documentation/imx500-converter?version=3.14.3&progLang=. Accessed 11 Dec 2024

13. Group, S.S.S.: Imx500 packager: Aitrios. https://developer.aitrios.sony-semicon.com/en/raspberrypi-ai-camera/documentation/imx500-packager?version=2024-11-21&progLang=. Accessed 11 Dec 2024
14. Humes, E., Navardi, M., Mohsenin, T.: Squeezed edge yolo: Onboard object detection on edge devices. arXiv preprint arXiv:2312.11716 (2023)
15. Intel: Intel® movidius™ myriad™ x vision processing unit (vpu). https://www.intel.com/content/www/us/en/products/details/processors/movidius-vpu/movidius-myriad-x/products.html
16. Jocher, G., Qiu, J., Chaurasia, A.: Ultralytics yolo (version 8.0.0) [computer software]. https://github.com/ultralytics/ultralytics
17. Kingma, D.P., Ba, J.: Adam: A method for stochastic optimization. arXiv preprint arXiv:1412.6980 (2014)
18. Lee, S.S., Nguyen, T.D., Meher, P.K., Park, S.Y.: Energy-efficient high-speed asic implementation of convolutional neural network using novel reduced critical-path design. IEEE Access **10**, 34032–34045 (2022)
19. Lin, T.-Y., et al.: Microsoft COCO: common objects in context. In: Fleet, D., Pajdla, T., Schiele, B., Tuytelaars, T. (eds.) ECCV 2014. LNCS, vol. 8693, pp. 740–755. Springer, Cham (2014). https://doi.org/10.1007/978-3-319-10602-1_48
20. Lodge, T., Crabtree, A., Brown, A.: Developing GDPR compliant apps for the edge. In: International Workshop on Data Privacy Management. Springer (2018)
21. Luxonis: Luxonis/depthai-python: Depthai python library. https://github.com/luxonis/depthai-python
22. Luxonis: OAK-FFC 4P. https://docs.luxonis.com/hardware/products/OAK-FFC%204P
23. Luxonis: Robotics vision core 2 (rvc2). https://docs.luxonis.com/hardware/platform/rvc/rvc2/#RVC2%20NN%20Performance
24. Murshed, M.S., Murphy, C., Hou, D., Khan, N., Ananthanarayanan, G., Hussain, F.: Machine learning at the network edge: a survey. ACM Comput. Surv. (CSUR) **54**(8), 1–37 (2021)
25. Núñez-Yáñez, J.L., Chouliaras, V.A.: Design and implementation of a high-performance and silicon efficient arithmetic coding accelerator for the h. 264 advanced video codec. In: 2005 IEEE International Conference on Application-Specific Systems, Architecture Processors (ASAP), pp. 411–416. IEEE (2005)
26. NVIDIA: Nvidia embedded systems for next-gen autonomous machines. www.nvidia.com/en-in/autonomous-machines/embedded-systems/
27. ONNX: Open neural network exchange. https://onnx.ai/
28. Redmon, J.: You only look once: unified, real-time object detection. In: Proceedings of the IEEE Conference on Computer Vision and Pattern Recognition (2016)
29. Reuther, A., Michaleas, P., Jones, M., Gadepally, V., Samsi, S., Kepner, J.: Survey and benchmarking of machine learning accelerators. In: 2019 IEEE High Performance Extreme Computing Conference (HPEC), pp. 1–9. Waltham, MA, USA
30. Sony: IMX500. https://developer.sony.com/imx500
31. Sony: Sony/model_optimization: Model compression toolkit (mct). https://github.com/sony/model_optimization. Accessed 11 Dec 2024
32. STMicroelectronics: STM32N6: Get a sneak peek at the future of AI-powered MCUs. https://blog.st.com/stm32n6/
33. Sun, K., Xiao, B., Liu, D., Wang, J.: Deep high-resolution representation learning for human pose estimation. In: Proceedings of the IEEE/CVF Conference on Computer Vision and Pattern Recognition, pp. 5693–5703 (2019)

34. Tan, M., Pang, R., Le, Q.V.: Efficientdet: Scalable and efficient object detection. In: Proceedings of the IEEE/CVF Conference on Computer Vision and Pattern Recognition, pp. 10781–10790 (2020)
35. TensorFlow: Tensorflow. https://www.tensorflow.org/
36. Tschand, A., et al.: Mlperf power: Benchmarking the energy efficiency of machine learning systems from microwatts to megawatts for sustainable ai (2024)
37. Tsoukas, V., Gkogkidis, A., Boumpa, E., Kakarountas, A.: A review on the emerging technology of tinyml. ACM Comput. Surv. **56**, 259–296 (2024)
38. Wang, Y., Li, M., Cai, H., Chen, W.M., Han, S.: Lite pose: efficient architecture design for 2d human pose estimation. In: Proceedings of the IEEE/CVF Conference on Computer Vision and Pattern Recognition, pp. 13126–13136 (2022)

Efficient Single Image Super-Resolution for Images with Spatially Varying Degradations

Simone Bianco(✉), Luca Cogo, Gianmarco Corti, and Raimondo Schettini

Department of Informatics, Systems and Communication, University of Milano-Bicocca, 20126 Milan, Italy
{simone.bianco,luca.cogo,gianmarco.corti,raimondo.schettini}@unimib.it

Abstract. Super-Resolution (SISR) is a computer vision task that aims to generate high-resolution images from their low-resolution counterparts. Typically, Super-Resolution methods use scaling factors of x2, x3, or x4 to uniformly enhance the resolution of the entire image. However, some acquisition devices, such as 360° cameras, produce images with non-uniform resolution across the frame. In this work, we propose to adapt state-of-the-art efficient methods for Single Image Super-Resolution to address the challenge of restoring images affected by spatially varying degradations. Specifically, we focus on the method that won the recent NTIRE 2024 Efficient Super-Resolution Challenge. For our experiments, synthetic images with different spatially varying types of degradation are generated, and the SISR method is specifically modified and trained to effectively handle such challenging scenarios. In addition to evaluating the developed method with traditional image quality metrics such as PSNR and SSIM, we also assess its practical impact on a downstream object detection task. The results on the WIDER FACE face-detection dataset, using the YOLOv8 object detection model, show that applying the proposed SISR approach to images with spatially varying degradations produces artifact-free outputs and enables object detectors to achieve superior performance compared to their application on degraded images.

Keywords: Single image super-resolution · Spatially varying degradation · Efficient · Deep learning

1 Introduction

Single Image Super-Resolution (SISR) is a computer vision task aimed at enhancing the resolution of low-quality images. It has various applications, including medical imaging, satellite image processing, and surveillance [31]. With the advent of deep learning, various methods based on Deep Neural Networks have been proposed and have surpassed traditional methods in various research areas.

E. Rodolà et al. (Eds.): ICIAP 2025, LNCS 16167, pp. 234–246, 2026.
https://doi.org/10.1007/978-3-032-10185-3_19

As a result, even in the research area of Super-Resolution (SR), new methods based on neural networks have been proposed, which are able to obtain better results than traditional techniques [29]. However, despite the progress made by deep learning models, many algorithms require considerable computational and memory capacity, making them difficult to adopt on a large scale or in contexts where resources are limited, as in the case of mobile devices and real-time applications. To this end, the NTIRE Efficient Super-Resolution Challenge [23] was organized, in which participants develop Super-Resolution methods that are as efficient as possible, where efficiency is evaluated in terms of metrics such as runtime, number of parameters, and computational complexity measured in FLOPs. These methods increase the resolution of their input images equally across the entire image; however, there are acquisition devices that generate images that do not have constant resolution. Certain sensors generate images that have spatially variable resolution, as in the case of 360° cameras, cameras with fish-eye lenses, and even acoustic imaging devices. In these scenarios, it can be useful to obtain images with constant resolution without losing information about the details of the content. This topic is still poorly addressed in the literature; in fact, when super-resolution is described, the focus is on the ability of the methods to increase resolution and image quality only in cases where degradation is constant, and these methods are not designed to work well in situations in which images have uneven resolution. The aim of this work is to evaluate the ability of Efficient Super-Resolution (ESR) methods to enhance images with spatially varying resolution. In order to do so, the winner of the NTIRE 2024 Efficient Super-Resolution challenge [23], SPAN [27], has been modified, trained and tested for this purpose. The model has been trained for such situations and tested on various types of degradation to observe whether it could improve the quality of the input images. Lastly, to test the impact of the application of these methods, SPAN was used in combination with YOLOv8 [9] on a face detection dataset to which spatially varying degradations were applied. Various tests were conducted to assess the YOLO model's ability on the face detection task when images present spatially varying resolution.

2 Related Works

Efficient Single Image Super-Resolution. Thanks to the recent rapid development of deep learning techniques, various Single Image Super-Resolution (SISR) methods based on neural networks have been proposed and have outperformed traditional approaches. SRCNN [4] is one of the first deep learning methods applied to the Super-Resolution task, and it is used to learn the mapping between a low-resolution image and its high-resolution version. A disadvantage of SRCNN is that it is computationally inefficient as it works in the space of high-resolution images, since the input image is resized before being processed by the network. In order to address this problem, ESPCN [26] proposed the use of a subpixel convolutional layer called PixelShuffle, which has been used by various methods to increase image resolution only at the end of the method. Other methods used a different approach, such as SRGAN [13], which is one of the first and most influential applications of the Generative Adversarial Networks [7] for SISR. Recently, SwinIR [14] has been proposed, which is based on Swin Transformers [16] and achieves state-of-the-art results while reducing the number of parameters. Despite the improvements, many methods still require intensive computations or many resources, which may hinder the ability to implement these models even on low-resource devices such as mobile devices. To this end, the NTIRE Efficient Super-Resolution Challenge [23] was organized with the aim of designing methods that are as efficient as possible while still maintaining good quality results. During the various editions of this challenge, many methods have been proposed that are capable of achieving impressive results despite being relatively small and without requiring many resources [10,15,27,30]. To do so, the various teams that participated in the challenge have employed various techniques such as reparameterization, knowledge distillation, and network pruning.

Spatially Varying Blur. Super-resolution can be conceptualized as a two-step process: image upsampling to increase spatial resolution, followed by deblurring to enhance details and recover high-frequency information. While modern approaches integrate these steps into a unified process, where both resolution enhancement and detail recovery are jointly optimized, image upsampling is a simple task and most of the complexity lies in the image deblurring task. Some approaches in the state of the art [5,6] address the combined problem of Super-Resolution and deblurring by explicitly breaking it down into the two stages previously described, i.e. upsampling and deblurring.

Image deblurring is a classic, well-studied computer vision problem. CNN-based models excel at learning data-driven representations and can adapt to a wide range of blur types without explicitly modeling the blur kernel. These methods have shown remarkable performance improvements over traditional techniques, particularly in handling real-world blurs [33]. However, most of the methods assume that the blur is uniform across the entire image. This task becomes more complex when the image presents a spatially varying blur. Several methods have been developed to address these challenges leveraging advancements in computational algorithms, optimization and deep learning. An early approach [20] approximates the blur contained in various regions with a spatially varying kernel using a combination of spatially invariant ones. Then the deblurring is applied to the single regions. Early deep learning deblurring methods were then implemented to remove uniform blur from images such as the method proposed by Schuler et al. [25] which is divided into two stages: the first estimates the blur kernel and the latent image from the blurry image, the second utilizes the blurry image and the latent one to predict the blur kernel. However, real-world images may contain spatially variable blur. In order to address this type of degradation, various methods have been proposed, especially to restore blur caused by moving objects [11,19,24]. Aljadaany et al. [2] proposed DrNet which does not require any estimate of the blur kernel and is further able to invert the effects of the blurring in blind image recovery tasks allowing it to obtain state-of-the-art results on the GoPro dataset [21] and results close to the state of the art on other datasets while maintaining a low inference time. Others address the problem of spatially varying blur in SR images, such as DMBSR [12], which is capable of achieving better results than other traditional SR methods on images with non-uniform blur.

3 Proposed Method

3.1 Degraded Images

Gathering real-world images that are characterized by spatially-varying resolution with their respective counterpart with uniform resolution is a very complex task, since it requires a pair of perfectly geometrically aligned images with two different resolutions in terms of spatial level of detail. For this reason, a synthetically generated dataset is employed in this work, in order to facilitate the experiments discussed in the following sections. To have more control over the type of degradation employed, and the images obtained, it is necessary to generate synthetically degraded images with higher resolution in certain parts of the image and lower in others. Given that super-resolution can be conceptualized as a two-step process, namely image upsampling and image deblurring, with the second one being the most critical one, we focus solely on the latter. Three different types of spatially varying degradation are here used: linear, radial, and random. To simulate a change in resolution, the Gaussian filter and a series of masks are used to make a gradual transition from the original image to the image with the filter applied. The masks used are images with values in the [0,1] range, such that, when applied to the original image, if the values are 1, the output will be the same as the original, while if the values are 0 the output will be the image with the applied Gaussian filter. This results in a blur-like effect that varies along the dimensions of the image itself. Let I_{orig} be the original image, $mask$ the mask used with values between [0,1] and I_{blur} the image on which the Gaussian filter was applied, then the degraded image I_{deg} is obtained as:

$$I_{deg} = I_{orig} \odot mask + I_{blur} \odot (1 - mask) \quad (1)$$

where $\odot$ denotes the element-wise multiplication. By doing so, it is possible to generate images that have a variable blur-like effect by simulating a change in resolution. Figure 1 shows an example of each degradation applied to a checkerboard image with the respective mask applied on it. In these cases, it is possible to observe the different spatially varying blur effects obtained using different degradations. In both linear and radial degradations the loss in detail is constant from the center of the image, where it is present a higher resolution, to the sides where a stronger blur is applied. The difference between the two cases is that in the linear case the transition occurs only along the horizontal axis, while in the radial degradation this takes place also along the vertical axis. Likewise to the first case, with the random degradation, the transition occurs only along the width of the image; however, it is not constant but random. The random mask applied to each image varies for each randomly degraded image. For this

(a) Ground truth (b) Linear (c) Radial (d) Random

Fig. 1. Images obtained by applying the different degradations. The degraded images are obtained by applying the Eq.(1) with the respective mask on the ground truth image. The masks are those in the top right corner of each degraded image, and a pixel value closer to white means a pixel closer to the original.

purpose, random value distributions are used to generate masks, allowing the simulation of random transitions from the original images to the images with the Gaussian filter applied.

3.2 Method Changes

During the experiments described in the next sections, two versions of SPAN [27] have been used. The difference between the two is the number of feature channels employed and therefore the number of parameters. For convenience, the smaller model and winner of the challenge is denoted SPAN-s, while the bigger one is referred to as SPAN. In the first case, the number of feature channels employed by the convolutional layers is 28 and in the second is 48. In both cases, the models were initially trained for 4× upsampling, with a 4× PixelShuffle as the final layer. However, since the image dimensions in our case remain unchanged, the models have been adapted to generate outputs with the same resolution as the input (i.e. the PixelShuffle layer was removed) while enhancing the level of detail. Then the models have been trained to handle the new spatially varying degradations. Both the versions have been trained with the same training configuration, that is: batch size of 16, initial learning rate of $1e^{-4}$ that is reduced every 10,000 iterations by a factor of 0.1, $L1$ loss function, Adam optimizer and 20,000 as the number of training iterations. The working environment employed during the following tests utilizes a Nvidia RTX 4070S GPU.

4 Experimental Results

4.1 ESR Methods for Spatially Varying Resolution

To assess the ability of Efficient Super-Resolution (ESR) models in handling images with spatially varying resolution, the SPAN models, modified as described in Sect. 3.2, have been trained from scratch on degraded images. The training was conducted using three versions of the DIV2K dataset [1], each modified to

Table 1. PSNR and SSIM obtained by applying the methods to the images in the test set. The best values contained in each column are listed in bold.

Model	BSD100					
	Linear		Radial		Random	
	PSNR↑	SSIM↑	PSNR↑	SSIM↑	PSNR↑	SSIM↑
Degraded	30.1696	0.9078	27.2973	0.8140	31.4426	0.9137
SPAN	30.9011	0.9210	**34.9025**	**0.9608**	32.4254	0.9306
SPAN-s	**31.3135**	**0.9310**	31.8414	0.9287	**32.5441**	**0.9409**

Model	Manga109					
	Linear		Radial		Random	
noentry	PSNR↑	SSIM↑	PSNR↑	SSIM↑	PSNR↑	SSIM↑
Degraded	28.8745	0.9450	25.7212	0.8878	29.2848	0.9368
SPAN	30.2921	0.9582	**35.4847**	**0.9797**	**30.9768**	**0.9548**
SPAN-s	**30.4650**	**0.9604**	31.5764	0.9633	30.8488	0.9547

Model	Set5					
	Linear		Radial		Random	
	PSNR↑	SSIM↑	PSNR↑	SSIM↑	PSNR↑	SSIM↑
Degraded	31.4673	0.9488	29.1126	0.9081	30.2307	0.9093
SPAN	32.6783	0.9588	**37.2752**	**0.9801**	31.6903	0.9276
SPAN-s	**33.2132**	**0.9638**	34.5581	0.9684	**32.5492**	**0.9457**

Model	Set14					
	Linear		Radial		Random	
	PSNR↑	SSIM↑	PSNR↑	SSIM↑	PSNR↑	SSIM↑
Degraded	29.9072	0.9192	27.5581	0.8575	31.9502	0.9425
SPAN	30.4842	0.9276	**35.2481**	**0.9691**	**32.6576**	0.9516
SPAN-s	**31.1062**	**0.9402**	32.4917	0.9540	32.4463	**0.9568**

Model	Urban100					
	Linear		Radial		Random	
	PSNR↑	SSIM↑	PSNR↑	SSIM↑	PSNR↑	SSIM↑
Degraded	22.6222	0.7307	21.9710	0.6920	22.4069	0.7141
SPAN	**25.4023**	**0.9381**	**24.8156**	**0.8189**	**25.6190**	**0.8388**
SPAN-s	24.6988	0.8159	24.5486	0.8114	24.4617	0.8052

incorporate a specific degradation: linear and radial degradations were applied using the same mask across all images, while for random degradation, a unique random mask was generated for each image. Similarly, for testing, three degraded versions were created for each of the BSD100 [17], Manga109 [18], Set5 [3], Set14 [32], and Urban100 [8] datasets.

Table 1 reports the PSNR and SSIM values, computed by comparing the degraded images and the outputs of the different methods against the original images. The results indicate that the models are able to effectively enhance image quality, making them more similar to the original counterparts and yielding

Table 2. PSNR value obtained by applying the SPAN-s model trained to handle the degradations present in the first column, on images degraded with the degradation present in the first row. The *Average* column contains the mean value obtained for each row.

Test				
Train	Linear	Radial	Random	Average
Linear	35.5975	30.2463	35.6630	33.8197
Radial	31.9280	36.5738	31.6145	33.5968
Random	35.4167	30.4630	35.5504	33.8725

higher PSNR and SSIM values than the degraded images. Moreover, despite the differences in model parameters, both versions are able to successfully improve degraded images achieving comparable PSNR and SSIM scores, with the only exception being SPAN-s trained on radially degraded images, which exhibits lower PSNR values compared to its larger counterpart.

Further experiments were conducted to assess the SPAN-s model's generalization ability on degradations it was not specifically trained for, with the corresponding PSNR values reported in Table 2. As expected, the highest PSNR values appear along the diagonal, corresponding to models tested on the same degradation type they were trained for. However, when applied to different degradation types, the models yield varying results. Notably, models trained on linear degradation perform well not only on their own degradation but also on random degradation, and vice versa. This is likely due to the fact that, unlike radial degradation, both linear and random cases involve resolution changes only along the image width. The radial case, on the other hand, exhibits uneven resolution even along the height of the image. Despite this, it still achieves better results on other degradation types when compared with the PSNR obtained by the other two methods in the radial degradation case. On average the models obtain similar results despite the differences in terms of degradations.

Figure 2 illustrates visual results for an image extracted from the test set. Notably, along the edges in both the linear and radial cases, certain details remain partially unrecovered. Therefore, even though the models are able to restore many of the details contained in the images, they still struggle in the areas where the degradation is stronger.

4.2 Efficient Super-Resolution for Face Detection

In order to evaluate the effectiveness of applying an Efficient Super-Resolution (ESR) method to images with spatially varying resolution, the previous discussed models were integrated into a face detection pipeline as preprocessing step. A subset of the WIDER Face dataset [28] was used for both training and evaluation, applying the degradations described in Sect. 3.1. The dataset was divided into 1,000 images for training, 200 for validation, and 500 for testing. Three different versions of YOLOv8 [9], with different sizes, were used as object detection

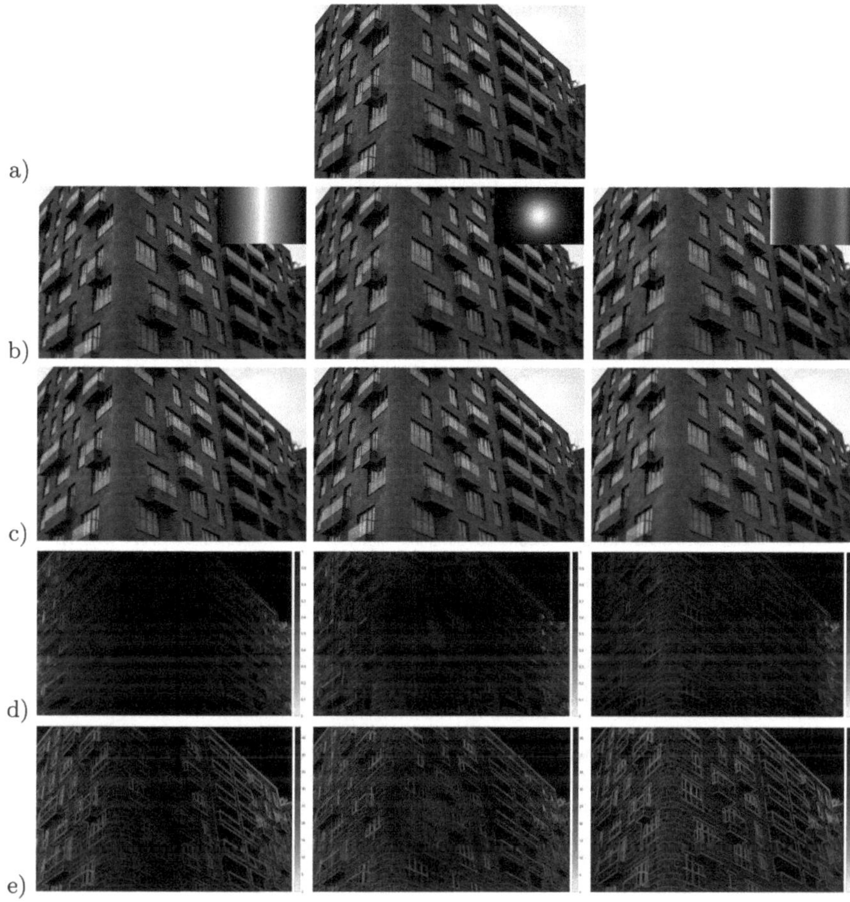

Fig. 2. Example result: a) original image; b) images degraded with their respective masks (linear, radial, and random); c) images restored with SPAN-s; d) SSIM value maps obtained by the original image and the restored one; e) the respective PSNR value maps. For all maps, a brighter color means larger errors, i.e. a lower level of details.

model to check the consistency of the results: Yolo8n, Yolo8s and Yolo8m. These models were fine tuned for 20 epochs using the same training configuration, with a learning rate of 0.01 and AdamW as the optimizer. Training was conducted on three different sets: the original images, the degraded images, and the images restored using the Super-Resolution methods. The top part of Table 3 reports the results obtained by the YOLO models on the restored test set,comparing their performance with models trained on the original (unmodified) images, degraded images, and restored images. The results are reported in terms of: Precision, indicating how many detections were correct; Recall, indicating the ratio of the instances identified by the model with respect to all instances of objects; mAP50,

Table 3. Results obtained by the different versions of YOLO on the considered WIDER FACE Test set. Top table: the *Original* rows contain the values obtained by training the different versions of YOLO on the original images (without the degradation); the *Degraded* rows contain the results obtained by the YOLO models trained on the degraded images; the *SPAN* and *SPAN-s* rows contain the results obtained by the YOLO models trained on the restored images using the respective model. Bottom table: results obtained by testing the versions of YOLO trained using the original images on the degraded images and on the images restored using the SPAN and SPAN-S models. In bold are marked the best results obtained by the models trained on the restored or degraded images, while the second best are underlined.

	Model	Linear				Radial				Random			
		P	R	mAP50	mAP50-95	P	R	mAP50	mAP50-95	P	R	mAP50	mAP50-95
Original	Yolo8n	**0.783**	**0.480**	**0.561**	**0.282**	0.783	**0.480**	**0.561**	**0.282**	0.783	**0.480**	**0.561**	**0.282**
Degraded	Yolo8n	0.740	0.455	0.518	0.255	0.737	0.450	0.519	0.252	**0.785**	0.439	0.518	0.255
SPAN	Yolo8n	0.756	0.471	0.544	0.271	**0.788**	0.463	0.544	0.268	0.774	0.476	0.546	0.265
SPAN-s	Yolo8n	0.767	0.453	0.530	0.259	0.772	0.447	0.530	0.263	0.778	0.459	0.533	0.259
Original	Yolo8s	**0.795**	**0.508**	**0.603**	**0.316**	0.795	**0.508**	**0.603**	**0.316**	0.795	**0.508**	**0.603**	**0.316**
Degraded	Yolo8s	0.779	0.460	0.546	0.282	0.797	0.460	0.555	0.281	**0.881**	0.452	0.550	0.283
SPAN	Yolo8s	0.782	0.494	0.586	0.301	**0.805**	0.499	0.594	0.306	0.789	0.490	0.582	0.300
SPAN-s	Yolo8s	0.790	0.488	0.574	0.295	0.797	0.487	0.583	0.296	0.771	0.481	0.565	0.291
Original	Yolo8m	**0.825**	**0.514**	**0.618**	**0.331**	**0.825**	**0.514**	**0.618**	**0.331**	**0.825**	**0.514**	**0.618**	**0.331**
Degraded	Yolo8m	0.786	0.486	0.571	0.298	0.807	0.466	0.568	0.299	0.770	0.477	0.564	0.292
SPAN	Yolo8m	0.794	0.505	0.589	0.309	0.806	0.507	0.604	0.317	0.804	0.502	0.590	0.306
SPAN-s	Yolo8m	0.783	0.496	0.591	0.311	0.814	0.489	0.593	0.312	0.779	0.493	0.580	0.303

	Model	Linear				Radial				Random			
		P	R	mAP50	mAP50-95	P	R	mAP50	mAP50-95	P	R	mAP50	mAP50-95
Original	Yolo8n	0.783	**0.480**	**0.561**	**0.282**	**0.783**	**0.480**	**0.561**	**0.282**	**0.783**	**0.480**	**0.561**	**0.282**
Degraded	Yolo8n	0.757	0.447	0.525	0261	0.762	0.426	0.506	0.251	0.756	0.437	0.517	0.255
SPAN	Yolo8n	**0.785**	0.463	0.549	0.274	0.779	0.479	0.553	0.276	0.768	0.479	0.550	0.275
SPAN-s	Yolo8n	0.749	0.474	0.544	0.271	0.760	0.470	0.543	0.270	0.767	0.456	0.537	0.266
Original	Yolo8s	0.795	**0.508**	**0.603**	**0.316**	0.795	**0.508**	**0.603**	**0.316**	0.795	**0.508**	**0.603**	**0.316**
Degraded	Yolo8s	0.816	0.440	0.551	0.285	0.772	0.431	0.535	0.275	0.761	0.447	0.542	0.276
SPAN	Yolo8s	**0.822**	0.477	0.587	0.306	**0.819**	0.486	0.594	0.308	**0.802**	0.488	0.587	0.304
SPAN-s	Yolo8s	0.806	0.475	0.577	0.298	0.766	0.492	0.579	0.299	0.774	0.472	0.564	0.290
Original	Yolo8m	**0.825**	**0.514**	**0.618**	**0.331**	**0.825**	0.514	**0.618**	**0.331**	**0.825**	**0.514**	**0.618**	**0.331**
Degraded	Yolo8m	0.807	0.472	0.565	0.298	0.794	0.464	0.559	0.290	0.799	0.459	0.559	0.291
SPAN	Yolo8m	0.812	0.501	0.603	0.321	0.797	**0.517**	0.611	0.324	0.797	0.504	0.601	0.316
SPAN-s	Yolo8m	0.810	0.488	0.593	0.313	0.793	0.502	0.596	0.313	0.821	0.465	0.577	0.305

i.e. mean average precision calculated at an intersection over union (IoU) threshold of 0.50; mAP50-95, i.e. the average of the mean average precision calculated at varying IoU thresholds, ranging from 0.50 to 0.95 [22]. Since we are interested in the overall assessment of model performance, the reference metric to consider is mAP. The obtained results show that, in all observed cases, models trained on images enhanced by a Super-Resolution method outperform those trained directly on degraded images.

More specifically, models incorporating the trained Super-Resolution methods to correct degradations achieve higher *mAP* values than those that do not, though their performance remains slightly below that of models trained on the original images. Figure 3 provides visual examples of face detections using SPAN-s for restoration and Yolo8n for detection. These examples highlight that applying restoration methods leads to more accurate detections and an increased number of detected faces. The bottom part of Table 3 also reports

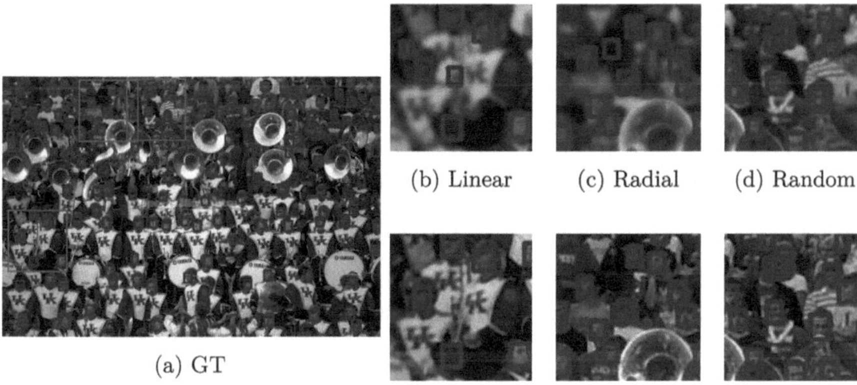

(a) GT (b) Linear (c) Radial (d) Random (e) Linear SPAN-s (f) Radial SPAN-s (g) Random SPAN-s

Fig. 3. Results obtained by using Yolo8n in the various analyzed cases. Where there is a stronger degradation (e.g., the areas in the red boxes) the methods that use the restored images detect more faces or are more accurate in localization. (Color figure online)

the results of YOLO models trained on original images and tested on both degraded and restored versions. In these cases, the models consistently achieve better performance on enhanced images than on degraded ones. This indicates that, even when trained solely on unmodified images, the models benefit from Super-Resolution preprocessing, improving detection performance despite not fully recovering all lost details.

5 Conclusions

This paper presents an investigation of the use of Efficient Super-Resolution (ESR) techniques to restore images with spatially varying degradations. By leveraging lightweight ESR models, we demonstrated their effectiveness in recovering image details and enhancing the visual quality of degraded images, particularly in scenarios with uneven resolution artifacts. Furthermore, we showcased the practical impact of ESR-restored images on downstream tasks, such as object detection, using the WIDER FACE dataset in conjunction with YOLOv8. The results underline how applying ESR methods significantly improves object detection performance compared to models operating on degraded images, though there is still a gap compared to models trained on pristine data.

While ESR models offer a computationally efficient solution, their performance is currently surpassed by more advanced methods, such as Transformer-based super-resolution models (e.g., SwinIR). Future work could focus on integrating lightweight Transformer-based architectures to better handle spatially varying degradations while maintaining efficiency. Another key limitation of this study is the reliance on synthetic datasets to simulate varying resolutions due to the lack of real-world datasets containing such degradations. Developing a dedicated real-world dataset, including images captured with devices prone to spatially varying degradations, such as 360° cameras, fish-eye lenses, or sonars, would provide a more robust basis for evaluation. As future work we plan to extend our method to other application domains, such as medical imaging, remote sensing or autonomous driving, which often encounter spatially varying resolution issues due to the nature of sensors used (e.g., sonars, radars, lidars, optical and electron microscopes), and where enhanced resolution may significantly improve interpretability and decision-making. We also plan to extend our method without making any assumptions about the degradation kernel thus permitting to experiment with real data even in absence of a higher resolution groundtruth.

Acknowledgments. This work has been carried out within the High-Performance Computing, Big Data and Quantum Computing National Research Center project, funded by the European Union - NextGenerationEU, under the National Recovery and Resilience Plan (NRRP) Mission 4 Component 2 Investment Line 1.4, project CN_00000013 / CN1, Innovation Grant SURE, CUP H43C22000520001.

References

1. Agustsson, E., Timofte, R.: Ntire 2017 challenge on single image super-resolution: dataset and study. In: Proceedings of the IEEE Conference on CVPR Workshops, pp. 126–135 (2017)
2. Aljadaany, R., Pal, D.K., Savvides, M.: Douglas-Rachford networks: learning both the image prior and data fidelity terms for blind image deconvolution. In: Proceedings of the IEEE/CVF Conference on CVPR, pp. 10235–10244 (2019)
3. Bevilacqua, M., Roumy, A., Guillemot, C., Alberi-Morel, M.L.: Low-complexity single-image super-resolution based on nonnegative neighbor embedding (2012)
4. Dong, C., Loy, C.C., He, K., Tang, X.: Learning a deep convolutional network for image super-resolution. In: Fleet, D., Pajdla, T., Schiele, B., Tuytelaars, T. (eds.) ECCV 2014. LNCS, vol. 8692, pp. 184–199. Springer, Cham (2014). https://doi.org/10.1007/978-3-319-10593-2_13
5. Esmaeilzehi, A., Ahmad, M.O., Swamy, M.: UPDCNN: a new scheme for image upsampling and deblurring using a deep convolutional neural network. In: 2019 IEEE International Conference on Image Processing (ICIP), pp. 2154–2158 (2019)
6. Esmaeilzehi, A., Ahmad, M.O., Swamy, M.: UPDRESNN: a deep light-weight image upsampling and deblurring residual neural network. IEEE Trans. Broadcast. **67**(2), 538–548 (2021)
7. Goodfellow, I., et al.: Generative adversarial networks. Commun. ACM **63**(11), 139–144 (2020)
8. Huang, J.B., Singh, A., Ahuja, N.: Single image super-resolution from transformed self-exemplars. In: Proceedings of the IEEE Conference on CVPR, pp. 5197–5206 (2015)
9. Jocher, G., Chaurasia, A., Qiu, J.: Ultralytics YOLOv8 (2023). https://github.com/ultralytics/ultralytics
10. Kong, F., Li, M., et al.: Residual local feature network for efficient super-resolution. In: Proceedings of the IEEE/CVF Conference on CVPR, pp. 766–776 (2022)

11. Kupyn, O., Budzan, V., Mykhailych, M., Mishkin, D., Matas, J.: DeblurGAN: blind motion deblurring using conditional adversarial networks. In: Proceedings of the IEEE Conference on CVPR, pp. 8183–8192 (2018)
12. Laroche, C., Almansa, A., Tassano, M.: Deep model-based super-resolution with non-uniform blur. In: Proceedings of the IEEE/CVF WACV, pp. 1797–1808 (2023)
13. Ledig, C., Theis, L., et al.: Photo-realistic single image super-resolution using a generative adversarial network. In: Proceedings of the IEEE Conference on CVPR, pp. 4681–4690 (2017)
14. Liang, J., Cao, J., et al.: Swinir: Image restoration using swin transformer. In: Proceedings of the IEEE/CVF ICCV, pp. 1833–1844 (2021)
15. Liu, J., Tang, J., Wu, G.: Residual feature distillation network for lightweight image super-resolution. In: Bartoli, A., Fusiello, A. (eds.) ECCV 2020. LNCS, vol. 12537, pp. 41–55. Springer, Cham (2020). https://doi.org/10.1007/978-3-030-67070-2_2
16. Liu, Z., et al.: Swin transformer: hierarchical vision transformer using shifted windows. In: Proceedings of the IEEE/CVF ICCV, pp. 10012–10022 (2021)
17. Martin, D., Fowlkes, C., Tal, D., Malik, J.: A database of human segmented natural images and its application to evaluating segmentation algorithms and measuring ecological statistics. In: Proceedings of the IEEE ICCV, vol. 2, pp. 416–423. IEEE (2001)
18. Matsui, Y., Ito, K., et al.: Sketch-based manga retrieval using manga109 dataset. Multimedia Tools Appl. **76**, 21811–21838 (2017)
19. Mustaniemi, J., Kannala, J., Särkkä, S., Matas, J., Heikkila, J.: Gyroscope-aided motion deblurring with deep networks. In: 2019 IEEE Winter Conference on Applications of Computer Vision (WACV), pp. 1914–1922. IEEE (2019)
20. Nagy, J.G., O'Leary, D.P.: Restoring images degraded by spatially variant blur. SIAM J. Sci. Comput. **19**(4), 1063–1082 (1998)
21. Nah, S., Hyun Kim, T., Mu Lee, K.: Deep multi-scale convolutional neural network for dynamic scene deblurring. In: Proceedings of the IEEE Conference on CVPR, pp. 3883–3891 (2017)
22. Padilla, R., Netto, S.L., Da Silva, E.A.: A survey on performance metrics for object-detection algorithms. In: 2020 International Conference on Systems, Signals and Image Processing (IWSSIP), pp. 237–242. IEEE (2020)
23. Ren, B., Li, Y., et al.: The ninth ntire 2024 efficient super-resolution challenge report. In: Proceedings of IEEE/CVF Conference on CVPR, pp. 6595–6631 (2024)
24. Ren, D., Zhang, K., Wang, Q., Hu, Q., Zuo, W.: Neural blind deconvolution using deep priors. In: Proceedings of IEEE/CVF Conference on CVPR, pp. 3341–3350 (2020)
25. Schuler, C.J., Hirsch, M., Harmeling, S., Schölkopf, B.: Learning to deblur. IEEE Trans. Pattern Anal. Mach. Intell. **38**(7), 1439–1451 (2015)
26. Shi, W., Caballero, J., et al.: Real-time single image and video super-resolution using an efficient sub-pixel convolutional neural network. In: Proceedings of the IEEE Conference on CVPR, pp. 1874–1883 (2016)
27. Wan, C., Yu, H., et al.: Swift parameter-free attention network for efficient super-resolution. In: Proceedings of the IEEE/CVF Conference on CVPR, pp. 6246–6256 (2024)
28. Yang, S., Luo, P., Loy, C.C., Tang, X.: Wider face: a face detection benchmark. In: Proceedings of the IEEE Conference on CVPR, pp. 5525–5533 (2016)
29. Yang, W., Zhang, X., et al.: Deep learning for single image super-resolution: a brief review. IEEE Trans. Multimedia **21**(12), 3106–3121 (2019)

30. Yu, L., Li, X., et al.: Dipnet: efficiency distillation and iterative pruning for image super-resolution. In: Proceedings of the IEEE/CVF Conference on CVPR, pp. 1692–1701 (2023)
31. Yue, L., Shen, H., Li, J., Yuan, Q., Zhang, H., Zhang, L.: Image super-resolution: the techniques, applications, and future. Sig. Process. **128**, 389–408 (2016)
32. Zeyde, R., Elad, M., Protter, M.: On single image scale-up using sparse-representations. In: Curves and Surfaces: 7th International Conference, pp. 711–730 (2012)
33. Zhang, K., et al.: Deep image deblurring: a survey. IJCV **130**(9), 2103–2130 (2022)

MADPOT: Medical Anomaly Detection with CLIP Adaptation and Partial Optimal Transport

Mahshid Shiri[1], Cigdem Beyan[1](✉), and Vittorio Murino[1,2]

[1] Department of Computer Science, University of Verona, Verona, Italy
{mahshid.shiri,cigdem.beyan,vittorio.murino}@univr.it
[2] AI for Good (AIGO) Research Unit, Istituto Italiano di Tecnologia, Genoa, Italy

Abstract. Medical anomaly detection (AD) is challenging due to diverse imaging modalities, anatomical variations, and limited labeled data. We propose a novel approach combining visual adapters and prompt learning with Partial Optimal Transport (POT) and contrastive learning (CL) to improve CLIP's adaptability to medical images, particularly for AD. Unlike standard prompt learning, which often yields a single representation, our method employs multiple prompts aligned with local features via POT to capture subtle abnormalities. CL further enforces intra-class cohesion and inter-class separation. Our method achieves state-of-the-art results in few-shot, zero-shot, and cross-dataset scenarios without synthetic data or memory banks. The code is available at https://github.com/mahshid1998/MADPOT

Keywords: Medical Anomaly Detection · Partial Optimal Transport · CLIP · Adapters · Learnable prompts · Few-shot · Zero-shot

1 Introduction

Medical anomaly detection (AD) aims to identify unusual patterns in medical data, a task made challenging by the absence of a universal definition of anomaly and the considerable variability in medical images, both across different modalities and anatomical regions [10,19,40]. Ideally, a model should exhibit versatility across diverse data types, regardless of variations in anomaly size, modality, or underlying anatomical structures. Additionally, it should be capable of handling tasks involving few-shot learning (where only a limited number of labeled examples are available for training, e.g., [12,18]) or zero-shot learning (where the model is trained on one set of modalities but tested on completely unseen modalities, e.g., [19]).

Recent advancements in large-scale pre-trained visual-language models (VLMs) have led to more robust and generalizable AD. Notably, CLIP [30] has been adapted by several studies, such as [7,10], for industrial defect detection. Similarly, in the field of medical AD, which involves anomaly classification

E. Rodolà et al. (Eds.): ICIAP 2025, LNCS 16167, pp. 247–259, 2026.
https://doi.org/10.1007/978-3-032-10185-3_20

(AC, i.e., binary classification of normal vs. abnormal) and anomaly segmentation (AS), CLIP-based architectures have also been effectively applied [19,41]. These studies [19,41] have demonstrated that relying solely on CLIP [30], which is pre-trained on natural images, is insufficient for medical AD. This limitation arises not only from the domain gap between natural and medical images, but also from CLIP's original training, which primarily focuses on aligning the class semantics of foreground objects. As a result, its ability to generalize and capture subtle visual abnormalities in medical AD is constrained. Consequently, CLIP can be leveraged through fine-tuning on medical images or by adapting the visual encoder, which primarily captures image semantics, to better align with the unique characteristics of medical data. For instance, MVFA [19] utilizes adapters, designed as linear layers, which are attached to different layers of the CLIP vision encoder, with training performed for both AC and AS tasks. That approach [19] is effective for both few-shot and zero-shot settings. In contrast, MediCLIP [41] uses convolutional layer adapters but focuses solely on few-shot AD, requiring a significant amount of synthesized medical image anomaly instances to achieve optimal performance. Another approach to adapting CLIP is through text-based adaptation via prompt engineering, where fixed hand-crafted prompts (e.g., "a black and white photo of a normal [brain]"), as applied in [19], can be used. Alternatively, prompt learning can be employed to align the textual representations more closely with task-specific language, i.e., medical terminology and anomaly-related descriptors. This approach is implemented through CoOP [42] in the few-shot medical AD pipeline of [41].

In this study, we combine both visual adapters and prompt learning. We argue that, since anomaly descriptions may share similarities across different datasets, incorporating textual information helps reduce reliance solely on visual data. This approach could enhance model performance, especially in data-scarce scenarios such as few-shot or zero-shot learning, where the test time modality is completely unseen. On the other hand, while learned prompts can potentially improve upon manually crafted ones, they may still fall short of fully representing a class that possesses multiple attributes and complex contextual relationships. For example, we can describe the class "Lung Tumor" from various perspectives, such as the size and shape of the tumor, its location within the lung, the density of surrounding tissue, and the presence of any vascular invasion or nearby organ involvement. This motivates us to learn multiple prompts. However, this is akin to matching the mean of the prompt features with the visual features, as all prompts are pulled toward a single point, resulting in them learning the same characteristic [9]. This undermines the goal of learning diverse and comprehensive prompts. To overcome this limitation, we target prompt learning with Optimal Transport (OT) [36], where OT is used to align local visual features with multiple textual prompts through a learned transport plan [9]. However, we argue that the classical OT problem may present two key limitations for medical AD. First, it imposes a balance condition, requiring the total "mass" (i.e., the summed probability) of the source and target distributions to be equal [23]. This constraint can be limiting when analyzing medical images, especially for detecting small, localized anomalies that constitute only a minor portion of the data. Second, the OT plan does not explicitly preserve local discriminative structures, such as groups or clusters of features, which we consider important for distinguishing between classes in classification tasks [38].

To address the first limitation, we employ Partial Optimal Transport (POT), which transports only a fixed fraction of the total mass. This allows us to guide the alignment toward the most informative image patches, typically small and subtle anomalies, while downplaying irrelevant or healthy regions. This selective focus is crucial in medical AD, where abnormalities often occupy only a minor portion of the image. To address the second limitation, we integrate contrastive learning (CL) with POT, enabling the model to explicitly learn local discriminative structures. While POT guides alignment toward informative regions, CL promotes the formation of well-separated clusters by pulling similar features closer and pushing dissimilar ones apart. This helps preserve meaningful intra-class and inter-class distinctions, which are essential for classification tasks in medical AD (see Fig. 1b). Our method, called MADPOT, demonstrates superior performance in medical AD across few-shot, zero-shot, and cross-dataset settings (i.e., training and testing occur on the same modality but different datasets), surpassing all state-of-the-art (SOTA) methods. Additionally, our pipeline eliminates the need for memory banks, as seen in e.g., [19], or additional synthetic data, as used in [41].

The main contributions of our work are: (1) The pioneering investigation of prompt learning with OT and POT and their novel integration with CL in medical AD, particularly in few-shot and zero-shot settings. This is the first exploration of these techniques in this context, highlighting their potential for reducing data dependency and enhancing model flexibility. **(2)** A comprehensive evaluation of OT, POT, CL, and vision adapters, alongside their various combinations, providing an in-depth analysis of how these components contribute to model performance across datasets with substantial modality and distribution differences. **(3)** Demonstration of strong generalization and noticeable performance improvements, validated through cross-dataset and cross-modality evaluations. Our method shows robustness across diverse medical modalities and anatomical regions, making it a possible reliable solution for real-world clinical applications.

2 Related Works

Anomaly Detection. The generic AD task is commonly addressed from two perspectives: (a) unsupervised methods (e.g., [16,31]), which model normal data distributions to detect deviations, and (b) supervised methods (e.g., [19,39,41]), which leverage labeled anomalies. For instance, PatchCore [31] compares test samples to a memory bank of normal embeddings, while CFLOW-AD [16] models normality using normalizing flows and Gaussian distributions. Although unsupervised methods benefit from large datasets, real-world settings often involve few labeled anomalies, motivating the development of few-shot AD approaches.

Yet, incorporating limited anomalies in training introduces reduced generalization. To mitigate this, DRA [12] learns disentangled representations, and BGAD [39] employs boundary-guided semi-push-pull contrastive learning. Recently, CLIP [30] has been leveraged for AD. April-GAN [10] maps CLIP-extracted visual features to the text feature space and uses multiple memory banks, which is an expensive setup with limited generalization. In medical AD, MVFA [19] extends this by applying multi-level CLIP adaptation with fixed prompts. MediCLIP [41], instead, uses learnable prompts [42] and synthetic abnormal data to improve generalization. However, a learnable prompt with OT/POT, as shown in our experiments, hinders performance, especially in certain modalities (e.g., histopathology). Our approach integrates POT and CL to enhance alignment and generalization across diverse medical data.

Optimal Transport. Originally proposed for minimizing resource allocation costs [36], OT has become an important tool in machine learning for comparing probability distributions as feature sets [29]. OT has been used in tasks such as domain adaptation [26], learning with noisy labels [14], causal discovery [35], and federated learning [27]. Partial OT extends classical OT by offering flexibility where exact mass preservation is impractical (See 3 for details). PLOT [9] is the most relevant study, using OT to learn local prompts with CoOP [42] for fine-grained vision-language matching. Motivated by PLOT, our approach uses POT instead and shows that POT alone is insufficient; CL is necessary for improved zero-shot and few-shot performance. This is the first study to apply OT and POT to medical AD while integrating CL, visual adapters, and learnable prompts for CLIP adaptation.

3 Method

In training, the data consists of triplets $\{x_i, c_i, M_i\}$, where each sample x_i is an image of size $\mathbb{R}^{h \times w \times 3}$, the label c_i takes a binary value, with 1 and 0 indicating normal and abnormal, respectively. When provided, the corresponding AS mask M_i is a binary map of dimensions $h \times w$. At inference, given a test image x_{test}, the model outputs both an anomaly classification (AC) decision and an AS mask prediction. Building on this setup, we propose MADPOT (see Fig. 1a), an innovative approach that leverages CLIP [30], incorporating visual features adaptation and projection alongside using POT for learning diverse text prompts. MADPOT learns separate multi-modal representations for normal and abnormal samples by simultaneously integrating CL and POT. While CL ensures that prompts for each class are distinguishable, POT promotes diversity within learnable prompts for each class, leading to better distinctions between normal and abnormal classes (see Fig. 1b). Below, we provide a detailed description of the components employed in the MADPOT.

Visual Feature Adapting. To adapt the pre-trained CLIP vision encoder [30] for the AC and AS tasks in medical imaging, MADPOT uses adapter and projection layers. This approach follows the findings of [19], which shows that it

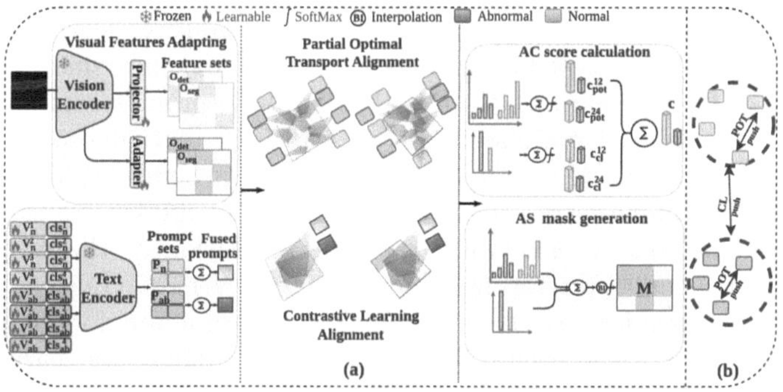

Fig. 1. (a) Overview of MADPOT: Visual features are extracted via an adapter and projection on the CLIP vision encoder. Separate prompt sets for normal and abnormal are learned and fused into class-specific prompts. POT produces multiple logits, and CL is applied to the visual features and fused prompts. The final score is the average of the logits from POT and CL. (b) POT enables the model to capture localized features, while CL encourages the prompts to differentiate between normal and abnormal classes. By combining CL and POT, we maximize multimodal similarity within each class and minimize it across classes, preserving the diversity of learned prompts within each class.

is preferable to traditional fine-tuning, as it helps avoid overfitting due to high model complexity and limited data. We append a small set of learnable layers to the visual branches of CLIP while keeping the backbone frozen.

In detail, for an input image x, we extract the i-th layer *feature* from the CLIP vision encoder, denoted as $f^i_{vis}(x) \in \mathbb{R}^{G \times d}$. Here, G represents the grid size, d is the feature dimension, and $i \in \{12, 24\}$. Learnable adapter $A(.)$ and projector $P(.)$ consist of two transformation stages. The first stage addresses the domain gap between natural and medical images. by introducing a shared linear transformation layer W^i_{shared} that maps CLIP features into a feature space better suited for medical AD, formulated as $F^i_{\text{shared}}(f^i_{vis}(x)) = ReLU(W^i_{\text{shared}} f^i_{vis}(x))$. To effectively handle both AC and AS, the second stage focuses on extracting features specialized for each task by applying two distinct linear transformation heads on top of the shared features: the first head, dedicated to AC, captures high-level characteristics essential for detecting anomalies, while the second head, highlights fine-grained spatial details for AS, expressed as $F^i_{\text{Det}}(f^i_{vis}(x)) = ReLU\Big(W^i_{\text{Det}}\, F^i_{\text{shared}}(f^i_{vis}(x))\Big)$ and $F^i_{\text{Seg}}(f^i_{vis}(x)) = ReLU\Big(W^i_{\text{Seg}}\, F^i_{\text{shared}}(f^i_{vis}(x))\Big)$. As a result, two feature sets, O^i_{Det} and O^i_{Seg}, are produced at each level. Both $A(.)$ and $P(.)$ follow the same transformation structure. However, $A(.)$ employs a residual connection as empirically found better-performing. Specifically, a constant value γ is introduced as a residual ratio to balance knowledge preservation and adaptation, formulated as $F^i_*(f^i_{vis}(x)) = \gamma F^i_{\text{shared}}(f^i_{vis}(x)) + (1-\gamma) f^i_{vis}(x)$, where F^i_* serves as input to the $i+1$ layer of the vision encoder.

Learnable Prompt Sets and Fused Prompts. Multiple learnable prompts (i.e., prompt sets) are used for each class (normal and abnormal) to support discriminative feature learning through POT. These prompt sets are fused to create class-discriminative embeddings via CL. The textual prompt for *normal* class is defined as $t_{\mathrm{n}}^k = \{v_1^k, v_2^k, \ldots, v_L^k, cls_{\mathrm{n}}^k\}$, where cls_{n}^k represents the k's normal class name (e.g., "flawless", "healthy") and $V_n^k = \{v_l^k\}_{l=1}^L$ consists of learnable tokens, each with the same dimension as the original word embedding, and L denotes the number of context words . Given t_{n}^k as input, the CLIP text encoder (f_{txt}) generates a set of prompts for class normal, $P_n = \{f_{txt}(t_{\mathrm{n}}^k)\}_{k=1}^K$, where each prompt is in $\mathbb{R}^d$. To enhance the learning of discriminative prompts across two classes of normal and abnormal, a fused prompt for the normal class is computed as $P_{\mathsf{fused},n} = \frac{1}{K}\sum_{k=1}^K P_n^k$. The same procedure is applied to the abnormal class (with class names such as pathological, with anomalies), yielding $P_{\mathrm{ab}} = \{f_{txt}(t_{\mathrm{ab}}^k)\}_{k=1}^K$, and the corresponding fused prompt $P_{\mathsf{fused},ab} = \frac{1}{K}\sum_{k=1}^K P_{\mathrm{ab}}^k$.

Multimodal Feature Alignment. Let us first formally describe classical and partial optimal transport (OT and POT, respectively), and then explain how we perform multimodal feature alignment using POT. OT is a constrained optimization problem that aims to find the most efficient way to transfer probability mass between two distributions [36]. In the discrete case, let α and β be probability simplex vectors, and let $C \in \mathbb{R}^{|\alpha|\times|\beta|}$ be the cost matrix. The objective of OT is to determine the optimal transport plan T by solving $disc_C(\alpha,\beta) = \min_{T\in U(\alpha,\beta)}\langle C,T\rangle$, where $\langle\cdot,\cdot\rangle$ denotes the Frobenius inner product, and $U(\alpha,\beta)$ represents the set of feasible transport plans: $U(\alpha,\beta) = \left\{T \in \mathbb{R}_+^{|\alpha|\times|\beta|} \mid T\mathbf{1}_{|\beta|} = \alpha, \quad T^\top\mathbf{1}_{|\alpha|} = \beta\right\}$. Directly solving the OT problem can be computationally expensive. To alleviate this, the Sinkhorn algorithm [33] introduces an entropic regularization term to accelerate the optimization process. The regularized OT formulation is given by: $disc_C(\alpha,\beta) = \min_{T\in U(\alpha,\beta)}\langle C,T\rangle + \lambda\langle T,\log T\rangle$, where $\lambda \geq 0$ is a regularization parameter. Under this formulation, the optimal transport plan T^* has a unique closed-form solution.

POT extends the classical OT formulation by relaxing the strict equality constraints on marginal distributions [15]. The objective function remains the same as OT formulation, but the feasible transport plan set $U(\alpha,\beta)$ is modified to allow for mass variations: $U(\alpha,\beta) = \left\{T \in \mathbb{R}_+^{|\alpha|\times|\beta|} \mid T\mathbf{1}_{|\beta|} \leq \alpha, \quad T^\top\mathbf{1}_{|\alpha|} = \beta\right\}$. Here, satisfy $\|\alpha\|_1 \geq \|\beta\|_1 = frac$ with $frac \in [0,1]$. For computational efficiency, as in OT, an entropic regularization term is added. The optimization problem can be reformulated as a Kullback-Leibler projection [5], and as introduced in [8], it can be solved in a few iterations by a fast implementation of Dykstra's Algorithm [13] by only performing matrix-vector multiplications, which is very similar to the Sinkhorn Algorithm [33]. In our implementation, by representing visual features and prompts as discrete distributions, POT maps prompts to relevant image patches. Unlike classical OT, which enforces strict alignment (mass conservation), POT relaxes this constraint. As a result, prompts are not forced to match every image patch, avoiding the inclusion of irrelevant information and improving focus on meaningful correspondences. In detail, to align visual feature sets O^i_{Det} and O^i_{Seg} with the prompt sets P_n and P_{ab} through POT, the cost matrix is defined on the basis of the cosine distance between the visual feature sets and the prompt sets. For class $j \in \{n, ab\}$, the cost matrices at each layer are given by: $C^i_{Det,j} = 1 - {O^i_{Det}}^\top P_j, \quad C^i_{Seg,j} = 1 - {O^i_{Seg}}^\top P_j$, where $C^i_{Det,j}$ and $C^i_{Seg,j} \in \mathbb{R}^{G \times k}$. For multimodal alignment, a two-stage optimization strategy (i.e. inner and outer loop) is employed. In the inner loop, the optimal transport plans $T^{i,*}_{Det,j}$ and $T^{i,*}_{Seg,j}$ are driven based on the defined cost matrices. This is achieved while keeping all learnable parameters fixed. It results in transport distances $dis^i_{Det,j}$ and $dis^i_{Seg,j} \in \mathbb{R}^{G \times k}$. In the outer loop, the transport plans are fixed, while the learnable parameters are optimized based on the distances as Eq. 1(left), where τ stands for the temperature of the softmax. To compute the image-level score at each layer by POT ($\hat{c}^i_{pot}$), we aggregate the patch-level distances by summing over all G number of patches and all k number of prompts for each class. To align fused prompts $P_{fused,n}$ and $P_{fused,ab}$ with visual feature sets through CL, the cosine similarity (sim) of the prompts and visual features is calculated, resulting in patch-level scores. To obtain image-level scores, the scores of each patch are summed as Eq. 1(right). The predicted AC score in the i-th feature layer is $\hat{c}^i_j = \hat{c}^i_{pot,j} + \hat{c}^i_{cl,j}$.

$$\hat{c}^i_{pot,j}(y = j|x) = \frac{\exp\big(\sum_{G\times k}(1 - dis^i_{Det,j})/\tau\big)}{\sum_{j'}\exp\big(\sum_{G\times k}(1 - dis^i_{Det,j'})/\tau\big)} \quad \hat{c}^i_{cl,j}(y = j|x) = \frac{1}{G}\sum_{G}\Big(\frac{\exp\big(sim(O^i_{Det}, P_{fused,j})/\tau\big)}{\sum_{j'}\exp\big(sim(O^i_{Det}, P_{fused,j'})/\tau\big)}\Big) \tag{1}$$

The anomaly mask at each layer ($\hat{M}^i_j$) is generated by aggregating ($dis^i_{Seg,j} \in \mathbb{R}^{G \times k}$) across all k prompts. The resulting logits from POT (i.e. $1 - dis^i_{Seg,j}$) and CL (i.e. $sim(O^i_{Seg}, P_{fused,j})$) are subsequently aggregated. Aggregation at this stage has been empirically found to be more effective for AS (up to +30% AD score), compared to the aggregation level similar to AC where aggregation is performed only at the final stage. Given that AS requires alignment with the input dimensions, (**BI**) modifies the anomaly logits to a size of $\sqrt{G} \times \sqrt{G}$ and subsequently resizes to the original input dimensions through bicubic interpolation (Eq. 2).

$$\hat{M}^i_j(y = j|x) = \frac{\exp\Big(\mathbf{BI}\big(\sum_k(1 - dis^i_{Seg,j}) + sim(O^i_{Seg}, P_{fused,j})\big)/\tau\Big)}{\sum_{j'}\exp\Big(\mathbf{BI}\big(\sum_k(1 - dis^i_{Seg,j'}) + sim(O^i_{Seg}, P_{fused,j'})\big)/\tau\Big)} \tag{2}$$

Loss Function. The loss function at feature level i is $Loss^i = \omega_1\,\mathrm{GDice}(\hat{M}^i, M) + \omega_2\,\mathrm{Focal}(\hat{M}^i, M) + \omega_3\,\mathrm{BCE}(\hat{c}^i, c)$. By default, all loss terms contribute equally ($\omega_1 = \omega_2 = \omega_3 = 1$). $GDice(.,.)$, $Focal(.,.)$, and $BCE(.,.)$ correspond to Generalized Dice, Focal, and Binary Cross Entropy losses. GDICE loss is specifically designed for highly imbalanced segmentation tasks, and Focal loss is beneficial in class imbalance scenarios, which is satisfied at the pixel level

AD, where anomalous pixels are significantly outnumbered by normal pixels. The overall loss is calculated as the sum of the losses at different feature levels: $Loss = \sum_i Loss^i$. During inference, $\hat{c}^i$ and $\hat{M}^i$ are averaged across feature layers to compute the final output of the model.

4 Experiments

We used the BMAD benchmark [3] in line with the latest SOTA: MVFA [19]. It consists of five modalities and six datasets: brain MRI [1,2,28], liver CT [6, 25], chest X-ray [37], retinal OCT (composed of two datasets; OCT17 [22], and RESC [17]), and digital histopathology (HIS) [4]. Brain, Liver, and RESC supply annotations for both AC and AS, while OCT17, Chest, and HIS are relevant only for AC. We report the results for AC and AS using the standard medical AD metric: Area Under the Receiver Operating Characteristic Curve (AUC).

Implementation Details. We use the CLIP [30] with the ViT-L/14 backbone in line with the recent SOTA, such as [10,19,41]. This CLIP model consists of 24 layers. We insert an adapter at the 12th layer and attach the projector to the 24th layer. This configuration is motivated by the ablation study in MVFA [19], which shows that using the 12th layer yields better performance compared to alternatives such as the 6th, 18th, or their combinations. The number of prompts in a prompt set is fixed to 4 as in PLOT [9]. Training is performed with the Adam optimizer with the learning rate of $1e^{-3}$, batch size 16, for 100 epochs. We set the adapter's residual ratio γ to 0.2 by default. Regarding the hyperparameters for solving POT, we fix the entropic regularization weight λ to 0.1, the maximum iteration number is set to 100, and the transport ratio *frac* is set to different values: 0.40.9 (step 0.1), while the final model uses 0.8. An early stopping is employed using a threshold of 0.001.

Table 1. Comparisons with SOTA in terms of AUC (%). Few-shot models use 16 samples per class. The results are reported in the AC/AS format. Best results are **bold**, second-best underlined.

	Method	HIS	Chest	OCT17	Brain	Liver	RESC	AVG
Unsup	CFLOWAD [16]	54.54	71.44	85.43	73.97/93.52	49.93/92.78	74.43/93.75	68.29/93.35
	RD4AD [11]	66.59	67.53	97.24	89.38/96.54	60.02/95.86	87.53/96.17	78.04/96.19
	PatchCore [31]	69.34	75.17	98.56	91.55/96.97	60.40/96.58	91.50/96.39	81.09/96.65
	MKD [32]	77.74	81.99	96.62	81.38/89.54	60.39/96.14	88.97/86.60	81.18/90.76
Few-shot	DRA [12]	79.16	85.01	99.87	82.99/80.45	80.89/93.00	94.88/84.01	87.13/85.82
	BGAD [39]	-	-	-	88.05/95.29	78.79/99.25	91.29/97.07	- /97.20
	APRIL-GAN [10]	81.16	78.62	99.93	94.03/96.17	82.94/99.64	95.96/98.47	88.77/98.09
	MediCLIP [41]	70.22	69.74	96.37	91.56/**98.08**	79.31/98.95	86.51/94.07	82.28/97.03
	MVFA [19]	82.62	85.72	99.66	94.40/97.70	83.85/99.73	97.25/99.07	90.58/98.83
	MADPOT (Ours)	**96.59**	**93.4**	**100**	**99.99**/97.97	**97.62/99.84**	**99.4/99.3**	**97.83/99.03**

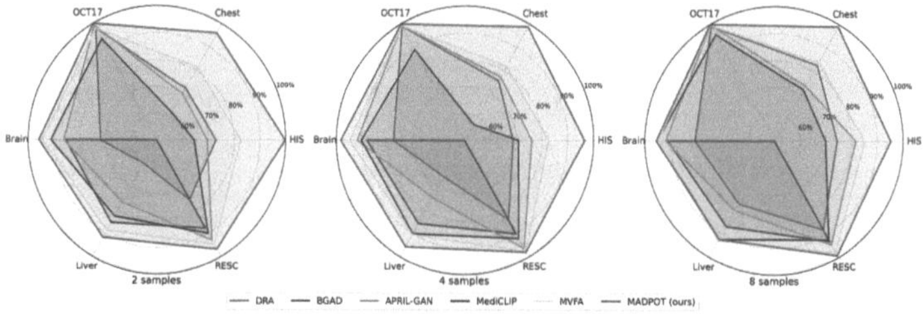

Fig. 2. Comparison with few-shot SOTA for 2, 4, and 8 shots per class (AUC %). For datasets with both AC and AS, the average of AC and AS is reported.

Comparing Few-Shot Performance with SOTA. Table 1 presents a comparison of MADPOT with SOTA methods. In that table, *Unsup* methods rely on large normal training datasets, while few-shot methods use 16 normal and 16 abnormal examples. MADPOT achieves the best AC performance across all six datasets, with an average AC score of 97.83%, surpassing the second-best (i.e., MVFA [19]) by 7.25%. Notably, on HIS and Chest, MADPOT achieves significant gains of 13.97% and 7.68%, respectively, showing its ability on the AC task. For AS, MADPOT achieves competitive performance, attaining near-perfect scores of 99.54% on Liver and 99.3% on RESC. On average, it outperforms all other methods, even though the second-best AS performance is close.

Moreover, Fig. 2 compares few-shot methods with varying numbers of normal and abnormal samples. As visualized by the outermost pink area in all plots, MADPOT consistently outperforms competing methods across all datasets, with particularly notable advantages in the Chest and HIS datasets. Although some methods, such as MVFA [19], show competitive performance in specific datasets, especially as the sample size increases to 8, none matches MADPOT across all. The consistent dominance of our approach confirms its value even when working with extremely limited training data.

Cross-dataset Analysis. MADPOT is evaluated against its strongest counterpart, MVFA [19], to assess its generalization performance in a cross-dataset setting. Each model was trained on 16 samples from the datasets described above and evaluated on unseen target datasets of the same modality (see Table 2). MADPOT consistently outperforms MVFA in all datasets, achieving an average performance gain of 5.75%, demonstrating its superior generalization.

Comparing Zero-Shot Performance with SOTA. Table 3 presents a comparison of MADPOT with zero-shot SOTA. MADPOT consistently achieves superior performance in AC, surpassing the second best across all evaluated datasets with a notable average gain of 7.99%. While our method's average AS score is marginally lower (by just 1.83% than the top performer), this slight difference is only due to the RESC dataset, where MADPOT lags by 6% AUC.

Ablation Studies. Several ablation studies were conducted to evaluate different components of our approach. This includes assessing the impact of prompt

Table 2. Cross-dataset evaluation. AC is reported for all datasets. Best are **bold**.

Source	Chest		Brain	OCT17	RESC	AVG
Target	NIHChest [34]	CheXpert [20]	ADNI [21]	OCTDL [24]		
MVFA [19]	61.91	80.41	53.94	88.47	93.94	75.73
MADPOT (Ours)	**64.9**	**90.69**	**54.63**	**99.21**	**97.95**	**81.48**

Table 3. Comparisons with SOTA in terms of AUC (%) in *zero-shot* scenario. The results are reported in the AC/AS format. Best results are **bold**, second-best <u>underlined</u>.

	Method	HIS	Chest	OCT17	Brain	Liver	RESC	AVG
Zero-shot	APRIL-GAN [10]	72.36	57.49	92.61	76.43/<u>91.79</u>	70.57/<u>97.05</u>	75.67/85.23	74.19/91.36
	MVFA [19]	<u>77.9</u>	<u>71.11</u>	<u>95.4</u>	<u>78.63</u>/90.27	<u>76.24</u>/**97.85**	<u>83.31</u>/**92.05**	<u>80.43</u>/**93.39**
	MADPOT (Ours)	**90.35**	**76.81**	**95.9**	**94.24/93.81**	**88.21**/96.27	**85.04**/<u>86.6</u>	**88.42**/<u>91.56</u>

learning strategies in the text branch (Table 4), evaluating the effectiveness of projection and adapter layers in the vision branch (Table 5), and the effect of POT parameter *frac*. Table 4 presents six variations based on the **inclusion or exclusion of CL, OT, POT, and their combinations** (e.g., CL with OT or CL with POT). Note that as POT is derived from OT, they cannot be used in combination. The case where all components are excluded (first row in the table) corresponds to using fixed, hand-crafted prompts, specifically adopting those from MVFA [19]. The case where only CL is enabled corresponds to using learnable prompts following the CoOP approach [42]. Our analysis reveals several key findings. *i)* Learnable prompts outperform fixed prompts, and the combination of CL and POT achieves the highest performance. *i)* OT or POT alone leads to significant performance degradation, demonstrating their inability to learn effective prompts without CL. *iii)* While CL alone improves AS (98.7% vs. 97.94%), the addition of POT provides the most balanced improvement, boosting AC by 4.93% compared to fixed prompts without compromising AS

Table 4. Ablation study evaluating the impact of different prompt learning strategies in *few-shot* scenario. Results are reported in the AC/AS format, with *few-shot* referring to 16 samples per class.

Scenario	CL	OT	POT	HIS	Chest	OCT17	Brain	Liver	RESC	AVG
few-shot	✗	✗	✗	<u>86.87</u>	<u>88.26</u>	99.71	93.91/95.73	92.35/99.62	96.33/98.46	92.90/97.94
	✓	✗	✗	85.61	89.11	99.47	94.13/<u>97.24</u>	92.35/<u>99.78</u>	97.45/<u>99.07</u>	93.02/<u>98.70</u>
	✗	✓	✗	9.8	46.55	75.43	15.75/51.9	55.01/53.05	52.59/52.05	42.52/52.33
	✗	✗	✓	67.54	48.25	83.34	20.36/49.76	61.09/49.87	52.49/50.42	55.51/50.02
	✓	✓	✗	79.05	88.25	**100**	<u>99.95</u>/95.90	<u>94.07</u>/99.15	**99.85**/98.87	<u>93.53</u>/97.97
	✓	✗	✓	**96.59**	**93.4**	**100**	**99.99/97.97**	**97.62/99.84**	<u>99.40</u>/**99.30**	**97.83/99.03**

Table 5. Ablation study evaluating the impact of adapter and projector. Results are reported in the AC/AS format, with *few-shot* referring to 16 samples per class.

Scenario	Adapter	Projector	HIS	Chest	OCT17	Brain	Liver	RESC	AVG
few-shot	✗	✗	79.72	85.74	<u>97.69</u>	88.57/96.33	67.25/98.05	97.12/<u>96.44</u>	86.01/<u>96.94</u>
	✓	✗	**97.71**	<u>86.26</u>	84.85	**100**/<u>96.62</u>	<u>96.57</u>/<u>98.95</u>	51.58/89.25	<u>86.16</u>/94.94
	✗	✓	58.87	52.52	70.09	99.59/91.85	86.8/96.24	<u>97.5</u>/<u>96.44</u>	77.56/94.84
	✓	✓	<u>96.59</u>	**93.4**	**100**	<u>99.99</u>/**97.97**	**97.62/99.84**	**99.40/99.30**	**97.83/99.03**

performance. *iv)* OT/POT shows clear benefits in AC, with gains also evident in AS. Table 5 evaluates **the impact of the adapter and projector**, showing that using the CLIP vision encoder without any adaptation (i.e., the first row) leads to suboptimal performance. This highlights the necessity of adaptation to better align the vision encoder with the AD task. Among the adaptation methods, the adapter was found to outperform the projector, while their combination achieved the best performance, with gains of 11.67% in AC and 4.19% in AS over the adapter-only setting. When considering the results in Table 4 and Table 5 together, it becomes clear that removing vision adaptation while retaining prompt learning with POT and CL yields lower results, particularly for AC, compared to the opposite configuration. **The impact of *frac*** on AC and AS was studied across three datasets: Brain, Liver, and RESC. Results showed sensitivity to this hyperparameter, with optimal performance between 0.5 and 0.8 (AC: 91.94%-99.00%, AS: 96.79%-99.03%). Extreme values (*frac* = 0.4 or 0.9) caused significant degradation (AC/AS: 81.61%/95.21% for 0.4, 82.76%/95.53% for 0.9). Peak performance was achieved at 0.8, suggesting that excluding 20% of the mass enhances relevant features.

5 Conclusion

We presented MADPOT, which adapts CLIP to medical AD by integrating learnable prompt tuning with POT and CL, and incorporating adaptive vision components for both semantic understanding and pixel-level segmentation in few/zero-shot settings. MADPOT outperforms baselines across diverse datasets, demonstrating robustness and generalization to varying modalities and distributions. Ablation studies show that combining learnable prompts with CL and POT is most effective in the text branch, while the adapterprojector pair drives vision performance, with notable gains. Their joint use further boosts results. Future work will extend MADPOT to support temporal data processing.

Acknowledgments. We acknowledge the financial support of the PNRR project FAIR - Future AI Research (PE00000013), under the NRRP MUR program funded by the NextGenerationEU.

Disclosure of Interests. The authors have no competing interests to declare that are relevant to the content of this article.

References

1. Baid, U., et al.: The RSNA-ASNR-MICCAI brats 2021 benchmark on brain tumor segmentation and radiogenomic classification. arXiv:2107.02314 (2021)
2. Bakas, S., et al.: Advancing the cancer genome atlas glioma MRI collections with expert segmentation labels and radiomic features. Sci. Data **4**, 1–13 (2017)
3. Bao, J., et al.: BMAD: benchmarks for medical anomaly detection. In: CVPR, pp. 4042–4053 (2024)
4. Bejnordi, B.E., et al.: Diagnostic assessment of deep learning algorithms for detection of lymph node metastases in women with breast cancer. JAMA **318**(22), 2199–2210 (2017)
5. Benamou, J.D., et al.: Iterative Bregman projections for regularized transportation problems. SIAM J. Sci. Comput. **37**(2), A1111–A1138 (2015)
6. Bilic, P., et al.: The liver tumor segmentation benchmark (LITS). Med. Image Anal. **84**, 102680 (2023)
7. Cao, Y., et al.: Adaclip: adapting clip with hybrid learnable prompts for zero-shot anomaly detection. In: ECCV, pp. 55–72 (2024)
8. Chang, W., et al.: Csot: curriculum and structure-aware optimal transport for learning with noisy labels. NeurIPS **36**, 8528–8541 (2023)
9. Chen, G., et al.: Plot: prompt learning with optimal transport for vision-language models. arXiv preprint arXiv:2210.01253 (2022)
10. Chen, X., et al.: A zero-/fewshot anomaly classification and segmentation method. arXiv:2305.17382 (2023)
11. Deng, H., Li, X.: Anomaly detection via reverse distillation from one-class embedding. In: CVPR, pp. 9737–9746 (2022)
12. Ding, C., et al.: Catching both gray and black swans: open-set supervised anomaly detection. In: CVPR, pp. 7388–7398 (2022)
13. Dykstra, R.L.: An algorithm for restricted least squares regression. J. Am. Stat. Assoc. **78**(384), 837–842 (1983)
14. Feng, C., et al.: Ot-filter: an optimal transport filter for learning with noisy labels. In: CVPR, pp. 16164–16174 (2023)
15. Frogner, C., et al.: Learning with a wasserstein loss. Adv. Neural Inf. Process. Syst. **28** (2015)
16. Gudovskiy, D., et al.: CFlow-ad: real-time unsupervised anomaly detection with localization via conditional normalizing flows. In: WACV, pp. 98–107 (2022)
17. Hu, J., et al.: Automated segmentation of macular edema in oct using deep neural networks. Med. Image Anal. **55**, 216–227 (2019)
18. Huang, C., et al.: Registration based few-shot anomaly detection. In: ECCV, pp. 303–319 (2022)
19. Huang, C., et al.: Adapting visual-language models for generalizable anomaly detection in medical images. In: CVPR, pp. 11375–11385 (2024)
20. Irvin, J., et al.: Chexpert: a large chest radiograph dataset with uncertainty labels and expert comparison. In: AAAI, pp. 590–597 (2019)
21. Jack, C.R., Jr., et al.: The Alzheimer's disease neuroimaging initiative (ADNI): MRI methods. J. Magn. Reson. Imag. **27**(4), 685–691 (2008)
22. Kermany, D.S., et al.: Identifying medical diagnoses and treatable diseases by image-based deep learning. Cell **172**(5), 1122–1131 (2018)
23. Koehl, P., et al.: Physics approach to the variable-mass optimal-transport problem. Phys. Rev. E **103**(1), 012113 (2021)

24. Kulyabin, M., et al.: OCTDL: optical coherence tomography dataset for image-based deep learning methods. Sci. Data **11** (2024)
25. Landman, B., et al.: MICCAI multi-atlas labeling beyond the cranial vault–workshop and challenge. In: MICCAI, vol. 5, p. 12 (2015)
26. Li, B., et al.: Unsupervised cross-domain image retrieval via prototypical optimal transport. In: AAAI, vol. 38, pp. 3009–3017 (2024)
27. Li, H., et al.: Global and local prompts cooperation via optimal transport for federated learning. In: CVPR, pp. 12151–12161 (2024)
28. Menze, B.H., et al.: The multimodal brain tumor image segmentation benchmark (brats). IEEE Trans. Med. Imag. **34**(10), 1993–2024 (2014)
29. Peyré, G.E.A.: Computational optimal transport: with applications to data science. Found. Trends Mach. Learn. **11**(5–6), 355–607 (2019)
30. Radford, A., et al.: Learning transferable visual models from natural language supervision. In: ICML, pp. 8748–8763 (2021)
31. Roth, K., et al.: Towards total recall in industrial anomaly detection. In: CVPR, pp. 14318–14328 (2022)
32. Salehi, M., et al.: Multiresolution knowledge distillation for anomaly detection. In: CVPR, pp. 14902–14912 (2021)
33. Sinkhorn, R.: Diagonal equivalence to matrices with prescribed row and column sums. Am. Math. Mon. **74**(4), 402–405 (1967)
34. Summers, R.: NIH Chest X-Ray Dataset of 14 Common Thorax Disease Categories. NIH Clinical Center, Bethesda, MD, USA (2019)
35. Tu, R., et al.: Optimal transport for causal discovery. arXiv:2201.09366 (2022)
36. Villani, C., et al.: Optimal Transport: Old and New, vol. 338 (2008)
37. Wang, X., et al.: Chestx-ray8: hospital-scale chest x-ray database and benchmarks. In: CVPR, pp. 2097–2106 (2017)
38. Wang, Y., et al.: Probability-polarized optimal transport for unsupervised domain adaptation. In: AAAI, vol. 38, pp. 15653–15661 (2024)
39. Yao, X., et al.: Explicit boundary guided semi-push-pull contrastive learning for supervised anomaly detection. In: CVPR, pp. 24490–24499 (2023)
40. Zhang, J., et al.: Viral pneumonia screening on chest x-rays using confidence-aware anomaly detection. IEEE TMI **40**(3), 879–890 (2020)
41. Zhang, X., et al.: MediClip: adapting clip for few-shot medical image anomaly detection. In: MICCAI, pp. 458–468 (2024)
42. Zhou, K., et al.: Learning to prompt for vision-language models. IJCV (2022)

Predicting Soccer Penalty Kick Direction Using Human Action Recognition

David Freire-Obregón(✉), Oliverio J. Santana, Javier Lorenzo-Navarro, Daniel Hernández-Sosa, and Modesto Castrillón-Santana

SIANI, Universidad de Las Palmas de Gran Canaria, Canaria, Spain
david.freire@ulpgc.es

Abstract. Action anticipation has become a prominent topic in Human Action Recognition (HAR). However, its application to real-world sports scenarios remains limited by the availability of suitable annotated datasets. This work presents a novel dataset of manually annotated soccer penalty kicks to predict shot direction based on pre-kick player movements. We propose a deep learning classifier to benchmark this dataset that integrates HAR-based feature embeddings with contextual metadata. We evaluate twenty-two backbone models across seven architecture families (MViTv2, MViTv1, SlowFast, Slow, X3D, I3D, C2D), achieving up to 63.9% accuracy in predicting shot direction (left or right)–outperforming the real goalkeepers' decisions. These results demonstrate the dataset's value for anticipatory action recognition and validate our model's potential as a generalizable approach for sports-based predictive tasks.

Keywords: Human Action Recognition · Action Prediction · Soccer · Vision Transformers · Penalty Kick

1 Introduction

Human action recognition (HAR) in video has gained significant attention in recent years [17], with action anticipation–predicting future actions from past observations–being one of its most challenging tasks [16]. Deep learning approaches are increasingly used for this purpose [13], but they require specialized datasets. In FIFA World Cups, shootouts decided three titles (1982, 1994, 2006), while UEFA Champions League finals saw penalties settle 30% of matches over 23 years. These moments highlight their significance in club and international soccer. This paper presents a dataset of 1,010 penalty kick clips from official soccer matches. It aims to train neural networks with HAR embeddings to predict shot direction. Model development is based on a representative subset of the data. Penalty kicks are pivotal in soccer, shaping match outcomes and tournament histories.

Integrating technology into soccer has significantly improved the analysis of game dynamics. The combination of video processing, data visualization, and

E. Rodolà et al. (Eds.): ICIAP 2025, LNCS 16167, pp. 260–272, 2026.
https://doi.org/10.1007/978-3-032-10185-3_21

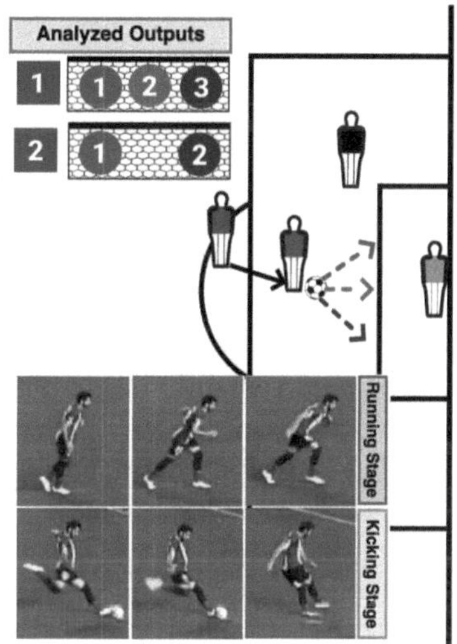

Fig. 1. Penalty-kick direction estimation. Our method predicts ball direction by combining metadata and HAR embeddings, using only kicker actions without ball trajectory data.

automation enables more profound insights into player behavior [12], tactical formations [18], and key moment detection [5]. These advances support applications such as strategy analysis and predictive modeling. In this context, penalty kick outcome prediction can benefit from incorporating factors like tempo, technique, and individual skills.

The motivation behind this work is to support goalkeeper decision-making through data-driven video analysis, both for offline tactical preparation and potentially for online support during matches, assuming future integration with a real-time tracking system. While personalized models could be valuable, our approach focuses on general motion cues that may reveal intent in real-time. The proposed method processes video clips using HAR encoders split into running and kicking phases (see Fig. 1), combined with contextual metadata in a two-stream classifier. We evaluate three-class (left, center, right) and two-class (left, right) formulations. The binary setup helps mitigate ambiguity in the "center" class, which can be visually and semantically unclear due to inconsistencies in human annotation and variations in camera angles. Results show that shot direction can be predicted with over 50% accuracy–exceeding 60% in the binary setting–outperforming the behaviors observed in real goalkeepers.

2 Related Work

Sports analysis has gained significant research interest, with technology transforming our understanding of sports and athlete performance. This research focuses on sports datasets and associated challenges, specifically within the context of ball sports, where the interaction between players and the ball is crucial in determining outcomes.

Sports datasets are generally categorized into still-image and video-sequence types. Still-image datasets, like the Leeds Sports Pose Dataset, focus on image classification and athlete poses. Video-sequence datasets, like the MTL-AQA diving dataset [21] and Fis-V skating dataset [28], provide dynamic scene insights for action recognition and player tracking, even under challenging conditions. Usually, the semantic structure of sports video content is organized into four layers: raw video, object, event, and semantic layers [23]. Raw video is processed to identify objects like players [11] and the ball [26]. The event layer captures actions and interactions, with research areas including action recognition [10], re-identification [1], facial expression recognition [22], and trajectory prediction [25]. The semantic layer summarizes the footage's content [4]. For this study, the semantic layer categorizes penalty kick outcomes based on the kicker's actions [2], excluding team dynamics to focus on individual actions and minimizing noise from other individuals present during the shot.

The goalkeeper's behavior during penalty kicks has been widely studied due to its key role in high-pressure, structured situations. Hunter et al. [15] proposed a model to improve goal likelihood by analyzing kicker and goalkeeper strategies. Earlier work [14] showed that goalkeepers anticipate shot direction using the kicker's cues, with kick speed being critical. Noël et al. [20] examined two strategies: keeper-independent (pre-decided shot direction) and keeper-dependent (reactive to the goalkeeper), finding the former more common (7886%) with similar success rates. Recent work highlights goalkeeper deception, showing that subtle misleading actions can reduce goal conversion, especially when kickers attend to the goalkeeper [30]. Our work proposes an automated approach, designing a deep learning model to predict the shot direction from kicker movements.

3 Method

3.1 Problem Formulation

Given a set of m of penalty kick samples, an observation $o^{(i)} = (F^{(i)}, y^{(i)}, \Gamma^{(i)})$ for $i = 1 \dots m$, is composed of annotated footage $F^{(i)}$, a label $y^{(i)}$ that corresponds to the direction of the shot, and contextual information $\Gamma^{(i)}$. The latter includes two distinct input variables known before the kicking action: the penalty-kick field side and the kicking foot. Both variables serve as binary indicators, distinguishing between left and right. Regarding the labels, two different scenarios are considered in this work: $y^{(i)} \in [left, center, right]$ and $y^{(i)} \in [left, right]$. Thus, the problem can be stated as finding the model parameter set θ of an end-to-end classifier that minimizes the log loss function:

$$J(\theta) = -\frac{1}{m}\sum_{i=1}^{m}\sum_{k=1}^{n} p(y^{(i)} = k)\log(p(\hat{y}(i) = k)) \tag{1}$$

where m is the number of penalty kick samples, n is the number of classes, $p(y^{(i)} = k)$ is the probability that the observation $o^{(i)}$ belongs to class k, and $p(\hat{y}^{(i)} = k)$ is the probability that the predicted class for the observation $o^{(i)}$ belongs to class k.

3.2 Multistage and Classifier Formulation

First, the original footage $F^{(i)}$ undergoes a pre-processing step to constrain the context, resulting in $F'^{(i)}$. Then, each footage $F'^{(i)}$ is manually divided into two sets of frames: 32 for the running stage and 16 for the kicking stage, values empirically chosen to cover typical pre-kick motions and maintain temporal balance across clips. Consequently, $F'^{(i)} = F'^{(i)}_{run} \cup F'^{(i)}_{kick}$.

Secondly, the feature extraction process involves passing the frames from the running and kicking stages through HAR backbones, grouped into chunks of frames, denoted as $\Omega(F'^{(i)}_{run}) = \{f^{i,1}_{run}, \ldots, f^{i,n^i_f}_{run}\}$ and $\Omega(F'^{(i)}_{kick}) = \{f^{i,1}_{kick}, \ldots, f^{i,n^i_f}_{kick}\}$ respectively, where $f^{i,j}_{run}$ and $f^{i,j}_{kick}$ represent the embedding of the j-th chunk of the footage $F'^{(i)}$, being n^i_f the number of chunks into which the footage was split. In order to obtain a single embedding per footage, a pooling operation is introduced as follows:

$$\begin{aligned} T^{(i)}_{run} &= pool(\Omega(F'^{(i)}_{run})) \\ T^{(i)}_{kick} &= pool(\Omega(F'^{(i)}_{kick})) \end{aligned} \quad (2)$$

where $pool \in [Average, Max]$ denotes the pooling operation. The footage embeddings, denoted as $T^{(i)}_{run}$ and $T^{(i)}_{kick}$, are then utilized as input for a subsequent classifier.

Consequently, if m represents the total number of penalty-kick instances in the dataset, the objective is to learn a mapping function C that predicts the kick direction based on the features extracted from the running and kicking stages, and contextual information:

$$C : (T^{(i)}_{\text{run}}, T^{(i)}_{\text{kick}}, \Gamma^{(i)}) \rightarrow y^{(i)} \quad (3)$$

where $T^{(i)}_{\text{run}}$ and $T^{(i)}_{\text{kick}}$ are the embeddings corresponding to the i-th penalty-kick instance, and $y^{(i)}$ is the direction of the shot, as previously defined. The classifier, C, integrates information from both the running and kicking stages along with contextual information ($\Gamma^{(i)}$).

4 Pipeline Description

This paper introduces a sequential pipeline with two core modules: embedding extraction and a classifier. Video pre-processing is manually executed before the pipeline. Visual representations are shown in Figs. 2 and 3.

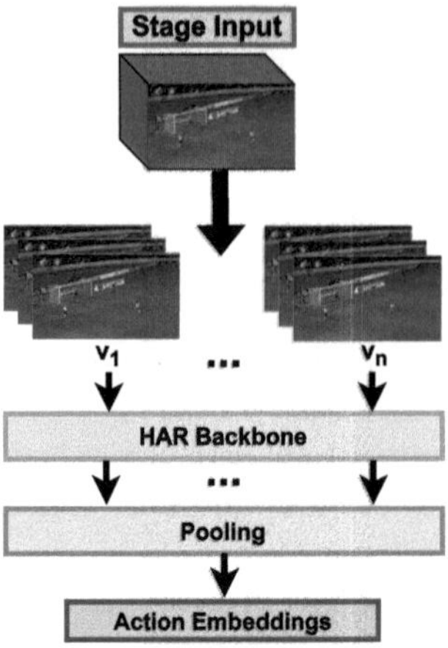

Fig. 2. Embeddings extraction module. The video is downsampled into n clips. A pre-trained HAR model extracts features, combined via pooling (average or max) to produce a final tensor for the classifier.

Step 1: Video Pre-processing: Context Constraint. To optimize embedding quality, extraneous elements (e.g., unrelated players, staff, supporters, referees) are removed from the input footage [10]. The primary subject, the kicker, is isolated using ByteTrack [29], a precise multi-object tracking network. Context-constrained pre-processing creates an ideal experimental setting.

For a kicker (i) at time (t) within interval ($[0, T]$), the bounding box ($BB^{(i)}(t)$) outlines the kicker's area in the frame. The pre-processed frame ($F'^{(i)}(t)$) is generated by superimposing $BB^{(i)}(t)$ onto the average frame ($\overline{f}$), computed from τ frames, where τ corresponds to the total number of frames in the considered sequence (e.g., $\tau = 48$ for a sequence composed of 32 running and 16 kicking frames):

$$F'^{(i)} = BB^{(i)}(t) \cup \overline{f} \tag{4}$$

The $\cup$ operation aligns the bounding box with the static background, producing footage with the kicker as the sole moving element. We first compute an average frame over the entire input sequence to represent the static background to achieve this. Then, for each frame in the sequence, the kicker's bounding box is extracted and superimposed onto this average frame. This results in a new sequence where the background remains fixed and only the kicker's motion is preserved, allowing the model to focus on the relevant dynamics. Since there is no camera movement during the analyzed stage of the penalty kick, this alignment is direct and does not require additional spatial transformations. Moreover, the footage was temporally cropped to begin at the start of the penalty sequence. In cases where the running phase is shorter than expected (i.e., 32 frames), temporal padding is applied to maintain a consistent input length.

Step 2: HAR Embeddings Extraction. Input footage $F'^{(i)}$ is downsampled into n_f^i overlapping clips, spaced one frame apart. A pre-trained HAR model extracts features consolidated via pooling (Eq. 2) into a single embedding. Two pooling methods (average and max) are evaluated. Twenty-two HAR backbones, ranging from MViT and Slowfast to C2D and X3D, are tested for kick direction challenges. The C2D model uses 2D CNNs for spatial feature extraction [24], while SlowFast employs a dual-pathway approach for temporal resolution [8]. Enhanced versions, SlowFastNLN and SlowNLN, integrate non-local operations for spatiotemporal dependencies [27] (NLN). Contrary, the I3D model processes video clips as 3D volumes [3], and its variant I3D_NLN includes non-local operations. The X3D model family (X3D-XS to X3D-L) adapts spatial and temporal dimensions [7]. Vision Transformers like MViTv1 [6] and MViTv2 [19] use multiscale feature hierarchies for improved recognition.

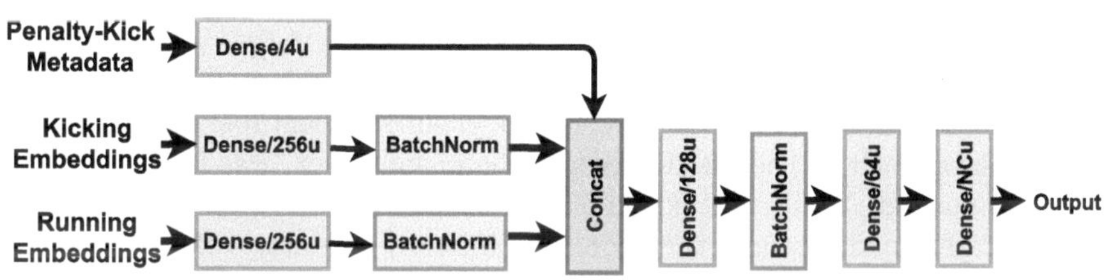

Fig. 3. The proposed classifier. It combines HAR backbone features and penalty-kick metadata for enhanced decision-making during running and kicking stages.

Step 3: Classification Module. The classifier integrates HAR backbone features and metadata, including the pitch side and the kicker's foot (Fig. 3). Inputs are processed via fully connected layers, concatenated, and passed through additional layers. Outputs classify ball position as *left*, *right*, or *center*, or as a binary *left/right* decision.

Fig. 4. Penalty-kick dataset samples. The dataset, sourced from the Internet, exhibits pose, scale, and lighting variability. Clips are edited to focus on the running and kicking stages.

5 Experiments and Results

This section covers dataset acquisition, experimental setup, and results. It begins with dataset details, including acquisition and cleaning, followed by methodology insights and data partitioning. Finally, it summarizes the experimental outcomes.

Dataset. No publicly available dataset for soccer penalty kicks exists. Our dataset, collected from online sources, exhibits variations in pose, scale, and lighting (see Fig. 4). Web scraping with keywords like "penalty-kick shootout" yielded 1010 clips, trimmed to include the run-up and kick result. The original videos are in 1280×720 resolution and last between 3 and 6 s. They capture the player from just before starting the run-up until the ball reaches the net, the goalkeeper, the post, or goes out of play. For our study, each clip was trimmed to 48 frames: 32 frames capturing the player's run-up, followed by 8 frames before and 8 frames after the kicking moment. This interval excludes visual information about the ball's trajectory or outcome, focusing solely on kick motion cues.

Camera viewpoint and video duration were critical. Clips with unsuitable perspectives (e.g., behind the goalkeeper) or fewer than 64 frames were excluded, reducing the dataset to 640 clips. ByteTrack ensured acceptable detection quality. Metadata, such as the pitch side and the kicking foot, was incorporated to enrich the dataset. While contextual features such as the kicking foot are manually annotated in this study, they could be inferred automatically in real-world settings using pose estimation or approach-angle analysis. Of the 640 clips, 395 penalties were on the right side of the pitch, 245 on the left, with 498 taken by right-footed players and 142 by left-footed players.

Table 1. Performance Comparison of HAR Architectures for Kick Estimation with Three Classes. This table compares the top backbones used in penalty-kick direction estimation across different architectures, along with the goalkeeper baseline (GK Baseline), corresponding to the actual direction chosen by the goalkeeper during each penalty kick. The table lists the architectures, optimal model variants, number of frames used for HAR embedding extraction, and additional performance metrics.

Architecture	Best Model	#Frames	Pooling	Accuracy	Precision	Recall	F1-Score
GK Baseline	-	-	-	46.0%	35.3%	44.2%	38.4%
C2D [24]	C2D_R50	8	Max	47.1%	33.1%	45.6%	38.4%
I3D [3]	I3D_R50	8	Average	45.0%	32.7%	46.3%	38.3%
Slow [9]	Slow8x8	8	Average	46.7%	28.9%	45.9%	35.5%
SlowFast [8]	SlowFast4x16	32	Average	46.2%	33.6%	46.4%	39.1%
NLN [27]	Slow_NLN_4x16	32	Max	45.1%	32.8%	47.8%	38.9%
X3D [7]	X3D_M	13	Max	45.9%	33.7%	45.3%	38.6%
MViTv1 [6]	MViT_CONV	32	Average	**51.9%**	34.4%	**47.1%**	39.7%
MViTv2 [19]	MViTv2_S	16	Average	51.6%	**35.5%**	45.8%	**40.1%**

Experimental Setup. We employed a 10-fold cross-validation strategy on our 640-sample dataset, with eight folds used for training, one for validation, and one for testing. Each fold contains 64 samples, roughly preserving the class distribution: ~30 right (47.3%), ~23 left (35.8%), and ~11 center (16.9%). Although class imbalance is present, we compensated for it by applying class weights during training. Underrepresented classes, such as *center*, were given higher importance in the loss function, helping the model learn from all classes effectively. While each fold includes 64 samples for testing and validation, cross-validation over 10 folds provides comprehensive coverage across the whole dataset, thereby mitigating the risks of bias and enhancing robustness. The dataset consists of 229 left, 303 right, and 108 center-directed penalty kicks, showing class imbalance.

Each architecture in Tables 1 and 2 includes the best-performing instance among several evaluated variants. The column #Frames indicates the number of frames processed per HAR input clip; shorter clips were temporally padded at the start. We evaluated average and max pooling strategies for embedding aggregation and selected the method that achieved the highest average validation performance during cross-validation.

Table 2. Performance Comparison of HAR Architectures for Kick Estimation with Two Classes. This table compares the top backbones used in penalty-kick direction estimation across different architectures, along with the goalkeeper baseline (GK Baseline), corresponding to the actual direction chosen by the goalkeeper during each penalty kick. The table lists the architectures, optimal model variants, number of frames used for HAR embedding extraction, and additional performance metrics.

Architecture	Best Model	#Frames	Pooling	Accuracy	Precision	Recall	F1-Score
GK Baseline	-	-	-	54.2%	53.3%	53.2%	53.3%
C2D [24]	C2D_R50	8	Max	58.3%	58.4%	55.8%	57.1%
I3D [3]	I3D_R50_IN1K	8	Max	55.4%	55.5%	54.2%	54.9%
Slow [9]	Slow4x16	4	Max	61.1%	**66.7%**	57.3%	61.6%
SlowFast [8]	SlowFast8x8	32	Max	60.5%	60.4%	57.7%	58.7%
NLN [27]	SlowFast8x8	8	Average	61.6%	60.3%	58.8%	59.5%
X3D [7]	X3D_M	13	Max	60.1%	59.1%	59.2%	59.2%
MViTv1 [6]	MViT_CONV	32	Average	61.8%	58.4%	58.1%	58.2%
MViTv2 [19]	MViTv2_S	16	Average	**63.9%**	64.9%	**60.2%**	**62.5%**

Table 1 compares HAR architectures for three-class kick estimation. Each row details an architecture, best model, frame count, pooling technique, and key performance metrics. MViTv1 and MViTv2 achieved the highest accuracy (51.9% and 51.6%), excelling in pattern recognition. Architectural choices, such as backbone selection and pooling strategies, significantly impact performance. SlowFast4x16 delivered competitive results, emphasizing tailored architectural design. The choice between max and average pooling influences outcomes, highlighting the need for strategic pooling selection. Table 2 evaluates HAR architectures for binary-class kick estimation (*right*, *left*). Accuracy is generally higher (55.9% to 63.9%), with MViTv2 leading at 63.9%. Slow4x16 achieved 66.7% precision, while MViTv2 balanced precision (64.9%), recall (60.2%), and F1-Score (62.5%). Pooling strategies significantly impact model performance.

We examined the classification outcomes for both the three-class and two-class tasks through confusion matrix analysis. In the three-class setting, the model correctly classified 60.4% of left shots and 61.3% of right shots, while center shots were correctly identified only 12.0% of the time, with the remaining misclassified as left (46.0%) or right (42.0%). This highlights the difficulty of reliably distinguishing the *center* class, likely due to its ambiguous visual definition. From the camera perspective, subtle differences between center and slightly off-center shots are challenging to detect, and the subjective nature of manual annotation may contribute to inconsistencies in labeling. This conceptual vagueness motivated our inclusion of a two-class setup, which removes the ambiguous center category and yields more reliable performance. In this regard, the two-class task showed improved and more stable performance, with left shots classified correctly 64.2% of the time and right shots 86.5%, supporting our decision to adopt this simplified setup.

As anticipated, the two-class models consistently outperform the three-class models across both tables, demonstrating higher accuracy. MViTv2 excels in kick direction estimation, achieving 51.6% accuracy in the three-class task, just behind MViTv1 (51.9%). However, it leads in binary classification with 63.9% accuracy, demonstrating its adaptability in simplifying classification while maintaining precision and recall. Its balanced performance–64.9% precision, 60.2% recall, and 62.5% F1-Score–ensures accurate predictions with minimal false positives, making it a reliable choice compared to models like Slow4x16, which, despite higher precision (66.7%), lacks the same recall and F1-Score.

MViTv2's superior performance stems from its use of average pooling, which provides a holistic view of temporal data rather than emphasizing only salient features like max pooling. This contributes to its ability to capture long-range dependencies and contextual relationships, crucial for analyzing dynamic actions such as penalty kicks. Consequently, transformer-based architectures like MViTv2 excel in interpreting sequential movements, setting them apart from traditional HAR models and solidifying their role in improving kick direction estimation. Metadata incorporation improved accuracy by 2–3% and F1-Score by 5%. Using two-stream embedding branches instead of one increased accuracy by 3–5%, highlighting the importance of integrating running and kicking phases.

6 Architecture Performance Analysis

We evaluated 22 HAR models selected to represent a broad range of state-of-the-art architectures. These 22 models correspond to variants across seven core architecture families, offering a representative sample of diverse HAR design strategies. The goal was to cover various HAR design principles and analyze their generalization and robustness in the penalty kick prediction task. While the previous section focused on the best-performing models, this section offers a broader analysis of each architecture's behavior across all its variants. Using boxplots, we highlight performance variability and consistency, helping to identify high-performing models and the most robust and stable architecture families. Figures 5 and 6 show the accuracy and F1-Score distributions for all evaluated models under the 2-class experiment configuration. To enhance clarity, NLN variants are grouped according to their source architectures, and green markers indicate the best accuracy achieved by each architecture, aligning with the top-performing instances reported in Table 2.

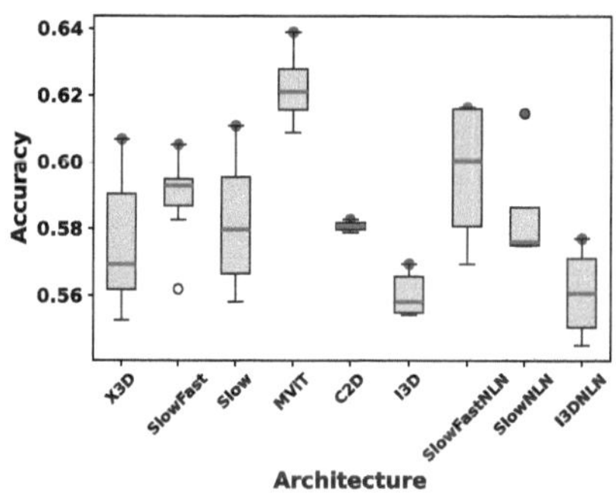

Fig. 5. Accuracy Analysis.

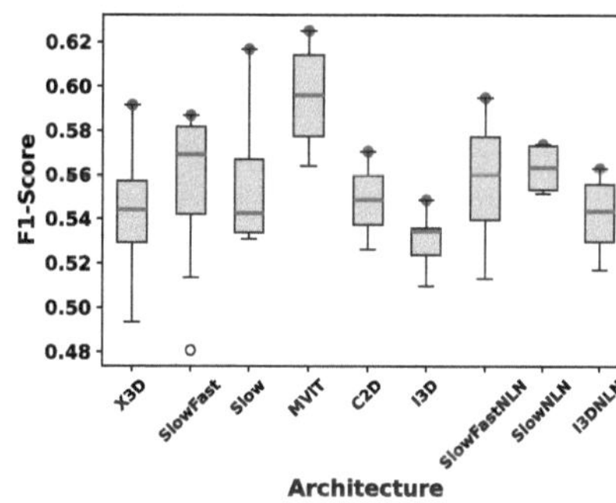

Fig. 6. F1-Score Analysis.

Figure 5 presents a boxplot analysis of accuracy distributions across architectures. MViT models show a broader performance range, with the best-reaching 63.9% accuracy. In contrast, I3DNLN and Slow models exhibit narrower distributions, indicating more consistent results. SlowFastNLN displays more significant variability, while SlowNLN includes a notable outlier with significantly higher accuracy, suggesting an exceptional configuration. Such outliers may arise from model design or data nuances. Overall, performance trends remain consistent between base models (SlowFast, Slow, I3D) and their NLN-enhanced versions.

Figure 6 visually captures the variation in F1-Score across the examined HAR architectures. Some architectures exhibit relatively consistent performance, as indicated by narrow interquartile ranges and medians near the center of the boxes. Notably, SlowFast and MViT demonstrate higher median F1-Scores, suggesting more reliable performance across different scenarios. Conversely, architectures like Slow display a wider interquartile range, indicating greater variability in F1-Score performance. This variability may be attributed to the sensitivity of these models to certain aspects of the input data or the complexity of the recognition task.

7 Conclusions

This study presents a penalty-kick dataset designed to train deep learning models for short-term human action anticipation. Building on this dataset, we propose a method for predicting penalty-kick direction using HAR embeddings. Our pipeline includes player tracking via ByteTrack, segmentation of running and kicking phases, and feature extraction. We evaluate 22 models across seven architectures–including NLN variants–identifying optimal configurations. Among them, MViTv1 and MViTv2 achieved 51.9% and 51.6% accuracy in three-class classification, while MViTv2 led in binary classification with 63.9%. Precision-recall trade-offs indicate balanced performance. Significantly, our best-performing models surpass the baseline accuracy of human goalkeepers, highlighting the potential of learned motion cues in decision-making under pressure. These results demonstrate that the dataset contains rich and representative HAR features, enabling effective anticipation of human actions. As such, it offers a valuable benchmark for testing new HAR models and advancing the state of the art. Despite the inherent difficulty of the task, the model's ability to generalize across subtle visual cues supports its practical application in real-time penalty analysis.

Acknowledgments. This work is partially funded by project PID2021-122402OB-C22/MICIU/AEI /10.13039/501100011033 FEDER, UE and by the ACIISI-Gobierno de Canarias and European FEDER funds under project ULPGC Facilities Net and Grant EIS 2021 04.

References

1. Akan, S., Varli, S.: Reidentifying soccer players in broadcast videos using body feature alignment based on pose. In: Int. Conf. on Computing, Networks and Internet of Things, pp. 440–444 (2023)
2. Artiles, J., Hernández-Sosa, D., Santana, O., Lorenzo-Navarro, J., Freire-Obregón, D.: Classifying soccer ball-on-goal position through kicker shooting action. In: Int. Conf. on Pattern Recognition Applications and Methods, pp. 79–90. SciTePress (2024)
3. Carreira, J., Zisserman, A.: Quo vadis, action recognition? a new model and the kinetics dataset. In: IEEE Conf. on Computer Vision and Pattern Recognition, pp. 4724–4733 (2017)
4. Cioppa, A., Deliège, A., Van Droogenbroeck, M.: A bottom-up approach based on semantics for the interpretation of the main camera stream in soccer games. In: IEEE Conf. on Computer Vision and Pattern Recognition Workshops, pp. 1846–184609 (2018)
5. Deliège, A., et al.: Soccernet-v2: a dataset and benchmarks for holistic understanding of broadcast soccer videos. In: The IEEE Conf. on Computer Vision and Pattern Recognition Workshops, June 2021
6. Fan, H., Xiong, B., Mangalam, K., Li, Y., Yan, Z., Malik, J., Feichtenhofer, C.: Multiscale vision transformers. IEEE Int. Conf. on Computer Vision, pp. 6804–6815 (2021)

7. Feichtenhofer, C.: X3D: expanding architectures for efficient video recognition. In: IEEE Conf. on Computer Vision and Pattern Recognition, pp. 200–210 (2020)
8. Feichtenhofer, C., Fan, H., Malik, J., He, K.: Slowfast networks for video recognition. In: IEEE Int. Conf. on Computer Vision (ICCV), pp. 6201–6210 (2018)
9. Feichtenhofer, C., Fan, H., Xiong, B., Girshick, R.B., He, K.: A large-scale study on unsupervised spatiotemporal representation learning. In: IEEE Conf. on Computer Vision and Pattern Recognition, pp. 3298–3308 (2021)
10. Freire-Obregón, D., Lorenzo-Navarro, J., Santana, O.J., Hernández-Sosa, D., Castrillón-Santana, M.: Towards cumulative race time regression in sports: I3D ConvNet transfer learning in ultra-distance running events. In: Int. Conf. on Pattern Recognition (ICPR), pp. 805–811 (2022)
11. Guo, T., Tao, K., Hu, Q., Shen, Y.: Detection of ice hockey players and teams via a two-phase cascaded cnn model. IEEE Access **8**, 195062–195073 (2020)
12. He, X.: Application of deep learning in video target tracking of soccer players. Soft. Comput. **26**(20), 10971–10979 (2022)
13. Hu, X., Dai, J., Li, M., Peng, C., Li, Y., Du, S.: Online human action detection and anticipation in videos: a survey. Neurocomputing **491**, 395–413 (2022)
14. Hunter, A.H., Murphy, S.C., Angilletta, M.J., Wilson, R.S.: Anticipating the direction of soccer penalty shots depends on the speed and technique of the kick. Sports **6**(3) (2018)
15. Hunter, A.H., Pavlic, T.P., Angilletta, M.J., Wilson, R.S.: Identifying the best strategy for soccer penalty success: a predictive model for optimising behavioural and biomechanical trade-offs. J. Biomech. **141**, 111208 (2022)
16. Jiyang Gao, Z.Y., Nevatia, R.: Red: Reinforced encoder-decoder networks for action anticipation. In: Proceedings of the British Machine Vision Conf. (BMVC), pp. 92.1–92.11, September 2017
17. Kong, Y., Fu, Y.: Human action recognition and prediction: a survey. Int. J. Comput. Vision **130**(5), 1366–1401 (2022)
18. Li, L., Zhang, T., Kang, Z., Zhang, W.H.: Design and implementation of a soccer ball detection system with multiple cameras. ArXiv abs/2302.00123 (2023)
19. Li, Y., Wu, C., Fan, H., Mangalam, K., Xiong, B., Malik, J., Feichtenhofer, C.: Mvitv2: improved multiscale vision transformers for classification and detection. In: IEEE Conf. on Computer Vision and Pattern Recognition, pp. 4794–4804 (2021)
20. Noël, B., Furley, P., van der Kamp, J., Dicks, M., Memmert, D.: The development of a method for identifying penalty kick strategies in association football. J. Sports Sci. **33**(1), 1–10 (2015)
21. Parmar, P., Morris, B.T.: What and how well you performed? a multitask learning approach to action quality assessment. In: IEEE Conf. on Computer Vision and Pattern Recognition, pp. 304–313. IEEE (2019)
22. Santana, O.J., Freire-Obregón, D., Hernández-Sosa, D., Lorenzo-Navarro, J., Sánchez-Nielsen, E., Castrillón-Santana, M.: Facial expression analysis in a wild sporting environment. Multimed. Tools Appl. **82**(8), 11395–11415 (2023)
23. Shih, H.C.: A survey of content-aware video analysis for sports. IEEE Trans. Circuits Syst. Video Technol. **28**(5), 1212–1231 (2018)
24. Simonyan, K., Zisserman, A.: Two-stream convolutional networks for action recognition in videos. ArXiv abs/1406.2199 (2014)
25. Teranishi, M., Fujii, K., Takeda, K.: Trajectory prediction with imitation learning reflecting defensive evaluation in team sports. In: IEEE Conf. on Consumer Electronics (GCCE), pp. 124–125 (2020)

26. Wang, S., Xu, Y., Zheng, Y., Zhu, M., Yao, H., Xiao, Z.: Tracking a golf ball with high-speed stereo vision system. IEEE Trans. Instrum. Meas. **68**(8), 2742–2754 (2019)
27. Wang, X., Girshick, R.B., Gupta, A.K., He, K.: Non-local neural networks. In: IEEE Conf. on Computer Vision and Pattern Recognition, pp. 7794–7803 (2017)
28. Xu, C., Fu, Y., Zhang, B., Chen, Z., Jiang, Y.G., Xue, X.: Learning to score figure skating sport videos. IEEE Trans. Circuits Syst. Video Technol. **30**(12), 4578–4590 (2020)
29. Zhang, Y., et al.: ByteTrack: multi-object tracking by associating every detection box. In: European Conf. on Computer Vision (2021)
30. Zheng, R., van der Zijden, B., Janssen, T., van der Kamp and, J.: Goalkeepers benefit from using deceptive actions in the soccer penalty kick. J. Sports Sci. **42**(13), 1224–1231 (2024)

AI-Driven Visual Monitoring of Industrial Assembly Tasks

Mattia Nardon[1], Stefano Messelodi[1], Antonio Granata[2], Fabio Poiesi[1], Alberto Danese[2], and Davide Boscaini[1](✉)

[1] Fondazione Bruno Kessler, Trento, Italy
dboscaini@fbk.eu
[2] Meccanica del Sarca s.p.a., Trento, Italy

Abstract. Visual monitoring of industrial assembly tasks is critical for preventing equipment damage due to procedural errors and ensuring worker safety. Although commercial solutions exist, they typically require rigid workspace setups or the application of visual markers to simplify the problem. We introduce ViMAT, a novel AI-driven system for real-time visual monitoring of assembly tasks that operates without these constraints. ViMAT combines a perception module that extracts visual observations from multi-view video streams with a reasoning module that infers the most likely action being performed based on the observed assembly state and prior task knowledge. We validate ViMAT on two assembly tasks, involving the replacement of LEGO components and the reconfiguration of hydraulic press molds, demonstrating its effectiveness through quantitative and qualitative analysis in challenging real-world scenarios characterized by partial and uncertain visual observations. Project page: https://tev-fbk.github.io/ViMAT.

Keywords: Visual monitoring · Assembly task · Object detection

1 Introduction

Assembly tasks, involving the procedural integration of atomic components to achieve a target configuration, are central to manufacturing [31]. In this work, we focus on the reconfiguration of hydraulic press molds, a critical process required when switching production between different products. Improper mold reconfiguration can damage the mold or the entire hydraulic press, incurring high repair costs and halting production for weeks or even months. Beyond financial loss, incorrectly assembled components can be ejected at high speed during molding, posing serious (potentially fatal) safety risks to nearby workers. As such, monitoring human-operated assembly tasks from video streams is essential. However, this is a challenging problem due to the partial visibility of both the worker's actions and the manipulated components, often caused by inherent occlusions. Industrial environments further complicate the task with cluttered workspaces and varying lighting conditions. To address these challenges, existing

E. Rodolà et al. (Eds.): ICIAP 2025, LNCS 16167, pp. 273–285, 2026.
https://doi.org/10.1007/978-3-032-10185-3_22

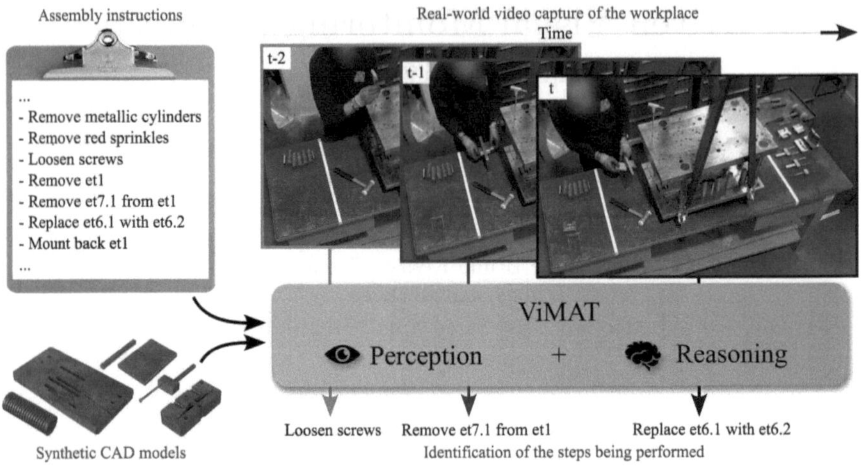

Fig. 1. We present ViMAT, a novel system for the real-time visual monitoring of industrial assembly tasks. Given prior knowledge on assembly instructions (top left) and synthetic CAD models of assembly components (bottom left), ViMAT integrates an AI-driven perception module, which extracts visual observations from real-world video streams (top right), with a probabilistic reasoning module that predicts the assembly state from these observations (bottom center-right).

commercial solutions [9] often simplify the problem by imposing rigid workspace setups, where assembly components are placed in predefined containers based on their category or role, or by applying visual markers to objects, containers, and the worker's wrists to facilitate action recognition. However, these approaches restrict worker flexibility, fail to generalize to unstructured work environments, and are prone to failure in real-world industrial settings, where visual markers can be easily compromised by dirt or debris.

In this work, we introduce ViMAT, a novel AI-driven system for real-time Visual Monitoring of industrial Assembly Tasks that operates without such constraints. We formalize the problem as a state-transition system, where states are specific configurations of assembly components, and transitions between them occur depending on the actions performed by the human operator. ViMAT combines a perception module, which extracts visual observations from multi-view video streams, with a probabilistic reasoning module that infers the most likely action being performed based on the observed assembly state and prior task knowledge (Fig. 1). The perception module leverages a deep learning-based object detector trained exclusively on synthetic data generated from digital twins that replicate the physical assembly components and eliminate the need for manual annotation of real-world data. The effectiveness of the resulting detector eliminates the need for rigid workspace organization and visual markers, while its robustness to lighting variations enables accurate detection in real-world industrial settings. To mitigate occlusions, we capture video footage from multiple viewpoints, perform object detection independently for each viewpoint, and then fuse the resulting predictions to produce spatially coherent observations. The probabilistic reasoner uses these observations to predict the full sequence of actions leading to the current configuration and demonstrates robustness to ambiguous actions.

We validate ViMAT on two distinct assembly tasks: replacing components in a LEGO-based system and reconfiguring a hydraulic press. Although both involve substituting certain components with others, the former is performed in a more controlled environment with simple LEGO bricks, whereas the latter occurs in a real manufacturing facility with actual mechanical components. In both scenarios, ViMAT achieves accurate real-time visual monitoring despite the challenges posed by real-world conditions. A comprehensive evaluation, including both quantitative and qualitative results, demonstrates the effectiveness of the proposed system and its advantages over baseline approaches.

In summary, our main contributions are:

- We formalize the problem of visual monitoring of assembly tasks as a state-transition system that infers the action being performed from partial visual observations on the assembly state and prior task knowledge.
- We enhance ViMAT's perception with a deep learning-based object detector trained exclusively on synthetic data generated from task-specific digital twins, eliminating the need for real-world annotations.
- We validate ViMAT on two distinct real-world assembly scenarios, demonstrating accurate real-time visual monitoring without relying on rigid constraints.

2 Related Works

Visual Monitoring of human-operated tasks aims to identify the actions performed by individuals from video recordings. In industrial environments, this capability is essential for ensuring adherence to standard operating procedures, promoting safe machinery usage, and safeguarding worker safety. Existing approaches typically interpret observed behaviour using either rule-based reasoning or probabilistic models. In this work, we focus on the latter and address real-world scenarios where the partial observability caused by inherent occlusions poses a significant challenge. Most competitors [6,27,31,32] rely on the interpretation of human actions, which can be ambiguous, deviate from standard procedures, or vary significantly across operators. In contrast, our approach centers on detecting assembly components and then infers the underlying action by identifying the predefined assembly instruction that is most likely to have resulted in the observed configuration.

Digital Twins are virtual replicas of physical objects that mirror their real-world counterparts either in appearance alone or also in functionality [21,24]. They are essential for enabling the analysis and simulation of complex systems or processes, particularly when direct inspection is impractical, costly, or hazardous. As a result, digital twins are playing an increasingly important role in manufacturing [8,30], robotics [14,26], and autonomous driving [19,34]. More recently, they have become valuable tools for training, testing, and monitoring computer vision systems. In particular, they have been used in combination with rendering engines [1,12] to generate synthetic datasets for training object detectors [18], pose estimators [22], and segmentation models [13]. A key advantage of using digital twins in this context is the automatic generation of accurate annotations, reducing the burden of manually labelling real-world data. In this work, we create digital twins of the assembly components, replicating both their shape and appearance. We extract precise 3D geometry from synthetic CAD models and approximate visual appearance using synthetic materials [2,4] that simulate photometric properties (e.g., colour, texture) and physical characteristics (e.g., reflectivity, signs of wear).

Object Detection involves identifying and localizing objects in an image by predicting their categories and corresponding bounding boxes. Key challenges include handling occlusions, scale variations, and ensuring temporal consistency when processing videos. The field has evolved from traditional approaches based on hand-crafted features [10,15] to data-driven approaches based on deep learning, which are generally classified into *single-stage* and *two-stage* methods. Single-stage detectors [25,28] perform object localization and classification jointly, offering faster inference. In contrast, two-stage detectors [17,29] first generate region proposals identifying areas where objects may be located and then classify them, usually achieving higher accuracy at the cost of increased latency. Models based on CNNs [25,28] have faster training convergence and lower inference times, while methods based on Transformers [7] enable end-to-end training and achieve better accuracy, but require higher computational costs. In this work, we adopt YOLOv8 [5,20], a CNN-based single-stage detector that offers a good trade-off between accuracy and efficiency. We finetune it on a synthetic dataset generated using task-specific digital twins and the BlenderProc [12] rendering engine.

3 Method

Assembly Tasks are formally defined by three components: *objects*, *configurations*, and *actions*. Objects include physical *elements*, which are manipulated during the assembly process, as well as *trays*, which are designated regions of the workspace associated with specific functions, such as storing elements before use, collecting unmounted elements, or holding the required tools. Note that we do not require trays to be physical containers. Configurations describe the relationships among objects and are defined by a set of Boolean predicates over one or more objects that hold true in a specific assembly state. Actions represent admissible interactions with objects that cause transitions between configurations by modifying the associated predicates. Every action is associated with a set of *preconditions*, i.e. predicates that must be satisfied for the action to be applicable.

We model an assembly task as a state-transition system, where states correspond to configurations and transitions to actions [11,16]. In this framework, a planner can identify the correct executions of an assembly task by finding ordered sequences of states transitioning from the input *initial configuration* C_{init} to the target *final configuration* C_{final} through valid actions. This system can be represented as a directed graph, where nodes correspond to configurations (with C_{init} as root and C_{final} as the sole leaf), edges represent actions, and paths connecting C_{init} to C_{final} define valid plans. Note that multiple such paths may exist, each corresponding to a different valid plan (see Fig. 3, bottom).

Visual Monitoring of an assembly task consists in estimating the current state of the system from visual data by observing its evolution during the execution of a plan. The task of visual monitoring is difficult for two main reasons: observations provide uncertain information, and the system is only partially observable. For this reason, we adopt a probabilistic approach [33] to find the sequence of states that maximizes the observations' probability. The observations on the working environment are provided by the perception module in the form of object detections. By processing these images, the system identifies and tracks the components of the assembly, enabling it to predict the probability distribution. Given the current view of the scene, the system updates the probability of being in a specific configuration based on the observed state of the components. By reasoning about the most probable configuration of the assembly, the system monitors the operator. This enables accurate tracking of the assembly progress and facilitates reliable execution of the process, ultimately guiding it to the desired target configuration C_{final}.

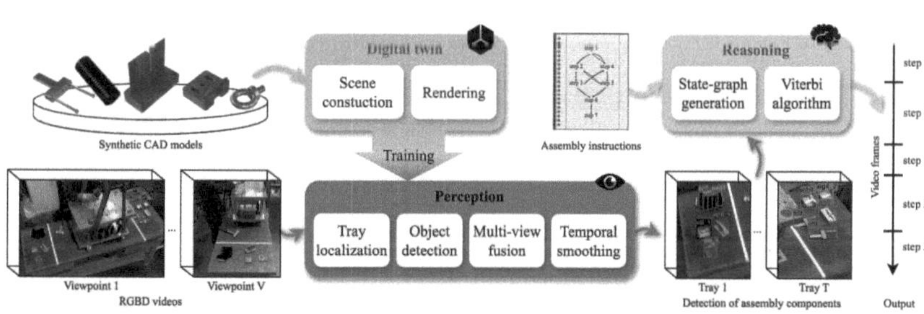

Fig. 2. Overview of ViMAT. Multi-view video frames are processed by the perception module (pink) to detect assembly components using a detector trained on a synthetic dataset generated by the digital twin module (green). These detections, along with prior task knowledge (assembly instructions), are passed to the probabilistic reasoning module (azure) to estimate the action being performed. (Color figure online)

System Overview. Figure 2 shows the architecture of ViMAT, comprising three main modules: *digital twin* (green), *perception* (pink), and *reasoning* (azure). The perception module processes multi-view video streams of an assembly task, crops predefined regions corresponding to various trays, and performs frame-level detection of assembly components within each region. To ensure accurate detection of task-specific components, we finetune the object detector on synthetic data generated by the digital twin module. Spatial and temporal coherence is enforced through multi-view fusion and temporal smoothing, respectively. The reasoning module then uses the resulting detections to infer the most likely action being performed, based on a predefined set of assembly instructions.

3.1 Digital Twin

Objects. Each element involved in the assembly process is modeled using its corresponding synthetic 3D representation from the input CAD models. Although geometric information (i.e., their 3D shape) is directly derived from these models, the visual appearance is obtained by assigning the appropriate textures and materials [2,4], designed to resemble the physical counterparts of the elements.

Scene Generation. To simulate realistic assembly environments, we procedurally generate 3D scenes by placing the elements on a planar surface. The placement is randomized but guided by prior knowledge and by preventing spatial conflicts among objects. This allows the generation of physically plausible and diverse scenes that reflect the variability observed in real assembly processes.

Synthetic Dataset Generation. We use a photorealistic rendering engine to generate images capturing partial views of the 3D scene [1,12]. To increase data variability, we perform random selection of camera viewpoints, distances from the scene, and lighting conditions. We generate photometric information by capturing RGB channels and depth information by capturing range distances from the sensor. In addition, we render the foreground object as a distinct layer to obtain segmentation masks, which are used to extract bounding boxes to supervise the training of our object detection model.

3.2 Perception

The perception module takes as input RGBD frames captured from multiple calibrated viewpoints and produces as output the assembly elements detection.
Tray Localization. The trays are positioned at pre-defined locations that do not change during the assembly process. While their positions are fixed, the elements within each tray are placed freely, without any specific ordering or categorization.

Object Detection. We train an object detector Φ_Θ on a synthetic dataset generated using a digital twin of the task-specific assembly environment (see Sect. 3.1). At inference time, Φ_Θ performs object detection frame-wise, producing 2D bounding boxes, class predictions, and confidence scores.

Multi-view Fusion. To improve the robustness and accuracy of Φ_Θ, we leverage multiple cameras through a multi-view fusion strategy. The process begins by selecting, for each viewpoint, the 2D detections that fall within each predefined tray region. This results in a set of per-tray, per-view detections with associated confidence scores. Every pixel inside the 2D detection is back-projected using depth, intrinsics, and the pinhole model, then transformed to the global frame via extrinsics, yielding a 3D point cloud per detection per view. To associate detections across viewpoints, we match these point clouds using a pairwise Point Cloud Intersection-over-Union (IoU) metric. This is based on a radius-based intersection strategy. Formally, given two point clouds $\mathcal{P}_1$ and $\mathcal{P}_2$, and a radius threshold r, we define the intersection set as: $\text{Intersection}(\mathcal{P}_1, \mathcal{P}_2, r) = \{p \in \mathcal{P}_2 \mid \exists q \in \mathcal{P}_1 \text{ such that } \|p - q\| < r\}$. The corresponding IoU value is computed using the usual formula. This method allows us to robustly identify overlapping detections across views, even in the presence of minor inaccuracies. Once the pairwise IoUs are computed and matching detections are identified, the results are aggregated across all views to obtain a consolidated observation, i.e., a vector representing the presence and confidence of each element in the workspace. The multi-view fusion helps mitigate occlusions and provides a more reliable perception of the scene.

Temporal Smoothing. To ensure consistency over time and reduce noise in the predictions, we apply a temporal smoothing strategy to the consolidated observation vectors. At each time step, the current observation vector is combined with the previously smoothed one using a weighted average. The weighting is asymmetric: if the current score of an element is higher than the previous one, it is given more weight to allow rapid adaptation to improvements; otherwise, the previous score is favored to dampen abrupt drops.

3.3 Reasoning

Assembly Instruction Formalization. Together with domain experts, we formalize the description of system configurations, the set of actions that can be performed and, for each action, the set of preconditions that have to hold in order to perform it. A configuration involves several objects and a set of predicates that can hold among them. In the LEGO scenario, for example, we identified 13 objects ($E_1, E_2, E_3, E_4, E_5, E_6, E_4', E_5', E_6', T_{\text{in}}, T_{\text{out}}, T_{\text{work}}, T_{\text{aux}}$) and 6 predicates (`is_joined`, `is_split`, `is_mounted`, `is_accessible`, `is_free`, `do_contain`).

A state (or configuration) is specified by the set of predicates that hold when the system is in that state. Actions describe the different modes the systems can be modified to pass from one configuration to another. Their effect is to update the list of predicates that are valid before the execution of the action. In the LEGO scenario, the set of elementary actions is $A = \{\texttt{join}, \texttt{split}, \texttt{mount}, \texttt{remove}, \texttt{put}, \texttt{take}\}$. In practice, the description of the reconfiguration process in terms of elementary actions often turns out to be redundant because some sequences of actions are always performed together and can be considered atomic. Therefore, we have described the process as a set of steps, each defined by a sequence of actions, and the preconditions that must hold in order to be executed. In this way, we obtain a more meaningful description of the activity to be performed by the operator, hiding too detailed and often obvious instructions. For example, we defined the step *Mount element* E_4 *on* E_3 which is composed by two actions (`take`(E_4, Tin) and `mount`(E_4, E_3)) and three preconditions (`do_contain`(Tin,E_4), `is_accessible`(E_3), `is_free`(E_3)). In the LEGO scenario we have identified 10 steps.

State-Graph Generation. The state-transition system can be used to find a plan (i.e. a sequence of steps) that, from some given initial configuration, permits to achieve an objective, typically specified as a goal state. By exploring the state-space starting from C_{init}, we automatically construct a directed graph $\mathcal{G}$ representing all feasible sequences of steps reaching C_{final} (Fig. 3, bottom).

Graph-Based Probabilistic Reasoning. The control of the execution of a given plan is performed by a controller module by observing the system evolution during the plan execution. In our case, the controller module does not take decisions, but operates as an external observer trying to estimate the current state of the system. Obviously, if it detects deviations from a valid plan, a warning message can be sent to the operator. Observations aim to test which predicates are valid in the observed configuration to provide the probability that the system is in that state. Because observations are intrinsically partial and uncertain, we adopt a probabilistic approach trying to estimate the most probable sequence of steps performed (or traversed configurations) given all the observations from the beginning to the current time. The Viterbi algorithm works by iteratively calculating the probabilities of being in each state at each time step, considering the previous state and the current observation:

$$V_{1,k} = P(y_1|s_k)\pi_k \qquad\qquad V_{t,k} = \max_{x\in S}(P(y_t|s_k)a_{x,s_k}V_{t-1,x})$$

where $S = \{s_1, \ldots, s_n\}$ are the states, initial probabilities $\{\pi_1, \ldots \pi_n\}$ are all zero except for the initial state (set to 1.0), y_t is the observation vector at time t, the transition matrix a is estimated from $\mathcal{G}$ and $P(y|s)$, the probability of observing y assuming the system is in state s, is estimated by comparing y with its expected value $E_y(s)$, i.e. $P(y|s) \propto \exp(-\|y - E_y(s)\|/\sigma)$. Estimates $V_{t,k}$ can be used to backtrack and find the most likely sequence of states (the Viterbi path).

4 Results

4.1 Experimental Setting

We conduct experiments in two distinct scenarios: a controlled LEGO scenario and an industrial scenario. In both scenarios, we use three RGB-D sensors for perception, and the communications between system modules are handled via sockets. We train as object detector Φ_Θ, the pretrained YOLOv8-X model (68.2M parameters) and finetune it for 100 epochs. We employ a cosine decay schedule, and the SGD optimizer. Input images are resized to 480×480 pixels with standard data augmentation techniques (random scaling, rotation, and HSV variations).

Datasets. For the LEGO scenario, we create a digital twin comprising the 13 unique objects. The 3D models are constructed using Stud.io [3] and Ldraw [4] to ensure accurate representation. We generate for this task a synthetic dataset of 10,000 images by using either Blender [1] or BlenderProc [12] as rendering engine. The industrial scenario uses a more complex digital twin with 20 distinct objects. These are modeled by extruding 2D blueprints to replicate the 3D geometry and textured using realistic materials from BlenderKit [2] to simulate surface appearance and signs of wear. A similar BlenderProc pipeline is employed to generate approximately 25,800 synthetic images. In both scenarios, real-world background images of the workspace are collected without any objects to detect. These images are included during the training of the object detector. This strategy helps to reduce false positive detections during inference.

Metrics. We evaluate object detection performance using mean average precision (mAP) on manually annotated real-world validation data.

4.2 Results

Quantitative Results. We present quantitative evaluations of ViMAT's object detection performance in both LEGO and industrial scenarios. Tables 1 and 2 detail the average mAP results for each setting, highlighting the impact of data augmentation (Aug.) and background images (BG) during training.

Table 1. Object detection performance in the LEGO setting.

	Data type	Aug.	BG	mAP
1	B, Synth.			60.4
2	B, Synth.	✓		71.8
3	BProc, Synth.	✓		76.4
4	BProc, Synth.	✓	✓	**79.7**
5	Real[1]			73.5
6	Real[1]	✓	✓	77.7

[1] Results on a subset of test sequences due to training data overlap.

LEGO scenario. Table 1 shows the average mAP across three test sequences. Our baseline model, trained solely on synthetic data generated with Blender (B) without augmentation or background, achieves 60.4% mAP. Applying data augmentation improves the average to 71.8%. Using BlenderProc (BProc) as the rendering engine further increases performance to 76.4% mAP. Incorporating background augmentation yields the best results, reaching 79.7% mAP overall. Models trained exclusively on real images outperform those trained solely on

Table 2. Object detection performance in the industrial scenario.

	Variant	Aug.	BG	Cam0	Cam1	Cam2	Avg.
1	Synth.	✓	✓	44.2	65.4	56.8	55.8
2	Synth.	✓	✓	41.4	72.9	59.5	58.4
3	Real			14.9	21.1	11.9	15.0
4	Real	✓		26.9	43.0	26.7	30.9
5	Real	✓	✓	17.3	29.8	29.1	26.3
6	S+R	✓	✓	56.5	60.5	61.5	60.0
7	S+R[2]	✓	✓	60.7	79.4	63.3	**66.7**
8	SAM-6D	-	-	20.2	19.9	10.2	16.7

[2] Model pretrained on Open Image V7

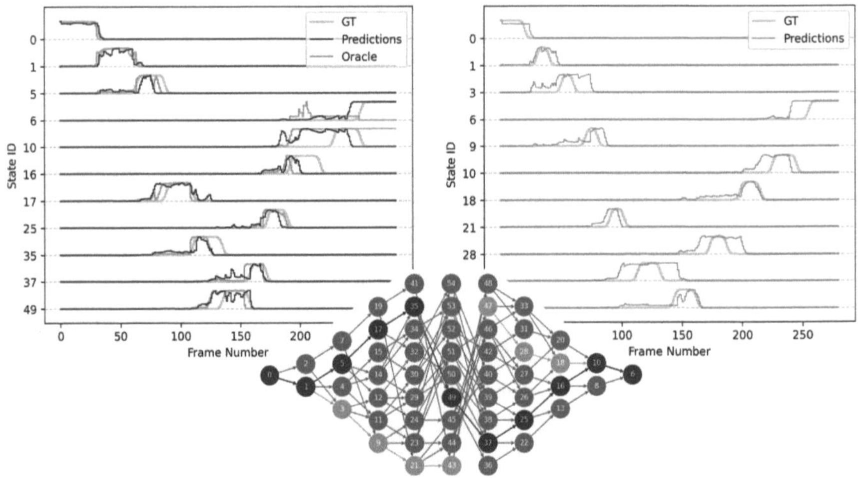

Fig. 3. Visual monitoring of the LEGO assembly: Prec. = 72.9, Recall = 77.9.

synthetic data. However, these real-image-only models exhibit reduced generalization on new sequences with different camera setups.

Industrial scenario. Table 2 shows the mAP results across the three cameras and their average. In this more challenging scenario, our method achieves a peak performance of 58.4%, reflecting the increased difficulty in detecting similar objects. The models trained using only real images perform worse than in the LEGO scenario, likely due to the limited data (only 8 images). Table 2 includes ablation studies, such as variations using synthetic and real data (S+R), and SAM-6D [23] performance. Combining synthetic and real data for training (S+R) yields improved results, highlighting the benefit of including real data when available. SAM-6D, a model-based zero-shot method, achieves lower performance, likely due to its poor calibration.

Qualitative Results. Figures. 3 and 4 show the state probabilities over time for different videos. Each graph compares the ground-truth states (pink) with ViMAT's predictions (blue or green) and with predictions using ground-truth object detections as observation vectors (azure). Precision and recall are computed by matching predicted and ground-truth active states at each time step.

LEGO scenario. At the bottom of Fig. 3, the complete assembly path is shown, illustrating the system's ability to track the progression from the initial (left) to the final configuration (right). ViMAT closely follows the ground-truth sequence, demonstrating strong temporal alignment. Notably, predictions from the trained object detector closely match those derived from oracle detections, indicating that the model, trained solely on synthetic data, generalizes well to the real-world and provides reliable input for downstream reasoning.

Industrial scenario. In Fig. 4, the state probabilities over time also follow the ground-truth trend, but with greater variability. This is expected due to the increased complexity of the environment and actions. Nevertheless, the system still captures the overall sequence of states effectively. In both scenarios, predictions often slightly anticipate the ground-truth annotations. This is likely due to the nature of many steps, where the configuration change (e.g., removing an object from a tray) occurs at the beginning of the step. Moreover, the industrial scenario yields lower precision and recall, likely reflecting the increased complexity and more challenging conditions compared to the controlled setting.

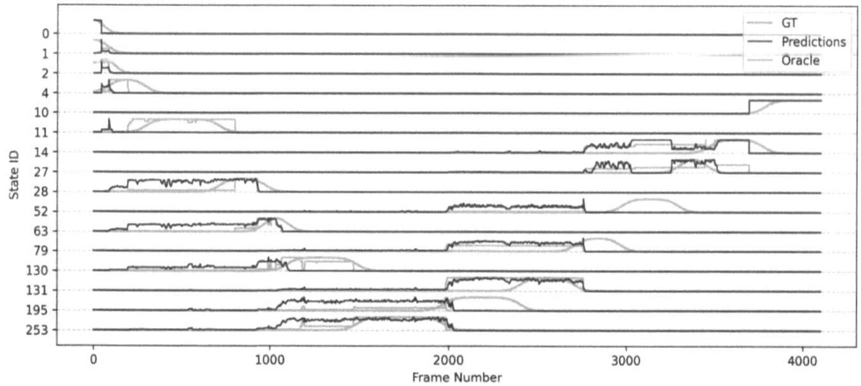

Fig. 4. Visual monitoring of the industrial assembly: Prec. = 43.1, Recall = 43.7.

Failure Cases and Limitations. A key limitation is the perception module, where missed or incorrect detections, especially for similar objects, impact downstream reasoning. The system also struggles to distinguish between actions with subtle differences.

5 Conclusions

We presented ViMAT, a real-time visual monitoring system for industrial assembly tasks that integrates AI-driven perception with probabilistic reasoning to infer assembly states from multi-view videos. ViMAT uses digital twins of the assembly components to finetune the object detector and predefined assembly instructions to guide reasoning. As future work, we aim to incorporate hand pose estimation for finer action recognition and detect deviations from correct assembly procedures.

Acknowledgments. This work has been partially funded by the Provincia Autonoma di Trento (Italy) under L.P. 6/99, as part of the NEXTMAG project.

References

1. Blender. https://www.blender.org/. Accessed Apr 2025
2. BlenderKit. https://github.com/BlenderKit/BlenderKit. Accessed Apr 2025
3. BrickLink Stud.io.. https://www.bricklink.com. Accessed Apr 2025
4. LDraw. https://www.ldraw.org/. Accessed Apr 2025
5. Ultralytics YOLOv8. https://docs.ultralytics.com. Accessed Apr 2025
6. Andrianakos, G., Dimitropoulos, N., Michalos, G., Makris, S.: An approach for monitoring the execution of human based assembly operations using machine learning. Procedia CIRP **86**, 198–203 (2019)
7. Carion, N., Massa, F., Synnaeve, G., Usunier, N., Kirillov, A., Zagoruyko, S.: End-to-end object detection with transformers. In: ECCV (2020)
8. Cimino, C., Negri, E., Fumagalli, L.: Review of digital twin applications in manufacturing. Comput. Ind. **113**, 103–130 (2019)
9. CSEM: Quality monitored assembly. https://www.youtube.com/watch?v=zBao3QunnGY&ab_channel=CSEMtechnologies. Accessed Apr 2025
10. Dalal, N., Triggs, B.: Histograms of oriented gradients for human detection. In: CVPR (2005)
11. Dean, T.L., Wellman, M.P.: Planning and control (1991)
12. Denninger, M. et al.: BlenderProc. arXiv preprint arXiv:1911.01911 (2019)
13. Dosovitskiy, A., Ros, G., Codevilla, F., Lopez, A., Koltun, V.: CARLA: An open urban driving simulator. In: CoRL (2017)
14. Feddoul, Y., Ragot, N., Duval, F., Havard, V., Baudry, D., Assila, A.: Exploring human-machine collaboration in industry: a systematic literature review of digital twin and robotics interfaced with extended reality technologies. Int. J. Adv. Manufact. Technol. **129**(5), 1917–1932 (2023)
15. Felzenszwalb, P.F., Girshick, R.B., McAllester, D., Ramanan, D.: Object detection with discriminatively trained part-based models. TPAMI **32**(9), 1627–1645 (2010)
16. Ghallab, M., Nau, D., Traverso, P.: Automated Planning: Theory and practice. Elsevier (2004)
17. Girshick, R., Donahue, J., Darrell, T., Malik, J.: Rich feature hierarchies for accurate object detection and semantic segmentation. In: CVPR (2014)
18. Hodaň, T., et al.: Photorealistic image synthesis for object instance detection. In: ICIP (2019)

19. Hu, X., Li, S., Huang, T., Tang, B., Huai, R., Chen, L.: How simulation helps autonomous driving: a survey of sim2real, digital twins, and parallel intelligence. Trans. Intell. Veh. **9**(1), 593–612 (2023)
20. Jocher, G., Qiu, J., Chaurasia, A.: Ultralytics YOLO (2023). https://github.com/ultralytics/ultralytics
21. Jones, D., Snider, C., Nassehi, A., Yon, J., Hicks, B.: Characterising the Digital Twin: a systematic literature review. CIRP-JMST **29**, 36–52 (2020)
22. Labbé, Y., et al.: MegaPose: 6D pose estimation of novel objects via render & compare. In: CoRL (2022)
23. Lin, J., Liu, L., Lu, D., Jia, K.: SAM-6D: Segment anything model meets zero-shot 6d object pose estimation. In: CVPR (2024)
24. Liu, M., Fang, S., Dong, H., Xu, C.: Review of digital twin about concepts, technologies, and industrial applications. J. Manuf. Syst. **58**, 346–361 (2021)
25. Liu, W., et al.: SSD: Single shot multibox detector. In: ECCV (2016)
26. Malik, A.A., Shao, G., Tarakhovsky, J.: Digital twins for robot systems in manufacturing. In: Digital Twins, Simulation, and the Metaverse: Driving Efficiency and Effectiveness in the Physical World through Simulation in the Virtual Worlds. Springer Nature Switzerland (2024)
27. Peixoto, P., Batista, J., Araujo, H.J.: Real-time human activity monitoring exploring multiple vision sensors. Robot. Auton. Syst. **35**(3), 221–228 (2001)
28. Redmon, J., Divvala, S., Girshick, R., Farhadi, A.: You Only Look Once: Unified, real-time object detection. In: CVPR (2016)
29. Ren, S., He, K., Girshick, R., Sun, J.: Faster R-CNN: Towards real-time object detection with region proposal networks. In: NeurIPS (2015)
30. Soori, M., Arezoo, B., Dastres, R.: Digital twin for smart manufacturing, A review. Sustainable Manufacturing and Service Economics **2** (2023)
31. Urgo, M., Tarabini, M., Tolio, T.: A human modelling and monitoring approach to support the execution of manufacturing operations. CIRP Ann. **68**(1), 5–8 (2019)
32. Urgo, M., Terkaj, W., Simonetti, G.: Monitoring manufacturing systems using AI: a method based on a digital factory twin to train CNNs on synthetic data. CIRP J. Manuf. Sci. Technol. **50**, 249–268 (2024)
33. Viterbi, A.: Error bounds for convolutional codes and an asymptotically optimum decoding algorithm. IEEE Trans. Inf. Theory **13**(2), 260–269 (1967)
34. Wang, K., Yu, T., Li, Z., Sakaguchi, K., Hashash, O., Saad, W.: Digital twins for autonomous driving: a comprehensive implementation and demonstration. In: ICOIN (2024)

Benchmarking Content-Based Puzzle Solvers on Corrupted Jigsaw Puzzles

Richard Dirauf[1,2](✉), Florian Wolz[2], Dario Zanca[1], and Björn Eskofier[1]

[1] Machine Learning and Data Analytics Lab, Department Artificial Intelligence in Biomedical Engineering, Friedrich-Alexander-Universität Erlangen-Nürnberg (FAU), 91052 Erlangen, Germany
{richard.dirauf,dario.zanca,bjoern.eskofier}@fau.de
[2] Siemens Healthineers AG, 91301 Forchheim, Germany
florian.wolz@siemens-healthineers.com

Abstract. Content-based puzzle solvers have been extensively studied, demonstrating significant progress in computational techniques. However, their evaluation often lacks realistic challenges crucial for real-world applications, such as the reassembly of fragmented artefacts or shredded documents. In this work, we investigate the robustness of State-Of-The-Art content-based puzzle solvers introducing three types of jigsaw puzzle corruptions: missing pieces, eroded edges, and eroded contents. Evaluating both heuristic and deep learning-based solvers, we analyse their ability to handle these corruptions and identify key limitations. Our results show that solvers developed for standard puzzles have a rapid decline in performance if more pieces are corrupted. However, deep learning models can significantly improve their robustness through fine-tuning with augmented data. Notably, the advanced *Positional Diffusion* model adapts particularly well, outperforming its competitors in most experiments. Based on our findings, we highlight promising research directions for enhancing the automated reconstruction of real-world artefacts.

Keywords: Jigsaw Puzzle Solvers · Robustness · Image Processing

1 Introduction

Real-world combinatorial optimization problems are NP-hard, complex, and time-consuming to solve. In addition to these challenges, automatic solvers must handle noisy data and non-optimal surrogate objectives, making robustness a critical requirement. One category of these real-world problems includes the reassembly of objects such as shredded documents [7] or fragmented historical artworks [9,10]. This challenge is in close relationship to solving jigsaw puzzles as both are defined similarly: Given a set of pieces, the goal is to reconstruct the original image or object, from which the pieces were created [6]. Throughout the years, many automated solvers for jigsaw puzzles were developed utilizing both

E. Rodolà et al. (Eds.): ICIAP 2025, LNCS 16167, pp. 286–298, 2026.
https://doi.org/10.1007/978-3-032-10185-3_23

heuristic and deep learning methods [6] with the hope of also being applicable to real-world problems.

Despite significant advancements, a recent study [17] demonstrates that nearly all existing jigsaw puzzle solvers are unsuitable to work on realistic archaeological fragments and those that can be applied perform poorly. This is because the fragments are eroded, irregularly shaped due to fractures, vary in size, or are missing due to weathering or human-induced damage. Such characteristics of a real-world archaeological puzzle are often not present in the datasets that are used to develop jigsaw puzzle solvers. Therefore, the authors identify a research gap between existing computational methods and the automated reconstruction of real-world artefacts.

To bridge this gap, it is essential to evaluate the robustness of existing solvers by systematically assessing their ability to handle these real-world challenges. Since isolating these characteristics from an archaeological dataset is impossible, a practical approach is to introduce controlled corruptions to standard jigsaw puzzles mimicking the complexities of real-world artefacts. These types of corruptions can be categorized into two groups: The first group involves modifications to the shape of a puzzle piece, such as irregular contours, multiple edges, or varying piece sizes [3,4]. The existing approaches to these challenges focus on fitting together the contours of neighbouring pieces, while mostly disregarding the pictorial content of the fragments. The second group includes corruptions that alter the visual content of a puzzle, such as erosion of texture and colour, the degradation of the edges of a piece, or the complete absence of multiple pieces. Most content-based jigsaw puzzle solvers are designed to reassemble uncorrupted, square-sized pieces [6] and their robustness to such modifications remains largely unexplored.

In this paper, we focus on content-modifying corruptions and systematically benchmark the State-Of-The-Art (SOTA) content-based jigsaw puzzle solvers to investigate their robustness to different types of corruptions. Using the *PuzzleWikiArts* dataset [14] we introduce three types of corruptions to simulate individual real-world fragment characteristics. We then evaluate the performance of selected solvers, comparing heuristic methods and deep learning models, to determine their effectiveness and limitations. Furthermore, we explore whether learning-based methods can be adapted to handle corruption more effectively. Based on our structured investigation, we derive suggestions for the most promising research directions for developing automated methods capable of reconstructing historical artefacts.

2 Related Work

When reviewing existing research, we focus on 2D content-based image puzzles with square-sized, non-rotated pieces, also known as Type 1 jigsaw puzzles [1], as it is the most frequently studied type in the literature [6]. We identify the SOTA puzzle solvers to test them in our benchmark by examining the comparative investigation by Giuliari *et al.* [2], which compare three heuristic solvers and four deep learning models, and Son *et al.* [13], which compiled the results reported for nine heuristic solvers.

2.1 Content-Based Heuristic Jigsaw Puzzle Solvers

Over the years, many heuristic algorithms have been proposed to automatically solve Type 1 jigsaw puzzles. Most of these methods consist of two key components: a metric to compute the pairwise compatibility between neighbouring puzzle pieces and a strategy to assemble the puzzle based on this metric. One of those pairwise compatibility measures, introduced by Gallagher [1], is the Mahalanobis Gradient Compatibility (MGC). It evaluates the local gradients at the boundaries of the jigsaw puzzle pieces, based on the assumption that neighbouring pieces should exhibit similar gradient distributions at their edges. Gallagher shows that MGC significantly outperforms previous compatibility metrics which rely on pixel intensity differences at piece boundaries. This is also confirmed by Giuliari *et al.* [2] and Son *et al.* [13]. Gallagher combines the MGC with a greedy, tree-based reassembly algorithm to solve different types of jigsaw puzzles.

The pairwise compatibility metric by Paikin and Tal [8] assesses the $L1$ norm of the boundary pixels between all pieces as a measurement of dissimilarity. Their deterministic heuristic identifies pairs of pieces with the smallest and most unique differences, as these are likely to be neighbours, and uses them as starting points to greedily reconstruct the original image. Their method achieves high reconstruction accuracies, outperforming Gallagher's solver [2,13].

Yu *et al.* [12] utilize MGC as their pairwise compatibility metric and combine it with a linear program, which exploits all pairwise matches simultaneously. According to the comparison by Son *et al.* [13], Yu *et al.*'s approach achieves the highest performance on two out of five jigsaw puzzle datasets and the second-highest performance on the remaining three datasets.

Overall, heuristic solvers show strong performance on the Type 1 jigsaw puzzle. Their main limitation lies in their pairwise compatibility metrics, which are not universally optimal for all possible variations of image puzzles. Among different approaches, the two most effective compatibility measurements are the MGC by Gallagher and the $L1$ norm by Paikin and Tal. Therefore, we select both solvers for our benchmark and additionally include the heuristic solver by Yu *et al.*due to its high performance. We refrain from analysing additional heuristic solvers in our experiments, as they either exhibit significantly lower performance compared to the selected approaches or lack open-source implementations, making it difficult to reproduce their reported results.

2.2 Content-Based Deep Learning Jigsaw Puzzle Solvers

In recent years, many researchers tried to apply deep learning models to the jigsaw puzzle problem. These methods follow a two-step process: first, visual features are extracted from the individual image pieces by a neural network. In the second step, the original permutation or the absolute positions of the puzzle pieces are predicted based on the processed features. Therefore, deep learning methods have the advantage that they can learn the reassembly task by leveraging both structure- and colour-based features across the entire image [6]. However, earlier deep learning approaches were limited to small puzzle sizes or a fixed number of possible input permutations.

Talon *et al.* [14,15] are the first who tested their model on puzzles up to 12×12 pieces but their performance is still behind those of the best heuristics. As one of the few papers in this field, they test how different approaches handle missing pieces, Gaussian noise, or missing borders in the puzzle. They report a performance drop for all investigated solvers but do not analyse the results further.

A recent paper by Giuliari *et al.* [2] introduced *Positional Diffusion*, a method based on Diffusion Probabilistic Models that is designed to solve ordering problems. For image puzzles, the method learns to predict the positions of the pieces in a normalized coordinate system by perturbing their position vectors during the diffusion process. The authors report that their method outperforms other solvers, for example, Paikin and Tal's approach, across most of the tested puzzle sizes. Notably, one of the compared deep learning solvers that is also developed by the same authors is a Transformer model that likewise predicts the pieces' coordinates but without the diffusion process.

In summary, research on applying deep learning models to the jigsaw puzzle problem is still limited, and only a few approaches achieve performance comparable to existing heuristic solvers. As a result, we include two deep learning models in our benchmark: the Transformer-based model and *Positional Diffusion*, both by Giuliari *et al.*.[1]

3 Methodology

This chapter presents our approach to benchmark the five selected, SOTA content-based jigsaw puzzle solvers on corrupted puzzles to determine their robustness. To test the performance of the solvers, we are analysing two established metrics [1]. The *Direct Comparison Metric* measures the percentage of

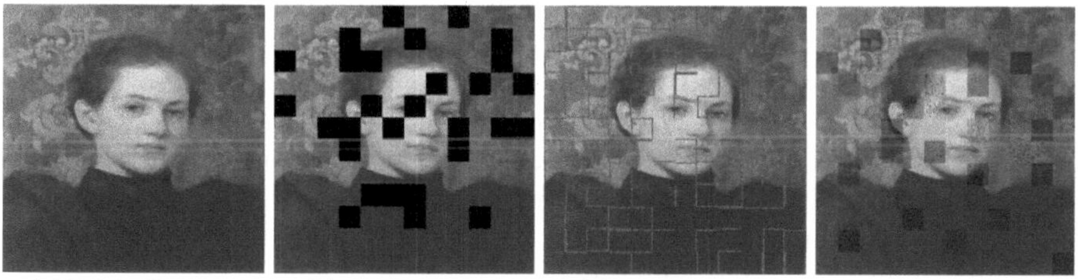

(a) Original Image (b) Missing Pieces (c) Eroded Edges (d) Eroded Contents

Fig. 1. Example of a 12×12 puzzle modified by each of the three corruption types. The original image is taken from the *PuzzleWikiArts* dataset [14].

[1] Links to implementations: Gallagher, Yu *et al.*, Paikin and Tal, Giuliari *et al.*.

pieces correctly placed in their absolute positions within the puzzle. The *Perfect Reconstruction Metric* is the percentage of jigsaw puzzles from a dataset that could be correctly reassembled into the original image. For our experiments, we define three types of corruptions inspired by real-world archaeological puzzles and describe their systematic application to unmodified jigsaw puzzles: missing pieces, eroded edges, and eroded contents. The corruptions are content-based only, as the selected solvers cannot handle more complex shape-based corruptions involving irregular fragments.

3.1 Types of Corruption for Jigsaw Puzzles

Missing Pieces. The first type of corruption includes missing pieces. Therefore, a percentage of the total number of pieces is removed from the puzzle and the solvers are tasked with reconstructing the original image as accurately as possible using the remaining pieces. The missing pieces are randomly selected, which can lead to holes in the image, as shown in Fig. 1b if the proportion is low. At higher percentages, solvers may need to assemble disconnected components of the image, further increasing the complexity of the task. In each experiment, the percentage of missing pieces is gradually increased and we limit the experiments to a maximum of 50%. Since all three heuristic solvers do not inherently support missing pieces, the selected pieces are replaced with black image patches of the same shape, with pixel values set to zero. This approach allows evaluation without adapting the original algorithms, ensuring their behaviour remains unchanged. In comparison, it is valid to remove the selected missing pieces from the input for both investigated deep learning models as they predict the absolute position of each puzzle piece. To assess the robustness of the puzzle solvers against this type of corruption, we analyse the two performance metrics. Since only the correct placement of the non-missing pieces should be considered, the computation of these metrics is adapted accordingly.

Eroded Edges. To simulate eroded edges on square-piece jigsaw puzzles in our benchmark, we adopt the approach of Huroyan *et al.* [5], who use this type of corruption to compare three heuristic solvers. Specifically, we randomly select any of the four edges of a puzzle piece with a fixed probability. For each selected edge, the two outermost pixel rows or columns are replaced with new values, which are uniformly sampled from the entire image. An example of this corruption is shown in Fig. 1c. While Huroyan *et al.*.,examine eroded edges with probabilities up to 14%, we extend this range by gradually increasing the probability up to 50%. At this maximum level, an average of two edges per piece are affected, severely limiting the puzzle solvers' ability to compare neighbouring edges and reconstruct the original image.

This type of corruption does not precisely replicate the irregularly shaped erosion seen in archaeological fragments. However, it serves as a valid approximation for Type 1 jigsaw puzzles. In real-world scenarios, eroded edges make it difficult to match fragments based on their boundaries. Similarly, our simulated erosion prevents solvers from using affected edges for the reassembly process.

As a result, the methods must rely on the remaining uncorrupted edges or the internal content of the pieces to reconstruct the puzzle.

Eroded Contents. The third category of corruption includes erosion suffered to the pictorial contents. As the causes and the severity of damage to an artwork differ for each real-world example, it is difficult to simulate all possibilities of decay. Therefore, we focus on four general effects that can impact historical paintings which are exposed to erosion, specifically loss in saturation, contrast and brightness as well as the complete loss of paint through flaking [11].

To control the amount of erosion to the contents of a jigsaw puzzle, we introduce the erosion factor $\mathcal{E} \in [0, 100]$ (in %) which controls the corruption of the pieces in two ways. First, we assume that in a realistic fragmented artwork, each fragment is affected differently by erosion, meaning that some parts could be unharmed while others are severely damaged. $\mathcal{E}$ controls the proportion of pieces that are modified by each of the four erosion effects. This means that a piece in the jigsaw puzzle can be either unmodified or modified by one or a combination of multiple corruption types. Secondly, $\mathcal{E}$ also controls the severity of the individual erosion effect. If a piece is selected for corruption, the proposition of the effect is sampled from a discrete list using a normal distribution centred around $\mathcal{E}$. As a result, the severity of the erosion is generally increased with higher $\mathcal{E}$, while each piece of a puzzle is affected differently by the four effects. A visual example of this type of corruption can be seen in Fig. 1d.

3.2 Content-Based Jigsaw Puzzle Dataset

We apply the selected corruptions on the *PuzzleWikiArts* dataset [14]. It is derived from *WikiArts* [16] and includes 63 thousand images of paintings, featuring a wide variety of contents and artistic styles. The images are commonly split into puzzles of 6×6 up to 12×12 pieces with each piece consisting of 32×32 pixels. For testing, the four puzzle sizes are uniformly distributed across the test set, resulting in approximately 3125 images per puzzle size.

3.3 Experimental Design

For all solvers, we use their standard implementations, with adjustments made only to the data loading and evaluation procedures to accommodate the specific types of corruption. Before benchmarking the solvers on corrupted puzzles, the SOTA solvers are first tested on standard Type 1 jigsaw puzzles without any corruption. This step is done to reproduce the performance of the four selected approaches as reported in [2]. Additionally, we compare the solvers based on their overall performance across the two evaluation metrics. This serves as a baseline before proceeding with the subsequent experiments, where we investigate the robustness of the jigsaw puzzle solvers towards missing pieces (**Experiment 1**), eroded edges (**Experiment 2**) and eroded contents (**Experiment 3**).

We extend the evaluation of the deep learning solvers by not only measuring the performance of the base models, which were trained on uncorrupted jigsaw puzzles but also by fine-tuning both models through data augmentation. By comparing the resulting performances, we assess how well the deep learning methods can adapt to this type of corruption by modifying only the input data while keeping the solver implementations unchanged.

4 Results

4.1 Baseline: Standard Jigsaw Puzzles

To establish a baseline for the subsequent experiments, we evaluate all five selected jigsaw puzzle solvers on standard Type 1 puzzles. The results are presented in Table 1. Comparing our results with those of Giuliari *et al.* [2] on the *Direct Comparison Metric*, we successfully reproduce the performance of three out of four solvers. Only for the solver by Paikin and Tal, we observe a significant divergence, with an average performance gap of 13.85% compared to the previously reported results. The reason for this discrepancy remains unclear, especially since the exact implementation used in the previous study is not specified.

For Type 1 jigsaw puzzles, the best-performing method is the heuristic by Yu *et al.*, which demonstrates high accuracy in reconstructing various puzzle sizes. In contrast, the other two heuristics show significantly worse performance. Among the deep learning models, the diffusion-based model outperforms the Transformer model, which shows a sharp decline as the puzzle size increases. While *Positional Diffusion* is competitive with the best heuristic methods in the *Direct Comparison Metric*, particularly for smaller puzzles, it struggles to accurately reconstruct larger puzzles.

4.2 Experiment 1: Corrupting Jigsaw Puzzles with Missing Pieces

In general, the performance of all jigsaw puzzle solvers declines as the number of missing pieces increases. This trend is evident in the evaluation metrics across

Table 1. Comparison of the solvers' performances in reconstructing standard Type 1 puzzles. The best-performing results are highlighted in **bold**, second-best results are <u>underlined</u>.

Metrics	*Direct Comparison*				*Perfect Reconstruction*			
Solvers	6×6	8×8	10×10	12×12	6×6	8×8	10×10	12×12
Gallagher	86.67	85.06	81.26	79.19	81.73	78.59	75.10	<u>71.65</u>
Paikin and Tal	82.71	81.77	80.53	80.81	74.99	70.92	65.55	63.83
Yu *et al.*	<u>98.73</u>	<u>98.09</u>	**96.83**	**95.63**	**97.78**	**95.73**	**93.25**	**91.57**
Transformer	98.24	96.12	90.49	80.23	90.74	72.59	36.88	6.45
Positional Diffusion	**99.14**	**98.32**	<u>96.79</u>	<u>93.62</u>	<u>96.88</u>	<u>91.08</u>	<u>81.53</u>	67.50

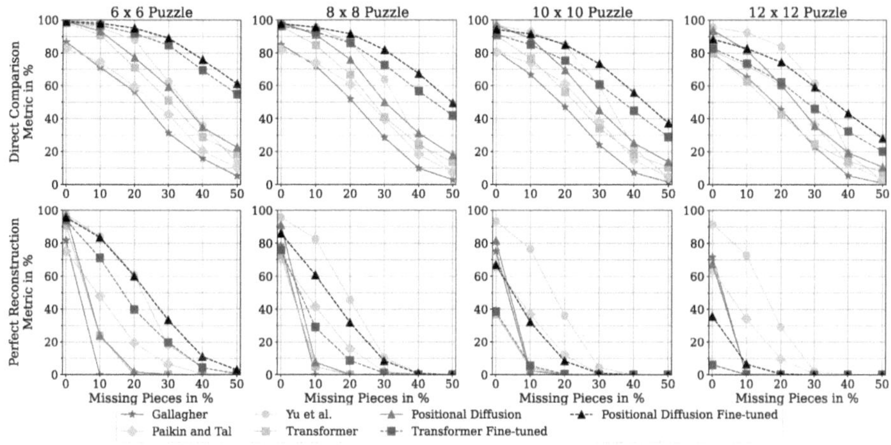

Fig. 2. Results of **Experiment 1** showing the performance of all solvers on puzzles with missing pieces for four different puzzle sizes.

all four puzzle sizes, as shown in Fig. 2. Among the heuristics, Yu *et al.*'s solver generally achieves the highest performance, followed by the methods of Paikin and Tal and then Gallagher. The Transformer and *Positional Diffusion* base models also perform worse than Yu *et al.*'s heuristic.

Fine-tuning the deep learning models on puzzles with missing pieces significantly improves their performance. Although the Transformer model still lags behind, *Positional Diffusion* outperforms Yu *et al.*'s heuristic on smaller puzzles with up to 8×8 pieces in the *Direct Comparison Metric*. However, for larger 12×12 puzzles, Yu *et al.*'s solver remains superior when 30% or fewer pieces are missing. Additionally, the heuristic reconstructs more jigsaw puzzles perfectly than *Positional Diffusion* for puzzles of size 8×8 or larger. The visual example in Table 2 presents the prediction of the five puzzle solvers on a jigsaw puzzle with 10% missing pieces. In this example, the three heuristics and *Positional Diffusion* can reconstruct the original image perfectly, while the fine-tuned Transformer model misplaces multiple pieces.

4.3 Experiment 2: Corrupting Jigsaw Puzzles with Eroded Edges

When jigsaw puzzles are corrupted with eroded edges, the performance of all three heuristics and both deep learning base models declines rapidly, as shown in Fig. 3. Among these, Yu *et al.*'s solver achieves the highest accuracies, followed by Paikin and Tal's and Gallagher's heuristic. The *Positional Diffusion* and Transformer base models are on par with Gallagher on the *Direct Comparison Metric*, but perform significantly worse on the other metric. However, as the level of corruption increases to 30% or more, the differences between the approaches are diminishing and all five methods fail to reconstruct most puzzles.

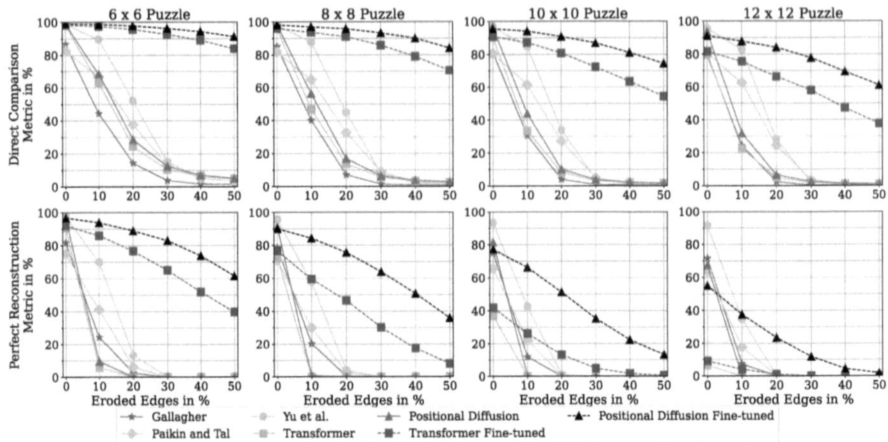

Fig. 3. Results of **Experiment 2** showing the performance of all solvers on puzzles with eroded edges for four different puzzle sizes.

Fine-tuning the deep learning models on this type of corruption again significantly improves their performance. Notably, *Positional Diffusion* becomes the best-performing method, outperforming both Yu *et al.*'s solver and the Transformer model in nearly all scenarios. However, it still faces challenges on larger puzzles, successfully reconstructing fewer than half of them when the edges are corrupted. The visual example of Table 2 confirms the previous findings. There, Gallagher's solver fails to reassemble a quadratic image while the other methods misplace multiple pieces that are severely affected by this type of corruption. The prediction by *Positional Diffusion* is almost perfect.

4.4 Experiment 3: Corrupting Jigsaw Puzzles with Eroded Contents

Figure 4 presents the performance of the jigsaw puzzle solvers across different levels of content erosion, controlled by $\mathcal{E}$, for all four puzzle sizes. Among the heuristics, Gallagher's solver performs the worst, followed by Paikin and Tal's method, with Yu *et al.*'s heuristic achieving the highest accuracies. However, the performance for all three heuristics declines rapidly and they struggle to reassemble most test images when faced with eroded contents. While the base models of the deep learning approaches follow a similar downward trend, their *Direct Comparison Metric* remains higher for smaller puzzles up to 8×8 pieces.

Fine-tuning the deep learning models on this type of corruption significantly improves their performance. *Positional Diffusion* remains the most effective solver, successfully reconstructing the highest number of puzzles in all scenarios. The Transformer model, while less effective than the diffusion model, still outperforms all other solvers by a considerable margin. A visual example is presented in Table 2 where the heuristics misplace every piece. The reconstructed image by

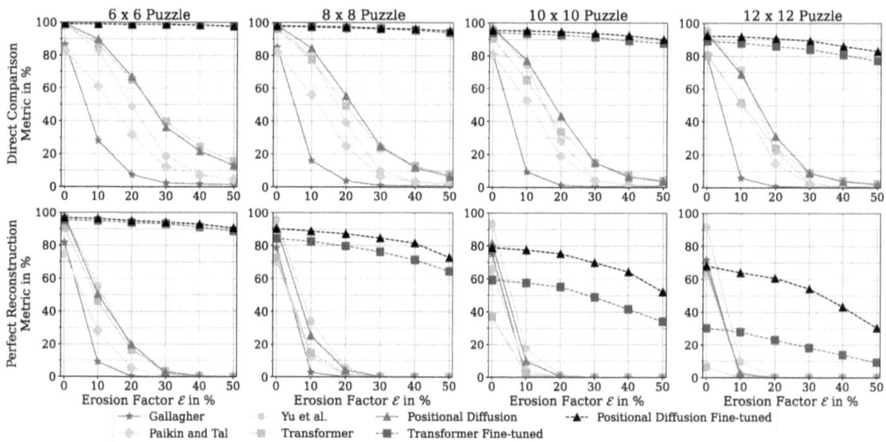

Fig. 4. Results of **Experiment 3** showing the performance of all solvers on puzzles with eroded contents for four different puzzle sizes.

the Transformer model shows more consistent structures while also misplacing many pieces. *Positional Diffusion* almost perfectly assembles the original image.

5 Discussion

Among the three heuristics, Yu *et al.*'s approach demonstrates the highest robustness across all corruption types. Not only is it the best-performing method on standard Type 1 puzzles, but it also consistently outperforms the approaches by Paikin and Tal and Gallagher across nearly all experiments and evaluation metrics. Since both methods by Gallagher and Yu *et al.*rely on the same pairwise compatibility metric for identifying neighbouring pieces, this suggests that Yu *et al.*'s linear programming approach is more robust compared to the greedy assembly strategies of the other heuristics. Nevertheless, all three heuristics degrade in performance if more corruption is applied to the puzzles. Corruptions affecting pixel values, such as the eroded edges and eroded contents, lead to a particularly sharp decline, as they directly impact the pairwise compatibility metric of the heuristics.

Similarly, the base models of the deep learning approaches are not robust to most types of corruption. However, both models demonstrate strong adaptability when trained on augmented data that includes corrupted puzzles. After fine-tuning, their performance improves significantly across all experiments compared to their base versions. The more complex diffusion model consistently outperforms the Transformer model, benefiting from its iterative refinement process during the multiple diffusion steps. Fine-tuning particularly enhances the robustness against eroded edges and eroded contents, as the model's embedding layers seem to become less sensitive to the specific type of corruption. Only if pieces

Table 2. Visual examples of the predictions from the three heuristics and two fine-tuned deep learning solvers on three types of corruptions. Each example is a 12 × 12 puzzle from the *PuzzleWikiArts* dataset [14].

Experiment Details	Gallagher	Paikin and Tal	Yu *et al.*	Transformer	*Positional Diffusion*
Missing Pieces (10%)	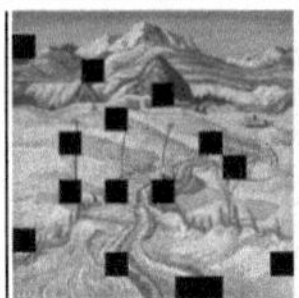		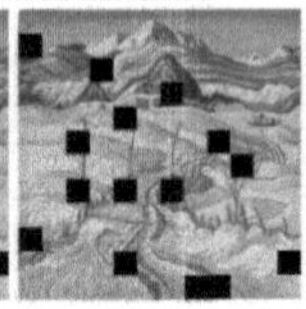	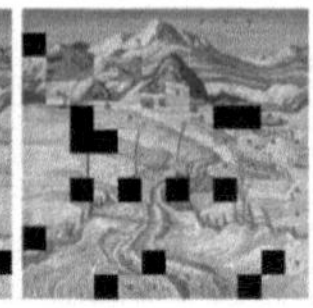	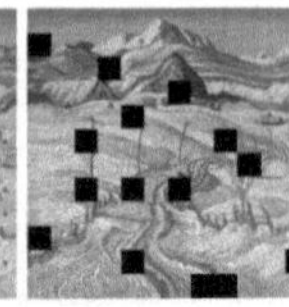
Eroded Edges (20%)					
Eroded Contents (30%)			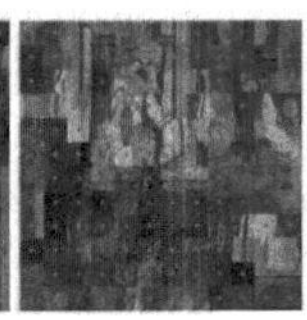	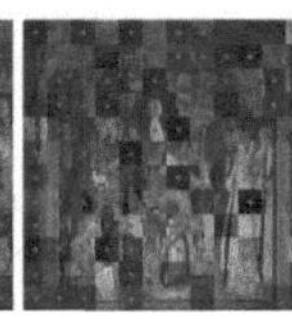	

are missing, both deep learning models struggle, which results in a lower number of perfectly reconstructed puzzles. Additionally, a general drawback of deep learning models is their declining performance as puzzle size increases and they require large amounts of training data.

6 Conclusion and Future Work

This paper explores the robustness and adaptability of content-based jigsaw puzzle solvers when they reassemble corrupted puzzles. Recent research has highlighted that most existing solvers cannot be used to reconstruct real-world fragmented artefacts. As a first step toward bridging this research gap, we benchmarked five SOTA puzzle solvers on three types of corruption, namely missing pieces, eroded edges, and eroded contents. The deep learning models demonstrated their adaptability, with the fine-tuned *Positional Diffusion* model emerging as the most robust solver in the majority of the experiments. It consistently outperformed the heuristic approaches, which rely on pairwise compatibility metrics that become unreliable in the presence of corrupted information. The diffusion model struggled only in cases with missing pieces in larger puzzles, where the method by Yu *et al.*demonstrated greater robustness.

Therefore, the development of more advanced deep learning models presents a promising direction for improving robustness. While this study focused on content-based solvers, a key challenge is the combination of content- and shape-based solvers to handle the irregularly shaped fragments that are typically found in real-world artefacts. Additionally, the creation of more realistic synthetic datasets is crucial. These datasets should incorporate more accurate forms of erosion, variable fragment sizes, and irregular shapes to better reflect real-world conditions and enhance model generalization. Another promising direction is the development of hybrid models that combine deep learning with heuristic approaches. Such advancements would bring us closer to fully automated and robust reconstruction methods capable of reassembling fragmented real-world objects.

Acknowledgments. The authors gratefully acknowledge the scientific support and HPC resources provided by the Erlangen National High Performance Computing Center (NHR@FAU) of the Friedrich-Alexander-Universität Erlangen-Nürnberg (FAU). The hardware is funded by the German Research Foundation (DFG).

Disclosure of Interests. This work was carried out in collaboration with the company Siemens Healthineers AG as part of a research agreement between Siemens Healthineers AG and Machine Learning and Data Analytics Lab, Friedrich-Alexander Universität Erlangen-Nürnberg (FAU). The authors acknowledge the support provided through this collaboration. All author affiliations are listed on the title page. The paper reflects the views of the scientists, and not the company.

References

1. Gallagher, A.C.: Jigsaw puzzles with pieces of unknown orientation. In: CVPR, pp. 382–389 (2012). https://doi.org/10.1109/CVPR.2012.6247699
2. Giuliari, F., et al.: Positional diffusion: graph-based diffusion models for set ordering. Pattern Recogn. Lett. **186**, 272–278 (2024). https://doi.org/10.1016/j.patrec.2024.10.010
3. Harel, P., Ben-Shahar, O.: Crossing cuts polygonal puzzles: models and solvers. In: CVPR, pp. 3084–3093 (2021). https://doi.org/10.1109/CVPR46437.2021.00310
4. Harel, P., Shahar, O.I., Ben-Shahar, O.: Pictorial and apictorial polygonal jigsaw puzzles from arbitrary number of crossing cuts. Int. J. Comput. Vision **132**(9), 3428–3462 (2024). https://doi.org/10.1007/s11263-024-02033-7
5. Huroyan, V., Lerman, G., Wu, H.T.: Solving jigsaw puzzles by the graph connection laplacian. SIAM J. Imag. Sci. **13**(4), 1717–1753 (2020). https://doi.org/10.1137/19M1290760
6. Markaki, S., Panagiotakis, C.: Jigsaw puzzle solving techniques and applications: a survey. Vis. Comput. **39**(10), 4405–4421 (2023). https://doi.org/10.1007/s00371-022-02598-9
7. Marques, M.A.O., Freitas, C.O.A.: Reconstructing strip-shredded documents using color as feature matching. In: ACM symposium on Applied Computing, pp. 893–894 (2009).https://doi.org/10.1145/1529282.1529475
8. Paikin, G., Tal, A.: Solving multiple square jigsaw puzzles with missing pieces. In: CVPR, pp. 4832–4839 (2015).https://doi.org/10.1109/CVPR.2015.7299116

9. Papaodysseus, C., Panagopoulos, T., Exarhos, M., Triantafillou, C., Fragoulis, D., Doumas, C.: Contour-shape based reconstruction of fragmented, 1600 BC wall paintings. IEEE Trans. Signal Process. **50**(6), 1277–1288 (2002). https://doi.org/10.1109/TSP.2002.1003053
10. Pintus, R., Pal, K., Yang, Y., Weyrich, T., Gobbetti, E., Rushmeier, H.: A survey of geometric analysis in cultural heritage. Comput. Graph. Forum **35**(1), 4–31 (2016). https://doi.org/10.1111/cgf.12668
11. Podany, J.C., Larson, J.H., Rosenthal, A.L., Brommelle, N.S., Insall, D.W., Zuccari, F.: Art conservation and restoration - wall paintings, preservation, restoration. Britannica (2023). https://www.britannica.com/art/art-conservation-and-restoration/Wall-paintings
12. Rui Yu, C.R., Agapito, L.: Solving jigsaw puzzles with linear programming. In: BMVC, pp. 139.1–139.12 (2016). https://doi.org/10.5244/C.30.139
13. Son, K., Hays, J., Cooper, D.B.: Solving square jigsaw puzzle by hierarchical loop constraints. IEEE Trans. Pattern Anal. Mach. Intell. **41**(9), 2222–2235 (2019). https://doi.org/10.1109/TPAMI.2018.2857776
14. Talon, D., Bue, A.D., James, S.: Ganzzle: Reframing jigsaw puzzle solving as a retrieval task using a generative mental image. In: ICIP, pp. 4083–4087 (2022).https://doi.org/10.1109/ICIP46576.2022.9897553
15. Talon, D., Del Bue, A., James, S.: GANzzle++: generative approaches for jigsaw puzzle solving as local to global assignment in latent spatial representations. Pattern Recogn. Lett. **187**, 35–41 (2025). https://doi.org/10.1016/j.patrec.2024.11.010
16. Tan, W.R., Chan, C.S., Aguirre, H.E., Tanaka, K.: Improved ArtGAN for conditional synthesis of natural image and artwork. IEEE Trans. Image Process. **28**(1), 394–409 (2019). https://doi.org/10.1109/TIP.2018.2866698
17. Tsesmelis, T., et al.: Re-assembling the past: The RePAIR dataset and benchmark for real world 2D and 3D puzzle solving. In: NeurIPS, vol. 37, pp. 30076–30105 (2024). https://proceedings.neurips.cc/paper_files/paper/2024/hash/3507ec8d7d6895eb9feb87a2098abe11-Abstract-Datasets_and_Benchmarks_Track.html

Successful 3D Reconstructions of Underwater World War II Wreckage

Jaren Gerdes[1(✉)], Jane Berk[1], Michael Hess[1], Shane Wechsler[2], Mark Bilinski[3], Raymond Provost[3], and Angaar Wasimi[3]

[1] Alutiiq LLC, San Diego, CA 92123, USA
{jaren.r.gerdes.ctr,jane.n.berk.ctr,michael.r.hess22.ctr}@us.navy.mil
[2] San Diego State University, San Diego, USA
shane.t.wechsler.ctr@us.navy.mil
[3] Naval Information Warfare Center Pacific, San Diego, CA 92152, USA
{mark.bilinski.civ,raymond.c.provost2.civ,angaar.wasimi.civ}@us.navy.mil

Abstract. Reconstruction of underwater wreckage scenes plays a critical role in maritime archaeology. Understanding the context and details of these submerged sites is essential for both preservation and research. Accurately representing these scenes digitally allows for the continued exploration of these sites without the need for extraneous expeditions. The Naval History and Heritage Command (NHHC), as managers of the Department of the Navy's sunken military craft, provided NIWC Pacific with hundreds of hours of deep-sea underwater video data of different World War II (WWII) shipwrecks collected by Vulcan, LLC. This data was released to us for the purposes of historical, archaeological, and educational research. Our team first pieced together individual video feeds using traditional photogrammetry techniques to create detailed three-dimensional representations of the wreck sites. We then successfully utilized two state-of-the-art frameworks for 3D reconstruction – Neural Radiance Fields (NeRFs) and Gaussian splatting – to create novel views for these comprehensive underwater scene reconstructions. In this paper we share qualitative and quantitative analysis of these techniques and explore the positives and negatives of each as they perform on our unstructured deep sea data.

Keywords: WWII · Underwater · 3D Reconstruction · Photogrammetry · NeRFs · Gaussian splatting

1 Introduction

Detailed representation of an underwater scene is a difficult task due to unique and complex environmental factors, especially sites such as wreckage. Light refraction, absorption, and scattering distort color and often cause objects to appear faded or blurry. This problem worsens as the depth increases, especially at the seafloor where the presence of suspended sand further limits visibility and clarity [12]. This loss in color, texture quality, and structure complicates

E. Rodolà et al. (Eds.): ICIAP 2025, LNCS 16167, pp. 299–311, 2026.
https://doi.org/10.1007/978-3-032-10185-3_24

scene renderings and often leads to inaccurate representations. Furthermore, the dynamic nature of an underwater environment also causes unpredictable changes, making data capture strenuous, and may result in flawed and distorted images. These factors make it difficult to accurately represent underwater scenes that require high fidelity and realism, which complicates our goal of accurate WWII wreckage site reconstruction.

Expeditions to underwater sites are often expensive and laborious; we aim to alleviate this by using existing remote-operated vehicle (ROV) video streams from previous trips to reconstruct sites of interest. Traditional photogrammetry techniques can accurately reconstruct a focus object as a point cloud, but struggle to visualize a scene in its entirety. We can build off of photogrammetry's base capabilities with state-of-the-art 3D reconstruction machine learning methods: Neural Radiance Fields and Gaussian splats. These innovative methods allow for the complete visual preservation of a wreckage site in 3D and facilitates further study of the site virtually, without a specialized scanning survey. Additionally, this enhanced representation offers a more complete picture of the wreck that is essential for the involvement of the general public [9]. The various available models for NeRFs and Gaussian splats offer unique benefits; we tested and compared a variety and selected the best for high-fidelity deep-sea renderings.

The paper is laid out as follows: Sect. 2 provides background information on the intricacies of underwater cultural heritage preservation, as well as the existing methods and models we will be testing; Sect. 3 discusses the data preprocessing performed prior to scene reconstruction; Sect. 4 explains the model generation processes; Sect. 5 presents our metrics and results. Finally, in Sect. 6 we discuss how the NeRFs and Gaussian splats performed on our data, and the steps we plan to take going forward.

2 Background

Neural Radiance Fields (NeRFs) [19] and 3D Gaussian splatting (3DGS) [16] are both highly effective rendering tools for novel-view synthesis of scenes that build off of the capabilities of photogrammetry. In this section, we give a brief overview of each of these techniques and their importance.

2.1 Photogrammetry

Photogrammetry has been around for many years but has become more widely adopted recently by countless domains due to advances in computer vision and compute processing hardware. Photogrammetry takes 2D images that are captured from different points of view and, when conducted properly, generates a 3D point cloud of the scene as well as a mapping of the underwater landscape with engineering accuracy [4,9]. Cultural heritage documentation has benefited greatly from photogrammetric adoption and advances due to its low upfront costs and low technical barrier to entry [14,15]. The stated challenges with underwater environments lead to difficulties in feature detection in the input images, which in turn leads to less desirable photogrammetry results. Even when the specific object of interest is reconstructed well, it often lacks context provided by a surrounding environment. This is where new techniques like NeRFs and Gaussian Splats can greatly enhance the 3D visual experience - especially for shipwreck expeditions which are expensive, difficult to conduct, and provide a very brief snapshot of the environment.

2.2 Neural Radiance Fields (NeRFs)

NeRFs are deep learning models that are used to generate and represent a 3D scene from a collection of 2D images or video frames and synthesize novel views [20]. They do so by representing a scene as a continuous 3D volumetric field instead of traditional meshes or point clouds. A scene is encoded using a neural network that takes in an (x, y, z) position coordinate and a viewing direction. The output is an estimated (r, g, b) color and density value for each point using ray casting, sampling, and accumulation. The only input required for this process is a set of images with known camera poses, typically calculated using structure-from-motion photogrammetry techniques. This allows for accurate modeling of how light interacts with a scene from various viewpoints, resulting in a highly realistic 3D representation.

Additionally, NeRF models can be trained to provide unique benefits depending on what kind of scene the user is trying to create. We compare each of the following three models to select the most effective one for our intended purposes.

Nerfacto is a model developed by the Nerfstudio [24] team consisting of a combination of previously existing models that achieves a balance between speed and quality [24]. It is comparable to other state-of-the-art methods while achieving a significant increase in processing time. This model first optimizes camera views, then generates ray bundles. The ray bundles are then fed into a proposal network sampler where density fields are calculated. These are then translated into the 3D scene where scene contraction, hash encoding, and appearance embedding construct the RGB, depth, and normal values for each of the pixels.

Instant-NGP permits primitive high-speed neural graphics renderings, which are usually known for being costly to train and evaluate [20]. Computer graphics primitives are fundamentally represented by mathematical functions that parameterize appearance, but by using multi-layer perceptrons (MLPs) and multi-resolution hash encodings these functions can help accurately and efficiently render high-fidelity NeRF reconstructions.

SeaThru-Lite is a model specifically designed for deep-sea underwater renderings that often have to deal with strongly influenced mediums; this lightweight variation of SeaThru-Nerf extends upon the original implementation by using advanced building blocks available in the Nerfstudio library [17]. This model deals with absorption and scattering by assigning separate color and density parameters to both the scene focus and medium. By separating the two, a clearer picture of the focus object in its undistorted state can be obtained and represented.

Despite their benefits, NeRF results are highly dependent on the quality of the input, and may suffer if the dataset is of low-quality, noisy, or does not contain comprehensive viewpoints of its central target. Furthermore, while NeRFs have the potential to render highly accurate and realistic scenes, this is achieved with an expensive to train neural network, and they do not easily allow real-time rendering without sacrificing fidelity [7,20,30].

2.3 Gaussian Splatting

Gaussian splatting is a radiance field rendering method that renders at the same high fidelity as NeRFs, but in real-time. Beginning in the same manner as NeRFs, the input is a set of images and their corresponding cameras calibrated via structure-from-motion and the resulting sparse point cloud. The scene is represented with 3D Gaussians that are defined by their mean, a covariance matrix, and opacity; while the color is represented via spherical harmonics to model view-dependent appearance [16]. The result is a real-time high-quality radiance field render that can be generated for a wide variety of scenes.

Several papers discuss improving reconstructed scenes by using depth regularization to remove floaters, especially in areas of the scene that are untextured [2,10,27]; however, the overall scale impact of depth map priors is scene-dependent. In some cases, the result was significant, and in others it was minimal, or even resulted in a decline in rendering quality [16]. Exposure variations between images can also be taken into account during the training process. This is done by optimizing a 3×4 affine transform for each image that is applied to the color of the rendered images. This addition by the Gaussian Splat authors is designed to greatly improve the reconstruction of "in the wild" captures. We include a combination of both depth map priors and exposure compensation during our training process in order to evaluate the effectiveness of their additions on our dataset.

Although the Gaussian splat technique offers promising results, it does still have some limitations. For example, artifacts, elongated, and "splotchy" Gaussians can be present in areas of the scene that are not well captured in the input dataset [13]. However, this is not necessarily limited to Gaussian splats alone; state-of-the-art NeRFs are also known to exhibit this behavior [25]. We do expect this to be a challenge in our 3D renders since our data was not captured for the explicit purpose of 3D reconstruction.

3 Data Preprocessing

We use two datasets for our training, inference, and method analysis: a USS WASP wreckage site (Aircraft) [11] and the sunken USS Indianapolis [26]. These datasets were both collected during underwater expeditions performed by Vulcan, LLC via an ROV deployed from the research vessel R/V Petrel, in 2019 and 2017 respectively. The datasets were collected over multiple dives resulting in hundreds of hours of video footage. Due to its simplicity, the in-depth

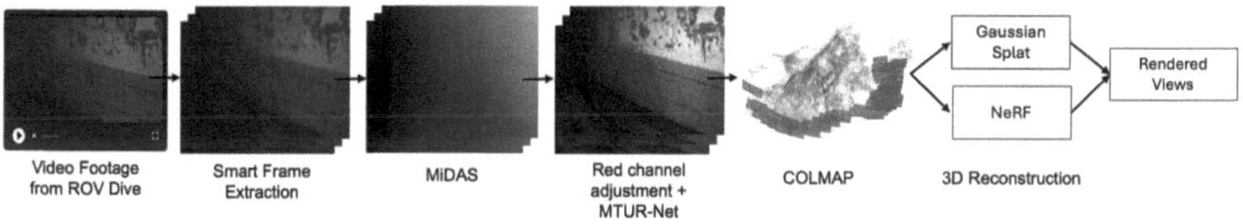

Fig. 1. Data Processing Workflow

hyperparameter and scenario testing will be performed on the Aircraft dataset first, and then extended to the more complex USS Indianapolis. The preprocessing methodologies discussed in Sects. 3.1, 3.2, and 3.3 were performed on both datasets. The full data processing pipeline is shown in Fig. 1.

3.1 Smart Frame Extraction

After we selected videos with the most comprehensive collection of views, we used an algorithm to reduce any duplicative frames. The first pass scrapes an even 2 frames per second (fps), then Meshroom [6] selects images where the camera shows greater than 5% pixel movement between frames. This process optimizes the number of images selected, while still prioritizing movement within the scene. Our final datasets consist of 222 Aircraft images and 257 USS Indianapolis images. After cropping out navigation details on the outer edges of the images, the final resolution is 1660×800.

3.2 Color Correction

The images were processed using a depth map guided red color cast removal algorithm. First, we use the monocular depth estimation algorithm, MiDAS, [21], to obtain the depth map for an image. The normalized red channel from the depth map is used to linearly reduce the red channel of the RGB image. The final step is to use a typical color correction algorithm - we use a medium transmission deep learning network called MTUR-Net [8,28]. Due to the extreme depths that this data was collected at, when only using MTUR-Net, the dark parts of the image were over compensated creating red color casting; this translated into

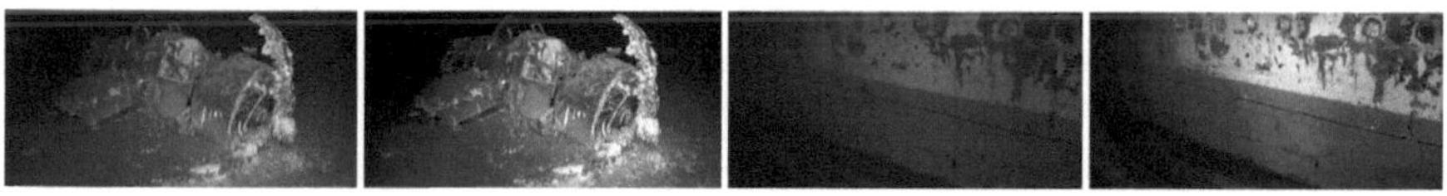

(a) Original (b) Color-Corrected (c) Original (d) Color-Corrected

Aircraft USS Indianapolis

Fig. 2. Original and Color-Corrected versions of the ground truth images used in the NeRF and Gaussian Splat model experiments.

noisy 3D reconstructions. Example images of our color correction method are shown in Fig. 2.

3.3 Point Cloud and Camera Positions

Both NeRFs and Gaussian splats require camera pose estimations as input; we leveraged the open source Structure-from-Motion (SfM) pipeline COLMAP [22, 23] to estimate the camera intrinsics, including the focal length, principal point, and sensor orientations, from a set of 2D images. Once these camera positions were calculated by identifying matching key points across the images, a 3D point cloud was constructed by solving the perspective-n-point problem. These camera positions and point clouds are the input to both our NeRF and Gaussian splat reconstructions.

4 3D Model Generation

We implemented a 90/10 training/testing split for each preprocessed dataset in order to validate each methods' performance. The split data was then fed into both NeRFs and Gaussian splats to construct the scenes, the details of which are described 4.1 and 4.2 respectively. Every tested variation of the NeRFs and Gaussian splats were trained for $30k$ epochs.

4.1 NeRFs

We use Nerfstudio, a modular NeRF development framework to train and render NeRF reconstructions. We experimented with Nerfacto, Instant-NGP, and SeaThru-Lite models. Since SeaThru-Lite is unique in its specific design for underwater reconstruction, we test this model with both color corrected and non-color corrected underwater images. Examples of each of the tested models are shown in Fig. 3. Nerfacto and SeaThru-Lite both have larger available models but did not perform as well when compared to their lightweight counterparts. Each of the tested models were trained with their default hyperparameters.

4.2 Gaussian Splatting

Four different input combinations were used to train the Gaussian splat: with and without depth maps, and with and without exposure compensation. The algorithm DepthAnythingV2 [29] was used to generate the depth maps priors. In addition, we used the exposure compensation hyperparameters suggested in [16]. An example of a resulting render from each of the combinations is shown in Fig. 4, while quantitative metrics for both methods are shown in Table 1.

5 Results

The 3D reconstruction methods were validated using the withheld image sets from our 90/10 splits. While the 3D reconstruction techniques are capable of generating novel points of view, for our results we specifically recreate the viewpoints with a ground truth counterpart. For each of these images three metrics were calculated to quantify our results. The first, Structural Similarity Index Measure (SSIM), considers the contrast, luminance, and structure of two images in order to determine how similar they are [5]. The second metric calculated is Peak Signal-to-Noise Ratio (PSNR), which compares an original image to a reconstructed image to determine how much of the original was lost in compression or reconstruction [1]. The final metric used is Learned Perceptual Image Patch Similarity (LPIPS), which utilizes deep learning models to evaluate the perceptual similarity between two images [3].

5.1 Aircraft

Instant-NGP scored highest in both SSIM and PSNR, while scoring just below Nerfacto in LPIPS. Inversely, Nerfacto came in second place in SSIM and PSNR, taking the top score for LPIPS. SeaThru-Lite without color correction lagged behind SeaThru-Lite with color correction by $\sim$ 5% for SSIM and $\sim$ 1% for PSNR and LPIPS respectively. SeaThru-Lite did not benefit from running on non-color corrected data as theorized; full details of each of the model's performance are available in Table 1. In the selected frame, Fig. 3b, SeaThru-Lite with color correction renders the most balanced image; however, when looking

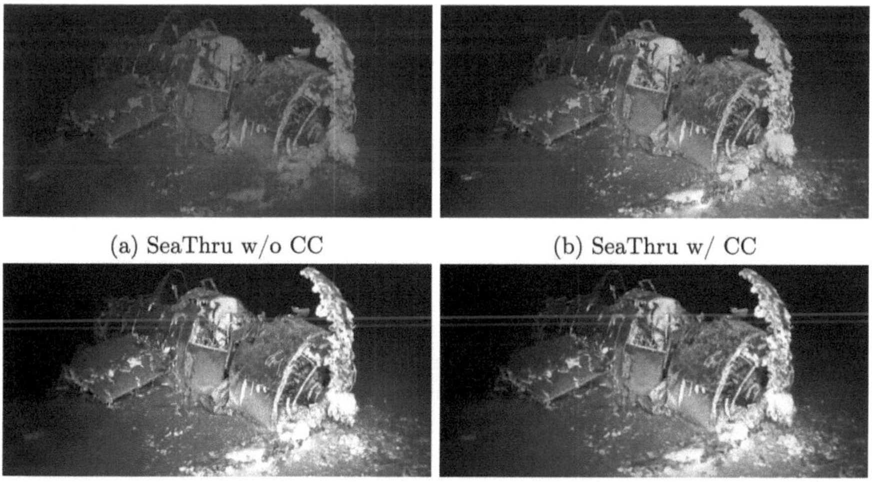

(a) SeaThru w/o CC (b) SeaThru w/ CC (c) Nerfacto (d) Instant-NGP

Fig. 3. Resulting renders from each of the tested NeRF models. Figure 2 used as ground truth.

at the entire validation set, it struggled to differentiate between the water and the object, causing distortion in some of the generated images. Instant-NGP, Fig. 3d, appears over-saturated, but overall resulted in very consistent visual clarity throughout the validation set, reinforcing its selection as the best performing model.

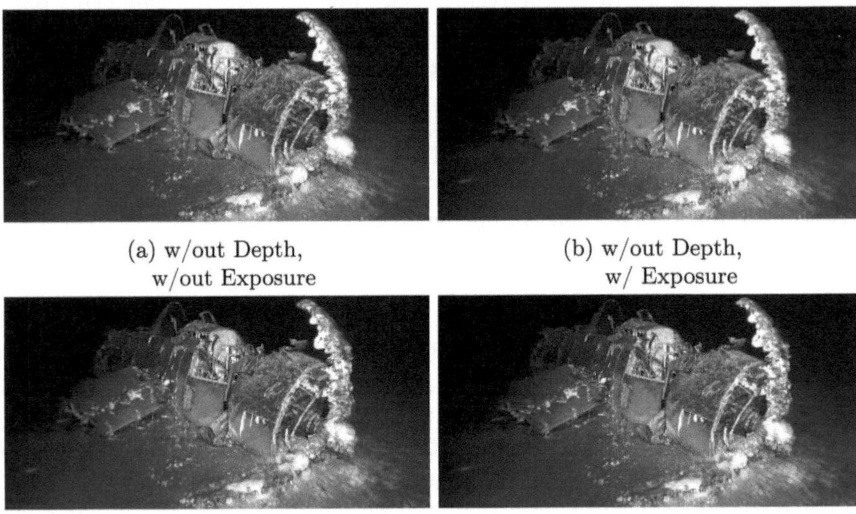

(a) w/out Depth, w/out Exposure

(b) w/out Depth, w/ Exposure

(c) w/ Depth, w/out Exposure

(d) w/ Depth, w/ Exposure

Fig. 4. Resulting renders from each of the tested Gaussian splat training combinations.

Next we consider the Gaussian splat experiments; shown in Fig. 4. Although all combinations create visually realistic views, there are some subtleties. By compensating for the discrepancies in exposure between frames during the training process (Figs. 4a and 4b), blurring is reduced, specifically around the propeller of the aircraft – this is also reflected in the quantitative metrics. Without considering the depth maps, the SSIM and PSNR increased $\sim$ 3%, while the LPIPS saw a decrease of $\sim$ 5%. If we incorporate the depth maps during training, we see the same reduction in blur and artifacts in Figs. 4c and 4d – this is again reflected in Table 1. With the inclusion of depth maps only, we see a $\sim$ 3% increase in SSIM, $\sim$ 4% increase in PSNR, and $\sim$ 5% reduction in LPIPS. Although subtle, this is visually evidenced by the clarity and lack of blurring around the propellor. Overall, the best combination is when depth and exposure are used, Fig. 4d. For easy comparison, Fig. 5 contains a ground truth, a novel view from Instant-NGP, as well as the same novel view for Gaussian Splat with depth maps and exposure in Rows 1, 2, and 3 respectively.

5.2 USS Indianapolis

After performing an in-depth analysis, we apply these techniques to the USS Indianapolis dataset. We used the training data from the 90/10 split to train an Instant-NGP model as well as a Gaussian splat with depth maps and exposure, and then rendered novel views. In Table 2, we see that the Gaussian Splat performs better than the NeRF for all three metrics; Fig. 5 presents the ground truth, a novel view from Instant-NGP, as well as the same novel view for Gaussian Splat with depth maps and exposure in Rows 1, 2, and 3 respectively.

Table 1. Aircraft: Quantitative metrics for NeRF and Gaussian splat models and training parameter combinations for the withheld validation set.

		SSIM↑	PSNR↑	LPIPS↓
NeRF	Nerfacto	0.73	20.40	0.26
	SeaThru-Lite	0.69	18.50	0.37
	SeaThru-Lite w/o CC	0.66	18.93	0.38
	Instant-NGP	0.79	23.25	0.28
Gaussian Splat	w/out depth, w/out exposure	0.74	20.51	0.41
	w/out depth, w/ exposure	0.76	21.05	0.39
	w/ depth, w/out exposure	0.76	21.31	0.39
	w/ depth, w/ exposure	0.77	21.33	0.39

Table 2. USS Indianapolis: Quantitative metrics for NeRF and Gaussian splat models for the withheld validation set. We used the best parameters and models from the Aircraft experiments.

	SSIM↑	PSNR↑	LPIPS↓
Instant-NGP	0.57	19.28	0.54
Gaussian Splat w/depth, w/exposure	0.71	22.54	0.31

5.3 Training and Rendering Time

We measured the time required to train and render images for each method. While Gaussian splats take longer to train than NeRFs, an average of 42 vs 17 minutes, they are significantly faster to render, an average of 0.15 vs. 0.42 seconds per image.

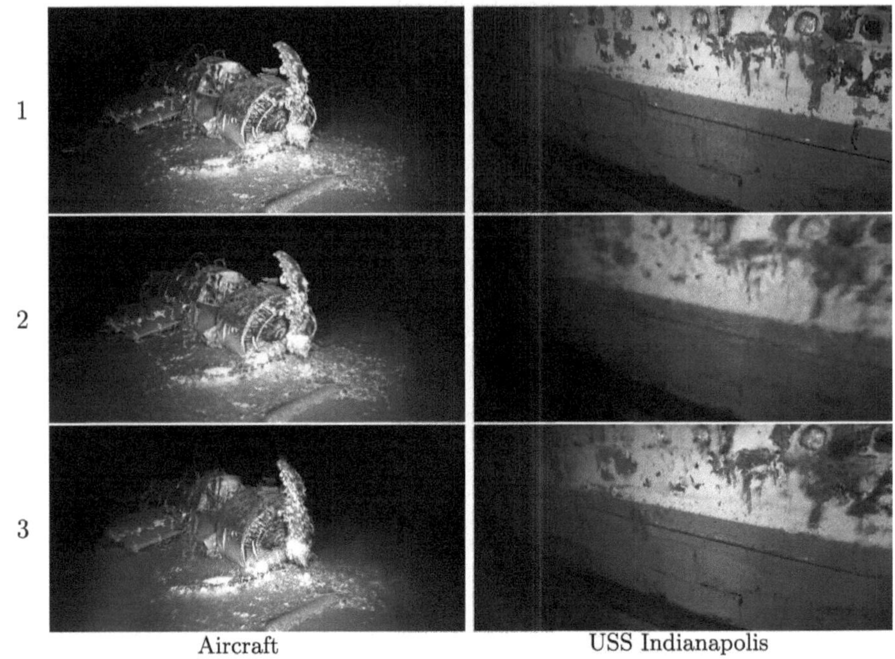

Fig. 5. From top to bottom. Row 1: Ground truth images used for NeRF and Gaussian Splat validation. Row 2: NeRF results rendered using the Instant-NGP model. Row 3: Example images rendered via Gaussian splat model trained with depth maps and with exposure.

6 Discussion and Conclusion

We evaluated multiple NERF and Gaussian splat reconstructions to identify the best method suited for 3D visualization of historical shipwreck sites. Instant-NGP scored the highest, with Nerfacto and SeaThru-Lite following shortly behind. We were surprised by the performance of SeaThru-Lite; though the model was designed specifically for shallow underwater reconstructions, it did not translate well to deep underwater scenes with inconsistent lighting such as ours.

Although including exposure compensation *or* depth maps during the Gaussian splat training process improved metrics overall, we saw the largest improvements when including both. However, the highest performing Gaussian splat lags behind the highest performing NeRF model for the Aircraft dataset, Instant-NGP, by $\sim$ 2.5% for SSIM, $\sim$ 8% in PSNR, and $\sim$ 28% for LPIPS. In all but LPIPS, the Gaussian splat is the second highest performing model overall. Since Gaussians can be rendered in near real time, there is still a choice to be made between the two methodologies depending on the use case.

When Instant-NGP and Gaussian splatting with depth maps and exposure are used for the challenge dataset, the USS Indianapolis, we see an improvement of $\sim 25\%$, $\sim 17\%$, and $\sim 43\%$ in SSIM, PSNR, and LPIPS respectively between Instant-NGP and Gaussian splatting. We theorize that this change in best performing model is due to the increase in the type of noise present in this dataset. Although both rendering methods are susceptible to noise, the extremely unstable lighting in our dataset has a greater impact on NeRFs.

In future work, we plan to continue to explore available reconstruction methods to achieve the highest quality deep underwater 3D renderings. New NeRF models are consistently being developed in order to account for a growing amount of use cases. For underwater specifically, UWNerf [31], builds off of SeaThru-Lite's methodology while considering the additional challenges, such as distance-dependent visibility and unstable illumination, that reconstructing underwater scenes creates. This model will be included in future testings of our work. There are also developments in Gaussian splats to more accurately render the medium as well as the geometry [18].

Finally, there is an increased importance in combating inconsistencies in data collected in the wild such as illumination, color, and camera positioning that are introduced by water depth distortion. We plan to continue color correction work so that the input to both NeRFs and Gaussian splats continues to be accurate and scene agnostic, resulting in 3D reconstructions with fewer artifacts.

Our research into both NeRFs and Gaussian splats have provided us with a thorough understanding of the difficulties that need to be overcome when reconstructing deep underwater scenes, and how best to approach this difficult task. Ultimately, we aim to render these WWII historical sites as accurately and with as much fidelity as possible – through either the adaptation of an existing approach, or the development of our own deep-underwater-aware reconstruction method – in order to ensure the continued preservation and study of these important sites for many years to come.

References

1. Fardo, F.A., et al.: A formal evaluation of PSNR as quality measurement parameter for image segmentation algorithms. arXiv:1605.07116 (2016)
2. Li, J., et al.: Dngaussian: Optimizing sparse-view 3D gaussian radiance fields with global-local depth normalization. arXiv preprint arXiv:2403.06912 (2024)
3. Zhang, R., et al.: The unreasonable effectiveness of deep features as a perceptual metric. arXiv:1801.03924 (2018)
4. Bolles, R.C., et al.: Epipolar-plane image analysis: an approach to determining structure from motion. Int. J. Comput. Vis. **1**, 7–55 (1987). https://api.semanticscholar.org/CorpusID:12598541
5. Wang, Z., et al.: Image quality assessment: from error visibility to structural similarity. IEEE Trans. Image Process. **13**(4), 600–612 (2004). https://doi.org/10.1109/TIP.2003.819861
6. AliceVision: Meshroom: A 3D reconstruction software. (2018). https://github.com/alicevision/meshroom
7. Barron, J.T., et al.: Mip-nerf: a multiscale representation for anti-aliasing neural radiance fields. In: Proceedings of the IEEE/CVF International Conference on Computer Vision, pp. 5855–5864 (2021)

8. Berk, J., et al.: Color correction of deep underwater images using depth maps (2025). [Manuscript submitted for publication]
9. Calantropio, A., et al.: Underwater cultural heritage documentation using photogrammetry. J. Mar. Sci. Eng. **12**(3) (2024). https://doi.org/10.3390/jmse12030413, https://www.mdpi.com/2077-1312/12/3/413
10. Chung, J., et al.: Depth-regularized optimization for 3D gaussian splatting in few-shot images. arXiv preprint arXiv:2311.13398 (2023)
11. Commander, U.T.F.: Wreckage of USS wasp cv-7 discovered in coral sea (2019). https://www.c7f.navy.mil/Media/News/Display/Article/1784781/wreckage-of-uss-wasp-cv-7-discovered-in-coral-sea/
12. Costa, E.: Survey and photogrammetry in underwater archaeological contexts at low visibility in the Venice lagoon. Digit. Appl. Archaeol. Cult. Heritage **24**, e00215 (2022). https://doi.org/10.1016/j.daach.2022.e00215
13. Foroutan, Y., et al.: Does gaussian splatting need SFM initialization? arXiv preprint arXiv:2404.12547 (2024)
14. Guarnieri, A., et al.: Digital photogrammetry and laser scanning in cultural heritage survey **35** (2004)
15. Ikeuchi, K., et al.: The great buddha project: modeling cultural heritage for VR systems through observation. In: ISMAR 2003, Proceedings of the 2nd IEEE/ACM International Symposium on Mixed and Augmented Reality, p. 7. IEEE Computer Society, USA (2003)
16. Kerbl, B., et al.: 3D gaussian splatting for real-time radiance field rendering. ACM Trans. Graph. **42**(4), 139–1 (2023)
17. Levy, D., et al.: Seathru-nerf: neural radiance fields in scattering media. arXiv:2304.07743v1 (2023)
18. Li, H., et al.: Watersplatting: fast underwater 3D scene reconstruction using gaussian splatting. arXiv preprint arXiv:2408.08206 (2024)
19. Mildenhall, B., et al.: Nerf: representing scenes as neural radiance fields for view synthesis. Commun. ACM **65**(1), 99–106 (2021)
20. Müller, T.: Instant neural graphics primitives with a multiresolution hash encoding. ACM Trans. Graph. (TOG) **41**(4), 1–15 (2022)
21. Ranftl, R., Lasinger, K., Hafner, D., Schindler, K., Koltun, V.: Towards robust monocular depth estimation: mixing datasets for zero-shot cross-dataset transfer. IEEE Trans. Pattern Anal. Mach. Intell. **44**(3), 1623–1637 (2022)
22. Schönberger, J.L., Zheng, E., Frahm, J.-M., Pollefeys, M.: Pixelwise view selection for unstructured multi-view stereo. In: Leibe, B., Matas, J., Sebe, N., Welling, M. (eds.) ECCV 2016. LNCS, vol. 9907, pp. 501–518. Springer, Cham (2016). https://doi.org/10.1007/978-3-319-46487-9_31
23. Schönberger, J.L., et al.: Structure-from-motion revisited. In: Conference on Computer Vision and Pattern Recognition (CVPR) (2016)
24. Tancik, M., et al.: Nerfstudio: a modular framework for neural radiance field development. arXiv:2302.04264v4 (2023)
25. Warburg, F., Weber, E., Tancik, M., Holynski, A., Kanazawa, A.: Nerfbusters: removing ghostly artifacts from casually captured nerfs. In: Proceedings of the IEEE/CVF International Conference on Computer Vision, pp. 18120–18130 (2023)
26. Werner, B.: Paul Allen finds lost world war II USS Indianapolis (2017). https://news.usni.org/2017/08/19/uss-indianapolis-wreckage-found
27. Xiong, H., et al.: Sparsegs: real-time 360° sparse view synthesis using gaussian splatting. Arxiv (2023)
28. Yan, K., et al.: Medium transmission map matters for learning to restore real-world underwater images. Appl. Sci. **12**(11), 5420 (2022)

29. Yang, L., et al.: Depth anything v2. arXiv preprint arXiv:2406.09414 (2024)
30. Yu, A., et al.: Plenoctrees for real-time rendering of neural radiance fields. In: Proceedings of the IEEE/CVF International Conference on Computer Vision, pp. 5752–5761 (2021)
31. Tang, Y., et al.: Neural underwater scene representation. In: CVPR (2024)

Explainable Machine Learning for Earthquakes: SHAP Interpretation of CNNs to Distinguish Seismic Spectrograms of Foreshocks and Aftershocks

Michele Magrini[1(✉)], Francesco Marrocco[1], Laura Laurenti[2], Gabriele Paoletti[3], Elisa Tinti[3], and Chris Marone[3,4]

[1] Department of Mathematics, Sapienza University of Rome, Rome, Italy
magrini.2066963@uniroma1.it
[2] Swiss Seismological Service (SED), ETH Zurich, Zurich, Switzerland
[3] Department of Earth Sciences, Sapienza University of Rome, Rome, Italy
[4] Department of Geosciences, Pennsylvania State University, Pennsylvania, USA

Abstract. Fault zone properties evolve during the seismic cycle and explainable machine learning can help us monitor these changes. We propose the innovative use of RGB spectrograms to represent 3-component seismic waveforms, facilitating both model training and interpretation. We apply a CNN to classify foreshocks and aftershocks from RGB spectrograms and use SHapley Additive exPlanations (SHAP) to interpret the model's decisions. Given the CNN's high accuracy (99.66%), we turn our attention to understanding its decision-making process through SHAP. SHAP reveals that the CNN distinguishes classes based on a narrow frequency band near 30 Hz, linked to elastic wave attenuation. The RGB format enables intuitive SHAP interpretation by preserving time-frequency structure across seismic components. Temporal SHAP analysis shows meaningful evolution in aftershocks, interpreted as fault healing, and highlights the complex and varied nature of foreshock behavior. Our work highlights SHAP as a robust and transparent method to extract physical insight from ML models and contributes to bridging data-driven approaches with seismological understanding.

Keywords: Explainability · Seismology · Earthquakes · AI · Machine Learning Interpretability · Seismic Sequence · SHAP · Foreshocks · Aftershocks · Spectrograms · Fault Zone Evolution

1 Introduction

Understanding and detecting earthquake precursors is one of the most challenging goals in geophysics. In recent years, the increasing availability of large-scale seismic datasets and advancements in computational power have paved the way

E. Rodolà et al. (Eds.): ICIAP 2025, LNCS 16167, pp. 312–323, 2026.
https://doi.org/10.1007/978-3-032-10185-3_25

for the application of Machine Learning (ML) techniques to seismology [1,2,7]. These methods have proven to be highly effective in automatically detecting and classifying seismic events by learning patterns and relationships that are challenging to identify using traditional analytical methods. Recent studies have moved beyond catalog development and developed Deep Learning (DL) methods to classify the state of the seismic cycle from raw seismic waveforms [3]. In particular, we build upon the work of Laurenti et al. [3], who classified seismic events as foreshocks (earthquakes that occur before the mainshock) or aftershocks (earthquakes that occur after the mainshock). Their research highlighted the potential of Convolutional Neural Networks (CNNs) for the classification of seismograms and for tracking the evolution of earthquake fault properties during the seismic cycle.

Despite major advances the widespread adoption of ML in scientific fields, and particularly in seismology, is often limited by the difficulty of interpreting model decisions. State-of-the-art neural networks function as "black boxes", providing high predictive accuracy but limited insight into the features that influence the decision-making process. This lack of interpretability poses a barrier to scientific validation, trust, and practical application, as it hinders the ability to link model outputs to established physical processes.

Here, we apply SHapley Additive exPlanations (SHAP) [4] to interpret the predictions of a CNN trained to identify the evolution of a seismic sequence, using 2D images of earthquake seismic spectrograms as input. Before the largest earthquake in the seismic sequence (called mainshock), smaller earthquakes named foreshocks may occur as fault stress increases toward failure. After the mainshock, there is always an increase in seismic activity as aftershocks (smaller earthquakes occurring after the mainshock) occur in response to fault stress redistribution and local dynamic stressing caused by the mainshock. The state of stress around the earthquake fault zone increases prior to the mainshock, and then decreases dramatically as fault slip occurs during the mainshock. Such differences in stress state cause changes in the elastic properties and seismic attenuation of the rock surrounding the fault. We train a CNN to detect these variations by classifying foreshocks and aftershocks.

We use SHAP techniques to quantify the contribution of input features for a given prediction. By applying SHAP to our trained model, we aim to identify which features of the seismic spectrograms are most influential for classification and how these features evolve over time relative to the mainshock event.

The goal of this work is to leverage SHAP to bridge the gap between high-performance ML models and physical understanding, ultimately enhancing the reliability and scientific rigor of ML applications in earthquake analysis.

2 Data and Methods

2.1 Dataset Details

We use seismic waveforms from the M6.5 2016 Norcia mainshock and seismic sequence in central Italy. The mainshock occurred near the town of Norcia on

October 30th 2016. In [3] the events were considered within a volume of a few kilometers around the Vettore fault, in the area that hosted the Norcia mainshock, where abundant seismicity was observed prior to and after October 30th. We followed the same approach by downloading seismic data from the NRCA station based on their criteria. The dataset consists of three-component seismic waveforms recorded at 100 Hz, corresponding to the North–South (NS), East–West (EW), and vertical (Z) components of ground motion. The training dataset is balanced, comprising 3,534 total seismic events, evenly split between foreshocks and aftershocks (1,767 events for each class). The test dataset consists of 1,165 events, with 738 foreshocks and 427 aftershocks. The dataset spans 10 weeks of foreshocks (from 2016-08-20 to 2016-10-26) and 8 weeks of aftershocks (from 2016-10-31 to 2016-12-20).

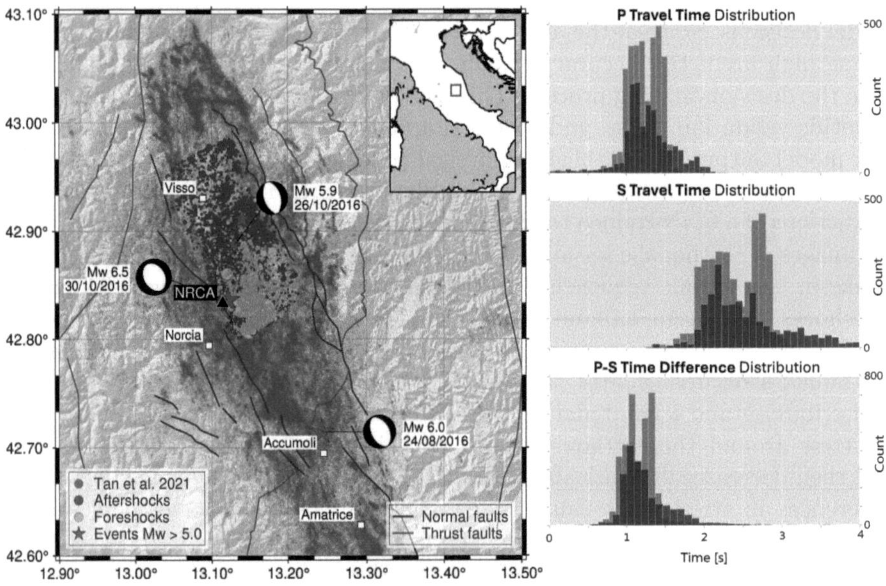

Fig. 1. Earthquake dataset based on the 2016 Norcia seismic sequence. Left panel shows earthquakes of the Tan *et al.* [8] catalog. Gray dots show all earthquakes (> 900k events) and red stars are M $\geq$ 5.0 earthquakes. The earthquakes selected for our experiments are highlighted in color: cyan symbols are foreshocks, and purple are aftershocks. We use data from the NRCA seismic station (black triangle). Right panels show the distributions of P- and S-Wave arrival times for our events. (Color figure online)

2.2 Preprocessing

Seismic waves generated by earthquakes include P-waves (fastest, compressional waves that arrive first), S-waves (slower, shear waves that arrive after P-waves), and surface waves (travel along the Earth's surface). For each earthquake in our dataset we extract a 13-second time window starting 5 s before the P-wave arrival. This window length was chosen to include the S-wave arrival, which for most of our events occurs within 3 to 5 s after the P-wave (Fig. 1). The 13-second window effectively captures the pre-P-wave signal and extends to approximately 5 s beyond most S-wave arrivals. By focusing on this time window, we aim to illuminate how the CNN model classifies earthquakes as foreshocks and aftershocks.

For each of the three-component seismograms, we computed the Short-Time Fourier Transform (STFT) of the input signal. This involves dividing the signal into small, overlapping segments – with a segment length of 64 points of the 100 Hz data and an overlap ratio of 0.8 in our specific setup. We then perform a Fourier transform on each segment. The result is a time-frequency representation that shows how the signal's energy is distributed across frequencies over time (Fig. 2). From this, we generate spectrograms, which are images constructed from the power values of the STFT, capturing the changes in frequency over time and providing a visual representation of the seismic signal's characteristics.

After computing the spectrograms for each channel of the input waveform (Fig. 2b), we convert the power values to a logarithmic scale and apply min-max normalization across the channels of each spectrogram, such that all values fall between 0 and 1. This normalization ensures that the model learns to focus on the temporal and frequency patterns, rather than variations in signal energy.

2.3 Image Creation

To integrate the three-component spectra into a format compatible with image-based models, we combine each spectrogram (Fig. 2b) into a single RGB image (Fig. 2c). Each component is assigned a color: Red for East-West, Green for North-South, and Blue for the Vertical component. This creates an RGB three-channel image that encapsulates multi-component temporal and spectral information in a compact and visually informative format.

Our image creation technique adapts a commonly used representation in Computer Vision, which we generalize to earthquake science. In seismology, the typical approach is to compute the spectrogram for each channel separately, and plot them as individual color-coded time-frequency maps, with amplitude or power (dB scale) represented by color intensity. Our approach for constructing RGB spectrograms allows the use of image-based DL models (here a CNN) on seismic data. This novel approach offers a practical way to integrate information from three seismic components and explore the potential of vision-based techniques in a domain where they are not yet widely adopted.

2.4 Training

The classification task is performed using a CNN designed to process the RGB spectrogram images generated from the three-channel seismic waveforms. The

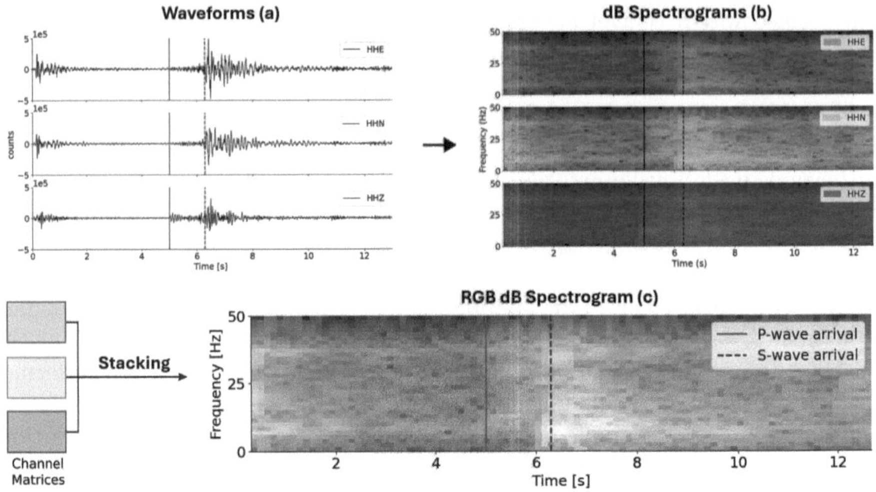

Fig. 2. RGB images from seismic spectrograms. (a) Three component seismograms for one earthquake showing E-W, N-S, and vertical components of ground velocity for 13 s (starting 5 s before the p-wave arrival). (b) Spectrograms are computed for each channel in a sliding time window of 0.64 s, with overlap of 0.5 s. (c) The RGB spectrograms are stacked to produce an RGB image corresponding to the seismic spectrogram.

model consists of three convolutional layers, each followed by batch normalization and ReLU activation functions. Max-pooling layers are introduced between convolutional blocks to progressively reduce the spatial dimensions and make the network focus on the most salient features. A dropout layer is applied after flattening to reduce overfitting. The convolutional blocks are followed by fully connected layers, culminating in an output layer with two units, corresponding to the two classes: foreshock and aftershock (Table 1).

Training Protocol. We employ a 7-fold cross-validation scheme to evaluate the model's performance and reduce variance associated with data partitioning. The final performance metrics correspond to the best-performing model across all folds. For optimization, we use the Adam optimizer with a learning rate of $1 \cdot 10^{-3}$. To prevent overfitting and avoid unnecessary training cycles, we apply early stopping, halting training when validation accuracy fails to improve for a predefined number of consecutive epochs.

2.5 Model Performance

Initial training and validation runs demonstrated stable convergence, with the model consistently achieving high accuracy across folds. The cross-validation results indicate that our 2D CNN slightly outperforms the 1D CNN proposed by Laurenti et al. [3] on the same NRCA traces (99.66% vs. 99.19% accuracy),

Table 1. CNN2D Model Summary

Layer Type	Output Shape	Param #
Conv2D + BatchNorm2D	(32, 33, 96)	896 + 64
MaxPool2D	(32, 16, 48)	0
Conv2D + BatchNorm2D	(64, 16, 48)	18.5K + 128
MaxPool2D	(64, 8, 24)	0
Conv2D + BatchNorm2D	(128, 8, 24)	73.9K + 256
Flatten + Dropout	(24576)	0
Linear	(128)	3.1M
Dropout	(128)	0
Linear	(2)	258
Total params:		3.2M
Model size:		37.1 MB

confirming both the reliability of our pipeline and the model's improved ability to capture discriminative features.

Quantitatively, the final selected model achieves a total accuracy of 99.66%, reflecting near-perfect classification performance across the dataset. Precision, recall, and F1 scores are all measured at 99.53%.

3 SHAP Results

SHAP. is a widely-used technique for explainable machine learning. The SHAP approach is based on a concept from cooperative game theory, where the contribution of each player to the overall outcome is quantified by considering all possible combinations in which that player could participate. In the context of ML, players represent features. As noted by Lundberg *et al.* [4], SHAP satisfies essential properties such as local accuracy and consistency, establishing it as a principled and robust method for evaluating model predictions. For our CNN model, SHAP allows us to measure the impact of each input feature (here, a region of pixels of the spectrogram) on the model's output probability. The method works by systematically perturbing input features and observing how the model's predictions change. By doing so across many testing spectrograms and recomputing the predictions, SHAP estimates the contribution of each section of the image to the final prediction score. The fundamental idea is to compute the marginal contribution of each feature by comparing the model output when that feature is present versus when it is absent, averaged over all possible subsets. By generating SHAP values for each pixel, we obtain visual explanations that highlight which regions of the spectrogram have the most influence on the model classification predictions.

Telea Inpainting. In order to compute SHAP values on image data, we generate perturbed images by masking certain regions (patches of pixels) corresponding to a given spectral frequency range. However, simply replacing masked regions with zeros or random noise would create unrealistic images that may mislead the model and distort the explanation. Instead, it is important to replace masked regions in a plausible way.

For this purpose, we use inpainting, a technique that fills in missing parts of an image by smoothly interpolating surrounding information. In this study, we adopt the Telea inpainting algorithm [9], based on the fast marching method. It reconstructs masked or missing areas by propagating pixel values from the mask boundary inward. The process uses both the gradient of known pixels and spatial weighting to ensure that the inpainted region transitions smoothly from the known data. The result is a natural in-fill of the masked segment that maintains the continuity of structures of the original image.

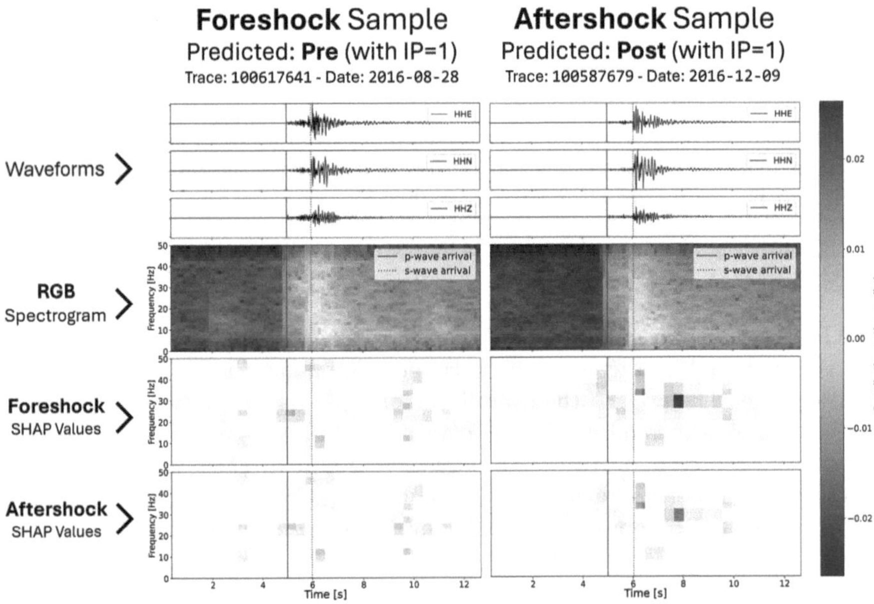

Fig. 3. Foreshock and Aftershock Samples with SHAP-Based Model Interpretations. The left panel presents a foreshock sample, while the right panel shows an aftershock sample. Each column includes (from top to bottom): three-component seismic waveforms, the corresponding RGB spectrogram, and SHAP value visualizations for both classes. The SHAP maps labeled *Foreshock* and *Aftershock* indicate the contribution of each time-frequency region toward the respective class prediction. Red regions reinforce the corresponding class, while blue regions indicate support for the alternative class. (Color figure online)

3.1 Average SHAP on Each Class

Averaging SHAP values across multiple correctly classified examples reveals class-level patterns in the model's decision-making process. For this analysis, 300 correctly predicted foreshock and aftershock samples are selected from the test set. To ensure that the results reflect meaningful model behavior rather than misclassification artifacts, only correctly classified events are included. SHAP values are computed for each sample using 5,000 impainting perturbations per spectrogram (Fig. 3).

Each SHAP computation produces a tensor of the same dimensions as the input spectrogram, where each pixel indicates its contribution to the model's predicted class probability. In Fig. 3, we show an example of one correctly classified foreshock and aftershock. Note that the SHAP values of the correct and wrong classes are visually opposite, even though numerically not complementary. This could represent real differences between the spectral content of elastic waves before (foreshock) and after (aftershock) the mainshock, but it could also arise from the way the model redistributes importance when masking and inpainting parts of the input. Rather than trying to extract physical meaning from each SHAP map, which could vary with signal noise and other factors, we averaged SHAP tensors across the selected samples (Fig. 4) to highlight robust features.

To further aid in interpretability of SHAP output, we separately analyzed temporal alignments for both P- and S-wave arrival times. The time between P and S depends on the distance between the hypocenter and the recording station. In the P-fixed mean SHAP tensor, spectrograms were aligned based on the P-wave arrival time, preserving temporal relationships relative to this reference. Conversely, in the S-fixed mean SHAP tensor, spectrograms were rearranged to align the S-wave arrival time, allowing a direct comparison of model explanations centered around this feature. These averaged SHAP maps provide insight into how the model distinguishes between foreshocks and aftershocks based on both spectral and temporal characteristics.

Observations for the Average SHAP Values. The average SHAP tensor maps show pixels (regions of spectral frequency and time) that contribute negatively (blue) or positively (red) to the model classification (Fig. 4). For the foreshock samples (top panels of Fig. 4), SHAP values indicate a positive contribution (approximately +0.006) from pixels in the 45–50 Hz range near the S-wave arrival. These frequencies are near the Nyquist for 100 Hz data sampling. In contrast, for aftershocks, the frequency band 25–30 Hz contributes strongly to the classification for the whole time period especially between the P and S-wave arrivals. These patterns are consistent for SHAP samples aligned on both P- and S-waves. This shows that the difference between foreshocks and aftershock seismograms is not an earthquake source effect, which would occur only after the P-wave, but is rather a seismic ray path effect. The SHAP results suggest that differences in frequency-dependent elastic wave attenuation, in particular in the 25–30 Hz band, are the primary source of the CNN performance (Fig. 4).

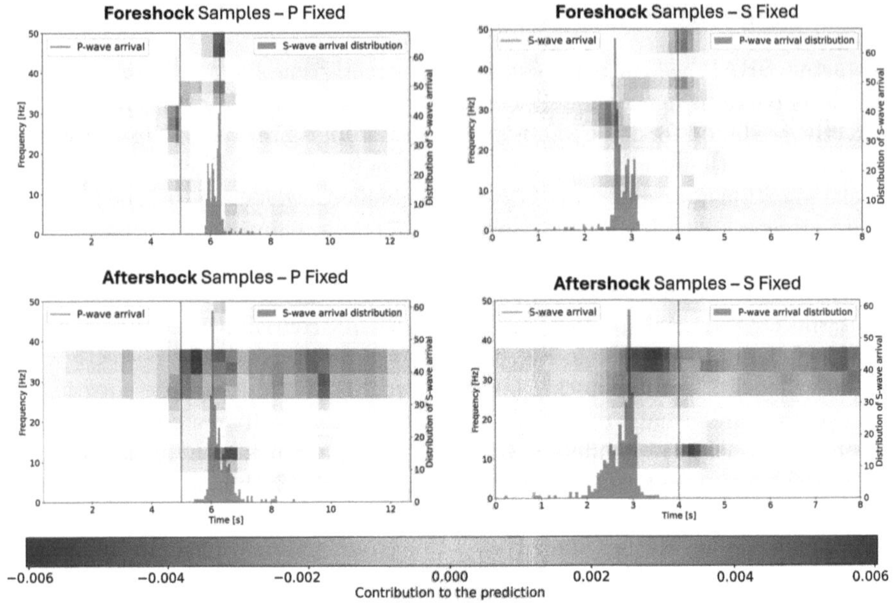

Fig. 4. Average SHAP analysis for correctly predicted foreshock and aftershock samples. In the first row, 300 correctly predicted foreshock samples are used to compute SHAP values. In the P-fixed Foreshock plot (top left), SHAP values are averaged while keeping the P-wave arrival aligned. In the S-fixed Foreshock plot (top right), the same 300 samples are rearranged to align the S-wave arrival before computing SHAP values. The second row follows the same procedure but with 300 correctly predicted aftershock samples. The grayscale histograms represent the distribution of arrival times for the opposite wave type in each case.

Physical Insights from the Analysis of Average SHAP Values. We find persistent patterns in the SHAP results for foreshocks and aftershocks. The frequency band from 25–30 Hz is consistently the most important factor in CNN performance classification. These results are consistent with expectations for wave attenuation, which is strongly dependent on fracture density and the evolution of fault zone damage before (foreshocks) and after (aftershocks) the mainshock. Our results are also consistent with previous works [3] that highlighted the role of frequency-dependent attenuation. One interpretation that can be made based on the SHAP results is that the model is very confident about distinguishing foreshocks and aftershocks because of the evolution of shear stress during the seismic cycle. Higher shear stress causes a higher degree of damage around the fault prior to mainshock failure and afterward. This is particularly apparent for waves that travel through the fault zone and damaged rock.

3.2 Temporal Evolution in SHAP Values

While average SHAP values for each class provide insight into the general patterns the model uses to distinguish foreshocks from aftershocks, seismic signals (and potentially the model's interpretation of them) are not static in time. In particular, seismic activity in the days or weeks leading up to and following a mainshock evolves, with changes in the ground's structure and resulting in different propagation of the seismic waves, with differences in frequency content and wave attenuation. A key question we address is whether the model's internal representations and feature importance also evolve as the temporal distance from the main event changes.

To investigate this, we analyzed the temporal evolution of SHAP values by grouping correctly classified seismic events based on their temporal proximity to the mainshock. Specifically, we divided the events into weekly bins, starting from eight weeks before the mainshock and continuing through eight weeks afterward. Within each bin, we computed the mean SHAP map following the same procedure described in the previous section. This allows us to visualize how the features that the model considers important shift as the main event approaches and recedes.

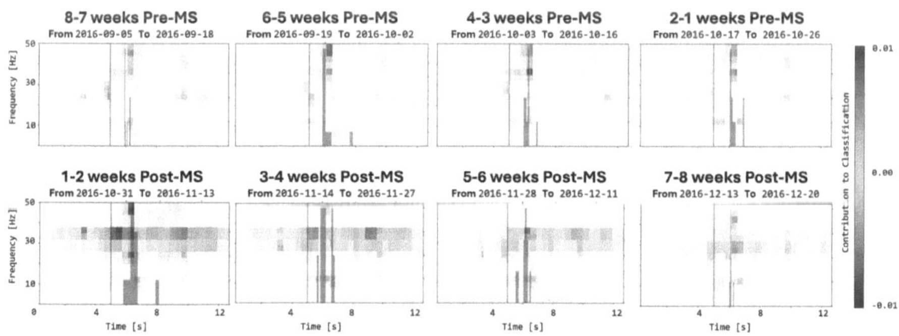

Fig. 5. Average SHAP values for correctly predicted samples across different time periods. The first row represents foreshocks, with SHAP tensors averaged over two-week intervals from August 5, 2016, to October 26, 2016. The second row shows aftershocks, with averages computed for the weeks spanning October 31, 2016, to December 20, 2016. We consider 20 samples for each week interval.

Observations for the Week SHAP Analysis. The SHAP value distributions across different time periods reveal distinct trends for both foreshock (first row of Fig. 5) and aftershock (second row of Fig. 5) samples.

In the foreshock samples, no clear temporal trend emerges across the analyzed weeks. We can observe a stronger contribution of the S wave with respect to the P wave; there are sporadic positive contributions in the 45–50 Hz range and negative contributions in the 25–30 Hz range. However, we cannot identify consistent patterns across time intervals since the intensity and localization vary and are sparse. The overall magnitude of SHAP values remains relatively low, and no evident shift in feature importance is discernible as the foreshock period progresses.

In the aftershock samples, SHAP values show a decreasing trend as the time progresses from the mainshock event. The stronger positive contribution in the 35–45 Hz band seems to weaken over time, suggesting a possible evolution of the fault properties, akin to healing processes, as the aftershock sequence advances. Moreover, the initial two aftershock weeks display a strong negative contribution in regions that had previously shown a strong positive effect in the foreshock samples, specifically around the 45–50 Hz range near the S-wave arrival. This suggests an inversion in feature importance between foreshock and aftershock classifications.

These temporal variations in SHAP distributions highlight systematic changes in the model's reliance on different frequency components over time, warranting further analysis of the underlying physical mechanisms driving these patterns.

Physical Insights from the Temporal Analysis of SHAP Values. The clearest temporal change in SHAP values occurs for aftershocks, where we see that the intensity of the contribution to classification decreases with time after the mainshock (Fig. 5). This change is consistent with previous works showing time dependent changes in wave speed during the postseismic period. It is also consistent with lab data showing time dependent changes associated with frictional healing and increased elastic wave speed following a failure event. While our SHAP data for foreshocks show only minor changes they are also consistent with previous works showing subtle changes in elastic wave speed and transmitted amplitude associated with progressive microcrack formation and damage as stress rises prior to failure.

4 Conclusions

In this study, we introduce a SHapley Additive exPlanations (SHAP) approach to interpret RGB images of earthquake spectrograms. By incorporating explainable ML our work provides an opportunity to significantly enhance the application of DL models in seismology. Our approach leverages CNNs to differentiate between earthquakes that occur before a mainshock (foreshocks) and those that occur afterwards (aftershocks), using RGB spectrograms generated from three-component seismic waveforms. The CNN not only achieves near-perfect classification performance but also, when coupled with SHAP analysis, offers valuable insights into the underlying physical mechanisms governing the seismic cycle.

The SHAP interpretation method allows us to quantify the contribution of different time-frequency components, revealing that high-frequency features near the S-wave arrival are particularly discriminative for foreshock classification, while broader frequency patterns in the 25–35 Hz range serve as key indicators in aftershock recognition. Additionally, the temporal analysis of SHAP values provides evidence of an evolving feature importance relative to the mainshock, which is consistent with the expected healing process of the fault zone. These findings build upon and extend the framework introduced in [3] and [4], bridging the gap between high-performance, data-driven methods and physical insights critical for seismic hazard assessment.

Overall, our results underscore the potential of integrating explainable ML models into earthquake analysis workflows. The ability to interpret model predictions in a physics-based context not only fosters trust in the model but also paves the way for real-time monitoring and early warning applications.

Acknowledgements. We thank the anonymous reviewers for their helpful and detailed feedback. This work was supported by European Research Council Advanced Grant 835012 (TECTONIC).

Data Availability Statement. All data analyzed during this study is available in the Zenodo repository at https://doi.org/10.5281/zenodo.15683047. All custom code used to preprocess the data, run the analyses, and generate the figures is available at our GitHub (https://github.com/mich1803/CNN-Explainability-on-Earthquakes) and Zenodo (https://doi.org/10.5281/zenodo.15747240) archives.

References

1. Beroza, G.C., Segou, M., Mousavi, S.M.: Machine learning and earthquake forecasting—next steps. Nat. Commun. **12**, 4761 (2021). https://doi.org/10.1038/s41467-021-24952-6
2. Kubo, H., Naoi, M., Kano, M.: Recent advances in earthquake seismology using machine learning. Earth Planets Space **76**(1), 36 (2024). https://doi.org/10.1186/s40623-024-01982-0
3. Laurenti, L., Paoletti, G., Tinti, E., Galasso, F., Collettini, C., Marone, C.: Probing the evolution of fault properties during the seismic cycle with deep learning. Nat. Commun. **15**, 10025 (2024). https://doi.org/10.1038/s41467-024-54153-w
4. Lundberg, S., Lee, S.I.: A unified approach to interpreting model predictions (2017). https://arxiv.org/abs/1705.07874
5. Magrini, M., Marrocco, F., Laurenti, L., Paoletti, G., Tinti, E., Marone, C.: Explainable machine learning for earthquakes - dataset. https://doi.org/10.5281/zenodo.15683047
6. Michele, M., Marrocco, F., Laurenti, L., Paoletti, G., Tinti, E., Marone, C.: Explainable machine learning for earthquakes - code. https://doi.org/10.5281/zenodo.15747240
7. Rouet-Leduc, B., Hulbert, C., Lubbers, N., Barros, K., Humphreys, C.J., Johnson, P.A.: Machine learning predicts laboratory earthquakes. Geophys. Res. Lett. **44**(18), 9276–9282 (2017). https://doi.org/10.1002/2017GL074677
8. Tan, Y.J., et al.: Machine-learning-based high-resolution earthquake catalog reveals how complex fault structures were activated during the 2016–2017 central Italy sequence. Seismic Rec. **1**(1), 11–19 (2021). https://doi.org/10.1785/0320210001
9. Telea, A.: An image inpainting technique based on the fast marching method. J. Graph. Tools **9** (2004). https://doi.org/10.1080/10867651.2004.10487596

RoadFusion: Latent Diffusion Model for Pavement Defect Detection

Muhammad Aqeel(✉), Kidus Dagnaw Bellete, and Francesco Setti

Department of Engineering for Innovation Medicine, University of Verona, Strada le Grazie 15, Verona, Italy
muhammad.aqeel@univr.it

Abstract. Pavement defect detection faces critical challenges including limited annotated data, domain shift between training and deployment environments, and high variability in defect appearances across different road conditions. We propose RoadFusion, a framework that addresses these limitations through synthetic anomaly generation with dual-path feature adaptation. A latent diffusion model synthesizes diverse, realistic defects using text prompts and spatial masks, enabling effective training under data scarcity. Two separate feature adaptors specialize representations for normal and anomalous inputs, improving robustness to domain shift and defect variability. A lightweight discriminator learns to distinguish fine-grained defect patterns at the patch level. Evaluated on six benchmark datasets, RoadFusion achieves consistently strong performance across both classification and localization tasks, setting new state-of-the-art in multiple metrics relevant to real-world road inspection.

Keywords: Pavement defect detection · Diffusion models · Road surface analysis

1 Introduction

Road infrastructure is a cornerstone of national development, underpinning mobility, economic activity, public safety, and territorial accessibility. The structural condition of pavement surfaces directly impacts vehicle performance, fuel efficiency, travel time reliability, and user safety. Poorly maintained roads lead to increased wear on vehicles and higher operating costs. In Europe, road maintenance alone can represent up to 40% of total transport infrastructure spending, emphasizing the importance of targeted and timely maintenance efforts [24].

For public administrations (PAs) managing extensive and aging road networks, early detection of surface anomalies is critical. Traditional visual inspection methods are labor-intensive, inconsistent, and lack scalability [15]. In Italy, for example, more than 250,000 km of roads require regular assessment. The national road agency, ANAS, allocates over 1.5 billion annually to pavement

E. Rodolà et al. (Eds.): ICIAP 2025, LNCS 16167, pp. 324–335, 2026.
https://doi.org/10.1007/978-3-032-10185-3_26

rehabilitation [22], yet resource constraints continue to limit large-scale, proactive maintenance. As a result, AI-driven inspection systems are gaining traction as cost-effective, scalable alternatives [18].

Recent advances in deep learning–particularly convolutional neural networks (CNNs)–have demonstrated strong performance in tasks such as crack classification, pothole detection, and texture anomaly recognition [29]. However, a key limitation of existing approaches is their primary focus on image-level classification, rather than on precise localization of defects. For real-world deployment, especially in public infrastructure management, simply knowing that a defect exists is insufficient. High-resolution, pixel-level localization is essential for prioritizing maintenance, estimating damage extent, and planning repairs efficiently.

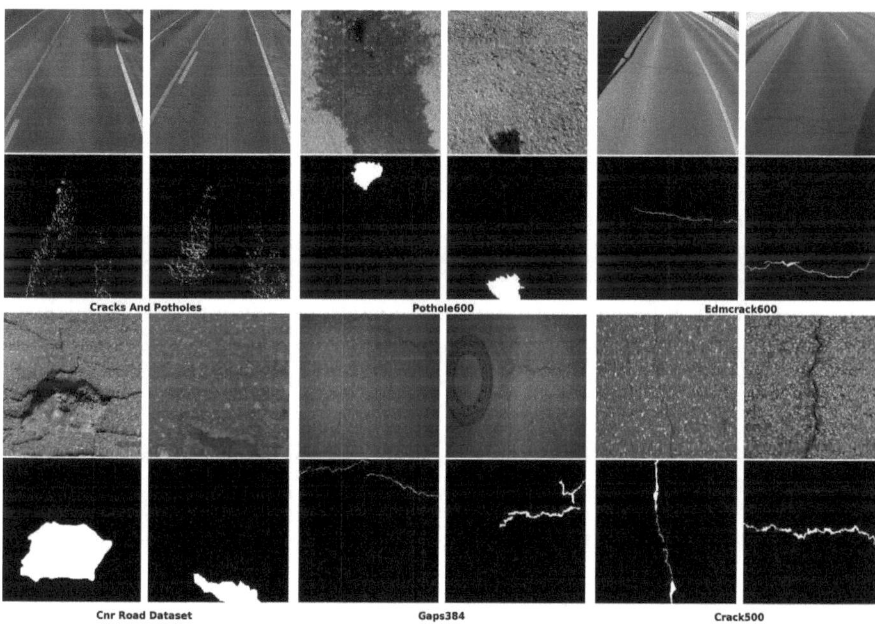

Fig. 1. Diverse examples of real-world pavement defects from six benchmark datasets used in our experiments. Each pair shows an input image (top) and its corresponding ground truth mask (bottom). The datasets include a range of defect types and appearances: surface cracking and patching in Cracks and Potholes, localized potholes in Pothole600, fine linear cracks in Edmcrack600 and Gaps384, irregular surface damage in CNR Road, and dense crack patterns in Crack500. This visual diversity highlights the challenges of consistent defect detection across datasets.

Additionally, several practical challenges persist. First, pre-trained models often struggle with domain shifts when applied to real pavement data. Second, the imbalance between abundant normal samples and limited defect samples reduces training effectiveness. Third, the visual diversity of pavement conditions–across materials, lighting, weather, and imaging perspectives–adds noise and complexity to feature learning. These issues contribute to false positives, missed detections, and unreliable predictions–outcomes that are costly and dangerous in practice [4,9]. Figure 1 illustrates the visual diversity of real-world pavement defects across several benchmark datasets, including cracks, potholes, surface wear, and texture anomalies. These examples highlight the complexity of the task and the need for models that can generalize well across different defect types, scales, and appearances.

To address these challenges, we propose RoadFusion, a novel framework that shifts the focus from simple defect classification to accurate spatial anomaly localization. The framework is designed to handle both the variability of real-world road conditions and the limitations of existing datasets. Our approach integrates synthetic anomaly generation using diffusion models with a dual-adaptor architecture that enhances feature learning for both normal and anomalous patterns.

Our main contributions are as follows:

- A dual-adaptor architecture that bridges the domain gap between pre-trained features and pavement-specific representations, using separate pathways for normal and anomalous samples to improve discriminative power.
- Integration of a latent diffusion model for generating diverse, realistic synthetic anomalies guided by text prompts and spatial masks, helping address the scarcity of annotated defect data.
- A streamlined inference pipeline that maintains computational efficiency while delivering high-resolution anomaly localization across challenging, real-world datasets.

2 Related Work

Pavement defect detection has evolved from rule-based image processing to deep learning-driven approaches. Early methods relied on handcrafted features like Gabor filters and morphological operations [15], but lacked robustness under real-world variations. Classical machine learning models (e.g., SVMs, Random Forests) improved performance by learning from labeled data, yet still depended on manual feature design.

Deep learning, particularly Convolutional Neural Networks (CNNs), marked a turning point by enabling end-to-end learning from raw imagery. Architectures such as U-Net [13] and CrackGAN [29] brought pixel-level precision to defect localization, while real-time detectors like YOLO [19] enabled practical deployment. More recently, hybrid models combining CNNs and transformers [14] have improved context modeling, benefiting detection of subtle or large-spanning defects.

Surface defect detection, now dominated by CNNs, remains critical across industrial applications. Recent work has focused on making these models more robust and generalizable under real-world conditions [1,4]. Self-supervised learning approaches [2,3] aim to improve defect detection without relying heavily on labeled datasets. By leveraging pretext tasks and unsupervised refinement strategies, these methods can identify subtle surface anomalies across varied domains.

Diffusion-based augmentation has shown promise in improving model performance under distribution shifts [5]. By generating realistic in-distribution samples, such approaches help mitigate overfitting and improve defect generalization, particularly relevant for surface inspection scenarios where data imbalance and domain variability are major obstacles.

Synthetic data generation has gained traction as a response to data scarcity in defect detection. GANs [16] were initially adopted for augmentation, while diffusion models have emerged as a more stable and expressive alternative for generating high-quality defect samples [28]. These approaches enable the creation of diverse training examples covering a wider range of defect appearances and environmental conditions.

In parallel, domain adaptation and transfer learning strategies have addressed the mismatch between training distributions and deployment scenarios [11]. These techniques help models maintain performance when faced with new pavement types, lighting conditions, or imaging systems not represented in the original training data. Collectively, these advancements represent a shift toward more data-efficient and adaptable solutions [10].

Despite these advances, challenges like domain shift, class imbalance, and scale variation persist. Our proposed framework, RoadFusion, builds on these insights by combining diffusion-based synthetic anomaly generation with a dual-adaptor architecture, offering improved performance in both classification and localization tasks under diverse real-world conditions.

3 RoadFusion Pipeline

The RoadFusion framework is introduced in this section. As illustrated in Fig. 2, RoadFusion consists of a Feature Extractor, dual Feature Adaptors (A and B), a Latent Diffusion-based Anomalous Image Generator, and a Discriminator. The framework operates with a streamlined single-flow architecture during inference. These modules will be described below in sequence.

3.1 Feature Extractor

The Feature Extractor obtains local features through a multi-scale approach as in [23]. We denote the training set and test set as $\mathcal{X}_{\text{train}}$ and $\mathcal{X}_{\text{test}}$. For any image $\mathbf{x}_i \in \mathbb{R}^{H\times W\times 3}$ in $\mathcal{X}_{\text{train}} \cup \mathcal{X}_{\text{test}}$, the pre-trained backbone network Φ extracts features from different hierarchical levels. We define $\mathcal{L}$ as the subset of selected hierarchical levels. The feature map from level $l \in \mathcal{L}$ is denoted as $\Phi_{l,i} \sim \Phi_l(\mathbf{x}_i) \in \mathbb{R}^{H_l\times W_l\times C_l}$, where H_l, W_l, and C_l represent the height, width, and channel dimensions. For an entry $\Phi_{l,i}^{h,w} \in \mathbb{R}^{C_l}$ at location (h,w), its neighborhood with patchsize p is defined as:

$$\mathcal{N}_p^{(h,w)} = \{(h',w') \mid h' \in [h-\lfloor p/2\rfloor,\dots,h+\lfloor p/2\rfloor],\ w' \in [w-\lfloor p/2\rfloor,\dots,w+\lfloor p/2\rfloor]\} \tag{1}$$

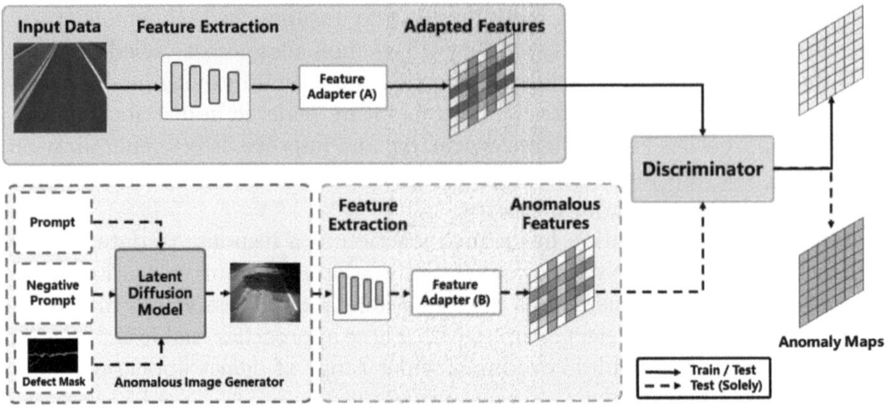

Fig. 2. Overview of the proposed RoadFusion architecture for pavement defect detection. The top pathway handles normal samples: defect-free road images are passed through a pre-trained feature extractor and then adapted via Feature Adapter (A) to generate domain-specific normal features. The bottom pathway generates synthetic anomalies using a latent diffusion model conditioned on prompts, negative prompts, and defect masks. These anomalous images are then processed through the same feature extraction pipeline and Feature Adapter (B) to obtain anomalous features. During training, the Discriminator learns to differentiate between normal and anomalous features. At test time, only the upper pathway is used to produce anomaly maps via the Discriminator.

Aggregating the features within the neighborhood $\mathcal{N}_p^{h,w}$ with aggregation function f_{agg} (using adaptive average pooling) yields the local feature $\mathbf{z}_{l,i}^{h,w}$:

$$\mathbf{z}_{l,i}^{h,w} = f_{\text{agg}}\left(\left\{\Phi_{l,i}^{h',w'} \mid (h',w') \in \mathcal{N}_p^{h,w}\right\}\right) \tag{2}$$

To combine features $\mathbf{z}_{l,i}^{h,w}$ from different hierarchies, all feature maps are linearly resized to the same dimensions (H_0, W_0). Concatenating the feature maps channel-wise produces the feature map $\mathbf{o}_i \in \mathbb{R}^{H_0 \times W_0 \times C}$:

$$\mathbf{o}_i = f_{\text{cat}}\left(\{\text{resize}(\mathbf{z}_{l',i}, (H_0, W_0)) \mid l' \in \mathcal{L}\}\right) \tag{3}$$

We define $\mathbf{o}_i^{h,w} \in \mathbb{R}^C$ as the entry of $\mathbf{o}_i$ at location (h, w) and simplify the expression as:

$$\mathbf{o}_i = \mathcal{F}_\Phi(\mathbf{x}_i) \tag{4}$$

where $\mathcal{F}_\Phi$ represents the Feature Extractor.

3.2 Feature Adaptors

To adapt features to the target domain of pavement surfaces, we employ two distinct Feature Adaptors. Feature Adaptor A, denoted as $\mathcal{G}_A$, processes features from normal images:

$$\mathbf{q}_{h,w}^i = \mathcal{G}_A(\mathbf{o}_{h,w}^i) \tag{5}$$

For the anomalous images generated by the Latent Diffusion Model, we utilize Feature Adaptor B, denoted as $\mathcal{G}_B$. After extracting features from the synthetic anomalous images using the same Feature Extractor:

$$\mathbf{o}_a^i = \mathcal{F}_\Phi(\mathbf{i}_a^i) \tag{6}$$

These features are processed through Feature Adaptor B:

$$\mathbf{q}_{h,w}^{i-} = \mathcal{G}_B(\mathbf{o}_a^{i,h,w}) \tag{7}$$

This separate adaptor pathway for anomalous features allows the framework to learn distinct representations for normal and defective pavement regions. Both Feature Adaptors A and B share the same architectural design, consisting of fully-connected layers, but maintain separate parameters to specialize in their respective domains. Experimental results demonstrate that this dual-adaptor approach more effectively differentiates between normal and anomalous features compared to using a single adaptor for both types of features.

3.3 Latent Diffusion-Based Anomalous Image Generator

The Latent Diffusion Model (LDM) [10,12] generates realistic pavement anomalies by leveraging a diffusion process that operates in a lower-dimensional latent space rather than directly in pixel space. In this paper, we generate images using DIAG [10] for its ability to adapt to new textures and to cope with different surface defects. To generate an anomalous image $\mathbf{i}_a$, the process begins with a defect-free pavement image, a textual anomaly description, and a location mask, forming the triplet $(\mathbf{i}_n, \mathbf{d}_a, \mathbf{m}_a)$. The text-conditioned LDM performs inpainting on image $\mathbf{i}_n$ using the mask $\mathbf{m}_a$.

Given a set of defect-free pavement samples $\mathcal{I}_n$, the framework incorporates textual descriptions $\mathcal{D}_a$ of pavement anomalies (cracks, potholes, raveling, etc.). Regions where these anomalies may realistically appear are designated through a set of binary masks $\mathcal{M}_a$. The LDM, conditioned on this information, inpaints plausible anomalies onto the defect-free samples. The output $\mathbf{i}_a$ represents an anomalous version of $\mathbf{i}_n$, with a realistic defect inpainted in the masked region $\mathbf{m}_a$. This process can be repeated with different parameters to generate a diverse set of anomalous images $\mathcal{I}_a$ for training.

3.4 Discriminator

The Discriminator $\mathcal{D}_\psi$ functions as a normality estimator, calculating a normality score at each spatial location (h, w). It processes both normal features $\{\mathbf{q}^i \mid \mathbf{x}_i \in \mathcal{X}_{\text{train}}\}$ from Feature Adaptor A and anomalous features $\{\mathbf{q}^{i-}\}$ from Feature Adaptor B during training. The Discriminator architecture employs a 2-layer MLP that outputs a scalar normality value $\mathcal{D}_\psi(\mathbf{q}_{h,w}) \in \mathbb{R}$.

3.5 Loss Function and Training

The training employs a truncated ℓ_1 loss formulation:

$$\ell_{h,w}^{i} = \max\left(0, \tau_{+} - \mathcal{D}_{\psi}(\mathbf{q}_{h,w}^{i})\right) + \max\left(0, -\tau_{-} + \mathcal{D}_{\psi}(\mathbf{q}_{h,w}^{i-})\right) \tag{8}$$

where τ_{+} and τ_{-} represent threshold values set to 0.5 and -0.5 respectively. The overall training objective is:

$$\mathcal{L} = \min_{A,B,\psi} \sum_{\mathbf{x}_i \in \mathcal{X}_{\text{train}}} \sum_{h,w} \frac{\ell_{h,w}^{i}}{H_0 \cdot W_0} \tag{9}$$

where A and B are the parameters of Feature Adaptors A and B respectively. The performance of this loss function is evaluated against standard cross-entropy loss in the experiments section.

3.6 Inference and Scoring Function

During inference, the Latent Diffusion-based Anomalous Image Generator and Feature Adaptor B are not used. For each test image $\mathbf{x}_i \in \mathcal{X}_{\text{test}}$, features are extracted through the Feature Extractor $\mathcal{F}_{\Phi}$ and adapted via Feature Adaptor A $\mathcal{G}_A$ to obtain features $\mathbf{q}_{h,w}^{i}$ as in Eq. (5). The anomaly score is calculated by the Discriminator $\mathcal{D}_{\psi}$:

$$\mathbf{s}_{h,w}^{i} = -\mathcal{D}_{\psi}(\mathbf{q}_{h,w}^{i}) \tag{10}$$

The anomaly map for localization is defined as:

$$\mathbf{S}_{AL}(\mathbf{x}_i) := \left\{\mathbf{s}_{h,w}^{i} \mid (h, w) \in H_0 \times W_0\right\} \tag{11}$$

This map is interpolated to match the input resolution and smoothed with a Gaussian filter ($\sigma = 4$). The final anomaly detection score for each image is determined by taking the maximum value from the anomaly map:

$$\mathbf{S}_{AD}(\mathbf{x}_i) := \max_{(h,w) \in H_0 \times W_0} \mathbf{s}_{h,w}^{i} \tag{12}$$

4 Experimental Results

4.1 Datasets

We evaluate our method on six public road damage datasets, each offering a unique combination of image characteristics and defect types:

- **Crack500** [30]: 500 high-resolution images (2000×1500) captured via smartphone, annotated for cracks.
- **GAPs384** [7]: 384 grayscale images (1920×1080) manually selected to focus on crack detection.

- **EdmCrack600** [20]: 600 pixel-level annotated images of road cracks from urban roads in Edmonton, Canada.
- **Pothole-600** [8]: 600 RGB images (400 × 400) annotated for potholes, with accompanying disparity maps and masks.
- **CPRID** [21]: 2,235 images (1024 × 640) from Brazilian highways, labeled for both cracks and potholes.
- **CNR Road** [25]: 20 high-resolution web images with detailed annotations of potholes.

These datasets span various regions, imaging methods, and damage types, forming a comprehensive benchmark for evaluating road surface anomaly detection.

4.2 Implementation Details

We implemented our model using the PyTorch framework and trained it on an NVIDIA RTX 4090 GPU for efficient training and inference. We use a pre-trained WideResNet-50 as the Feature Extractor, with features extracted from multiple intermediate layers and aggregated into a 1536-dimensional representation. Feature Adaptors A and B are fully connected layers without bias, sharing architecture but using separate parameters. Anomalous images are generated using a latent diffusion model guided by text prompts and spatial masks, then passed through the same feature pipeline to obtain anomalous features. The Discriminator is a two-layer MLP with batch normalization and leaky ReLU (slope 0.2). We train the model using Adam with learning rates of 0.0001 for the adaptors and 0.0002 for the discriminator, a weight decay of 0.00001, batch size 16, and 60 training epochs.

4.3 Evaluation Metrics

We evaluate performance using a comprehensive set of metrics to assess both classification and localization capabilities. For classification, we report Precision, Recall, Macro-F1, and AUROC–including both image-level (I-AUROC) and pixel-level (P-AUROC) variants. For localization and segmentation quality, we include mean Average Precision (mAP), Intersection over Union (IoU), and Average Precision (AP) where applicable. These metrics provide a balanced view of the model's accuracy, generalization, and ability to precisely identify defect regions.

4.4 Quantitative Results

RoadFusion demonstrates superior performance across all six benchmark datasets as shown in Table 1, confirming its robustness to diverse pavement types and defect characteristics. On Crack500, it achieves the highest Macro-F1 (0.91) and Recall (0.90), with a strong P-AUROC of 0.73–critical for early detection where missing subtle cracks can accelerate infrastructure deterioration. For CNR

Table 1. Performance comparison of different methods across multiple datasets. Best results for each metric are highlighted in **bold**.

Dataset	Method	P.	R.	M.-F1	mAP	IoU	AP	I/A	P/A
CNR Road	Eisenbach [7]	0.85	**0.91**	0.85	0.79	0.70	**0.85**	0.77	0.81
	RoadFusion	**0.89**	0.87	**0.88**	**0.83**	**0.76**	0.81	**0.84**	**0.87**
Crack500	Liu [17]	0.85	0.85	0.86	0.83	0.74	—	0.73	0.71
	Yang [26]	**0.93**	0.85	0.88	**0.90**	0.73	—	0.74	0.75
	RoadFusion	0.91	**0.90**	**0.91**	0.88	**0.79**	**0.89**	**0.79**	**0.81**
Cracks & Potholes	Maeda [19]	**0.89**	0.80	0.84	**0.84**	0.68	**0.85**	0.79	0.78
	RoadFusion	0.87	**0.89**	**0.88**	0.82	**0.74**	0.83	**0.82**	**0.80**
EDM Crack	Zhang [27]	**0.86**	0.78	0.82	**0.83**	0.67	**0.84**	0.75	0.74
	RoadFusion	0.82	**0.83**	**0.84**	0.79	**0.72**	0.80	**0.76**	**0.81**
GAPS384	Eisenbach [7]	0.86	**0.91**	**0.88**	**0.88**	0.71	**0.89**	0.80	**0.82**
	RoadFusion	**0.90**	0.88	0.87	0.85	**0.78**	0.86	**0.84**	0.78
Pothole600	Dhiman & Klette [6]	**0.89**	0.78	0.83	**0.83**	0.67	**0.84**	0.72	0.75
	RoadFusion	0.88	**0.86**	**0.87**	0.81	**0.75**	0.82	**0.79**	**0.83**

Road, RoadFusion reports superior Precision (0.89) and Macro-F1 (0.88), along with the highest I-AUROC (0.82), indicating reliable discrimination between normal and defective pavements. When handling the multi-defect Cracks & Potholes dataset, RoadFusion outperforms baselines in both Recall (0.89) and Macro-F1 (0.88), while delivering the highest P-AUROC (0.80), demonstrating adaptability to scenes with mixed damage categories. On the challenging EDM Crack dataset, characterized by fine-scale crack structures, RoadFusion improves IoU from 0.67 to 0.72 compared to baselines and achieves a P-AUROC of 0.81.

For GAPS384, RoadFusion attains the highest IoU (0.78) and mAP (0.85), confirming accurate defect localization even in visually complex road surfaces. On Pothole600, containing large, irregular defects, it maintains high P-AUROC (0.83) with an IoU of 0.75, alongside balanced Precision and Recall metrics resulting in a strong Macro-F1 of 0.87. The consistent excellence in both detection metrics (Macro-F1, AUROC) and localization metrics (IoU, mAP) across all datasets demonstrates RoadFusion's superior generalization capabilities, from fine cracks to extensive potholes, making it a robust solution for real-world pavement monitoring applications.

4.5 Synthesized Anomalies

Figure 3 shows a selection of anomalous pavement images generated by our diffusion-based anomaly synthesis pipeline. These samples were created by inpainting realistic defects, such as cracks, patches, and surface damage, onto clean road images using textual prompts and binary location masks. The visual diversity in shape, texture, and scale mirrors real-world defect patterns, which helps the model generalize effectively during training. These images serve as the input for extracting anomalous features via the backbone network and are subsequently processed through Feature Adaptor B in our training pipeline.

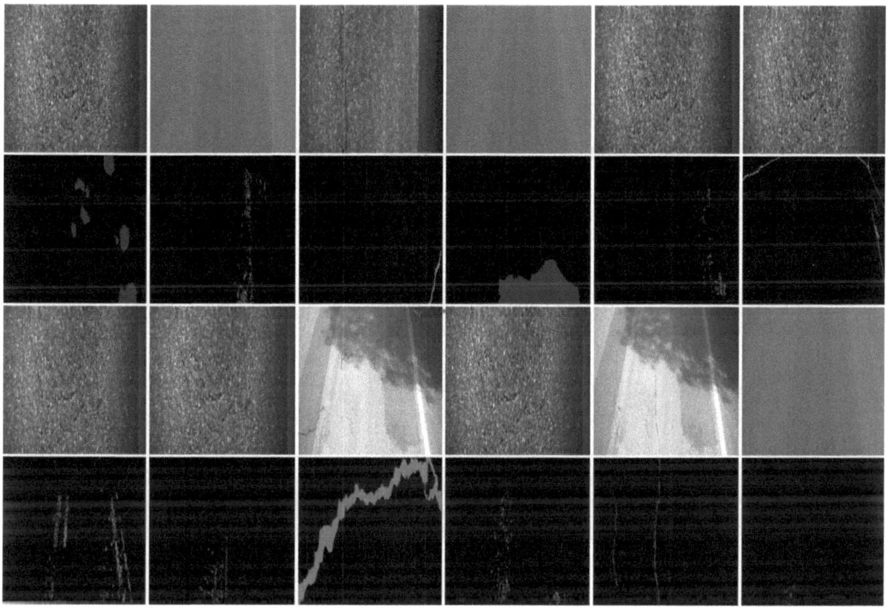

Fig. 3. Examples of synthesized pavement anomalies generated by the latent diffusion model. The samples show a range of defect types–including cracks, surface erosion, and patch damage–inpainted onto clean road images using textual prompts and spatial masks. These synthetic anomalies are used during training to extract anomalous features for the discriminator.

5 Conclusion

We introduced RoadFusion, a diffusion-based framework for pavement defect detection that combines anomaly generation with dual-adaptor feature learning. By leveraging latent diffusion to synthesize diverse pavement anomalies, our approach addresses the limited availability of labeled defect data. The dual feature adaptors enable domain-specific feature alignment, improving defect localization and classification. Experiments across six benchmark datasets demonstrate that RoadFusion consistently outperforms existing methods in both detection and localization tasks. Results confirm strong generalization across various road surfaces and defect types, while qualitative samples validate the realism of synthesized anomalies. RoadFusion provides an effective solution for pavement monitoring when annotated data are limited or diverse defect types are expected.

Acknowledgements. This study was carried out within the PNRR research activities of the consortium iNEST (Interconnected North-Est Innovation Ecosystem) funded by the European Union Next-GenerationEU (Piano Nazionale di Ripresa e Resilienza (PNRR) Missione 4 Componente 2, Investimento 1.5 D.D. 1058 23/06/2022, ECS_00000043).

References

1. Aqeel, M., Sharifi, S., Cristani, M., Setti, F.: Meta learning-driven iterative refinement for robust anomaly detection in industrial inspection. In: European Conference on Computer Vision, pp. 445–460. Springer (2024). https://doi.org/10.1007/978-3-031-92805-5_28
2. Aqeel, M., Sharifi, S., Cristani, M., Setti, F.: Self-supervised learning for robust surface defect detection. In: International Conference on Deep Learning Theory and Applications (2024)
3. Aqeel, M., Sharifi, S., Cristani, M., Setti, F.: Self-supervised iterative refinement for anomaly detection in industrial quality control. In: International Joint Conference on Computer Vision, Imaging and Computer Graphics Theory and Applications (2025)
4. Aqeel, M., Sharifi, S., Cristani, M., Setti, F.: Towards real unsupervised anomaly detection via confident meta-learning. In: Accepted to Proceedings of the IEEE/CVF International Conference on Computer Vision (2025)
5. Capogrosso, L., et al.: Diffusion-based image generation for in-distribution data augmentation in surface defect detection. In: International Joint Conference on Computer Vision, Imaging and Computer Graphics Theory and Applications (2024)
6. Dhiman, A., Klette, R.: Pothole detection using computer vision and learning. IEEE Trans. Intell. Transp. Syst. **21**(8), 3536–3550 (2019)
7. Eisenbach, M., et al.: How to get pavement distress detection ready for deep learning? A systematic approach. In: International Joint Conference on Neural Networks (IJCNN) (2017)
8. Fan, R., Wang, H., Bocus, M.J., Liu, M.: We learn better road pothole detection: from attention aggregation to adversarial domain adaptation. In: European Conference on Computer Vision Workshops (2020)
9. Garilli, E., Roncella, R., Hafezzadeh, R., Giuliani, F., Autelitano, F.: UAV photogrammetry for monitoring the cold asphalt patching pothole repairs. In: International Conference on Maintenance and Rehabilitation of Pavements (2024)
10. Girella, F., Liu, Z., Fummi, F., Setti, F., Cristani, M., Capogrosso, L.: Leveraging latent diffusion models for training-free in-distribution data augmentation for surface defect detection. In: International Conference on Content-Based Multimedia Indexing (CBMI) (2024)
11. He, Y., et al.: Pavement surface defect detection using mask region-based convolutional neural networks and transfer learning. Appl. Sci. **12**(15), 7364 (2022)
12. Ho, J., Jain, A., Abbeel, P.: Denoising diffusion probabilistic models. Adv. Neural. Inf. Process. Syst. **33**, 6840–6851 (2020)
13. Jenkins, M.D., Carr, T.A., Iglesias, M.I., Buggy, T., Morison, G.: A deep convolutional neural network for semantic pixel-wise segmentation of road and pavement surface cracks. In: European Signal Processing Conference (EUSIPCO) (2018)

14. Jiang, T.Y., Liu, Z.Y., Zhang, G.Z.: YOLOV5S-road: road surface defect detection under engineering environments based on CNN-transformer and adaptively spatial feature fusion. Measurement **242**, 115990 (2025)
15. Koch, C., Georgieva, K., Kasireddy, V., Akinci, B., Fieguth, P.: A review on computer vision based defect detection and condition assessment of concrete and asphalt civil infrastructure. Adv. Eng. Inform. **29**(2), 196–210 (2015)
16. Kyslytsyna, A., Xia, K., Kislitsyn, A., Abd El Kader, I., Wu, Y.: Road surface crack detection method based on conditional generative adversarial networks. Sensors **21**(21), 7405 (2021)
17. Liu, Y., Yao, J., Lu, X., Xie, R., Li, L.: DeepCrack: a deep hierarchical feature learning architecture for crack segmentation. Neurocomputing **338**, 139–153 (2019)
18. Loprencipe, G., de Almeida Filho, F.G.V., de Oliveira, R.H., Bruno, S.: Validation of a low-cost pavement monitoring inertial-based system for urban road networks. Sensors **21**(9), 3127 (2021)
19. Maeda, H., Sekimoto, Y., Seto, T., Kashiyama, T., Omata, H.: Road damage detection and classification using deep neural networks with smartphone images. Comput. Aided Civ. Infrastruct. Eng. **33**(12), 1127–1141 (2018)
20. Mei, Q., Gül, M., Azim, M.R.: Densely connected deep neural network considering connectivity of pixels for automatic crack detection. Autom. Constr. **110**, 103018 (2020)
21. Passos, B.T., Cassaniga, M.J., Fernandes, A.M.R., Medeiros, K.B., Comunello, E.: Cracks and potholes in road images (2020)
22. Pompigna, A., Mauro, R.: Smart roads: a state of the art of highways innovations in the smart age. Eng. Sci. Technol. Int. J. **25**, 100986 (2022)
23. Roth, K., Pemula, L., Zepeda, J., Schölkopf, B., Brox, T., Gehler, P.: Towards total recall in industrial anomaly detection. In: IEEE/CVF Conference on Computer Vision and Pattern Recognition (2022)
24. Schroten, A., et al.: Overview of transport infrastructure expenditures and costs. Publications Office of the European Union, Luxembourg, vol. 870 (2019)
25. Thompson, E.M., et al.: SHREC 2022: pothole and crack detection in the road pavement using images and RGB-d data. Comput. Graph. **107**, 161–171 (2022)
26. Yang, F., Zhang, L., Yu, S., Prokhorov, D., Mei, X., Ling, H.: Feature pyramid and hierarchical boosting network for pavement crack detection. IEEE Trans. Intell. Transp. Syst. **21**(4), 1525–1535 (2019)
27. Zhang, A., et al.: Automated pixel-level pavement crack detection on 3d asphalt surfaces using a deep-learning network. Comput. Aided Civ. Infrast. Eng. **32**(10), 805–819 (2017)
28. Zhang, H., Chen, N., Li, M., Mao, S.: The crack diffusion model: An innovative diffusion-based method for pavement crack detection. Remote Sens. **16**(6), 986 (2024)
29. Zhang, K., Zhang, Y., Cheng, H.D.: CrackGAN: pavement crack detection using partially accurate ground truths based on generative adversarial learning. IEEE Trans. Intell. Transp. Syst. **22**(2), 1306–1319 (2020)
30. Zhang, L., Yang, F., Zhang, Y.D., Zhu, Y.J.: Road crack detection using deep convolutional neural network. In: IEEE International Conference on Image Processing (ICIP) (2016)

Historical Postcards Classification Combining Visual Content and Text Description

Matthieu Pelingre and Salvatore Tabbone(✉)

Université de Lorraine, CNRS LORIA UMR 7503, Campus Scientifique, Vandoeuvre-lès-Nancy, France
{matthieu.pelingre,salvatore.tabbone}@univ-lorraine.fr

Abstract. This article introduces an end-to-end pipeline to index a collection of 4,294 digitized French historical postcards spanning the years 1899 to 1920. These postcards are visually rich documents with both pictures and texts. We propose to classify these postcards according to their visual content and following their topics. Therefore, the postcards are annotated with keywords defined from the transcription of the printed text set on them. The keyword annotation is then automatically extended to postcards without text description.

1 Introduction

Historical documents represent an invaluable source of information and constitute an essential part of our cultural heritage. Therefore, their preservation for future generations is of utmost importance. In this perspective, extensive collections of historical documents are accessible to archives, museums, and libraries, offering valuable perspectives on the past for both historians and the broader public. Despite their availability, historical documents are very difficult to index and interpret manually as analyzing their contents manually can be very demanding and time-consuming, especially for extensive collections.

Postcards are of great interest to different researchers and especially historians. Postcards have historical value, and they help us to see what life was like at the time they were printed (e.g., design or photographs of cities, common clothing styles, or architecture of the time). Historians agree that, in the early twentieth century, postcards were a revolutionary way of combining short messages with an image and have left an indelible imprint on the history of human communication [9,17].

For computer vision research, postcards pose various new challenges as they vary considerably in content, which can be grouped into different categories (town, landscape, monument, people, object, etc.). Some postcards have no text description, which is problematic for text-based analysis. In addition, the text descriptions of some postcards do not correspond to the main instance. Search methods based on the image instance become necessary to prevent these cards

E. Rodolà et al. (Eds.): ICIAP 2025, LNCS 16167, pp. 336–348, 2026.
https://doi.org/10.1007/978-3-032-10185-3_27

from being forgotten. In addition, postcards have different sizes and photometry. Some are in black-and-white or sepia photographs, while others are colorized (see Fig. 1). Analyzing these postcards using recent deep learning techniques [1] offers a valuable opportunity to retrieve significant information, thereby enriching the historical value of this heritage. Thanks to automatic deep analysis models, these precious sources of information are more accessible, while simplifying their storage and archiving.

Among the great number of works to analyze historical documents [2,3,5, 12,20] few studies are dedicated to historical postcards. Recent works trained a deep neural network (i.e., YOLOv5x object detection models) to automatically detect persons and classify them as male or female in a collection of historical postcards [19]. In the context of card classifications, mention can also be made of García et al. [8] who used semantic segmentation deep neural networks to classify postcards image pixels into different classes (e.g., stamps, postmarks, handwritten text or illustrations, among others). Smits et al. [22] proposed a pipeline combining Computer Vision segmentation, Handwritten Text Recognition, and Large Language Models to extract and disambiguate address information from a collection of 102K historical postcards.

This paper introduces the development and validation of methods that aim to index a collection of 4,294 digitized historical postcards spanning the years 1899 to 1920. This collection allows to discover the history and heritage of the eastern France region. These postcards are visually rich documents with both pictures and texts. They contain details about the events, places, and individuals of the time, as well as aspects of culture and arts, among other things. Our contribution is to present an end-to-end pipeline dedicated to classify these postcards according to their visual content and following their topics. To annotate these images by keywords, we propose a transcription of the printed text included on the postcards. Moreover, we automatically extend this keyword annotation to images without text descriptions.

Fig. 1. Samples of Historical Postcards from the Grand Est region in France.

2 Proposed Approach

Our approach focuses on non-exclusive multi-label classification based on topic representations and keywords. Topics are defined following historians' knowledge, after a preliminary study on a part of the corpus: 'urban', 'rural', 'destruction', 'landscape', 'monuments', 'person', 'religion', 'work', 'public building', 'art'.

As postcards are visually rich documents, containing images and text we use the texts to define keywords and therefore to annotate the postcards. According to image multi-label classification, using the 10 topics, we will be able to identify related keywords for each topic. Furthermore, we can then use these keywords to refine our classification, extending the annotation to postcards having no text description.

The overall approach is as follow (Fig. 2): firstly, keywords are defined by our text extraction pipeline and the corresponding images will be tagged using a multi-label image classifier. The embedded images having keywords will be considered in our process by feeding a set of Support-Vector Machine (SVM) classifiers. Finally, the embeddings of images where no keywords could be extracted will be compared with images sharing the same topics. Keywords of these latter images are used to predict the annotation for images without text description.

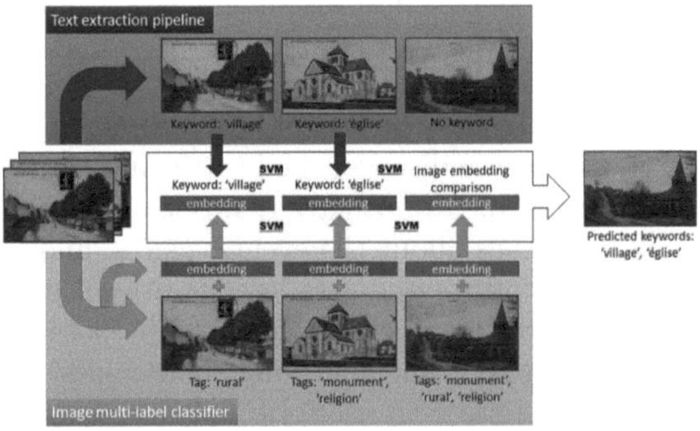

Fig. 2. Proposed global pipeline for image classification

2.1 Text Extraction Pipeline

In order to extract textual information from postcards, we need to feed up a deep neural model and, to define the ground truth. Therefore, the first part of the extraction pipeline consists of detecting bounding boxes. As printed texts can have different orientations within the same postcard, text orientation is detected before extraction using an OCR to recognize the content. Since manual information extraction is extremely time-consuming, iterative annotation processes are used to define this ground truth.

Printed Text Ground Truth Building: Text areas on postcards can be categorized into three main classes: printed, handwritten, and scene text (Fig. 3a). Printed text may contain information on the content of the postcard image, as well as on the photographer, illustrator, and/or publisher. It is also possible that a printed text zone contains other miscellaneous information (poem, quotation, ...). Handwritten text may contain additional information about the context of the postcard image (anecdotes about the location, description of photographed people, ...) or about the lives of the postcard's contemporaries. Finally, the scene text, or scene text, generally corresponds to placards, road signs, store signs, billboards, and advertisements present in the picture, mainly in urban environments. This last class is particularly confusing, as it can easily be misidentified as printed text. Indeed, similar fonts can be used in both classes. All these text zones can be in different orientations on the same postcard (Fig. 3b).

Fig. 3. a) Printed (pink), handwritten (yellow) and scene (cyan) texts; b)Postcard with printed text oriented at 0°, 90° and 270° (Color figure online)

EasyOCR [11] is set up to perform an initial recording of both the text, the bounding box containing it, as well as its orientation and confidence score. The languages set up in the EasyOCR model are French, English and German (indeed, we can find First World War postcards with these three languages in our dataset). This pipeline was applied to the 4294 images in the dataset. The bounding boxes were then transposed to the Yolo detection dataset format, then imported into CVAT [6] for manual validation on the first 700 postcards. A significant number of false positives could then be noticed, in particular due to the texts of postage and date stamps. The rest of the data extracted by EasyOCR was, in fact, hardly usable to build a suitable ground truth. To avoid this problem, specific bounding boxes for postmarks were added. In this way, the data set obtained could be used to train detection models on 6 classes: "postage stamp", "date stamp", "other marks", "handwritten", "printed" and"scene text". A first YOLOv11 model [13] is fine-tuned on these 700 postcards and used to generate bounding box predictions, which facilitated the annotation for another batch of 330 postcards. After manual validation on CVAT, the process was reiterated on 350 postcards, enhancing the efficiency in time of the annotation process, to obtain a dataset of 1,375 annotated postcards with 6 bounding box classes.

Detection: Instead of using conventional OCR algorithms, fine-tuning a detection model allows different processing methods to be applied depending on the predicted class. Therefore, a 5-fold cross validation is performed, by fine-tuning YOLOv11 over 50 epochs on the 1,375 annotated postcards dataset. Obtained scores are shown in Table 1.

Fine-tuned YOLOv11 models perform well with stamps and printed text, but are less efficient in the "handwritten" and "scene text" classes. This can

Table 1. 5-fold cross validation scores for text and stamp classification

	Images	Instances	P	R	mAP@50	mAP@50-95
All	275 ± 0	1164 ± 0	0.893 ± 0.027	0.788 ± 0.005	0.836 ± 0.008	0.664 ± 0.013
Postage stamp	66 ± 0	68 ± 0	0.985 ± 0.006	0.997 ± 0.007	0.995 ± 0.000	0.914 ± 0.008
Date stamp	70 ± 0	79 ± 0	0.966 ± 0.020	0.920 ± 0.021	0.957 ± 0.009	0.847 ± 0.011
Other marks	46 ± 0	47 ± 0	0.931 ± 0.033	0.751 ± 0.032	0.836 ± 0.021	0.675 ± 0.018
Handwritten	40 ± 0	113 ± 0	0.822 ± 0.045	0.771 ± 0.025	0.834 ± 0.034	0.577 ± 0.041
Printed	270 ± 0	611 ± 0	0.927 ± 0.023	0.931 ± 0.017	0.966 ± 0.004	0.724 ± 0.016
Scene text	61 ± 0	246 ± 0	0.727 ± 0.095	0.356 ± 0.018	0.429 ± 0.035	0.250 ± 0.019

be explained by the lack of instances of the "handwritten" class in the dataset, but also by the variability of these classes. Indeed, each person has his own handwriting and scene text has a lot of different sizes, fonts, orientations, or styles. However, printed text detection gives very good results.

Printed Text Transcription: Bounding boxes obtained during annotation are also used to extract specifically printed text areas, to evaluate different OCR algorithms. EasyOCR is used to extract text in those areas as well as their orientation. 222 text areas, belonging to 110 postcards, are then manually corrected, replacing characters, words, or even whole paragraphs, depending on EasyOCR's difficulty in reading those text areas. These data are then used as ground truth to evaluate different OCR models and pre-processing for text orientation detection and text recognition.

To extract printed text using OCR, five reliable open-source libraries are evaluated: Tesseract OCR [21], EasyOCR [11], docTR [14], Calamari-OCR [23] and PaddleOCR [16]. Tesseract OCR, EasyOCR and docTR are evaluated for angle detection and all of them for text recognition, with and without Otsu binarization. Only available pre-trained models from each library are used, with language parameters set to French, English and German when available, for Tesseract, EasyOCR and PaddleOCR. Architectures used for docTR are ResNet 50 for detection and PARSeq for recognition, as this combination give the best results on their benchmarks. Finally, for Calamari-OCR, an ensemble of "historical_french" pre-trained models are used for recognition. As orientation can only take 4 discrete values (0, 90, 180 and 270°), weighted precision, recall and f1-score are used as metrics (Table 2). It should be noted, however, that over 90% of texts are 0°-oriented. Tesseract OCR achieves best results over our data, with no impact from binarization. The second is EasyOCR with a negative impact from binarization. Finally, docTR achieves worst results, especially on 0° oriented texts, with a good precision but bad recall, and with a positive impact of binarization.

Text transcription is evaluated, by Levenshtein normalized distance, Jaccard similarity and CER. A first evaluation is performed on rotated images, to obtain only 0°-oriented texts. Obtained results are shown in Table 3. Except for Cala-

Table 2. OCR evaluation for text orientation detection

	Precision	Recall	f1-score
Tesseract	0.8934	0.9276	0.9102
EasyOCR	0.9290	0.7104	0.8011
docTR	0.9823	0.3077	0.4421
Binarized Tesseract	0.8934	0.9276	0.9102
Binarized EasyOCR	0.9477	0.6697	0.7730
Binarized docTR	0.9651	0.4751	0.6328

mari and docTR, binarization has a negative impact over text recognition. In fact, Calamari does not work on unbinarized images. For text recognition, EasyOCR gives best performances, without binarisation. Regarding these results, Tesseract OCR was then used for text orientation detection and EasyOCR for text transcription.

Table 3. OCR evaluation for text recognition on rotated images

	Levenshtein	Jaccard	CER
Tesseract	0.8483	0.7708	0.1694
EasyOCR	0.9046	0.7842	0.1166
docTR	0.8015	0.7393	0.2620
PaddleOCR	0.7517	0.6370	0.3207
Calamari	0.2568	0.0289	0.8305
Binarized Tesseract	0.8312	0.7475	0.1866
Binarized EasyOCR	0.8472	0.6584	0.1827
Binarized docTR	0.8059	0.7228	0.2778
Binarized PaddleOCR	0.6137	0.5024	0.4597
Binarized Calamari	0.8248	0.6217	0.2267

2.2 Image Multi-label Classifier

Ground Truth Building for Image Classification: 1,375 postcards are tagged, exploiting the 10 topics defined previously where a postcard can have several topics, depending on its content. In the same manner, to improve the efficiency defining the ground truth these postcards are pre-annotated using a CLIP ViT-B/32 model in zero-shot mode [18]. A topic is considered as True, if its probability, given by CLIP, is greater than 10%. This threshold is set empirically based on a pre-annotation process on a random set of twenty postcards. Obtained results are then manually corrected to build a ground truth. The CLIP zero-shot results on the 1,375 images are then compared with their manually corrected annotations (Table 4).

Table 4. Evaluation of CLIP ViT-B/32 in zero-shot on 1,375 postcards.

	precision	recall	f1-score	Instances
All	0.6620	0.5872	0.5405	2965
urban	0.8145	0.3141	0.4534	573
rural	0.5843	0.8255	0.6843	424
destruction	0.7424	0.3798	0.5026	129
landscape	0.5689	0.9489	0.7113	509
monuments	0.4568	0.7124	0.5567	386
person	0.5769	0.2027	0.3000	74
religion	0.8936	0.1795	0.2989	234
work	0.9762	0.2908	0.4481	141
public building	0.4983	0.7487	0.5984	394
art	0.9999	0.1089	0.1964	101

According to these results, CLIP in zero-shot mode is accurate in precision for the topics 'urban', 'destruction', 'religion', 'work' and 'art'. However, recall on these topics is rather low, suggesting a high false-negative rate. This can be partially explained by CLIP's training on the ImageNet dataset [7], which mostly contains contemporary images. It has, for example, been observed that postcards in urban areas featuring horses are generally classified by CLIP as 'rural', also explaining the lower precision for this class. Similarly, the majority of postcards tagged as 'work' depicted industrial work, whereas the false negatives for this topic are mostly rural, with small-scale agricultural work. Obtained distribution of topics in the annotated dataset and number of topics per image distribution are shown on Fig. 4. As we can remark, most of the postcards are tagged with two topics.

Text Keywords: To define text keywords, the texts from all 4294 images are extracted using the textual pipeline with the fine-tuned YOLOv11 as a detector, Tesseract OCR for orientation detection and EasyOCR for text recognition. The words are then tokenized and the stop words removed using the Natural Language Toolkit library [4]. Finally, remaining words are lemmatized, using spaCy library [10] to find 20,160 lemmas with 1,545 unique values. Using those lemmas, texts containing information on editors, illustrators and photographers have been separated from texts containing information on image content. Indeed, the former generally contains words whose roots can be easily found (e.g. 'éditeur', 'imprimerie', 'photographe' ...). Only the lemmas in text zones containing information about the image content are thus selected reducing the number of unique lemmas to 1,482. They are then grouped by postcard. For each lemma in each

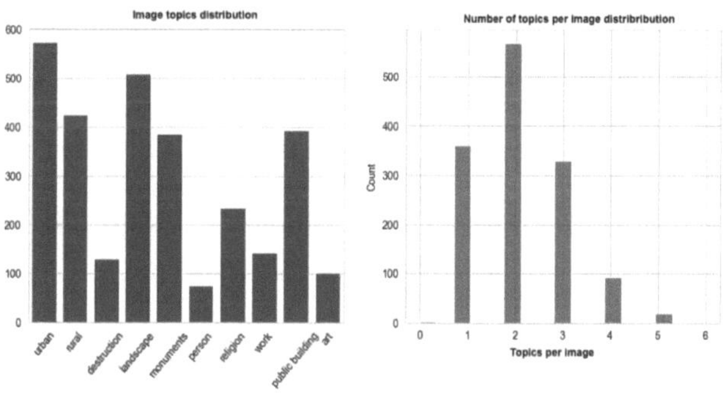

Fig. 4. Data distribution topics in the obtained annotated dataset.

postcard, a term frequency-inverse document frequency score (TF-IDF) is calculated. Subsequently, to further reduce the number of lemmas to study, only the words with a TF-IDF above the last quartile value of all TF-IDF scores, are selected, i.e. 268 words. These words are manually sorted and selected, according to our 10 topics and the knowledge of the historians. Finally, 67 keywords are kept as shown in Fig. 5.

Fig. 5. WordCloud of the 67 keywords found in the postcards.

3 Experimental Results

The ground truth thus obtained is used to train and evaluate the first models constituting the image classification pipeline proposed in this paper.

Image Multi-label Classification: Cross validations are performed on CLIP and DINOv2 [15] models, using the 1,375 annotated postcards dataset with 10 classes, with 10% kept as the same test set for all the models. The following data augmentations are used for fine-tuning: random horizontal and vertical flip, random Colour Jitter and random Greyscale. Each model is fine-tuned on 50 epochs, using Adam optimizer and Binary Cross Entropy With Logits Loss (Sigmoid + BCE) for Multi-label Classification. Indeed, Binary Cross Entropy is used instead of Categorical Cross-Entropy as our problem can be summarized as whether or not the image belongs to each class independently (One-vs-rest classification). The learning rate is also divided by 5, every time the validation loss reaches a plateau for 5 epochs and the training is stopped if the plateau is held for 10 epochs, keeping the best model weights. The starting learning rates are $5e^{-5}$ for CLIP and $5e^{-6}$ for DINOv2 models. Models are then evaluated using precision, recall and f1-score, a class being considered as True if its probability was greater than 50%. Best 5-fold cross-validation results are obtained with DINOv2 ViT-B/14. Its scores are shown in Table 5.

Table 5. 5-fold cross validation scores for image multi-label classification by DINOv2 ViT-B/14

	precision	recall	f1-score	support
All	0.875 ± 0.013	0.775 ± 0.012	0.817 ± 0.010	535 ± 7
urban	0.864 ± 0.027	0.839 ± 0.042	0.850 ± 0.018	104 ± 6
rural	0.868 ± 0.033	0.857 ± 0.031	0.862 ± 0.019	76 ± 8
destruction	0.887 ± 0.082	0.518 ± 0.088	0.645 ± 0.070	22 ± 5
landscape	0.935 ± 0.008	0.895 ± 0.027	0.914 ± 0.016	93 ± 4
monuments	0.867 ± 0.052	0.749 ± 0.034	0.802 ± 0.023	68 ± 8
person	0.704 ± 0.188	0.448 ± 0.166	0.540 ± 0.178	13 ± 4
religion	0.875 ± 0.062	0.775 ± 0.063	0.818 ± 0.026	42 ± 7
work	0.911 ± 0.032	0.686 ± 0.052	0.782 ± 0.043	24 ± 3
public building	0.828 ± 0.040	0.705 ± 0.081	0.759 ± 0.054	71 ± 5
art	0.779 ± 0.144	0.505 ± 0.088	0.598 ± 0.063	19 ± 5

As we can see from this table, the model performs well, except for classes 'destruction', 'person' and 'art' with a recall below, or close to, 0.5. Topics of these classes are too general and therefore highly variable which can add confusion in the classification.

Best models are then evaluated on the same test sets as well as CLIP in zero-shot with classes being considered as True above 10 and 50% probability. To compare with common CNN architectures used in image classification, a 5-fold cross-validation is performed on ResNet50 and EfficientNetV2 M pre-trained models, under the same conditions as before, but with an initial learning rate of $1e^{-4}$. Obtained scores are shown in Table 6. DINOv2 with the ViT-B/14 encoder achieves the best performances and is then used in the pipeline for tagging the rest of the dataset i.e. 2919 images.

Keywords Prediction: In Sect. 2, text keywords were defined and could be retrieved after text lemmatization. However, 2587 postcards out of 4294 did not receive any keywords. This may be due to various factors. A card may have no text (95 postcards). The only text it contains may be related only to the editor and not to the image content (120 postcards). Finally, the postcard text may not contain a keyword or, to a lesser extent, the potential keyword may have

Table 6. Evaluation of all best models on the same test set

	Model type	precision	recall	f1-score	Instances
zero-shot CLIP > 0.1	ViT-B/32	0.5929	0.5347	0.4793	288
zero-shot CLIP > 0.5	ViT-B/32	0.4571	0.2361	0.2956	288
fine-tuned InceptionV3	CNN	0.7368	0.5208	0.5839	288
fine-tuned ResNet50	CNN	0.7667	0.5694	0.6302	288
fine-tuned EfficientNetV2 M	CNN	0.7776	0.6701	0.7094	288
fine-tuned CLIP	ViT-B/32	0.8260	0.7257	0.7612	288
fine-tuned DINOv2	ViT-S/14	0.8816	0.7465	0.7997	288
fine-tuned DINOv2	ViT-B/14	0.9138	0.7743	0.8320	288

been poorly transcribed. It is also possible that the text might be in another language than French.

In order to link these keywords to the image topics, based on fine-tuned DINOv2 predictions across the entire dataset, the co-occurrence frequency is computed between each found keyword and the predicted topics. A keyword is considered as relevant to a given topic if more than one third of the images, containing that keyword, are classified under that topic. To avoid over- or under-representation of certain classes within each topic, keywords with an image count deviating significantly from the mean (outside the range of mean ± one standard deviation) are discarded. Additionally, keywords exhibiting high visual variability (e.g., "moulin", which may refer to windmills, watermills, street names, city names, or geographic locations) are filtered out to improve generalization. Selected keywords for each topic can be found in Table 7 and as we can remark at the end the set is reduced to 37 unique keywords.

Table 7. Selected keywords per topic

urban	rural	destruction	landscape	monuments	person	religion	work	public building
fontaine	col	guerre	bois	cathédrale	monument	abbaye	cave	caserne
hôpital	lac	mort	canal	château	statue	cathédrale	marché	hôtel
lycée	montagne	ruine	cascade	porte		tombeau	usine	lycée
poste	village		col	statue		église	vin	pont
préfecture			hiver	église				préfecture
quai			jardin					
square			lac					
théâtre			neige					
			parc					
			pont					

As only a few dozen postcards are available for each keyword, the use of SVMs seems appropriate, as they are well suited to small datasets. Thus, a set of one-vs-rest multi-class SVM are trained to predict keywords based on the image embeddings. Each multi-class SVM corresponds to a specific topic and trained only on the image embeddings provided by DINOv2 for that topic (dimension of a vector: 768). The number of classes is set following the different keywords belonging to the corresponding topic. Finally, hyperparameter tuning for each SVM is performed using randomized search 5-fold cross-validation, with search spaces defined over reciprocal distributions for the regularization parameter (C) between 1 and $1e^3$ and the kernel coefficient (gamma) between $1e^{-6}$ and $1e^{-3}$. Kernel types include polynomial (degree 3), radial basis function (RBF), and sigmoid. To ensure that predicted keywords are relevant, false positives must be minimized, and therefore precision maximized. Best SVM hyperparameters for every topic can be found in Table 8 as well as their scores on each test set.

Table 8. SVM hyper-parameters and results on test set per image class

	C	gamma	kernel	precision	recall	f1-score	Instances
urban	299	3.83e-4	rbf	0.819	0.536	0.626	28
rural	56	5e-6	sigmoid	0.710	0.700	0.689	10
destruction	45	1.09e-4	rbf	1.000	0.500	0.656	16
landscape	4	2.69e-4	rbf	0.715	0.539	0.589	76
monuments	515	8.93e-4	rbf	0.884	0.457	0.591	46
person	45	2.19e-4	rbf	0.654	0.563	0.602	16
religion	4	4.48e-4	poly	0.728	0.722	0.721	18
work	178	2e-6	sigmoid	0.917	1.000	0.952	9
public building	135	5.75e-4	rbf	0.904	0.640	0.714	50

Out of 2587 images processed automatically to predict their keywords, 1 keyword was predicted for 842 images, 2 keywords for 80 and 3 keywords for 10 images. From these 932 images, an evaluation on four sets of 200 images randomly drawn is shown Table 9.

Table 9. Postcards annotation evaluation

	Images	predicted keywords	true positives	precision
Sample 1	200	221	181	0.819
Sample 2	200	219	179	0.817
Sample 3	200	221	180	0.814
Sample 4	200	228	192	0.842

Most of the time, errors in keyword predictions can be explained as, in the majority of cases, they occur on images very similar to those that could define the keyword when training the corresponding SVM (e.g. predicting 'church' for a synagogue, 'hotel' for a baroque castle or 'canal' for rivers).

4 Conclusion

In this article, an end-to-end pipeline dedicated to the classification of historical postcard images is proposed, as well as the creation of the ground truth required to train and evaluate the models used. The aim is to index the dataset enabling to easily retrieve images on their visual content or keywords. As some postcards have no text description, keywords are automatically extended to these images.

Acknowledgements. We would like to thank Image'Est for making historical postcards data available and the Grand Est Region, France who supported this work.

References

1. Amri, A., Tabbone, S.: Historical postcards retrieval through vision foundation models. In: Gouet-Brunet, V., Kosti, R., Weng, L. (eds.) Proceedings of the 6th Workshop on the Analysis, Understanding and Promotion of Heritage Contents, SUMAC 2024, Melbourne, VIC, Australia. ACM (2024)
2. Belay, B.H., et al.: A historical handwritten dataset for Ethiopic OCR with baseline models and human-level performance. In: ICDAR (2024)
3. Beshirov, A., Dobreva, M., Dimitrov, D., Hardalov, M., Koychev, I., Nakov, P.: Post-OCR text correction for Bulgarian historical documents. Int. J. Digit. Libr. **26**(4) (2025)
4. Bird, S., Klein, E., Loper, E.: Natural Language Processing with Python: Analyzing Text with the Natural Language Toolkit. O'Reilly Media, Inc. (2009)
5. Constum, T., Preel, L., Larcher, T., Paquet, T., Tranouez, P., Bree, S.: End-to-end information extraction in†handwritten documents: understanding Paris marriage records from†1880 to†1940. In: ICDAR (2024)
6. CVAT.ai Corporation: Computer Vision Annotation Tool (CVAT) (2023). https://github.com/cvat-ai/cvat
7. Deng, J., Dong, W., Socher, R., Li, L.J., Li, K., Fei-Fei, L.: ImageNet: a large-scale hierarchical image database. In: CVPR 2009 (2009)
8. García, B., Moreno, B., Vélez, J.F., Sánchez, Á.: Deep layout extraction applied to historical postcards. In: International Work-Conference on the Interplay Between Natural and Artificial Computation, pp. 346–355. Springer (2022)
9. Gillen, J.: The Edwardian Picture Postcard as a Communications Revolution: A Literacy Studies Perspective. Routledge (2023)
10. Honnibal, M., Montani, I., Van Landeghem, S., Boyd, A.: spaCy: Industrial-strength Natural Language Processing in Python (2020). https://doi.org/10.5281/zenodo.1212303
11. JaidedAI: JaidedAI/EasyOCR (2025). https://github.com/JaidedAI/EasyOCR. Original-date: 2020-03-14T11:46:39Z

12. Jian, C., Jin, L., Liang, L., Liu, C.: HisDoc R-CNN: robust Chinese historical document text line detection with dynamic rotational proposal network and iterative attention head. In: ICDAR (2023)
13. Jocher, G., Qiu, J., Chaurasia, A.: Ultralytics YOLO (2023). https://github.com/ultralytics/ultralytics
14. Mindee: doctr: Document text recognition (2021). https://github.com/mindee/doctr
15. Oquab, M., et al.: Dinov2: learning robust visual features without supervision (2024)
16. PaddlePaddle: Paddleocr (2020). https://github.com/PaddlePaddle/PaddleOCR/
17. Pyne, L.: Postcards: The rise and fall of the world's first social network. Reaktion Books (2021)
18. Radford, A., et al.: Learning transferable visual models from natural language supervision (2021)
19. Schuerkamp, R., Barrett, J., Bales, A., Wegner, A., Giabbanelli, P.J.: Enabling new interactions with library digital collections: automatic gender recognition in historical postcards via deep learning. J. Acad. Librariansh. **49**(4), 102736 (2023)
20. Shi, Y., Liu, C., Peng, D., Jian, C., Huang, J., Jin, L.: M5hisdoc: a large-scale multi-style Chinese historical document analysis benchmark. In: Oh, A., Naumann, T., Globerson, A., Saenko, K., Hardt, M., Levine, S. (eds.) Advances in Neural Information Processing Systems, vol. 36, pp. 78483–78495. Curran Associates, Inc. (2023)
21. Smith, R.: An overview of the tesseract OCR engine. In: ICDAR (2007)
22. Smits, T., Haverals, W., et al., L.V.: Greetings from! Extracting address information from 100, 000 historical picture postcards. In: Workshop on Computational Humanities Research (2023)
23. Wick, C., Reul, C., Puppe, F.: Calamari - a high-performance tensorflow-based deep learning package for optical character recognition. Digit. Humanit. Q. **14**(1) (2020)

VidCLearn: A Continual Learning Approach for Text-to-Video Generation

Luca Zanchetta[1(✉)], Lorenzo Papa[2], Luca Maiano[1], and Irene Amerini[1]

[1] Sapienza University of Rome, Rome, Italy
zanchetta.1848878@studenti.uniroma1.it, {maiano,amerini}@diag.uniroma1.it
[2] ESA philab, Frascati, Italy
lorenzo.papa@esa.int

Abstract. Text-to-video generation is an emerging field in generative AI, enabling the creation of realistic, semantically accurate videos from text prompts. While current models achieve impressive visual quality and alignment with input text, they typically rely on static knowledge–making it difficult to incorporate new data without retraining from scratch. To address this limitation, we propose **VidCLearn**, a continual learning framework for diffusion-based text-to-video generation. **VidCLearn** features a student-teacher architecture where the student model is incrementally updated with new text-video pairs, and the teacher model helps preserve previously learned knowledge through generative replay. Additionally, we introduce a novel temporal consistency loss to enhance motion smoothness and a video retrieval module to provide structural guidance at inference. Our architecture is also designed to be more computationally efficient than existing models retaining a satisfactory generation performance. Experimental results show **VidCLearn** superiority over baseline methods in terms of visual quality, semantic alignment, and temporal coherence.

Keywords: Text-to-Video Generation · Continual Learning · Knowledge Distillation · Generative Replay

1 Introduction

Text-to-Video (T2V) generation is a cutting-edge computer vision task that involves generating dynamic video sequences starting from a textual description; it aims at filling the existing gap between natural language understanding and video synthesis. Key aspects of this task are the following: *i)* semantic alignment of generated videos with respect to the input textual description; *ii)* temporal consistency between generated frames, and *iii)* high-quality generated frames, possibly avoiding visual artifacts. However, text-to-video generative models have to face several challenges, including the need for high-quality large-scale datasets, huge computational resources and a large amount of time to ensure a proper training. Another crucial limitation of existing text-to-video generative models

E. Rodolà et al. (Eds.): ICIAP 2025, LNCS 16167, pp. 349–361, 2026.
https://doi.org/10.1007/978-3-032-10185-3_28

is the following: their knowledge is *static*, in the sense that it's not easy to maintain the knowledge of such models up to date, because it would require a new training from scratch process each time, resulting in a huge waste of time and computation.

In order to overcome these limitations, we propose **VidCLearn**: a generative replay continual learning approach for training a text-to-video generative model in such a way that its knowledge increases over time. Our work builds upon Tune-A-Video [1], a text-to-video generative model that is able to generate a novel video starting from a single text-video pair as input and an edited prompt that guides the generation. Tune-A-Video preserves the motion information at inference time thanks to a DDIM inversion operation performed on the input video, which produces latent noisy frames providing structural guidance during the denoising procedure. Our proposed method consists of a knowledge distillation-like paradigm, where a *student* model is fine-tuned on the currently available text-video pair, while a *teacher* model is in charge of preserving as much as possible the student's previously acquired knowledge over time. Additionally, in order to provide structural guidance to the model at inference time, we propose to integrate a *retrieval mechanism* based on a cosine similarity score computed between the inference prompt and all the training prompts the model has seen so far. Our experiments demonstrate the superiority of the proposed approach with respect to the vanilla Tune-A-Video model and several baseline strategies, which lack of generalization capabilities. We achieve encouraging results that prove the effectiveness of our approach, without significantly increasing the computational requirements for proper training. In fact, our approach leverages only one NVIDIA A100 GPU, resulting in minimal energy consumption and resource-intensive computations, thanks to the almost lightweight core Tune-A-Video architecture. In summary, we propose a continual learning framework for text-to-video generation, introducing a temporal consistency loss for motion coherence, a cosine similarity-based retrieval mechanism for structural guidance, and demonstrate its effectiveness through extensive experiments.

2 Related Works

The Text-to-Video generation field has evolved rapidly in the past few years, addressing complex challenges like temporal and spatial coherence of the generated video frames. Models like *Sync-DRAW* [2] leverage attention mechanisms for enhancing temporal consistency, while other models like *ModelScopeT2V* [3] propose to adapt a pre-trained Stable Diffusion model [4] to the temporal domain. The *Video Diffusion Model* proposed by Ho et al. [5] employs diffusion models for generating videos in the pixel space, while other models like *Pixel-Dance* [6] and *MagicVideo* [7] propose to generate videos leveraging the latent space, which is less costly. Hybrid models, such as *Show-1* [8], combine pixel-based and latent-based video diffusion models, in order to exploit the advantages of both approaches while reducing their drawbacks. There exist also text-to-video generative models based on autoregressive approaches, such as *NUWA-Infinity*

[9], *CogVideo* [10], and *Phenaki* [11]. Transformer-based approaches, such as *NUWA* [12] and *GenTron* [13], are able to generate high-quality videos, too. The current state-of-the-art in text-to-video synthesis is represented by *Sora* [14], a diffusion transformer-based model that has demonstrated a remarkable ability to generate even one minute-long videos while maintaining high visual quality, unlike earlier works, that are able to generate only short clips. Efficient models such as *VideoLCM* [15], which is able to achieve high-fidelity video synthesis with only 4~6 sampling steps, or *Text2Video-Zero* [16], which is able to leverage the power of existing text-to-image synthesis methods in order to make them suitable for the video domain in a "zero-shot" fashion, allow to perform video generation from text with significantly reduced computational cost. There exist also efficient models that are able to significantly limit the amount of needed training data, such as *Make-A-Video* [17] or the previously mentioned *Tune-A-Video*, that have revolutionized the text-to-video generation field.

Continual learning strategies in generative models, particularly diffusion models, address the challenge of catastrophic forgetting. Most of them are based on a generative replay approach; among them, *GPPDM* [18] leverages the concept of class prototype in order to capture the core characteristics of images belonging to a given class. A key contribution is that of *DSG* [19] and *Generative Distillation* [20] models, which employ knowledge distillation for transferring the knowledge in an effective way; the strategies proposed in these two works have inspired our proposed method. Other approaches, such as *STAMINA* [21] and *C-LoRA* [22], involve low-rank adaptation mechanisms for enhancing knowledge retention. There exist also class incremental learning approaches, such as *DiffClass* [23], trying to overcome limitations like privacy issues or domain gaps between synthetic data and real data by means of a multi-distribution-matching technique for fine-tuning diffusion models.

Continual learning, despite its potential to enable dynamic knowledge expansion, has been largely unexplored in text-to-video generation. In this work, we take a step toward addressing this gap by proposing a generative replay-based continual learning strategy for text-to-video synthesis.

3 Methodology

Consider a setting in which text-video pairs become available over time, one at a time. Our goal is to train a text-to-video generative model in such a way that its knowledge can be updated over time, without the need to re-train the model from scratch each time a new text-video pair becomes available.

We propose **VidCLearn**: a novel approach employing a continual learning technique for text-to-video generation. Specifically, our strategy consists of applying a generative replay continual learning approach to the core Tune-A-Video model, so that its knowledge is both preserved and increased over time. We achieve this by means of a knowledge distillation-like paradigm: the core Tune-A-Video model acts as the student, while a duplicate of the student with frozen weights acts as the teacher. The pipeline of our approach is shown in

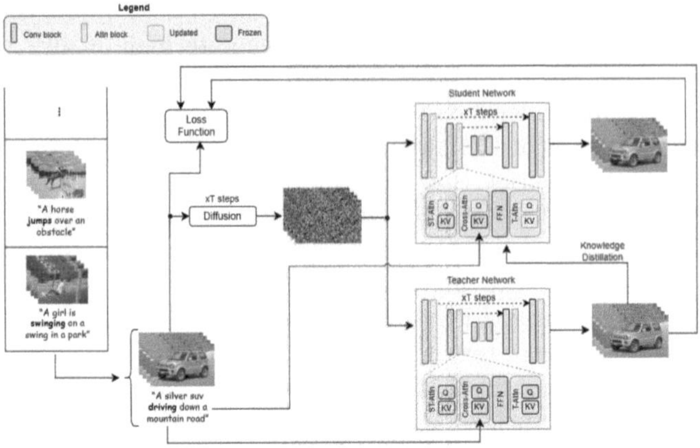

Fig. 1. VidCLearn pipeline.

Fig. 1. When the very first text-video pair becomes available, there is no prior knowledge to retain; therefore, the student model is tuned by means of the one-shot video tuning strategy of the core Tune-A-Video model. In particular, a latent representation of the input video is created by means of the underlying Stable Diffusion's variational autoencoder; then, a standard forward diffusion process is performed on the latent representation of each video frame, thus producing noisy latents. During the reverse diffusion process, the student model is asked to predict the amount of noise to be removed from each noisy latent representation, in order to get back the original ones; a target amount of noise is defined based on the noise scheduler of the model, and a reconstruction loss is computed at each training step between the prediction of the model and the predefined target amount of noise. As soon as a new text-video pair becomes available, before tuning is actually performed, the student model is duplicated, and the weights of the duplicate are frozen; this ensures that, during the training process, the duplicate retains the knowledge the student has acquired so far. The student model is now free to learn; the original one-shot video tuning strategy is applied once again, but now the duplicate contributes to the learning process of the student by means of a *distillation loss*, essentially acting as a teacher. First of all, a Mean Squared Error (MSE) is computed between the output of the student network (i.e., the amount of noise to be removed from the current denoising step) $\hat{y}_i$ and the target noise y_i to be removed for the i-th element in the noisy video sequence:

$$L_s = \frac{1}{N} \sum_{i=1}^{N} (y_i - \hat{y}_i)^2 . \tag{1}$$

Here, N denotes the total number of frames, while L_s denotes the student's reconstruction loss. The teacher model is then asked to distill its knowledge by means of a KL-Divergence loss:

$$L_{KL} = T^2 \cdot D_{KL}\left(Softmax\left(\frac{\mathbf{t}}{T}\right) \middle\| \log\left(Softmax\left(\frac{\mathbf{s}}{T}\right)\right)\right). \tag{2}$$

Here, T is a temperature parameter used to control the softness of the output distributions $\mathbf{t}$ and $\mathbf{s}$ produced by the teacher and the student networks, respectively. The distillation loss can be thus obtained by means of a linear combination between the two terms L_{KL} and L_s [24]:

$$L_{distill} = \alpha \cdot L_{KL} + (1 - \alpha) \cdot L_s, \tag{3}$$

where α is a crucial parameter that balances knowledge retention with the learning of new knowledge. Notice that, when the model is being tuned on the very first text-video pair, we have $L_{distill} = L_s$, without any α coefficient. Additionally, in order to enhance the temporal consistency between the generated frames, we propose to integrate a *temporal consistency loss* considering the difference on the temporal dimension between consecutive frames for both the prediction of the student model and the target noise that should be predicted. In particular, let us denote with $\boldsymbol{P} \in \mathbb{R}^{B \times C \times T \times H \times W}$ the model's prediction, and with $\boldsymbol{N} \in \mathbb{R}^{B \times C \times T \times H \times W}$ the target noise the model's prediction has to resemble. A squared difference between the temporal changes of both the predictions of the model and the target noise, averaged over all the frames M, is then computed as follows:

$$L_t = \frac{1}{M} \sum_{i=1}^{M} \left(\left(\boldsymbol{P}_i^{(t+1)} - \boldsymbol{P}_i^{(t)} \right) - \left(\boldsymbol{N}_i^{(t+1)} - \boldsymbol{N}_i^{(t)} \right) \right)^2 \tag{4}$$

Therefore, considering both the distillation loss and the temporal consistency loss we have previously defined, our final proposed loss is defined as follows:

$$L_{tot} = \gamma \cdot L_{distill} + \lambda \cdot L_t, \tag{5}$$

where γ and λ are additional coefficients used for scaling the contribution of the two loss components with respect to the total loss function. Additionally, we propose to integrate a *retrieval mechanism* in order to choose the training video that should be used for providing structural guidance to the model at inference time. Our proposed approach, depicted in Fig. 2, is based on Sentence-BERT (sBERT) [25]: a modification of the BERT model [26] using siamese and triplet networks that is able to derive semantically meaningful sentence embeddings from input sentences. We feed sBERT with the inference prompt and each prompt used for training the model, one by one, and we let sBERT compute a similarity score between them. The highest similarity score is then used to retrieve the training prompt resembling the most the inference prompt, and the corresponding training video will in turn be used for providing structural guidance to the model at inference time. Finally, a *DDIM inversion* operation is performed on the retrieved video, resulting in an inverted latent noise, that will in turn be used as the starting point for a *DDIM sampling* operation. The latent representation resulting from this latter DDIM sampling operation is then decoded, thus providing the generated video.

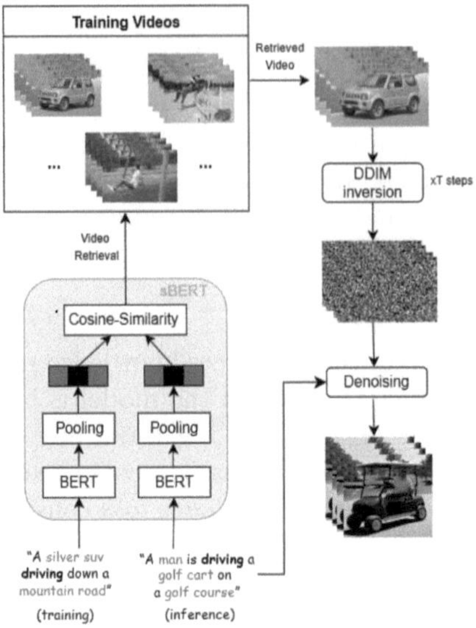

Fig. 2. VidCLearn inference procedure.

4 Experiments and Results

In this section, we evaluate the effectiveness of our proposed continual learning approach for text-to-video generative models, with the aim of maximizing temporal consistency and generation quality while minimizing knowledge degradation over time.

Dataset. We evaluate our method on the DAVIS dataset [27], as done by the authors of Tune-A-Video. Since the plain DAVIS dataset does not provide paired text-video data, we employ a pre-trained BLIP-2 model [28] for generating the textual descriptions of each video frame. The caption representing each single video is determined by the highest CLIP Score [29] value resulting from each group of frames, along with their corresponding descriptions. This setup guarantees the availability of 90 text-video pairs for training and 30 text-video pairs for evaluation, to be provided to the model over time.

Evaluation Metrics. In order to evaluate different aspects of our model's performance, we rely on some of the most commonly used evaluation metrics. We evaluate the visual quality of the model's output by means of the the *Fréchet Video Distance* (FVD) [30], the *Fréchet Inception Distance* (FID) [31], and the *Inception Score* (IS) [32] metrics. We also evaluate the semantic alignment between the generated content and the corresponding prompt by means of the *CLIP Score* metric. Finally, we evaluate the knowledge retention's degree of our model by means of the *Forward Transfer* (FWT) and the *Backward Transfer* (BWT) metrics [33].

Implementation Details. At training time, we sample 20 uniform frames at resolution of 512×512 from each input video, and we fine-tune the student model for 500 steps with a learning rate η of 3×10^{-5}, a temperature T of 4, a distillation weight γ of 1 and a temporal weight λ of 10. Moreover, the balancing coefficient α of the distillation loss (Eq. 3) is set to 0.8. At inference time, we perform DDIM inversion with 100 inverted latent steps, and then we generate the output video with 150 inference steps. Each tuning on the current text-video pair takes slightly more than one hour on an NVIDIA A100 GPU, and requires around 37 MiB of GPU RAM to be performed; at inference time, each generation takes about one minute to be performed and requires around 7 MiB of GPU RAM.

4.1 Quantitative Evaluation

Table 1. Naive Approach: visual quality and semantic alignment.

Videos	FVD ↓	FID ↓	IS ↑	CLIP ↑
5	865,824	9,344	10,272	32,058
10	935,043	19,915	9,298	31,002
25	1224,906	20,105	6,043	26,870
50	954,725	11,570	6,891	27,677
75	952,074	11,538	7,318	28,187
90	617,230	7,112	6,328	26,743

Since there is no prior work employing continual learning techniques for solving the text-to-video generation task, the reference baseline against which we compare the performance of our method consists on iteratively fine-tuning the Tune-A-Video model on all the training text-video pairs, provided sequentially over time. We measure the behavior of the baseline model over time, as more text-video pairs become available; the results, summarized in Table 1, show an overall decrease of the model's performance, as the number of text-video pairs increases over time. This is a clear sign of the catastrophic forgetting issue arising.

One of the simplest strategies to avoid the catastrophic forgetting issue is the *Elastic Weight Consolidation (EWC)* technique [34]: this technique selectively protects important parameters of the model as its knowledge is expanded over time. By applying the EWC technique, as summarized in Table 2, we notice a slight increase in performance with respect to the naive baseline: we obtain a reduction of 15,42% in FID, an increase of 20,93% in IS and an increase of 5,46% in CLIP Score. However, these slight variations show that the behavior of the model remains almost the same as before: the catastrophic forgetting issue still persists, as the performance of the model decreases over time.

Table 2. EWC Approach: visual quality and semantic alignment.

Videos	FVD ↓	FID ↓	IS ↑	CLIP ↑
5	1096,049	10,155	10,820	31,839
10	1210,202	25,408	9,589	30,278
25	1593,644	18,167	6,751	28,508
50	1264,515	12,656	6,942	28,200
75	844,169	7,481	7,091	25,641
90	901,032	6,162	8,003	28,288

Our proposed approach significantly reduces the catastrophic forgetting issue arising in the baseline, while improving both visual quality and temporal consistency of the generated video frames. Compared to the baseline, as shown in Table 3, experiments performed with our proposed method demonstrate an overall improvement of the model's performance. Specifically, since we obtain a 18,97% increase in FVD, a 125,42% reduction in FID, and a 102,78% increase in IS, the visual quality of the generated videos improves. Additionally, since we obtain a 23,03% increase in CLIP Score, also the semantic alignment between the generated content and the inference prompt improves. We can draw similar conclusions by comparing our approach with the EWC strategy: in this latter case, we obtain a 22,7% reduction in FVD, a 95,31% reduction in FID and a 37,63% increase in IS, demonstrating better performance in terms of visual quality of generated videos. In addition, we also obtain a 14,02% increase in the CLIP score, showing an improvement in performance also in terms of semantic alignment.

Table 3. VidCLearn Approach: visual quality and semantic alignment.

Videos	FVD ↓	FID ↓	IS ↑	CLIP ↑
5	740,210	8,505	12,002	32,800
10	820,430	15,704	10,885	31,754
25	980,112	14,002	9,123	29,932
50	852,764	10,200	9,005	30,125
75	765,821	6,512	10,304	29,342
90	734,320	3,155	12,832	32,903

Our proposed approach also shows a significant performance improvement in terms of knowledge retention. In fact, when compared with the Naive approach, our proposed VidCLearn approach leads to a 215,17% increase in FWT (1,002 Naive vs 3,158 VidCLearn) and a 158,73% increase in BWT (0,538 Naive vs 1,392 VidCLearn); moreover, when compared with the EWC strategy, our approach leads to a 30,62% improvement in FWT (2,191 EWC vs 3,158 VidCLearn) and a 27,23% improvement in BWT (1,013 EWC vs 1,392 VidCLearn). All the experiments were conducted after the knowledge of the model was expanded with 90 training videos.

4.2 Qualitative Evaluation

Table 1 shows, from a quantitative point of view, that iteratively fine-tuning the Tune-A-Video model on all the training text-video pairs that become available sequentially over time leads to catastrophic forgetting. Let's see what happens visually when we ask the model to generate a video in response to the inference prompt "*a man is driving a golf cart on a golf course*"; Fig. 3 shows its behavior. Specifically, the first row shows the last training video on which the model was tuned; the second row, instead, shows the ground truth video, while the last three rows show the actual generated video with the different methodologies. In the third row we can see that with the Naive approach the model responds to the inference prompt by generating a video that highly resembles the last video seen during training. A very similar behavior is observed when applying the *Elastic Weight Consolidation (EWC)* technique: quantitatively, the performance of the model decreases over time, despite a slight overall increase with respect to the naive baseline. Let's see again what happens visually when we ask the model to generate a video in response to the same inference prompt as before; Fig. 3 (see row number four) shows its behavior. Even the application of the EWC technique results in the model generating a video that highly resembles the last video seen during training, in response to the given inference prompt; however, in this case, the visual quality of the generated frames appears to be better than before. Table 3, on the other hand, shows that the application of our proposed approach leads to a significant increase in the performance of the model: the generated video frames have better visual quality and temporal consistency, and the degree of knowledge retention is higher. Let's see again what happens visually when we ask the model to generate a video in response to the same inference prompt as before; Fig. 3 (last row) shows its behavior. With our approach, the model responds to the inference prompt by generating a video that highly resembles the corresponding ground truth video, as an additional proof of the superiority of our proposed method.

4.3 Ablation Study

To evaluate the contribution of each component in our proposed approach, we have conducted an ablation study by systematically removing or altering individual components, and analyzing their impact on model performance. We evaluate the performance of our model after learning the entire training partition of the DAVIS dataset using metrics such as *FVD*, *FID*, *IS*, and *CLIP score*. Therefore, we investigated the impact of the temporal consistency loss (Eq. 4) on model performance by increasing the temporal importance factor λ. Table 4 summarizes the results of our experiments. Clearly, the presence of this component benefits

Fig. 3. Qualitative evaluation: first row shows the frames of the last training video; second row shows the ground truth with the textual prompt "*a man is driving a golf cart on a golf course*"; third row shows the video frames generated by the Naive approach, the fourth with the EWC approach and the last row with the proposed **VidCLearn** approach.

the overall performance of the model; however, the temporal importance factor λ should not be set too high, otherwise performance starts to degrade. For instance, as reported in Table 4, we can see that as the temporal importance factor λ increases, the performance of the model in terms of the semantic alignment between its input and its output decreases: in fact, the CLIP score decreases from 32,903 with $\lambda = 10$, to 32,783 with $\lambda = 20$, to 32,651 with $\lambda = 30$.

Table 4. Ablation study on the temporal loss.

α	γ	λ	FVD ↓	FID ↓	IS ↑	CLIP ↑
0,7	1	0	791,997	4,007	12,613	32,699
0,7	1	1	774,002	4,249	12,437	32,771
0,8	1	10	734,320	3,155	12,832	32,903
0,8	1	20	699,161	3,051	12,165	32,783
0,8	1	30	660,582	3,090	13,217	32,651

We also investigated the impact of our proposed retrieval mechanism for structural guidance on model performance. Table 5 summarizes the results of our experiments; here, "*last*" means that at inference time, structural guidance was provided by the last training video, while "*retrieval*" means that structural guidance was provided by the training video retrieved by our proposed

Table 5. Ablation study on the retrieval mechanism.

Guidance	α	γ	λ	FVD ↓	FID ↓	IS ↑	CLIP ↑
Last	0,7	1	0	782,928	3,541	10,874	31,490
Retrieval	0,7	1	0	791,997	4,007	12,613	32,699
Last	0,7	1	1	758,124	3,322	10,832	31,631
Retrieval	0,7	1	1	774,002	4,249	12,437	32,771
Last	0,8	1	10	728,164	2,899	11,324	31,898
Retrieval	0,8	1	10	734,320	3,155	12,832	32,903
Last	0,8	1	20	709,237	3,111	11,200	32,080
Retrieval	0,8	1	20	699,161	3,051	12,165	32,783
Last	0,8	1	30	684,854	3,169	10,871	32,159
Retrieval	0,8	1	30	660,582	3,090	13,217	32,651

retrieval mechanism. Our model benefits from the dynamic structural guidance provided by such a retrieval mechanism, leading to an overall increase in model performance.

5 Conclusions

This work presents a continual learning approach for text-to-video generation, addressing issues like catastrophic forgetting and motion consistency. Existing models are almost static and cannot adapt to new data without full retraining. We tackle this by applying continual learning techniques to enable dynamic knowledge expansion over time. To reduce computational costs, we adopt a lightweight architecture inspired by Tune-A-Video. Our method achieves promising results, preserving prior knowledge while maintaining spatial and temporal coherence. However, challenges remain, including difficulty generating unseen motion patterns and improving visual quality. Future work could explore perceptual losses, memory replay strategies, and model compression techniques like pruning or quantization to address these limitations more efficiently.

Acknowledgments. This study has been partially supported by SERICS (PE00000014) under the MUR National Recovery and Resilience Plan funded by the European Union - NextGenerationEU.

References

1. Wu, J.Z., et al.: Tune-a-video: one-shot tuning of image diffusion models for text-to-video generation. In: Proceedings of the IEEE/CVF International Conference on Computer Vision (2023)

2. Mittal, G., Marwah, T., Balasubramanian, V.N.: Sync-draw: automatic video generation using deep recurrent attentive architectures. In: Proceedings of the 25th ACM International Conference on Multimedia (2017)
3. Wang, J., et al.: ModelScope text-to-video technical report. arXiv preprint arXiv:2308.06571 (2023)
4. Rombach, R., et al.: High-resolution image synthesis with latent diffusion models. In: Proceedings of the IEEE/CVF Conference on Computer Vision and Pattern Recognition (2022)
5. Ho, J., et al.: Video diffusion models. Adv. Neural. Inf. Process. Syst. **35**, 8633–8646 (2022)
6. Zeng, Y., et al.: Make pixels dance: high-dynamic video generation. In: Proceedings of the IEEE/CVF Conference on Computer Vision and Pattern Recognition (2024)
7. Zhou, D., et al.: MagicVideo: efficient video generation with latent diffusion models. arXiv preprint arXiv:2211.11018 (2022)
8. Zhang, D.J., et al.: Show-1: Marrying pixel and latent diffusion models for text-to-video generation. Int. J. Comput. Vis., 1–15 (2024). https://doi.org/10.1007/s11263-024-02271-9
9. Liang, J., et al.: NUWA-infinity: autoregressive over autoregressive generation for infinite visual synthesis. Adv. Neural. Inf. Process. Syst. **35**, 15420–15432 (2022)
10. Hong, W., et al.: CogVideo: large-scale pretraining for text-to-video generation via transformers. arXiv preprint arXiv:2205.15868 (2022)
11. Villegas, R., et al.: PHENAKI: variable length video generation from open domain textual descriptions. In: International Conference on Learning Representations (2022)
12. Wu, C., et al.: Nüwa: Visual synthesis pre-training for neural visual world creation. In: European Conference on Computer Vision. Springer Nature Switzerland, Cham (2022). https://doi.org/10.1007/978-3-031-19787-1_41
13. Chen, S., et al.: GenTron: diffusion transformers for image and video generation. In: Proceedings of the IEEE/CVF Conference on Computer Vision and Pattern Recognition (2024)
14. Liu, Y., et al.: Sora: a review on background, technology, limitations, and opportunities of large vision models. arXiv preprint arXiv:2402.17177 (2024)
15. Wang, X., et al.: VideoLCM: video latent consistency model. arXiv preprint arXiv:2312.09109 (2023)
16. Khachatryan, L., et al.: Text2video-zero: text-to-image diffusion models are zero-shot video generators. In: Proceedings of the IEEE/CVF International Conference on Computer Vision (2023)
17. Singer, U., et al.: Make-a-video: text-to-video generation without text-video data. arXiv preprint arXiv:2209.14792 (2022)
18. Doan, K., et al.: Class-prototype conditional diffusion model for continual learning with generative replay. arXiv preprint arXiv:2312.06710 (2023)
19. He, J., et al.: Continual learning with diffusion-based generative replay for industrial streaming data. In: 2024 IEEE/CIC International Conference on Communications in China (ICCC). IEEE (2024)
20. Masip, S., et al.: Continual learning of diffusion models with generative distillation. arXiv preprint arXiv:2311.14028 (2023)
21. Smith, J.S., et al.: Continual diffusion with STAMINA: STack-And-Mask INcremental Adapters. In: Proceedings of the IEEE/CVF Conference on Computer Vision and Pattern Recognition (2024)
22. Smith, J.S., et al.: Continual diffusion: continual customization of text-to-image diffusion with C-LORA. arXiv preprint arXiv:2304.06027 (2023)

23. Meng, Z., et al.: DiffClass: Diffusion-based class incremental learning. European Conference on Computer Vision. Springer, Cham (2025). https://doi.org/10.1007/978-3-031-73021-4_9
24. Hinton, G.: Distilling the knowledge in a neural network. arXiv preprint arXiv:1503.02531 (2015)
25. Reimers, N.: Sentence-BERT: sentence embeddings using Siamese BERT-Networks. arXiv preprint arXiv:1908.10084 (2019)
26. Devlin, J.: BERT: pre-training of deep bidirectional transformers for language understanding. arXiv preprint arXiv:1810.04805 (2018)
27. Pont-Tuset, J., et al.: The 2017 DAVIS challenge on video object segmentation. arXiv preprint arXiv:1704.00675 (2017)
28. Li, J., et al.: Blip-2: bootstrapping language-image pre-training with frozen image encoders and large language models. In: International Conference on Machine Learning. PMLR (2023)
29. Hessel, J., et al.: ClipScore: a reference-free evaluation metric for image captioning. arXiv preprint arXiv:2104.08718 (2021)
30. Unterthiner, T., et al.: FVD: a new metric for video generation. DGS@ICLR 2019 (2019)
31. Heusel, M., et al.: GANs trained by a two time-scale update rule converge to a local NASH equilibrium. In: Advances in Neural Information Processing Systems, vol. 30 (2017)
32. Salimans, T., et al.: Improved techniques for training GANs. In: Advances in Neural Information Processing Systems, vol. 29 (2016)
33. Lopez-Paz, D., Ranzato, M.A.: Gradient episodic memory for continual learning. In: Advances in Neural Information Processing Systems, vol. 30 (2017)
34. Kirkpatrick, J., et al.: Overcoming catastrophic forgetting in neural networks. Proc. Nat. Acad. Sci. **114**(13), 3521–3526 (2017)

Ensemble-Enhanced Semantic Segmentation for Efficient Waste Recycling

Mohammadhossein Akbari Moafi(✉) and Horst Possegger

Institute of Visual Computing, Graz University of Technology, Graz, Austria
mohammadhossein.akbarimoafi@studenti.unipd.it, possegger@tugraz.at

Abstract. Efficient and accurate annotation workflows are crucial for advancing semantic segmentation, especially in waste recycling applications, where detailed labeling is time-consuming and prone to errors. This work proposes a new pipeline to accelerate the annotation process by combining the ensemble of three semantic segmentation models (UperNet, OCRNet, and DeepLabv3+) with the Segment Anything Model (SAM), for instance segmentation. Within this pipeline, we investigate two categories of approaches: (1) purely semantic segmentation-based methods–Majority Voting (MV), Max Logits (ML), and SDF Mask Retrieval (SDF-MR)–and (2) hybrid methods that integrate both semantic and instance segmentation, namely Max Logits with SAM Refinement (ML-SAM) and SDF Mask Retrieval with SAM Refinement (SDF-SAM). The work is carried out on household waste data that have been annotated with 11-class masks. The experiments demonstrated significant improvements in both annotation accuracy and efficiency compared to individual models. Integrating SAM into the system also resulted in substantially better boundary predictions and an increase in mIoU of 4.97% compared to the best-performing individual model. This approach demonstrates the potential to enhance waste recycling solutions by efficiently generating the necessary labels through the combination of semantic and instance segmentation models.

Keywords: Semantic Segmentation · Waste Recycling · Ensemble Learning · Annotation Workflow

1 Introduction

Waste recycling is essential for sustainable development, as it reduces landfill use, lowers pollution, and conserves finite resources. However, the pace of urbanization and the growth of the global population have led to an unprecedented increase in waste generation, projected to reach 3.4 billion tonnes by 2050 [8]. Waste recycling systems often cannot adequately meet demand, and cities are

M. Akbari Moafi—Work done during internship at TU Graz.

E. Rodolà et al. (Eds.): ICIAP 2025, LNCS 16167, pp. 362–374, 2026.
https://doi.org/10.1007/978-3-032-10185-3_29

struggling to manage this influx. Manual, low-tech recycling methods are labor intensive and susceptible to errors, often leading to contamination and reducing the value of recyclable materials. Such inefficiencies hinder the wider implementation of recycling schemes, especially in densely populated urban environments, where the volume and variety of waste can overwhelm current systems.

State-of-the-art computer vision approaches, particularly those based on deep learning, have enabled automated systems that enhance recycling efficiency and support intelligent waste management. However, their effectiveness is highly dependent on large volumes of high-quality annotated data. Semantic segmentation classifies images at the pixel level to accurately separate material types, making it especially useful for optimizing the recycling process. Nonetheless, its practical deployment is hindered by the time-consuming and error-prone process of generating detailed annotations [12], particularly in the presence of overlapping objects, varied textures, or poor lighting. These issues limit scalability and often introduce label noise. This highlights the need for advanced annotation workflows that reduce manual effort while improving segmentation accuracy across diverse and cluttered waste categories.

This paper presents a novel annotation pipeline in which ensemble strategies effectively address key challenges by leveraging the complementary strengths of three advanced semantic segmentation models, namely UperNet [17], OCRNet [18], and DeepLabv3+ [4], sharing the same ResNet-101 backbone [7], coupled with an advanced instance segmentation model SAM [9] to reduce the time and effort required for annotation. Within this pipeline, SAM focuses solely on detecting objects in an image and refining their boundaries at the instance level, while the pixel-accurate semantic segmentation models predict the corresponding material classes. Due to the high cost and complexity of capturing high-resolution waste images, datasets in this domain are extremely small. Consequently, the proposed pipeline leverages pre-trained models with proven performance to ensure robust generalization in low-data scenarios. By generating high-quality pre-labels, it reduces manual effort and supports efficient human correction. Addressing issues such as dataset types, annotation errors, and classification concerns, this method aims to transform management processes and make it easier to deploy intelligent recycling systems and data operations.

2 Related Work

The application of computer vision in waste recycling has advanced significantly through the adoption of deep learning techniques such as convolutional neural networks, which improve the accuracy of classifying and segmenting recyclable materials in complex scenes [5,11,14]. However, creating high-quality pixel-level annotations remains labor intensive, especially in cluttered environments where objects overlap or have similar appearances.

Recent studies have investigated annotation workflows that employ ensemble learning to reduce manual effort while enhancing label quality [6,16]. These systems often aggregate the outputs of multiple segmentation models to generate accurate pseudo-labels. Depending on the quality of the output, these masks can be used either directly or optionally refined by human annotators. This ensemble-based strategy significantly accelerates the labeling process while maintaining annotation quality, particularly beneficial in low-data scenarios.

Further advances integrate both semantic and instance segmentation models to improve boundary precision and object separation. Such hybrid approaches enable models to not only classify material types, but also distinguish individual object instances, which is critical in complex domains [2,10]. Building on this, our study combines ensemble learning with SAM and multiple semantic segmentation models to enhance annotation efficiency. This approach reduces reliance on large-scale manual labeling while supporting scalable, high-quality pre-labeling in complex scenes.

3 Methodology

This work employs a combination of semantic segmentation models, which are trained on RGB images, in order to assign a label to each pixel from the set $\{1, \ldots, C\}$, generating a segmentation mask that matches the resolution of the input image. Additionally, SAM [9] is incorporated to enhance instance-level segmentation, leveraging its zero-shot segmentation capabilities to handle diverse and unseen object categories effectively. The selection of models was guided by several key considerations. First, each model was chosen for its ability to achieve high accuracy on semantic segmentation benchmarks, particularly on datasets with complex and diverse structures. Second, the models differ in design, employing various architectures and strategies to generate diverse perspectives on segmentation tasks, thereby enhancing overall robustness. Additionally, the selected models demonstrated adaptability to fine-grained boundary requirements and the heterogeneous characteristics of domain-specific datasets, such as those in the recycling context.

3.1 Ensemble Strategies

This study investigates sophisticated ensemble strategies, such as majority voting as a baseline, alongside tailored approaches such as max logits aggregation and retrieved SDF masks, in order to handle the unique challenges of semantic segmentation and refine them using SAM. These strategies are designed to leverage the complementary strengths of semantic and instance segmentation techniques to address complex object scenarios and improve boundary refinement. Figure 1 illustrates the overall pipeline of the proposed framework.

Majority Vote (MV): The majority voting is a straightforward yet effective ensembling technique where the most frequently predicted label across all individual models determines the final class label of each pixel. Let $y_i(p)$ be the label

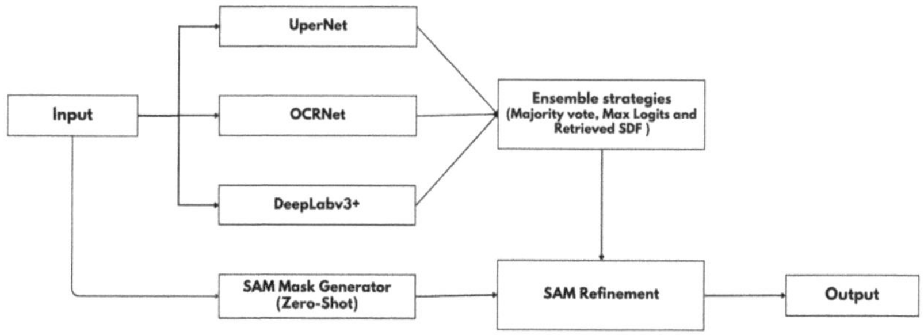

Fig. 1. Overview of the framework pipeline, including models, ensemble strategies, and SAM Refinement.

that the i-th model predicts for pixel p given N segmentation models. The final predicted label $Y(p)$ is given by:

$$Y(p) = \arg\max_{c \in C} \sum_{i=1}^{N} \mathbf{1}(y_i(p) = c), \tag{1}$$

where C is the set of all possible classes and $\mathbf{1}(\cdot)$ is the indicator function that equals 1 when the condition is true and 0 otherwise.

Max Logits (ML): By aggregating models' logits, this method is designed to take advantage of the probabilistic output of semantic segmentation models. Unlike majority voting, which operates on discrete class predictions, this approach leverages the raw model logits, averaging them across models to create a unified probabilistic representation. Mixing the confidence levels of different models produces smoother and more consistent segmentation boundaries.

Formally, let $s_i(p, c)$ denote the logit value assigned by model i for class c at pixel p. Given N models, the final logits are computed as:

$$S(p, c) = \frac{1}{N} \sum_{i=1}^{N} s_i(p, c). \tag{2}$$

The final class label is then determined by applying the argmax function over the averaged logits. This strategy effectively smooths inconsistencies by considering the confidence distribution rather than relying on hard labels. Consequently, it reduces abrupt transitions and enhances segmentation coherence.

SDF Mask Retrieval (SDF-MR): The signed distance function (SDF) mask retrieval (MR) strategy is a novel adaptation inspired by the use of SDFs in binary medical image segmentation, where precise boundary delineation is essential [1]. In this work, we extend the concept to multi-class segmentation and integrate it into an ensemble-based annotation pipeline. To the best of our knowledge, this is the first use of class-specific SDFs to aggregate predictions from multiple segmentation models for annotation refinement. SDF-based ensembling, in contrast to traditional pixel-wise voting techniques, depicts object borders as continuous scalar fields, where the value of each pixel indicates how far away it is from the closest boundary. This allows for better boundary refinement, particularly in cases with overlapping objects or uncertain segmentation edges. The contribution lies in replacing conventional binary/multi-label masks with class-specific SDFs for aggregating predictions from multiple models, improving annotation quality in complex scenes.

The SDF of a segmentation mask M can be computed using the Euclidean distance function for each pixel p. The implicit truncated SDF $d_c(p)$ for class c is defined as:

$$d_c(p) = \begin{cases} -\min\left(\min_{q\in\partial M_c} \|q-p\|_2, \tau\right), & \text{if } M(p) = c \\ \min\left(\min_{q\in\partial M_c} \|q-p\|_2, \tau\right), & \text{if } M(p) \neq c \\ 0, & \text{if } p \in \partial M_c \end{cases} \tag{3}$$

where ∂M_c represents the boundary of class c, and τ is a threshold that ensures the SDF remains bounded, avoiding the influence of distant background regions. Negative values indicate regions inside an object, positive values represent background, and boundary pixels are set to zero. The final mask is obtained by aggregating SDFs from multiple models and applying a zero threshold:

$$Y(p) = \arg\max_{c\in C} \mathbf{1}(\bar{d}_c(p) \leq 0) \cdot c, \tag{4}$$

where $\bar{d}_c(p)$ is the average SDF computed over models. This approach refines boundaries and reduces inconsistencies by leveraging geometric properties instead of raw pixel classifications.

3.2 SAM Refinement

To improve segmentation accuracy, SAM is employed for mask refinement by enhancing object boundaries, resolving inconsistencies, and enabling instance-aware segmentation. This refinement is applied to both ML and SDF-MR outputs, aligning the initial segmentation masks with SAM-generated masks to ensure greater consistency and update regions with high model agreement.

Given an initial segmentation mask M, SAM-generated masks S are used to refine object boundaries. The refinement process consists of two main steps: (1) Mask alignment and region merging based on Intersection over Union (IoU), and (2) Updating segmentation labels using a region of interest (ROI) approach. Let M_c denote the binary mask for class c in the base segmentation mask and S_i the corresponding SAM-generated mask. If the IoU between the masks exceeds a predefined threshold θ, the SAM mask is merged with the base mask to refine the object boundaries:

$$M_c^{\text{refined}} = M_c \cup S_i, \quad \text{if } \text{IoU}(M_c, S_i) > \theta. \tag{5}$$

To ensure controlled refinement, an ROI mask is used to limit modifications to areas where SAM provides high-confidence segmentations. Given a binary ROI mask R, the final refined mask is:

$$M^{\text{final}}(p) = \begin{cases} M^{\text{refined}}(p), & \text{if } R(p) = 1 \\ M(p), & \text{otherwise} \end{cases} \tag{6}$$

Leveraging SAM helps to resolve ambiguities in overlapping objects and helps the final predictions capture even finer details, thus making the object segmentation masks more accurate and consistent.

3.3 Models Within the Ensemble

To achieve high-quality segmentation outputs, this study utilizes a combination of semantic and instance segmentation models. Semantic segmentation models give class-level pixel annotation, while SAM refines these outputs for more object-level details. This integration is considered an approach to address waste recycling challenges, given the precision in classification, adaptation to unseen objects, and boundary delineation required in an efficient annotation framework.

For semantic segmentation, UperNet [17], OCRNet [18], and DeepLabv3+ [4] are utilized, each with distinct architectural designs. UperNet integrates a Pyramid Pooling Module and Feature Pyramid Network to capture both local and global context effectively. Its ability to extract rich hierarchical characteristics at multiple scales is made possible by the ResNet-101 backbone, which makes it especially useful for recognizing a variety of waste categories, even in cluttered environments like garbage bins. OCRNet improves segmentation accuracy by explicitly modeling contextual information at the object-level and calculates the representation of each object and its surrounding context by extracting feature maps using the ResNet-101 backbone. This ensures accurate segmentation in situations where waste items overlap or share similar appearances. It is resilient to changes in item sizes and shapes due to its attention-based process, which improves concentration on important areas. DeepLabv3+, by adding an encoder-decoder structure, expands on DeepLabv3 and allows for more precise spatial segmentation. While the decoder improves spatial resolution, the Atrous Spatial Pyramid Pooling module in the encoder records multi-scale characteristics. Its proven capability in boundary delineation on Cityscapes and PASCAL VOC datasets justified its application in common waste recycling tasks, where objects often overlap and have complex boundaries.

For instance segmentation, SAM [9] is employed, as it is designed to generalize across diverse segmentation tasks. SAM is trained on an extremely massive dataset containing more than 1 billion masks over 11 million images, enabling strong generalization across various domains. Leveraging zero-shot capabilities, SAM can segment objects without prior task-specific training, making it highly adaptable for various applications. This is specifically more useful in applications like waste recycling, where training data is limited and accurate identification and separation of materials such as plastic, metals, and paper are essential.

4 Experimental Setup

4.1 Dataset

The dataset used in this study consists of a total of 401 images, with 321 training images and 80 test images. These images are annotated at the pixel level across 11 classes. Additionally, an 'Ignore' category is available to the human annotators as an option during training and evaluation to exclude ambiguous pixels, ensuring cleaner annotation results. The input images have dimensions of 2464 × 2056 pixels, capturing high-resolution details essential for precise semantic segmentation. Both the percentage coverage and the number of each class in the images are displayed in Table 1 for the training and testing datasets, respectively.

Preprocessing: To adapt the dataset for the semantic segmentation models–UperNet, OCRNet, and DeepLabv3+–a unified preprocessing pipeline is applied with minor variations for each model. The general preprocessing pipeline consists of resizing all images to 1024 × 1232 pixels while maintaining their aspect ratio; then the cropping is applied, where the images are randomly cropped to 500 × 500 pixels for UperNet and 512 × 512 pixels for DeepLabv3+ and OCRNet. To enhance generalization, augmentation techniques, such as random flipping with probability of 50% and photometric distortion, are applied to simulate various real-world conditions. The preprocessing pipeline ensures that underrepresented classes are sampled sufficiently during training by enforcing a maximum category ratio during cropping in order to resolve class imbalance.

For testing, augmentations like cropping and photometric distortion are excluded to maintain consistency. The testing pipeline focuses on resizing and flipping to ensure a fair evaluation.

Table 1. Class Distribution in the Dataset.

a) Training Dataset

Class	% Coverage	Count
Background	60.73	321
Residual Waste	7.76	289
Organic Waste	5.00	186
Biodegradable Bag	1.18	44
Glass	0.62	23
Metal	3.71	138
Paper	5.50	205
Cardboard	0.94	35
Plastic	6.44	240
PET	1.72	64
Other Garbage Bag	6.40	238

b) Testing Dataset

Class	% Coverage	Count
Background	54.51	80
Residual Waste	8.79	69
Organic Waste	5.73	45
Biodegradable Bag	1.15	9
Glass	1.91	15
Metal	4.08	32
Paper	6.49	51
Cardboard	1.15	9
Plastic	7.14	56
PET	1.27	10
Other Garbage Bag	7.78	61

4.2 Training Details

The experiments in this study are carried out using the MMSegmentation Framework [3], an open-source PyTorch-based toolbox for semantic segmentation, extended with custom components to support ensemble strategies and SAM Refinement.

All segmentation models in this study were initialized with pre-trained weights from the ImageNet dataset. The models were then fine-tuned on the waste dataset using an NVIDIA RTX 2070 GPU and an Intel(R) Core(TM) i7-8700 CPU (6 cores, 12 threads) with varying batch sizes and iteration counts tailored to each architecture. UperNet and OCRNet were trained using a batch size of 4, while DeepLabv3+ utilized a batch size of 2 due to its higher memory requirements. The total number of training iterations was set to 150k for UperNet, 100k for OCRNet, and 200k for DeepLabv3+. The initial learning rate was set to 0.001 and decayed using a poly-schedule, and the optimization was performed using the Stochastic gradient descent optimizer. During training, data augmentation techniques such as random resizing, cropping, flipping, and photometric distortions are used, as described in Sect. 4.1, to enhance model robustness. The training setup partially follows the baseline from [15], incorporating selected hyperparameters while adapting other settings to better fit the methodology. For evaluation, the best-performing weights were selected at 140k iterations for UperNet, 50k iterations for OCRNet, and 140k iterations for DeepLabv3+.

5 Results and Discussion

5.1 Performance Analysis

The performance of the individual models and the ensemble strategies was measured using mean Intersection-over-Union (mIoU), pixel accuracy (ACC), and class-wise IoU. These metrics are widely used in segmentation tasks and are particularly suitable for assessing the accuracy of predictions in diverse and complex datasets, such as waste recycling. Table 2 summarizes the performance comparison of individual semantic segmentation models and ensemble strategies.

Regarding the ensemble results, the pipeline showed a notable improvement over individual models in terms of mIoU and ACC. To aid interpretation, Fig. 2 presents a comparison between the individual models and the ensemble outputs. The Majority Vote (MV) strategy moderately improved overall performance by partially offsetting individual model biases. The Max Logits (ML) approach further enhanced accuracy by aggregating raw model outputs, achieving an mIoU of 51.11%. The integration of SAM into the ensemble strategies resulted in the highest mIoU improvements. The SDF Mask Retrieval with SAM Refinement (SDF-SAM) method, for example, achieved a mIoU of 52.17% due to the additional refinement of SAM and the clear boundary delineation of the SDF masks. Similarly, the Max Logits with SAM Refinement (ML-SAM) approach achieved the highest mIoU of 52.65%, demonstrating the capacity of SAM to resolve boundary ambiguity and produce a more accurate prediction.

Table 2. Performance Comparison of Individual Models and Ensemble Strategies.

Model/Strategy	mIoU (%)	ACC (%)	Inference Speed (s/image)
UperNet	46.20	91.33	0.51
OCRNet	47.68	91.64	0.77
DeepLabv3+	44.95	91.25	**0.26**
MV	49.33	93.66	11.37
SDF-MR	50.32	93.59	8.94
ML	51.11	93.70	1.41
SDF-SAM	52.17	93.77	19.15
ML-SAM	**52.65**	**93.79**	10.63

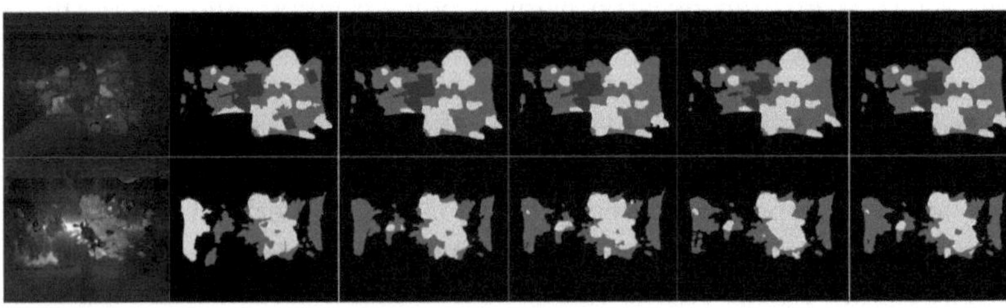

Fig. 2. Qualitative comparison of individual models and ensemble outputs. Each row (left to right) shows the input image, ground truth, outputs of UperNet, OCRNet, DeepLabv3+, and the corresponding ensemble method (top: ML, bottom: SDF-MR). (Color figure online)

5.2 Impact of SAM Refinement

By reducing the manual effort needed for boundary refinement, the proposed framework simplifies the annotation process. The visual examples in Figs. 3 and 4 demonstrate the effect of SAM Refinement on the outputs of the SDF-MR and ML strategies, revealing improvements in boundary delineation and segmentation quality. Interestingly, class-wise IoU analysis reveals that SAM Refinement significantly improved segmentation in *Cardboard*, *Paper*, and *PET* classes, which have complex shapes and varied manifestations. Figure 5 shows the five most improved classes, where *Cardboard* and *Paper* achieved the highest IoU gains of 21.50% and 8.12%, respectively, highlighting the effectiveness of SAM Refinement in enhancing segmentation accuracy.

The inference speed measurements in Table 2 demonstrate the computational cost introduced by the combination of ensemble techniques and SAM Refinement. While SDF-MR took 8.94 s/image, the SDF-SAM approach increased the inference speed to 19.15 s/image. Similarly, ML-SAM, which had the highest accuracy, required 10.63 s/image. Although iterative region-based refinement

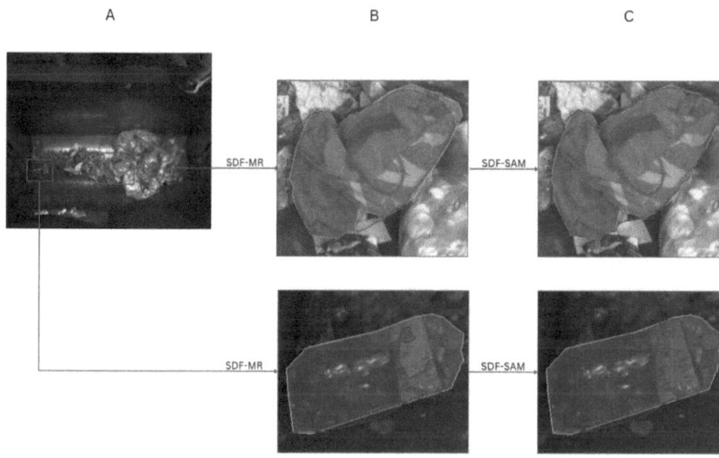

Fig. 3. Impact of SAM on boundary refinement in the SDF-MR strategy. Column A shows the input image with red boxes indicating selected areas. Columns B and C present zoomed-in predictions from the SDF-MR and SDF-SAM strategies, respectively, with blue overlays indicating predicted regions and yellow overlays indicating ground truth boundaries. (Color figure online)

using SAM introduces additional computational cost, the total inference time remains under 20 s/image, still significantly faster than manual labeling. This trade-off is well justified in annotation tasks where real-time performance is not required and high boundary precision is crucial, particularly in cluttered datasets like waste or plant imagery.

While fine-tuning SAM could potentially improve segmentation performance, it was deliberately employed in zero-shot mode to preserve generalizability, avoid task-specific dependencies, and reduce training complexity. This strategy aligns with the intended use of the framework in low-data annotation scenarios, where efficiency and minimal setup are critical, and enables rapid integration into scalable offline annotation workflows. Additionally, although various annotation-assistance frameworks have emerged in recent literature, they are not directly applicable in this context due to differing input modalities, task formulations, limited availability for benchmarking, or the requirement for extensive training or retraining. Given these considerations, the proposed pipeline emphasizes modularity and practicality by combining pre-trained semantic segmentation models with zero-shot SAM refinement. By eliminating the need for retraining, the pipeline remains lightweight and broadly applicable, which is particularly valuable in fast prototyping scenarios and practical deployment settings. The framework can also be integrated into annotation platforms such as CVAT [13], an open-source tool for image and video labeling, where the predicted masks serve as editable annotations and reduce manual correction effort.

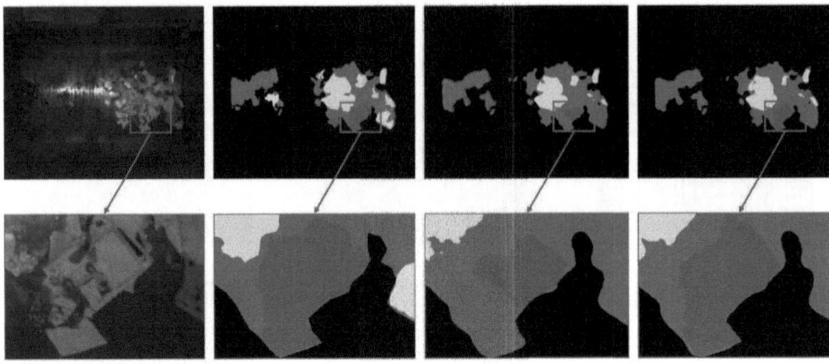

Fig. 4. Effect of SAM Refinement on segmentation using the ML strategy. The top row (left to right) shows the input image, ground truth, ML output, and its SAM-refined version. The bottom row presents zoomed-in views of the selected region. (Color figure online)

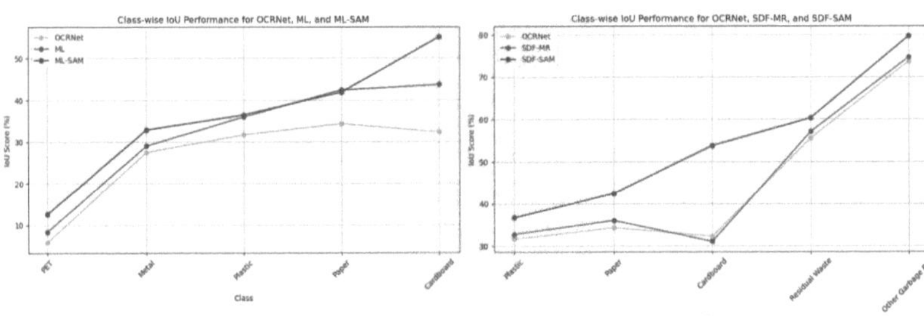

Fig. 5. IoU comparison of ML (left) and SDF (right) SAM Refinements with OCRNet. (Color figure online)

6 Conclusion

By integrating the advantages of three semantic segmentation models–UperNet, OCRNet, and DeepLabv3+–and SAM for refinement, this study presented a novel pipeline for effective semantic segmentation in waste recycling.

The pipeline improved accuracy and boundary precision using ensemble strategies such as Majority Vote, Max Logits, and SDF Mask Retrieval, with SAM-based refinement helping resolve ambiguities and improve object-level clarity, achieving high ACC (approximately 93%) and solid mIoU performance (50–52%). The ensemble methods outperformed individual models; Max Logits combined with SAM Refinement achieved the best mIoU of 52.65%. Despite the limited dataset size, the framework performed robustly and is well-suited for low-data scenarios, where pre-trained models and refinement strategies help minimize manual labeling. By streamlining the annotation process, it reduces human effort and facilitates high-quality dataset creation for waste management, making it suitable for offline workflows and potentially integrable with real-world labeling tools.

Challenges remain with underrepresented or visually similar classes, which could be addressed through refinement techniques, adaptive weighting, or targeted augmentation. While SAM and ensemble methods increase memory usage, this is justified by improved quality, and lighter variants or reduced models could help mitigate cost. This hybrid framework advances intelligent waste management and shows potential for generalization to other domains such as medical imaging and industrial automation. Future work will explore its scalability to larger or more diverse datasets. Active or semi-supervised learning could further reduce reliance on full annotations, supporting sustainable recycling.

References

1. Bogensperger, L., Narnhofer, D., Ilic, F., Pock, T.: Score-based generative models for medical image segmentation using signed distance functions. In: Pattern Recognition, pp. 3–17 (2024)
2. Buakum, B., Kosacka-Olejnik, M., Pitakaso, R., Srichok, T., Khonjun, S., Luesak, P., Nanthasamroeng, N., Gonwirat, S.: Two-stage ensemble deep learning model for precise leaf abnormality detection in centella asiatica. AgriEngineering **6**(1), 620–644 (2024) P., Nanthasamroeng, N., Gonwirat, S .: Two-stage ensemble deep learning model for precise leaf abnormality detection in centella asiatica. AgriEngineering **6**(1), 620–644 (2024)
3. Chen, K., Wang, J., Pang, J., et al.: MMDetection: open MMLab detection toolbox and benchmark. CoRR arXiv:1906.07155 (2019)
4. Chen, L.C., Zhu, Y., Papandreou, G., Schroff, F., Adam, H.: Encoder decoder with atrous separable convolution for semantic image segmentation. In: European Conference Computer Vision, pp. 833–851 (2018)
5. Choi, J., Lim, B., Yoo, Y.: Advancing plastic waste classification and recycling efficiency: integrating image sensors and deep learning algorithms. Appl. Sci. **13**(18), 10224 (2023)
6. Dang, T., Nguyen, T.T., McCall, J., Elyan, E., Moreno-García, C.F.: Two-layer ensemble of deep learning models for medical image segmentation. Cogn. Comput. **16**(3), 1141–1160 (2024)
7. He, K., Zhang, X., Ren, S., Sun, J.: Deep residual learning for image recognition. In: IEEE Conference Computer Vision Pattern Recognition, pp. 770–778 (2016)
8. Kaza, S., Yao, L.C., Bhada-Tata, P., Woerden, F.V.: What a Waste 2.0: a global snapshot of solid waste management to 2050. World Bank (2018)
9. Kirillov, A., et al.: Segment anything. In: International Conference Computer Vision, pp. 3992–4003 (2023)
10. Lema, D.G., Usamentiaga, R., García, D.F.: Enhancing automated inspection in metal industries: zero-shot segmentation of surface defects using bounding box prompts. Meas. Sci. Technol. **35**(8), 085604 (2024)
11. Mao, W.L., Chen, W.C., Wang, C.T., Lin, Y.H.: Recycling waste classification using optimized convolutional neural network. Resour. Conserv. Recycl. **164**, 105132 (2021)

12. Rasmussen, C.B., Kirk, K., Moeslund, T.B.: The challenge of data annotation in deep learning-a case study on whole plant corn silage. Sensors **22**(4), 1596 (2022)
13. Sekachev, B., Manovich, N., Zhiltsov, M., et al.: Computer Vision Annotation Tool (CVAT) (2024). https://doi.org/10.5281/zenodo.3497105
14. Sirimewan, D., Harandi, M., Peiris, H., Arashpour, M.: Semi-supervised segmentation for construction and demolition waste recognition in-the-wild: adversarial dual-view networks. Resour. Conserv. Recycl. **202**, 107399 (2024)
15. Waltner, G., Jaschik, M., Rinnhofer, A., Possegger, H., Bischof, H.: An intelligent scanning vehicle for waste collection monitoring. In: International Conference Image Anal. Process, pp. 38–50 (2022)
16. Weder, S., Blum, H., Engelmann, F., Pollefeys, M.: LabelMaker: automatic semantic label generation from RGB-d trajectories. In: International Conference 3D Vision, pp. 334–343 (2024)
17. Xiao, T., Liu, Y., Zhou, B., Jiang, Y., Sun, J.: Unified perceptual parsing for scene understanding. In: European Conference Computer Vision, pp. 432–448 (2018)
18. Yuan, Y., Chen, X., Wang, J.: Object-contextual representations for semantic segmentation. In: European Conference Computer Vision, pp. 173–190 (2020)

FSMDet: Vision-Guided Feature Diffusion for Fully Sparse 3D Detector

Tianran Liu(✉), Morteza Mousa Pasandi, and Robert Laganiere

VIVA Lab, Department of EECS, University of Ottawa, Ontario, Canada
{tliu157,mmous029,laganier}@uottawa.ca

Abstract. Fully sparse 3D detection has attracted an increasing interest in recent years. However, the sparsity of the features in these frameworks challenges the generation of proposals because of the limited diffusion process. In addition, the quest for efficiency has led to only a few works on vision-assisted fully sparse models. In this paper, we propose FSMDet (Fully Sparse Multi-modal Detection), which uses visual information to guide the LiDAR feature diffusion process while still maintaining the efficiency of the pipeline. Specifically, most fully sparse works focus on complex customized center fusion diffusion/regression operators. However, we observed that if the adequate object completion is performed, even the simplest interpolation operator leads to satisfactory results. Inspired by this observation, we split the vision-guided diffusion process into two modules: a Shape Recover Layer (SRLayer) and a Self Diffusion Layer (SDLayer). The former uses RGB information to recover the shape of the visible part of an object, and the latter uses a visual prior to further spread the features to the center region. Experiments demonstrate that our approach successfully improves the performance of previous fully sparse models that use LiDAR only and reaches SOTA performance in multimodal models. At the same time, thanks to the sparse architecture, our method can be up to 5 times more efficient than previous SOTA methods in the inference process.

Keywords: RGB-LiDAR fusion · Fully sparse 3D detection · Shape recovery

1 Introduction

The 3D detection of objects using LiDAR has made great progress in the past few years. However, the vast majority of dense detector, whether they be anchor-based [2,4,15,16,21,24,25,41] or anchor-free [36,42] works on dense Bird's Eye View (BEV) representations which are costly to process. For instance, for every doubling of the detection range, the computational cost will increase by 2–4 times [7]. To overcome this issue, fully sparse detectors have witnessed an increased interest in recent years. Different from point-based methods [22,23,26,27], which perform time-consuming point neighborhood aggregation, fully sparse models

E. Rodolà et al. (Eds.): ICIAP 2025, LNCS 16167, pp. 375–389, 2026.
https://doi.org/10.1007/978-3-032-10185-3_30

tend to introduce a voxel or a point-wise segmentation to significantly reduce the computation cost early in the network.

However, sparse models come also with their issues. Two main questions in fully sparse detectors are center feature missing and image signal fusion. Specifically, for the detectors that adopt a dense detection head on the BEV map, the normal pipeline first feeds the voxelized lidar signal to backbone, which always consists of a stack of submanifold sparse convolution layers [10]. However, this step won't change the occupancy of voxels. After the backbone, the z-axis(height) will be flattened to obtain the BEV map, which will still be sparse. Normally, the 2D CNN will process the BEV maps to diffuse the features to the empty voxels and then the detection head can generate the proposals for every non-empty voxel. The feature located in the voxel at the object's center will be the main candidate to regress the predicted box of the object. However, this method won't solve the issues created by missing voxel center feature in a fully sparse pipeline. Many fully sparse pipelines propose modules that vote the center of objects directly [7] or use the voted center as virtual voxels [8] to further aggregate features. However, this is a complex strategy that requires several hand-craft parameter settings which makes the model far from satisfactory in terms of deployment and speed of inference. It is worth noting that the recently proposed SAFDNet [37] aims at achieving to achieve this goal through a foreground-only feature diffusion approach, but this parameter-free operation makes it difficult to ensure not only the accuracy of the diffusion but also the quality of the diffused features.

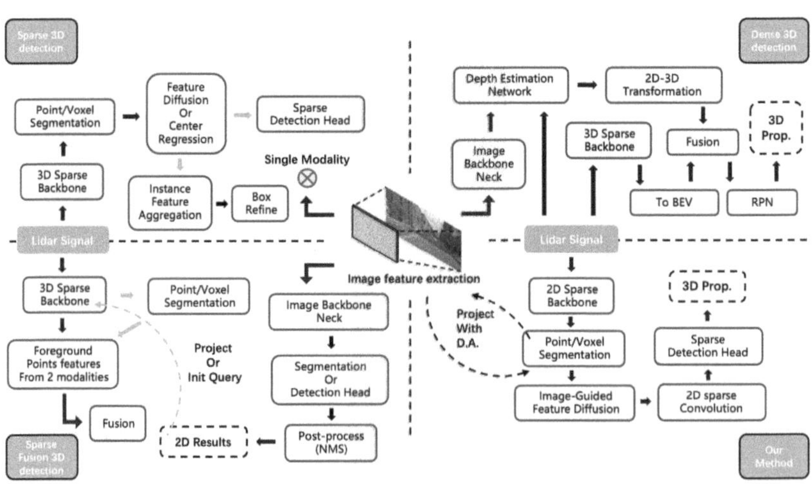

Fig. 1. A vision-centric view for LiDAR-RGB fusion in a fully sparse framework. Our proposed FSMDet integrates image features without any parameters for 2D tasks. The feature diffusion is guided by RGB features. Prop. and D.A. stand for proposals and deformable attention. The orange arrows represent the optional steps for models in the same categories.

In this work, we demonstrate that vision information can be used to guide the feature diffusion in a 3D detection framework. Only a few works [5,18,33] tried to introduce the RGB features into a fully sparse pipeline. However, most of them need a 2D segmentation or detection head to localize the objects from the image signal. These modules which include RPN and NMS operations limit the overall inference speed of the 3D detection framework. In other words, 2D detection is only intended to be used as an assist to the subsequent 3D detection results. Such a simple merged network tends to introduce a large amount of parameter redundancy. This inefficiency in RGB fusion approaches in fully sparse networks led us to revisit how to introduce RGB information while maintaining architectural sparsity. This paper introduces the FSMDet network (Fully Sparse Multi-modal Detection), a straightforward and efficient multi-modal 3D detector, that only keeps the necessary parameters for the image-branch and retains the sparsity of the structure.

As shown in Fig 1, our proposed method is as follows: the voxelized LiDAR features interact with RGB features through deformable attention operations to select the foreground RGB features. The foreground voxel features are diffused in 2 ways: using the shape-recovery layer (SRLayer) and the self-diffusion layer (SFLayer). Specifically, considering that the diffusion process changes the occupancy, we believe that the process of an ideal feature diffusion should essentially be like shape completion. From an idealized test, we experimentally proved that for a fully sparse model, even if there is no specific design for the center feature diffusion, we can still get nearly 100% accuracy with the completed objects. This conclusion can also be extended to most point-based methods. However, since the full shape of objects is not always available in the image, it is not a trivial task to make a network predicting the full shape of objects. Therefore, we further tested the visible part completion with color information. The result shows that this is even better for small objects with almost the same performance in other categories compared with the full-shape competition.

The experiment shows that a shape recovery operation at the early stage of the network is essential for accuracy. Inspired by this observation, the SRLayer is used at an early stage of the network to recover the shape of the visible part with features from the camera and the LiDAR modalities. The self-diffusion layer (SDLayer) is performed before the sparse detection head, which expands the feature at the boundary of objects to the center, thus solving the center feature estimation problem mentioned before.

Our proposed FSMDet network efficiently and effectively introduces visual information into a fully sparse detector: on a desktop GPU, without any CUDA kernel or deployment optimization, our method achieve a competitive accuracy comparing with SOTA fusion-based solutions with up to 5.3x speedup.

2 Related Work

Fully sparse 3D detection. A premise that has sometimes been overlooked in past 3D detection methods [4,19,24] is the different importance of differentiating foreground objects from the background which significant impact computing resources. FSD [6] first proposed a solution with a segmentation head to distinguish objects from background signals and a voting network to regress the center of objects. Subsequently FSDv2 [8] improves the performance by introducing virtual voxels and a multi-stage pipeline. Later, the FSD++ [9] utilized multi-frame information to improve the segmentation process. Considering the complexity of the pipeline, VoxelNeXt [3] adopt a relatively simple structure with extra down-sampling steps, which get better results with lower latency. The SAFDNet [37] proposed another strategy that used a 2D sparse convolution layer named AFD to diffuse the foreground lidar signal. Although most of the works achieved the SOTA performance, none of the mentioned works introduced multi-modal signal analysis into the fully sparse 3D detection pipeline.

Object Completion in 3D detection. The concept of object completion starts from lidar-based 3D detection: With a sub-network recovering the missing signal of distant or occluded objects, the performance of 3D detection can be boosted. Specifically, BtcDet [34] used cylinder coordination to voxelise the space and then try to recover the missing signal by predicting the occupancy of voxels. SPG [35] tries to use an unsupervised method to expand the foreground voxels. Both PG-RCNN [14] and PC-RGNN [38] tried to densify the foreground region after proposing the region in the first stage. Sparse2Dense [30] proposed a new way to complete the object in the hidden space by an explicit optimization. This same idea is behind many lidar-RGB fusion solutions. SFDNet [32] and VirConv [31] introduce dense pseudo points to complete the objects and the latter further proposed different modules to remove possible noise. Distinct from these 2 works, WYSIWYD [20] introduces a mesh deformation method that only completes the depth of foreground parts visible in the RGB image.

3 FSMDet

3.1 Preliminary

Previous works in fully sparse detection [7,8,37] attribute its performance flaw to center feature aggregation, and then proposed different methods to improve the performance of this module. In this section, we introduce a series of ideal experiments to show that another important factor that affects the performance of a sparse 3D detector is the density of the foreground objects point cloud representation. In this first experiment, our objective is to test the performance of sparse models with plain center feature aggregation modules under different lidar inputs. Note that when choosing the models to be tested, it is important to select ones that do not change the center voxels' occupancy. This way, the change in performance can be attributed to the different lidar inputs.

With a 3D ground truth box B_i of object i, we can obtain the lidar point set in this box, denoted by L_i. Using the full shape (FS) object point completion methods described in [20,34], we can obtain a densified object, denoted by $\widetilde{L_i}$, as shown in Fig 3. Our experiment, under ideal condition, consists then in completing the full shape of all ground-truth foreground objects in a scene. For this experiments, we use 3 classic models: IA-SSD [39], Point-RCNN [26], and FSD [7]. As shown in the FS line of Table 1, even for these models without a modern center feature aggregation module, good performance can be obtained if the objects are completed. This suggests that even the most basic center regression method is sufficient for well-completed objects to give us good results in 3D detection.

However, although full-shape completion can lead to a good results, it is still a non-trivial task to recover the full shape either from the image or the lidar signal. Therefore, we further use the obtained $\widetilde{L_i}$ to generate the visible part of objects, denoted by $\hat{L_i}$. Specifically, for every single L_i in the dataset, a unique 2D instance I_i on the image can be identified. First, we adopted a surface reconstruction from the Possion [13] method, and then placed the obtained hull(reconstructed surface), denoted by $\mathcal{M}_i$ in 3D space. Based on the known camera intrinsic matrix, extrinsic matrix, and the pixel 2D coordinates of pixels in the region I_i, a ray casting model can be established. Note that since the rays from the pixel on the boundary sometimes miss the object, a volume expansion coefficient δ is used here to make sure there is always a bijection from the pixel set to the depth set. This process is illustrated in Fig. 2.

Table 1. The performance of different sparse models with completed objects on the KITTI validation set. FS and VP stand for full shape and Visible Part completion respectively.

Models	Comp	Car 3D AP_{R40}			Car BEV AP_{R40}			Ped 3D AP_{R40}			Ped BEV AP_{R40}		
		Easy	Mod.	Hard	Easy	Mod.	Hard	Easy	Mod.	Hard	Easy	Mod.	Hard
PointRCNN	VP	99.9	99.6	97.2	99.9	99.6	97.2	85.3	77.9	86.2	68.1	80.9	71.2
	FS	97.2	97.2	97.1	97.2	97.2	97.1	77.8	72.8	70.3	80.3	74.8	71.1
IA-SSD	VP	99.9	99.7	99.4	99.9	99.3	99.3	75.5	71.3	66.4	77.5	75.1	70.3
	FS	99.7	99.4	99.4	99.6	99.2	99.4	71.9	69.7	86.2	74.8	74.3	72.8
FSD-V1	VP	99.9	99.7	99.7	99.9	99.7	99.7	87.2	80.1	72.1	89.9	82.7	73.6
	FS	99.9	99.9	99.9	99.9	99.9	99.9	81.7	77.9	71.1	83.3	77.1	72.9

With the equation of line T_c and $\mathcal{M}_i$, the distance traveled by a ray of light from point c when it arrives on the mesh can be expressed as $|\delta\mathcal{M}_i \cap T_c - c_i|$. Considering the slope of T_C, the depth of the point on the mesh hit by the light to the camera plane is d_i. Here $\cdot$ refers to scalar multiplication.

$$d_c = |\delta\mathcal{M}_i \cap T_c - c_i| \cdot |\boldsymbol{x_c}| \tag{1}$$

After estimating the depth of every pixel in the region I_i, we get a dense visible surface of the object, denoted by $\hat{L_i}$, as shown in the left most sub-figure of Fig 3. Replacing the original $\widetilde{L_i}$ with $\hat{L_i}$, we repeated the same experiment on the selected models in VP lines of Table 1, show that even with only visible part completion, the upper boundary of these models is still high enough. A

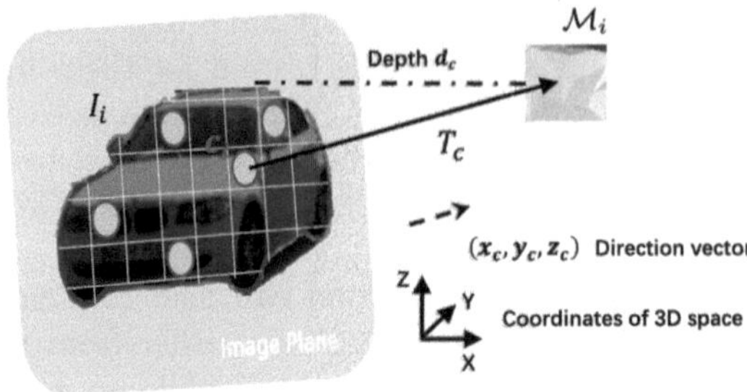

Fig. 2. The ray casting model for visible part ground truth generation. All calculations are performed in LiDAR (3D space) coordination. Here c_i stand for the location of pixel c in 3D space.

counter-intuitive fact that also needs to be mentioned is that VP is even more suitable for small objects such as pedestrians.

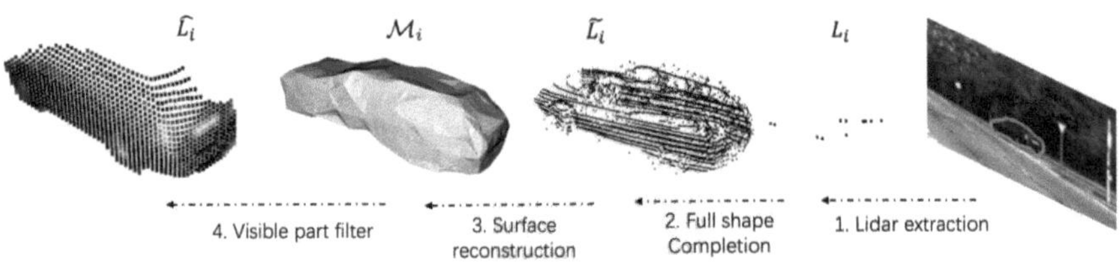

Fig. 3. From sparse lidar signal to visible part ground truth: we first obtain the full shape objects and then reconstruct the surface. With a ray-cast model, we can paint the color information onto the object's surface and filter out the non-visible part.

During the training, the completion of objects causes a change of occupancy, or in other words, it diffuses the point features. This experiment shows that the detection results are satisfactory when the object shapes are adequately completed, even with a vanilla design for the center regression or center feature aggregation modules. This inspired us to split the feature diffusion into two steps: the shape recovery diffusion and the diffusion of features to the object center. Since only the visible part of objects is available on the image, here we use VP to supervise the first-step diffusion.

3.2 Overall Structure

In Sect. 3.1, we experimentally demonstrated that the visible part completion of the foreground object in the LiDAR signal improve the detection results despite the diffculty of center prediction. Based on this result, we introduce the shape recovery module, which diffuses the features to occupy near voxels, at an early layer of the network. Our experiments will also demonstrate that it's beneficial to do this step at an early stage.

Specifically, as shown in Fig 4, we use SRLayer (Shape Recover Layer) and SDLayer (Self Diffusion Layer) to do the two steps mentioned respectively. After the voxelization of the space, Submanifold sparse convolution [11] with residual design was used as the basic block to process the LiDAR data. As in work [37], we call these stacked sparse convolution block **SRB** (Sparse Residual Block). We choose $\frac{1}{4}$ of the original size to perform shape recovery and color features aggregation as a good balance between efficiency and performance. With the coordinates of points centroid in the different voxels and the calibration information, we fuse the voxel and image features with a deformable attention operator and replace the original feature. These augmented features will be further process with SRBs, and then a voxel classification head will be applied to get the foreground parts of the scene. The fused features and the foreground masks are used as the input of our SRLayer. SRLayers use 2D VP object completion as the supervision signal.

The output of SRLayers will go through the SRBs for further downsampling. Finally, the BEV transformation squeezes the feature map to get a 2D representation of the space, the SDLayer will be used to deliver the feature to the center. The output size of the proposed cross-modality backbone will be $\frac{1}{8}$ of the width and length of the original input.

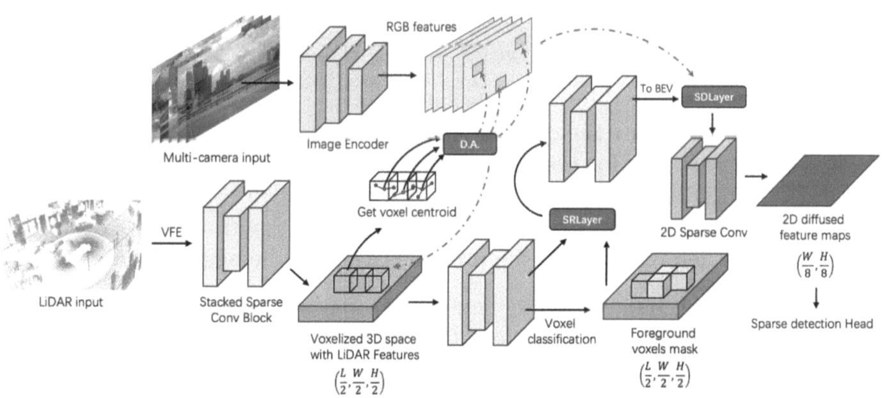

Fig. 4. Proposed pipeline: We complete the diffusion of features in two steps, the SRLayer uses RGB features to recover the visible part of the contours of objects while the SDLayer diffuses the features to the center of objects. D.A. stand for the deformable attention. The operation related to color features is marked in red.

3.3 Proposed Method

Deformable attention for feature fusion. Let $V = \{V_i, V_{f_i}, c_i\}_{i=1}^{|V|}$ represent all of the non-empty voxels in a sparse feature map, V_i is the indices of the voxel i, V_{f_j} stand for the features in the voxel and c_i is the centroid of the LiDAR points in this voxel. With the camera intrinsic and extrinsic matrix, the 3D-2D transformation can be denoted as $\mathcal{P}$, to obtain the 2D-pixel coordinates $\mathbf{p}_i$, as shown in Eq. 2.

$$\mathbf{p}_i = \mathcal{P}(c_i) \tag{2}$$

Normally, we will interpolate the image features to the original size and acquire the image feature for points in 2D coordinates. However, as pointed out in previous lidar segmentation work [40], the calibration of different modalities is not always reliable. Therefore the correspondency between semantic information in down-sampled image feature maps and voxel centroid point is not guaranteed. For this reason, we introduce deformable attention in this step to allow the network to freely select features near $\mathbf{p}_i$, following the design of LoGoNet [17] with a minor modification.

Specifically, V_{f_i} will be used as query of the attention calculation. The image features sampled from the pixels around $\mathbf{p}_i$, denoted by $\hat{R}_i$, will be used as Key and Value. The image features located at $\mathbf{p}_i$ and all of its 3-order neighbour will be noted as $R_{\mathbf{p}_i}$ and $R^3_{\mathbf{p}_i}$, with the image map itself denoted by R. With M attention head and K sample points, we can obtain $\hat{R}_i$ from Eq. 3.

$$\begin{aligned} \Delta\mathbf{p}_{mki} &= MLP(MLP(R^3_{\mathbf{p}_i}) \cdot V_{f_i}) \\ R_{mi} &= \mathcal{G}(R, \mathbf{p}_i + \Delta\mathbf{p}_{mki}) \\ \hat{R}_i &= \sum_{m=1}^{M} W_m \left[\sum_{k=1}^{K} A_{mik} \cdot (W'_m R_{mi}) \right] \end{aligned} \tag{3}$$

The $\mathcal{G}(a, b)$ represents the operation of obtaining the features of a at location b. The W'_m, W_m stands for the trainable weight, and A_{mik} stands for attention weight. In the original design, the $\Delta\mathbf{p}_{mki}$ was only decided by the LiDAR feature V_{f_i}, however, here we add an interaction between V_{f_i} and all features around $R_{\mathbf{p}_i}$, which will help to identify the proper features to be aggregated.

Shape Recover Layer. The augmented features will replace the original one and is sent to the next SRBs. Note that at this stage, we keep the original feature map size. The obtained map is then sent to a classification head to split the foreground voxels from the background information. Considering the large space difference between the various classes of objects, we set a category number $\mathcal{S}$ for the classification. Using $\mathcal{M}$ to represent the index of foreground voxels, the $\mathcal{M}$ and processed voxels $V' = \left\{V_i, V'_{f_i}, c_i\right\}_{i=1}^{|V|}$ will be sent to the SRLayer.

With the foreground voxels that are denoted by $V'_o = \{V_i, V'_{f_i}, c_i | V_i \in \mathcal{M}\}_{i=1}^{|\mathbf{M}|}$, we first need to calculate the possible direction of diffusion. A possible direction should ensure that the voxel in that direction is visible on the image and secondly that the next two voxels in that direction are not occupied. As shown in the left part of Fig 5, all voxels in V'_o will have 4 candidate directions to diffuse, but here only 2 of them are possible for diffusion (marked with red). Specifically, for a

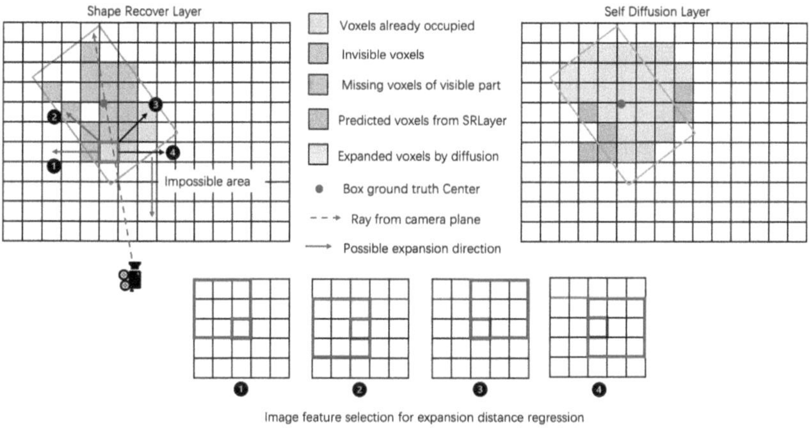

Fig. 5. Illustration of shape recover layer and self diffusion layer. We use the sparse signal to recover the shape of objects. After Shape recovery, we expand the boundary voxels along the ray direction, the center point will be naturally located in this area.

specific voxel in V_o', denoted by $\left\{V_t, V_{f_t}', c_t\right\}$, which can be projected to a pixel located at $\mathbf{p}_t$, we use $V_\varnothing$ to denote all invisible voxels. The possible diffusion direction J_t can be described in Eq. 4. Here V_j is the nearest non-empty voxels on direction J_t.

$$J_t = \{J_{ti} | J_{ti} \cap V_\varnothing, \|V_t - V_j\| < 2\}_{i=1}^{4} \quad (4)$$

For every J_{ti}, we use V_{f_t}' and the direction-specific image features (as shown in the bottom of Fig. 5), denoted by $R_{\mathbf{p}_t}^i$, to regress the expand distance d_t. For the empty voxel in range d_t alone the direction J_t, denoted by $\left\{V_{t+d}, V_{f_{t+d}}', c_{t+d}\right\}$, the $V_{f_{t+d}}'$ will be obtain by Eq. 5.

$$V_{f_{t+d}}' = MLP\Bigg(\bigcup_{\mathbf{N}(p_{t+d},d)} R_{\mathbf{p}_{t+d}}^i\Bigg) \quad (5)$$

Here $\mathbf{N}(x, l)$ and $\bigcup$ stand for the l-order neighborhood of pixel x and the concatenation operation of the neighboring features. With the new occupied voxels, the V' will be sent to the next SRB to process the features further.

The Shape Recover Layer needs supervision from the generated VP ground truth. Considering the computational complexity, we only apply the supervision on the BEV plane. Specifically, after voxelizing the VP ground truth, we can get the ground truth for the shape recovery, denoted by V_{gt}. Using $\hat{V}$ to represent the score of predicted recovery voxels, the loss can be written as Eq. 6.

$$\mathcal{L} = \sum_{i=1}^{\mathbf{S}} \alpha_j \sum_{j=1}^{3} focal\_loss(\hat{V}_{ij}, V_{gt}) \quad (6)$$

Note that the boundary voxels only make up a small portion. Here we use a multi-class focal loss to calculate if the predicted occupied voxel cannot cover the boundary ground truth. If it is located in the 2D box, it won't be treated as negative samples. Here we further split the voxel predicted for different size objects into 3 categories: the predicted voxels located at the boundary correctly, the predicted voxels not situated at the boundary but in the box, and the others. Specific α_j will be used to scale these different cases here.

Self Diffusion Layer. Before being sent to the detection head, the sparse signal needs to be further diffused to the center of objects. Note that since we have identified the invisible voxels $V_{\varnothing}$, we can directly diffuse the features along the ray across the occupied voxels. As we have shown in the right part of Fig. 5, in most cases, the center can be naturally covered. Considering the size of different objects, the scope parameter σ_0 of this diffusion depends on the result of the classification head. Following SAFDNet [37], we also adopt an adaptive method in which for larger objects, the diffusion distance alone ray casting will be larger than that of small objects. However, when the objects are occluded, the diffusion along the ray-casting direction will miss the center. Therefore, considering the size of voxels here, we also set a parameter σ_1 to control the diffusion of features perpendicular to the direction of the ray casting.

4 Experiments

4.1 Dataset and Metrics

We conducted experiments on the nuScenes [1] dataset for a comparison. The nuScenes dataset was collected with 6 cameras, a radar sensor, and a 32-beam LiDAR sensor under different weather and illumination conditions. The model proposed was evaluated by the official evaluation script with NDS and MAP as index.

4.2 Implementation Details

We used the OpenPCDet [28] as the framework to implement the mentioned design. All experiments were performed on RTX3090, we used 8 as batch size on 4 GPUs for training and 1 batch size for inference speed test. Training in all experiments lasted for 20 epochs. During training, in addition to random rotation and translation of the point cloud signals and random cropping and normalization of the image signals, we change the GT-sampling commonly used in the lidar-based model to the multimodal GT-sampling proposed in PointAugmenting [29]. We use ResNet [12] pre-trained on NuImage as the image backbone. The volume expansion parameter δ was set to 1.15 to ensure all rays from every pixel wouldn't miss the object. d in Eq. 5 was set to 2. Considering the size of different objects occupied, we set $\sigma_0 = 6$ and $\sigma_1 = 4$. We set the $\alpha_0 = \alpha_1 = 0.5$, $\alpha_2 = 1$ to scale the loss mentioned in Eq. 6 (Fig. 2).

4.3 Main Results

We compared the inference speed with recent SOTA models. Since we did not need any parameters for tasks on the image plane, our proposed method takes only 25% (as a minimum) of the time the SOTA solutions use to take while maintaining a decent detection performance. As shown in Tables 2 and 3, our proposed method obtained 70.1 mAP and 72.1 NDS in nuScenes Dataset. This performance is only 2.4% to 3.8% less than the SOTA but we only used 18% to 47% of the inference time that is needed for SOTA models.

Table 2. Performance comparsion with SOTA models on **nuScences validation set**. The best results are shown in bold. ↑ higher is better, ↓ lower is better.

Methods	mAP↑	NDS↑	mATE↓	mASE↓	mAOE↓	mAVE↓	mAAE↓
FSDv2	65.4	70.8	0.270	0.248	0.272	0.207	0.187
SAFDNet	68.3	72.3	0.251	0.242	0.311	0.258	0.127
SparseFusion	71.0	73.1	0.277	0.247	0.270	0.253	0.188
MSMDFusion	69.2	72.0	0.283	0.254	0.282	0.252	0.185
BEVFusion	67.3	70.7	0.286	0.255	0.313	0.251	0.186
FSMDet	70.0	72.2	0.285	0.251	0.296	0.254	0.179

Table 3. Efficiency comparison, our proposed model achieves a balance point between inference speed and performance.

Model	Modality	Sparsity	Image result	FPS	mAP	NDS
SAFDNet	LiDAR	✓	N/A	15.7	68.3	72.3
SparseFusion	LiDAR+RGB	✓	✓	5.3	71.0	73.1
MSMDFusion	LiDAR+RGB	×	×	2.1	71.4	73.9
BEVFusion	LiDAR+RGB	×	×	8.1	67.3	70.7
FSMDet	LiDAR+RGB	✓	×	11.2	70.0	72.2

4.4 Ablation Study

In Table 4, we break down the performance boost in the proposed model. Here the D.A Color and Proj Color stand for the two different methods to introduce RGB color features: the deformable attention or direct projection. As shown in the table, the parameter-based solution benefits from the contribution of these models. In experiments on Table 4, if we do not apply the SDLayer, the ADF in SAFDNet [37] is used.

Table 4. The verification of the effect of each module on the model on the final accuracy

Models	D.A Color	Proj Color	SRLayer	SDLayer	mAP	NDS
FSMDet	✓	–	–	–	67.9	70.4
FSMDet	-	✓	–	–	67.6	70.8
FSMDet	✓	–	✓	–	69.8	71.5
FSMDet	✓	–	✓	✓	70.0	72.2

Here in Table 5, we compare the effect of different positions of the SRLayer on accuracy. If we insert the shape recovery operation too early, the shallow representation of LiDAR features can barely help the expansion regress. On the other hand, if the SRLayer is too close to the SDLayer, the feature of the newly occupied layer also cannot aggregate the nearby features very well.

Table 5. Different stages to insert the shape recover layer

Modules	Block-1	Block-2	Block-3	Block-4	mAP	NDS
SRLayer	✓	–	–	–	61.2	63.6
SRLayer	–	✓	–	–	70.0	72.2
SRLayer	–	- -	✓	–	68.7	70.9
SRLayer	–	–	–	✓	63.1	67.2

5 Conclusion

In this work, we proposed a cross-modality method for the RGB-LiDAR fusion in a fully sparse framework. Starting from the idealized experiment which reveals that the core of center regression for detection models should be the visible part completion, we designed the SRLayer to achieve this step with the guidance from the image features. This way, the diffusion alone of the ray from the image plane can naturally cover the center of objects in most cases. As the experiments prove the potential of our method, we hope the FSMDet can promote the research in data fusion for fully sparse models.

References

1. Caesar, H., Bankiti, V., Lang, A.H., Vora, S., Liong, V.E., Xu, Q., Krishnan, A., Pan, Y., Baldan, G., Beijbom, O.: nuScenes: a multimodal dataset for autonomous driving (2020). arxiv:1903.11027
2. Chen, Y., Li, Y., Zhang, X., Sun, J., Jia, J.: Focal sparse convolutional networks for 3D object detection. In: Proceedings of the IEEE/CVF Conference on Computer Vision and Pattern Recognition, pp. 5428–5437 (2022)

3. Chen, Y., Liu, J., Zhang, X., Qi, X., Jia, J.: Voxelnext: fully sparse voxelnet for 3d object detection and tracking. In: Proceedings of the IEEE/CVF Conference on Computer Vision and Pattern Recognition, pp. 21674–21683 (2023)
4. Deng, J., Shi, S., Li, P., Zhou, W., Zhang, Y., Li, H.: Voxel R-CNN: towards high performance voxel-based 3D object detection. In: Proceedings of the AAAI Conference on Artificial Intelligence, vol. 35, pp. 1201–1209 (2021)
5. Erabati, G.K., Araujo, H.: Srfdet3d: sparse region fusion based 3D object detection. Neurocomputing **593**, 127814 (2024). https://doi.org/10.1016/j.neucom.2024.127814
6. Fan, L., Wang, F., Wang, N., Zhang, Z.X.: Fully sparse 3D object detection. Adv. Neural. Inf. Process. Syst. **35**, 351–363 (2022)
7. Fan, L., Wang, F., Wang, N., Zhang, Z.: Fully sparse 3D object detection (2022). arxiv:2207.10035
8. Fan, L., Wang, F., Wang, N., Zhang, Z.: Fsd v2: improving fully sparse 3D object detection with virtual voxels (2023). arXiv preprint arXiv:2308.03755
9. Fan, L., Yang, Y., Wang, F., Wang, N., Zhang, Z.: Super sparse 3D object detection. IEEE Trans. Pattern Anal. Mach. Intell. **45**(10), 12490–12505 (2023)
10. Graham, B., Engelcke, M., van der Maaten, L.: 3D semantic segmentation with submanifold sparse convolutional networks. In: CVPR (2018)
11. Graham, B., van der Maaten, L.: Submanifold sparse convolutional networks (2017). arxiv:1706.01307
12. He, K., Zhang, X., Ren, S., Sun, J.: Deep residual learning for image recognition. In: Proceedings of the IEEE Conference on Computer Vision and Pattern Recognition, pp. 770–778 (2016)
13. Kazhdan, M., Bolitho, M., Hoppe, H.: Poisson surface reconstruction. In: Proceedings of the Fourth Eurographics Symposium on Geometry Processing. vol. 7 (2006)
14. Koo, I., Lee, I., Kim, S.H., Kim, H.S., Jeon, W.J., Kim, C.: Pg-RCNN: semantic surface point generation for 3D object detection. In: Proceedings of the IEEE/CVF International Conference on Computer Vision, pp. 18142–18151 (2023)
15. Lang, A.H., Vora, S., Caesar, H., Zhou, L., Yang, J., Beijbom, O.: Pointpillars: fast encoders for object detection from point clouds. In: Proceedings of the IEEE/CVF Conference on Computer Vision and Pattern Recognition, pp. 12697–12705 (2019)
16. Li, J., Luo, C., Yang, X.: Pillarnext: rethinking network designs for 3D object detection in lidar point clouds (2023). arxiv:2305.04925
17. Li, X., Ma, T., Hou, Y., Shi, B., Yang, Y., Liu, Y., Wu, X., Chen, Q., Li, Y., Qiao, Y., He, L.: Logonet: towards accurate 3D object detection with local-to-global cross-modal fusion (2023). arxiv:2303.03595
18. Li, Y., Fan, L., Liu, Y., Huang, Z., Chen, Y., Wang, N., Zhang, Z.: Fully sparse fusion for 3D object detection. IEEE Trans. Pattern Anal. Mach, Intell (2024)
19. Liang, T., Xie, H., Yu, K., Xia, Z., Lin, Z., Wang, Y., Tang, T., Wang, B., Tang, Z.: Bevfusion: a simple and robust lidar-camera fusion framework. Adv. Neural. Inf. Process. Syst. **35**, 10421–10434 (2022)
20. Liu, T., Zhang, Z., Pasandi, M.M., Laganiere, R.: What you see is what you detect: towards better object densification in 3D detection (2023). arxiv:2310.17842
21. Mao, J., Xue, Y., Niu, M., Bai, H., Feng, J., Liang, X., Xu, H., Xu, C.: Voxel transformer for 3D object detection. In: Proceedings of the IEEE/CVF International Conference on Computer Vision, pp. 3164–3173 (2021)
22. Qi, C.R., Litany, O., He, K., Guibas, L.J.: Deep Hough voting for 3D object detection in point clouds. In: Proceedings of the IEEE International Conference on Computer Vision (2019)

23. Qi, C.R., Yi, L., Su, H., Guibas, L.J.: Pointnet++: deep hierarchical feature learning on point sets in a metric space (2017). arxiv:1706.02413
24. Shi, S., Guo, C., Jiang, L., Wang, Z., Shi, J., Wang, X., Li, H.: Pv-RCNN: point-voxel feature set abstraction for 3d object detection. In: Proceedings of the IEEE/CVF Conference on Computer Vision and Pattern Recognition, pp. 10529–10538 (2020)
25. Shi, S., Jiang, L., Deng, J., Wang, Z., Guo, C., Shi, J., Wang, X., Li, H.: Pv-RCNN++: point-voxel feature set abstraction with local vector representation for 3D object detection. Int. J. Comput. Vision **131**(2), 531–551 (2023)
26. Shi, S., Wang, X., Li, H.: Pointrcnn: 3D object proposal generation and detection from point cloud (2019). arxiv:1812.04244
27. Shi, W., Ragunathan, R.: Point-GNN: graph neural network for 3D object detection in a point cloud (2020). arxiv:2003.01251
28. Team, O.D.: Openpcdet: an open-source toolbox for 3D object detection from point clouds (2020). https://github.com/open-mmlab/OpenPCDet
29. Wang, C., Ma, C., Zhu, M., Yang, X.: Pointaugmenting: cross-modal augmentation for 3D object detection. In: Proceedings of the IEEE/CVF Conference on Computer Vision and Pattern Recognition (CVPR), pp. 11794–11803 (2021)
30. Wang, T., Hu, X., Liu, Z., Fu, C.W.: Sparse2dense: learning to densify 3D features for 3D object detection. Adv. Neural. Inf. Process. Syst. **35**, 38533–38545 (2022)
31. Wu, H., Wen, C., Shi, S., Wang, C.: Virtual sparse convolution for multimodal 3D object detection. In: CVPR (2023)
32. Wu, X., Peng, L., Yang, H., Xie, L., Huang, C., Deng, C., Liu, H., Cai, D.: Sparse fuse dense: towards high quality 3D detection with depth completion. In: Proceedings of the IEEE/CVF Conference on Computer Vision and Pattern Recognition, pp. 5418–5427 (2022)
33. Xie, Y., Xu, C., Rakotosaona, M.J., Rim, P., Tombari, F., Keutzer, K., Tomizuka, M., Zhan, W.: Sparsefusion: fusing multi-modal sparse representations for multi-sensor 3D object detection. In: Proceedings of the IEEE/CVF International Conference on Computer Vision, pp. 17591–17602 (2023)
34. Xu, Q., Zhong, Y., Neumann, U.: Behind the curtain: learning occluded shapes for 3D object detection (2021). arXiv preprint arXiv:2112.02205
35. Xu, Q., Zhou, Y., Wang, W., Qi, C.R., Anguelov, D.: SPG: unsupervised domain adaptation for 3D object detection via semantic point generation. In: Proceedings of the IEEE/CVF International Conference on Computer Vision, pp. 15446–15456 (2021)
36. Yin, T., Zhou, X., Krahenbuhl, P.: Center-based 3D object detection and tracking. In: Proceedings of the IEEE/CVF conference on computer vision and pattern recognition, pp. 11784–11793 (2021)
37. Zhang, G., Chen, J., Gao, G., Li, J., Liu, S., Hu, X.: SAFDNet: a simple and effective network for fully sparse 3D object detection (2024)
38. Zhang, Y., Huang, D., Wang, Y.: Pc-RGNN: point cloud completion and graph neural network for 3D object detection. In: Proceedings of the AAAI Conference on Artificial Intelligence, vol. 35, pp. 3430–3437 (2021)
39. Zhang, Y., Hu, Q., Xu, G., Ma, Y., Wan, J., Guo, Y.: Not all points are equal: learning highly efficient point-based detectors for 3D lidar point clouds (2022). arxiv:2203.11139
40. Zhang, Z., Zhang, Z., Yu, Q., Yi, R., Xie, Y., Ma, L.: Lidar-camera panoptic segmentation via geometry-consistent and semantic-aware alignment (2023). arxiv:2308.01686

41. Zhou, Y., Tuzel, O.: Voxelnet: end-to-end learning for point cloud based 3D object detection. In: Proceedings of the IEEE Conference on Computer Vision and Pattern Recognition, pp. 4490–4499 (2018)
42. Zhou, Z., Zhao, X., Wang, Y., Wang, P., Foroosh, H.: Centerformer: center-based transformer for 3D object detection (2022). arxiv:2209.05588

Efficient Odd-One-Out Anomaly Detection

Silvio Chito, Paolo Rabino(✉), and Tatiana Tommasi

Politecnico di Torino, Turin, Italy
{silvio.chito,paolo.rabino,tatiana.tommasi}@polito.it

Abstract. The recently introduced odd-one-out anomaly detection task involves identifying the odd-looking instances within a multi-object scene. This problem presents several challenges for modern deep learning models, demanding spatial reasoning across multiple views and relational reasoning to understand context and generalize across varying object categories and layouts. We argue that these challenges must be addressed with efficiency in mind. To this end, we propose a DINO-based model that reduces the number of parameters by one third and shortens training time by a factor of three compared to the current state-of-the-art, while maintaining competitive performance. Our experimental evaluation also introduces a Multimodal Large Language Model baseline, providing insights into its current limitations in structured visual reasoning tasks. The project page can be found at https://silviochito.github.io/EfficientOddOneOut/.

Keywords: Anomaly Detection · Multi-view Analysis · 3D learning

1 Introduction

The concept of anomaly is both fundamental and inherently ambiguous. While in general anomalies signal critical deviations from expected patterns, their very definition is context-dependent as the singularity of a sample hinges on the framework in which it occurs. In the computer vision literature, anomaly detection and segmentation have been formalized mainly focusing on semantic novelty with the identification of previously unseen categories [1,10], or searching for defects as localized structural variations from standard reference data [9,11,23,48]. However, the concept of anomaly is broader and recent works have started to investigate logical inconsistencies [4,20] and to consider multi-modal frameworks including depth [34,46] and language [17,19]. Still, most of these works address settings with isolated objects, while realistic applications usually involve scenes with multiple objects in complex layouts and call for anomaly detection methods capable of handling clutter while dealing with multi-view analysis to elaborate on 3D object geometry and appearance. Moreover, every single scene may individually drive its own meaning of normalcy on the basis of cross-object comparisons and relative anomaly criteria. This requires models

E. Rodolà et al. (Eds.): ICIAP 2025, LNCS 16167, pp. 390–402, 2026.
https://doi.org/10.1007/978-3-032-10185-3_31

with an advanced level of relational reasoning to detect the *odd-looking* instances while generalizing to scenes including objects (both categories and instances) not observed during training.

To push research in the described challenging direction and closer to real world requirements, a new framework has been recently introduced which formalizes anomaly detection as an odd-one-out task [7]. Given multi-view images, this problem consists of identifying the samples in a scene that deviate from the others. It is inspired by standard tests in cognitive science, often used to assess intelligence and to provide valuable insights into an agent's ability to manage high-level understanding of perceptual similarity.

When designing new models to solve this problem, it is crucial to prioritize efficiency and avoid unnecessarily complex architectures. Indeed, such multi-object scenarios including outliers are common in industrial production, where minimizing inference time is critical, along with reducing re-training efforts by improving generalization. In this work we align with these objectives by introducing streamlined approaches that leverage the internal representations and extensive semantic knowledge of pre-trained large-scale models. Our contributions are summarized as follows:

- we explore the use of Multimodal Large Language Models (MLLM) to tackle the anomaly detection task in the odd-one-out setting. We present a pipeline to define a baseline approach and assess its performance;
- building on previous work [7], we show how the self-supervised representation extracted by DINOv2 can be used in a direct and effective manner by introducing an approach that reduces by one third the number of model parameters and cuts training time by a factor of three while incorporating a module that enhances relational reasoning and advances state-of-the-art results.

2 Related Works

The elusive nature of the anomaly concept has fostered an extensive body of research, aiming to frame it in a wide range of settings accompanied by CNN and Transformer-based solutions. Given access to annotated outlier data, the problem of anomaly detection can be formalized as a supervised task [30]. However, anomalies are generally rare: despite some efforts to define image collections of industrial objects with labeled scratches, dents, or missing parts [3], their annotation is costly and anyway limited, leading to models that may not be proficiently used in practice. Indeed, most of the literature has focused on unsupervised approaches that have access only to normal samples during training, either in abundance or in limited quantity, with the latter setting known as few-shot anomaly detection [27,43]. Usually the problem is restricted to a binary task, defining separate models for each object category [15,18], but in the last years the literature has been moving towards a more natural unified multi-class framework [13,48]. Over time, research has expanded beyond early approaches based on pixel reconstruction which assume that a network trained on normal images can accurately reconstruct anomaly-free regions while failing on anomalous areas [32,47]. This strategy has been extended to feature reconstruction [15,31] and knowledge distillation [5,36], under the hypothesis that a student network trained to mimic a teacher exposed only to normal data will underperform on anomalies. More recent works leverage the rich representations of large-scale pre-trained vision models by computing feature distances [12,34], though these methods often involve computationally expensive memorization and matching steps. Another line of research revisits the supervised setting by synthesizing anomalies through noise injection, image mixing [22], or generative models [25,38]. Finally, the rise of large-scale multimodal language-vision models has enabled zero-shot anomaly detection via tailored prompting and the use of textual descriptions [17,19].

The overview provided above covers literature focusing on 2D data. Anomaly detection has been mainly studied on images despite real world applications clearly need 3D spatial reasoning. The MVTech3D dataset was one of the first proposed for 3D anomaly benchmarks [6] and has been followed by a few others with growing cardinality and resolution as well as different 3D formalization (point cloud, multi-view) [24,41]. Still, they all capture isolated instances and the proposed methods inherit standard strategies from the 2D literature. Only recently this niche has been extended by considering different settings such as semantic novelty detection to recognize unknown categories [2,33], conditional anomaly detection to compare a query image to a reference shape [8], and odd-one-out contextual anomaly detection involving multi-object scenes [7].

The method introduced in [7] employs a 2D CNN architecture to extract visual features, which are then projected into 3D space to represent the observed instances. These representations are subsequently mapped back to the 2D domain to align with the internal feature space of DINOv2 [29], enabling a distillation strategy that captures object geometry while injecting open-world knowledge. However, this two-step mapping introduces unnecessary complexity, which could be mitigated by directly extracting DINOv2 features from the multi-view input images. In the final stage, the method compares object-centric feature volumes by performing voxel-level part matching. We argue that this process can be streamlined and integrated with a strategy that explicitly encodes the relative distinctiveness of each instance with respect to the average normalcy observed in the scene. We detail our proposed approach in the following section.

3 Method

We formalize the odd-one-out task by following [7] and considering a scene containing N instances of an object category $o_{i=1}^{N}$. The goal is to identify which of these instances is odd, meaning that they present significant individual variations from the others that instead share a high visual similarity. We observe the scene from M points of view corresponding to $\mathbf{I}_{t=1}^{M}$ RGB images with their camera projection matrices $\mathbf{P}_t$. Starting from these inputs and knowing the object locations in the scene, we need a model able to encode the appearance of separate instances and assign them a binary label $y_i \in \{0, 1\}$, where 1 identifies the

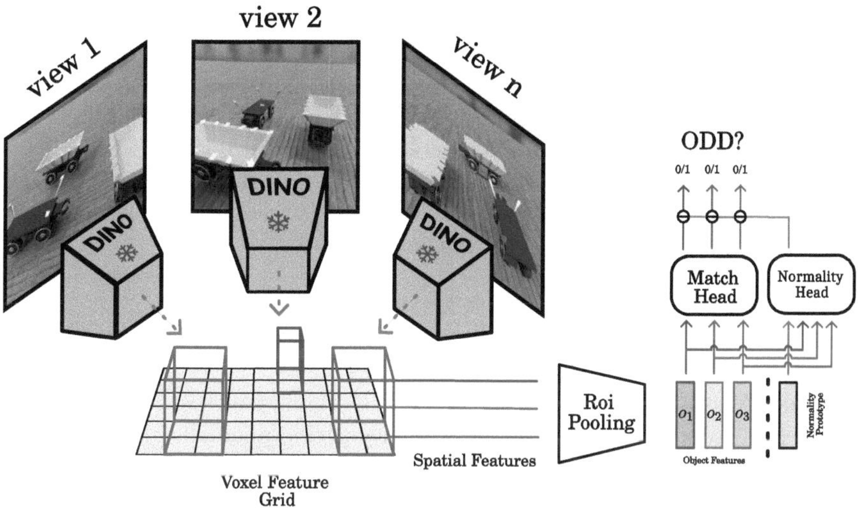

Fig. 1. Schematic representation of our approach. It takes as input M views of the scene, which are processed using the DINOv2 encoder. The resulting features are projected onto a voxel grid and associated with each of the $N = 3$ objects. A subsequent pooling step yields per-object representations. These are further refined by the context and residual anomaly heads, which encode both object-to-object similarity and the relative deviations of each object from the scene-specific average normalcy.

anomaly. We target an efficient method that leverages the embedding space of pre-trained models and works with a minimal memory footprint (see Fig. 1).

Feature Encoder. Several existing foundation models have shown impressive zero-shot performance in a variety of vision tasks thanks to their highly informative features that support strong generalization. In particular, DINOv2 [29] captures rich global and contextual cues, which are essential for accurately representing objects within complex 3D environments. We choose this model as a feature encoder for the multi-view scene images of our odd-one-out task. Thus, we obtain the features $\mathbf{F}_t = \texttt{DINOv2}(\mathbf{I}_t) \in \mathbb{R}^d$ for each input view, and project them into the 3D voxel space using the known camera parameters. The feature volumes are then averaged over all input views as in [28,39] and fed to a 3D CNN-based network for refinement, obtaining the final representation $\mathbf{G} = \texttt{3DCNN}(aggr(\{\Pi_{proj}(\mathbf{F}_t, \mathbf{P}_t)\}_{t=1}^{M})) \in \mathbb{R}^{d \times v_x \times v_y \times v_z}$.

Context Match Head. From the obtained voxel feature grid, our goal is to extract the individual object instance features by using their 3D bounding boxes. However, working directly with 3D crops within the voxel grid poses several challenges: it is computationally expensive and can introduce artifacts due to the projection of high-dimensional DINOv2 features into 3D space, especially when object sizes vary. To mitigate these issues, we first normalize all object crops to a fixed $d \times 8 \times 8 \times 8$ size using ROI pooling [16], then apply average pooling to squeeze their spatial representations into compact d dimensional embeddings

$\mathcal{O} = \{\mathbf{o}_1, \dots, \mathbf{o}_N\}$, $\mathbf{o}_i \in \mathbb{R}^d$. These instance representations are interpreted as tokens and provided as input to a standard transformer encoder of L layers, allowing the model to reason about the relative appearance of objects in the same scene. The output embeddings $\mathcal{Z} = \{\mathbf{z}_1, \dots, \mathbf{z}_N\}$, $\mathbf{z}_i \in \mathbb{R}^q$ reflect both the intrinsic features of individual objects and their relationships with others in the scene. We apply a simple linear layer to produce the final anomaly score, representing the probability p that a given input is anomalous. The network is trained by minimizing the binary cross-entropy loss

$$\mathcal{L}_{\text{BCE}} = -\sum_{i=1}^{N} \left[y_i \log(p_i) + (1 - y_i) \log(1 - p_i) \right] . \tag{1}$$

Residual Anomaly Head. Besides relying on pair-wise cross-instance matching, we can boost the identification of scene-specific anomalies by guiding the model to focus on how much an object deviates from the average normality. This deviation can be encoded into discriminative features via a residual anomaly module. We propose to devise it by adding to the network a 1-layer transformer encoder that takes as input the objects' representations $\mathcal{O}$ together with a learnable token $\mathbf{t}_0 \in \mathbb{R}^d$. The latter attends to all object embeddings, and the corresponding output $\mathbf{z}_0 \in \mathbb{R}^q$ is the only token retained for further processing. Specifically, it acts as a proxy for the scene-specific normality prototype and is supervised through a mean squared error loss

$$\mathcal{L}_{\text{normality}} = \|\bar{\mathbf{z}} - \mathbf{z}_0\|_2^2 , \tag{2}$$

where $\bar{\mathbf{z}} = \frac{1}{|\{i|y_i=0\}|} \sum_{i:y_i=0} \mathbf{z}_i$ is the average embedding of the ground-truth normal objects. Moreover, the residual anomaly module refines each object embedding via a projection head $f(\mathbf{z}_i)$ (composed of linear layer, LayerNorm, and ReLU) to then calculate the difference between the obtained object representations and the learned centroid $\mathbf{r}_i = f(\mathbf{z}_i) - \mathbf{z}_0$.

Finally, a linear layer elaborates on $\mathbf{r}_i$ to produce the anomaly score, representing the probability that a given input is anomalous. Thus, when the residual anomaly head is active, the network is trained by minimizing $\mathcal{L}_{\text{BCE}} + \mathcal{L}_{normality}$.

4 The MLLM Baseline

Recent advances in Multimodal Large Language Models (MLLMs) have demonstrated remarkable reasoning capabilities. By leveraging textual knowledge, these models significantly enrich visual understanding and enable a wide range of downstream tasks in a zero-shot setting (*i.e.*without task-specific training), including anomaly detection [17,27]. While MLLMs have proven effective in identifying anomalies based on the global appearance of objects, they still exhibit limitations in visual grounding [14]. Furthermore, their ability to reason about fine-grained anomalous details is currently an open question [44], especially considering that most existing benchmarks are restricted to single-object scenes observed from a single viewpoint.

To explore the performance of MLLMs on the odd-one-out task, we present a tailored pipeline that leverages the Set-of-Mark (SoM) prompting method [45]. This strategy supports spatial understanding by detecting the objects in the scene and overlaying visual marks directly on the images, which are then provided as input to the MLLM. Specifically, we use gemini-flash 2.0 [40] and feed it the following prompt: *Here is a list of images taken from multiple view-points of the same scene, each object is annotated with an index and the bounding box, one or more objects are different from the majority of all other objects, reply only with a list of indices of the odd objects.* The annotated images are concatenated to the prompt by following [45] as shown in Fig. 2. The model outputs the annotated index of the odd objects from which we can evaluate the prediction accuracy.

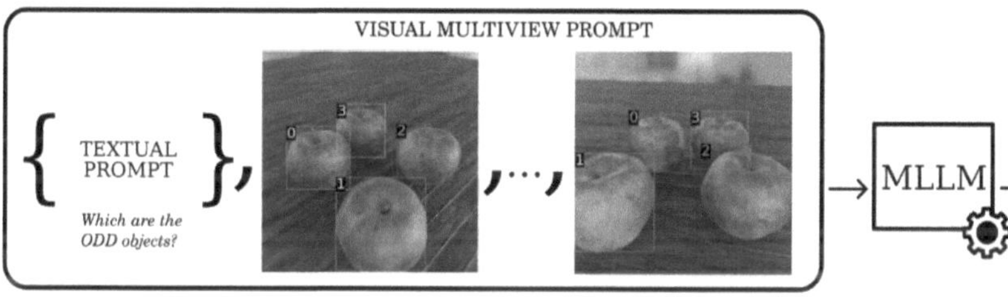

Fig. 2. Representation of the Set-of-Mark [45] prompting strategy for the MLLM baseline. Note that the color of the bounding boxes and of the numerical indexes are randomized according to the SoM strategy.

5 Experimental Setup

Datasets. We adopt the same experimental setup of [7], considering three tracks from two datasets containing objects with synthetically generated anomalies. *Toys* is composed of 8K scenes with 3–6 objects from the 51 categories of the Toys4D dataset [37] rendered in 20 views. The collection is divided into a training set with 5K scenes from 39 categories and two disjoint test sets: one (*seen*) with 1K scenes from the seen categories but with unseen object instances, and the other (*unseen*) containing 2K scenes from the remaining 12 novel categories. *Parts* was defined from the ABC dataset [21] mostly containing mechanical parts shapes. The collection contains 15K scenes, each consisting of 3–12 objects rendered from 20 viewpoints, and is divided into train and test splits of respectively 12K and 3K scenes, with the latter including only *unseen* shapes. As the datasets are provided[1] with the anomalous objects always listed as first in the set, we ensured all labels were shuffled to avoid accidental information leakage and preserve the integrity of the evaluation.

[1] https://github.com/VICO-UoE/OddOneOutAD.

Baselines. Our main baseline is the method that introduced the odd-one-out task in [7], which we refer to as *OOO*. It included two multi-view object detection methods that we also adopt as reference: *ImVoxelNet* [35] and *Detr3D*. In addition, we use our newly defined zero-shot baseline that we indicate as *MLLM*.

Metrics. As in [7], we evaluate the anomaly prediction performance via two metrics. The Area Under the ROC curve (*AUC*) measures the ability of a model to discriminate odd from normal objects regardless of the chosen confidence threshold. The *Accuracy* considers an operating scenario with a threshold equal to 0.5. In our analysis we feed the models with the ground truth location of the objects and focus only on the odd identification performance. This is justified by observations of the previous work [7] about the negligibility of the localization error due to the relative simplicity of the setting. Note that for the zero-shot MLLM baseline we collect the Accuracy, but we do not have access to the confidence score so we omit the AUC.

Implementation Details. All the models consider $M = 5$ views as input, each with resolution 256×256. For our approach we use DINOv2 ViT-S/14 as feature encoder and a four-scale Encoder-Decoder-based 3D CNN as the 3D backbone. While projecting the raw DINOv2 features (with DINOv2 frozen parameters) over the 3D voxel grid with dimension $96 \times 96 \times 16$, we use a bilinear interpolation for augmenting the height and width of the feature maps and then reduce channels from 384 to 32 with a 1×1 convolution resulting in a final voxel grid of dimension $32 \times 96 \times 96 \times 16$ (each voxel represents 4 cm). The object crops of dimensions $32 \times 8 \times 8 \times 8$ are flattened along the spatial dimensions through average pooling and projected into a 256 dimensional space via a linear layer. The Transformers employed by the *Context Match Head* and *Residual Anomaly Head* use a multi-head attention mechanism with 8 different 32-dimensional heads. We train our method on one NVIDIA V100 GPU for 50 epochs using a learning rate of 2×10^{-4}, batch size 20, along with a cosine annealing scheduler and AdamW [26] optimizer.

Table 1. Main results for the three benchmark tracks. Ours presents comparable performance to that of the leading competitor (OOO) on Toys and largely outperforms it on the challenging Parts dataset.

Model	Toys Seen		Toys Unseen		Parts	
	AUC	Accuracy	AUC	Accuracy	AUC	Accuracy
ImVoxelNet [35]	78.13	65.55	73.19	60.12	72.80	64.34
DETR3D [42]	79.16	67.37	74.60	62.98	74.49	65.11
OOO [7]	**91.78**	83.21	**89.15**	**81.57**	86.12	79.68
MLLM baseline	–	52.23	–	53.35	–	60.73
Ours	89.49	**84.86**	85.57	81.43	**89.72**	**88.81**

6 Results

Benchmark. The results of our experimental analysis are presented in Table 1. Our method performs similarly to OOO on Toys Seen and slightly worse on Toys Unseen, while showing a significant advantage over the leading competitor on Parts Unseen. We remark that the difference between these two datasets is in the geometric nature of the comprised objects: Toys includes free-form shapes with high inter-class variability, while Parts consists of mechanical components characterized by low semantic diversity and predominantly rigid angular geometries. Besides being more challenging due to the higher object count in the scenes and fine-grained shape differences, the limited categorical diversity of the latter dataset better reflects industrial quality inspection scenarios and aligns well with our approach, which is specifically designed to prioritize efficiency. We acknowledge that the OOO modules, which learn a task-specific 2D representation for each view, combine them in 3D and subsequently enhance them by passing through a differentiable rendering to distill DINOv2 features, offer greater flexibility in handling class variability. However, this comes at a high cost in terms of model complexity, including increased training time and a large number of parameters, due to the sparse voxel attention employed in the matching head.

The MLLM baseline falls short on the Toys tracks, and on Parts it gets accuracy results close to those of the detection methods ImVoxelsNet and DETR3D. Overall, MLLM shows a bias towards normal data that might have been part of its large scale pre-training and, similarly to the detection methods, it recognizes large cracks and fractures but struggles when intra-group comparisons over multiple views are necessary. These limitations align with what was discussed in [44] and call for new tailored modules to guide generalist models when dealing with tasks requiring multi-view consistency.

Detailed analysis and Robustness. We show a breakdown of the Accuracy of our approach across different anomaly types in Fig. 3 (a), together with success and failure qualitative results in Fig. 4. We can see from the breakdown that our model is particularly good at detecting material issues, bumps, and fractures that usually consist of darker regions of the object, while it struggles with 3D specific anomalies like missing parts, translations, and deformations that require a more global 3D reasoning.

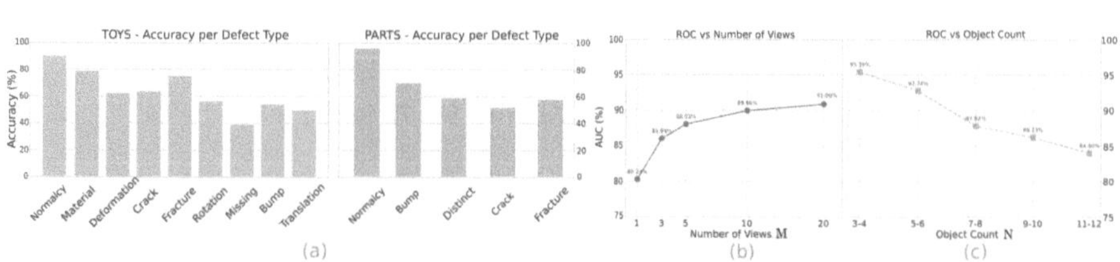

Fig. 3. We report (a) accuracy across different anomaly types and normalcy data on Toys Unseen and Parts Unseen; AUC on Parts Unseen when changing (b) the number of views at inference time and (c) the number of objects in the scene.

Furthermore, the right part of Fig. 3 shows the AUC of our model when varying the number of input views during testing (b) and the performance when changing the object count (c). Both results are obtained on the Parts Unseen dataset and highlight the advantage provided by a growing amount of images as well as the relative robustness across scenes becoming progressively more difficult with a range of 4% point variation around the main result (Accuracy 88.81 in Table 1).

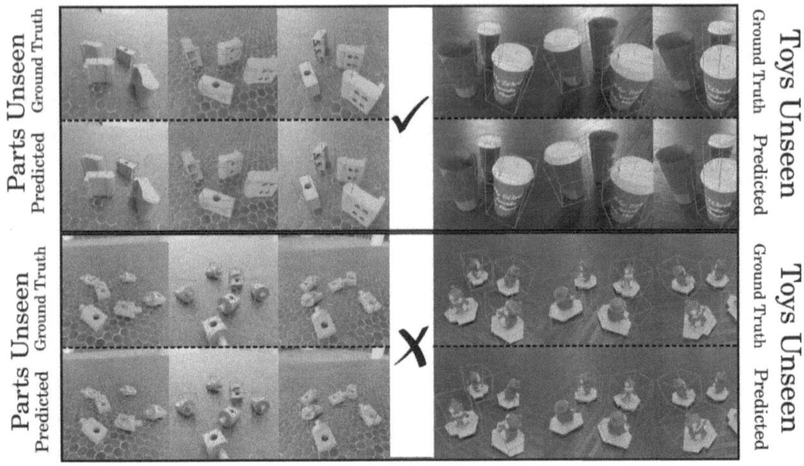

Fig. 4. Qualitative results of our approach on Toys Unseen and Parts Unseen. We show three views for each scene. The green boxes indicate normal objects while the red ones indicate anomalous objects. The ✓ and ✗ indicate respectively correct and wrong predictions. (Color figure online)

Table 2. Ablation results across the three benchmark tracks for various versions of our model's head.

Ours Head	Toys Seen		Toys Unseen		Parts		Memory	Inf. time
	AUC	Acc.	AUC	Acc.	AUC	Acc.	(GB)	(ms)
Sparse Voxel Attn.	86.11	78.79	85.48	77.32	86.32	84.06	1.23	337
Context	89.18	**85.42**	85.09	81.27	89.14	88.65	**0.36**	**271**
Context + Residual	**89.49**	84.86	**85.57**	**81.43**	**89.72**	**88.81**	0.54	286

Ablations. Among the architectural design choices of our method, there is the *dense attention* on the pooled representations adopted within the Context Match Head. This is the standard choice for Transformers, but the original OOO approach exploited a different *sparse voxel attention*. Building on voxel features certainly provides highly detailed information, but comes with the need for sparsity to avoid severe inefficiency. Thus, the cross-instance matching in OOO was performed considering only geometrically corresponding voxel locations. To explore the potential of the sparse voxel attention, we integrated it into our Context Match Head by replacing the original dense attention mechanism. For this experiment we keep the same configuration proposed [7] by running our model for 50 epochs with batch size 4 and two different fixed learning rates for Toys (2×10^{-5}) and Parts (2×10^{-6}).

In Table 2 we compare different versions of our model, also ablating the Residual Anomaly Head. From the obtained results we can conclude that the sparse voxel attention performs worse than the other versions on average. Moreover, it consumes significantly more memory and is considerably slower. Adding the Residual Anomaly Head slightly boosts the performance with a minimal change in memory and inference time. The sparse version of the model also uses significantly more parameters (72M vs. 48M) and requires much longer training time, lasting about 6h on Toys and 22h on Parts (our models take around 2.5h on Toys and 5.5h on Parts).

7 Conclusions

In this work we focused on the odd-one-out anomaly detection task that poses several challenges by moving anomaly detection closer to real world conditions. It involves multi-object scenes with scene-specific anomaly criteria that can only be inferred by cross-instance comparisons. This problem calls for models able to manage spatial reasoning to elaborate on multiple scene views, along with relational reasoning to interpret the context and generalize over different object categories and layouts. Considering the relevance of the task for industrial applications, efficiency is also a key requirement. By focusing on this aspect, we presented a model that, despite the significantly low capacity (number of parameters) and memory needs, gets competitive performance compared to the state of the art. Several in-depth analyses confirmed the effectiveness of the adopted architectural choices as well as the overall model's robustness.

To test the zero-shot effectiveness of Multimodal Large Language Models on the odd-one-out task, we also introduced a baseline that leverages the Set-of-Mark prompting method to support visual grounding. The low experimental results revealed the limitations of this approach, signaling the need for new solutions to leverage the extensive internal knowledge of generalist models when performing fine-grained comparisons in complex multi-view settings.

We believe that future solutions for the odd-one-out task should continue to build on the capabilities of pre-trained foundation models, with efficiency remaining a key objective. This would enable the transition from anomaly detection to full 3D anomaly localization, while allowing models to articulate the rationale behind the predictions through natural language. Furthermore, the task could be extended to encompass logical and functional aspects, bridging perception and robotics, and evolving into a true intelligence test for autonomous agents.

Acknowledgments. This study was carried out within the FAIR - Future Artificial Intelligence Research and received funding from the European Union Next-GenerationEU (PIANO NAZIONALE DI RIPRESA E RESILIENZA (PNRR) - MISSIONE 4 COMPONENTE 2, INVESTIMENTO 1.3 - D.D. 1555 11/10/2022, PE00000013). This manuscript reflects only the authors' views and opinions, neither the European Union nor the European Commission can be considered responsible for them.

References

1. Ahmed, F., Courville, A.: Detecting semantic anomalies. In: AAAI (2020)
2. Alliegro, A., Borlino, F.C., Tommasi, T.: 3DOS: Towards 3d open set learning - benchmarking and understanding semantic novelty detection on point clouds. In: NeurIPS (2022)
3. Bergmann, P., Batzner, K., Fauser, M., Sattlegger, D., Steger, C.: The mvtec anomaly detection dataset: a comprehensive real-world dataset for unsupervised anomaly detection. IJCV **129**(4), 1038–1059 (2021)
4. Bergmann, P., Batzner, K., Fauser, M., Sattlegger, D., Steger, C.: Beyond dents and scratches: logical constraints in unsupervised anomaly detection and localization. IJCV **130**, 947–969 (2022)
5. Bergmann, P., Fauser, M., Sattlegger, D., Steger, C.: Uninformed students: student-teacher anomaly detection with discriminative latent embeddings. In: CVPR (2020)
6. Bergmann, P., Jin, X., Sattlegger, D., Steger, C.: The mvtec 3d-ad dataset for unsupervised 3d anomaly detection and localization. In: IJCV (2022)
7. Bhunia, A., Li, C., Bilen, H.: Odd-one-out: anomaly detection by comparing with neighbors. In: CVPR (2025)
8. Bhunia, A.K., Li, C., Bilen, H.: Looking 3d: anomaly detection with 2d-3d alignment. In: CVPR (2024)
9. Bogdoll, D., Nitsche, M., Zöllner, J.M.: Anomaly detection in autonomous driving: a survey. In: CVPR (2022)
10. Chan, R., et al.: Segmentmeifyoucan: a benchmark for anomaly segmentation. In: NeurIPS (2021)
11. Deecke, L., Ruff, L., Vandermeulen, R.A., Bilen, H.: Transfer-based semantic anomaly detection. In: ICML (2021)
12. Defard, T., Setkov, A., Loesch, A., Audigier, R.: Padim: a patch distribution modeling framework for anomaly detection and localization. In: ICPR (2021)
13. Ding, C., Pang, G., Shen, C.: Catching both gray and black swans: open-set supervised anomaly detection. In: CVPR (2022)
14. El Banani, M., et al.: Probing the 3d awareness of visual foundation models. In: CVPR (2024)
15. Fang, Z., Wang, X., Li, H., Liu, J., Hu, Q., Xiao, J.: Fastrecon: few-shot industrial anomaly detection via fast feature reconstruction. In: ICCV (2023)
16. Girshick, R., Donahue, J., Darrell, T., Malik, J.: Rich feature hierarchies for accurate object detection and semantic segmentation. In: CVPR (2014)
17. Gu, Z., Zhu, B., Zhu, G., Chen, Y., Tang, M., Wang, J.: Anomalygpt: detecting industrial anomalies using large vision-language models. In: AAAI (2024)
18. Huang, C., Guan, H., Jiang, A., Zhang, Y., Spratling, M., Wang, Y.F.: Registration based few-shot anomaly detection. In: ECCV (2022)

19. Jeong, J., Zou, Y., Kim, T., Zhang, D., Ravichandran, A., Dabeer, O.: Winclip: zero-/few-shot anomaly classification and segmentation. In: CVPR (2023)
20. Kim, S., An, S., Chikontwe, P., Kang, M., Adeli, E., Pohl, K.M., Park, S.H.: Few shot part segmentation reveals compositional logic for industrial anomaly detection. In: AAAI (2024)
21. Koch, S., et al.: Abc: a big cad model dataset for geometric deep learning. In: CVPR (2019)
22. Li, C.L., Sohn, K., Yoon, J., Pfister, T.: Cutpaste: self-supervised learning for anomaly detection and localization. In: CVPR (2021)
23. Li, H., Chen, Z., Xu, Y., Hu, J.: Hyperbolic anomaly detection. In: CVPR (2024)
24. Liu, J., et al.: Real3d-ad: a dataset of point cloud anomaly detection. In: NeurIPS (2023)
25. Liu, Z., Zhou, Y., Xu, Y., Wang, Z.: Simplenet: a simple network for image anomaly detection and localization. In: CVPR (2023)
26. Loshchilov, I., Hutter, F.: Decoupled weight decay regularization. In: ICLR (2019)
27. Lv, W., Su, Q., Xu, W.: One-for-all few-shot anomaly detection via instance-induced prompt learning. In: ICLR (2025)
28. Murez, Z., van As, T., Bartolozzi, J., Sinha, A., Badrinarayanan, V., Rabinovich, A.: Atlas: end-to-end 3d scene reconstruction from posed images. In: ECCV (2020)
29. Oquab, M., et al.: DINOv2: learning robust visual features without supervision. In: TMLR (2024)
30. Pang, G., Shen, C., Cao, L., Van Den Hengel, A.: Deep learning for anomaly detection: a review. ACM Comput. Surv. **54**(2) (2021)
31. Patra, S., Taieb, S.B.: Revisiting deep feature reconstruction for logical and structural industrial anomaly detection. In: TMLR (2024)
32. Pirnay, J., Chai, K.: Inpainting transformer for anomaly detection. In: ICIAP, pp. 394–406 (2022)
33. Rabino, P., Alliegro, A., Tommasi, T.: 3d semantic novelty detection via large-scale pre-trained models. IEEE Access **12**, 135352–135361 (2024)
34. Roth, K., Pemula, L., Zepeda, J., Schölkopf, B., Brox, T., Gehler, P.: Towards total recall in industrial anomaly detection. In: CVPR (2022)
35. Rukhovich, D., Vorontsova, A., Konushin, A.: Imvoxelnet: image to voxels projection for monocular and multi-view general-purpose 3d object detection. In: WACV (2022)
36. Salehi, M., Sadjadi, N., Baselizadeh, S., Rohban, M.H., Rabiee, H.R.: Multiresolution knowledge distillation for anomaly detection. In: CVPR (2021)
37. Stojanov, S., Thai, A., Rehg, J.M.: Using shape to categorize: Low-shot learning with an explicit shape bias. In: CVPR (2021)
38. Sträter, L.P., Salehi, M., Gavves, E., Snoek, C.G., Asano, Y.M.: Generalad: anomaly detection across domains by attending to distorted features. In: ECCV (2025)
39. Sun, J., Xie, Y., Chen, L., Zhou, X., Bao, H.: NeuralRecon: real-time coherent 3d reconstruction from monocular video. In: CVPR (2021)
40. Team, G., et al.: Gemini: a family of highly capable multimodal models. arXiv:2312.11805 (2023)
41. Wang, C., et al.: Real-iad: a real-world multi-view dataset for benchmarking versatile industrial anomaly detection. In: CVPR (2024)
42. Wang, Y., Guizilini, V.C., Zhang, T., Wang, Y., Zhao, H., Solomon, J.: DETR3d: 3d object detection from multi-view images via 3d-to-2d queries. In: CORL (2021)
43. Xie, G., Wang, J., Liu, J., Jin, Y., Zheng, F.: Pushing the limits of fewshot anomaly detection in industry vision: Graphcore. In: ICLR (2023)

44. Xu, J., Lo, S.Y., Safaei, B., Patel, V.M., Dwivedi, I.: Towards zero-shot anomaly detection and reasoning with multimodal large language models (2025)
45. Yang, J., Zhang, H., Li, F., Zou, X., Li, C., Gao, J.: Set-of-mark prompting unleashes extraordinary visual grounding in gpt-4v. arXiv:2310.11441 (2023)
46. Zavrtanik, V., Kristan, M., Skočaj, D.: Cheating depth: enhancing 3d surface anomaly detection via depth simulation. In: WACV (2024)
47. Zavrtanik, V., Kristan, M., Skočaj, D.: Reconstruction by inpainting for visual anomaly detection. Pattern Recogn. **112**, 107706 (2021)
48. Zhao, Y.: Omnial: a unified cnn framework for unsupervised anomaly localization. In: CVPR (2023)

Physics-Based 3D Simulation for Synthetic Data Generation and Failure Analysis in Packaging Stability Assessment

Samuel Seligardi[1], Pietro Musoni[1(✉)], Eleonora Iotti[1], Gianluca Contesso[2], and Alessandro Dal Palù[1]

[1] Department of Mathematical, Physical and Computer Sciences, University of Parma, Parma, Italy
pietro.musoni@unipr.it
[2] ACMI S.p.A., Fornovo di Taro, Parma, Italy

Abstract. The design and analysis of pallet setups are essential for ensuring safety of packages transportation. With rising demands in the logistics sector, the development of automated systems utilizing advanced technologies has become increasingly crucial. Moreover, the widespread use of plastic wrapping has motivated researchers to investigate eco-friendly alternatives that still adhere to safety standards. We present a fully controllable and accurate physical simulation system capable of replicating the behavior of moving pallets. It features a 3D graphics-based virtual environment that supports a wide range of configurations, including variable package layouts, different wrapping materials, and diverse dynamic conditions. This innovative approach reduces the need for physical testing, cutting costs and environmental impact while improving measurement accuracy for analyzing pallet dynamics. Additionally, we train a deep neural network to evaluate the rendered videos generated by our simulator, as a crash-test predictor for pallet configurations, further enhancing the system's utility in safety analysis.

Keywords: 3D graphics modeling · Physical simulations · Digital Twins

1 Introduction

In logistics safety, packages must withstand road or rail transport, safeguarding truck workers and preventing product loss or damage. The rigidity and strength of the load plays a critical role in transportation efficiency: permanent deformations and excessive movements of the load during transit directly impact the stability of the truck and the durability of the external wrapping, which may become deformed or torn.

Existing safety regulations require companies to conduct evaluations on every type of product they plan to transport, ensuring compliance with stability standards throughout the transportation process. They require specific tolerances

E. Rodolà et al. (Eds.): ICIAP 2025, LNCS 16167, pp. 403–415, 2026.
https://doi.org/10.1007/978-3-032-10185-3_32

for loaded pallets, when subjected to force caused by acceleration in a specific direction (more details in Sec. 2). Standardized crash tests packaging setups typically involve a physical moving platform capable of applying both positive and negative accelerations to simulate real-world transport conditions. It is worth noting that each package layout must be tested individually. As a consequence, the main drawback of such a testing approach lies in its costs, in terms of energy consumption, but also in their environmental impact.

Automating the control and evaluation of these physical tests is not always straightforward. Regulations impose fine-grained requirements about the amount of elastic deformations of packages and the resulting condition of wrapping materials after significant acceleration changes. The analysis of a video recording of a test can detect the complete collapse of the pallet or the displacement of packages outside the pallet, by using common computer vision systems. However, current safety standards require more precise assessments and finer techniques.

Our work overcomes the limitations of the physical approach thanks to a 3D Digital Twin of the whole testing process. The main advantages are: (i) the ability to maintain full control over packaging configurations and testing parameters; (ii) accurate measurements of all objects during the simulation and execution of large-scale experiments; (iii) our digital replica drastically reduces both economic and environmental costs associated with physical testing machinery, thus allowing packaging industries to provide larger coverage of simulated conditions and the development of optimized packaging configurations to increase safety.

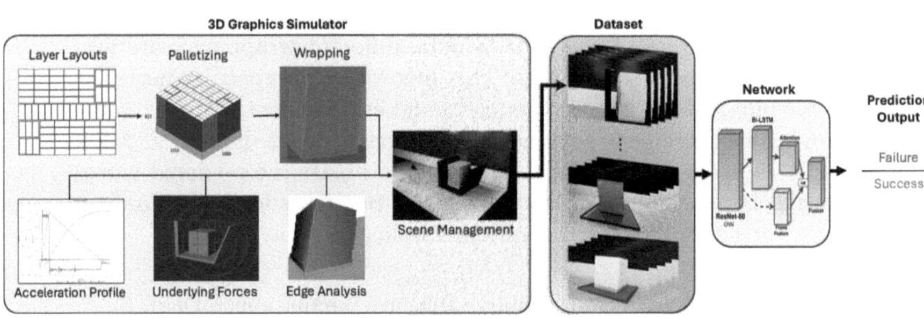

Fig. 1. System's pipeline.

Our system's pipeline, depicted in Fig. 1, is composed of the following elements: a 3D graphics virtual crash test, based on rigid body dynamics combined with cloth simulation; a simulation generator (simulator) to produce synthetic renderings; an AI-powered pipeline that leverages synthetic data to perform success/failure classification and give feedback to automation engineers.

This paper presents our modeling of pallets and packages, as well as the acceleration bench, as rigid bodies subjected to forces and accelerations. We also represent the external envelope as a mesh with cloth dynamics (see Sec. 2). Cloth simulation ensures control over many material parameters, such as elasticity, and permits the mesh to shrink and adhere to the underlying bodies. Then, a simulation engine, detailed in Sec. 3, is developed to automatically generate virtual replicas of various packaging configurations and setups, as well as acceleration conditions, with rendered videos as similar as possible to real videos obtained from the physical acceleration bench. We feed a deep neural network, detailed in Sec. 4, that returns the classification of success or failure cases to the final users, who can adjust the configuration to improve the predicted safety levels. Sec. 5 draws the conclusions.

2 Background

Our virtual simulator engine aims at generating models that precisely replicate the physical testing environment. To the best of our knowledge, this is the first attempt to develop digital simulations that address safety in the transport of palletized units by focusing on their stability and rigidity under stress tests.

However, the idea of creating a 3D physical simulation to model challenging conditions is not new [7,20,21]. Digital Twin modeling [16,18] is one of the ways this problem is addressed. The acceleration test bench that we model and simulate as a Digital Twin is a machinery produced by the ESTL Company [3] (from now on, the ESTL machine), designed for the packaging industry. But, differently from conventional Digital Twins, that perfectly reproduce one single scenario, our simulator also offers a high degree of flexibility in comparison to the ESTL machine, such as changes in configurations, types of motion, and viewpoints.

Other approaches include 3D scene reconstruction, both with standard methods and deep learning [10,14], text-to-3D generation [5], diffusion models [13], and Neural Radiance Field (NeRF) [8]. In our case, such methods cannot be employed due to the lack of multi-view quality data of the testing machinery. In fact, the ESTL machine cannot be modified due to safety protocols: its system provides only accelerometer data and high-speed recordings from a single fixed lateral view, capturing motion in the perpendicular direction.

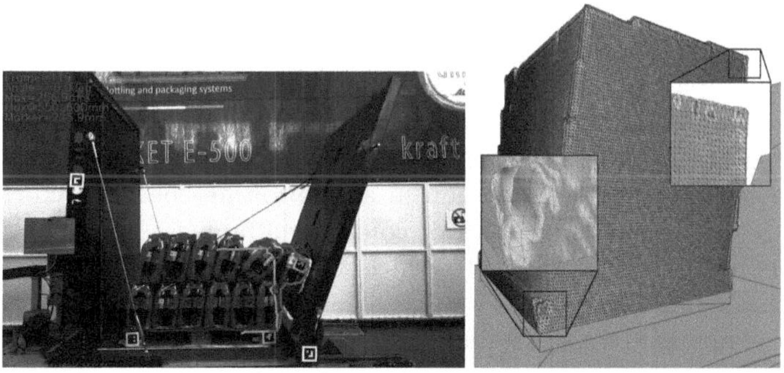

Fig. 2. Left: Video analysis of an experimental crash-test without wrapping, showing elastic deformations (red lines) during motion. Right: Cloth simulation highlighting mesh edge stretching, with stress areas marked in red.

Safety Standards. We refer to European regulations, but regulations from different countries can handled similarly by our system, having the possibility to simulate and measure any physical property involved (e.g., accelerations and displacements). The European Road Worthiness Directive [1] introduces the basis for the EUMOS 40509 standard [2], which specifies how to perform palletized units crash tests and how to interpret the resulting measures as successful tests or failed tests.

Following EUMOS 40509, the test should provide an acceleration impulse lasting $0.5s$ followed by a constant deceleration, until the unit stops. The acceleration impulse to be supported is $0.5g$, but generally tests are performed in a range from $0.3g$ to $0.8g$ accelerations, in step of $0.1g$, to actually stress the loaded unit. Two types of deformations may occur: elastic, which are temporary and observed only during the testing window, and permanent, which represent the residual changes in the load unit after the test is completed. A passed test requires that: (i) the permanent displacement of any part of the test load unit must not exceed 5% of the total height of the load unit; (ii) within the lowest $20cm$ of the test load unit, the permanent displacement on the wooden pallet must be less than $4cm$; (iii) during testing, the elastic displacement of all parts of the test load unit must not exceed 10% of its total height; (iv) no visible structural damage or product leakage should occur by the end of the test. The ESTL machine consists of a body, called *sleight*, which loads the palletized unit. The sleight moves with the tested acceleration patterns and transmits it to the loaded unit, while a high-speed recording camera records the motion. A proprietary software applies computer vision elaborations to the video and outputs the measurement of the elastic and permanent deformations, as specified by the EUMOS 40509 standard, as shown in Fig. 2 on the left.

Cloth Physical Simulation. The modeling of the wrapping material is a critical aspect in simulating the behavior of a moving pallet. Some works use particle simulations [17], Virtual Element Method (VEM) [6], and neural networks [15] to model and simulate elasto-plastic materials. These approaches typically consider a single material, whereas our problem involves frequent variation in materials such as LLDPE, heat-shrink wrap, and kraft paper. LLDPE, in various configurations of stretch, thickness, and layering, is the most used. Heat-shrink wrap has specific properties that make it more rigid than classic elasto-plastic. Kraft paper is recyclable and sustainable, but more prone to tearing. In our work, to model the behavior of the wrapping around packages, we employ the physical simulation of cloth. Cloth simulation in animation has been a well-established research area since the early days of computer graphics. A seminal contribution in this field is the work of [4], which laid solid groundwork for physically based cloth simulation. In recent years, the rapid growth in computational power and the development of novel techniques have significantly enhanced the realism and performance of cloth simulation systems, attracting growing interest from both researchers and industries, notably the fashion industry. Cloth simulation in 3D animation aims to digitally approximate physical behavior and continuum properties. Various modeling approaches have been explored, including mass-spring systems [12], continuum-based methods [19], and yarn-level simulations

[11]. The parameters of these systems strongly influence the dynamics of the simulated cloth, making their selection a critical aspect of the process. Consequently, parameter estimation and mapping between simulation parameters and real-world material properties remain active areas of research. Some 3D software platforms, such as Maya[1] and Blender[2], provide preset configurations for various types of fabrics. Moreover, such presets are typically only rough starting points and require manual fine-tuning.

3 The Simulation Engine

The core component of our pipeline is the physical simulation system (simulator). Among many alternatives, we choose the 3D modeling software Blender, for the wide variety of customizations it enables with easy applicability, its open-source nature, and the possibility to run scripts to automatize the 3D scene generation.

Our simulation engine allows us to automate scene configuration, perform comprehensive validation checks, and render videos of packaging units subjected to various acceleration conditions. By leveraging Blender's Python API, we created a system that automatically generates simulations of the physical behavior of various layouts of palletized loads during transport.

3.1 Simulation System Architecture

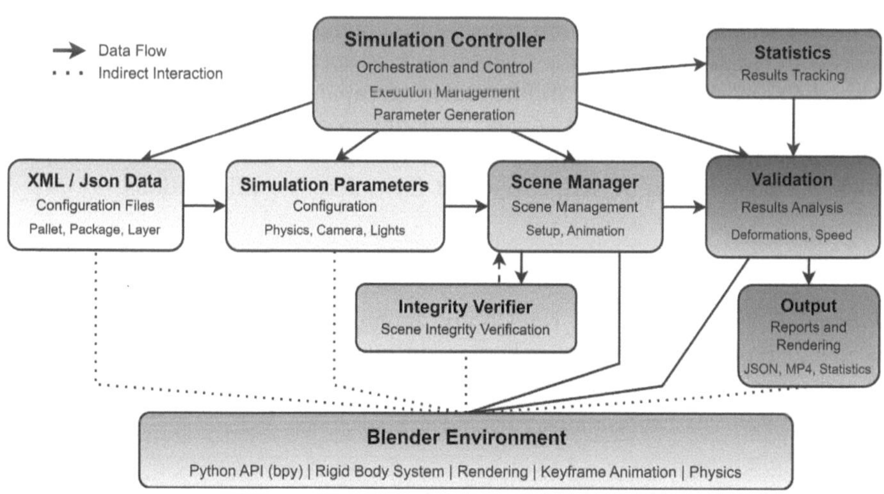

Fig. 3. Architecture diagram of the simulation system showing component relationships and data flow.

[1] https://www.autodesk.com/it/products/maya/
[2] http://www.blender.org/

Our simulator follows a modular architecture design to ensure flexibility, maintainability, and extensibility. The architecture comprises several specialized components that interact to create, execute, and validate the simulations, as shown in Fig. 3.

The architecture gives a clear separation of tasks, with distinct components for configuration parsing, parameter management, scene setup, physics simulation, and result analysis. Data flow from configuration files through parameter setting, scene construction, simulation execution, and finally to validation and output generation. In detail, the system is structured around the following core components: *the Simulation Controller* is the central component that manages execution flow, coordinates the overall simulation process, and handles parameter generation; *the Statistics module* tracks simulation outcomes, success and failure rates, and provides reports for multiple simulation runs; *the XML Data* contains the structured configuration information including pallet specifications, package dimensions, and stacking patterns loaded from external files; *the Simulation Parameters* stores and validates all simulation parameters, including physical properties, camera settings, lighting configurations, and animation timelines; *the Scene Manager* handles all interactions with Blender environment, including object creation, material assignment, physical property configuration, and animation generation; *the Validation* performs deformation analysis and speed monitoring to determine if a simulation represents a success or failure case according to the EUMOS 40509 standard; *the Integrity Verifier* ensures that all essential objects and their properties are correctly configured before and during simulation execution, preventing potentially invalid simulations; *the Output module* manages the generation of simulation results, including rendered videos, JSON reports, and statistical data; *the Blender Environment* is the underlying 3D platform providing physics simulation, rendering capabilities, and Python API for automation. The environment comes pre-configured with essential scene objects (ESTL machine, pallet, wrapping mesh, force fields, camera, package sample), which can be modified by script to adapt for different configurations.

3.2 Physical Modeling

We employ a combination of rigid body dynamics, force fields, and cloth simulation to replicate real-world behaviors in our simulator. The fusion of these three aspects for the 3D industrial simulation in this scenario is an important contribution of our work.

Rigid Body Dynamics. The core of our physical simulation utilizes Blender's rigid body system to model packages and pallets. The *pallet* and each *package* are defined as a discrete rigid body with box collision shape. The ith-package P_i (similarly, the pallet P), has the following properties.

- Dimensions extracted from configuration files: $(d_x^{(i)}, d_y^{(i)}, d_z^{(i)})$;
- Accurate mass values that influence physical behavior: m_i;

- Collision boundaries that define interaction surfaces: $[l_i, u_i]$;
- Friction coefficients that replicate surface interactions: φ_i.

The rigid body system calculates the motion of packages under external forces while handling collisions between adjacent packages. By the control scene system, we can access the ith-package center of mass (x_i, y_i, z_i). This allows us to observe, and gather data, on how packages shift, tilt, or fall during acceleration and deceleration phases.

Perimeter Force Fields. One of the key innovations in our simulation approach is the use of perimeter force fields to replicate the restraining effect of wrapping materials. Rather than directly modeling the material's physical properties, we simulate its effect through strategically positioned force fields around the cargo. The kth-*force fields* F_k, with k from 1 to 4, are configured with the following characteristics:

- Strength values proportional to cargo mass and acceleration: $F_v^{(k)}$;
- Positioned at a calculated distance from the cargo perimeter: d_k;
- Height-dependent influence that matches packaging behavior: σ_h;
- Appropriate falloff patterns to simulate realistic restraint

In particular, the strength value $F_v^{(k)}$ of the kth-force field is distributed over the cargo height: $F_v^{(k)} = F_{\text{base}}^{(k)} \cdot \sigma_h$. The punctual strength $F_{\text{base}}^{(k)}$ is obtained from the chosen acceleration a times the total mass Σ_{m_i} of the palletized unit, also considering the part of the tension coefficient T impacting on the force field: $F_{\text{base}}^{(k)} = \frac{T}{4} \cdot \Sigma_{m_i} \cdot a$. Both T and σ_h are calibrated parameters that adjust the restraining force based on empirical observations of packaging behavior.

Cloth Simulation for Wrapping Representation. While the force fields provide the physical effects of the material of the external envelope, representing the wrapping is important for both analytical and visual presentation purposes. We use Blender's cloth simulation system to create the wrapping material that envelops the pallet load. The calculated physics of the cloth simulation allow us to manage mesh edges and monitor the critical tear points in the wrapping. As shown in Fig. 2 on the right, with the cloth simulation we can visualize stress points in the wrapping material, providing additional insights for potential failures. The simulation calculates the variation of the length of each individual mesh edge during the simulation, highlighting areas where the packaging experiences the greatest tension during transportation.

In detail, the wrapping W, modeled as cloth: (i) is a surface mesh that dynamically adapts to the cargo geometry; (ii) responds to the movement of the packages; (iii) provides indicators of packaging stress points; (iv) enables the identification of potential tearing locations.

The combination of rigid bodies, force fields, and cloth simulation captures both the macro-level stability of the entire load and the micro-level stress patterns in the packaging material.

3.3 Parameters and Configurability

A key feature of our simulator is its high degree of configurability, which allows for extensive experimentation with different package arrangements, physical properties, and testing conditions. The system's parametric approach supports both testing of specific scenarios and wide analysis through randomized configurations. The main parameter categories include *Physical Parameters*, *Testing Conditions*, *Visual Parameters*, *Validation Thresholds*. Physical Parameters refer to package dimensions, weights, pallet characteristics, and stacking patterns. Stacking patterns are often called *palletizing schema*, and consist of one or more layers of packages, each of these layers can be disposed with different layouts (relative positions). Layers' layouts give the major contribution to pallets stability, and our results empirically confirm this insight. The proposed setups (layouts, packages dimensions, number of layers) are real-world setups provided by manufacturing companies. The manufacturing companies use softwares that help and guide the creation of configurations, and outputs the physical parameters as XML files. Our system supports these XML files as inputs in the simulator architecture, as shown in Fig. 3. Testing Conditions consist of acceleration values ($0.3g$ to $0.8g$), acceleration duration ($0.35s$ to $0.5s$), and deceleration rates, while Visual Parameters include camera settings, lighting configurations, and rendering options. Finally, Validation Thresholds are defined by elastic deformation limits, permanent deformation tolerances, and bottom layer specific constraints.

Fig. 4. Top row: Comparison of the same palletizing scheme under varying acceleration conditions. Bottom row: Different stacking patterns demonstrate how palletizing design impacts unit stability under identical test conditions. Left examples are successful; right ones show failures.

This comprehensive parameterization enables our simulator to generate a diverse range of scenarios. As shown in Fig. 4, even small variations in parameters can yield to significantly different outcomes, highlighting the importance of thorough testing across the parameter space.

Our implementation allows for two complementary approaches to parameter selection, i.e., *deterministic configuration* and *random parameter generation*. Using the deterministic configuration approach, parameters are loaded from predefined JSON files for precise reproduction of specific test cases, useful for validation against physical tests or investigating known problematic scenarios. Usually, the random parameter approach could be used to automatically create parameter sets within valid ranges, in order to perform massive generation, and to enable broad exploration of the parameter space, discovering unforeseen failure patterns and edge cases. The second row of Fig. 4 illustrates how different palletizing patterns, one of many configurable aspects, can significantly impact load stability under identical acceleration conditions. This variability allows packaging engineers to identify optimal stacking configurations before physical implementation.

4 Experiments

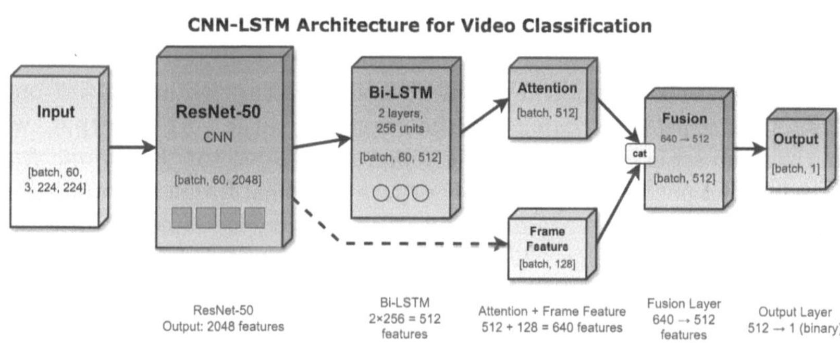

Fig. 5. The architecture of the deep neural network to perform video classification.

Dataset. To evaluate the usability of our simulator in a real-world context, we generated a dataset $D = \{(v_j, o_j)\}$ of labeled videos, by rendering a diverse set of simulations across various configurations. From real palletizing schema provided by the packaging company's software, in the form of XML configurations, we obtained package dimensions and layers layouts. The simulator varied these configurations by rotating layouts, and expanding the number of layers from 4 to N, where N is the original number of layers of the palletized schema. Then, the simulator tests these augmented palletized units from $0.3g$ to $0.8g$ accelerations in step of $0.1g$, and with acceleration impulses lasting from $0.35s$ to $0.5s$. Other variations include changing the camera position and its focal length. The final number of rendered simulations is 1500. Each video v_j consists of about $60 \div 70$

frames, with 1920 × 1080 resolution. The simulator report for v_j metadata is used for labeling, where $o_j \in \{\text{success}, \text{failure}\}$, but also carries other relevant information. For example, in case of failure, it provides details on where the violation (whether permanent or elastic deformation) occurred. These details are used to preprocess input data to enhance elaboration. About 1/3 of these videos represent successful tests, while the others 2/3 are failed tests.

This dataset is then split in D_{train} and D_{test} to train and validate a deep neural network classifier, capable of analyzing videos and predicting potential failure cases in virtual crash tests.

Architecture of the Network. The network architecture is inspired by most used and efficient models in literature, such as [9]. The architecture of the network, illustrated in Fig. 5, is as follows. Our method starts with *input reshaping* to process each video frame individually using a Convolutional Neural Network (CNN). The video input, consisting of N_f frames, each a 3-channel color image at 1920 × 1080 resolution, is reshaped depending on the batch size which is multiplied by the number of frames. Additionally, the resolution is reduced to 224 × 224. Next, during *feature extraction*, the ResNet-50 model processes the reshaped input, producing 2048 features. These are leveraged for temporal sequence analysis by a Long Short-Term Memory (LSTM) model. In *temporal sequence analysis*, the LSTM processes the input feature vector, combining 256 units from each direction in its bi-directional setup, obtaining 512 output features. An *attention mechanism* is then applied to the LSTM output, which summarizes the sequence into a single representative vector. For *frame feature fusion*, a parallel pathway is used to incorporate critical frame information. The 2048 critical frame features are reduced to 128 and concatenated with the attention context vector, forming $512 + 128 = 640$ features. A fusion layer refines this to 512. Finally, in the *output prediction stage*, the fused features are passed through an output layer to generate predictions, representing the desired outcome based on the video data.

Results. For the training process, we used different hyperparameters and optimization techniques to ensure efficient learning and prevent overfitting. The chosen *loss function* is a weighted Binary Cross-Entropy Loss, where successful test cases have a higher weight to counterbalance the ratio 1 : 2 between success and failure labels. The *optimizer* is AdamW, with a learning rate of $5 \cdot 10^{-4}$ and weight decay of 10^{-5} for better regularization. The model was trained for a maximum of 100 epochs with an early stopping patience of 10 epochs to prevent overfitting. We used a batch size of 8 with gradient accumulation steps of 2, effectively simulating a batch size of 16 while maintaining memory efficiency. Automatic mixed precision training was enabled to accelerate the training process. To further augment data, random transformations including resizing, color jitter, horizontal flips, and rotations were applied with a probability of 0.5 during training.

Our experiments demonstrate the effectiveness of the CNN-LSTM architecture for classifying pallet simulation videos. The model achieved its best performance at epoch 22, with Validation Loss of **0.2296**, Accuracy of **92.66%**, Precision of **86.36%**, Recall of **95.00%**, F1 Score of **90.48%**.

Fig. 6. An example of style transfer of our simulation. First image is the style reference, a real test with paper wrapping, second image is a frame of the simulation (a configuration with plastic wrapping), third image the result created with https://lensgo.ai/.

5 Conclusions and Future Work

In this paper, we presented the design and development of a novel virtual simulator for crash tests of load units, based on 3D graphics of cloth. The simulator was proved capable of generating a wide variety of synthetic renderings that adhere to the real physical simulation with the acceleration bench machinery. The obtained renderings are fully controllable, and the automatically calculated metadata give precious information about test results. Moreover, the flexibility of such an instrument provides us with the ability to create different configurations, enriching the typologies of crash tests. Finally, the pipeline was completed by training a neural network that classifies success and failure cases, allowing for a complete labeling of test videos. All European packaging companies are required to adhere to EUMOS 40509, and have great interest in cutting costs and environmental impact of these crash tests. Our simulator is the first proposal targeted at the industry world that deals with pallet stability and safety in a completely virtual 3D environment.

As observed during simulations, the layout of the packages on the pallet and the positioning of each layer significantly affect the overall stability of the load. As future development, our tool could be integrated in an automatic system to support packaging optimization and transportation planning during pallet assembly. Our system does not allow the elastic forces of the cloth simulation to influence the motion of rigid bodies. A more flexible tool, such as NVIDIA Warp, PhysX, or Bullet, would allow us to implement this feature from scratch, improving the controllability of our system and bringing the simulation closer to real-world cloth behavior. Moreover, conditions like tearing or high strain of the external envelope could be treated separately, by measuring the tension between edges of the mesh, and detecting critical point and implemented with this tools. Finally, despite the rendered results of our simulations show high fidelity to the real physical behavior of these systems, the application shown by training a neural network for real-world test predictions depends on the visual similarity between the simulation videos and environmental setups. In the future, we aim to explore generative models for style transfer, which would allow us to modify our renders to better match the visual characteristics of real-world videos. A preliminary result can be seen in Fig. 6. This approach opens to new challenges in style transfer including computational costs and time, the use of open-source architectures, and better identification of pallet, machine, and package materials. This would provide full control over the physics while achieving a visual result that more closely resembles the actual footage.

References

1. European road worthiness directive. https://eur-lex.europa.eu/eli/dir/2014/45/oj/eng (2014), Accessed 14 April 2025
2. Eumos, the european safe logistics association. quality standards. https://eumos.eu/quality-standards/ (2020), Accessed 14 April 2025
3. Engineering & solutions for transport & logistic nv (estl nv), wafelstraat 46, 8540 deerlijk, belgium. www.estl.be Accessed 14 April 2025
4. Baraff, D., Witkin, A.: Large steps in cloth simulation. In: Proceedings of Conference on Computer Graphics and Interactive Techniques, pp. 43–54 (1998)
5. Chen, R., Chen, Y., Jiao, N., Jia, K.: Fantasia3d: disentangling geometry and appearance for high-quality text-to-3d content creation. In: Proceedings of the IEEE/CVF International Conference on Computer Vision, pp. 22246–22256 (2023)
6. Cihan, M., Hudobivnik, B., et al.: 3d mixed virtual element formulation for dynamic elasto-plastic analysis. Comput. Mech. **68**, 1–18 (2021)
7. Felsenstein, C., Benedict, K., Baldauf, M.: Maritime safety and security challenges-3d simulation based training. TransNav, Int. J. Marine Navig. Saf. od Sea Transp. **7**(3), 327–336 (2013)
8. Gao, K., Gao, Y., He, H., Lu, D., Xu, L., Li, J.: Nerf: neural radiance field in 3d vision, a comprehensive review. arXiv preprint arXiv:2210.00379 (2022)
9. Ge, H., Yan, Z., Yu, W., Sun, L.: An attention mechanism based convolutional lstm network for video action recognition. Multimedia Tools Appl. **78**(14), 20533–20556 (2019). https://doi.org/10.1007/s11042-019-7404-z
10. Guo, H., et al.: Neural 3d scene reconstruction with the manhattan-world assumption. In: Proceedings of the IEEE/CVF Conference on Computer Vision and Pattern Recognition, pp. 5511–5520 (2022)
11. Kaldor, J.M., James, D.L., Marschner, S.: Efficient yarn-based cloth with adaptive contact linearization. In: ACM SIGGRAPH 2010 Papers, pp. 1–10 (2010)
12. Kwang, J.C.: Stable but responsive cloth. In: Siggraph 2002, pp. 604–611 (2002)
13. Liu, Z., Feng, Y., Black, M.J., et al.: Meshdiffusion: Score-based generative 3d mesh modeling. Preprint arXiv:2303.08133 (2023)
14. Lyu, X., Chang, C., Dai, P., Sun, Y.T., Qi, X.: Total-decom: decomposed 3d scene reconstruction with minimal interaction. In: Proceedings of the IEEE/CVF Conference on Computer Vision and Pattern Recognition, pp. 20860–20869 (2024)
15. Malik, A., Abendroth, M., Hütter, G., Kiefer, B.: A hybrid approach employing neural networks to simulate the elasto- plastic deformation behavior of 3d-foam structures. Adv. Eng. Mater. **24**(2), 2100641 (2022)

16. Pylianidis, C., Osinga, S., Athanasiadis, I.N.: Introducing digital twins to agriculture. Comput. Electron. Agric. **184**, 105942 (2021)
17. Shi, H., Xu, H., Huang, Z., Li, Y., Wu, J.: Robocraft: learning to see, simulate, and shape elasto-plastic objects in 3d with graph networks. Int. J. Robot. Res. **43**(4), 533–549 (2024)
18. Tao, F., Xiao, B., Qi, Q., Cheng, J., Ji, P.: Digital twin modeling. J. Manuf. Syst. **64**, 372–389 (2022)
19. Wang, H., O'Brien, J.F., Ramamoorthi, R.: Data-driven elastic models for cloth: modeling and measurement. ACM Trans. Graphics (TOG) **30**(4), 1–12 (2011)
20. Wang, Z., Han, K., Tiwari, P.: Digital twin simulation of connected and automated vehicles with the unity game engine. In: 2021 IEEE 1st International Conference on Digital Twins and Parallel Intelligence (DTPI), pp. 1–4. IEEE (2021)
21. Zhao, D., Lucas, J.: Virtual reality simulation for construction safety promotion. Int. J. Inj. Contr. Saf. Promot. **22**(1), 57–67 (2015)

Segmenting Anything in Architecture: A Tailored Approach to Segmenting Floor Plan Documents

Claudio Piciarelli, Silvia Zottin(✉), Axel De Nardin, and Gian Luca Foresti

Department of Mathematics, Computer Science and Physics,University of Udine, Udine, Italy
{claudio.piciarelli,silvia.zottin,axel.denardin, gianluca.foresti}@uniud.it

Abstract. Room segmentation in architectural floor plan documents is crucial for tasks such as property valuation, space optimization, and automated design analysis. Unlike natural images, floor plans consist of abstract and sparse visual elements that pose unique challenges for segmentation models. In this work, we investigate the applicability of the Segment Anything Model (SAM), a general-purpose foundation model, to room-level segmentation in floor plans. We evaluate SAM on two heterogeneous datasets: a private collection and the public R3D dataset. Our results demonstrate that SAM consistently outperforms other semantic segmentation state-of-the-art approaches, enabling accurate surface area computation with minimal user input. We further analyze the impact of prompt quantity and training source variability, confirming SAM's robustness and practical utility for real-world applications in architectural analysis.

Keywords: Floor Plan Segmentation · Segment Anything Model · Document Analysis

1 Introduction

Document analysis is a fundamental challenge in computer vision and machine learning, encompassing the automated extraction, classification, and interpretation of information from both handwritten [4,6,27] and printed documents [2,15]. Over the years, various techniques have been proposed to process different types of documents, including sacred texts [28,29], business reports [10,19], novels [5,7], scientific articles [9], and floor plans [21,23]. Each category presents distinct challenges, requiring specialized methodologies for effective analysis.

Among the many tasks in document analysis, such as optical character recognition, layout analysis, handwriting recognition, and symbol detection, we focus on floor plan segmentation, a problem with significant practical implications. Floor plans serve as essential blueprints in architecture, real estate, and urban

E. Rodolà et al. (Eds.): ICIAP 2025, LNCS 16167, pp. 416–428, 2026.
https://doi.org/10.1007/978-3-032-10185-3_33

planning, yet their automated interpretation remains challenging due to the coexistence of hand-drawn and digital drawings, stylistic variations, and architectural conventions that have evolved over decades. Additional complexities arise from scanning artifacts such as noise, distortions, and missing information, which can hinder accurate analysis.

In this work, we specifically address room segmentation, the task of precisely delineating rooms within a floor plan at the pixel level. Accurate segmentation is crucial for applications such as space optimization, automated design modifications, and property valuation, facilitating informed decision-making for architects, surveyors, and real estate professionals. To this end, we investigate the Segment Anything Model (SAM) [11], a state-of-the-art segmentation framework leveraging advanced deep learning techniques to provide robust and generalizable segmentation capabilities.

Our study investigates SAM's effectiveness in segmenting architectural floor plans, with a specific focus on its ability to accurately delineate rooms across a wide range of layouts and styles.

We conduct a comprehensive evaluation on two datasets: a private dataset provided by a public institution, and R3D [25], a publicly available benchmark that includes both hand-drawn and digital architectural floor plans. These datasets feature substantial stylistic variability and cover different historical and architectural contexts, providing a robust basis for assessing SAM's generalization capabilities. By analyzing SAM's performance across this diverse collection, we aim to shed light on its strengths and limitations for room segmentation, and to evaluate its practical viability as a tool for real-world applications in architecture, construction, and real estate.

2 Related Work

Floor plan analysis has emerged as a key area of research in computer vision, driven by the growing availability of digitized architectural documents and the need for robust spatial understanding in indoor environments [16]. A wide range of tasks has been explored in this domain, including boundary detection, object recognition, and room analysis.

Among the foundational tasks is boundary detection, which involves identifying key structural elements such as walls and openings [1,17]. These components are essential for reconstructing the topology of indoor spaces. Closely related is the task of binarization, particularly relevant for hand-drawn or degraded floor plans. Suh *et al.* [18] recently proposed a two-step binarization pipeline using Gaussian Mixture Modeling and Random Forests to convert noisy inputs into clean binary maps suitable for further processing.

At the semantic level, object detection techniques have been developed to recognize architectural elements within floor plans. For instance, ArchNetv2 [22] can detect thirteen types of common elements, such as windows, doors, and stairs, enabling more detailed floor plan understanding.

A crucial aspect of semantic interpretation is room segmentation, which aims to isolate and label individual rooms. Traditional methods can be categorized into vector-based and raster-based approaches. Early vector-based techniques, such as Raster-to-Vector [13], relied on geometric primitives and integer programming to reconstruct room layouts. More recent methods like VectorFloorSeg [24] employ graph neural networks to model spatial relationships between structural components. While these approaches produce structured outputs ideal for applications like 3D modeling, they often depend on handcrafted heuristics and extensive post-processing, making them less robust across diverse floor plan styles.

To overcome these limitations, raster-based deep learning approaches have emerged, operating directly on pixel-level data. Deep Floor Plan [25] proposed a multi-task architecture that jointly segments rooms and boundaries using a boundary-guided attention mechanism. Similarly, Offset-Guided Attention [20] introduced an offset-aware module to enhance boundary-region consistency. Despite their effectiveness, these methods often struggle with spatial ambiguities, especially in open layouts, and still require vectorization steps for practical usage.

Most existing methods focus on semantic room segmentation, where the aim is to assign each region a room type. However, many real-world applications such as real estate analytics, regulatory compliance, and architectural planning, require not only segmentation, but geometrically precise room partitioning to enable reliable computation of surface areas.

In this work, we shift the focus from semantic classification to room segmentation, presenting an in-depth analysis of the Segment Anything Model applied to architectural floor plans. Our goal is to provide a practical and generalizable tool for extracting precise room-level masks from rasterized floor plans, enabling automatic computation of habitable surface areas. This represents a crucial step toward integrating floor plan understanding into scalable, real-world workflows in architecture, construction, and urban planning.

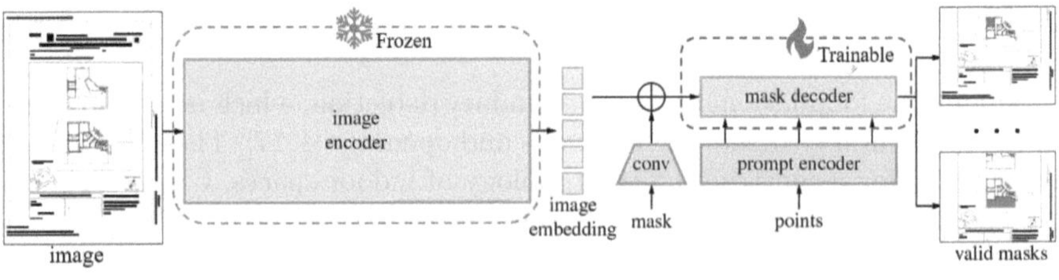

Fig. 1. Overview of the Segment Anything Model fine-tuned for a floor plan room segmentation task. The architecture consists of a ViT-based image encoder (green section), a prompt encoder for user or automatic point inputs (violet sections), and a lightweight mask decoder (orange section) producing room segmentation outputs. (Color figure online)

3 Methods and Experiments

In this section, we describe the methodology adopted to adapt and evaluate the Segment Anything Model for the task of room segmentation in architectural floor plan documents. We detail the model architecture, the datasets used, the experimental setup, and the evaluation metrics employed to assess segmentation performance.

3.1 Model Overview

Segment Anything Model (SAM) [11] is a state-of-the-art segmentation model designed to generate high-quality segmentation masks for a wide range of images with minimal user input. SAM is trained on the SA-1B dataset, which contains over 1 billion masks across 11 million general-purpose images. An overview of SAM's architecture is presented in Fig. 1. Its architecture consists of three main components:

- Image Encoder: A Vision Transformer (ViT) backbone processes the input image $I \in \mathbb{R}^{H \times W \times 3}$ into a grid of dense visual embeddings $E_I \in \mathbb{R}^{h \times w \times d}$, where $h = H/p$, $w = W/p$, and p is the patch size (typically 16 or 32). SAM supports different ViT variants, including ViT-B (91M parameters), ViT-L (308M parameters), and ViT-H (636M parameters).
- Prompt Encoder: User or automatically generated prompts (e.g., points, bounding boxes, or masks) are embedded into the same latent space using a lightweight multilayer perceptron and positional encodings, producing a prompt embedding E_P. In our work, we employ point-based prompts for guiding the segmentation process.
- Mask Decoder: A two-way attention mechanism fuses the image and prompt embeddings to produce a set of binary segmentation masks $\{M_1, M_2, ..., M_k\} \in \mathbb{R}^{H \times W}$, each accompanied by a confidence score.

In this study, we fine-tune SAM for the specific task of room segmentation in architectural floor plans, adapting it to the structural and semantic characteristics of architectural drawings. Fine-tuning is achieved by freezing both the image and prompt encoders, and retraining the mask decoder only. This approach also allows to pre-compute the image and prompt embeddings at training stage, allowing a very fast training procedure.

3.2 Datasets

For our experiments, we used a private dataset consisting of 105 floor plan images provided by a public entity. This collection includes both hand-drawn and digital diagrams of a building's layout, exhibiting significant variability in style and structure. The dataset spans from the 1950 s to the present day, encompassing a wide range of architectural representations. The ground truth annotations were manually created by expert surveyors, who segmented the images at a pixel-precise level. The segmentation task involved identifying two classes: rooms and background. Unfortunately, at this time, the dataset cannot be shared due to privacy issues; however, we are negotiating the possibility to publish a properly anonymized version.

Additionally, we extended our experiments to the publicly available R3D dataset [25]. The R3D dataset consists of 232 floor plan images, each with a resolution of 512×512 pixels. In this dataset, we considered only room segmentation, treating all other segmentation classes as background.

In Fig. 2, we present sample images from the private dataset and R3D dataset along with their corresponding ground truth room annotations highlighted in distinct colors.

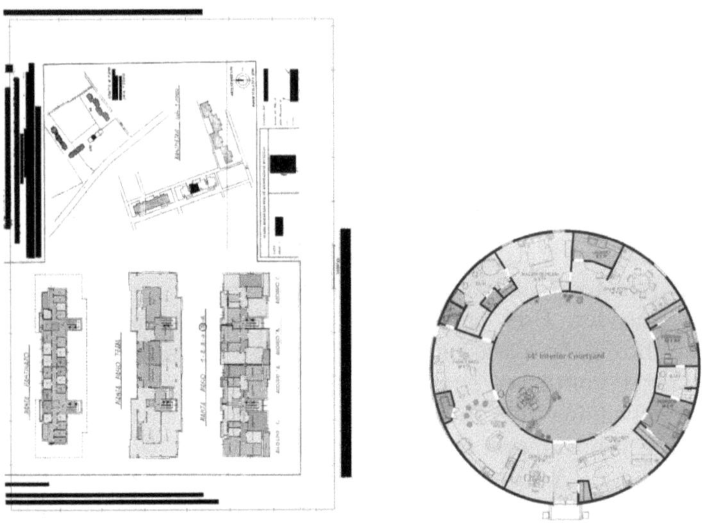

(a) Private dataset **(b)** R3D dataset

Fig. 2. Original images and corresponding GT masks from the private dataset and R3D dataset [25]. Each color represents a region describing a different room in the original image.

3.3 Experimental Setup

The original images from the private dataset have all been rescaled to the same size of 2896×2048 pixels, thus retaining the original A4 width-to-height ratio. The ground truth data is composed of a separate binary mask for each room in the floor plan, thus each floor plan image is typically associated with several masks: in total, the dataset consists of 1246 image/mask pairs. Each image/mask pair was then cropped to a random 1024×1024 square fully enclosing the room, since this is the maximum input size supported by SAM without rescaling.

Since SAM is a prompt-based model, the dataset must also include proper prompts, which can be points, bounding boxes, or textual prompts. We chose to use point prompts since we plan to use the model in a real-world application where the user will click on specific rooms to compute their surface area. In the next experiments, we will always use a single-point prompt unless specified otherwise. In this case, the point is chosen as the mask centroid (or a manually selected point in case the centroid falls outside the mask, as could happen, for example, in L-shaped rooms). In the experiments with multiple-point prompts, we randomly selected both positive (inside the mask) and negative (outside the mask) points.

The final dataset, composed of cropped versions of the original images, their masks, and the corresponding prompt points, was randomly split into train, validation, and test sets with a 6:2:2 ratio.

A grid search of the hyperparameter space showed that the best results could be achieved with a learning rate $lr = 2 \times 10^{-5}$ and a block size $bs = 16$ using a Dice loss. As stated before, training involved only the mask generation part of the network. Early stopping was used to stop the training after ten epochs without validation loss improvements.

3.4 Evaluation Criteria

We evaluate the performance of SAM using several standard metrics for pixel-wise image segmentation, including Intersection over Union, Recall, Precision, and Dice score. Intersection over Union (IoU), also known as the Jaccard index, quantifies the overlap between the predicted segmentation mask and the ground truth mask by computing the ratio of their intersection to their union. It is formulated as.

$$\text{IoU} = \frac{TP}{TP + FP + FN} \tag{1}$$

where TP (true positives) represents the number of correctly identified pixels, FP (false positives) accounts for pixels wrongly classified as positive, and FN (false negatives) denotes pixels that should have been classified as positive but were not.

Recall measures the model's effectiveness in detecting relevant objects and is expressed as:

$$\text{Recall} = \frac{TP}{TP + FN} \tag{2}$$

Precision indicates the accuracy of positive classifications and is computed as:

$$\text{Precision} = \frac{TP}{TP + FP} \tag{3}$$

Lastly, the Dice score, which serves as the harmonic mean of precision and recall, provides a balanced assessment of model performance. It is given by:

$$\text{Dice} = 2 \times \frac{\text{Precision} \times \text{Recall}}{\text{Precision} + \text{Recall}} \tag{4}$$

4 Results

As a first test, we evaluated the performance of the standard, pre-trained SAM model on the private dataset in the three variants *base*, *large* and *huge*. The three variants differ only in the image encoder, which is one of the three encoders described in the original Vision Transformer paper [8]. From the analysis of the first three results shown in Table 1, two main aspects can be deduced:

- the standard SAM model performs poorly on the task of floor plan segmentation. This is probably due to the lack of specific domain knowledge, which is especially needed when two rooms should be considered separated even if they visually appear connected, e.g., due to doors, as shown in Fig. 3;
- the base image encoder achieves better results than the larger ones, probably because the relatively simple graphical structure of floor plans does not require an overly complex encoder to be described.

The last three rows of Table 1 show the results of the fine-tuned model, clearly showing an evident performance boost compared to the standard models. Also, in this case, the best results are achieved using the base image encoder. For this reason, all the results presented in the rest of this section were obtained with the base encoder.

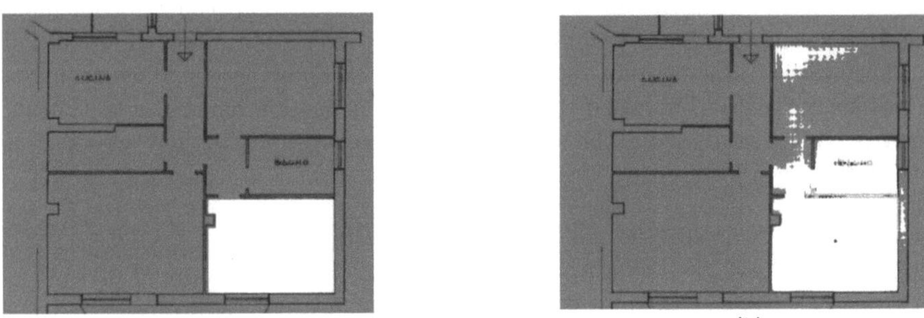

(a) (b)

Fig. 3. Segmentation failure of the pre-trained models. (a) expected result; (b) segmentation computed by the pre-trained *sam-vit-base* model. Red dot shows the prompt point.

A second test aimed at analyzing how the model performance is affected by the number of prompt points. SAM can accept prompts with a variable number of points, both positive (inside the ground-truth mask) and negative (outside the mask). Results shown in Table 1 were computed using a single positive prompt point. We considered four scenarios *s1*–*s4*:

- *s1*: 1 positive point (mask centroid);
- *s2*: 2 random positive points;
- *s3*: 3 random positive points;

Table 1. Performance evaluation of the original pre-trained SAM model and the proposed fine-tuned one, with different ViT image encoders (base, large, huge).

Model	IoU	Dice	Precision	Recall
Pre-trained, base	0.311	0.405	0.724	0.563
Pre-trained, large	0.235	0.318	0.640	0.570
Pre-trained, huge	0.248	0.331	0.653	0.577
Fine-tuned, base	**0.856**	**0.913**	**0.937**	**0.910**
Fine-tuned, large	0.821	0.889	0.929	0.879
Fine-tuned, huge	0.816	0.885	0.915	0.887

– *s4*: 3 random positive points and up to 3 random negative points.

The number of negative points can potentially be less than 3 if some points fall outside the cropping area. Table 2 shows the results for all the possible combinations of the four scenarios, both in the training phase and in the testing phase. The table reports only the IoU performance metric, as the other metrics (Dice, Precision, Recall) have similar patterns. Results show that the model generally performs better when the number of prompt points used for testing is at least the same as or greater than the number of points used for training. The best overall result is reached in the s4/s4 configuration; however, it is worth noting that the performance difference compared to the s1/s1 case is not exceedingly large. This result could be relevant for systems where the input points must be manually inserted by human operators: in this case, a small performance degradation could be acceptable in exchange for faster interaction with the system because fewer points need to be inserted.

Table 2. IoU values with varying numbers of prompt points both in train phase and test phase. In bold, the best score for each column.

		Test			
	Point	s1	s2	s3	s4
Train	s1	**0.856**	**0.824**	0.828	0.816
	s2	0.607	0.822	0.834	0.826
	s3	0.473	0.801	**0.846**	0.837
	s4	0.479	0.762	0.795	**0.861**

As discussed in Sect. 3.2, the main dataset on which the model was tested is currently private. For the sake of experiment reproducibility and in order to promote the production of comparative results by the research community, we also share experimental results on the public R3D dataset [25]. The model was therefore trained on the Private Dataset (PD), on R3D, and on a combination of the two datasets. The three resulting models were tested on the same three different scenarios, always using a single prompt point. As in the previous experiment, we report only the IoU metric, as the remaining ones showed similar patterns. The results, reported in Table 3, show that the best performance is always achieved when the model is trained on both datasets.

Table 3. IoU values with different datasets both in train phase and test phase. In bold, the best score for each column.

		Test		
	Dataset	PD	R3D	PD+R3D
Train	PD	0.856	0.677	0.744
	R3D	0.731	0.838	0.826
	PD+R3D	**0.903**	**0.894**	**0.843**

Results shown in Table 3 highlight that model performance can be improved by increasing the variability of the training dataset, e.g. by adding data from R3D. We investigated whether similar results could be achieved by training on our private dataset alone, but with data augmentation. The adopted augmentation strategy involved the random use of image and point manipulations like vertical and horizontal flips, rotations of multiples of 90°, and random rotations in the range of ±20°. Experimental results (see Table 4) show that dataset augmentation provided a slight benefit, but not as significant as the introduction of R3D. Moreover, without data augmentation, the training procedure is relatively fast, since the image encoder is not fine-tuned and thus image embeddings can be pre-computed. With data augmentation, this is no longer possible, since images randomly change at each epoch, thus requiring the computation of image embeddings at each use. This dramatically increased the training time from 4 s/epoch using a single NVIDIA H100 GPU to roughly 20 s/epoch even when using four H100 GPUs in parallel. For this reason, we did not investigate the use of data augmentation further.

Table 4. Performance metrics of the model, trained with an augmented version of the private dataset.

Dataset	IoU	Dice	Precision	Recall
PD	0.867	0.921	0.945	0.913
R3D	0.714	0.806	0.893	0.806
PD+R3D	0.761	0.839	0.906	0.841

Finally, we performed a comparison of the proposed model with some relevant state-of-the-art image segmentation models: Pyramid Attention Network

(PAN) [12], Fully Convolutional Network (FCN) [14], Pyramid Scene Parsing Network (PSPNet) [26], and DeepLabv3+ [3]. All the models were trained and tested on both the private dataset and R3D, without data augmentation. However, these are promptless semantic segmentation models that detect all rooms at once, without the ability to extract each room separately. In order to compare with these models, our fine-tuned SAM was used to segment every single room with only one prompt point; the resulting masks were then stacked together to form a full segmentation map (see Fig. 4 for an example) that could be compared with other models. Results are shown in Table 5, the proposed fine-tuned SAM model outperformed the other models on the private dataset, and performed better or at par with competitors on R3D.

Table 5. Comparative results of segmentation models on the private dataset and R3D dataset. In bold the best score for each metric and each dataset.

	Private dataset				R3D			
Models	IoU	Dice	Precision	Recall	IoU	Dice	Precision	Recall
PAN [12]	0.558	0.622	0.788	0.581	0.804	0.891	0.884	0.901
FCN [14]	0.817	0.890	0.924	0.861	0.927	**0.962**	0.962	0.963
PSPNet [26]	0.773	0.857	0.855	0.860	0.912	0.953	0.952	0.953
Deeplabv3+ [3]	0.833	0.901	0.929	0.877	0.919	0.958	**0.968**	0.955
Fine-tuned SAM	**0.906**	**0.950**	**0.960**	**0.942**	**0.928**	**0.962**	0.956	**0.970**

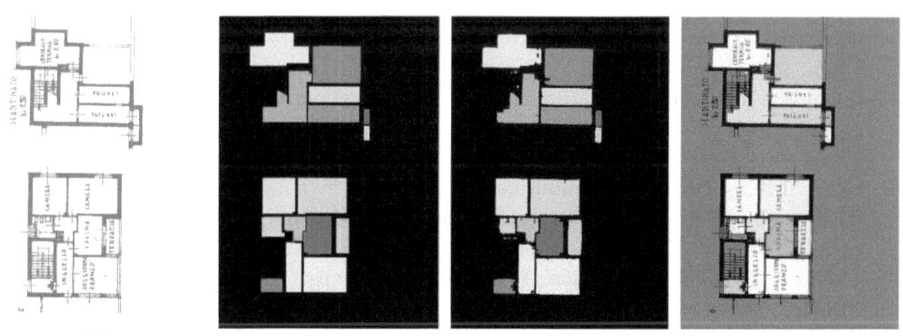

Fig. 4. Qualitative example on the private dataset. (a) detail of floor plan map; (b) the corresponding ground truth; (c) segmentation of each room; (d) segmentation results overlaid to the original map.

5 Conclusion

In this study, we evaluated the Segment Anything Model for room segmentation in architectural floor plans, demonstrating its strong performance across diverse datasets. Fine-tuning SAM led to significant improvements in segmentation accuracy, with the model achieving high IoU, Dice, Precision, and Recall scores. Our results showed that the model performed better when trained on a combination of two datasets, highlighting the importance of dataset diversity. Additionally, SAM's performance was robust even with minimal user input, requiring only a single point prompt to achieve effective segmentation.

For future work, we plan to integrate additional information, such as scale reduction factors, to enable the calculation of total habitable area. This would enhance the model's applicability in real-world architectural applications, such as property valuation and space optimization, making it a valuable tool for professionals in architecture and real estate.

Acknowledgement. Partial financial support was received from: PNRR DD 3277 del 30 dicembre 2021 (PNRR Missione 4, Componente 2, Investimento 1.5) - iNEST; from Strategic Departmental Plan on Artificial Intelligence, DMIF, University of Udine; dall'Unione europea - Next Generation EU, Missione 4 Componente 2 CUP G23C24000790006 Progetto COVERT.

References

1. Ahmed, S., Weber, M., Liwicki, M., Langenhan, C., Dengel, A., Petzold, F.: Automatic analysis and sketch-based retrieval of architectural floor plans. Pattern Recogn. Lett. **35**, 91–100 (2014). https://doi.org/10.1016/j.patrec.2013.04.005, frontiers in Handwriting Processing
2. Chau, P.P.M., Bakkali, S., Doucet, A.: Docsum: domain-adaptive pre-training for document abstractive summarization. In: Proceedings of the Winter Conference on Applications of Computer Vision (WACV) Workshops, pp. 1303–1312 (February 2025)
3. Chen, L.-C., Zhu, Y., Papandreou, G., Schroff, F., Adam, H.: Encoder-Decoder with Atrous Separable Convolution for Semantic Image Segmentation. In: Ferrari, V., Hebert, M., Sminchisescu, C., Weiss, Y. (eds.) ECCV 2018. LNCS, vol. 11211, pp. 833–851. Springer, Cham (2018). https://doi.org/10.1007/978-3-030-01234-2_49
4. De Nardin, A., Zottin, S., Colombi, E., Piciarelli, C., Foresti, G.L.: Is imagenet always the best option? an overview on transfer learning strategies for document layout analysis. In: Foresti, G.L., Fusiello, A., Hancock, E. (eds.) Image Analysis and Processing - ICIAP 2023 Workshops, pp. 489–499. Springer Nature Switzerland, Cham (2024). https://doi.org/10.1007/978-3-031-51026-7_41
5. De Nardin, A., Zottin, S., Paier, M., Foresti, G.L., Colombi, E., Piciarelli, C.: Efficient few-shot learning for pixel-precise handwritten document layout analysis. In: Proceedings of the IEEE/CVF Winter Conference on Applications of Computer Vision (WACV), pp. 3680–3688. Waikoloa, Hawaii (2023). https://doi.org/10.1109/WACV56688.2023.00367

6. De Nardin, A., Zottin, S., Piciarelli, C., Colombi, E., Foresti, G.L.: A one-shot learning approach to document layout segmentation of ancient arabic manuscripts. In: 2024 IEEE/CVF Winter Conference on Applications of Computer Vision (WACV), pp. 8112–8121 (2024). https://doi.org/10.1109/WACV57701.2024.00794
7. De Nardin, A., Zottin, S., Piciarelli, C., Foresti, G.L., Colombi, E.: In-domain versus out-of-domain transfer learning for document layout analysis. Int. J. Doc. Anal. Recogn. (IJDAR) (2024). https://doi.org/10.1007/s10032-024-00497-4
8. Dosovitskiy, A., et al.: An image is worth 16×16 words: transformers for image recognition at scale. arXiv preprint arXiv:2010.11929 (2020)
9. Gemelli, A., Marinai, S., Pisaneschi, L., Santoni, F.: Datasets and annotations for layout analysis of scientific articles. Int. J. Doc. Anal. Recogn. (IJDAR) **27**(4), 683–705 (2024). https://doi.org/10.1007/s10032-024-00461-2
10. Kavasidis, I., et al.: A Saliency-Based Convolutional Neural Network for Table and Chart Detection in Digitized Documents. In: Ricci, E., Rota Bulò, S., Snoek, C., Lanz, O., Messelodi, S., Sebe, N. (eds.) ICIAP 2019. LNCS, vol. 11752, pp. 292–302. Springer, Cham (2019). https://doi.org/10.1007/978-3-030-30645-8_27
11. Kirillov, A., et al.: Segment anything. In: Proceedings of the IEEE/CVF International Conference on Computer Vision (ICCV), pp. 4015–4026 (October 2023)
12. Li, H., Xiong, P., An, J., Wang, L.: Pyramid attention network for semantic segmentation (2018). https://arxiv.org/abs/1805.10180
13. Liu, C., Wu, J., Kohli, P., Furukawa, Y.: Raster-to-vector: revisiting floorplan transformation. In: 2017 IEEE International Conference on Computer Vision (ICCV), pp. 2214–2222 (2017). https://doi.org/10.1109/ICCV.2017.241
14. Long, J., Shelhamer, E., Darrell, T.: Fully convolutional networks for semantic segmentation. In: 2015 IEEE Conference on Computer Vision and Pattern Recognition (CVPR), pp. 3431–3440 (2015). https://doi.org/10.1109/CVPR.2015.7298965
15. Ngo, P.: Digital line segment detection for table reconstruction in document images. In: Sclaroff, S., Distante, C., Leo, M., Farinella, G.M., Tombari, F. (eds.) Image Analysis and Processing - ICIAP 2022, pp. 211–224. Springer International Publishing, Cham (2022)
16. Pokuciński, S., Mrozek, D.: Methods and applications of space understanding in indoor environment—a decade survey. Appl. Sci. **14**(10) (2024). https://doi.org/10.3390/app14103974, https://www.mdpi.com/2076-3417/14/10/3974
17. Ryall, K., Shieber, S., Marks, J., Mazer, M.: Semi-automatic delineation of regions in floor plans. In: Proceedings of 3rd International Conference on Document Analysis and Recognition. vol. 2, pp. 964–969 vol.2 (1995). https://doi.org/10.1109/ICDAR.1995.602062
18. Suh, H., Kim, H., Yu, K.: Machine learning-based binarization technique of hand-drawn floor plans. Int. J. Doc. Anal. Recogn. (IJDAR) (2025). https://doi.org/10.1007/s10032-025-00514-0, https://doi.org/10.1007/s10032-025-00514-0
19. Thomas, E., et al.: Raptor: refined approach for product table object recognition. In: Proceedings of the Winter Conference on Applications of Computer Vision (WACV) Workshops, pp. 1333–1342 (2025)
20. Wang, Z., Sun, N.: Offset-guided attention network for room-level aware floor plan segmentation. IEEE Access **11**, 63667–63677 (2023). https://doi.org/10.1109/ACCESS.2023.3288598
21. Wei, L., Lai, C.: Wall segmentation in house plans: fusion of deep learning and traditional methods. Vis. Comput. **40**(9), 6015–6031 (2024). https://doi.org/10.1007/s00371-023-03150-z

22. Xu, Z., Jha, N., Mehadi, S., Mandal, M.: Multiscale object detection on complex architectural floor plans. Autom. Constr. **165**, 105486 (2024). https://doi.org/10.1016/j.autcon.2024.105486
23. Xu, Z., Yang, C., Alheejawi, S., Jha, N., Mehadi, S., Mandal, M.: Automatic floor plan analysis using a boundary attention-based deep network. Int. J. Doc. Anal. Recogn. (IJDAR) **28**(1), 19–30 (2025). https://doi.org/10.1007/s10032-024-00487-6
24. Yang, B., Jiang, H., Pan, H., Xiao, J.: Vectorfloorseg: two-stream graph attention network for vectorized roughcast floorplan segmentation. In: Proceedings of the IEEE/CVF Conference on Computer Vision and Pattern Recognition (CVPR), pp. 1358–1367 (2023)
25. Zeng, Z., Li, X., Yu, Y.K., Fu, C.W.: Deep floor plan recognition using a multi-task network with room-boundary-guided attention. In: Proceedings of the IEEE/CVF International Conference on Computer Vision, pp. 9096–9104 (2019)
26. Zhao, H., Shi, J., Qi, X., Wang, X., Jia, J.: Pyramid scene parsing network. In: 2017 IEEE Conference on Computer Vision and Pattern Recognition (CVPR), pp. 6230–6239 (2017). https://doi.org/10.1109/CVPR.2017.660
27. Zottin, S., et al.: Exploring few-shot text line segmentation approaches in challenging ancient manuscripts. In: CEUR Workshop Proceedings. vol. 3937 (2025)
28. Zottin, S., De Nardin, A., Colombi, E., Piciarelli, C., Pavan, F., Foresti, G.L.: U-diads-bib: a full and few-shot pixel-precise dataset for document layout analysis of ancient manuscripts. Neural Comput. Appl. **36**(20), 11777–11789 (2024). https://doi.org/10.1007/s00521-023-09356-5
29. Zottin, S., De Nardin, A., Foresti, G.L., Colombi, E., Piciarelli, C.: ICDAR 2024 competition on few-shot and many-shot layout segmentation of ancient manuscripts (sam). In: Barney Smith, E.H., Liwicki, M., Peng, L. (eds.) Document Analysis and Recognition - ICDAR 2024, pp. 315–331. Springer Nature Switzerland, Cham (2024)

Automated Instance Label Generation for Traffic Domain to Enhance YOLACT-YOLO

Dick Scholte[1,2](✉), Matthijs H. Zwemer[2], Peter H.N. de With[1], and Egor Bondarev[1]

[1] Department of Electrical Engineering, Eindhoven University, Eindhoven, Netherlands
d.scholte@tue.nl
[2] ViNotion B.V., Eindhoven, The Netherlands

Abstract. This paper presents a novel approach to generate accurate instance masks for challenging traffic surveillance scenes to allow retraining YOLACT-YOLOv9 model for surveillance domain. We propose a two-stage method as follows. First, we automate the generation of instance segmentation labels by applying the Segment Anything Model using 2D bounding boxes as prompts to generate instance masks. Three specific adaptations are proposed to improve the segmentation quality: adopting multiple masks per object, incorporating foreground/background knowledge for each object, and filtering of challenging instance masks. Second, we introduce a novel multi-scale soft loss function for training YOLACT-YOLOv9 to handle missing instance masks based on consistency regularization. Experimental results show that detection and segmentation accuracy is significantly improved, particularly in challenging traffic scenes. The final model obtains the detection and instance segmentation performance of 92% mAP and 84% mAP, respectively.

Keywords: Instance Segmentation · Object Detection · Traffic Surveillance · Semi-Supervised · YOLOv9 · Automated Label Generation

1 Introduction

Nowadays automated systems for visual traffic surveillance are widely employed, supporting a range of tasks. In these systems, live video streams are interpreted by automated procedures to measure the movements of traffic participants. In order to analyze movements correctly, it is essential to detect, localize, and follow all traffic participants accurately. Object localization commonly uses 2D bounding boxes to represent object locations. A more accurate representation would be to describe the contours of the traffic participants, which can be done by instance segmentation. For instance, the road location of a large elongated

CONVOLVE - EU Horizon 2020 (Grant nr. 101070374).

E. Rodolà et al. (Eds.): ICIAP 2025, LNCS 16167, pp. 429–441, 2026.
https://doi.org/10.1007/978-3-032-10185-3_34

object differs enormously if estimated by a bounding box or instance segmentation (if all camera parameters are known), as depicted in Fig. 1 by the red dot.

Fig. 1. A scene containing a large truck, where its location is represented by a either a 2D bounding box or a instance mask, and the estimated road location as red dot. (Color figure online)

The development of instance segmentation models for traffic monitoring applications poses several challenges. The main challenge is the lack of datasets containing ground-truth for instance segmentation in traffic surveillance scenes. A second challenge is to maintain real-time inference while incorporating instance segmentation into a detection model. A third challenge is that traffic participants in surveillance scenarios can be heavily occluded. The final challenge is to ensure high performance in all types of lighting and weather conditions.

With the appearance of the newest state-of-the-art instance segmentation models, we are advancing toward the goal of automated label generation for the traffic surveillance domain. However, there are multiple approaches for generating pseudo-ground truth using Box2Mask [10], Dinov2 [13], or Segment Anything Model (SAM) [9]. Unsupervised and self-supervised learning approaches have shown effectiveness in standard image domains, for example, indoor scenes and autonomous driving. However, these methods often struggle with traffic surveillance imagery due to challenges such as low image quality and under-segmentation. Alternatively, some generation procedures leverage SAM. The generation procedure [15] has shown promising results, but the generation procedure did not incorporate highly occluded objects, and as a result, the instances split into multiple different regions were neglected. In addition, the generation procedure tries to create instance masks for all bounding boxes. However, it can fail to generate an instance mask in difficult situations, e.g. high occlusions, or bad weather. During training, these missing instance masks were ignored.

The focus of this work is on a two-fold approach in order to face the challenges of occlusions, adverse conditions, and to improve the methodology of [15]: (1) enhance the instance mask generation procedure with additional properties from traffic surveillance scenes and (2) enforce the real-time YOLACT-YOLOv9 [1] model with an additional loss function to better cope with missing instance masks. The extension of the instance mask generation procedure specifically addresses issues related to occluded objects and aims at improving segmentation performance under poor lighting conditions. First, it improves the segmentation quality for partly occluded traffic participants. Second, the procedure solves mask errors due to inter-object occlusions. Third, and last, it removes erroneous segmentation masks of hard conditions such as rain, fog, snow, and poor lighting. Moreover, the YOLACT-YOLOv9 model should be extended with an additional loss function to handle (sparsely) missing instance segmentation ground-truth labels. As a result, the two-fold approach allows to generate instance masks as (pseudo) ground-truth labels, relieving the burden of manual annotation, while utilizing the computationally efficient YOLACT-YOLOv9 model for retraining.

2 Related Work

A. Weakly Supervised Instance Segmentation. Weakly supervised instance segmentation is an approach in which an algorithm learns to segment individual objects in an image using only limited or weak annotations. These methods reduce the cost and effort of labeling large datasets while leveraging additional techniques like unsupervised learning or pseudo labels to enhance instance segmentation accuracy. Pseudo labels help to refine the training process but remain imperfect approximations of true labels.

The impact of imperfect annotations and performance is explored extensively [8]. Jiménez *et.al.* found that noise in the bounding box significantly affects segmentation performance, while noise in the segmentation also impacts bounding-box estimation performance. Furthermore, pretraining models with similar data improves predictive performance. Rizve *et.al.* [14] experimented with training with a large number of incorrect pseudo labels and therefore introduced a pseudo label selection process to reduce noise. In addition, Galvez *et.al.* [4] demonstrated that the removal of noisy annotations from the data has a positive impact on model performance. On the other hand, Shen *et.al* [16] proposed box-level supervision [10], which simplifies training compared to image-level supervision. However, these models are computationally too complex for edge devices and do not achieve state-of-the-art performance in the traffic domain [15].

B. Generating Pseudo Ground-Truth. Besides the box-supervised method to generate pseudo instance labels, there are several other methods. One of the methods, U2Seg [12], is based on Dinov2 [13], which provides strong feature embeddings but requires additional processing for instance labels. The model captures semantic information, but does not inherently distinguish between separate objects of the same class. This can lead to under-segmentation in traffic scenes. On the other hand, Labelmaker [19], which employs a student-teacher network to generate instance labels, which requires already annotated data. Moreover, the WSCOS model [5] and the generation approach [15] employ SAM [9] which uses prompts as guidance for segmentation and achieves state-of-the-art performance. Given that SAM is not error-free, pseudo ground-truth labels generated by SAM may contain inaccuracies, as noted in [18]. These errors

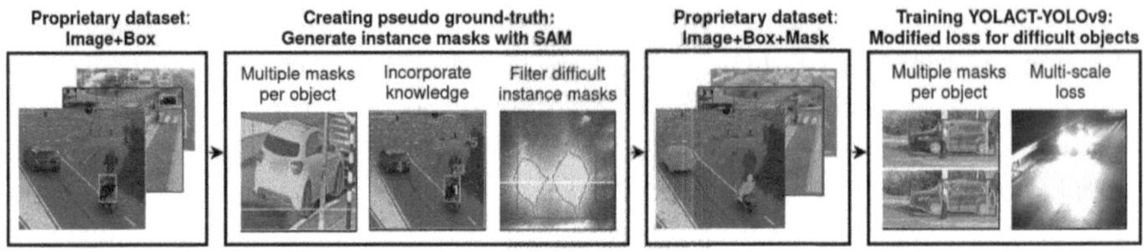

Fig. 2. Proposed approach for a dataset with bounding boxes only. First, generation of pseudo-ground truth with SAM. Second, training of YOLACT-YOLOV9 model.

are expressed in several forms, including (1) incomplete segmentation, (2) over-segmentation, and (3) incorrect labeling, often due to object confusion in complex scenarios. For example, SAM frequently generates multiple disconnected masks for occluded objects, as highlighted in [6]. Moreover, its performance is significantly lower in low contrast scenes and under low light or night conditions, as discussed in [7].

C. Consistency Regularization. Semi-supervised methods exist to improve supervised learning while using partially unlabeled data [3]. A form of semi-supervised learning is consistency regularization, where the discrepancies are penalized between the predictions of the modified inputs. By enforcing the consistency in the network response, it helps the model to generalize better and become more robust to variations and noise in the data. STAC [17] extends semi-supervised methods by augmentation-driven consistency regularization and gains increased performance. In addition, DFCPS [2] introduced cross-pseudo-supervision, combining coherent learning with self-training. This approach leverages pseudo labels from multiple augmentation angles to better exploit unlabeled data, enhancing performance and generalization. The refined pseudo labels and filtering strategies further improve the quality of the labels and the robustness.

3 Methods

This section elaborates on a proposed two-stage approach to train the YOLACT-YOLOv9 model on the proprietary dataset, as shown in Fig. 2. First, in Sect. 3.1 the instance masks for training are generated by SAM, since the YOLACT-YOLOv9 model requires these instance segmentation annotations for training. Second, in Sect. 3.2 the YOLACT-YOLOv9 model is trained on the dataset with the generated ground truth, for creating instance masks. To this end, the YOLACT-YOLOv9 model is adapted to cope with missing labels.

3.1 Creating Pseudo Ground-Truth Data for Instance Segmentation

This section describes the first step that uses the existing dataset containing images and bounding-box annotations as input and then generates masks for each annotated bounding box by utilizing SAM (see left-hand side of Fig. 2). Each bounding box corresponding to an object in the dataset (together with the image) is used as a prompt for SAM to predict an instance mask for this specific object. We propose two modifications to handle (partially) occluded objects. First, we compute the predictions of multiple instance masks per object. Second, we incorporate the knowledge about other nearby located objects, causing occlusions between objects. Both modifications are discussed in detail below.

A. Prediction of Multiple Masks per Object. Typically, each object is represented by only one (associated) instance mask. These instance masks do overlap per image, resulting in multiple classes per pixel on occasion. The indexes of these masks determine which object is foreground. In the traffic surveillance use case, multiple partially occluded objects commonly occur. Therefore, multiple instance masks are often required and applied. An example of an occlusion is depicted in Fig. 2, where a lamppost divides the delineation of the car into two regions. Therefore, we allow SAM to predict multiple instance masks per object, such that we obtain more accurate instance masks. We have empirically determined that up to a maximum of five mask predictions should be merged to obtain a robust segmentation mask (SAM may predict less than five masks).

B. Incorporating Knowledge About Foreground Objects. If two bounding boxes have a large overlap, the estimation of instance masks with SAM by using both boxes individually as a prompt, results in the same instance segmentation for both persons, i.e. the mask is predicted for the person in the foreground. To solve this issue, four additional processing steps per image are proposed and also shown in Fig. 3. *First*, a sorted list based on bottom locations of all bounding boxes is created to order the objects, according to their distances to the camera.*Second*, an empty mask image (called negative mask) is initialized (Row 2, Column 1 in Fig. 3). This negative mask is used to filter out pixel areas that have been employed in instance masks of other objects, since each pixel in the image can only correspond to a single object. *Third*, the first bounding box is selected from the sorted list and used as a prompt for SAM to generate a mask prediction. The areas that are high in the negative mask image are then subtracted from the mask prediction to obtain the segmentation mask for the currently prompted object. *Fourth*, the resulting mask is added to the negative mask. These pixels are excluded for the remaining bounding boxes in the sorted list. The third and fourth step are repeated until instance masks are estimated for all bounding boxes in the image.

C. Filtering of Inaccurate Instance Masks. SAM is not always able to create instance masks, due to low-lighting conditions, low-quality images or objects being too heavily occluded. An example, where the headlights of a car oversaturate the image, is shown in Fig. 2. The predicted instance masks mainly cover the bright headlights of the car. Such erroneous masks should be avoided,

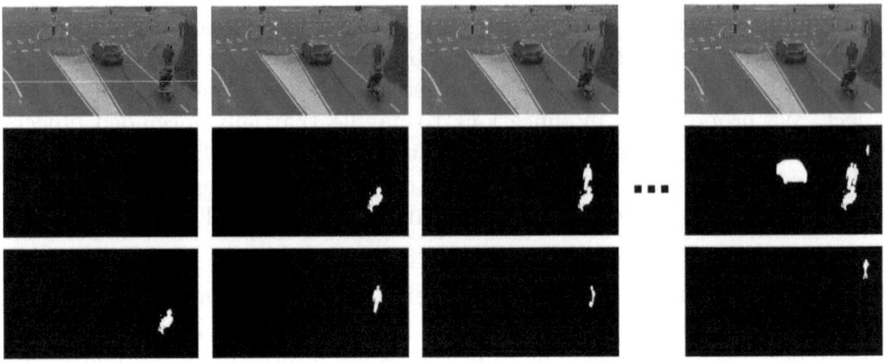

Fig. 3. Each column is a new object for segmentation and consists of (Row 1) the input image with bounding box and (Row 2) the negative mask for SAM (combined from previous objects and cannot be used again). The last row shows the resulting mask.

since these masks are applied to supervise the YOLACT-YOLOv9 model during training. Therefore, faulty instance masks should be removed from the dataset. However, manually removing all these instance masks is time-consuming. Therefore, we propose to avoid prediction of segmentation masks based on a heuristic approach. Typically, SAM has problems with estimating masks when parts of an object are over-saturated in the image. Therefore, we decide to skip the prediction of instance masks if the average pixel-intensity value of the pixels in the object area (defined by the bounding box in the dataset) is higher than 200. This method may remove actual samples, but also ensures that no falsely predicted masks become part of the dataset without costly manual effort. With this, the YOLACT-YOLOv9 model should be adapted to cope with (sparsely) missing ground-truth labels for learning the instance segmentation task.

3.2 Training YOLACT-YOLOv9 on Pseudo Ground-Truth Data

The second step is training the YOLACT-YOLOv9 model with the pseudo-ground truth. Normally, the model accepts only a single segmentation mask per bounding box. Therefore, the YOLACT-YOLOv9 data loading process is adjusted to accept multiple instance segmentations for a single object and then combines them into a single mask. Furthermore, we propose to modify the model such that it can be trained with missing instance labels.

A. Training with Partially Missing Ground-Truth for Instance Masks. The original loss function during the training of the YOLACT-YOLOv9 model is composed of four individual loss terms, namely the objectness, class, box, and mask losses. The work of Scholte *et al.* [15] has introduced weighting of the mask loss by the number of instance masks that are actually present in the batch. The loss is weighting a batch more if objects lack instance labels and is defined by:

$$\mathcal{L} = \lambda_{obj}\mathcal{L}_{\text{obj}} + \lambda_{\text{cls}}\mathcal{L}_{\text{cls}} + \lambda_{\text{box}}\mathcal{L}_{\text{box}} + \lambda_{\text{mask}}\alpha_{\text{mask}}\frac{gt_{\text{all}}}{gt_{\text{mask}}}\mathcal{L}_{\text{mask}}\,, \tag{1}$$

where λ_{obj}, λ_{cls}, λ_{box} and λ_{mask} denote scalars to weight the respective loss functions, gt_{mask} is the number of ground-truth masks within the image, gt_{all} is the total number of ground-truth annotations within the image, and α_{mask} is a normalization factor that indicates the fraction of objects in the dataset that contain ground-truth masks, to ensure that the mask loss is not negligible in case a batch contains only a few instance labels.

B. Multi-Scale Loss for Missing Instances. Neglecting all samples that lack mask annotations causes confusion during training. Therefore, we propose to utilize a novel multi-scale soft loss function, which is based on a multi-scale assigner that encourages the model to make invariant consistent predictions. The proposed multi-scale soft loss ($\mathcal{L}_{\text{soft}}$) enforces consistency for mask prediction by utilizing only bounding boxes as ground-truth labels.

The multi-scale soft loss $\mathcal{L}_{\text{soft},o}$ for object o is calculated by comparing the predicted instance masks $\hat{M}_{t,i}$ with a constructed ground-truth mask $M_{t,o}$ by

$$\mathcal{L}_{\text{soft},o} = \frac{1}{P_o}\sum_{i=0}^{P_o}\text{BCE}\left(M_{t,o}, \hat{M}_{t,i}\right), \tag{2}$$

where P_o denotes the number of positive anchors that are matched to object o, and BCE$(\cdot)$ represents the binary cross-entropy function. Parameter $M_{t,o}$ is constructed for the single object o with a missing instance mask. The task-aligned assigner that is incorporated by the YOLOv9 model generates a dynamic number of positive anchors for each level in the feature pyramid. An instance mask is predicted for each positive anchor j by the YOLACT-YOLOv9 head in the current iteration t. From the predicted masks for all anchors that are matched to the same object, together with the same combination of previous iterations on the same object/images during training, we define the updated mask as:

$$M_{t,o} = M_{\text{neg}}\left(\lambda_{\text{hist}}M_{t-1} + \frac{1-\lambda_{\text{hist}}}{P_o}\sum_{j=0}^{P_o}\hat{M}_{t,j}\right), \tag{3}$$

where $t-1$ indicates the previous and t the current iteration of the same object during training, λ_{hist} is a weight factor that determines how much information from the mask historically predicted is included. The sum over $M_{t,j}$ forms a new mask that is normalized over the prediction masks over all positive anchors P for the current object. All available instance masks present in the same image are treated as negatives and combined into one mask, which is converted and used as a negative ground-truth label, called M_{neg}. The pixel-by-pixel multiplication of M_{neg} and $M_{t,i}$ yields an instance mask that does not overlap with any of the other available instance masks and is called the constructed ground-truth

Fig. 4. Example images of the proprietary dataset and manual annotated test set.

mask. Averaging over the historical masks is of interest due to the variation of augmentations during the training process.

Both mask losses are combined with addition, where both utilize the same weight factor to balance the amount of ground-truth data with and without instance masks. The combination of Equation (2) and Equation (1) results in:

$$\mathcal{L} = \lambda_{obj}\mathcal{L}_{\text{obj}} + \lambda_{\text{cls}}\mathcal{L}_{\text{cls}} + \lambda_{\text{box}}\mathcal{L}_{\text{box}} + \frac{\lambda_{\text{mask}}\alpha_{\text{mask}}gt_{\text{all}}}{gt_{\text{mask}}}(\mathcal{L}_{\text{mask}} + \frac{1}{N}\sum_{o=0}^{N}\mathcal{L}_{\text{soft},o}), \quad (4)$$

where the $\mathcal{L}_{\text{soft},o}$ per object o is accumulated over the number of objects in a batch N in the image.

4 Experiments and Results

This section first describes the proprietary traffic surveillance dataset and the utilized metrics. Then, the experiments and corresponding results are presented in three subsequent Sects. 4.1 through 4.3.

A. Traffic Surveillance Dataset. The experiments are performed with the proprietary dataset and the COCO2017 dataset [11]. The proprietary dataset with bounding boxes only, contains 135,000 images (shown in Fig. 4) of traffic surveillance in a broad variety of different scenes, such as busy traffic scenes, crowded pedestrian areas, and bicycle roads. The proprietary dataset is annotated with 6 traffic classes defined as follows: pedestrian, cyclist, motorcyclist, car, bus, and truck. Annotated bounding-box labels cover the entire object, even if occluded parts are within the considered bounding box. Testing a trained model on the proprietary dataset directly is impossible because of the lack of ground-truth instance masks. Therefore, two test sets are created manually, as shown in at right-hand side of Fig. 4. The 'proprietary daytime test set', consists of 100 images containing 1,129 pedestrians, 77 cyclists, 11 motorcyclists, 370 cars, 21 trucks and 21 buses. The 'proprietary nighttime test set', consists of 100 images containing 142 cars and 7 trucks. The test sets are separated to show the differences in evaluation. The remaining proprietary dataset, is split into a training set (123,829 images) and a validation set (13,127 images).

Table 1. Results of the original SAM method and the proposed generation procedure for the proprietary dataset.

Test set	Method	Mask performance				
		$mAP_{0.50-0.95}$	$mAP_{0.50}$	$mAP_{0.75}$	$mAP_{0.85}$	$mAP_{0.95}$
Daytime	SAM	77.0	94.5	84.3	69.6	29.1
Daytime	Proposed	81.8	94.9	88.5	77.6	39.0
Nighttime	SAM	54.2	90.1	59.0	33.4	0.0
Nighttime	Proposed	54.3	90.5	59.0	33.4	0.0

B. Evaluation Metrics. To evaluate the performance, the Average Precision (AP) metric is used. The AP is an approximation of the area under the precision-recall curve and is computed similarly to the COCO benchmark.

4.1 Creation of Pseudo Labels Employing SAM

This first experiment evaluates the quality of the pseudo labels generated by SAM. This experiment measures the difference between the original SAM-based generation from the work of Scholte*et al.* [15] and the proposed generation procedure that allows multiple masks, incorporates foreground/background knowledge and adds extra filtering of the mask (see Sect. 3.1). To this end, both generation procedures are tested on both proprietary test sets.

The results of the experiment are depicted in Table 1. The proposed generation procedure outperforms the original results by a margin of 4.8% for $mAP_{0.50-0.95}$. Moreover, a significant improvement of 8% and 9.9% can be observed for $mAP_{0.85}$ and $mAP_{0.95}$, respectively. Figure 5 visualizes the importance of allowing multiple instance masks and incorporating knowledge about the foreground causing the major improvements. The results on the proprietary nighttime test set are completely different, both procedures achieve poor performance. Performance is only high for $mAP_{0.50}$. The $mAP_{0.75}$ to $mAP_{0.95}$ drop significantly to 59.0%, 33.4% and 0.0%, respectively. We can hypothesize that SAM underestimates the objects as shown in Fig. 5, such that the overlap thresholds are not satisfied. Low-quality images do not contain clear edges, thereby making it harder to distinguish individual objects.

Based on the results, it can be concluded that incorporating additional information in the generation procedure improves the quality of the pseudo-ground-truth annotations and it mainly improves segmentation quality of (highly) occluded objects. Furthermore, it can be concluded that the nighttime results of SAM contain inaccuracies, especially for poor lighting conditions.

4.2 Training of YOLOACT-YOLOv9 Model with Fractions of Masks

The second experiment measures whether both bounding box and instance segmentation annotations are required for *every* object in a dataset, or if instance

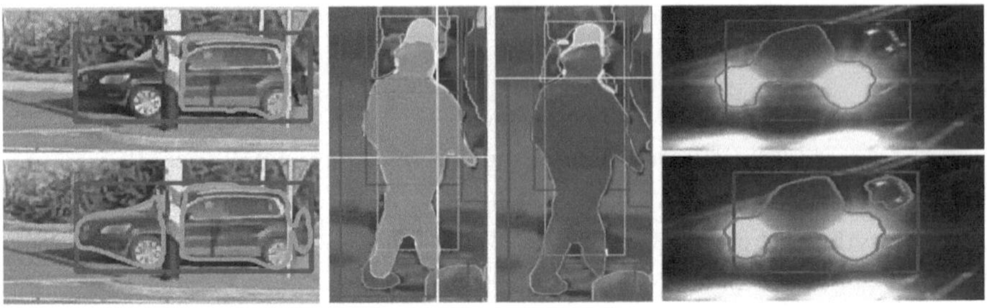

Fig. 5. Generated masks by naive method and proposed method. Left: difference in amount of instance masks for a box. Middle: incorporating foreground, overlapping is shown as orange. Right: poor results for both methods for a nighttime image. (Color figure online)

segmentation annotations for only part of the data would be sufficient for training the YOLACT-YOLOv9 model. This knowledge significantly reduces the required manual annotation effort and/or allows training with an imperfect dataset.

The experiment evaluates the effect of training the YOLOv9-YOLACT model when only a fraction of instance masks is available and is performed on the COCO2017 dataset [11]. The amount of available ground-truth segmentation is varied from 30% to 100% of the objects. Note that for all experiments, the bounding-box ground-truth is always available and not discarded, only the amount of available instance masks is different. The experiment is carried out for the 100%, 70%, 50% and 30% splits. The training of the YOLOv9-YOLACT model is first performed with the standard loss function as described in Eq. (1). This standard loss function completely ignores bounding boxes without instance masks during training. Then the training is performed with the proposed multi-scale loss (see Equation (4)).

Table 2. Results of the training with a limited amount of instance masks available during training, but all bounding boxes are available.

		Box performance				Mask performance			
Loss function	**Metric**	**Split**				**Split**			
		100%	**70%**	**50%**	**30%**	**100%**	**70%**	**50%**	**30%**
Standard (Eq. (1))	$mAP_{0.50}$	69.2	67.3	65.8	59.6	66.1	63.3	62.5	57.0
	$mAP_{0.50-0.95}$	52.1	50.1	48.2	42.7	43.0	40.5	39.1	34.9
Multi-scale (Eq. (4))	$mAP_{0.50}$	69.2	67.5	67.2	62.9	66.1	64.4	63.3	59.7
	$mAP_{0.50-0.95}$	52.1	50.3	50.1	45.1	43.0	41.4	40.5	36.6

The results are shown in Table 2. It can be observed that the training with the multi-scale loss outperforms the standard loss training for all splits. The segmentation performance with the standard loss drops with 9.1% for $mAP_{0.5}$ and 8.1% for $mAP_{0.5-0.95}$ when only 30% of the instance labels are available. For the multi-scale loss training, the segmentation performance only drops with 6.4% for both $mAP_{0.5}$ and $mAP_{0.5-0.95}$. Besides that, the bounding box performance drops evenly with the segmentation performance and the impact of missing segmentation data is notable. Moreover, the margin increases with the number of missing instance masks. At a training split of 70% the margin is only 0.2% for both the $mAP_{0.5}$ and $mAP_{0.5-0.95}$, while the margin increases to 2.7% for $mAP_{0.5}$ and 1.7% for $mAP_{0.5-0.95}$ with a split of 30%. It can be concluded that training with the multi-scale loss increases the performance when training with partial availability of segmentation labels. Furthermore, the drop in performance is negligible when small parts of the objects lack segmentation labels.

4.3 Train YOLACT-YOLOv9 Model While Removing Hard Annotations Containing Bad Weather and Lighting Conditions

This experiment aims to assess whether it is necessary to remove incorrect instance labels from the proprietary dataset. The proprietary dataset contains several low-quality images in terms of contrast and saturation due to poor weather conditions and low-lighting conditions. The filtering approach explained in Sect. 3.1 removes instance masks for overexposed images, resulting in a dataset with (sparsely) missing instance labels. In total, there are 483,078 ground-truth bounding boxes with 482,617 instance labels generated with the proposed generation procedure, where 1,443 instance labels are additionally removed by the filtering approach. Note that the wrong instance labels are removed to prevent learning on the erroneous masks. Therefore, we hypothesize that removing these samples actually improves performance. This is supported by the results of the previous experiment, which has shown that limited missing of instance masks has negligible impact on the recognition performance when the multi-scale loss is utilized during training. In this experiment, we validate the YOLACT-YOLOv9 model by training on either the unfiltered dataset, or by training on the filtered dataset and testing on the proprietary nighttime test set.

The results of the models trained with and without filtering of the incorrect instance masks are presented in Table 3. For the proprietary nighttime test set, a small performance increase can be observed. Note that training on the filtered proprietary dataset yields an improvement of 0.7% and 0.5% in $mAP_{0.50-0.95}$ on the nighttime test set for detection and segmentation, respectively. Since the proprietary dayttime test set does not contain poor-quality images, there is no difference in the performance, as expected. It can be concluded that removing such poor-quality images affect the results positively together with the implementation of the multi-scale loss. Furthermore, for the daytime test set, the YOLACT-YOLOv9 model achieves a high final performance of 92.7% and 84.1% in $mAP_{0.50}$ for detection and segmentation, respectively.

Table 3. Detection and segmentation performance of YOLACT-YOLOv9 models trained on the proprietary dataset with and without applying the filtering.

	Proprietary daytime test set				**Proprietary nighttime test set**			
Filter	**Box**	**Mask**	**Box**	**Mask**	**Box**	**Mask**	**Box**	**Mask**
	$mAP_{0.50}$		$mAP_{0.50-0.95}$		$mAP_{0.50}$		$mAP_{0.50-0.95}$	
✗	92.7	84.1	69.1	54.1	99.2	97.0	83.5	66.8
✓	92.7	84.1	69.2	54.2	99.5	97.4	84.2	67.3

5 Conclusions

In this research, we have proposed and evaluated several methods to enhance the joint training of object detection and instance segmentation for traffic surveillance applications. First, we have improved an automated label-generation procedure for datasets with bounding boxes only. The segmentation accuracy for occluded objects is improved by enabling the generation of multiple masks per object. Specific handling of overlapping objects by integrating knowledge of foreground/background results in a more robust instance segmentation generation procedure. The segmentation performance reaches an $mAP_{0.5}$ score of 94.9%. Second, we have extended the training of the YOLACT-YOLOv9 model by introducing a novel loss function to cope with missing ground-truth instance masks. The proposed multi-scale loss, combined with consistency regularization techniques, enables the model to handle sparse and imperfect instance segmentation labels without sacrificing performance. Third, we have applied a filtering method to exclude low-quality or erroneous segmentation masks from training. This filtering mechanism prevents learning from poorly lit or over-saturated images, enhancing the models ability to generalize across diverse traffic conditions. In conclusion, the proposed automated label generation method combined with modifications to the YOLACT-YOLOv9 architecture and advanced loss functions, have resulted in the ability to train a robust detection and segmentation model suited for real-time analysis in traffic surveillance applications. The final performance of the trained YOLACT-YOLOv9 model obtains a detection performance of 92.7% mAP and a segmentation performance 84.1% mAP.

References

1. Chang, H.S., Wang, C.Y., Wang, R.R., Chou, G., Liao, H.Y.M.: Yolor-based multitask learning. In: arXiv:2309.16921 (2023)
2. Chen, Y., et al.: Semi-supervised medical image segmentation method based on cross-pseudo labeling leveraging strong and weak data augmentation strategies. In: 2024 IEEE International Symposium on Biomedical Imaging (ISBI), pp. 1–5 (2024)
3. Fan, Y., Kukleva, A., Schiele, B.: Revisiting consistency regularization for semi-supervised learning. In: Int. J. Comput. Vis. 131, pp. 626–643 (2023)

4. Galvez Jimenez, L., Franzin, A., Decaestecker, C.: Training data selection to improve multi-class instance segmentation in digital pathology. In: Proceedings of the 2023 10th International Conference on Bioinformatics Research and Applications, pp. 27–33. ICBRA '23, New York, NY, USA (2024)
5. He, C., et al.: Weakly-supervised concealed object segmentation with sam-based pseudo labeling and multi-scale feature grouping. In: Advances in Neural Information Processing Systems. vol. 36, pp. 30726–30737. Curran Associates, Inc. (2023)
6. Ji, G.P., Fan, D.P., Xu, P., Zhou, B., Cheng, M.M., Van Gool, L.: Sam struggles in concealed scenes — empirical study on segment anything. Sci. China Inf. Sci. **66**(12) (2023)
7. Ji, W., Li, J., Bi, Q., Liu, T., Li, W., Cheng, L.: Segment anything is not always perfect: an investigation of sam on different real-world applications. Mach. Intell. Res. **21**(4), 617–630 (2024)
8. Jiménez, L.G., Decaestecker, C.: Impact of imperfect annotations on CNN training and performance for instance segmentation and classification in digital pathology. Comput. Biol. Med. **177**, 108586 (2024)
9. Kirillov, A., et al.: Segment anything. In: Proceedings of the IEEE/CVF ICCV, pp. 4015–4026 (2023)
10. Li, W., et al.: Box2mask: box-supervised instance segmentation via level-set evolution. IEEE Trans. Pattern Anal. Mach. Intell. **46**(7), 5157–5173 (2024)
11. Lin, T.Y., et al.: Microsoft coco: common objects in context
12. Niu, D., Wang, X., Han, X., Lian, L., Herzig, R., Darrell, T.: Unsupervised universal image segmentation. In: Proceedings of CVPR, pp. 22744–22754 (2024)
13. Oquab, M., et ak.: Dinov2: learning robust visual features without supervision. In: arXiv:2304.07193 (2024)
14. Rizve, M.N., Duarte, K., Rawat, Y.S., Shah, M.: In defense of pseudo-labeling: an uncertainty-aware pseudo-label selection framework for semi-supervised learning. In: International Conference on Learning Representations (2021)
15. Scholte, D., Urselmann, T., Zwemer, M., Bondarev, E., De With, P.H.: Automated generation of instance segmentation labels for traffic surveillance models. In: Proceedings of the 19th International Joint Conference on Computer Vision, Imaging and Computer Graphics Theory and Applications - Volume 3, pp. 350–358 (2024)
16. Shen, W., et al.: A survey on label-efficient deep image segmentation: bridging the gap between weak supervision and dense prediction. IEEE Trans. Pattern Anal. Mach. Intell. **45**(8), 9284–9305 (2023)
17. Sohn, K., Zhang, Z., Li, C.L., Zhang, H., Lee, C.Y., Pfister, T.: A simple semi-supervised learning framework for object detection. In: arXiv:2005.04757 (2020)
18. Wang, Y., Zhao, Y., Petzold, L.: An empirical study on the robustness of the segment anything model (sam). Pattern Recogn. **155**, 110685 (2024)
19. Weder, S., Blum, H., Engelmann, F., Pollefeys, M.: Labelmaker: Automatic semantic label generation from RGB-d trajectories. In: 2024 International Conference on 3D Vision (3DV), pp. 334–343 (2024)

Accelerating YOLO with EOBranch: An Early Exit Approach for Adaptive Object Detection

Dick Scholte[1,2(✉)], Matthijs H. Zwemer[2], and Egor Bondarev[1]

[1] Eindhoven University, Department of Electrical Engineering, Eindhoven, The Netherlands

[2] ViNotion B. V., Eindhoven, The Netherlands

d.scholte@tue.nl

Abstract. This paper presents an efficient early-out branch (EOBranch) integrated into YOLO-based object detection architectures to address the computational challenges inherent in edge-based traffic surveillance applications. By enabling the early exit for background images, the proposed EOBranch significantly reduces processing time and computational load without sacrificing detection accuracy. We evaluated our approach in YOLOv6 and YOLOv9, through a series of experiments that examine training strategies, optimal branch placement, and extended branch architectures. Experimental results reveal that, while a finetuning strategy delivers high early-exit performance, optimal branch placement is critical. Deeper placements significantly improved average precision, but are computationally more expensive. Placing an extended EOBranch earlier in the backbone achieved early-exit APs of 97.4% and 98.4% for YOLOv6 and YOLOv9, respectively. In particular, these optimized configurations led to reductions of processing time up to 46% for 24 h of traffic scene processing, highlighting the potential for substantial energy savings and enhancing the efficiency of traffic surveillance systems.

Keywords: Early Exit · Branch · YOLO · Object Detection · Dynamic Neural Networks · Edge Computing · Traffic Surveillance · Efficiency

1 Introduction

Modern urban environments increasingly rely on automated visual traffic surveillance systems to monitor and improve traffic flow. These systems support a variety of tasks by analyzing live video streams to monitor and interpret the movements of traffic participants. Edge cameras can be deployed in areas without access to a fixed power connection and therefore require battery-powered

CONVOLVE - EU Horizon 2020 (Grant nr. 101070374).

E. Rodolà et al. (Eds.): ICIAP 2025, LNCS 16167, pp. 442–455, 2026.
https://doi.org/10.1007/978-3-032-10185-3_35

Table 1. Traffic presence in three typical urban surveillance scenes over a 24 h period. A large percentage the scenes are empty and do not contain any objects.

Bus stop	Railway Crossing	City Center

59%	45%	20%

solutions, which makes energy efficiency a critical consideration. This not only extends battery life, but also reduces environmental impact, even where power availability is not a constraint. Research on dynamic neural networks presents a promising approach to optimize performance under these limitations.

Dynamic neural networks offer a promising roadmap as they adapt their structures or parameters during inference based on input complexity. Traffic surveillance systems could benefit from this, by optimizing the process efficiency while maintaining localization. Table 1 illustrates the object sparsity over a 24 h period of three typical traffic videos, where from 20% to 59% of the frames have no objects present in the scene. Therefore, the integration of dynamic neural networks into edge-based surveillance has the potential to significantly improve computational resource management and reduce overall energy consumption.

To address these challenges, several innovative methods have been proposed in recent years. BranchyNet [12] introduces early-exit strategies that allow intermediate layers of a network to output predictions when the prediction confidence is high, thus saving unnecessary computation. Other approaches employ dynamic techniques that adjust the computational load based on the input data complexity of the input [3,14]. Furthermore, research of Matsubara *et al.* [8] employs a neural filter indicating whether an image contains an object or not, decreasing energy consumption together with compression of bottleneck layers.

Building upon these advancements, our research contributes an early-exit mechanism specifically tailored for YOLO-based object detection models. More specifically, we target the YOLOv6 [6] and YOLOv9 [13] architectures for industrial deployment. The early-exit branch, called EOBranch, is integrated into different YOLO models to assess its impact and if the EOBRanch can be applied for different feature extraction methods. This comparative analysis allowed us to identify how the early-exit placement influences the different feature extraction capabilities. Our work investigates the impact on overall system performance, detailing on how it can significantly reduce processing time without degrading detection accuracy. To that end, we introduce a high-accuracy background estimation method designed to efficiently distinguish between scenes containing relevant objects and empty scenes. Additionally, the paper offers an in-depth overview of various training approaches to identify the training method yielding the best performance, and investigates the optimal layer placement.

2 Related Work

Several classes of techniques exist to reduce the inference time of static neural networks, such as pruning to remove redundant parameters, quantization to reduce computation precision, and knowledge distillation to allow training of lighter models [2,4,10]. Although such methods can be effective, they require input data processing through all layers of the model, regardless of the content of the input data. Dynamic architectures with side branches can adaptively exit earlier for specific data inputs, thus reducing unnecessary computation.

A. Early Out Dynamics in Models. Early out dynamics involve a range of strategies that vary in structure, location, and number of side branches, as well as in their exit policies [9,11]. BranchyNet [12] integrates side branches into intermediate layers of a deep neural network to accelerate inference. This early-exit branch not only enhances performance, but also reduces computational costs. In addition, early-exit dynamics addresses critical training challenges, such as counteracting the vanishing gradients or mitigation of overthinking [5].

B. Early Out in Object Recognition. Recent research, such as ESNO [3], has applied early-exit branches to object detection, enabling significant computational savings while maintaining a good balance between accuracy and efficiency. On the other hand, AdaDet [14] employs stochastic variables along with an entropy-based criterion to assess the detection uncertainty, allowing high-confidence samples to exit early, while more complex images are processed by the entire network. On the other hand, Matsubara *et al.* [8] integrate a novel approach based on early exiting, which they call a neural filter. They attempt to filter out empty images (without objects of interest) before they are transmitted from a local edge device to an edge server to reduce the data transfer between these resources. The neural filter responsible for image-empty decisions is placed early in the network and achieves a 91.9% ROC-AUC score. The neural filter consists of two adaptive average pools, three convolutions, and a linear layer.

C. Different Training Methods. Various training strategies have been proposed to optimize neural networks with early-exit branches, each with distinct advantages and challenges. One common approach is joint training, where losses from all exit branches are treated separately or combined and are trained simultaneously (as used in AdaDet [14]). Although this simplifies the implementation process, it can lead to increased training complexity. A second approach is separate training, where each side branch functions as an independent classifier [8]. This approach is particularly beneficial when different exit points are designed to capture different features or levels of abstraction, allowing for targeted optimization of each branch. Another method, known as two-stage training, first involves optimizing the backbone network prior to freezing its parameters and then training the exit branches. This strategy is highly effective when leveraging pre-trained models, enabling rapid experimentation with detachable side

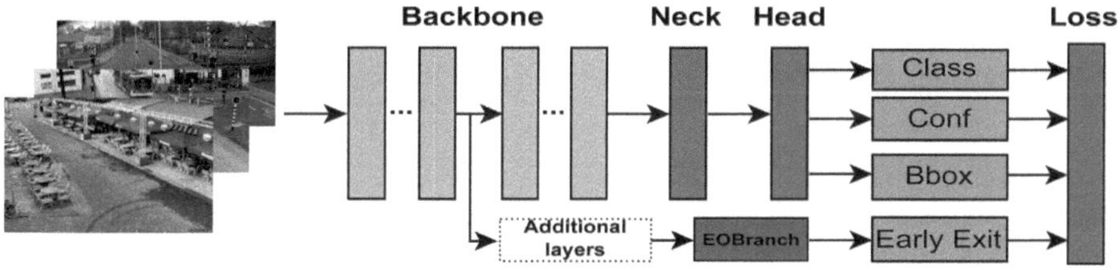

Fig. 1. Overview of the proposed system with one EOBranch attached to the backbone of a YOLO-model. The EOBranch can be extended with additional layers.

branches while reducing overall training time. However, the fixed nature of the backbone during training of a specific branch can result in suboptimal performance. Other approaches involve iterative training or distillation-based training, as described in the reviews [9,11]. These are effective for multiple early-exit branches, but for computational complexity reasons we restrain to one early exit.

3 Method

This section elaborates on the addition of an early-exit branch to the YOLOv6 [6] and YOLOv9 [13] detection models. The proposed Early Out Branch (EOBranch) method is designed for computational efficiency. It serves as an auxiliary component to standard YOLO models and can be seamlessly integrated as a side branch (see Fig. 1). The model processes the input image in a conventional layer-by-layer manner, until it reaches the EOBranch. At this stage, the EOBranch evaluates whether the image contains objects. If the confidence score exceeds a predefined threshold, inference is terminated early, which indicates the absence of objects. In contrast, if the confidence score is below the threshold, inference continues, and all features are computed for further object detection.

3.1 Design Considerations

Two critical design considerations arise to make the use of early-exit branches beneficial to reduce computational load. First, the computational efficiency of the early-exit branch itself is important. Since the branch adds additional operations during inference, its design must balance the computational cost with the efficiency gains it provides. Second, the point location where the early-exit branch is attached to the feature layers of the detection model is crucial. Although an earlier placement can yield greater computational savings by terminating unnecessary processing earlier, the branch must maintain a high level of classification accuracy. Wrong decisions by the early-exit branch will result in false negatives, where images that contain objects are incorrectly discarded as background. The following two subsections discuss our proposed architecture for the EOBranch and elaborate on the placement of the EOBranch in the YOLO-models.

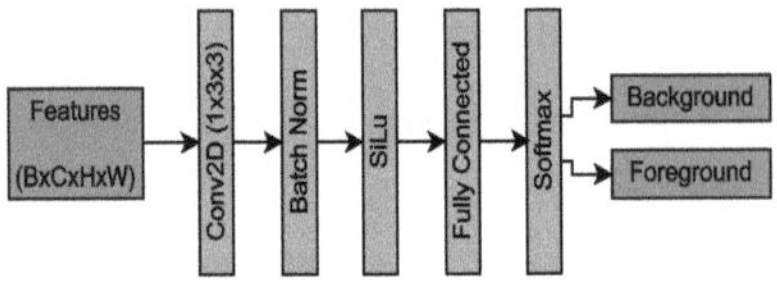

Fig. 2. Structure of the proposed EOBranch. Conv2D compresses the features. The fully connected layer classifies between foreground and background.

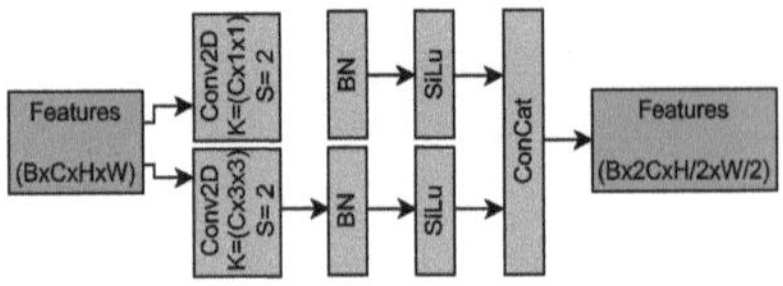

Fig. 3. Structure of the proposed FastADown (F). It applies a Conv2D (3x3) for spatial feature extraction and another Conv2D (1x1) for efficient downsampling.

3.2 Structure of the EOBranch

The EOBranch architecture consists of one convolutional layer and one fully connected layer, as illustrated in Fig. 2. The Conv2D layer aggregates the input channels into a single output channel corresponding to the width and height of the feature map and continues with a BatchNorm layer and a SiLu activation layer. Subsequently, the fully connected layer produces an output consisting of two neurons, activated by a Softmax function to classify the image. Since the EOBranch is small, it cannot significantly enrich the features, inserting additional layers in front can help to ensure that the intermediate features become more discriminative and descriptive. In addition, they can improve the backpropagation of earlier layers, leading to better overall training.

The layers to extend the EOBranch that are used in this paper are spatial resolution reducing layers. Besides the standard YOLO Conv layer (C) and the YOLOv9 ADown layer (A), a FastADown layer is proposed. The Conv layer consists of a convolutional layer, a batch norm layer, and a swish activation layer. The ADown layer processes the input via two parallel branches (average pooling and max pooling with 1×1 convolution) and then concatenates its results. One branch focuses on smoother, context-aware features, while the other captures the strongest responses in the feature map. Both layers reduce the spatial resolution while increasing the number of input channels, capturing a richer set of features.

The FastADown layer is proposed as an optimized version of the ADown layer, improving speed while retaining benefits of ADown. The structure is illustrated in Fig. 2. FastADown achieves greater efficiency by eliminating the channel splitting and separate pooling operations. Instead, it directly applies a 3×3 convolution for spatial feature extraction and a lightweight 1×1 convolution for efficient downsampling. This approach reduces computational overhead while enhancing representational capacity with minimal additional cost.

3.3 Placement of the EOBranch

To keep computational complexity low, we propose to incorporate only one EOBranch. The placement location of the EOBranch is crucial. While earlier placement saves time, feature maps in the early stages of the network do not

Table 2. Timings for each layer of the backbone, where the entire model is 100%.

YOLOv6-M				YOLOv9-C			
Layer	**Time**	**Sum**	**Out shape**	**Layer**	**Time**	**Sum**	**Out shape**
Stem_1	3.8%	3.8%	48x320x320	Conv_1	9.8%	9.8%	64x320x320
ERBlock_2	13.4%	17.2%	96x160x160	Conv_2	1.7%	11.5%	128x160x160
ERBlock_3	12.7%	29.9%	192x80x80	REPNCSPELAN4_3	10.8%	22.3%	256x160x160
ERBlock_4	17.3%	47.2%	384x40x40	ADown_4	5.5%	27.8%	256x80x80
ERBlock_5	10.6%	57.8%	768x20x20	REPNCSPELAN4_5	7.8%	35.6%	512x80x80
				...			
				REPNCSPELAN4_9	2.8%	49.0%	512x20x20

contain rich features that are suitable for determining if an object is present in a scene. More abstract and informative features are required to enable robust decision-making in an early-exit branch. To compare backbone computational loads in YOLOv6-M and YOLOv9-C, we measured inference times per block as shown in Table 2. For YOLOv6-M, each additional ERBlock takes a significant amount of time. The ERBlock_2 already accounts for 13.4% of the entire model. Therefore, we propose evaluating the placement EOBranch after the first, second, or third layer. For YOLOv9-C, the REPNCSPELAN4 blocks take the longest time in the backbone. Note that, the difference between Conv_1 and Conv_2 is only 1.7%. Given that our additional layers in front of the EOBranch incur computational cost, it is more efficient to place the branch after YOLOv9-C-2, where richer features are available. Thus, we propose to evaluate the EOBranch after the Conv_2, REPNCSPELAN4_3, and ADown_4 layers.

3.4 Updated Loss Function

In order to train the early-exit branch, an additional loss function and added to the original loss function of the YOLO models. The early out loss function $\mathcal{L}_{\text{early}}$ is combined with the loss function of the YOLO models as indicated by

$$\mathcal{L} = \lambda_{\text{obj}}\mathcal{L}_{\text{obj}} + \lambda_{\text{cls}}\mathcal{L}_{\text{cls}} + \lambda_{\text{box}}\mathcal{L}_{\text{box}} + \lambda_{\text{early}}\mathcal{L}_{\text{early}}, \quad (1)$$

where λ_{obj}, λ_{cls}, λ_{box} and λ_{early} are scalars to weight the respective loss functions. The objectness loss $\mathcal{L}_{\text{obj}}$ and the class loss $\mathcal{L}_{\text{cls}}$ are calculated by the binary cross entropy (BCE) function and the box loss $\mathcal{L}_{\text{box}}$ is calculated by the complete intersection over union loss (CIoU) function, similar to the original implementation. The introduced early-exit loss function $\mathcal{L}_{\text{early}}$ for the EOBranch is calculated by the BCE function to learn the one-hot encoded background versus non-background classification task. Overview shown in Fig. 1.

3.5 Training Methods of the Early-Exit Branch

This section describes three different training methods implemented to train the detection model. *Freezing during training* involves training of the EOBranch only. Initially, the standard detection model was trained. Subsequently, all layers except the early-out branch are frozen, allowing only the early-out branch to be trained. This approach ensures that the detection performance remains unaffected while optimizing the early-out branch. *Finetune training* begins after convergence of the baseline detection model, which is trained without any additional branches. At this stage, the EOBranch is attached, and the model is further trained for several epochs without freezing any layers. This allows all parameters to adjust jointly, benefiting from the pre-trained weights. *Joint training* trains the detection model in an end-to-end manner, without pre-training or freezing any layers. This represents the most straightforward approach as it eliminates the need for additional training steps or modifications.

4 Experimental Setup and Experiments

A. NuImages Dataset. Since the proprietary traffic surveillance dataset is private due to privacy regulations, the NuImages dataset [1] is deployed for evaluation due to its diversity in class distribution, spatiotemporal settings, and weather conditions. Except for the autonomous viewpoint of the NuImages dataset, this set resembles the proprietary dataset.

To evaluate our method, background images are required. To ensure a balanced distribution between background images and those containing objects, the NuImages dataset is split based on object categories. Images containing at least one annotation of a car, bus, or truck are classified as object-containing, while the remaining images are considered backgrounds. The background subset contains 23,167 images, and an equal number of object-containing images are selected to ensure that each scene type contributes equally. The resulting dataset is divided into a training set of 32,327 images, a validation set of 4,669 images, and a test set of 9,338 images.

B. Evaluation Metrics. To evaluate the performance of the early out branches, the following metrics are used: precision (P), recall (R) and average precision (AP) as an approximation of the area under the precision-recall curve. For evaluating detection performance, the metrics $mAP_{0.5}$ and $mAP_{0.5:0.95}$ are commonly used. These metrics, adopted from the COCO dataset benchmark [7], not only compute the average precision across all object classes but also average the precision across ten Intersection over Union (IoU) thresholds ranging from 0.5 to 0.95 in increments of 0.05. For efficiency evaluation, the number of Giga Floating Point Operations (GFLOPs) is considered as a key metric. This factor provides insight into the computational complexity and resource requirements.

Table 3. The three training procedures from Sect. 3.5 evaluated with the EOBranch attached after ERBLOCK_3 of YOLOv6-M and after ADown_4 of YOLOv9-C.

YOLOv6-M				YOLOv9-C			
Training	**Early-exit**		**Detection**	**Training**	**Early-exit**		**Detection**
procedure	**AP**	**P@R$_{90/95}$**	**mAP$_{.5/.5-.95}$**	**procedure**	**AP**	**P@R$_{90/95}$**	**mAP$_{.5/.5-.95}$**
Baseline	-	-	78.8/61.0	Baseline	-	-	78.6/61.3
Joint	99.7	99.9/99.3	76.3/56.1	Joint	98.7	97.8/92.6	77.6/60.6
Finetune	99.6	99.8/99.0	78.2/60.3	Finetune	98.6	97.4/91.7	78.4/61.0
Freeze	98.7	97.9/93.1	78.8/61.0	Freeze	97.9	94.6/87.1	78.6/61.3

C. Training Details. We evaluate EOBranch using YOLOv6-M and YOLOv9-C, which have comparable computational complexities of 85.74 and 101.96 GFLOPS, respectively. This selection aligns with practical deployment constraints, limiting models to around 100 GFLOPS for efficient inference. For all hyper-parameters, the default values are used. We only exclude mosaic augmentation in order to have a balanced set of background and non-backgrounds samples during training. The baseline detection performance was obtained with a batch-size of 40 and finetuning of 40 epochs on 2 RTX 3090 cards. For all experiments, λ_{early} is set to 2.5 such that it is proportional to other losses.

4.1 Experiment: Comparison of Training Procedures

The first experiment evaluates different training procedures as detailed in Sect. 3.5. The primary objective is to achieve high accuracy for the early-exit branch, while maintaining the overall detection performance. For this experiment, the EOBranch is implemented without additional layers and tested on YOLOv6-M and YOLOv9-C. The *Joint Training* approach trains the detector and EOBranch simultaneously for 60 epochs. The *Finetuning approach* starts with 40 epochs of training the detection model without the EOBranch loss, followed by 20 epochs of joint finetuning for detection and early exit. The *Freezing approach* first fully trains the detection model and then trains the early-exit branch for 20 epochs, while keeping the weights of the detection model frozen. Hence, all models undergo a total of 60 training epochs. Since the backbones of YOLOv6 and YOLOv9 differ, the EOBranches are placed after the ERBLOCK_3 and ADown_4 layers, respectively. The results of the different training procedures are presented in Table 3. The evaluation targets are divided into the early-exit and the detection performance.

Impact on object detection performance. Freezing the detection model while training the EOBranch does not affect detection performance with respect to the original implementation as expected. The detection performance slightly decreases when finetuning the detection model with an EOBranch. Joint training results in the lowest detection performance, where the decrease is most significant for the YOLOv6-M model (1.9% $\text{mAP}_{0.50}$ and 4.2% $\text{mAP}_{0.50-0.95}$).

Table 4. EOBranch evaluated placed after different layers for YOLOv6 and YOLOv9.

YOLOv6-M				**YOLOv9-C**			
Layer	**Early-exit**		**Detection**	**Layer**	**Early-exit**		**Detection**
	AP	**P@R$_{90/95}$**	**mAP$_{.5/.5-.95}$**		**AP**	**P@R$_{90/95}$**	**mAP$_{.5/.5-.95}$**
Stem_1	82.8	65.6/60.3	78.3/60.4	Conv_1	77.3	61.7/57.2	78.6/61.1
ERBlock_2	98.2	95.0/88.9	78.5/60.4	Conv_2	84.5	65.3/58.8	78.5/61.1
ERBlock_3	99.6	99.8/99.0	78.3/60.2	REPNCSPELAN4_3	97.3	92.5/84.3	78.4/61.0
				REPNCSPELAN4_4	98.6	97.4/91.7	78.4/61.0

Performance of EOBranch. The finetuning approach yields the best balance between early-exit performance and overall detection performance. For example, on the YOLOv6-M model, finetuning achieves an AP of 99.6%, outperforming the freezing approach with an AP of 98.7%, without having a significant detection performance drop-off. Similarly, on the YOLOv9-C model, finetuning resulted in AP of 98.6%, compared to an AP score of 97.9% for the freezing approach. Although the freezing method maintained competitive detection accuracy, its EOBranch performance is inferior to the finetuning and joint training methods. The difference in EOBranch accuracy between the finetuning and joint training approaches is negligible. These results suggest that finetuning is the most efficient training procedure among those evaluated.

4.2 Experiment: Optimal Position of EOBranch

The second experiment evaluates the optimal position of EOBranch and investigates the impact on performance when placing EOBranch earlier or later in the backbone. To that end, the performance of the EOBranch is evaluated when attached after the Stem_1, ERBlock_2, and ERBlock_3 layers for the YOLOv6-M model; and after the Conv_1, Conv_2, REPNCSPELAN4_3, and REPNCSPELAN4_4 layers for the YOLOv9-C model.

The results in Table 4 highlight the performance impact of the EOBranch placement. For YOLOv6-M, positioning the EOBranch too early (after the Stem_1 layer) yields an early-out AP of 82.8% and a low P@R$_{90}$ of 65.6%. This indicates that the feature maps contain limited information. When the EOBranch is placed later in the network (after ERBlock_1 or ERBlock_2), the early-out P@R$_{90}$ scores improve significantly to 95.0% and 99.8%, respectively. Similarly, for YOLOv9-C, an early branch placement (after Conv_2) results in a early-out AP of 84.5% and an early-out P@R$_{90}$ of 65.3%, while subsequent placements after the REPNCSPELAN4_3 or REPNCSPELAN4_4 layers improve the early-out scores to 97.3% and 98.6% AP, respectively. These findings highlight that aligning the EOBranch with feature maps extracted in later stages of the backbone is essential to maximize the EOBranch performance. However, this comes at the cost of more computations before the early-out decision can be made.

Table 5. The extended EOBranches employing the Conv (C), ADown (A) and FastADown (F) in front of the EOBranch for YOLOv6.

YOLOv6-M after Stem_1				**YOLOv6-M after ERBlock_2**			
EOB	**Early-exit**		**Detection**	**EOB**	**Early-exit**		**Detection**
config	**AP**	**P@R$_{90/95}$**	**mAP$_{.5/.5-.95}$**	**config**	**AP**	**P@R$_{90/95}$**	**mAP$_{.5/.5-.95}$**
C	88.8	73.0/65.5	78.6/60.7	C	98.6	96.8/91.8	78.4/60.4
A	89.0	74.0/66.0	78.5/60.6	A	95.3	84.9/76.1	78.3/60.4
F	87.9	72.1/64.8	78.7/60.6	F	98.6	97.3/91.7	78.3/60.2
A-C	93.2	79.9/71.2	78.6/60.5	C-C	95.6	86.0/78.0	78.3/60.4
A-A	89.3	71.8/63.4	78.5/60.5	C-F	98.9	98.2/93.7	78.1/60.2
A-C-C	91.2	75.3/67.1	78.7/60.7	F-C	98.8	98.1/93.8	78.0/60.1
A-C-A	96.6	90.3/80.9	78.3/60.3	F-F	98.9	98.3/94.1	78.2/60.3
A-C-A-C	97.4	93.0/85.7	78.9/60.8				
A-C-A-A	95.1	85.3/76.2	78.6/60.7				

4.3 Experiment: Increasing Complexity of the EOBranch

The third experiment aims to extend the EOBranch to improve the early-exit performance for early placements, balancing between improved performance and additional computational complexity. This experiment uses the extended EOBranch with the additional layers in front as described in Sect. 3.3. This will add some computational complexity to the early-out branch, but will increase the early-exit performance. Additional higher-level features can be computed in the branch itself, not influencing the features of the detection model itself. Aiming to optimize the early-exit performance after the Stem_1 and ERBlock_2 layers of YOLOv6-M and after the Conv_2 layer of YOLOv9-C. The computed results are described in Table 5 and Table 6.

Performance of extended EOBranch on YOLOv6-M. The EOBranch extended with one or two layers after the Stem_1 layer resulted in suboptimal performance. Each additional block from A to A-C to A-C-A to A-C-A-C contributes to a rise in early-exit performance in terms of AP (89.0% to 97.4%), as well as P@R$_{95}$ (66.0% to 85.7%). This demonstrates the significant impact of extending the EOBranch. However, positioning the EOBranch after the ERBlock_2 yields superior results. The lowest AP is observed in A (95.3%) with an increase to an AP of 98.9% for F-F. This confirms that deeper placements offer better feature utilization, ensuring higher efficiency in early exits. However, mAP scores remained relatively unchanged, with the best configuration reaching 78.4% in mAP$_{0.50}$ and 60.4% for mAP$_{0.50-0.95}$.

Performance of extended EOBranch on YOLOv9-C. For YOLOv9-C, EOBranches extended with three layers show better early-exit performance as shown in Table 6. The best performing configuration is the F-C-F setup with an AP of 98.4%, P@R$_{90}$ of 96.3%, and P@R$_{95}$ of 90.9%. The difference between two and three additional layers is mainly visible in the P@R$_{95}$ with an

Table 6. The extended EOBranches employing the Conv (C), ADown (A) and FastADown (F) in front of the EOBranch for YOLOv9.

YOLOv9-C after Conv_2							
EOB	**Early-exit**		**Detection**	**EOB**	**Early-exit**		**Detection**
config	**AP**	**P@R$_{90/95}$**	**mAP$_{.5/.5-.95}$**	**config**	**AP**	**P@R$_{90/95}$**	**mAP$_{.5/.5-.95}$**
C	94.4	81.5/72.9	78.3/61.0	F-F	97.5	92.5/84.6	78.1/61.0
A	92.4	76.5/67.8	77.0/60.3	C-C-C	98.4	95.7/89.3	78.2/61.3
F	96.2	86.9/77.4	77.3/60.5	C-C-F	97.2	91.6/83.6	78.1/61.1
C-C	97.7	93.7/85.8	78.3/61.4	F-C-C	98.4	96.1/90.0	78.4/61.3
C-F	97.6	93.3/85.0	78.4/61.1	F-C-F	98.4	96.3/90.9	78.4/61.1
F-C	97.7	93.4/85.0	78.3/61.3				

increase of more than 5% to 90.9%. The mAP scores have only small deviations between the various configurations, with a small decrease for the single extended EOBranches.

Overall, the findings suggest that the placement of the EOBranch later in the network, particularly after higher-level feature extraction layers, provides the best balance between early-out and detection performance. The addition of extensions (C, A, F) to the EOBranch further enhances early-exit performance.

4.4 Experiment: Computational Complexity

The fourth experiment evaluates the complexity of various EOBranch configurations. While prior experiments demonstrate that extending EOBranch improves early-exit performance without compromising detection accuracy, this experiment focuses on minimizing computational cost. Therefore, different EOBranch placements are compared, assessing both complexity and early-exit latency. Timings, averaged over 100 images on an Orin Nano, are normalized to a baseline model without early exit to emphasize relative inference overhead.

Table 7 presents the complexity analysis of the extended EOBranch using three key metrics. The EOB column (%) indicates the additional computational time of EOBranch, normalized to the full model's inference time. BEOB (%) reflects the total processing time with early exit, also normalized to the standard model. The GFLOPS column reports the computational cost.

Results on inference time and GFLOPS. The results indicate substantial differences in computational overhead across configurations. For example, YOLOv6's A-C-A-C setup after Stem_1 yields the highest EOBranch overhead (22.49% processing time, 5.44 GFLOP), whereas the F-F configuration after ERBlock_2 significantly reduces overhead (3.89% processing time, 2.38 GFLOPS) and achieves a more efficient early-exit inference time of 22.78%. In YOLOv9-C, the F-C-F configuration after Conv_2 offers a balanced trade-off, incurring only 4.26% overhead and saving 81.44% per background image. These results highlight the importance of strategic EOBranch placement.

Table 7. The timings of the best (extended) EOBranches are shown, normalized to the branchless model's processing time. EOB is the processing time of the (extended) EOBranch, BEOB represents the time from input to output of the EOBranch.

Ext.	Place	EOB	BEOB	GFLOPS	Ext.	Place	EOB	BEOB	GFLOPS
	YOLOv6-M			85.74		**YOLOv9-C**			101.96
A-C-A-C	Stem_1	22.49%	29.48%	91.18	F-C-F	Conv_2	4.26%	18.56%	109.96
F-F	ERBlock_2	3.89%	22.78%	88.12	F	REPNC..._3	4.47%	29.76%	110.42
-	ERBlock_3	1.08%	33.08%	85.76	-	ADown_4	0.88%	31.59%	102.00

Table 8. The (extended) EOBranches evaluated on traffic surveillance streams of 24 h. The results present the relative time saved over 24 h in percentage.

Config	Place	City	Bus	Railway	Config	Place	City	Bus	Railway
YOLOv6-M					**YOLOv9-C**				
A-C-A-C	Stem_1	-4.96	16.68	30.70	F-C-F	Conv_2	14.42	**33.63**	**46.08**
F-F	ERBlock_2	14.12	32.22	43.94	F	REPNCS..._3	11.89	28.62	29.46
-	ERBlock_3	14.19	29.32	29.12	-	ADown_4	**14.76**	30.14	40.11

4.5 Experiment: Efficiency in Traffic Surveillance Applications

As outlined in the Introduction, this work aims to reduce processing time and computational complexity through early-exit strategies in real-time traffic surveillance. In this application-focused experiment, inference time savings are evaluated across three representative surveillance scenes (Table 1). YOLOv6-M and YOLOv9-C were finetuned on a proprietary traffic dataset with integrated EOBranches using the optimal configurations identified earlier. EOBranch processing time was recorded per frame, depending on whether an early exit occurred. The proprietary dataset contains 135,000 images of traffic surveillance in a broad variety of different scenes, such as busy traffic scenes, crowded pedestrian areas, and bicycle roads. The proprietary dataset has around 10 % of background images and is annotated with 6 traffic classes defined as follows: pedestrian, cyclist, motorcyclist, car, bus, and truck.

Traffic surveillance results. The results of Table 8 highlight that model complexity and EOBranch placement directly influence efficiency. For YOLOv6-M, placing the F-F EOBranch after ERBlock_2 achieved up to 43.94% time savings at a railway crossing, 32.22% at a bus stop, and 14.12% in the city center. For YOLOv9-C, the most efficient setup was the F-C-F EOBranch after Conv_2, reducing processing time by up to 46.08% at a railway crossing, 33.63% at a bus stop, and 14.42% in the city center, performing well across all scenarios.

Overall, this experiment demonstrates that extended EOBranches are a viable approach to reduce computational overhead in edge-based traffic surveillance. The F-C-F EOBranch after the Conv_2 layer in YOLOv9-C saves the most inference time with a significant reduction of 46.08% in the processing time of an ordinary railway crossing.

5 Conclusions

In this work, we have introduced an early-exit strategy specifically tailored for YOLO-based object detection architectures to improve computational efficiency in edge-based traffic surveillance applications. We have evaluated the performance of the early-exit branch across multiple YOLO model architectures, integrating it into the YOLOv6 and YOLOv9 models. We have investigated the impact on overall early-exit performance and detection performance.

Experiments were conducted in a series of stages, starting with comparing training procedures. It shows that a finetuning training approach outperforms other while balancing the early exit AP and detection performance. In the second experiment, optimal branch placement was investigated, showing that better developed feature maps are necessary to maximize the early-exit performance. The third experiment focused on increasing the branch complexity such as A-C-A-C in YOLOv6 and F-C-F in YOLOv9, to enhance the early exit AP to 97.4% and 98.4%, respectively. The fourth experiment analyzed the computational complexity and processing time. For YOLOv6 and YOLOv9 the best configurations, incurred only 3.89% and 4.26% additional overhead, resulting in a total early-exit inference time of 22.78% and 18.56% relative to the entire model, respectively.

Finally, an application-based evaluation on traffic surveillance streams demonstrated that these optimized configurations reduce inference time by up to 46%, confirming that deeper, well-optimized EOBranch placements significantly improve overall efficiency while maintaining high detection accuracy. The empirical findings emphasize the potential of early-exit strategies to improve inference speed, extend battery life in edge devices, and promote sustainable systems.

References

1. Caesar, H., et al.: nuscenes: a multimodal dataset for autonomous driving. In: Proceedings of the IEEE/CVF Conference on CVPR (2020)
2. Han, S., Pool, J., Tran, J., Dally, W.J.: Learning both weights and connections for efficient neural networks. In: Advances in Neural Information Processing Systems. vol. 28. Curran Associates, Inc. (2015)
3. Hector, R., Umar, M., Mehmood, A., Li, Z., Bhattacharyya, S.: Scalable object detection for edge cloud environments. Front. Sust. Cities **3** (2021)
4. Jacob, B., et al.: Quantization and training of neural networks for efficient integer-arithmetic-only inference. In: Proceedings of the IEEE Conference on CVPR (2018)
5. Kaya, Y., Hong, S., Dumitras, T.: Shallow-deep networks: understanding and mitigating network overthinking. In: Proceedings of the 36th International Conference on Machine Learning. Proceedings of Machine Learning Research, vol. 97, pp. 3301–3310. PMLR (2019)
6. Li, C., et al.: Yolov6 v3.0: a full-scale reloading. In: arXiv:2301.05586 (2023)

7. Lin, T.Y., et al.: Microsoft COCO: common Objects in Context. Lecture Notes in Computer Science (including subseries Lecture Notes in Artificial Intelligence and Lecture Notes in Bioinformatics) **8693 LNCS**(PART 5), 740–755 (5 2014)
8. Matsubara, Y., Levorato, M.: Neural compression and filtering for edge-assisted real-time object detection in challenged networks. CoRR **abs/2007.15818** (2020)
9. Matsubara, Y., Levorato, M., Restuccia, F.: Split computing and early exiting for deep learning applications: survey and research challenges. ACM Comput. Surv. **55**(5), 1–30 (2022)
10. Phuong, M., Lampert, C.: Distillation-based training for multi-exit architectures. In: 2019 IEEE/CVF ICCV, pp. 1355–1364 (2019)
11. Rahmath P, H., Srivastava, V., Chaurasia, K., Pacheco, R.G., Couto, R.S.: Early-exit deep neural network - a comprehensive survey. ACM Comput. Surv. **57**(3) (Nov 2024)
12. Teerapittayanon, S., McDanel, B., Kung, H.T.: Branchynet: fast inference via early exiting from deep neural networks. CoRR **abs/1709.01686** (2017)
13. Wang, C.Y., Yeh, I.H., Mark Liao, H.Y.: Yolov9: learning what you want to learn using programmable gradient information. In: Computer Vision – ECCV 2024, pp. 1–21. Springer Nature Switzerland, Cham (2025)
14. Yang, L., Zheng, Z., Wang, J., Song, S., Huang, G., Li, F.: Adadet: an adaptive object detection system based on early-exit neural networks. IEEE Trans. Cognitive Develop. Syst. **16**(1), 332–345 (2024)

Mamba-OTR: A Mamba-Based Solution for Online Take and Release Detection from Untrimmed Egocentric Video

Alessandro Sebastiano Catinello(✉), Giovanni Maria Farinella, and Antonino Furnari

Department of Mathematics and Computer Science, University of Catania, Catania, Italy
ale.catinello.c@gmail.com, {giovanni.farinella,antonino.furnari}@unict.it

Abstract. This work tackles the problem of Online detection of "Take" and "Release" (OTR) of an object in untrimmed egocentric videos. This task is challenging due to severe label imbalance, with temporally sparse positive annotations, and the need for precise temporal predictions. Furthermore, methods need to be computationally efficient in order to be deployed in real-world online settings. To address these challenges, we propose Mamba-OTR, a model based on the Mamba architecture. Mamba-OTR is designed to exploit temporal recurrence during inference while being trained on short video clips. To address label imbalance, our training pipeline incorporates the focal loss and a novel regularization scheme that aligns model predictions with the evaluation metric. Extensive experiments on EPIC-KITCHENS-100, the comparisons with transformer-based approach, and the evaluation of different training and test schemes demonstrate the superiority of Mamba-OTR in both accuracy and efficiency. These findings are particularly evident when evaluating full-length videos or high frame-rate sequences, even when trained on short video snippets for computational convenience. The proposed Mamba-OTR achieves a noteworthy mp-mAP of 45.48 when operating in a sliding-window fashion, and 43.35 in streaming mode, versus the 20.32 of a vanilla transformer and 25.16 of a vanilla Mamba, thus providing a strong baseline for OTR. We will publicly release the source code of Mamba-OTR to support future research.

Keywords: Online Action Detection · Take/Release Action Detection · Egocentric Untrimmed Video Analysis

1 Introduction

Wearable devices provided with cameras are able to capture visual information from a first-person perspective, enabling the development of personalized, context-aware assistive technologies to support user daily activities [10]. A key requirement for such systems is the ability to detect fine-grained, atomic

E. Rodolà et al. (Eds.): ICIAP 2025, LNCS 16167, pp. 456–468, 2026.
https://doi.org/10.1007/978-3-032-10185-3_36

actions—such as *take* and *release* of an object—which are essential for downstream tasks like intention prediction, object interaction tracking, and anomaly detection during goal-directed human behavior. To address this challenge, action recognition algorithms must operate on streaming video data in an online fashion while maintaining temporal coherence by emitting a single, unambiguous prediction per action instance. We refer to this task as "Online Take-Release detection" (OTR). Previous work has explored various strategies, including detecting the *starting* frame of an action [12] and identifying *contact* frames [11]. However, recent findings suggest that predicting the *ending* frame of an action yields superior performance, as it reduces the uncertainty caused by partial observations and premature predictions [3], while inducing a small delay in the predictions, which is acceptable for most applications. Figure 1 illustrates this setup, originally discussed in [3]. Despite the advantages of this formulation, OTR still faces multiple challenges, notably, the extreme class imbalance between positives (last frame of a take/release action) and negatives (any other frame), the need to suppress multiple detections, and the requirement of online and computationally efficient processing in order to support deployment to real-world scenarios.

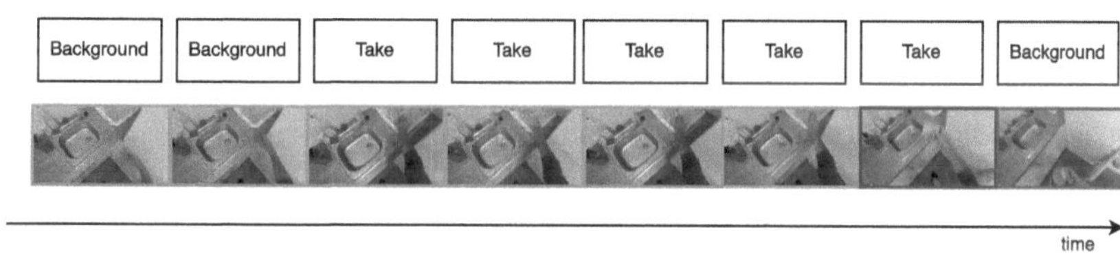

Fig. 1. Algorithms are tasked to process the video online and output a single prediction corresponding to the last frame of the take/release action (blue frame), while avoiding predictions for background (red frame) or any other frame (green ones). (Color figure online)

We propose a novel approach based on the Mamba architecture [8]. Our method uses the focal loss during training and incorporates tailored regularization techniques that align the model's behavior with the evaluation metric, encouraging precise and temporally consistent predictions. We evaluated our model on the EPIC-KITCHENS-100 dataset, focusing exclusively on the *take* and *release* verbs, and conduct extensive experiments across various architectural configurations. The proposed Mamba-OTR has been compared with respect to different approaches based on Tranformers and Mamba-based models, showing the advantages of each introduced optimization. Our results demonstrate that, although the task remains challenging, it becomes significantly more tractable when approached with an appropriate training strategy. Furthermore, we show that Mamba's inherent recurrence enables efficient training on short clips while allowing inference over full-length videos. This training-inference decoupling leads to substantial improvements in inference speed and enhances predictive performance. Our optimized Mamba-OTR model achieves an mp-mAP of 51.76

and a mean inference time of $0.14s$, versus the mp-mAP of 20.32 and the inference time of $0.28s$ of a standard Transformer module.

In sum, our main contributions are as follows: 1) A training pipeline incorporating regularization strategies to address severe annotation imbalance of OTR; 2) Comprehensive benchmarking against existing Transformer- and Mamba-based methods; 3) The Mamba-OTR model, which is a strong baseline to support research in this area. We plan to release our code.

2 Related Work

Online Detection of Action Start/End. The considered OTR problem is closely related to previous investigations aiming to detect action start or end frames in an online fashion. The Online Detection of Action Start (ODAS) task [12] aims to identify the exact moment an action begins in a streaming setting, emphasizing low latency and temporal precision. Several methods have approached this challenge using third-person video with 3D convolutional networks [12], LSTM-based models with reinforcement learning [6], and weak supervision [7]. More recently, quasi-online strategies tailored explicitly for OTR have relaxed strict causality by introducing short buffer windows [11]. While similar in evaluation and setup, the Online Detection of Action End (ODAE) task [3] proposes predicting the frame where the action concludes, leveraging the full temporal extent of the action and reducing the ambiguity present in early-stage predictions. For applications involving real-time monitoring, end-point predictions provide a sufficiently clear signal for downstream tasks.

In this work, we consider the ODAE formulation proposed in [3] to address the OTR problem, given its reduced ambiguity and practical relevance.

Deep Learning Architectures for Online Video Processing. Research in online video processing aims to balance the temporal accuracy of detections with computational efficiency. Transformer-based models, such as OadTR [13], TeSTra [12] or LSTR [15] have been proposed. In particular, these two last methods employ dual-memory mechanisms to integrate both short- and long-term temporal dependencies, but their limited scalability poses challenges in real-time applications. To address these limitations, alternative architectures based directly on recurrent neural networks [1,14] or Mamba [4,8] architectures have been proposed. These latter models performs sequence modeling in linear time using selective state spaces, offering competitive or superior performance compared to Transformers while reducing computational complexity in both time and space.

In this work, we adopt the Mamba architecture due to its ability to parallelize training and operate in a recurrent manner at inference time, making it particularly well-suited for efficient online video processing.

Class imbalance. Class imbalance is a persistent challenge in deep learning, particularly in object detection, where background regions vastly outnumber foreground objects. Solutions like *focal loss* [9] prioritize hard examples, while *class-balanced loss* [5] and *LDAM loss* [2] adjust for skewed label distributions via re-weighting and margin tuning. Although developed for image-based tasks, these methods are applicable to temporal problems such as action detection, where background frames dominate and relevant actions are short.

In this paper, we show that adapting imbalance-aware losses is crucial for improving performance in online action detection.

3 Problem Definition and Evaluation Metrics

Following the work in [3], we define OTR as the task of identifying whether the current frame at time t' marks the end of a *take* or *release* action, using only video frames observed up to t'. Each ground truth action is defined as $a = (c, t)$, where c is the action class and t is the end timestamp. Predictions are expressed as tuples $\hat{a} = (\hat{c}, \hat{t}, s)$, including the predicted class $\hat{c}$, predicted timestamp $\hat{t}$, and confidence score s. To evaluate our models, we adopt the point-level mean Average Precision (p-mAP) metric, as introduced in [12]. A predicted action $\hat{a} = (\hat{c}, \hat{t}, s)$ is matched to a ground truth action $a = (c, t)$ if the following conditions are met: 1) The predicted and ground truth action classes match ($\hat{c} = c$); 2) The temporal offset $\delta = |\hat{t} - t|$ is less than or equal to a threshold ϕ. Matching is performed greedily based on descending confidence scores. Each prediction and each ground truth can be matched at most once. Matched predictions are counted as true positives; unmatched predictions are false positives, and unmatched ground truth actions are false negatives. Take/Release mAP is computed based on these true and false positives, considering confidence scores s. To account for different levels of temporal precision, we report the mean point-level mAP (mp-mAP), computed by averaging the p-mAP values over a range of temporal thresholds ϕ ranging from 1 to 10 s (with 1-second intervals).

4 Architecture and Techniques

The proposed Mamba-OTR is obtained by combining a streamlined Mamba architecture with the focal loss and different regularization techniques. For comparison, we assess the effect of the proposed techniques on both Transformer-based and Mamba-based architectures. In the following, we first describe base architectures, then provide details on the training loss and proposed regularization techniques.

4.1 Base Architectures

All base architectures take as input visual features extracted from input frames and output an action prediction for each frame. Possible actions are “take”, “release” or “background”. Given this output, we extract $(\hat{c}, \hat{t}, s)$ tuples from each video.

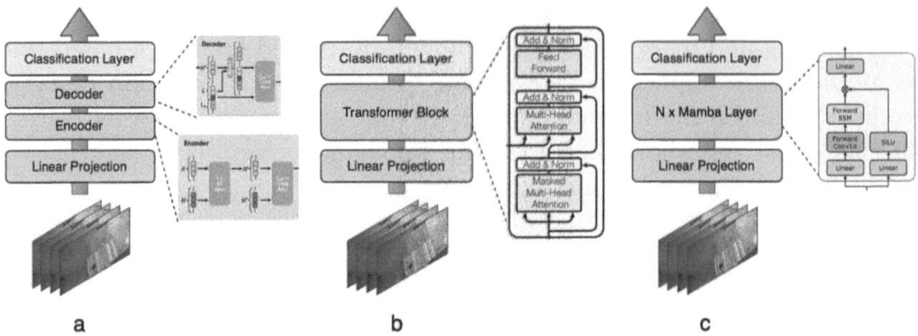

Fig. 2. Models architecture overview. a) TeSTra uses an encoder for long-term memory and a decoder to combine long- and short-term information. b) Transformer processes only short-term memory using standard attention blocks. c) MAMBA-OTR employs one or more Mamba layers for efficient temporal modeling.

TeSTra. We consider the TeSTra architecture [16] as a state-of-the-art Transformer-based model optimized for online video processing. TeSTra leverages a dual-memory structure, where an encoder builds a long-term memory from past observations, which is then fused with a short-term memory via cross-attention in a decoder to produce frame-level predictions. See Fig. 2(a).

Transformer. We also consider a simplified version of TeSTra where the long-term memory module is removed, retaining only the self-attention mechanism over a short temporal window. This leads to a standard transformer module operating over the video in a sliding-window fashion. We train this model based on the TeSTra code-based, hence taking advantage of its data-loading and pre-processing pipeline. See Fig. 2(b).

Mamba. We finally consider a Mamba [8] architecture as a base-model. This architecture ingest linearly projected visual features extracted from frames and stacks different Mamba layers followed by a classification head. See Fig. 2(c).

4.2 Training Loss and Regularization Techniques

We train our model using the following loss function:

$$\mathcal{L} = \sum_{i}^{m} FL(h(x_i)) + \lambda R \tag{1}$$

where FL is the focal loss calculated on the output of the model, given a specific dataset example $h(x_i)$ where h is the used model, and R is the regularization term.

The focal loss [9], originally introduced for object detection, is a widely used technique for addressing class imbalance by down-weighting easy examples and focusing the learning on hard, misclassified ones. In our case, when each action is represented by a single positive frame among thousands of negatives, this imbalance is particularly severe. We show that the focal loss plays a crucial role in guiding the model to concentrate on these precise endpoints, rather than learning from potentially misleading earlier frames and overfitting to the majority class.

The regularization term R is designed to reduce redundant detections around the same ground truth "take" or "release" action while maintaining model confidence. Specifically, we propose three different approaches to implement this regularization, which are described in the following: Entropy Minimization, Sliding Window Regularization, and Fixed Window Regularization.

Entropy Minimization.Entropy minimization offers a simple way to promote sparsity by encouraging confident predictions and suppressing uncertain ones. The loss is defined as:

$$\mathcal{R}_{entropy} = -\sum_{i} p_i \log(p_i) \tag{2}$$

where p_i is the predicted probability at frame i. While effective at reducing multiple low-confidence outputs, this approach lacks control over where predictions are concentrated in time.

Sliding Window. To refine the prediction distribution, we use a sliding window approach. Given a window size w, the loss sums the predictions within each window:

$$\mathcal{R}_{SW} = \sum_{f} \left(\sum_{i=f-\lfloor w/2 \rfloor}^{f+\lfloor w/2 \rfloor} p_i \right) \tag{3}$$

where f stands for each frame in the output vector.

This encourages the model to output a single high-confidence prediction per ground truth by suppressing excessive activations within each segment, as shown in Fig. 3 (a). However, since the approach treats both background and action-relevant frames equally, it may penalize non-action regions.

Fixed Window. To directly address the issue of multiple predictions around ground truth instances, we refine the sliding window approach by centering the window exclusively on the ground truth frames. Instead of applying the penalty over the entire sequence, we define a fixed window of size w around each ground truth instance and minimize the sum of predictions within this region:

$$\mathcal{R}_{FW} = \sum_{g \in G} \left(\sum_{i=g-\lfloor w/2 \rfloor}^{g+\lfloor w/2 \rfloor} p_i \right) \tag{4}$$

where G represents the set of ground truth action-ending frames (See Fig. 3 (b).

5 Experiments

In this section we consider the dataset introduced in [3], which is a curated subset of EPIC-KITCHENS-100, containing only two action classes: *take* (e.g., "get",

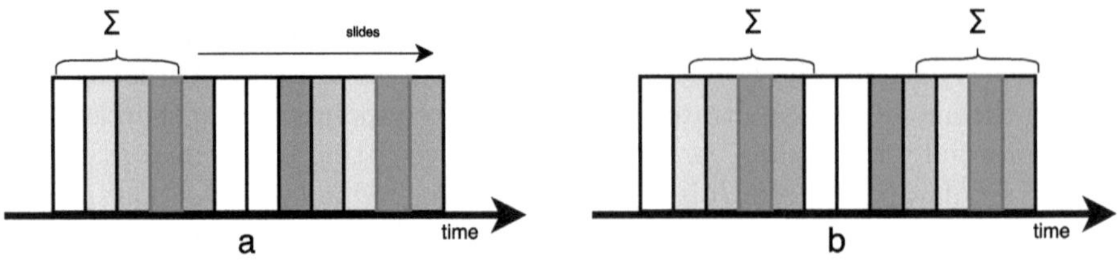

Fig. 3. (a) Sliding Window: a window slides on all frames, encouraging sparse predictions. (b) Fixed Window: we place windows only around ground truth actions encouraging sparse predictions only in those parts of the video.

"fetch") and *release* (e.g., "put", "leave-on"). Experiments aim to assess the different design choices leading to the definition of the final Mamba-OTR architecture. Due to the wide range of experimental settings, results are presented progressively, with only top-performing settings carried forward. All models were trained using high-level features from [12], sampled at 4 fps, and evaluated using the mp-mAP metric, with numbers reported in percentage.

5.1 Baseline Performance and Effect of Using the Focal Loss

Table 1 reports the performance of the three base architectures when trained using standard cross entropy or focal loss. The top section of the table reproduces the baseline models from [3]: *TeSTra* includes a 2-second long memory (8 frames) alongside short-term memory; *Transformer* removes the long memory, using a single transformer block with short memory; and *Mamba* replaces the transformer block with two Mamba layers. All models use 1 s (4 frames) for the short memory.[1] As can be observed, Transformer architectures struggle leveraging long-range information, with TeSTra and a standard Transformer performing on par. In these settings, Mamba already brings a better performance.

Table 1. Comparison of different models with and without Focal Loss.

Model	Long Memory	Focal Loss	mp-mAP↑
TeSTra	8 frames	–	20.10
Transformer	–	–	20.32
Mamba	–	–	25.16
TeSTra	8 frames	✓	**36.59**
Transformer	–	✓	**38.48**
Mamba	–	✓	**41.01**

[1] See [3] for more ablations on these models, for which we only report optimal configurations.

The lower section shows the same models trained with focal loss. The results demonstrate a substantial improvement across all models, with *Mamba-OTR* achieving the best performance (25.16 $\rightarrow$ 41.01). These findings confirm the critical role of focal loss in handling the extreme class imbalance arising in this task. We use focal loss for training in all subsequent experiments.

5.2 Number of Mamba Layers and Input Frames During Training

Table 2 presents the effect of varying the number of layers in the Mamba architecture. All models were trained with 4 frames (1 s) as input with the focal loss. Our results show that increasing the number of layers from 2 to 3 led to a small improvement in performance, from 41.01 to 41.63, while further increasing the number of layers to 4 resulted in a performance degradation, with the model achieving even worse results than the 2-layer configuration, all while significantly increasing the model size. Since the configuration with 3 Mamba layers resulted in improved performance, it will be used in the subsequent experiments.

Table 2. Performance of Mamba when varying the number of layers.

# Mamba layers	mp-mAP↑
2	41.01
3	**41.63**
4	40.52

Table 3. Performance of Mamba when varying the number of frames fed to the model during training.

Input frames	mp-mAP↑
4 (1 s)	41.63
8 (2 s)	41.84
12 (3 s)	42.47
20 (5 s)	**42.98**
40 (10 s)	41.77

Table 3 demonstrates how varying the number of input frames impacts the performance of the Mamba model. As observed, increasing the number of input frames improves performance, with an increase from 4 frames (1 s) to 20 frames (5 s), where the mp-mAP rises from 41.63 to 42.98. However, when the number of input frames is increased to 40 (10 s), the improvement over the 4-frames model is marginal and not as significant when compared to the 20-frames (5 s) configuration. This is coherent with the observation that in the considered dataset each action lasts around 8 frames ($\sim 2s$) in average. Thus, a window of 20 frames allows to capture both the action and enough context on previous relevant interactions.

For the purposes of the subsequent experiments, we will use 20-frames model for subsequent experiments.

It is worth noting that, while processing longer sequences is challenging for transformer architectures, it is much more natural for Mamba models, due to their recurrent nature. Nevertheless, we see benefits in keeping the input size restricted to 20 frames (5 s) at training time, while we will show benefits when extending this window at inference time, leveraging the generalization ability of the Mamba architecture.

5.3 Regularization

Given that the point-level mAP metric associates each ground truth instance with the closest prediction in time and penalize all other predictions made in its vicinity, the ideal model behavior is to produce a single, high-confidence prediction for each ground truth event. The regularization techniques we introduce aim to encourage this behavior during training by guiding the loss to penalize multiple nearby predictions appropriately. Throughout all experiments, we used a regularization weight $\lambda = 0.01$.

Table 4. Performance of MAMBA-OTR with different regularization techniques. Both the window-based approach use a 4 frames window size.

Regularization	mp-mAP↑
None	42.98
Entropy	42.70
Sliding Window	43.42
Fixed Window	**45.48**

Table 5. Performance of MAMBA-OTR with the Fixed Window regularization when varying the window size.

Window size	mp-mAP↑
4 frames	**45.48**
12 frames	44.05
20 frames	44.40

Table 4 reports the impact of different regularization strategies on the Mamba architecture with 3 layers and a input frames length of 20 frames (5 s). Notably, entropy regularization alone does not improve model performance. This is likely because it promotes general sparsity without enforcing any locality or temporal structure, thus failing to capture the intended prediction behavior. Better results arise from the window-based regularization approaches, which operate in the neighborhood of ground truth annotations. The *Sliding Window* technique yields a modest improvement of 0.44 mp-mAP, increasing performance from 42.98 to 43.42. More significantly, the *Fixed Window* method–where regularization is applied strictly around each positive (either take or release) ground truth annotation–produces a substantial improvement of 2.06 over the *Sliding Window* approach and 2.5 over the baseline model, achieving a final mp-mAP of 45.48.

Table 5 ablates the effectiveness of the best-performing regularization strategy, *Fixed Window*, when considering different window sizes. While the performance differences are relatively small, it is worth noting that even the lowest-performing configuration (12 frames, 44.05 mp-mAP) still outperforms the *Sliding Window* regularization from Table 4 by 0.63 mp-mAP. With a 20 frame

Table 6. Improvements in performance brought by the Fixed Window regularization when used with TeSTra and Transformer.

Model	Long memory	Short memory	Regularization	Window size	mp-mAP↑
TeStra	8 frames	4 frames	–	–	36.59
Transformer	–	4 frames	–	–	38.48
TeStra	8 frames	4 frames	Fixed Window	4 frames	**37.06**
Transformer	–	4 frames	Fixed Window	4 frames	**38.96**

window, the improvement increases to 0.98 mp-mAP over *Sliding Window*. Nevertheless, the optimal result is obtained with a window size of 4 frames (1 s), achieving an mp-mAP of 45.48, consistent with the results previously reported in Table 4. This is coherent with our previous observation that an take and release actions have a short duration, lasting in average 8 frames. Our fixed window regularization suppresses duplicate predictions around ground truth locations, encouraging the model to make more sparse and accurate predictions at test time, which results in an improved point level mAP. We refer to this model as Mamba-OTR.

Table 6 further shows the effect of the fixed window regularization technique on the two transformer-based models. Even if by small margins, regularization is beneficial with these architectures as well.

Table 7. Inference performance of Mamba-OTR and Transformer when processing videos using a sliding window and in streaming settings at test time.

	Time ↓		mp-mAP↑	
Model	**Video**	**Frame**	**Sliding↑**	**Streaming↑**
Mamba-OTR	**0.14 s**	**8ns**	**45.48**	43.35
Transformer	**0.14 s**	**8ns**	38.96	0.04

5.4 Extension to Longer Input Sequences at Test Time

Restricting input sequences to 20 frames (5 s) at training time enables a controlled and regularized learning, leading to best results. Nevertheless, processing videos in short chunks at test time is impractical for recurrent architectures such as Mamba, leading to slow inference due to the need of re-setting the hidden state of the model at every chunk or, worse, re-processing the same frame multiple time when a sliding window scheme is considered. In this section, we show that, while trained on fixed-length chunks, Mamba-OTR can generalize to longer sequences at inference time, achieving faster processing and even increased performance. On the contrary, transformer-based architectures do not possess this ability, likely due to their explicit modeling of temporal relations with positional embeddings.

Table 7 compares Mamba-OTR with the best-performing Transformer variant when tested using a sliding window approach (as in previous experiments) and in streaming mode, processing the full video one frame at a time, without relying on chunking or sliding window approaches. Note that this latter setting allows to deploy models more naturally, without requiring storing a buffer of previous observations. For each model, we also report mean processing time per video and per frame. We leave out the time required for feature extraction from this analysis. Both models were trained using the focal loss and the *Fixed Window* regularization strategy with a window size of 4 frames. The Transformer model is trained with an input size of 1 second (4 frames), no long memory and with the fixed window regularization with a window size of 4 frames, which proved optimal among the Transformer variants. The Mamba-OTR configuration includes 3 Mamba layers, an input frames length of 20-frames (5 s) and a fixed window regularization with a window size of 4 frames. We note that both architectures introduce minimal computational overhead beyond the backbone, both requiring only around 0.14 s per video and 8 nanoseconds for each frame. mp-mAP values confirms the inability of the Transformer-based model to deal with the shift between training-time and test-time processing. Mamba-OTR is robust and retains almost the same performance in streaming mode, obtaining an mp-mAP of 43.35, a small decrease from the original 45.48.

5.5 Qualitative Results

To further demonstrate the impact of Mamba-OTR on this task, we present a qualitative comparison in Fig. 4, where we evaluate Mamba without Focal Loss, Mamba with Focal Loss, and our proposed Mamba-OTR with regularization across two video examples exhibiting comparable levels of difficulty. As illustrated, the model trained without Focal Loss produces numerous predictions without any clear correlation with the ground truth. In contrast, both Mamba with Focal Loss and Mamba-OTR show improved behavior, with the latter displaying a marked advantage. Specifically, in the top example of Fig. 4, Mamba-OTR consistently generates a single, well-localized prediction for each ground truth event from the beginning to the midpoint of the video, a property not observed in the other models. A similar trend is observed in the bottom example of Fig. 4, where Mamba with Focal Loss produces a dense cluster of predictions in the middle portion of the video, whereas Mamba-OTR successfully condenses this into a single take prediction.

Fig. 4. Qualitative comparison between different regularization techniques.

6 Conclusion

We introduced Mamba-OTR, a Mamba-based model for the online detection of take and release actions in egocentric video. By combining focal loss with a fixed-window regularization approach, our method effectively addresses the challenges posed by class imbalance while aligning with the specific evaluation metric that penalizes multiple predictions around the same ground truth. Through extensive experiments, we shown that Mamba-OTR achieves state-of-the-art performance in terms of accuracy with a minimal overhead over the feature extractor, allowing to process full-length videos in real-time.

Acknowledgments. This research has been funded by the European Union - Next Generation EU, Mission 4 Component 1 CUP E53D23008280006 - Project PRIN 2022 EXTRA-EYE, and FAIR PNRR MUR Cod. PE0000013 - CUP: E63C22001940006.

References

1. An, J., Kang, H., Han, S.H., Yang, M.H., Kim, S.J.: Miniroad: Minimal RNN framework for online action detection. In: Proceedings of the IEEE/CVF International Conference on Computer Vision (2023)
2. Cao, K., Wei, C., Gaidon, A., Arechiga, N., Ma, T.: Learning imbalanced datasets with label-distribution-aware margin loss. In: Advances in Neural Information Processing Systems (NeurIPS) (2019)
3. Catinello, A.S., Farinella, G.M., Furnari, A.: Online detection of end of take and release actions from egocentric videos. In: International Conference on Computer Vision Theory and Applications (2025)
4. Chen, G., et al.: Video mamba suite: State space model as a versatile alternative for video understanding. arXiv preprint arXiv:2403.09626 (2024)
5. Cui, Y., Jia, M., Lin, T.Y., Song, Y., Belongie, S.: Class-balanced loss based on effective number of samples. In: Proceedings of the IEEE/CVF Conference on Computer Vision and Pattern Recognition (CVPR) (2019)
6. Gao, M., Xu, M., Davis, L.S., Socher, R., Xiong, C.: Startnet: online detection of action start in untrimmed videos. In: ICCV (2019)
7. Gao, M., Zhou, Y., Xu, R., Socher, R., Xiong, C.: Woad: weakly supervised online action detection in untrimmed videos. In: Proceedings of the IEEE/CVF Conference on Computer Vision and Pattern Recognition (2021)

8. Gu, A., Dao, T.: Mamba: linear-time sequence modeling with selective state spaces. arXiv preprint arXiv:2312.00752 (2023)
9. Lin, T.Y., Goyal, P., Girshick, R., He, K., Dollár, P.: Focal loss for dense object detection. In: IEEE International Conference on Computer Vision (2017)
10. Plizzari, C., et al.: An outlook into the future of egocentric vision. International Journal of Computer Vision (2024)
11. Scavo, R., Ragusa, F., Farinella, G.M., Furnari, A.: Quasi-online detection of take and release actions from egocentric videos. In: International Conference on Image Analysis and Processing (2023)
12. Shou, Z., et al.: Online detection of action start in untrimmed, streaming videos. In: ECCV (2018)
13. Wang, X., et al.: Oadtr: online action detection with transformers. In: Proceedings of the IEEE/CVF International Conference on Computer Vision (2021)
14. Xu, M., Gao, M., Chen, Y.T., Davis, L.S., Crandall, D.J.: Temporal recurrent networks for online action detection. In: ICCV (2019)
15. Xu, M., et al.: Long short-term transformer for online action detection. NeurIPS (2021)
16. Zhao, Y., Krähenbühl, P.: Real-time online video detection with temporal smoothing transformers. In: European Conference on Computer Vision (2022)

Zero-Shot Fibers Detection in Scanning Electron Micrographs

Pierluigi Carcagnì[1](✉), Marco Del Coco[1], Gennaro Gentile[2], Mariacristina Cocca[2], Melania Paturzo[3], and Marco Leo[1]

[1] Institute of Applied Sciences and Intelligent Systems, National Research Council of Italy, via Monteroni snc, 73100 Lecce, Italy
{pierluigi.carcagni,marco.delcoco,marco.leo}@cnr.it

[2] Institute for Polymers, Composites and Biomaterials, Italian National Research Council, Via Campi Flegrei 34, 80078 Pozzuoli, Napoli, Italy
{gennaro.gentile,mariacristina.cocca}@cnr.it

[3] Institute of Applied Sciences and Intelligent Systems, National Research Council of Italy, Via Campi Flegrei 34, 80078 Pozzuoli, Napoli, Italy
melania.paturzo@cnr.it

Abstract. The release of natural and artificial microfibers during laundering is considered one of the major sources of pollution in oceans. Quantifying microfibers is therefore necessary, and microscopy is the most used technique for any detection purpose. Scanning Electron Microscopy (SEM) can detect micro/nano scale objects, but related image analysis has proven more challenging than in traditional optical microscopy. Traditionally, trained experts count microfibers during washing cycles, but this task is tedious, subjective, and challenging. The development and application of automated techniques for extracting relevant image features has become necessary, at least to support the final human counting. Unfortunately, this area of research has been poorly investigated so far since the microfibers aspect has a large variability. This makes it difficult to transfer knowledge from human to machine. In this paper, a large AI model is applied, to the best of our knowledge, for the first time for fibers detection in SEM images. Its validity has been tested on a set of micrographs of wastewater from 4 washing cycles of cotton fabrics and 1 washing cycle of polyamide fabrics. Besides, the code and the annotated dataset of micrographs are also provided for research purposes.

Keywords: Foundation Models · Fibers Counting · Scanning Electron Microscopy · Zero-Shot Image Segmentation

1 Introduction

In recent years, the release of natural and artificial microfibers has become a hot topic [8,9]. Textile washing is in fact considered to be one of the main sources of microplastic pollution in the oceans [4]. While natural fibers such as cotton are

E. Rodolà et al. (Eds.): ICIAP 2025, LNCS 16167, pp. 469–480, 2026.
https://doi.org/10.1007/978-3-032-10185-3_37

not plastic, they behave similarly to microplastics in the environment, persisting for weeks to months before breaking down. They accumulate in soil, waterways and oceans where they can be ingested by wildlife, potentially causing blockages or chemical exposure [18]. Consequently, there has been an increasing urgency to understand these pollutants better and develop strategies to prevent them from entering the environment, where they pose a health risk [17]. This process starts from counting the microfibers released after standardized washing procedure and then finding adequate filtration methods (e.g. Wastewater Treatment Optimization) [10]. The collected data will serve as a foundation for designing targeted solutions to reduce fiber emissions and their environmental footprint. To quantify microfiber release, different analytical procedures have been adopted. However, the comparison among the results reported in the literature is quite difficult due to discrepancies in the used approaches [15]. A primary source of variation comes from the different laundering conditions, the washing machines (domestic, industrial or simulated laundering apparatus), the washing conditions and detergents, the filtration methods, and the filter types. Additionally, differences in the results are also due to the different analytical methods used for the microfiber quantification such as morphological, spectroscopic, thermal, and gravimetric approaches [6,10,23,26]. Microscopy is the most commonly used technique for several detection purposes [2]. A critical review of processing and classification techniques for images and spectra in microplastic research has been provided in [7]. Scanning Electron Microscopy (SEM) is a highly flexible method for obtaining precise surface information and high-resolution images of materials. It achieves far higher resolution than optical microscopy by scanning a specimen's surface with a focused electron beam. SEM equipment has resolutions ranging from less than one nanometer to multiple nanometers. Hence, it is possible to exploit SEM analysis to enable the detection of objects having micro- and nano-dimensions. On the other hand, compared to optical microscopy, electron microscopy image analysis has proven to be more difficult. Furthermore, the grey-scale nature of electron microscopy images means there is naturally less information to discern details [24]. The most common approach is to handle fragments mounted directly on the conductive adhesive. In this way, the fibers can be altered and contaminated and only the largest ones can be observed. On the other hand, they are well contrasted against the uniform background of the support. The best practice should be to directly observe the filters by scanning electron microscope (SEM) under actual laundry conditions. This method is currently the only reliable approach for detecting very small-sized fibers and that, additionally, ensures the preservation of fiber integrity, reducing the risk of damage or alteration during analysis. By maintaining the fibers in their original state, this technique also minimizes the potential for contamination, leading to more accurate and consistent results. Furthermore, it enables the assessment of filter efficiency under real-world laundry conditions, offering valuable insights into performance during actual use [5]. Unfortunately, observing the filter complicates the counting process because the filter's porosity creates a highly noisy background, unlike the common cases in which fragments are placed on the conductive adhesive.

Traditionally, trained experts count microfibers during washing cycles, but this is tedious, subjective, and challenging, especially in visually analyzing SEM images. Developing and applying automated techniques for extracting relevant picture features has become necessary [21]. Unfortunately, there are no previous automatic approaches working on SEM images acquired in a filter-based setup.

This paper presents the first known application of cutting edge technique, namely foundation models, for a fully automatic (zero-shot) fibers detection in filter-based method. Experiments were carried out on a dataset of 84 annotated SEM micrographs of wastewater from 4 washing cycles of Cotton fabrics and 21 annotated SEM micrographs of wastewater from 1 washing cycle of polyamide fabrics. Quantitative results demonstrated the ability of foundation models to detect candidate fiber regions, highlighting the effectiveness of foundation models in assisting fiber counting and reducing both the processing time and the subjectivity. Besides, the annotated dataset of micrographs is also provided for research purposes. The rest of the paper is organized as follows: Sect. 2 reports related work whereas Sect. 3 describes the algorithmic pipeline implemented to detect the fibers in SEM micrographs. Then, Sect. 4 reports experimental setup and results with both qualitative and quantitative analysis of the pipeline outcomes on a set of 84 + 21 = 105 micrographs, with a discussion about its strengths and weaknesses. Finally, Sect. 5 concludes the paper.

2 Related Work

Recently, it has been demonstrated that cutting-edge deep learning techniques can be exploited for electron microscopy image segmentation with great speed and accuracy [19].

It is worth noticing that previous attempts to use an automatic approach for microfiber segmentation rely on fragments mounted directly on the conductive adhesive. Even if this setup allows a uniform background, widely used approaches in computer vision for semantic segmentation tasks (R-CNN, U-NET) are prone to errors when they come to the analysis of fiber-shaped objects [13]. These methods require a combination with object-aware pixel embedding to perform instance segmentation on densely packed and tangled fibers for further quantification [25]. This additional process requires human supervision to provide shape and size knowledge to the model. Besides, fibers shape and textures change over different observations and then it became challenging to train AI models to be able to generalize over different fiber types and acquisition conditions [12]. Although some datasets were built for training deep neural networks, the experiments suffered from high costs and operational complexities. Current datasets are limited in size, geographically biased, and lack the diversity needed to train robust AI models [20,27].

A game change in this kind of situation has been introduced by foundation models, also known as large AI models, which have gained relevance in many fields of automatic processing. They are machine learning or deep learning models trained on extremely broad data such that they can achieve astonishing generalization capabilities, making them suitable for a wide range of use cases [3] without any specific tuning being required. They were introduced for natural language processing, but now they serve as the backbone of computer vision tasks such as image classification, object detection, and image segmentation. These models are trained on massive, diverse datasets to capture visual features that are universal to many different domains. They can then be used to perform specific tasks without the data needed to train a bespoke model from scratch. Among foundation models, the Segment Anything Model (SAM) is specifically addressed to RGB image segmentation [22]. The application of vision foundation models in microscopy is in its infancy, laying the groundwork for solving image analysis tasks in this domain with a small set of powerful deep learning models [1]. Recently in [14] SAM has been exploited to perform automated analysis of isolated aluminum nanoparticles.

3 Automatic Candidate Fibers Detection

The automatic detection of fibers in SEM micrographs is mainly an image segmentation problem but, due to its greater complexity, it requires multi-stage assessments to get accurate outcomes. The proposed solution exploits a SAM architecture for obtaining a preliminary zero-shot segmentation and then a refinement stage by including context knowledge-based processing.

The Segment Anything Model (SAM) is the most advanced, versatile model designed to segment any type of object within an image. It represents a significant leap forward in the field of image segmentation due to its flexibility, accuracy, and ability to generalize across a wide range of objects and scenarios [22]. The SAM Architecture leverages an encoder-decoder structure and it is trained on vast and diverse datasets that include various object types, environments, and conditions. This extensive pretraining enables the model to generalize well to new, unseen images. Unlike traditional models which may be tailored for specific object classes, SAM is designed to segment any object, regardless of its category. It can produce high-resolution segmentation masks, crucial for applications requiring detailed object delineation, and can exemplify the trend towards more generalizable and adaptable models in computer vision, leveraging advancements in deep learning, multi-modal integration, and large-scale pretraining to achieve state-of-the-art performance across diverse segmentation tasks.

The proposed algorithmic pipeline is schematically represented in Fig. 1.

Firstly, a zero-shot segmentation (no training required) by SAM model is applied to the acquired micrographs. The exploited SAM architecture has three components: an image encoder, a flexible prompt encoder, and a fast mask decoder. The encoder is an MAE [16] pre-trained Vision Transformer (ViT) [11]. There are two types of prompts: sparse which includes points, boxes and text and dense which includes a mask. Point is represented as a sum of positional encoding of the points' location and one of the two learned embeddings

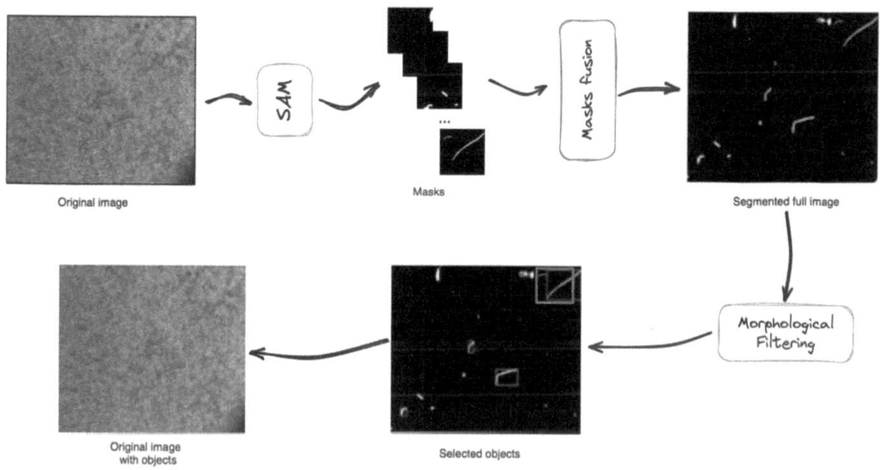

Fig. 1. The introduced algorithmic pipeline that highlights candidates for fiber occurrences in the input micrograph.

to indicate either a foreground point or a background point. Boxes are represented by an embedding pair. The pre-trained model involves predicting a set of "valid masks" for a given prompt to the model. The generated image embeddings and prompt embeddings are then mapped to the final mask, segmenting the objects in the scene. The model was trained on over 11 million licensed, privacy-preserving images with over 1 billion masks.

For fiber candidates' segmentation, SEM raw images are given as input to the SAM architecture to segment objects present in the acquired micrograph in a completely unsupervised way (no fine-tuning of the model has been previously performed due to the absence of ground truth mask annotations). According to the above description, in this step, the output of the SAM model consists of a set of different segmentation mask images (of the same resolution as the input micrograph), each one related to a hypothetical semantically recognized object. The masks are then bitwise merged to obtain a single final segmentation image. At this stage, every object identified by the model is returned, whether a fiber or not.

To filter out non-fiber occurrences, a second stage has been introduced in the pipeline exploiting morphological filtering relying on shape descriptors and unsupervised machine learning methodologies. As detailed in Fig. 2, a contour extraction procedure is first performed on the objects segmented by the SAM model. For each contour, two different shape descriptors are computed, and these scores are finally clustered using a k-means ($k = 2$) approach to isolate geometric shapes belonging to the fiber class.

Among several shape descriptors, the two that experimentally maximize the performance of the clustering procedure to best discriminate fiber-like sizes and

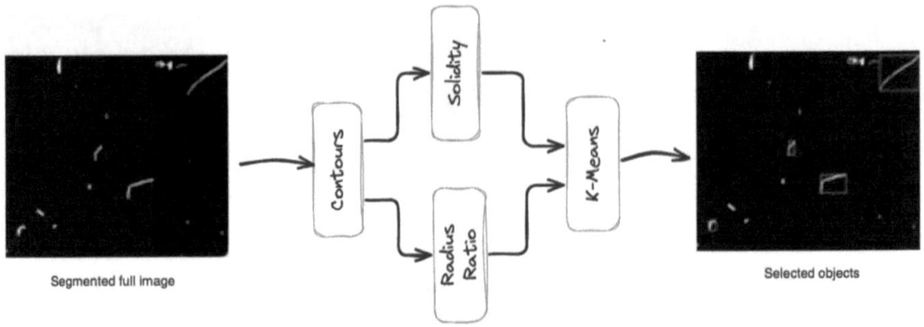

Fig. 2. Schematic representation of the morphological filtering step of the introduced pipeline.

shapes have been chosen, i.e. $Solidity = \frac{ContourArea}{ConvexHullArea}$ and $RadiusRatio = \frac{InscribedCircle}{CircumscribedCircle}$.

A graphic description of solidity and radius ratio shape descriptors is reported in Fig. 3.

Fig. 3. Graphic representation of Solidity and Radius Ratio descriptors in pictures a) and b) respectively.

Qualitative outcomes of segmented mask processing related to fiber and non-fiber detections are reported in Fig. 4. The top row concerns an actual fiber region, whereas the bottom row concerns a segmented region not belonging to a fiber. The leftmost column depicts the binary regions, whereas the central and rightmost column shows the solidity (the convex hull area in blue and the contour area in green) and radius ratio (inscribed circle in blue and circumscribed circle in green) outcomes.

4 Experimental Setup and Results

Cotton (CO) and 100% woven polyamide-6,6 (PA) fabrics were cut to a size of 9×9 cm^2, thermo-sealed edges and then were washed by using a lab scale Gyrowash apparatus (James H. Heal & Co, UK). Tests were performed at 40°C

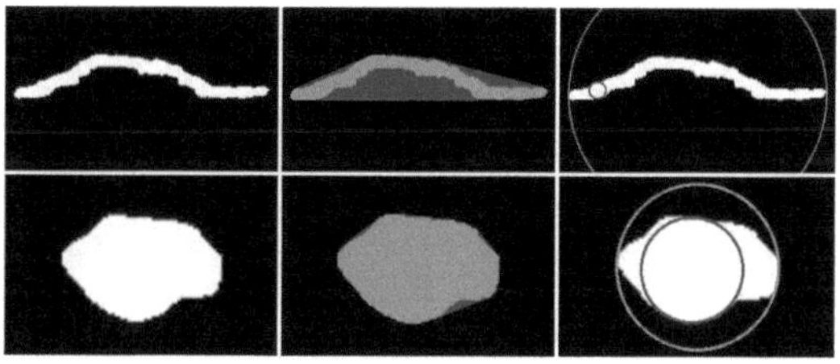

Fig. 4. Fiber Shape Analysis related to an actual fiber region (top row) and to a region not belonging to an actual fiber (bottom row).

using milli-Q water plus a commercial liquid detergent for 45 min. The washing water was filtered through polyvinylidene fluoride (PVDF) filters (Durapore, Merck Millipore) with an average pore width of 5 μm. The filter surfaces were analyzed by using a scanning electron microscope, SEM Quanta 200FEG (FEI, The Netherlands) in low vacuum mode (PH_2O = 0.7 torr), with a large field detector (LFD) and an accelerating voltage of 30 kV. Four cotton samples (CO_1,CO_2,CO_3, CO_4), and 1 polyamide sample (PA) were tested. At the end of the acquisition phase, 21 micrographs were available for each washing cycle and then 105 SEM micrographs were acquired in total. The resolution of each micrograph was 1024 × 882 pixels. The actual number of microfibers in each micrograph was determined by visual observation by one expert operator who provided annotations in terms of rectangular bounding boxes, making this way available the ground truth for assessing algorithms for fibers counting. Each image can either contain fibers or not. A total of 387 occurrences of Cotton fibers (83 in CO_1, 89 in CO_2 and CO_3 and 126 in CO_4) and 66 occurrences of PA fibers were annotated. This was considered the ground truth for the experiments described in the following sections.

In the experimental phase, all the acquired micrographs were processed using the algorithmic pipeline described in Sect. 3.

Figure 5 refers to a CO_4 micrograph: the outcome of the segmentation performed by the SAM model (top row) and of the following refinement process (bottom row) leveraging shape descriptors are reported. Green and blue bounding boxes are over-imposed on the input micrograph, where green bounding boxes indicate regions with shape and size compliant with those of fibers, and blue bounding boxes refer to discarded regions.

The discriminative capabilities of the two exploited descriptors are highlighted in Fig. 6, where clustering outcomes for each set of acquired fiber images are depicted.

It displays the clusters generated by the K-means algorithm, with the axes representing the selected parameters for shape characterization. After clustering, K-means identifies two unlabeled clusters. Following this, the label "non-fibers" is automatically assigned to the cluster with the centroid farthest from the origin of the axes, while the label "fibers" is applied to the remaining cluster. In Table 1,

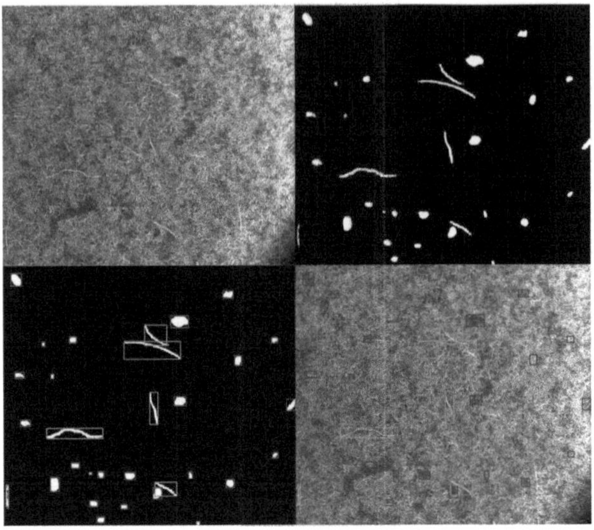

Fig. 5. Segmentation by zero-shot SAM (top row) and refinement (bottom row) outcomes in a CO_4 micrograph. See the text for more explanation.

the mean centroids and range values of the exploited descriptors in each washing cycle are reported.

The validity of the above assumption about clusters' labelling was finally provided by the quantitative evaluation described in the following.

The proposed algorithmic pipeline was then quantitatively evaluated by comparing the automatically extracted bounding boxes with those provided by the experts who annotated the acquired micrographs. In Table 2 the quantitative evaluation of the capability of the proposed pipeline to detect fibers is reported.

Detections have been considered correct if there is a minimum overlap of 25% between the bounding boxes provided by the manual annotation procedure with the ones built around the contours of the objects retained after the morphological analysis described in Sect. 3.

As qualitatively and quantitatively demonstrated, the proposed pipeline works very well in the majority of tested cases. This can help to speed up the tedious task of manually counting fibers, also reducing the subjectivity bias. It is worth noting that the pipeline works at its best in case of fibers sparse in the micrographs i.e. providing separate blobs as in Fig. 7a where red and green bonding boxes indicate fibers automatically detected and manually annotated respectively.
On the contrary, some difficulties arose in some rare cases where the fibers appear as in a cluster such as in Fig. 7b. In such cases, not all the fibers were automatically detected, and then their counting can lead to some underestimation with a decrease in quantitative performance in Table 2. This is the reason why the scores for the CO_1 set of micrographs are the worst among all the set (i.e. the set

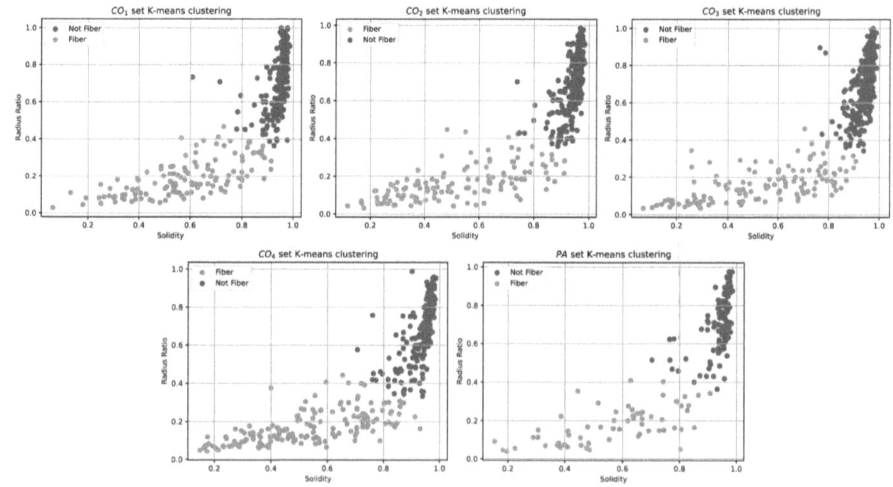

Fig. 6. Clustering exploiting Solidity and Radius Ratio for each segmented region in the 4 CO and 1 PA washing cycles.

Table 1. Clusters mean centroids and range values of the exploited descriptors for the 5 considered washing cycles.

Tested Sample	Class	Min Solidity	Max Solidity	Min Radius Ratio	Max Radius Ratio	Centroid
CO_1	Fiber	0.063	0.91	0.028	0.47	0.58, 0.19
	Non Fiber	0.61	0.99	0.36	1.0	0.94, 0.73
CO_2	Fiber	0.12	0.91	0.03	0.45	0.52, 0.18
	Non Fiber	0.74	0.98	0.36	0.98	0.94, 0.70
CO_3	Fiber	0.07	0.99	0.033	0.46	0.54, 0.16
	Non Fiber	0.76	0.99	0.34	1.0	0.94, 0.71
CO_4	Fiber	0.15	0.93	0.043	0.44	0.53, 0.18
	Non Fiber	0.71	0.99	0.33	0.99	0.93, 0.67
PA	Fiber	0.15	0.91	0.039	0.41	0.57, 0.18
	Non Fiber	0.70	0.99	0.36	0.98	0.94, 0.72

Table 2. Quantitative fibers detection results.

Tested Sample	TP	FP	FN	Precision	Recall	F1-Score
CO_1	58	52	25	0.53	0.70	0.60
CO_2	69	55	20	0.57	0.78	0.65
CO_3	70	57	19	0.55	0.79	0.65
CO_4	95	41	31	0.70	0.75	0.73
PA	51	20	15	0.72	0.77	0.74

contains more micrographs with clustered fibers). Anyway, the proposed pipeline in all the considered micrographs detected all the regions containing the cotton fibers. It fails only in counting every single fiber, but this is a refinement step that can be delegated to further post-processing steps, either automatic or manual if required by the specific application.

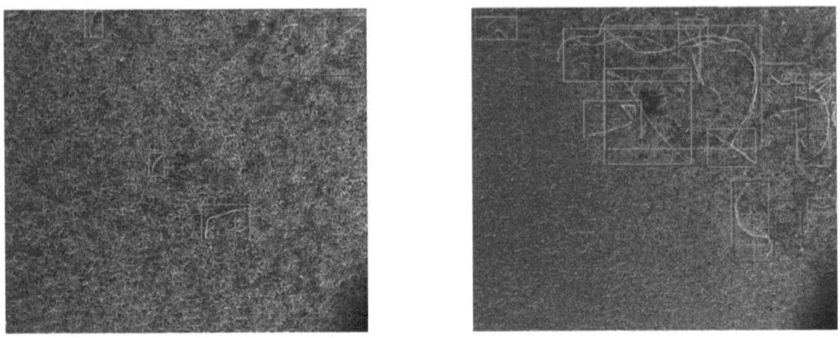

(a) (b)

Fig. 7. a) A simple scenario (belonging to the CO_4 set of micrographs), and b) a complex scenario (belonging to the CO_1 set of micrographs). See text for explanation.

5 Conclusions

This paper presents the first application of a foundation model to assist in counting micro- and nanofibers in SEM images of wastewater from washing cycles. In particular, the considered model can detect fibers with a zero-shot approach, so it does not need to be adjusted when the acquisition conditions and fiber types change. This is a relevant result, as the proposed pipeline can assist human experts in fiber counting, speeding up the process and reducing subjectivity in fiber detection. Besides, unlike existing supervised approaches, it does not require any preliminary training on annotated images, which are very difficult to obtain, especially when filters are directly observed. In these cases, the underlying porosity makes the background very noisy, and the appearance of the background and foreground is not stable under different setup conditions, making it impossible to generalize training knowledge. Future work will deal with these challenging cases and with the extension of experiments on different types fibers and, subsequently, introducing an automatic classification approach.

Acknowledgements. This research was partially supported by the project Future Artificial Intelligence Research—FAIR CUP B53C220036 30006 grant number PE0000013 and partially supported by the project "Co-creating strong uptake of REMEDIES for the future of our oceans through deploying plastic litter valorisation and prevention pathways" funded by the European Union's HORIZON EUROPE Innovation program under grant agreement No. 101093964.

References

1. Archit, A., et al.: Segment anything for microscopy. Nature Methods, pp. 1–13 (2025)
2. Baruah, A., Sharma, A., Sharma, S., Nagraik, R.: An insight into different microplastic detection methods. Int. J. Environ. Sci. Technol. **19**(6), 5721–5730 (2022)
3. Bommasani, R., et al.: On the opportunities and risks of foundation models. arXiv preprint arXiv:2108.07258 (2021)
4. Boucher, J., Friot, D.: Primary microplastics in the oceans: a global evaluation of sources, vol. 10. Iucn Gland, Switzerland (2017)
5. Conkle, J.L., Báez Del Valle, C.D., Turner, J.W.: Are we underestimating microplastic contamination in aquatic environments? Environ. Manage. **61**(1), 1–8 (2018)
6. Corami, F., Rosso, B., Bravo, B., Gambaro, A., Barbante, C.: A novel method for purification, quantitative analysis and characterization of microplastic fibers using micro-ftir. Chemosphere **238**, 124564 (2020)
7. Cowger, W., et al.: Critical review of processing and classification techniques for images and spectra in microplastic research. Appl. Spectrosc. **74**(9), 989–1010 (2020)
8. De Falco, F., Cocca, M., Avella, M., Thompson, R.C.: Microfiber release to water, via laundering, and to air, via everyday use: a comparison between polyester clothing with differing textile parameters. Environ. Sci. Technol. **54**(6), 3288–3296 (2020)
9. De Falco, F., Di Pace, E., Cocca, M., Avella, M.: The contribution of washing processes of synthetic clothes to microplastic pollution. Sci. Rep. **9**(1), 6633 (2019)
10. De Falco, F., et al.: Evaluation of microplastic release caused by textile washing processes of synthetic fabrics. Environ. Pollut. **236**, 916–925 (2018)
11. Dosovitskiy, A., et al.: An image is worth 16x16 words: transformers for image recognition at scale. arXiv preprint arXiv:2010.11929 (2020)
12. Elahi, M., Afolaranmi, S.O., Martinez Lastra, J.L., Perez Garcia, J.A.: A comprehensive literature review of the applications of ai techniques through the lifecycle of industrial equipment. Discover Artif. Intell. **3**(1), 43 (2023)
13. Frei, M., Kruis, F.E.: Fiber-cnn: expanding mask r-cnn to improve image-based fiber analysis. Powder Technol. **377**, 974–991 (2021)
14. Gaines, T.B., Boyle, C., Keller, J.M., Maschmann, M.R., Price, S., Scott, G.J.: Scanning electron microscope image segmentation with foundation ai vision model for nanoparticles in autonomous materials explorations. In: 2024 IEEE Conference on Artificial Intelligence (CAI), pp. 1266–1271. IEEE (2024)
15. Hazlehurst, A., Tiffin, L., Sumner, M., Taylor, M.: Quantification of microfibre release from textiles during domestic laundering. Environ. Sci. Pollut. Res. **30**(15), 43932–43949 (2023)
16. He, K., Chen, X., Xie, S., Li, Y., Dollár, P., Girshick, R.: Masked autoencoders are scalable vision learners. In: Proceedings of the IEEE/CVF Conference on Computer Vision and Pattern Recognition, pp. 16000–16009 (2022)
17. Hegarty, J., dos Reis, R., Dravid, V.P.: Laundry to laboratory: Automated image analysis for the characterization of fibrous microplastics. ACS ES&T Water (2024)
18. Henry, B., Laitala, K., Klepp, I.G.: Microfibres from apparel and home textiles: prospects for including microplastics in environmental sustainability assessment. Sci. Total Environ. **652**, 483–494 (2019)

19. Horwath, J.P., Zakharov, D.N., Mégret, R., Stach, E.A.: Understanding important features of deep learning models for segmentation of high-resolution transmission electron microscopy images. npj Computational Materials **6**(1), 108 (2020)
20. Iida, Y., Watanabe, K., Ominami, Y., Toyoguchi, T., Murayama, T., Honda, M.: Development of rapid and highly accurate method to measure concentration of fibers in atmosphere using artificial intelligence and scanning electron microscopy. J. Occupational Health **63**(1), e12238 (2021)
21. Jacobs, R.: Deep learning object detection in materials science: Current state and future directions. Comput. Mater. Sci. **211**, 111527 (2022)
22. Kirillov, A., et al.: Segment anything. In: Proceedings of the IEEE/CVF International Conference on Computer Vision, pp. 4015–4026 (2023)
23. Peñalver, R., Arroyo-Manzanares, N., López-García, I., Hernández-Córdoba, M.: An overview of microplastics characterization by thermal analysis. Chemosphere **242**, 125170 (2020)
24. Shah, A., Schiller, J.A., Ramos, I., Serrano, J., Adams, D.K., Tawfick, S., Ertekin, E.: Automated image segmentation of scanning electron microscopy images of graphene using u-net neural network. Materials Today Commun. **35**, 106127 (2023)
25. Shi, B., et al.: Automatic quantification and classification of microplastics in scanning electron micrographs via deep learning. Sci. Total Environ. **825**, 153903 (2022)
26. Volgare, M., et al.: Washing load influences the microplastic release from polyester fabrics by affecting wettability and mechanical stress. Sci. Rep. **11**(1), 19479 (2021)
27. Xu, J., Wang, Z.: Efficient and accurate microplastics identification and segmentation in urban waters using convolutional neural networks. Sci. Total Environ. **911**, 168696 (2024)

7Bench: A Comprehensive Benchmark for Layout-Guided Text-to-Image Models

Elena Izzo[1(✉)], Luca Parolari[1], Davide Vezzaro[1,2], and Lamberto Ballan[1]

[1] Department of Mathematics, University of Padova, Padova, Italy
{elena.izzo,luca.parolari,davide.vezzaro.4}@studenti.unipd.it, lamberto.ballan@unipd.it

[2] Brain Technologies Innovation Division, Brain Technologies srl, Torino, Italy

Abstract. Layout-guided text-to-image models offer greater control over the generation process by explicitly conditioning image synthesis on the spatial arrangement of elements. As a result, their adoption has increased in many computer vision applications, ranging from content creation to synthetic data generation. A critical challenge is achieving precise alignment between the image, textual prompt, and layout, ensuring semantic fidelity and spatial accuracy. Although recent benchmarks assess text alignment, layout alignment remains overlooked, and no existing benchmark jointly evaluates both. This gap limits the ability to evaluate a model's spatial fidelity, which is crucial when using layout-guided generation for synthetic data, as errors can introduce noise and degrade data quality. In this work, we introduce 7Bench, the first benchmark to assess both semantic and spatial alignment in layout-guided text-to-image generation. It features text-and-layout pairs spanning seven challenging scenarios, investigating object generation, color fidelity, attribute recognition, inter-object relationships, and spatial control. We propose an evaluation protocol that builds on existing frameworks by incorporating the layout alignment score to assess spatial accuracy. Using 7Bench, we evaluate several state-of-the-art diffusion models, uncovering their respective strengths and limitations across diverse alignment tasks. The benchmark is available at https://github.com/Elizzo/7Bench.

Keywords: Layout-guided Models · Diffusion Models · Benchmark

1 Introduction

Layout-guided text-to-image models extend the standard text-to-image generation paradigm by introducing explicit layout control. Among the various layout representations, bounding boxes have emerged as the most commonly adopted format due to their simplicity and ease of annotation. This format enables users

E. Izzo, L. Parolari, and D. Vezzaro—These authors contributed equally to this work and are listed in alphabetical order.

E. Rodolà et al. (Eds.): ICIAP 2025, LNCS 16167, pp. 481–493, 2026.
https://doi.org/10.1007/978-3-032-10185-3_38

to intuitively specify the position, scale, and appearance of objects, offering a more controllable generation process. Such fine-grained control over the image content fosters new opportunities for leveraging synthetic data, consequently increasing the demand and usage of layout-guided diffusion models in other computer vision tasks, *e.g.* synthetic dataset generation [11,17].

Despite the high potential of these tools, recent studies have identified several limitations [6]. In fact, models often struggle to generate images that accurately adhere to the given layout and textual prompt, especially when given uncommon spatial arrangements [26]. Additionally, they inherit fundamental issues from diffusion models, including catastrophic neglect, attribute binding, and attribute leaking, which affect both text and layout alignment.

Quantifying and identifying errors in text-image and layout-image faithfulness is crucial to properly evaluate and compare different layout-guided diffusion models. However, benchmarks collected for evaluating textual *and* layout alignment are still missing. As a result, the strategies adopted by the researchers to evaluate layout-guided generative models are heterogeneous and mainly based on traditional text-to-image benchmarks [1,9,23]. These include tests for text-image alignment in scenarios involving colors, attributes, or relations, but either lack any dedicated evaluation of layout alignment or suffer from inconsistencies among different evaluation protocols. For instance, layout alignment is assessed using a random subset of annotated datasets for object detection or grounding [3,13,19] or by researchers collecting their own sets of box-and-text pairs [26]. A recent effort [4] introduces the evaluation of spatial skills (*e.g.* position, size or shape) but overlooks semantic evaluation. This lack of a unified approach leads to a significant misalignment among studies in terms of evaluation metrics, the number of samples investigated, and the specific layout skills being tested.

In this paper, we propose 7Bench, a comprehensive benchmark for layout-guided models characterized by pairs of textual prompts and sets of bounding boxes to evaluate models on 7 different scenarios. The aim is to investigate models' abilities in objects' generation, colors, attributes, relationships, and spatial control. We design a structured prompt collection annotated with bounding boxes, collecting a dataset with 224 samples equally distributed in the 7 scenarios. We introduce an evaluation method based on two complementary measures: the text-alignment and the layout-alignment scores. Our evaluation protocol extends an existing framework that focuses on text-image alignment by introducing an assessment of layout accuracy. We systematically evaluate state-of-the-art layout-guided text-to-image diffusion models using our benchmark, comparing both fine-tuned models on grounding information and training-free approaches. Results obtained on 7Bench demonstrate room for improvement in all models, validating the scientific rationale of our approach.

In summary, our main contributions include: (i) 7Bench, a novel benchmark for layout-guided text-to-image models, designed to evaluate both text-image and layout-image alignment across seven diverse and challenging scenarios; (ii) a meticulous collection of 224 text and bounding boxes pairs, ensuring a broad coverage of various semantic and spatial arrangements; (iii) an evaluation protocol based on two complementary measures: text-alignment and layout-alignment scores, which assess the alignment of generated images with respect to both textual descriptions and spatial layouts; and (iv) a comprehensive evaluation of four state-of-the-art layout-guided text-to-image diffusion models using 7Bench, providing insights into the strengths and weaknesses of different approaches, including a comparison between fine-tuned and training-free models.

2 Related Works

Benchmarks for Text-to-Image Generation. Text-to-image generation is the process of creating images from textual descriptions. The rise of diffusion models [8] revolutionised this task, outperforming traditional GAN-based approaches [21] in generating high-quality and diverse images. Despite the outstanding performance, these models often struggle to produce images that accurately align with the description, failing to bind attributes or specific details to objects [6]. As a result, benchmarks and evaluation metrics have been introduced to measure the faithfulness of a generated image to its text input on various challenging scenarios: colors, attributes, counting, relations, or objects [1,5,9,10,23,25]. However, with the introduction of layout-guided text-to-image models, all these benchmarks miss two key elements: the input layouts of the object described in the textual prompt and an evaluation metric to quantify the layout alignment.

Layout-Guided Text-to-Image Diffusion Models. Layout-guided text-to-image diffusion models incorporate predefined spatial layouts for finer control over image generation. Generally, these methods inject grounding information into pre-trained models by adding new trainable layers or tokens [13,27], or manipulating cross-attention layers at inference time to steer the generation process without requiring additional training [3,19,26]. Evaluating these models requires assessing both text and layout alignment, but the absence of comprehensive benchmarks has led researchers to develop their own evaluation strategies.

Evaluation of Layout-Guided Diffusion Models. The evaluation of layout-guided models is often a two-step framework assessing text and layout faithfulness separately. Text alignment is evaluated by benchmarks from the traditional text-to-image domain. Instead, layout alignment is often assessed by generating images from bounding boxes and captions sampled from COCO2014 [15] or Flickr30K Entities [20] validation sets. The generation quality is then quantified with metrics like FID [7] and YOLO score [14]. However, this evaluation is heterogeneous among papers and does not follow a shared protocol. For instance, in [13] a subset of 30k samples are generated while in [19] only 5k. Moreover, in [26], the authors point out the need for a distinct evaluation dataset, independent of COCO images, since it is commonly used to train the models. In [4], a benchmark for layout-guided image generation for spatial control skills is introduced. However, it focuses only on spatial abilities without considering all the other challenging scenarios for generative models. In this work, we introduce a comprehensive benchmark for layout-guided models, aiming to test them on various skills, from colors to spatial layouts.

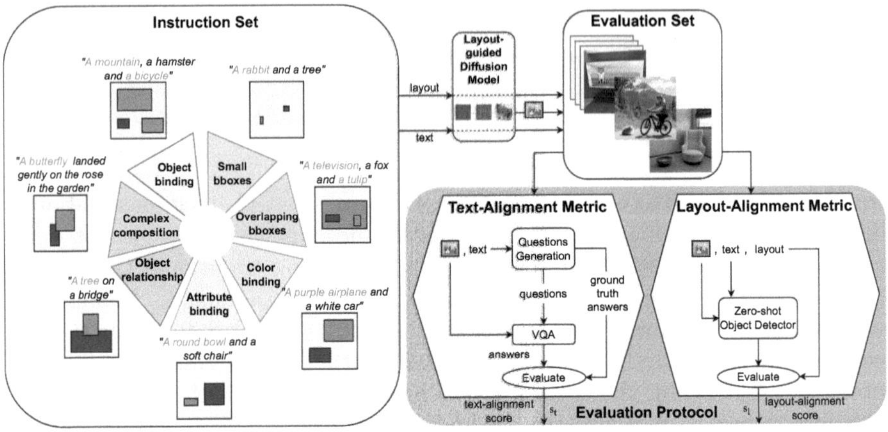

Fig. 1. 7Bench overview. Our benchmark contains prompts from 7 different scenarios based on semantic and spatial arrangement characteristics. The evaluation protocol to apply on the set of generated images is based on two metrics: text- and layout-alignment scores to evaluate text and layout faithfulness, respectively.

3 7Bench Benchmark

In this section, we present 7Bench, a benchmark specifically designed to evaluate the performance of layout-guided text-to-image diffusion models. It assesses how faithfully the generated image adheres to both the input text and the specified layout by employing an evaluation protocol that measures alignment along both dimensions. The benchmark comprises 224 diverse text prompts, each paired with a set of bounding boxes that define the intended spatial arrangement of objects in the image. Samples are carefully curated to probe the capabilities of layout-aware generative models under a range of challenging conditions. Overall, the benchmark spans 7 distinct scenarios, categorized based on the complexity and structure of the textual descriptions and associated layouts. An overview of the benchmark is illustrated in Fig. 1.

3.1 Prompt and Bounding-Box Collection

We created a benchmark of 224 samples. Each sample is a pair made by a textual prompt and a set of bounding boxes (one for each object in the prompt). We divided the 224 samples into seven scenarios based on the characteristics of their textual prompts and bounding boxes: object binding, small bboxes, overlapping bboxes, color binding, attribute binding, object relationship, and complex composition. We collected 32 prompts for each scenario to carefully balance the distribution and do not under/over-represent one specific case.

The sets of textual prompts are generated following a template whose formalism is inspired by the disentangle representation theory [24]. Each prompt contains N objects, each of which can be qualified by an attribute or be in relation to another object. Hence, the object at position i in a prompt is a token belonging to the set $\mathcal{O}$, it can be qualified by an attribute a_i in the set $\mathcal{A}$ or in relation r_{ij} to another object at position j, where the relation is in the set $\mathcal{R}$.

For example, for a prompt with $N = 2$ objects, the template t is:

$$t = "det(o_1, a_1)a_1 o_1 rel(and, r_{12}) det(o_2, a_2)a_2 o_2" \tag{1}$$

where, $o_i \in \mathcal{O}$, $a_i \in \mathcal{A}$, $det(o_i,a_i)$ is a determinant that depends on the object or attribute if present, and $rel(and, r_{ij})$ is the coordinating conjunction *and* or a relation $r_{ij} \in \mathcal{R}$ if present. We vary the number of objects N in $\{1, 2, 3, 4\}$. Figure 2 shows the objects' distribution across all collected prompts.

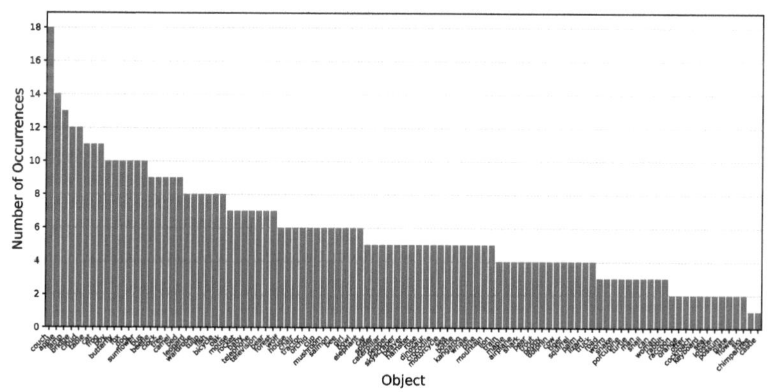

Fig. 2. Number of occurrences by object computed across all prompts in 7Bench.

In order to define the layout requirements, the position of each object o_i in a textual prompt is defined by a bounding box $bbox_i$. In particular, the desired position of the object o_i in the generated image is defined by a set of coordinates $bbox_i = [x^i_{min}, y^i_{min}, x^i_{max}, y^i_{max}]$, where $0 \leq x_{min} < x_{max} \leq 1$ and $0 \leq y_{min} < y_{max} \leq 1$. Unless otherwise specified, each bounding box is manually collected by arranging a realistic layout for each prompt that maintains proportions between objects and does not contain overlaps.

Hereafter, we illustrate each scenario investigated in our benchmark, describing the characteristics of their textual prompts and bounding boxes.

Object Binding. This scenario aims to study the catastrophic neglect in the layout-guided models and the right localization of each generated object. Catastrophic neglect is a phenomenon that characterizes the text-conditioned diffusion models and consists in not generating or mixing together the elements described in the prompt. For this scenario, we collect a set of samples varying the number of objects N in $\{1, 2, 3, 4\}$. We do not add any attribute or relation. The resulting textual prompt is constructed following the template:

$$t_{obj} = "det(o_1)o_1, det(o_2)o_2, det(o_3)o_3 and det(o_4)o_4". \quad (2)$$

Small Bboxes. The use of small bounding boxes as the input represents a critical challenge for layout-guided text-to-image diffusion models. Due to their limited spatial extent, small bounding boxes often fail to provide sufficient conditioning signals during the denoising process, leading to inaccurate object placement or complete omission [18]. In order to investigate generative abilities under such conditions, we introduce the *small bboxes* scenario.

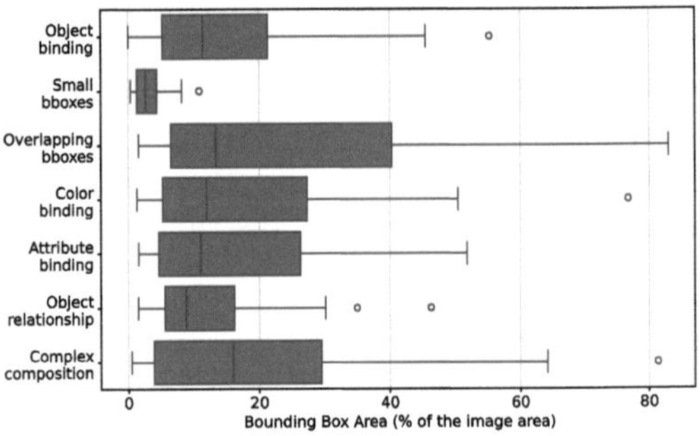

Fig. 3. Distribution of bounding boxes area per scenario. Values are displayed percentage with respect to the image size.

It is characterized by a spatial layout where the bounding boxes's area is between 3% and 10% of the total image area. Mathematically, given a bounding box $bbox_i = [x^i_{min}, y^i_{min}, x^i_{max}, y^i_{max}]$, its area A_i is defined as $A_i = (x^i_{max} - x^i_{min}) \cdot (y^i_{max} - y^i_{min})$. Given an $w \times h$ image and its area $A_{image} = w \cdot h$, then A_i is randomly chosen following the constraint $0.03 \cdot A_{image} \leq A_{image} \cdot A_i \leq 0.1 \cdot A_{image}$. Figure 3 shows the distribution of the areas of the bounding box collected in this scenario in comparison with the other ones. The textual template is t_{obj} from the *object binding* scenario.

Overlapping Bboxes. Similar to the *small bboxes* scenario, another critical and relevant aspect to investigate in a layout-guided diffusion model is the use of overlapping bounding boxes as guidance. Those can be used to specify background and foreground layers or to specify details on an object. As previously done, we follow the textual template t_{obj}, that represent 1-4 objects without attributes or relations. All the bounding boxes in a sample overlap at least one other bounding box in the same sample. Specifically, $\forall i \in \{1, ..., N\} \quad \exists j_{\in\{1,...,N\}} \neq i | A_i \cap A_j > 0$.

Color Binding. This scenario aims to investigate the adherence of color-based guidance on depicted objects. Despite color being the most commonly used attribute for describing objects in scenes, state-of-the-art text-conditioned diffusion models still demonstrate issues in applying the correct color to the target object. For studying this characteristic in layout-guided text-to-image diffusion models, we define a set $\mathcal{C}$ of eleven universal basic colors based on the findings of Berlin and Kay [2]: *black, blue, brown, gray, green, pink, purple, red, white, yellow, orange.* The textual template report a color attribute for each object described in the prompt:

$$t_{color} = "det(c_1)c_1o_1, det(c_2)c_2o_2, det(c_3)c_3o_3 and det(c_4)c_4o_4". \tag{3}$$

Attribute Binding. This scenario investigates the capacity of a generative model to correctly bind an object and its attribute. It extends the *color binding* scenario by adding several attributes related to various types: color, shape, material, appearance, and dimension of the object. In total, we investigate 30 different attributes setting $\mathcal{A} = \{$ *aggressive, black, blue, bright, clean, crowded, dark, fast, fluffy, fuzzy, green, happy, large, pink, red, rotten, rough, shiny, short, silver, small, smooth, snowy, soft, tall, warm, white, wooden, yellow*$\}$. For this scenario, we avoid the use of relations and we vary the number of objects N in $\{1, 2, 3, 4\}$:

$$t_{attr} = "det(a_1)a_1o_1, det(a_2)a_2o_2, det(a_3)a_3o_3 and det(a_4)a_4o_4". \tag{4}$$

Object Relationship. The relationships between objects are crucial when describing complex scenes, but is also a difficult task for generative models. For this reason, we introduce the *object relationship* scenario. It is characterized by a textual prompt with 2 or 4 objects correlated to each other by a relation, and does not use attributes. Inspired by studies on spatial cognition [12], we define a set $\mathcal{R}$ of 11 relations: $\mathcal{R} = \{$*above, below, beside, far from, near, next to, on, over, to the left of, to the right of, under* $\}$. The textual template is defined as:

$$t_{rel} = "det(o_1)\mathfrak{r}_{12} det(o_2)o_2 and det(o_3)\mathfrak{r}_{34} det(o_4)o_4". \tag{5}$$

Complex Composition. In order to test layout-guided models with more open-world and challenging compositional textual prompts and layout, we collect 32 samples in which to combine all the previous investigated scenarios. Hence, in the *complex composition* scenario, the textual prompts can contain from 1 to 4 objects. Each object is qualified by at least one attribute that describes its color, shape, material, appearance, or dimension, and it is in relation to another object. The bounding boxes associated with each object can have an area smaller than 10% of the generated image and can overlap each other.

3.2 Evaluation Protocol

To evaluate the alignment between the generated images and the input prompt, we introduce two complementary measures: the text-alignment score s_{text} and the layout-alignment score s_{layout}. These scores quantify the image adherence to both the semantic and spatial constraints specified in the prompt.

Text Faithfulness. We define the text-alignment score $s_{text} \in [0, 1]$ as the TIFA score [9], a widely used metric for evaluating semantic alignment in text-to-image models. This score represents the proportion of correct answers produced by a Vision Question Answering (VQA) model when analyzing the generated images. The evaluation process involves generating a set of questions and corresponding answers based on the input prompt. These are created by a Large Language Model (LLM), ensuring that the assessment remains independent of the image generation process. The VQA model is then queried with these questions while processing the generated image, and its responses are compared against the expected answers from the LLM. The final score s_{text} is computed as the accuracy of the VQA model's answers with respect to the reference answers provided by the LLM. A higher s_{text} value indicates strong semantic consistency between the image and the prompt.

Layout Faithfulness. We define the layout-alignment score $s_{layout} \in [0, 1]$ as the Area Under Curve (AUC) of accuracy@k values computed over a range of Intersection over Union (IoU) thresholds. The s_{layout} score captures the spatial accuracy of objects' placement within the generated images.

Specifically, given a generated image $\hat{I}$, the set of objects $\{o_i\}_{i=1}^{N}$ and the set of target bounding boxes representing the ground truth position of objects $\{t_i\}_{i=1}^{N}$, we apply an object detector (*e.g.* OWL-ViT [16]) to obtain a set of M detections associated with their corresponding label and confidence score: $D = \{(d_j, l_j, c_j)\}_{j=1}^{M}$. For each target object o_i, we filter the set of detections by matching the label with the object, thus $\hat{D}_i = \{(d_j, l_j, c_j) \in D \mid l_j = o_i\}$. Among the filtered detections $\hat{D}_i$, we select the bounding box with the higher score $\hat{d}_i = d_{j^*}$ with $j^* = \arg\max_{j:(d_j,l_j,c_j)\in\hat{D}_i} c_j$. We then compute the IoU between the selected detection $\hat{d}_i$ and the target bounding box t_i, denoted as $\text{IoU}_i = \text{IoU}(t_i, \hat{d}_i)$. Subsequently, these IoU values are thresholded at multiple levels $k \in \{0.1, 0.2, \ldots, 0.9\}$ to calculate

$$accuracy@k = \frac{1}{N}\sum_{i=1}^{N} \mathbf{1}[\text{IoU}_i \geq k]. \tag{6}$$

Then, the final layout-alignment score s_{layout} is obtained as the area under the resulting accuracy@k curve.

4 Experiments

4.1 Experimental Setup

In this section, we present a systematic evaluation of 4 popular layout-guided text-to-image diffusion models on our 7Bench benchmark. The models are all open source and have been accurately chosen to explore a wide range of methods and techniques. In particular, we test GLIGEN (G) [13], that is trained with grounding information, and three training-free approaches: Attention Refocusing (G_AR) [19], BoxDiff (G_BD) [26], and Cross Attention Guidance (SD_CAG) [3]. The first two are built on top of GLIGEN, the last one uses Stable Diffusion as the underlying model. Finally, we include Stable Diffusion v1.4 (SD) [22] in the analysis for a comparison in terms of textual alignment.

Given a model, we generate 16 images for each sample in 7Bench, varying the seed from 1 to 16. We get a total of 17.920 synthetic images with dimension 512×512, equally distributed in the five models under investigation.

We evaluate the generated images using the evaluation protocol described in Sect. 3.2. We use pre-trained weights for TIFA and OWL-ViT[1],[2].

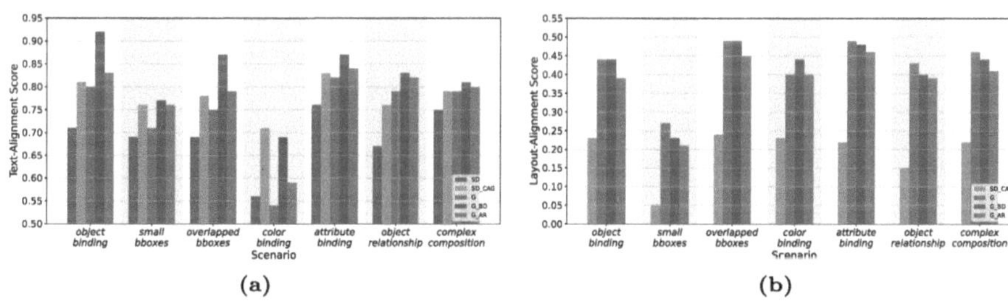

Fig. 4. Performance per scenario. Figure 4a and Fig. 4b show textual and layout alignment, respectively.

4.2 Quantitative and Qualitative Evaluation

We first present an evaluation average over scenarios (Fig. 4), reporting both text-alignment (Fig. 4a) and layout-alignment (Fig. 4b) results. We recall that the two metrics are complementary measures that aim to quantify text and layout alignment, respectively. Both scores take values in $[0, 1]$, where 1 means perfect alignment and 0 means complete misalignment.

Depending on the model and scenario, the text alignment varies approximately from 0.55 to 0.9, and the layout alignment from 0.05 to 0.5, showing room for improvement in the current state-of-the-art models in all the scenarios in our benchmark. In terms of layout-alignment score, the curves show that

[1] https://github.com/Yushi-Hu/tifa.
[2] https://huggingface.co/docs/transformers/model_doc/owlvit.

methods significantly struggle to generate objects that respect a provided small bounding box. Instead, they seem to not struggle in *overlapping bboxes*, often finding creative solutions to overlap objects as requested. Overall, models based on a trained layout-guided model (GLIGEN, BoxDiff, and Attention Refocusing) obtain superior performance with respect to Cross Attention Guidance (based on SD). This is not surprising considering that GLIGEN gets the best results in the majority of the scenarios.

As concerns the text-alignment (Fig. 4a), there is more scores' fluctuation both among models and scenarios. Overall, training-free approaches are good allies and can significantly boost the model to which they are added improving its concept knowledge. In particular, BoxDiff generally outperforms the other models, and SD_CAG gets results comparable to those of the other training-free approaches despite being built on top of SD instead of GLIGEN. Interestingly, all layout-guided models converge to the same score in *complex composition*, suggesting their comparable abilities when open-world and challenging compositional textual and layout are provided.

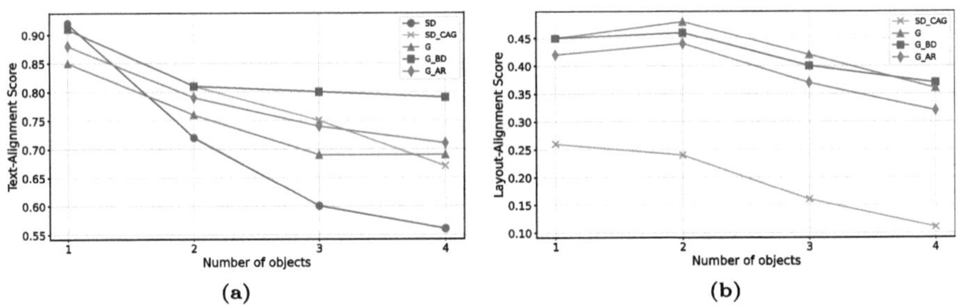

Fig. 5. Performance per number of objects in the prompt. Figure 5a and Fig. 5b show textual and layout alignment, respectively.

Secondly, we present the results averaged over the number of objects in the textual prompt (Fig. 5). Both text-alignment (Fig. 5a) and layout-alignment (Fig. 5b) curves confirm the expected trend: a progressive degradation in performance increasing the number of objects in the prompt. In addition, models based on GLIGEN exhibit a lower degradation factor with respect to Cross Attention Guidance and Stable Diffusion.

Finally, Fig. 6 shows some qualitative results of all the layout-guided models under investigation on a subset of randomly selected prompts. In the images, we show the input bounding box of each object in the textual prompt and the filtered bounding box obtained by applying our Layout Faithfulness pipeline.

Fig. 6. Qualitative results. Ground-truth object locations are indicated by dashed bounding boxes, while detected objects are outlined with solid bounding boxes. Colors identify different target objects.

5 Conclusion

We introduced 7Bench, a benchmark tailored for evaluating layout-guided text-to-image diffusion models across diverse and challenging scenarios. Unlike prior evaluations that focus solely on text alignment or adopt inconsistent strategies for layout assessment, 7Bench provides a unified protocol to measure both text-image and layout-image faithfulness. Our results highlight the limitations of training-free approaches and the importance of fine-tuning for effective spatial control. We believe that 7Bench will support more rigorous comparisons, drive the development of more controllable generative models, and enable progress in downstream tasks where structured generation is essential.

Acknowledgments. We acknowledge the CINECA award under the ISCRA initiative, for the availability of high performance computing resources and support.

References

1. Bakr, E.M., Sun, P., Shen, X., Khan, F.F., Li, L.E., Elhoseiny, M.: Hrs-bench: holistic, reliable and scalable benchmark for text-to-image models. In: ICCV (2023)
2. Berlin, B., Kay, P.: Basic color terms: their universality and evolution. Univ of California Press (1991)
3. Chen, M., Laina, I., Vedaldi, A.: Training-free layout control with cross-attention guidance. In: WACV (2024)
4. Cho, J., Li, L., Yang, Z., Gan, Z., Wang, L., Bansal, M.: Diagnostic benchmark and iterative inpainting for layout-guided image generation. In: CVPRW (2024)
5. Feng, W., et al.: Training-free structured diffusion guidance for compositional text-to-image synthesis. In: ICLR (2023)
6. Grimal, P., Borgne, H.L., Ferret, O., Tourille, J.: TIAM - A metric for evaluating alignment in text-to-image generation. In: WACV (2024)
7. Heusel, M., Ramsauer, H., Unterthiner, T., Nessler, B., Hochreiter, S.: Gans trained by a two time-scale update rule converge to a local nash equilibrium. In: NeurIPS (2017)
8. Ho, J., Jain, A., Abbeel, P.: Denoising diffusion probabilistic models. In: NeurIPS (2020)
9. Hu, Y., Liu, B., Kasai, J., Wang, Y., Ostendorf, M., Krishna, R., Smith, N.A.: TIFA: accurate and interpretable text-to-image faithfulness evaluation with question answering. In: ICCV (2023)
10. Huang, K., Sun, K., Xie, E., Li, Z., Liu, X.: T2i-compbench: a comprehensive benchmark for open-world compositional text-to-image generation. In: NIPS (2023)
11. Kupyn, O., Rupprecht, C.: Dataset enhancement with instance-level augmentations. In: ECCV (2024)
12. Landau, B., Jackendoff, R.: Whence and whither in spatial language and spatial cognition? Behav. Brain Sci. **16**(2), 255–265 (1993)
13. Li, Y., et al.: GLIGEN: open-set grounded text-to-image generation. In: CVPR (2023)
14. Li, Z., Wu, J., Koh, I., Tang, Y., Sun, L.: Image synthesis from layout with locality-aware mask adaption. In: ICCV (2021)
15. Lin, T., et al.: Microsoft COCO: common objects in context. In: ECCV (2014)
16. Minderer, M., et al.: Simple open-vocabulary object detection with vision transformers. In: ECCV (2022)
17. Parolari, L., Izzo, E., Ballan, L.: Harlequin: color-driven generation of synthetic data for referring expression comprehension. In: ICPR (2024)
18. Patel, Z., Serkh, K.: Enhancing image layout control with loss-guided diffusion models. In: WACV (2025)
19. Phung, Q., Ge, S., Huang, J.: Grounded text-to-image synthesis with attention refocusing. In: CVPR (2024)
20. Plummer, B.A., Wang, L., Cervantes, C.M., Caicedo, J.C., Hockenmaier, J., Lazebnik, S.: Flickr30k entities: Collecting region-to-phrase correspondences for richer image-to-sentence models. In: ICCV (2015)
21. Reed, S.E., Akata, Z., Yan, X., Logeswaran, L., Schiele, B., Lee, H.: Generative adversarial text to image synthesis. In: ICML (2016)

22. Rombach, R., Blattmann, A., Lorenz, D., Esser, P., Ommer, B.: High-resolution image synthesis with latent diffusion models. In: CVPR (2022)
23. Saharia, C., et al.: Photorealistic text-to-image diffusion models with deep language understanding. In: NeurIPS (2022)
24. Trager, M., Perera, P., Zancato, L., Achille, A., Bhatia, P., Soatto, S.: Linear spaces of meanings: Compositional structures in vision-language models. In: ICCV (2023)
25. Wu, X., Yu, D., Huang, Y., Russakovsky, O., Arora, S.: Conceptmix: a compositional image generation benchmark with controllable difficulty. In: NeurIPS (2024)
26. Xie, J., et al.: Boxdiff: text-to-image synthesis with training-free box-constrained diffusion. In: ICCV (2023)
27. Yang, Z., et al.: Reco: region-controlled text-to-image generation. In: CVPR (2023)

Efficient Attention on Microcontrollers via Linearized Transformer Blocks

Alberto Ancilotto(✉), Edoardo Castagnini, and Elisabetta Farella

Fondazione Bruno Kessler, Trento, Italy
aancilotto@fbk.eu

Abstract. Vision transformers (ViTs) have become the go-to solution in many computer vision tasks due to their ability to capture long-range dependencies in data and, consequently, achieve remarkable performance in dense prediction tasks. However, their high computational and memory requirements make these architectures unsuitable for embedded and low-power devices. This work introduces a novel linear ViT architecture optimized for resource-constrained embedded hardware. Our model operates with less than 1MB of parameters and 400KB of RAM, relying only on operations compatible with most embedded runtimes. We propose a linear attention mechanism based solely on fully connected layers, enabling execution on platforms that do not support typical transformer operations. To ease deployment, we apply a Hardware-Aware Scaling (HAS) strategy that enables one-shot network scaling to meet specific hardware constraints, without requiring an expensive Network Architecture Search (NAS). Using just three hyperparameters, HAS supports independent control over model size, memory usage, and computational cost. Our network architecture shows better generalization performance and up to 80% lower latency compared to current alternatives for embedded devices for Imagenet-1K classification. When coupled with a DETR detection head, we achieve state-of-the-art mAP/latency tradeoffs and up to a 90% reduction in latency compared to existing approaches, thanks to better optimization for the lightweight CPUs and small accelerators typical of tiny, resource-constrained devices.

1 Introduction

The advent of Vision Transformers (ViTs) has impacted the computer vision field significantly. The ability to model long-range dependencies allows these architectures to achieve higher performance than their previous convolutional counterparts, leading to widespread adoption in many fields [5,9,33]. However, ViTs generally require higher computational capabilities than classical convolutional architectures, both in terms of complexity and memory requirements, making them unsuitable for deployment on microcontrollers (MCUs) and other resource-constrained devices. Recent transformer-based models [4,24] have made progress toward efficient execution on some embedded platforms (e.g., Nvidia Jetson) by reducing parameter counts, memory usage, and latency. However,

E. Rodolà et al. (Eds.): ICIAP 2025, LNCS 16167, pp. 494–506, 2026.
https://doi.org/10.1007/978-3-032-10185-3_39

these architectures are still incompatible with highly constrained devices such as 32-bit MCUs. This limitation is due to two main factors: (i) the extreme memory constraints of MCUs (usually < 2MB in storage and RAM) and (ii) the lack of support in mainstream deployment toolchains [18] for core operations used in attention blocks (i.e. batch matrix multiplication, layer normalization and GeLU activations).

Despite these challenges, transformer-based architectures have shown greater generalization capabilities than CNNs [2,25,38], thus making them particularly appealing for MCUs. In real-world applications on these devices, neural networks often operate with lower-quality cameras and simplified preprocessing steps than those used with public datasets during training, leading to a domain shift and, inevitably, to considerable performance degradation. Domain adaptation and continual learning techniques that could mitigate this performance drop are challenging to implement on these devices [3,31] - therefore, architectures with stronger out-of-the-box generalization capabilities are particularly valuable in this setting.

This work proposes a novel ViT architecture that targets compatibility and efficiency for MCUs and tiny embedded devices. The contribution of our work is twofold: first, we introduce a hybrid network based on a novel scalable backbone coupled with linear attention blocks that maximize compatibility with embedded hardware platforms; second, we propose the Hardware-Aware Scaling (HAS) technique for scaling our ViT to different computational constraints. Specifically, we propose a multi-head linear attention (MLA) block that relies on a sequence of weight-sharing linear layers. As described in Sect. 3, this block implements a linear attention operation, providing broader support to consumer runtimes and devices, therefore addressing the lack of support in standard libraries for commonly used operations, such as Batch Matrix Multiply (BMM), LayerNorm, and GeLU. To the best of our knowledge, this is the first ViT implementation fully supported by tinyML consumer frameworks such as `STM32Cube.ai` [29], `nncase` [16], `GAPflow` [37], `Tflite-micro` [8].

The paper is structured in four main sections. After an overview of the existing literature on CNN and ViT compression, we describe the proposed architecture in Sect. 3. In 3.1, we present the optimizations to the multi-head linear attention block for tiny devices, analyzing the effects that scaling the different block tensors has on memory usage, parameter count, and operations. In Sect. 3.2, we present our proposed attention block based on linear layers and in Sect. 3.3, we detail the final attention block design and the proposed network. We analyze the effects of the proposed scaling method in Sect. 3.4 and compare the classification and object detection performance against existing architectures in Sect. 4.

2 Related Works

Several techniques to reduce the computational complexity of deep learning approaches have been studied in recent years. MobileNet [13] and Xception [6] were the first architectures to rely on depthwise-separable approximation of standard 2D convolutions to reduce the number of parameters. MobileNetV2 [34], MobileNetV3 [12] and MobileNetV4 [32] improved on this idea by using inverted residual blocks and automated network architecture search (NAS) to optimize the parameter/performance trade-off. EfficientNet [35] introduced the concept of compound scaling, where the three main dimensions of a neural network (resolution, depth, and width) are jointly scaled to increase performance. ShuffleNet V2 [27] introduced practical guidelines for optimizing not only the number of parameters but also the operation/latency trade-offs in modern ConvNets. MCUNet [21], MCUNet V2 [20], and V3 [22] are specifically targeted at small microcontroller units (MCUs) while still using the inverted residual block architecture to optimize the number of parameters. More recently, PhiNets [30] and XiNets [1] introduced hardware-aware scaling, allowing disjoint optimization of memory, parameters, and latency in an MCU-targeted network.

Since the advent of ViT architectures [7], optimizing the resources required by these models has been an active field of research. The Linformer [39] architecture was the first to propose a low-rank approximation for the attention operation, achieving linear complexity with respect to the input sequence length. However, this architecture still relies on softmax dot-product attention, which is an expensive operation to perform on MCUs due to the exponentials involved. In contrast, [15] first proposed using a linearized attention block based on a ReLU nonlinearity (lite MLA). ConvFormer [9] optimized efficiency using a hybrid architecture comprising both convolutional and ViT blocks. EfficientFormer [19] brought significant speedups by downsampling the Q, K, and V maps used in multi-head self-attention. Liu et al. [24] started optimizing the accuracy/latency trade-offs in ViTs; however, their work still targets GP-GPUs and larger embedded devices. Recently, Cai et al. [4] extended the lite MLA mechanism of [15] to create efficient transformer architectures targeting embedded devices. Nonetheless, networks presented in [4] are still too large for MCU-based implementations. MobileViT [28] and MobileNetV4 [32] worked towards bringing multi-head self-attention to mobile devices with significant improvements in performance/latency tradeoffs.

2.1 Hardware Aware Scaling

Traditional network scaling methods, such as width scaling [13,34] and compound scaling [35], focus on optimizing a single tradeoff (e.g. accuracy vs. operations or latency). While suitable for powerful edge devices like smartphones, they often underutilize resources on ultra-constrained platforms such as microcontrollers. In such settings, three key hardware limits must be jointly considered: MACs (latency/energy), FLASH (model size), and RAM (intermediate tensor size). Classical scaling typically addresses only one, yielding suboptimal results. Hardware Aware Scaling (HAS) [30] introduces three independent hyperparameters to scale different network parts in alignment with these constraints, maximizing resource utilization. Unlike compound scaling, HAS tunes each dimension separately to match specific hardware constraints, maximizing resource usage.

2.2 Multi-Head Linear Attention

Figure 2 shows a typical MLA block computing self-attention from the *keys* $\mathbf{K}'_h \in \mathbb{R}^{WH \times d_k}$, *queries* $\mathbf{Q}'_h \in \mathbb{R}^{WH \times d_q}$, and *values* $\mathbf{V}_h \in \mathbb{R}^{WH \times d_v}$ tensors. The linear attention operation for head h can be expressed (before scaling) as:

$$\mathbf{O}_h = \left(\Phi(\mathbf{Q}'_h)\Phi(\mathbf{K}'_h)^\top\right)\mathbf{V}_h, \tag{1}$$

where $\Phi(x) = \max\{0, x\}$ denotes the ReLU activation function. As in [15], using $\mathbf{K} = \Phi(\mathbf{K}')$ and $\mathbf{Q} = \Phi(\mathbf{Q}')$ we can rewrite the above equation as:

$$\mathbf{O}_h = \mathbf{Q}_h(\mathbf{K}_h^\top \mathbf{V}_h). \tag{2}$$

This re-parametrization reduces the memory footprint of the self-attention block. However, the original implementation relies on $d_q = d_k = d_v$, thus limiting the scalability of the proposed MLA block. The following sections will analyze the tradeoffs that can be achieved by varying the ratios between these dimensions.

3 Architecture Overview

In the following sections, we outline the main optimizations of our attention block compared to existing approaches. We first analyze the effects of varying the size of the keys, queries, and values vectors - d_k, d_q, d_v, respectively. A similar approach is presented in [19,24]; unlike previous works, however, we consider the joint effects of different feature maps dimensionality on memory, parameter count, and operations (Sect. 3.1). In Sect. 3.2, we detail our attention mechanism, which relies on linear layers to maximize compatibility. We show that this is mathematically equivalent to the original MLA formulation presented in [15]. We present the final attention block design in Sect. 3.3 and propose a scaling strategy for implementing Hardware Aware Scaling (HAS). Finally, in Sect. 3.4, we analyze the HAS implementation for our network.

3.1 Scaling Feature Maps in MLA

Vision transformer architectures are derived from networks used for NLP tasks, and generally use the same sizes for the $\mathbf{K}, \mathbf{Q}$, and $\mathbf{V}$ vectors [4,39]. This choice, dictated by the formulation of the attention operation (Eq. 2), may not be optimal in tiny architectures, especially given the large number of operations involving these vectors. In fact, from Eq. 2, it follows that $d_{qk} := d_q = d_k$ should hold, while no particular constraint is imposed on d_v.

To hit the Pareto-optimal tradeoff between performance and computational cost, we modeled the impact of the dimensionality of the keys, queries, and value vectors analytically. We tested different block configurations, varying the ratio d_v/d_{qk} while keeping their product constant. This allows us to keep the memory requirement of the feature map constant (i.e. proportional to $d_v \times d_{qk}$). Figure 1 shows, as the ratio d_v/d_{qk} varies, the effects on overall accuracy, number of parameters, RAM needed for computation, and number of operations (Multipy-Accumulate - MAC) in the attention block. In the following, we present the analytical model for the computational requirements of the MLA block.

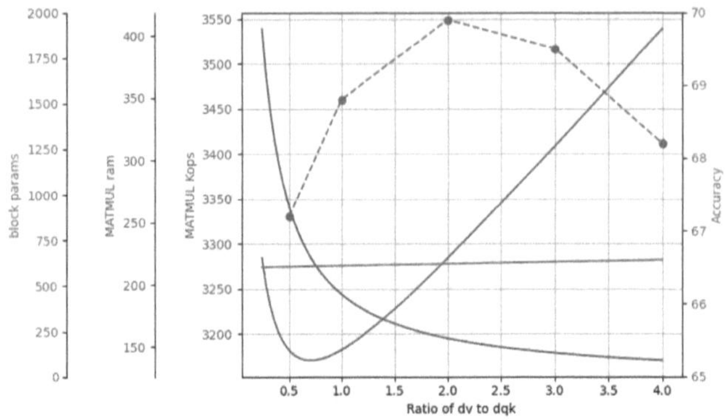

Fig. 1. Effects of varying the d_v/d_{qk} ratio on performance (gray), number of parameters (green), required RAM (light blue), and number of operations (red)(Color figure online)

Peak RAM Usage. It refers to the amount of RAM needed to store the input tensor composed of V, K and Q; it can be computed as a function of the number of heads h and spatial dimensions of the feature map W, H, following:

$$\text{RAM}_{\text{peak}} = h(\overbrace{WHd_v}^{\text{RAM(V)}} + \overbrace{WHd_q}^{\text{RAM(Q)}} + \overbrace{WHd_k}^{\text{RAM(K)}}) \tag{3}$$

Operation Count. It considers the number of operations required for the h matrix-matrix multiplications and the ReLU activation applied on $\mathbf{K} = \Phi(\mathbf{K}')$ and $\mathbf{Q} = \Phi(\mathbf{Q}')$.

$$\text{MAC}_{\text{att}} = \overbrace{2hWHd_vd_{qk}}^{\text{MAC(BMM)}} + \overbrace{2hWHd_{qk}}^{\text{ReLU(K')} + \text{ReLU(Q')}} \tag{4}$$

Parameter Count. It takes into account the number of parameters in the block, which consists of a pointwise convolution that expands the channels to $h(d_v + 2d_{qk})$ and a second pointwise convolution that projects the attention output to the number of input channels C

$$\text{Params}_{\text{block}} = \overbrace{Ch(d_v + 2d_{qk})}^{\text{Params(PW1)}} + \overbrace{Chd_v}^{\text{Params(PW2)}} \tag{5}$$

Our analysis, summarized in Fig. 1, shows that a ratio between 1 ($dv = d_{qk}$) and 2 is the best compromise to minimize the number of operations and amount of RAM required, with lower values optimizing the latter and higher values beneficial to the former. To simplify the search space of our network, we base the remainder of the paper and the final architecture on a d_v/d_{qk} ratio of 2. This provides slightly higher accuracy while still keeping the peak RAM usage of the attention layers under 250 kB for a 224×224 image.

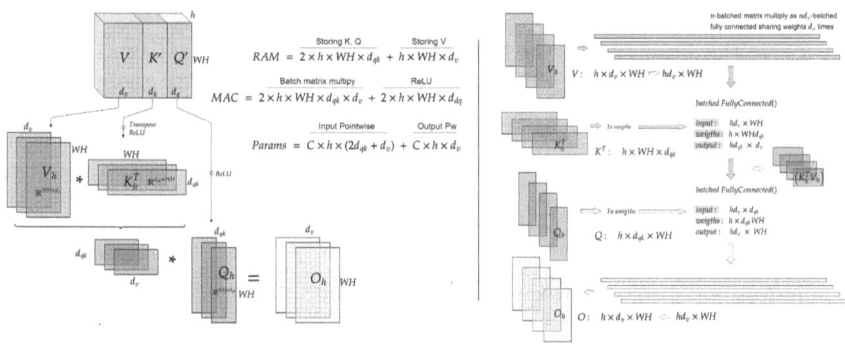

Fig. 2. Left: the linear attention block used in our architecture, with the different operations detailed. Right: the linearized attention computation, realizing h BMM operations via hd_v fully connected layers, reusing weights across d_v iterations.

3.2 An MCU-Compatible Attention Mechanism

Linear attention is typically implemented using batched matrix multiplications (BMMs), computing $O_h = \mathbf{Q}_h\mathbf{K}_h^\top\mathbf{V}_h$ for each head. This is expressed across all heads as:

$$\mathbf{O} = \text{BMM}(\mathbf{Q}, \text{BMM}(\mathbf{K}^\top, \mathbf{V})) \in \mathbb{R}^{h\times WH\times d_v} \tag{6}$$

where $\mathbf{Q} \in \mathbb{R}^{h\times WH\times d_{qk}}$, $\mathbf{K}^\top \in \mathbb{R}^{h\times d_{qk}\times WH}$, and $\mathbf{V} \in \mathbb{R}^{h\times WH\times d_v}$.

While efficient on GPUs, this BMM-based implementation is unsupported on many embedded runtimes. Instead, we propose a mathematically equivalent formulation using standard linear layers, which are widely supported. Rewriting Eq. 2, we express attention as:

$$\mathbf{O}_h = \begin{bmatrix} \mathbf{Q}_h(\mathbf{K}_h^\top \mathbf{v}_{h,1}) \\ \mathbf{Q}_h(\mathbf{K}_h^\top \mathbf{v}_{h,2}) \\ \vdots \\ \mathbf{Q}_h(\mathbf{K}_h^\top \mathbf{v}_{h,d_v}) \end{bmatrix} \tag{7}$$

This reformulation converts each head into d_v linear layers, resulting in $h \cdot d_v$ total layers with shared weights per head. This enables post-training conversion without accuracy loss, using efficient BMMs during training and linear layers during deployment.

3.3 Attention Block Design

The proposed architecture follows a hybrid ViT-convolutional approach, similar to [9], with 4 downsampling stages. The first two use purely convolutional blocks (from PhiNet [30]), while the last two introduce our novel attention blocks to manage complexity. Each attention block includes a pointwise expansion convolution projecting the input to C_{exp} channels:

$$C_{\text{exp}} = E_0 \beta (d_v + 2d_{qk}) h \tag{8}$$

where E_0 is the base expansion factor (default 4), β controls attention width, and h is the number of attention heads. This is followed by a linearized attention layer (see Sect. 3.2) and a final pointwise projection back to the input dimension.

To link C_{exp} with the input channels C_{in}, we define α such that:

$$C_{\text{exp}} = E_0 \alpha \beta C_{\text{in}}, \quad h = \frac{\alpha C_{in}}{d_v + 2d_{qk}} \tag{9}$$

This removes one degree of freedom and ties attention complexity to β:

$$\frac{C_{\text{exp,att}}}{C_{\text{exp,conv}}} = \beta \tag{10}$$

This formulation enables the implementation of hardware-aware scaling. Since modulating β scales the complexity only of the attention blocks, HAS can be used to reduce the parameter count independently when FLASH is the limiting factor, without much influence on RAM or MAC usage (FIg. 3).

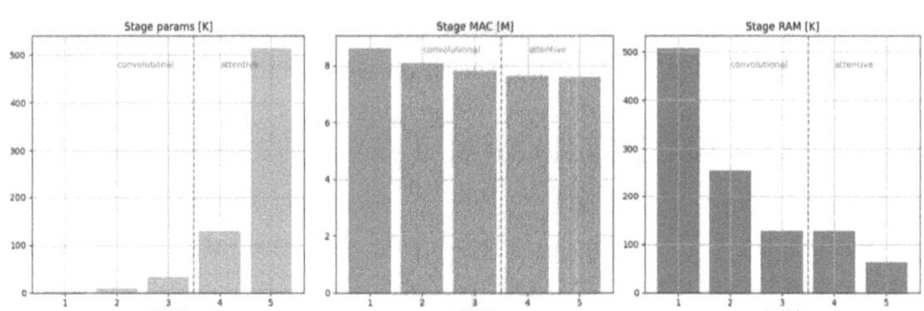

Fig. 3. Resources required for the different network stages. The number of parameters increases for blocks at the end of the network, while peak memory usage decreases.

3.4 HAS Implementation

As introduced in Sect. 2.1, HAS relies on three tunable hyperparameters targeting distinct constraints of embedded systems: RAM, operations, and parameter count. Table 1 summarizes the effect of each.

Table 1. Effect of each scaling factor on relevant network metrics and accuracy.

Factor	Scale	MAC [M]	Params	RAM [kB]	Acc
	1.0	40.1	1.39	530	70.3%
α	0.85	28.4	0.95	451	68.4%
	0.75	23.0	0.72	398	66.5%
	1.0	40.1	1.39	530	70.3%
β	0.75	32.5	0.95	530	69.8%
	0.5	25.9	0.70	530	67.6%
	4	40.1	1.39	530	70.3%
E_0	3	37.6	1.38	398	69.9%
	2	34.9	1.37	265	67.8%

Table 2. Comparison with other efficient CNNs and ViTs for ImageNet classification performance, generalization accuracy, and on-device latency

Model	Type	Params [M]	MAC [M]	ImageNet	ImageNet-A	ImageNet-C	Latency (RPi4)
PhiNet [30]	CNN	1.3	120	64.0%	5.59%	23.8%	31.1
MCUNet in3 [21]	CNN	0.74	82	62.2%	2.33%	12.3%	69.1
MCUNet in4 [21]	CNN	1.7	126	68.4%	3.03%	16.1%	99.5
MobileNet V3 small [12]	CNN	2.5	56	68.7%	5.42%	24.2%	34.2
ShuffleNet V2 × 1.0 [27]	CNN	2.2	140	69.4%	6.05%	28.9%	32.9
MobileNet V2 0.75 [34]	CNN	1.9	177	70.1%	4.85%	23.2%	40.7
MIT - EfficientViT B0 [4]	ViT	3.7	115	67.8%	5.71%	27.6%	58.2
MSRA - EfficientViT M0 [24]	ViT	2.3	39.5	63.2%	6.31%	29.3%	14.8
Ours (2M)	**ViT**	**2.0**	**40.1**	**70.3%**	**6.41%**	**29.3%**	**21.4**

The width factor α linearly scales all feature map widths, quadratically increasing operations and latency. In contrast, β adjusts the expansion ratio of attention vs. convolution blocks, primarily affecting parameter count with minimal memory impact. Lastly, E_0 controls the size of early-stage feature maps and directly impacts peak memory usage. These results show that when a specific constraint (e.g., memory or parameter count) dominates, scaling β or E_0 instead of α leads to more efficient accuracy-resource trade-offs. For instance, at a parameter budget of 0.7M, β scaling achieves 1.2% higher accuracy than α; at a 400kB RAM limit, E_0 scaling yields a 3.4% improvement.

4 ImageNet Classification

As previously discussed, one of the strengths of a ViT is its superior generalization capabilities compared to a CNN. To validate this propriety for our approach, we also tested our approach on ImageNet-A [11] (a set of natural adversarial examples, difficult-to-classify images that challenge most classifiers) and ImageNet-C [10] (containing corrupted versions of images from the original test set, e.g. by applying different levels of blur or noise). These benchmarks are widely used to assess the real-world robustness and generalization of models. Table 2 compares the performance of our approach to other state-of-the-art architectures for embedded devices on ImageNet, ImageNet-A and ImageNet-C classification accuracy and resulting latency.

Results show how the proposed network outperforms alternative approaches in terms of both classification performance and generalization capabilities. Furthermore, we verify how, even in embedded scenarios, ViTs provide better generalization capabilities than convolutional counterparts, providing significant real-world benefits for tiny devices. However, while ViTs have seen great developments in terms of efficiency, they still prove to be slower than CNNs with a comparable number of operations (e.g. MobileNet V2 can reach 4.34 MMAC/ ms and ShuffleNetV2 4.25 MMAC/ ms, while the fastest transformer architecture [24] only reaches 2.7 MMAC/ ms when tested on device).

5 COCO Object Detection

We also tested the viability of our approach in an object detection task. Specifically, we developed a transformer-based object detector based on RT-DETR [40], given the excellent tradeoffs between mAP and latency already demonstrated by this architecture compared to other lightweight object detectors [26]. Specifically, we use the same framework presented in [40], using our proposed network as the backbone. In addition, we modified the encoder and decoder of RT-DETR to use our linear transformer. This modification allows us to take full advantage of many embedded accelerators that do not normally support the execution of standard ViT. In fact, while recent state-of-the-art detectors allow excellent performance with a low number of parameters, older architectures are still common in embedded environments, despite lower performance, because they rely on common operations more widely supported by target devices [20,21,34]. Our architecture fills the gap between modern high-performance detectors and architectures supported on MCUs and embedded devices. By replacing the transformer blocks in the detection head of RT-DETR with our linear transformer, we achieve significant latency reduction on small devices (Table 3). These often have to rely on CPU for operations that require exponentials (e.g. softmax and GeLU), even when equipped with accelerators. For a minimal accuracy drop (easily recovered by scaling), our detector allows for massive speedup on these heavily constrained platforms (Results in Tables 3 and 4).

Table 3. Our linear transformer allows execution leveraging simple embedded accelerators. On an `STM32MP2`, this results in almost 90% latency reduction.

Transformer	Latency (STM32MP2)	mAP	Params [M]
Standard	2648	33.8	1.95
Linear	218	32.1	1.95
Linear (scaled)	279	33.9	2.21

Table 4. Our network outperforms state-of-the-art object detectors for mAP/latency tradeoff on embedded devices, while also guaranteeing compatibility with most common embedded runtimes and accelerators.

Model	Latency	mAP	Params [M]
RT-DETR (ResNet18) [26]	116	45.5	22.0
YoloV8 XS (custom)	78	28.0	1.9
YoloV8 N [14]	131	37.3	3.5
YoloV8 S [14]	216	44.9	11.2
YoloV11 XS (custom)	76	32.5	1.9
YoloV11 N [17]	106	39.5	2.6
YoloV11 S [17]	140	47.0	9.4
MCUNet-H7 [20]	80	26.0	2.0
XiNet a50 [1]	28	28.0	1.3
XiNet a60 [1]	48	32.5	1.7
XiNet a70 [1]	87	36.0	2.0
SSDlite MBNetV3 [23]	61	25.4	3.4
SSDlite MBNV2 FPNl [23]	98	28.2	4.3
EfficientDet D0 [36]	125	33.6	3.9
EfficientDet D1 [36]	168	38.4	6.6
EfficientDet D2 [36]	216	41.8	8.1
Ours (2M)	33	32.1	2.0
Ours (6M)	52	40.1	6.5
Ours (15M)	92	45.6	15.0

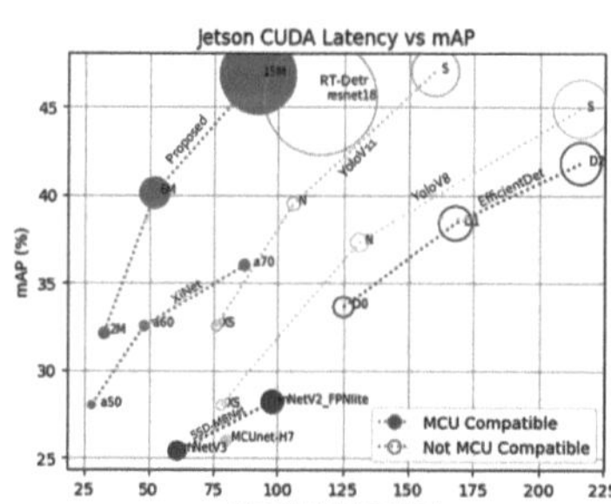

6 Conclusion

This paper describes a novel ViT architecture specifically designed for low-power embedded devices and microcontrollers. The architecture is based on the ReLU MLA mechanism and is carefully optimized to meet the stringent memory, operations, and parameter count requirements typical of these platforms. We propose a novel attention computation that relies on batched linear operations, allowing compatibility with most embedded toolchains. Our experiments show that our scaling approach can outperform previous state-of-the-art approaches under the severe memory, operation, and parameter count constraints of tiny embedded devices. In classification tasks, ViTs proved useful on these devices thanks to their higher generalization capabilities. However, these architectures still require higher latency than ConvNets. Instead, for object detection tasks, this optimized architecture fills the gap between modern high-performance models while ensuring compatibility with most embedded runtimes and accelerators.

Disclosure of Interests. The authors have no competing interests to declare that are relevant to the content of this article.

References

1. Ancilotto, A., Paissan, F., Farella, E.: XiNet: efficient neural networks for tinyML. In: Proceedings of the IEEE/CVF International Conference on Computer Vision (ICCV), pp. 16968–16977 (2023)
2. Bai, Y., Mei, J., Yuille, A.L., Xie, C.: Are transformers more robust than CNNs? Adv. Neural. Inf. Process. Syst. **34**, 26831–26843 (2021)
3. Cai, H., Gan, C., Zhu, L., Han, S.: Tiny transfer learning: towards memory-efficient on-device learning. arXiv preprint arXiv:2007.11622 (2020)
4. Cai, H., Li, J., Hu, M., Gan, C., Han, S.: Efficientvit: lightweight multi-scale attention for high-resolution dense prediction. In: Proceedings of the IEEE/CVF International Conference on Computer Vision, pp. 17302–17313 (2023)
5. Carion, N., Massa, F., Synnaeve, G., Usunier, N., Kirillov, A., Zagoruyko, S.: End-to-end object detection with transformers. In: European Conference on Computer Vision, pp. 213–229. Springer (2020)
6. Chollet, F.: Xception: Deep learning with depthwise separable convolutions. In: Proceedings of the IEEE Conference on Computer Vision and Pattern Recognition, pp. 1251–1258 (2017)
7. Dosovitskiy, A., et al, L.B.: An image is worth 16x16 words: transformers for image recognition at scale. In: International Conference on Learning Representations (2021)
8. Google: tflite-micro. https://github.com/tensorflow/tflite-micro
9. Gu, P., Zhang, Y., Wang, C., Chen, D.: Convformer: combining CNN and transformer for medical image segmentation. In: 2023 IEEE 20th International Symposium on Biomedical Imaging (ISBI), pp. 1–5 (2022)
10. Hendrycks, D., Dietterich, T.: Benchmarking neural network robustness to common corruptions and perturbations. In: International Conference on Learning Representations (2019). https://openreview.net/forum?id=HJz6tiCqYm
11. Hendrycks, D., Zhao, K., Basart, S., Steinhardt, J., Song, D.: Natural adversarial examples. In: Proceedings of the IEEE/CVF Conference on Computer Vision and Pattern Recognition, pp. 15262–15271 (2021)
12. Howard, A., Sandler, M., et al.: Searching for mobilenetv3. In: Proceedings of the IEEE/CVF International Conference on Computer Vision (2019)
13. Howard, A.G.E.A.: Mobilenets: efficient convolutional neural networks for mobile vision applications. arXiv preprint arXiv:1704.04861 (2017)
14. Jocher, G., Qiu, J., Chaurasia, A.: Ultralytics YOLO (2023). https://github.com/ultralytics/ultralytics
15. Katharopoulos, A., Vyas, A., Pappas, N., Fleuret, F.: Transformers are RNNS: fast autoregressive transformers with linear attention. In: International Conference on Machine Learning, pp. 5156–5165. PMLR (2020)
16. kendryte: nncase. https://github.com/kendryte/nncase

17. Khanam, R., Hussain, M.: Yolov11: an overview of the key architectural enhancements (2024). https://api.semanticscholar.org/CorpusID:273532028
18. Lai, L., Suda, N., Chandra, V.: CMSIS-NN: efficient neural network kernels for arm cortex-m CPUS. arXiv preprint arXiv:1801.06601 (2018)
19. Li, Y., et al.: Efficientformer: vision transformers at mobilenet speed. In: Proceedings of the 2022 Conference on Neural Information Processing Systems arXiv:2206.01191 (2022). https://api.semanticscholar.org/CorpusID:249282517
20. Lin, J., Chen, W.M., Cai, H., Gan, C., Han, S.: Mcunetv2: memory-efficient patch-based inference for tiny deep learning. In: Neural Information Processing Systems (2021)
21. Lin, J., Chen, W.M., Lin, Y., Cohn, J., Gan, C., Han, S.: Mcunet: tiny deep learning on IoT devices. arXiv preprint arXiv:2007.10319 (2020)
22. Lin, J., Zhu, L., Chen, W.M., Wang, W.C., Gan, C., Han, S.: On-device training under 256kb memory. Adv. Neural. Inf. Process. Syst. **35**, 22941–22954 (2022)
23. Liu, W., et al.: SSD: single shot multibox detector. arXiv preprint arXiv:1512.02325 (2016)
24. Liu, X., Peng, H., Zheng, N., Yang, Y., Hu, H., Yuan, Y.: Efficientvit: memory efficient vision transformer with cascaded group attention. In: Proceedings of the IEEE/CVF Conference on Computer Vision and Pattern Recognition, pp. 14420–14430 (2023)
25. Liu, Z., et al.: Swin transformer: hierarchical vision transformer using shifted windows. In: Proceedings of the IEEE/CVF International Conference on Computer Vision, pp. 10012–10022 (2021)
26. Lv, W., Zhao, Y., Chang, Q., Huang, K., Wang, G., Liu, Y.: Rt-detrv2: improved baseline with bag-of-freebies for real-time detection transformer. arXiv preprint arXiv:2407.17140 (2024)
27. Ma, N., Zhang, X., Zheng, H.T., Sun, J.: Shufflenet v2: practical guidelines for efficient CNN architecture design. In: Proceedings of the European Conference on Computer Vision (ECCV) (2018)
28. Mehta, S., Rastegari, M.: Separable self-attention for mobile vision transformers. Trans. Mach. Learn. Res. (2023). https://openreview.net/forum?id=tBl4yBEjKi
29. ST Microelectronics: X-cube-ai. https://www.st.com/en/embedded-software/x-cube-ai.html
30. Paissan, F., Ancilotto, A., Farella, E.: Phinets: a scalable backbone for low-power AI at the edge. ACM Trans. Embed. Comput. Syst. (2022). https://doi.org/10.1145/3510832
31. Paissan, F., et al.: Structured sparse back-propagation for lightweight on-device continual learning on microcontroller units. In: Proceedings of the IEEE/CVF Conference on Computer Vision and Pattern Recognition (CVPR) Workshops (2024)
32. Qin, D., et al.: Mobilenetv4 - universal models for the mobile ecosystem. In: Euro pean Conference on Computer Vision (ECCV) arXiv:2404.10518 (2024)
33. Radford, A., Kim, J.W., et al.: Learning transferable visual models from natural language supervision. In: International Conference on Machine Learning (2021)
34. Sandler, M., Howard, A., Zhu, M., Zhmoginov, A., Chen, L.C.: Mobilenetv2: inverted residuals and linear bottlenecks. In: Proceedings of the IEEE Conference on Computer Vision and Pattern Recognition, pp. 4510–4520 (2018)
35. Tan, M., Le, Q.: Efficientnet: rethinking model scaling for convolutional neural networks. In: International Conference on Machine Learning, pp. 6105–6114. PMLR (2019)

36. Tan, M., Pang, R., Le, Q.V.: Efficientdet: scalable and efficient object detection. In: 2020 IEEE/CVF Conference on Computer Vision and Pattern Recognition (CVPR) (2019)
37. Technologies, G.: Automated design intelligence with gapflow. https://greenwaves-technologies.com/automated-design-intelligence-with-gapflow-overview-and-benchmarks-on-gap8/
38. Touvron, H., Cord, M., Douze, M., Massa, F., Sablayrolles, A., Jégou, H.: Training data-efficient image transformers & distillation through attention. In: International Conference on Machine Learning, pp. 10347–10357. PMLR (2021)
39. Wang, S., Li, B.Z., Khabsa, M., Fang, H., Ma, H.: Linformer: self-attention with linear complexity. arXiv preprint arXiv:2006.04768 (2020)
40. Zhao, Y., et al.: DETRS beat YOLOS on real-time object detection (2023)

VS-Sim: A Synthetic Dataset for Viewpoint Shift Robustness

Riccardo Turra[1], Matteo Simoncini[1], Henrique Piñeiro Monteagudo[1,2](✉), Aurel Pjetri[1,3], Samuele Salti[2], and Leonardo Taccari[1]

[1] Verizon Connect, Florence, Italy
{riccardo.turra,matteo.simoncini,henrique.pineiro.monteagudo,aurel.pjetri,leonardo.taccari}@verizonconnect.com
[2] University of Bologna, Bologna, Italy
samuele.salti@unibo.it
[3] University of Florence, Florence, Italy
https://henriquepm.github.io/VS-SIM

Abstract. We release VS-Sim, a synthetic dataset of road scene images that can be used to study the robustness to viewpoint shift of Computer Vision models for several tasks. Our dataset includes images from a frontal camera in different positions, and annotations for tasks both in frontal view (depth, semantic segmentation) and Bird's Eye View (semantic segmentation). The paper also includes an analysis of the robustness of a Bird's Eye View segmentation model to viewpoint shifts. Experiments indicate that viewpoint shift significantly degrades the performance of a model trained on data from a single viewpoint, and suggest that using multiple viewpoints during training helps to mitigate the performance drop across different scenarios.

Keywords: Synthetic Dataset · Domain Shift · Computer Vision

1 Introduction

Computer Vision models for tasks related to Autonomous Driving (AD) and driving scene understanding, such as object detection, depth estimation, Bird's Eye View reconstruction, are expected to work reliably in different domains and on a long tail of challenging conditions.

In many practical scenarios, due to either the lack of a sufficient amount of data or to the difficulties and costs associated to data annotation, it is common to use *off-the-shelf* models trained on public datasets, possibly fine-tuned on a small dataset for the target domain. However, the target domain can be significantly different from the one on which models have been developed and evaluated, and the performance gap can be wide.

All the largest and most popular datasets, such as KITTI [8], nuScenes [2], Waymo [20], Cityscapes [3], are acquired either from a single vehicle with a very specific camera setup, or from multiple vehicles with a camera setup which is as

E. Rodolà et al. (Eds.): ICIAP 2025, LNCS 16167, pp. 507–519, 2026.
https://doi.org/10.1007/978-3-032-10185-3_40

close as possible to a single pre-defined one. This single-setup acquisition introduces an intrinsic Domain Shift problem when switching to applications where the installation setup of a camera, and, thus, the *viewpoint*, may be significantly different from the one in existing public datasets on which models are typically trained.

Indeed, in a real-world application, a frontal camera could be installed in several different positions: if it is an integrated camera, it is often behind the rear-view mirror, while dashcams can be installed virtually anywhere on the windshield or on the dashboard. The installation height is also directly related to the type of vehicle: comparing passenger cars with particularly tall vehicles, such as heavy duty trucks, the difference between the camera height could be in the order of 2 m. Another source of variability is the camera angle, that can vary depending on the vehicle (cameras on tall vehicles need to point down to capture footage in front of the vehicle) and the quality of installation, that might be suboptimal *e.g.*, for aftermarket devices like dashcams. As a result, different camera setups can have different yaw, pitch and roll angles, height from the ground, and horizontal displacement, as shown in Fig. 1.

Fig. 1. Video frames from real-world frontal dashcams with different viewpoints.

Although other domain shifts on driving scenes, such as adverse weather condition or image corruption, have been studied in the scientific literature [6], the robustness of neural network models to camera viewpoint shift is, to the best of our knowledge, an understudied problem in the literature, with few notable exceptions [12]. Building annotated datasets with data from multiple viewpoints covering most of the scenarios that might come up in practice is difficult, costly and time consuming. Such difficulties, however, could be easily overcome with synthetic data, that easily allows us to control for confounders (other sources of uncertainty of domain shift), isolate the effect of the viewpoint shift, and streamline evaluation.

In this paper, with the aim of raising awareness on the problem and speeding up the scientific research toward *viewpoint robustness*, we introduce VS-Sim, a dataset of synthetic driving scenes acquired from different viewpoints. Moreover, we estimate the impact of the viewpoint shift in driving scenes on the problem of Bird's Eye View (BEV) semantic segmentation (SS), a popular task in road scene understanding in which the position of the camera with respect to the scene is crucial. Due to the inherent geometric nature of the task, even small viewpoint changes are likely to impact the performance of a model more than for other tasks, such as semantic segmentation or object detection in perspective view. We show that a BEV SS model trained on a single viewpoint transfers poorly to different ones, especially if the angle with the ground changes, and that simply training on multiple viewpoints yields a model that generalizes better also on viewpoints that were not used during training, indicating that adding viewpoint augmentations during training can be useful also for tasks that typically require an accurate camera calibration.

2 Related Work

2.1 Domain Generalization in Road Scenes

Domain adaptation and domain generalization are well-known problems in autonomous driving and road scene understanding. Indeed, driving scenes in the wild are known to have a wide range of conditions and contain a long tail of non-nominal events that may introduce a significant domain shift with respect to popular datasets: examples are adverse weather conditions [15], rare safety-critical scenarios [21], as well as different scene geometry [12,23]. A simple yet effective approach is to accurately curate datasets so that they better align to the data distribution of the target domain, as well as using data augmentation techniques to cope with the scarcity of such data [13,15]. This is not always possible or sufficient, and there is a wide body of research on techniques to either adapt to a target domain [14,19] or to train models with better generalization capabilities [24].

2.2 Datasets with Multiple Viewpoints

Compared to other sources of domain shift, viewpoint shift comes with a substantial change in the geometry of the scene, that is challenging to obtain from real-world sensors. Most autonomous driving datasets, such as nuScenes [2], Waymo [20], Cityscapes [3], contain data acquired from rigs with several viewpoints from 360 surrounding cameras. However, this data cannot be directly used to study robustness to small viewpoint shifts of cameras that have the same function (e.g., frontal view).

There are few real-world datasets with driving scenes from multiple frontal viewpoints. A notable example is Mapillary Vistas [16], that contains road images from an extremely diverse set of cameras across the world, with very diverse geographies, environmental conditions and viewpoints. The only available annotations are for semantic segmentation. BDD100K [26] includes 100k annotated images captured by mobile phones used as dashcams. The dataset contains annotations for object detection and some semantic segmentation tasks. Note that in both BDD100K and Mapillary Vistas, camera extrinsics are not known.

Several synthetic datasets in the literature include data acquired from different viewpoints, although robustness to viewpoint shifts is usually not the main focus of the dataset. Virtual KITTI [7] is a synthetic replica of KITTI [8], that includes some variations of the orientation of frontal camera ($\pm 15°$ in the yaw angle). IDDA [1] has data from 5 different vehicles. The main difference is the camera height and the shape of the vehicle hood that is visible in the image. SYNTHIA [18] is another large-scale sythetic dataset focused on semantic segmentation of urban scenes. It includes 2 frontal cameras with a horizontal displacement, and images from other 6 side/rear cameras. A notable exception is the dataset described in [12], with synthetic data for a total of 36 frontal viewpoints, with perturbations in pitch, yaw, height, and depth.

2.3 Viewpoint Robustness

Several articles have shown that tasks such as Object Detection or Semantic Segmentation are generally quite robust to variability in camera pose, although there might be some degradation of performance to the different appearance that some objects might have from orientations not seen in the training set. [7] reports experiments on Virtual-KITTI around Multiple Object Tracking with 2 non-standard camera orientations ($\pm 15°$ in the yaw angle), showing a degradation of ~5% in MOTA. The work in [1] reports experiments on IDDA about the impact of viewpoint shifts on Semantic Segmentation of the frontal view image. Their results show that SS predictions for an unseen viewpoint are mainly affected by the presence of a different vehicle hood, whose pixels are misclassified. Overall, models do not appear to be heavily affected by the viewpoint shift (much less than other perturbations such as weather), nor do Domain Adaptation methods improve much (up to 4%). The work in [22] shows that Semantic Segmentation does not degrade much, on average around 3% of mIOU, also in very different viewpoints from what was seen during training. The work in [23] proposes a domain adaptation method to reduce the gap due to viewpoint shift for the detection of pedestrians in different real-world datasets (Cityscapes [3] and MOT [4]) in which they appear under significantly different perspectives.

Arguably, viewpoint shift is even more relevant for tasks that require some kind of 3D understanding of the world, such as depth estimation or Bird's Eye View reconstruction. In the depth estimation community, much effort in the last few years has gone towards foundation models that work zero-shot to estimate depth across multiple domains and geometries [9,11,25], though we know of no work that attempts to directly measure the effect of different viewpoints. In tasks where the aim is to reconstruct a Bird's Eye View representation, to the best of our knowledge the only relevant existing work is [12], in which the authors measure the impact of viewpoint shift in BEV reconstruction of surrounding vehicles. On their synthetic dataset, they train a model on a standard "source" viewpoint and test it on data from several different viewpoints, unseen during training. They show how a small change in the viewpoint significantly affects BEV reconstruction. Moreover, they propose a viewpoint augmentation technique involving novel view synthesis to transform images to a specific target viewpoint.

In this paper we propose to further expand the work of [12] on *viewpoint robustness* by measuring the performance drop due to viewpoint shift in several realistic positions for camera installations. In addition to [12] we include the case of a non-negligible roll angle, and consider dense BEV semantic segmentation rather then just the detection of surrounding vehicles. Moreover, we look at the problem through the lens of domain generalization, rather than only of domain adaptation as in [12], thus under the assumption that the target domain is not known and, potentially, challenging. We show that a naive approach, training a BEV model on a mix of different viewpoints, already helps to obtain a model with better generalization capabilities.

3 The VS-Sim Dataset

The VS-Sim dataset is generated using CARLA [5], a popular open-source simulator for autonomous driving research based on the Unreal engine. To acquire the raw RGB data and the annotations, we use three types of virtual sensors: the RGB camera, acting as a road-facing dashcam to collect frontal view images; the depth camera, that generates depth masks corresponding to the RGB images; the semantic segmentation (SS) camera, used to generate SS ground truth both for frontal views and for the Bird's Eye View. An example of RGB frame and corresponding annotations from the other virtual sensors is presented in Fig. 2.

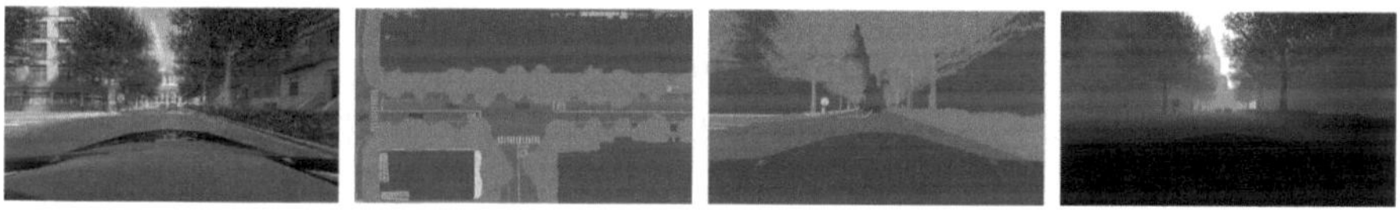

Fig. 2. Frontal RGB image and pixel-level ground truth for BEV semantic segmentation, semantic segmentation, depth.

We place frontal view cameras inside the cabin of the ego-vehicle, in positions that commonly occur in real dashcam installations: one in a default position (top-center, above the rear-view mirror), one on the top-left corner of the windshield, and one at the bottom touching the dashboard. Moreover, we include some common angle perturbations: a roll by 5° to the right and four different pitch rotations (−10°, −5°, 5°, 10°) with respect to the default position. For each position, we capture RGB, depth and semantic segmentation. All the virtual cameras simultaneously record the same scene. The available viewpoints are summarized in Fig. 3 and Table 1.

Fig. 3. Positions of the cameras on the windshield and reference system.

Table 1. Summary of the available viewpoints.

Name	Position	Roll	Pitch	Yaw
Default	Top-center of windshield (A)	0	0	0
Bottom	Bottom center of windshield (C)	0	0	0
Top-Left	Top-left of windshield (B)	0	0	0
Roll	Top-center of windshield (A)	+5°	0	0
Pitch^{--}	Top-center of windshield (A)	0	−10°	0
Pitch^{-}	Top-center of windshield (A)	0	−5°	0
Pitch^{+}	Top-center of windshield (A)	0	+5°	0
Pitch^{++}	Top-center of windshield (A)	0	+10°	0

We configure the intrinsic parameters of the frontal cameras to resemble a commercial dashcam, with a resolution of 1280×720 and a horizontal field of view of 120°. In Fig. 4, we report a sample set of images with the various viewpoints. To capture BEV ground truth, we position a SS camera with the same resolution and a FOV of 10° at a height of 650 m above the scene. With this configuration, we aim to minimize perspective distortion and approximate an orthographic camera, which is not available in CARLA. Semantic segmentation masks contain pixel-level information for a total of 11 classes: road, sidewalk, building, wall, vegetation, terrain, occlusion, person, two-wheeler, car, and truck, analogous to categories in similar real-world datasets [2,8].

Fig. 4. Viewpoints included in the dataset for each location. Horizon highlighted by a dashed line.

To provide sufficient diversity, data are captured in seven different CARLA maps, namely Town 1, 2, 3, 4, 5, 7, and 10. For each map, we generate 100 different 20-second trajectories, varying the position of the vehicle in the map and the surrounding objects. In total, using a sensor sampling rate of 1 Hz, the dataset consists of a total of 14,000 samples for each of the 8 viewpoints, each one including RGB images and frontal view annotations (depth and semantic segmentation), and a mask with BEV semantic segmentation ground truth.

4 Experimental Evaluation

For a fair representation of model performance, all experiments are carried out on spatially separated train/test sets, splitting each town in an area used for training and one used for testing, so that images extracted in the same part of the map belong to the same split. The train set is composed of a total of 10466 images per position while the test set is composed of 1491 images per position. In the following sections training is always performed on the aforementioned train split of the specified position and evaluation is performed on the test split.

4.1 BEV Semantic Segmentation

The Bird's Eye View (BEV) is a popular representation of the environment surrounding the ego-vehicle for assisted and autonomous driving. It has several desirable properties: the scale is constant regardless of the viewpoint (contrary to perspective view images), distances are metric if the ground is flat, and it is compact, containing most of the relevant information for road scene understanding in a single two-dimensional raster. The task of BEV semantic segmentation from a camera consists in obtaining a class value for each pixel in this top-down representation taking as input a RGB monocular image.

We use SimpleBEV [10] as the BEV semantic segmentation network for our experiments. SimpleBEV is a minimal yet high performance architecture composed of three stages. First, an image encoder processes the RGB image. Then, a parameter-free 3D lifting mechanism fills a 3D feature volume by bilinearly sampling the encoder features in positions obtained by projecting the voxel centers using the camera instrinsics. Finally, those features are collapsed along the vertical direction and fed to a UNet-like network with a semantic segmentation head that generates the final BEV semantic segmentation. We slightly modify the network as in [17] for our setting by switching to a multiclass segmentation task from the original monoclass paradigm and getting rid of the auxiliary task heads.

4.2 Measuring Performance Drop from Viewpoint Shift

To estimate the effect of the viewpoint shift on a BEV model for dense semantic segmentation, we train what we call a "baseline model" on the Default position. Then, we measure the quality of the predicted BEV on acquisitions from other viewpoints. The results are presented in Table 2.

Compared to the performance on the default viewpoint that the model was trained on, the results show a drop in performance for all viewpoints, although with very different levels of degradation. On the Bottom and Top-Left positions, despite the ego-vehicle occluding a bigger part of the frames (see Fig. 4), the

Table 2. Performance on BEV SS of a baseline model trained on the Default position, and tested with different camera viewpoints. Best is in bold, second best underlined.

Test viewpoint	Road	Sidew.	Build.	Wall	Veget.	Terrain	Car	Truck	mIoU	Δ
Default	**81.20**	**61.61**	**63.45**	**3.67**	**66.31**	**42.45**	**42.45**	**21.81**	**48.23**	-
Bottom	79.27	55.96	61.60	2.45	65.03	38.99	30.84	12.62	43.22	−10.4%
Top-Left	78.81	56.25	62.25	3.18	65.71	39.79	27.19	8.57	42.72	−11.4%
Roll	58.56	36.86	39.99	1.31	52.68	26.79	31.63	10.90	32.34	−32.9%
Pitch^{--}	34.31	25.10	32.97	0.49	44.19	22.3	1.45	0.18	20.12	−58.3%
Pitch^{-}	55.48	22.00	21.21	0.37	37.94	15.99	1.38	0.71	19.38	−59.8%
Pitch^{+}	30.85	23.43	26.50	0.35	33.56	21.91	0.09	0.00	17.09	−64.6%
Pitch^{++}	41.73	14.38	10.97	0.09	16.16	10.05	0.13	0.00	11.69	−75.8%

network demonstrates quite good generalization, resulting in a modest decrease in mIoU (around 10%). The largest drop occurs for classes with rather small objects (car, truck), where even small localization errors have a huge impact on the IoU.

The Rolled viewpoint has a significant impact on the performance, impacting the reconstruction especially on pixels that are outside the central portion of the BEV (i.e., far from the vanishing point). This is quite evident looking at Fig. 5, that compares the prediction of the baseline model on the RGB input from the default viewpoint and the prediction on the rolled RGB input.

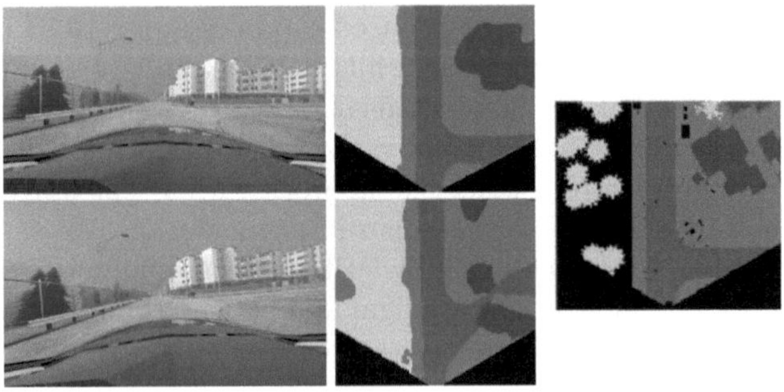

Fig. 5. Example of BEV reconstructions for a non-default viewpoint. On the left, the RGB image from the default position and the rolled viewpoint, respectively. In the middle, prediction of the baseline model on the 2 images. On the right, the BEV ground truth.

Finally, the largest decrease occurs when the viewpoint is pitched. BEV models, such as SimpleBEV, typically construct a 3D feature volume aligned with the camera reference system. If the camera optical axis is not parallel to the ground plane – which is the case when there is pitch – reducing the vertical dimension of this volume results in a plane which is tilted with the respect to the ground plane. This causes the output BEV to be deformed and explains the degradation in the results. In Fig. 6 it is easy to appreciate that even slight rotations along the pitch axis cause a drastic degradation of the quality of the prediction.

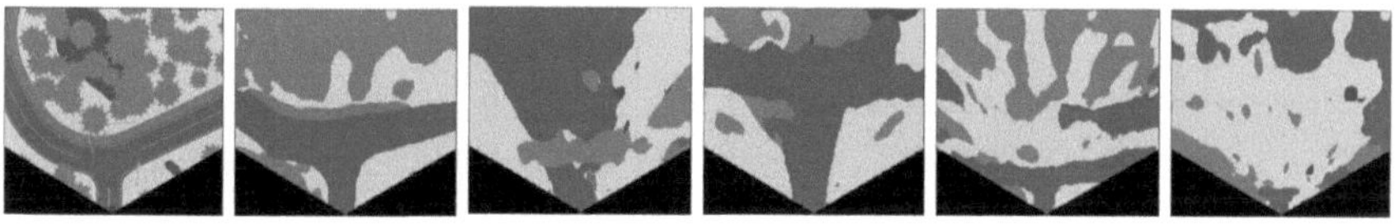

Fig. 6. Example of BEV maps predicted by the baseline model on viewpoints with different pitch.

4.3 Increasing Viewpoint Robustness on Multiple In-Domain Viewpoints

We run a second set of experiments by training the same SimpleBEV architecture on data acquired from a mix of positions. We refer to it as *mixed model*. We hypothesize that augmenting the training dataset with different viewpoints could enhance the model generalization and mitigate the degradation due to viewpoint shifts. Our aim with these experiments is to validate this hypothesis.

We train the model on a dataset composed of a mix of viewpoints (Default, Top-Left, Bottom, Rolled) and evaluate on a test set that includes only positions seen at training time. From the results in Table 3, we observe that the mixed model can perform almost equally well across all those viewpoints. The only major difference is the class "truck", which drops to almost 0 for all viewpoints. However, note that the number of trucks in the dataset is very small.

Table 3. Performance on BEV SS of a model trained on a mixture of viewpoints (Default, Bottom, Top-Left, Roll) and tested on those (in-domain) viewpoints.

Test viewpoint	Road	Sidewalk	Building	Wall	Veget.	Terrain	Car	Truck	mIoU
Default	82.83	60.67	66.01	1.85	69.29	45.23	33.00	1.51	45.05
Bottom	80.96	57.01	62.21	1.20	68.68	43.38	29.46	0.29	42.90
Top-Left	82.61	60.64	65.16	2.01	69.03	44.41	32.55	1.40	44.73
Roll	82.85	60.78	66.16	1.90	69.38	45.10	31.07	1.60	44.86

In Fig. 7 we compare the results obtained on all viewpoints of the baseline model (trained only on the Default viewpoint) and of the mixed model (trained on Default, Bottom, Top-Left, Roll). The mixed model does significantly better

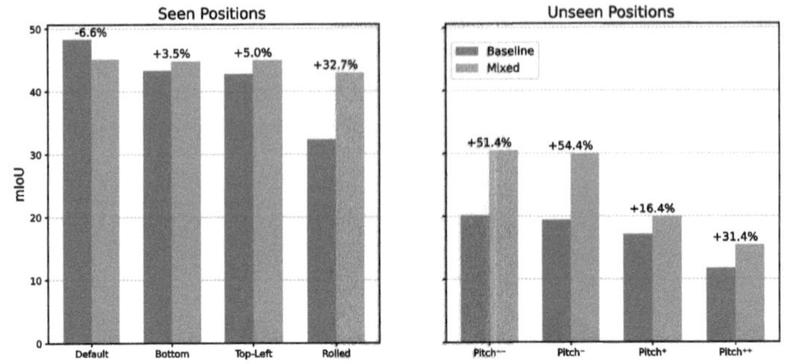

Fig. 7. Comparison of the results of the baseline model (in blue, trained on Default only) and of the mixed model (in orange, trained on Default, Bottom, Top-Left, Roll). Results on viewpoints seen (left) and unseen (right) during training. (Color figure online)

on all positions compared to the baseline model, except for the default viewpoint, indicating that targeting a single viewpoint is preferable, if possible. Still, this experiment shows that a BEV reconstruction model can learn to perform its task on multiple viewpoints, if they do not differ too much and they are provided at training time.

4.4 Increasing Viewpoint Robustness on Unseen Viewpoints

Our last experiment intends to measure if training on multiple viewpoints can help generalize to viewpoints that were *not seen* at training time. We consider again the mixed model trained on a mix of viewpoints (Default, Top-Left, Bottom, Rolled) and evaluate on a test set of unseen viewpoints (Pitch^{--}, Pitch^{-}, Pitch^{+}, Pitch^{++}). The results are reported in Table 4.

Table 4. Performance on BEV SS of a model trained on a mixture of viewpoints (Default, Bottom, Top-Left, Roll) and tested on unseen viewpoints (Pitch^{--}, Pitch^{-}, Pitch^{+}, Pitch^{++}).

Test viewpoint	Road	Sidewalk	Building	Wall	Veget.	Terrain	Car	Truck	mIoU
Pitch^{--}	63.80	40.25	49.26	0.48	56.62	32.79	0.57	0.00	30.47
Pitch^{-}	66.02	34.90	49.40	0.66	56.62	29.40	2.18	0.33	29.94
Pitch^{+}	38.11	27.95	35.64	0.19	36.64	20.58	0.02	0.00	19.89
Pitch^{++}	48.62	16.79	18.92	0.06	26.97	11.41	0.14	0.00	15.36

The performance is not competitive with the results obtained by the same model on viewpoints that were seen in training (Table 3). The very low scores on small objects, especially car, show that localization remains very imprecise.

However, in Fig. 7 we can observe that the training on multiple viewpoints was able to increase the viewpoint robustness, even for unseen perturbations. More specifically, the results for negative pitch perturbations are rather good, and show that the mixed model generalized significantly better than the Baseline. We hypothesize that this result is due to the similarity between the appearance of objects in the bottom camera pose (seen in training) and this viewpoint, reducing the domain shift between the train and test sets.

5 Conclusion

With VS-Sim, we release a synthetic dataset of road scene images that can be used to study the robustness to viewpoint shifts of modern Computer Vision methods. We also provide the first analysis of viewpoint robustness for Bird's Eye View dense semantic segmentation from monocular camera.

The experimental results show that a BEV model trained on data in a single position is significantly affected even by small variations in viewpoint. Among the camera positions we studied, translations turned out to affect performance the least, while changes of orientation have a much larger impact. Even small positive rotations along the pitch axis (*i.e.*, lower horizon) greatly reduce mIoU scores across all classes, and completely destroy the ability to localize smaller objects.

We plan on extending the dataset to include additional viewpoints and more driving scenarios. Future research can exploit the dataset for analysis on other tasks such as depth estimation, frontal-view semantic segmentation, and object detection. The dataset could also be used for research on data augmentations that introduce different points of view (*e.g.*, using novel view synthesis) to improve viewpoint robustness.

Acknowledgments. Henrique Piñeiro Monteagudo is supported by the SMARTHEP project, funded by the European Union's Horizon 2020 research and innovation programme, call H2020-MSCA-ITN-2020, under Grant Agreement n. 956086.

References

1. Alberti, E., Tavera, A., Masone, C., Caputo, B.: IDDA: a large-scale multi-domain dataset for autonomous driving. IEEE Rob. Autom. Lett. **5**(4), 5526–5533 (2020)
2. Caesar, H., et al.: nuScenes: a multimodal dataset for autonomous driving. In: CVPR (2020)
3. Cordts, M., et al.: The cityscapes dataset for semantic urban scene understanding. In: Proceedings of the IEEE Conference on Computer Vision and Pattern Recognition, pp. 3213–3223 (2016)
4. Dendorfer, P., et al.: MOT20: a benchmark for multi object tracking in crowded scenes. arXiv preprint arXiv:2003.09003 (2020)
5. Dosovitskiy, A., Ros, G., Codevilla, F., Lopez, A., Koltun, V.: CARLA: an open urban driving simulator. In: Conference on Robot Learning, pp. 1–16. PMLR (2017)

6. Drenkow, N., Sani, N., Shpitser, I., Unberath, M.: A systematic review of robustness in deep learning for computer vision: mind the gap? (2022)
7. Gaidon, A., Wang, Q., Cabon, Y., Vig, E.: Virtual worlds as proxy for multi-object tracking analysis. In: CVPR (2016)
8. Geiger, A., Lenz, P., Urtasun, R.: Are we ready for autonomous driving? The KITTI vision benchmark suite. In: 2012 IEEE Conference on Computer Vision and Pattern Recognition, pp. 3354–3361. IEEE (2012)
9. Guizilini, V., Vasiljevic, I., Chen, D., Ambruş, R., Gaidon, A.: Towards zero-shot scale-aware monocular depth estimation. In: Proceedings of the IEEE/CVF International Conference on Computer Vision, pp. 9233–9243 (2023)
10. Harley, A.W., Fang, Z., Li, J., Ambrus, R., Fragkiadaki, K.: Simple-BEV: what really matters for multi-sensor BEV perception? In: 2023 IEEE International Conference on Robotics and Automation (ICRA), pp. 2759–2765 (2023)
11. Ke, B., Obukhov, A., Huang, S., Metzger, N., Daudt, R.C., Schindler, K.: Repurposing diffusion-based image generators for monocular depth estimation. In: Proceedings of the IEEE/CVF Conference on Computer Vision and Pattern Recognition (CVPR) (2024)
12. Klinghoffer, T., et al.: Towards viewpoint robustness in bird's eye view segmentation. In: Proceedings of the IEEE/CVF International Conference on Computer Vision, pp. 8515–8524 (2023)
13. Kong, L., Xie, S., Hu, H., Ng, L.X., Cottereau, B.R., Ooi, W.T.: RoboDepth: robust out-of-distribution depth estimation under corruptions. In: Thirty-Seventh Conference on Neural Information Processing Systems Datasets and Benchmarks Track (2023)
14. Man, Y., Gui, L.Y., Wang, Y.X.: DualCross: cross-modality cross-domain adaptation for monocular BEV perception. In: IROS (2023)
15. Michaelis, C., et al.: Benchmarking robustness in object detection: autonomous driving when winter is coming (2020)
16. Neuhold, G., Ollmann, T., Rota Bulo, S., Kontschieder, P.: The mapillary vistas dataset for semantic understanding of street scenes. In: Proceedings of the IEEE International Conference on Computer Vision, pp. 4990–4999 (2017)
17. Monteagudo, H.P., Taccari, L., Pjetri, A., Sambo, F., Salti, S.: RendBEV: semantic novel view synthesis for self-supervised bird's eye view segmentation. In: 2025 IEEE/CVF Winter Conference on Applications of Computer Vision (WACV) (2025)
18. Ros, G., Sellart, L., Materzynska, J., Vazquez, D., Lopez, A.M.: The SYNTHIA dataset: a large collection of synthetic images for semantic segmentation of urban scenes. In: Proceedings of the IEEE Conference on Computer Vision and Pattern Recognition, pp. 3234–3243 (2016)
19. Shenaj, D., et al.: Learning across domains and devices: style-driven source-free domain adaptation in clustered federated learning. In: Proceedings of the IEEE/CVF Winter Conference on Applications of Computer Vision, pp. 444–454 (2023)
20. Sun, P., et al.: Scalability in perception for autonomous driving: Waymo open dataset. In: Proceedings of the IEEE/CVF Conference on Computer Vision and Pattern Recognition, pp. 2446–2454 (2020)
21. Taccari, L., et al.: Classification of crash and near-crash events from dashcam videos and telematics. In: 2018 21st International Conference on Intelligent Transportation Systems (ITSC), pp. 2460–2465. IEEE (2018)

22. Testolina, P., Barbato, F., Michieli, U., Giordani, M., Zanuttigh, P., Zorzi, M.: SELMA: semantic large-scale multimodal acquisitions in variable weather, daytime and viewpoints. IEEE Trans. Intell. Transp. Syst. **24**(7), 7012–7024 (2023)
23. Vidit, V., Engilberge, M., Salzmann, M.: Learning transformations to reduce the geometric shift in object detection. In: Proceedings of the IEEE/CVF Conference on Computer Vision and Pattern Recognition, pp. 17441–17450 (2023)
24. Wang, S., et al.: Towards domain generalization for multi-view 3D object detection in bird-eye-view. In: 2023 IEEE/CVF Conference on Computer Vision and Pattern Recognition (CVPR), pp. 13333–13342 (2023)
25. Yang, L., Kang, B., Huang, Z., Xu, X., Feng, J., Zhao, H.: Depth anything: unleashing the power of large-scale unlabeled data. In: CVPR (2024)
26. Yu, F., et al.: BDD100K: a diverse driving dataset for heterogeneous multitask learning. In: Proceedings of the IEEE/CVF Conference on Computer Vision and Pattern Recognition, pp. 2636–2645 (2020)

Few-Shot Anomaly Classification by Learning to Inpaint Nominal Images

Rubin Carkaxhia[1], Musawar Ali[1(✉)], Nicola Fioraio[2], Luigi Di Stefano[1], and Samuele Salti[1]

[1] University of Bologna, Bologna, Italy
rubin.carkaxhia@studio.unibo.it,
{musawar.ali2,luigi.distefano,samuele.salti}@unibo.it
[2] Eyecan.ai, Bologna, Italy
nicola.fioraio@eyecan.ai
https://www.eyecan.ai/

Abstract. Automated inspection is a crucial aspect in modern industrial manufacturing. Despite its importance, the methods used to perform it suffer from the data scarcity intrinsic to the problem, where only a few anomalous samples are usually available. Recently, fine-tuning of foundation inpainting models has been proposed as a solution. The fine-tuned model is then used to inpaint nominal images where areas corresponding to defective masks have been removed. While effective, random pairing sometimes applies the mask on background or logically unfeasible areas. To counteract this phenomenon, we experiment with generating high-resolution defective images inpainting the few available real defects. Since the resulting images would show limited variance in the non-defective parts, we propose to fine-tune another inpainting model to change nominal parts of generated images. Experimental results on the MVTec-AD dataset demonstrate that our method generates images with complementary properties with respect to those produced by the baseline and training on an ensemble of generated data produces a new state of the art result.

Keywords: Anomaly classification · Diffusion models · ControlNet

1 Introduction

Automated anomaly inspection has become a crucial aspect in industrial manufacturing, with anomaly classification being one of the anomaly inspection tasks with most practical usefulness. However, anomaly inspection is made hard by the intrinsic shortage of data: while nominal samples are usually abundant, examples of anomaly are scarce. Although image classification is well studied topic for which effective methods are available, such shortage limits the performance that can be obtained in a real scenario. Several approaches have been proposed to exploit the abundance of nominal samples. Some of them are

E. Rodolà et al. (Eds.): ICIAP 2025, LNCS 16167, pp. 520–532, 2026.
https://doi.org/10.1007/978-3-032-10185-3_41

based on unsupervised methods relying on training ML models only on nominal images [3,10,34,37], which however cannot handle anomaly classification. Other approaches relying on generative models like GANs [14] have also been proposed [12,26,44]. However, they still require a large amount of anomaly data at training time, hence failing to generate realistic defects in real scenarios. Several other approaches based on diffusion models [18] have been proposed. In particular, Anomaly Diffusion [20] achieves state-of-the-art performance in anomaly classification. In this method, only a limited number of real images are available to train the image generator, while the task-specific models are trained exclusively on images sampled from the generator. However, the generated images often fail to closely resemble real defective samples. To overcome this limitation, a recently proposed framework named AnomalyControl [1] suggested to use Stable Diffusion Inpainting [32] to generate defects, by filling defective masks applied to nominal images. The idea behind this framework is that fine-tuning a foundational inpainting model like Stable Diffusion Inpaint is possible even with limited available data, both nominal and defective. Indeed, the authors were able to generate high resolution defective images effectively bypassing the data scarcity problem, outperforming existing anomaly generation methods in anomaly classification. However, the main limitation of AnomalyControl is the random pairing of defective masks and nominal images. Since the pairs are formed randomly, the mask can end up being applied on the background or in areas where a defect cannot logically appear as shown in Fig. 1. In this paper, we tackle this limitation by generating high resolution defective images by inpainting multiple times the few available real images of anomalies, thereby ensuring by construction that the mask and the image are properly aligned. To cope with the scarcity of anomalous images we create a Variance Generator by fine-tuning another Stable Diffusion Inpainting model to generate non-defective parts of images. Thus, we can increase the variance of the defective images generated. Purposely, we leverage the publicly available stable diffusion inpainting model trained on LAION-400M [38] as the backbone while training ControlNet [46] conditioned on masks created by a Perlin noise generator and text prompts describing the object to synthesize. We show that this procedure generates better defective images for the objects in the MVTec AD dataset [2] where AnomalyControl struggles due to the main limitation discussed above, which results in higher classification accuracy for those objects. Since the images generated by our method have complementary properties with respect to those generated by AnomalyControl, training on a mixture of data generated by both models is shown to give the best results and attain the new state of the art in anomaly classification for the MVTec AD dataset.

2 Related Work

2.1 Anomaly Inspection

Anomaly inspection usually involves anomaly detection, segmentation and classification [12]. Due to data scarcity of industrial anomaly images, the first two tasks are addressed with the help of unsupervised approaches where the models are

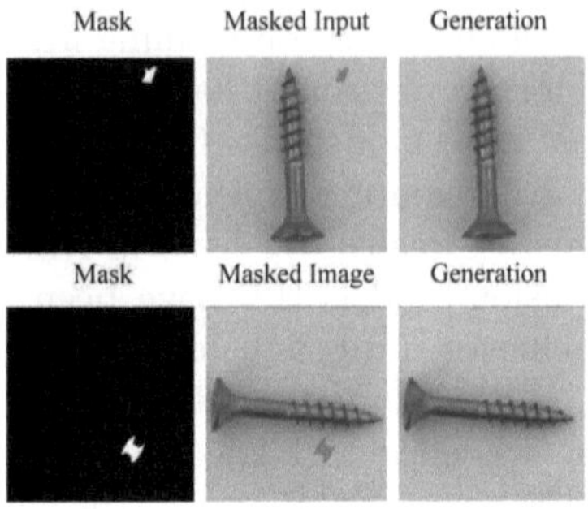

Fig. 1. Two examples showing that if the mask applied to a nominal image ends up in the background the model proposed in [1] does not generate a defective image.

trained solely on nominal samples. Anomalies are detected based on the distribution gap between predictions and nominal samples. Unsupervised approaches fall into two categories: (1) feature-based methods using deep models [3,5,9–11,15,23,25,29,30,36,41,42], and (2) reconstruction-based methods leveraging autoencoders [4,19,31,43], inpainting [27], or diffusion models [40]. These models fail to accurately reconstruct defects, enabling per-pixel anomaly localization. More recently, zero-shot and few-shot methods [6–8,16,21,39,47] have achieved encouraging results. However, anomaly classification remains limited by data scarcity, and this can not be tackled by unsupervised paradigms.

2.2 Generative Models

Image generation lies at the core of our approach. While GANs [14] have shown promising results due to their min-max loss, they are notoriously difficult to train. Diffusion models [18], especially Latent Diffusion Models (LDMs) [32], have recently set new benchmarks by operating in compressed latent spaces and enabling text-conditioned outputs. LDMs also support image inpainting by extending input channels to include masked latents and downsampled masks, allowing localized modifications. AnomalyControl leverages this capability to generate defects on nominal samples, enhancing anomaly detection, and to introduce variability through random mask-based re-generation. Fine-tuning methods like Textual Inversion [13] and DreamBooth [35] enable concept learning from limited data by embedding images into the LDM's textual space. Additionally, ControlNet [46] facilitates spatial conditioning for more precise control of frozen LDMs.

2.3 Anomaly Generation

To address the scarcity of real defective images, various model-free approaches [23,24,43,45] have been proposed, typically generating defects by either pasting textures within the same image or transferring them across nominal images.

However, these methods often lack realism. Model-based approaches like DFM-GAN [12] and SDGAN [26], which rely on GANs [14], also struggle to produce convincingly defective samples. Recent diffusion-based methods have shown promise: AnomalyDiffusion [20] uses few-shot learning with LDMs, conditioning Stable Diffusion using Textual Inversion. In contrast, AnomalyControl [1] also uses few-shot learning with LDMs, but conditions Stable Diffusion Inpainting model using ControlNet for more realistic anomaly generation. All these techniques focus on generating plausible defects under limited data. In this work, we propose: (1) a method to introduce variability by inpainting nominal textures on the images, and (2) enhancing AnomalyControl by using defective samples to generate new, foreground-aligned defects.

3 Method

3.1 Background

Diffusion models [18] are a recent approach to image generation that provides the best quality, representing a paradigm shift. They consist of two key processes: the forward process, where noise is incrementally added to training images according to the noise schedule $\beta_1, ..., \beta_T$, and the reverse process, where noise is progressively removed to synthesize new images. The forward process can be defined as:

$$q(x_t|x_0) = \sqrt{\bar{\alpha}_t}x_0 + \sqrt{1-\bar{\alpha}_t}\epsilon, \qquad \epsilon \sim \mathcal{N}(0, I) \tag{1}$$

where an image x_0 is transformed to a noisy image x_t, with t uniformly sampled from $1, ..., T$, ϵ drawn from $\mathcal{N}(0, I)$ and $\bar{\alpha_t} := \prod_{i=0}^{t} \alpha_i = \prod_{i=0}^{t}(1 - \beta_i)$ with $\beta_i \in (0, 1)$. Directly computing the posterior of the reverse diffusion process is intractable. To approximate it, a neural network is trained to learn to predict the noise that was added to the input image at time t, so the neural network predicts x_{t-1} given x_t. This is usually done by learning a U-Net [33] that models the noise predictor $\epsilon_\theta(x_t; t)$. Given the learned noise predictor, during the inference stage $\hat{x}_{t-1}$ is sampled from $\hat{x}_t$ using:

$$\hat{x}_{t-1} = \frac{1}{\sqrt{\alpha_t}}(\hat{x}_t - \frac{1-\alpha_t}{\sqrt{1-\bar{\alpha}_t}}\epsilon_\theta(\hat{x}_t; t)) + \sigma_t\epsilon \tag{2}$$

where $\epsilon \sim \mathcal{N}(0, I)$. The process starts at $\hat{x}_T \sim \mathcal{N}(0, I)$ and, in an iterative way, progresses from $\hat{x}_T$ to $\hat{x}_T - 1$ until it reaches $\hat{x}_0$.

3.2 Stable Diffusion Inpainting

Stable Diffusion [32] is a text-to-image diffusion model that accelerates generation by operating in a lower-dimensional latent space, produced via a VQ-VAE encoder [22]. An input image x_0 is encoded as $z_0 = E(x_0)$, then gradually noised to z_T through a forward diffusion process. The model is trained to predict the added noise ϵ_θ, conditioned on a text prompt c. At inference, a random latent

$\hat{z}_T \sim \mathcal{N}(0, I)$ is denoised into $\hat{z}_0$, and the decoder reconstructs the final image $\hat{x}_0 = D(\hat{z}_0)$. Stable Diffusion Inpainting is a fine-tuned variant that fills masked regions with contextually coherent content while preserving unmasked areas. It follows the same training scheme, with the difference that the U-Net [33] receives additional inputs: the noisy latent z_t, the masked latent z_0^m, and the downsampled binary mask $m_\downarrow$.

3.3 ControlNet

ControlNet [46] is a neural network architecture that provides more control over image generation by adding extra conditioning. In particular, ControlNet augments the U-Net [33] architecture by freezing the parameters of the original encoder and middle block, and then creating a trainable copy of both. These trainable copies process the control image and produce outputs that are added to the corresponding skip connections of the frozen U-Net decoder. When used for inpainting the ControlNet receives as inputs the noisy latent, the original mask, and the text embedding.

3.4 Proposed Method

As pointed out in Sect. 1, our approach deploys the idea that generating defective images by the inpainting model proposed in [1] while starting from the few available real defects ensures alignment between images and masks, thereby mitigating the main problem of AnomalyControl [1]. To increase variance we fine-tune another Stable Diffusion Inpainting [32] model to replace good parts with other good parts. Before training the model, several masks are randomly generated using the Perlin noise generator provided by the authors of [20].

3.5 Model Training

Since Stable Diffusion Inpainting [32] is trained on LAION-400M [38] rather than industrial data, we adopt ControlNet [46] to fine-tune it on the nominal samples of the MVTec AD dataset. We refer to this fine-tuned model as the Variance Generator. Figure 2 illustrates the training pipeline. For each nominal image x, a randomly sampled mask m is used to create a masked version x^m. These are encoded into the latent space as $z_0 = E(x)$ and $z_0^m = E(x^m)$. Gaussian noise is added to z_0 at timestep t to obtain z_t. The U-Net [33] receives as input the concatenation of z_t, z_0^m, and a downsampled mask $m_\downarrow$. ControlNet is conditioned on z_t and the full-resolution mask m. Both the U-Net and ControlNet are further conditioned via a text embedding c, produced by a CLIP encoder [28] from prompts like "A photo of a [*]", where [*] denotes an object from the MVTec AD dataset (e.g., hazelnut, bottle, ...). The ControlNet is trained using the standard MSE loss of diffusion models:

$$L = \mathbb{E}_{(c,x,m),t}\left[\|\epsilon - \epsilon_\theta(z_t, t, c, m, z_0^m, m_\downarrow)\|_2^2\right] \quad (3)$$

As shown in Fig. 3, the masked regions in the generated images (4th column) differ from those in the originals (3rd column), indicating the model's ability to produce diverse and realistic variations of good (i.e. without anomalies) objects' parts.

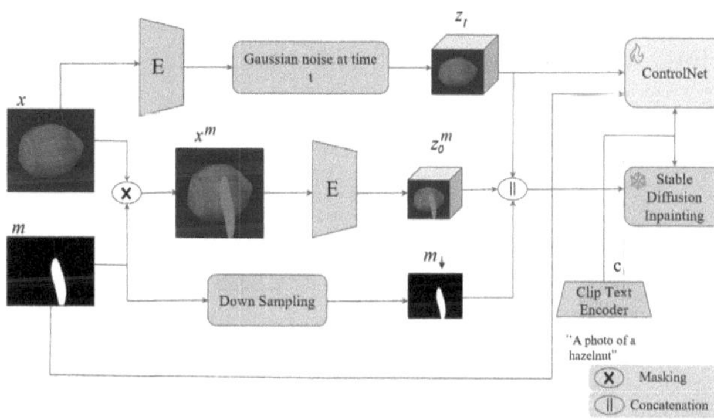

Fig. 2. Training pipeline of the Variance Generator.

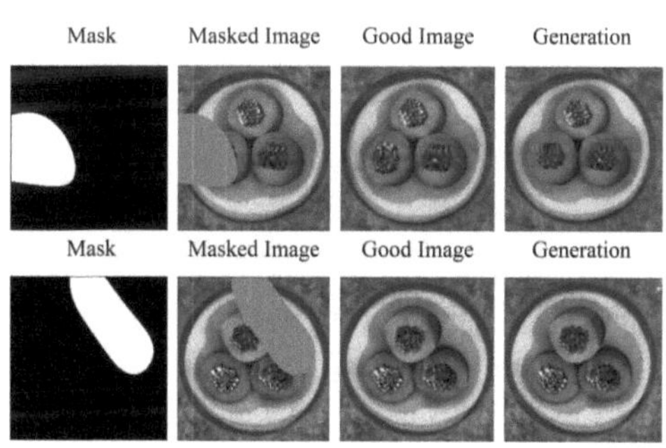

Fig. 3. Two examples showing how the Variance Generator replaces good parts with other good parts.

3.6 Image Generation Pipeline

Figure 4 shows our image generation pipeline, which includes two models: AnomalyControl [1], to generate initial defective images, and our proposed Variance Generator, to increase diversity. For each defective image–mask pair, random augmentations are first applied to both components to enhance variability. These are fed into AnomalyControl, which synthesizes a defective image conditioned on a defect-specific prompt (e.g., *"A photo of a color stain"* for a metal nut anomaly from the MVTec-AD [2] dataset). Next, a Perlin noise-based random mask is generated, ensuring no overlap with the existing defect mask. If a non-overlapping mask cannot be found after 10 attempts, the image from AnomalyControl is kept as-is without applying an additional mask and we proceed to the next image–mask pair. Otherwise, an object-specific prompt is sampled (e.g., *"A photo of a metal nut"*). Finally, the image generated by AnomalyControl, the Perlin noise mask and the prompt are passed to the Variance Generator, which produces the final defective image with enhanced variation.

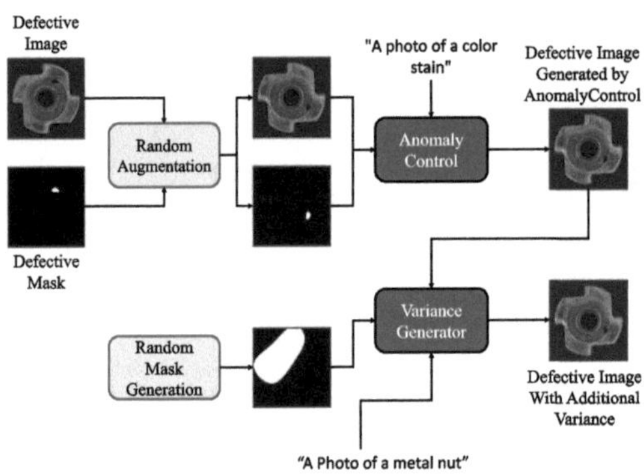

Fig. 4. Image Generation Pipeline.

4 Experiments

4.1 Dataset

The MVTec-AD dataset [2], used throughout our experiments, consists of 15 objects captured at resolutions ranging from 700 × 700 to 1024 × 1024 pixels. For defective samples, the dataset provides both pixel-level anomaly masks and image-level labels that indicate the specific anomaly class associated with each object. Following the protocol established in AnomalyControl, we split the images of each anomaly class into a training set (1/3) and a test set (2/3), using the exact same splits as reported in their paper [1].

4.2 Implementation Details

We train our generative models on an NVIDIA GeForce RTX 2080 Ti GPU. For each object, training takes approximately 11 to 13 h, and models are trained for a total of 20,000 steps. Training the classifier by the same GPU according to the evaluation protocol proposed in [1], which deploys 500 generated images per defect class and requires running 10 trials, takes about 1 h per object. We generate the images using an NVIDIA TITAN V GPU, which requires about 2 s per image.

4.3 Evaluation Protocol

The evaluation protocol adopted in [1] deals with generating 500 synthetic images for each defect class and object. We conduct three separate experiments based on the size of the masks used within the Variance Generator. In particular, we will denote as small, medium and large masks those covering between 5%–10%, 11%–18% and 11%–20% of the image area, respectively. Akin to [1], to address anomaly classification we utilize the generated defective images to train a linear classifier on top of a frozen ResNet-34 [17] backbone pre-trained on ImageNet. To perform a comparison with the results reported in AnomalyControl [1], we evaluate the models on the test set. Each model is trained 10 times, and we report the average performance across the 10 trials. For all experiments, we use the final training checkpoint. On Fig. 5 we show some examples of generated defective images.

Table 1. Ablation study of mask sizes

Object	Baseline	Large	Medium	Small
bottle	73.72 ± 2.76	78.84 ± 3.52	**80.47** ± 3.15	78.14 ± 2.98
cable	73.75 ± 4.18	**76.88** ± 5.27	75.78 ± 5.15	76.72 ± 2.92
capsule	35.07 ± 3.87	**42.53** ± 3.12	41.07 ± 2.97	35.07 ± 4.22
carpet	56.45 ± 6.96	**65.65** ± 4.79	62.26 ± 3.76	62.74 ± 6.90
grid	**39.00** ± 6.54	29.50 ± 3.50	35.75 ± 3.72	37.00 ± 3.50
hazelnut	83.13 ± 4.70	83.75 ± 3.70	84.38 ± 4.39	**84.79** ± 3.96
leather	64.29 ± 4.72	69.68 ± 8.20	**70.95** ± 7.14	67.78 ± 5.90
metal nut	**71.88** ± 2.10	**71.88** ± 2.71	70.63 ± 1.68	69.84 ± 2.22
pill	37.40 ± 2.11	**48.02** ± 5.11	47.19 ± 4.09	44.69 ± 3.62
screw	51.36 ± 3.99	**53.70** ± 2.72	52.47 ± 3.63	49.63 ± 4.71
tile	**79.12** ± 5.05	77.72 ± 6.18	75.97 ± 3.60	76.84 ± 3.40
transistor	68.21 ± 3.37	71.79 ± 6.28	**72.86** ± 3.98	68.21 ± 5.64
wood	**58.57** ± 5.24	56.43 ± 6.99	54.76 ± 5.22	53.10 ± 5.00
zipper	67.68 ± 3.50	62.81 ± 5.30	**70.73** ± 4.60	68.78 ± 2.13
Average	61.40 ± 4.22	63.51 ± 4.81	**63.95** ± 4.07	62.38 ± 4.07

4.4 Anomaly Classification

Table 1 compares the performance of classifiers trained on images generated using our method with masks of varying size, starting from real defective images. As a reference, we include the results of the pre-trained AnomalyControl model, applied on the same images (referred to as Baseline in the table). Results show how using medium-sized masks,i.e., masks covering between 11% and 18% of the image area, achieves the best performance. They also highlight how AnomalyControl struggles to create effective images when applied on real defective images, which validates the importance of the variance generator we propose.

Table 2. Anomaly classification accuracy between AnomalyControl, Ensemble and our model.

Object	Ours	AnomalyControl	Ensemble
bottle	80.47 ± 3.52	79.76 ± 3.47	**81.86** ± 2.28
cable	**75.78** ± 5.27	69.37 ± 4.35	66.41 ± 2.90
capsule	41.07 ± 3.12	**60.79** ± 2.82	57.73 ± 7.11
carpet	62.26 ± 4.79	51.44 ± 2.98	**64.03** ± 2.61
grid	35.75 ± 3.50	38.75 ± 6.50	**48.50** ± 4.50
hazelnut	**84.38** ± 3.70	80.20 ± 4.25	80.42 ± 2.67
leather	70.95 ± 8.20	67.61 ± 4.62	**85.24** ± 4.26
metal nut	**70.63** ± 2.71	66.36 ± 1.61	68.28 ± 2.43
pill	47.19 ± 5.11	**63.08** ± 4.31	61.46 ± 2.87
screw	**52.47** ± 2.72	29.25 ± 4.93	36.17 ± 6.10
tile	75.97 ± 6.18	70.70 ± 2.74	**78.25** ± 4.91
transistor	72.86 ± 6.28	**79.64** ± 4.41	73.21 ± 8.03
wood	54.76 ± 6.99	**73.52** ± 3.06	70.95 ± 3.66
zipper	**70.73** ± 5.30	66.92 ± 4.79	69.27 ± 4.49
Average	63.95 ± 4.81	64.89 ± 3.91	**67.27** ± 4.20

Table 2 reports a comparison between our model with medium size images and the original results from AnomalyControl (i.e. when applied on nominal images only). Our model performs particularly well on objects where Anomaly-Control struggles. Since the images generated by our method exhibit complementary properties to those produced by AnomalyControl, training on a mixture of data generated by both models (i.e., 50% from each), referred to as the Ensemble approach in the Table, yields the best results. This approach, indeed, achieves the new state-of-the-art in anomaly classification on the MVTec AD [2] dataset.

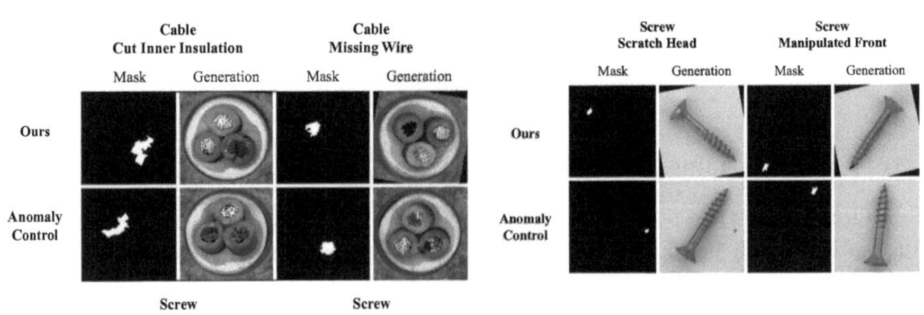

Fig. 5. Qualitative comparison between AnomalyControl and our method using similar masks.

5 Conclusion

In this paper, we proposed a new strategy to generate defective images starting from already existing anomaly images with aligned masks, thereby mitigating one of the common errors in AnomalyControl [1]. However, due to data scarcity, when starting with only a few anomaly images, the nominal parts of the generated images tend to be very similar. To address this limitation, the Variance Generator is trained and used to replace the nominal parts of the generated images, thereby increasing their variance. The experimental results demonstrate that the adoption of the Variance Generator within the image generation pipeline can significantly improve the results on objects where AnomalyControl struggles, and that training the classifier on a mixture of images generated by our best model and AnomalyControl leads to the new state-of-the-art in anomaly classification.

References

1. Ali, M., Fioraio, N., Salti, S., Di Stefano, L.: AnomalyControl: few-shot anomaly generation by controlnet inpainting. IEEE Access (2024)
2. Bergmann, P., Fauser, M., Sattlegger, D., Steger, C.: MVTec AD–a comprehensive real-world dataset for unsupervised anomaly detection. In: Proceedings of the IEEE/CVF Conference on Computer Vision and Pattern Recognition, pp. 9592–9600 (2019)
3. Bergmann, P., Fauser, M., Sattlegger, D., Steger, C.: Uninformed students: student-teacher anomaly detection with discriminative latent embeddings. In: Proceedings of the IEEE/CVF Conference on Computer Vision and Pattern Recognition, pp. 4183–4192 (2020)
4. Bergmann, P., Löwe, S., Fauser, M., Sattlegger, D., Steger, C.: Improving unsupervised defect segmentation by applying structural similarity to autoencoders. arXiv preprint arXiv:1807.02011 (2018)
5. Cao, Y., Wan, Q., Shen, W., Gao, L.: Informative knowledge distillation for image anomaly segmentation. Knowl.-Based Syst. **248**, 108846 (2022)
6. Cao, Y., et al.: Segment any anomaly without training via hybrid prompt regularization. arXiv preprint arXiv:2305.10724 (2023)
7. Chen, X., Han, Y., Zhang, J.: A zero-/few-shot anomaly classification and segmentation method for CVPR 2023 VAND workshop challenge tracks 1&2: 1st place on zero-shot ad and 4th place on few-shot ad. arXiv preprint arXiv:2305.17382 (2023)
8. Chen, X., et al.: CLIP-AD: a language-guided staged dual-path model for zero-shot anomaly detection. arXiv preprint arXiv:2311.00453 (2023)
9. Chiu, L.L., Lai, S.H.: Self-supervised normalizing flows for image anomaly detection and localization. In: Proceedings of the IEEE/CVF Conference on Computer Vision and Pattern Recognition, pp. 2926–2935 (2023)
10. Defard, T., Setkov, A., Loesch, A., Audigier, R.: PaDiM: a patch distribution modeling framework for anomaly detection and localization. In: International Conference on Pattern Recognition, pp. 475–489. Springer (2021)
11. Deng, H., Li, X.: Anomaly detection via reverse distillation from one-class embedding. In: Proceedings of the IEEE/CVF Conference on Computer Vision and Pattern Recognition, pp. 9737–9746 (2022)

12. Duan, Y., Hong, Y., Niu, L., Zhang, L.: Few-shot defect image generation via defect-aware feature manipulation. In: Proceedings of the AAAI Conference on Artificial Intelligence, vol. 37, pp. 571–578 (2023)
13. Gal, R., et al.: An image is worth one word: personalizing text-to-image generation using textual inversion. arXiv preprint arXiv:2208.01618 (2022)
14. Goodfellow, I., et al.: Generative adversarial nets. In: Advances in Neural Information Processing Systems, vol. 27 (2014)
15. Gudovskiy, D., Ishizaka, S., Kozuka, K.: CFLOW-AD: real-time unsupervised anomaly detection with localization via conditional normalizing flows. In: Proceedings of the IEEE/CVF Winter Conference on Applications of Computer Vision, pp. 98–107 (2022)
16. He, H., et al.: A diffusion-based framework for multi-class anomaly detection. In: Proceedings of the AAAI Conference on Artificial Intelligence, vol. 38, pp. 8472–8480 (2024)
17. He, K., Zhang, X., Ren, S., Sun, J.: Deep residual learning for image recognition. In: Proceedings of the IEEE Conference on Computer Vision and Pattern Recognition, pp. 770–778 (2016)
18. Ho, J., Jain, A., Abbeel, P.: Denoising diffusion probabilistic models. arXiv preprint arxiv:2006.11239 (2020)
19. Hou, J., Zhang, Y., Zhong, Q., Xie, D., Pu, S., Zhou, H.: Divide-and-assemble: learning block-wise memory for unsupervised anomaly detection. In: Proceedings of the IEEE/CVF International Conference on Computer Vision, pp. 8791–8800 (2021)
20. Hu, T., et al.: AnomalyDiffusion: few-shot anomaly image generation with diffusion model. In: Proceedings of the AAAI Conference on Artificial Intelligence, vol. 38, pp. 8526–8534 (2024)
21. Jeong, J., Zou, Y., Kim, T., Zhang, D., Ravichandran, A., Dabeer, O.: WinCLIP: zero-/few-shot anomaly classification and segmentation. In: Proceedings of the IEEE/CVF Conference on Computer Vision and Pattern Recognition, pp. 19606–19616 (2023)
22. Kingma, D.P., Welling, M., et al.: Auto-encoding variational bayes (2013)
23. Li, C.L., Sohn, K., Yoon, J., Pfister, T.: CutPaste: self-supervised learning for anomaly detection and localization. In: Proceedings of the IEEE/CVF Conference on Computer Vision and Pattern Recognition, pp. 9664–9674 (2021)
24. Lin, D., Cao, Y., Zhu, W., Li, Y.: Few-shot defect segmentation leveraging abundant defect-free training samples through normal background regularization and crop-and-paste operation. In: 2021 IEEE International Conference on Multimedia and Expo (ICME), pp. 1–6. IEEE (2021)
25. Massoli, F.V., Falchi, F., Kantarci, A., Akti, Ş., Ekenel, H.K., Amato, G.: MOCCA: multilayer one-class classification for anomaly detection. IEEE Trans. Neural Netw. Learn. Syst. **33**(6), 2313–2323 (2021)
26. Niu, S., Li, B., Wang, X., Lin, H.: Defect image sample generation with GAN for improving defect recognition. IEEE Trans. Autom. Sci. Eng. **17**(3), 1611–1622 (2020)
27. Pirnay, J., Chai, K.: Inpainting transformer for anomaly detection. In: International Conference on Image Analysis and Processing, pp. 394–406. Springer (2022)
28. Radford, A., et al.: Learning transferable visual models from natural language supervision. In: International Conference on Machine Learning, pp. 8748–8763. PMLR (2021)

29. Reiss, T., Cohen, N., Bergman, L., Hoshen, Y.: PANDA: adapting pretrained features for anomaly detection and segmentation. In: Proceedings of the IEEE/CVF Conference on Computer Vision and Pattern Recognition, pp. 2806–2814 (2021)
30. Rippel, O., Mertens, P., Merhof, D.: Modeling the distribution of normal data in pre-trained deep features for anomaly detection. In: 2020 25th International Conference on Pattern Recognition (ICPR), pp. 6726–6733. IEEE (2021)
31. Ristea, N.C., et al.: Self-supervised predictive convolutional attentive block for anomaly detection. In: Proceedings of the IEEE/CVF Conference on Computer Vision and Pattern Recognition, pp. 13576–13586 (2022)
32. Rombach, R., Blattmann, A., Lorenz, D., Esser, P., Ommer, B.: High-resolution image synthesis with latent diffusion models. In: Proceedings of the IEEE/CVF Conference on Computer Vision and Pattern Recognition, pp. 10684–10695 (2022)
33. Ronneberger, O., Fischer, P., Brox, T.: U-Net: convolutional networks for biomedical image segmentation. In: Medical Image Computing and Computer-assisted Intervention–MICCAI 2015: 18th International Conference, Munich, Germany, 5–9 October 2015, Proceedings, Part III, vol. 18, pp. 234–241. Springer (2015)
34. Roth, K., Pemula, L., Zepeda, J., Schölkopf, B., Brox, T., Gehler, P.: Towards total recall in industrial anomaly detection. In: Proceedings of the IEEE/CVF Conference on Computer Vision and Pattern Recognition, pp. 14318–14328 (2022)
35. Ruiz, N., Li, Y., Jampani, V., Pritch, Y., Rubinstein, M., Aberman, K.: DreamBooth: fine tuning text-to-image diffusion models for subject-driven generation. In: Proceedings of the IEEE/CVF Conference on Computer Vision and Pattern Recognition, pp. 22500–22510 (2023)
36. Salehi, M., Sadjadi, N., Baselizadeh, S., Rohban, M.H., Rabiee, H.R.: Multiresolution knowledge distillation for anomaly detection. In: Proceedings of the IEEE/CVF Conference on Computer Vision and Pattern Recognition, pp. 14902–14912 (2021)
37. Schlegl, T., Seeböck, P., Waldstein, S.M., Schmidt-Erfurth, U., Langs, G.: Unsupervised anomaly detection with generative adversarial networks to guide marker discovery. In: International Conference on Information Processing in Medical Imaging, pp. 146–157. Springer (2017)
38. Schuhmann, C., et al.: LAION-400M: open dataset of CLIP-filtered 400 million image-text pairs. arXiv preprint arXiv:2111.02114 (2021)
39. Wei, S., Wei, X., Ma, Z., Dong, S., Zhang, S., Gong, Y.: Few-shot online anomaly detection and segmentation. arXiv preprint arXiv:2403.18201 (2024)
40. Wyatt, J., Leach, A., Schmon, S.M., Willcocks, C.G.: AnoDDPM: anomaly detection with denoising diffusion probabilistic models using simplex noise. In: Proceedings of the IEEE/CVF Conference on Computer Vision and Pattern Recognition, pp. 650–656 (2022)
41. Yi, J., Yoon, S.: Patch SVDD: patch-level SVDD for anomaly detection and segmentation. In: Proceedings of the Asian Conference on Computer Vision (2020)
42. Yoa, S., Lee, S., Kim, C., Kim, H.J.: Self-supervised learning for anomaly detection with dynamic local augmentation. IEEE Access **9**, 147201–147211 (2021)
43. Zavrtanik, V., Kristan, M., Skočaj, D.: DRAEM-a discriminatively trained reconstruction embedding for surface anomaly detection. In: Proceedings of the IEEE/CVF International Conference on Computer Vision, pp. 8330–8339 (2021)
44. Zhang, G., Cui, K., Hung, T.Y., Lu, S.: Defect-GAN: high-fidelity defect synthesis for automated defect inspection. In: Proceedings of the IEEE/CVF Winter Conference on Applications of Computer Vision, pp. 2524–2534 (2021)

45. Zhang, H., Wu, Z., Wang, Z., Chen, Z., Jiang, Y.G.: Prototypical residual networks for anomaly detection and localization. In: Proceedings of the IEEE/CVF Conference on Computer Vision and Pattern Recognition, pp. 16281–16291 (2023)
46. Zhang, L., Rao, A., Agrawala, M.: Adding conditional control to text-to-image diffusion models. In: Proceedings of the IEEE/CVF International Conference on Computer Vision, pp. 3836–3847 (2023)
47. Zhou, Q., Pang, G., Tian, Y., He, S., Chen, J.: AnomalyCLIP: object-agnostic prompt learning for zero-shot anomaly detection. arXiv preprint arXiv:2310.18961 (2023)

AGCD-Net: Attention Guided Context Debiasing Network for Emotion Recognition

Varsha Devi[1(✉)], Amine Bohi[2(✉)], and Pardeep Kumar[3]

[1] CESI LINEACT Laboratory, UR 7527, Nice 06200, France
vdevi@cesi.fr

[2] CESI LINEACT Laboratory, UR 7527, Dijon 21800, France
abohi@cesi.fr

[3] ACENTAURI Team, Inria Centre at Université Côte d'Azur, Sophia Antipolis 06560, France
pardeep.kumar@inria.fr

Abstract. Context-aware emotion recognition (CAER) enhances affective computing in real-world scenarios, but traditional methods often suffer from context bias-spurious correlation between background context and emotion labels (e.g. associating "garden" with "happy"). In this paper, we propose **AGCD-Net**, an Attention Guided Context Debiasing model that introduces *Hybrid ConvNeXt*, a novel convolutional encoder that extends the ConvNeXt backbone by integrating Spatial Transformer Network and Squeeze-and-Excitation layers for enhanced feature recalibration. At the core of AGCD-Net is the Attention Guided - Causal Intervention Module (AG-CIM), which applies causal theory, perturbs context features, isolates spurious correlations, and performs an attention-driven correction guided by face features to mitigate context bias. Experimental results on the CAER-S dataset demonstrate the effectiveness of AGCD-Net, achieving state-of-the-art performance and highlighting the importance of causal debiasing for robust emotion recognition in complex settings.

Keywords: Human emotion recognition · Context awareness · Bias reduction · Causal intervention module

1 Introduction

Emotion recognition is a key task in artificial intelligence, with applications including but not limited to health [2,3], human-robot interaction [24], and social robotics [6]. It involves classifying input into categories such as *angry*, *sad*, *happy*, etc. Traditional methods rely on single modalities (e.g., facial analysis [4], body postures [1], or acoustic cues [20]), or on multimodal fusion [25], but they often fail in unconstrained environments due to pose variations, occlusions, or overreliance on facial cues [2,7,24].

E. Rodolà et al. (Eds.): ICIAP 2025, LNCS 16167, pp. 533–545, 2026.
https://doi.org/10.1007/978-3-032-10185-3_42

Context-Aware Emotion Recognition (CAER) has emerged to address these issues by jointly leveraging facial and contextual cues [13,15,16,29]. Most approaches adopt a two-stream architecture to extract face and context features before fusing them for emotion prediction. Despite improvements, these models suffer from *context bias*, where spurious correlations between background context and emotion (e.g., associating "hospital" with "sadness") lead to misclassifications.

Recent works tackle this bias using causal frameworks such as CCIM [26] and CLEF [27]. CCIM build a computationally expensive confounder dictionary by clustering contexts (e.g., "hospital", "beach") to reduce context bias, but it applies uniform adjustments across each representative context, potentially suppressing fine-grained emotional cues within similar-looking but emotionally diverse contexts. CLEF avoids the dictionary but performs context debiasing after face-context fusion, using two parallel pipelines for factual and counterfactual, limiting its ability i) to refine context representations in isolation and ii) model the complex interactions between facial and contextual cues.

To overcome these limitations, we propose the **Attention Guided Context Debiasing Network (AGCD-Net)**, which performs instance-level correction of context features before fusion. AGCD-Net introduces a *Hybrid ConvNeXt* encoder with integrated Spatial Transformer Networks (STN) and Squeeze-and-Excitation (SE) blocks for robust and aligned feature extraction. It further incorporates a dual-stream attention mechanism and the novel **Attention Guided - Causal Intervention Module (AG-CIM)**, which isolates and suppresses spurious context features under guidance from face embeddings. Unlike CCIM [26], AG-CIM performs targeted refinements instead of uniform corrections; unlike CLEF [27], it adjusts context representations before fusion, avoiding post-hoc bias suppression. The key contributions of our work are as follows:

1. **Attention-based Dual Encoding Mechanism**: Enhances recognition accuracy by independently encoding face and context features using a robust Hybrid ConvNeXt architecture.
2. **Causal Intervention Module**: Dynamically adapts and debiases context features based on facial information, reducing spurious correlations.
3. **End-to-End Framework**: Integrates encoding, causal intervention, and feature fusion for seamless optimization.
4. **State-of-the-Art Performance**: Demonstrates superior accuracy of the proposed AGCD-Net compared to existing methods.

2 Related Work

Facial Emotion Recognition (FER) has evolved from early methods using handcrafted features like FACS [23] and LBPs [21] to more advanced deep learning approaches such as CNNs, RNNs, and 3D-CNNs for capturing spatial and temporal information in static FER [5] and dynamic FER [11,12]. While these traditional techniques were effective in controlled settings, they struggled with real-world challenges such as pose variations, lighting conditions, occlusions and limited contextual awareness (overlooking the broader context surrounding the subject) [2,4].

To address these limitations, Context-Aware Emotion Recognition (CAER) emerged, integrating facial and contextual features for more robust emotion analysis in diverse settings. Early work by *Kosti et al.* [13] introduced CAER using a two-stream Convolutional Neural Network (CNN) for encoding face and context features. Later models enhanced this framework using attention mechanisms [15,28] or multimodal fusion with body cues and scene understanding [17]. Despite these advancements, CAER models often face challenges with *context bias*, where over-reliance on background context leads to misinterpretation of emotional states, underscoring the need for methods to address this issue.

To tackle this problem, causal theory-based frameworks have been introduced. These approaches aim to reduce contextual bias by identifying and separating meaningful context features from confounding factors. The Contextual Causal Intervention Module (CCIM) [26] utilizes a confounder dictionary to learn this separation. This dictionary is built using an unsupervised learning method to identify the clusters of context feature by masking out the face-specific attributes in images. However, this framework requires significant computational resources, increasing model complexity for the generation of the confounder dictionary. Moreover, this method is sub-optimal, as the predefined intervention treats all contexts uniformly, making it difficult to distinguish beneficial cues from harmful ones. While CLEF [27] adopts a parallel structure involving factual (face+context) and counterfactual (context-only) branches. The framework then compares the predictions from *factual* branch with the counterfactual predictions to isolate the harmful bias and debias final emotion predictions. Such a late-stage intervention limits the model's capacity to disentangle contextual bias at a granular level and makes it harder to remove spurious contextual influences without also affecting useful emotional cues.

In contrast to these methods, our proposed AGCD-Net addresses context bias through early face-guided correction of context features, before fusion, ensuring that misleading correlations are removed while preserving semantically rich cues. This dynamic and attention-driven strategy overcomes the limitations of rigid prototypes and post-hoc adjustments, offering a more flexible and interpretable solution for real-world emotion recognition.

3 Preliminary

To build a strong foundation for the proposed model, we begin by introducing essential concepts from *causal inference theory*, which aims to analyze the affect of change in one variable (light or context) on final output (classification) while keeping other variables (subject/face) constant. To get this understanding, we look at two scenarios:

1. **Factual**: This scenario represents what is actually observed (face) in a specific setting (context/background).

2. **Counterfactual**: This represents "what if" scenario. The face placed in a different context to assess the effect of background changes. The aim is to generate counterfactual to eliminate spurious correlations in data [19].

Comparing these scenarios, reveals how much the context biases emotion recognition. A large shift suggests a *bias* rooted in the context rather than the facial expression. To address this, we apply causal adjustment only to the biased variable (i.e. context features) while keeping facial features unchanged to isolate context influence. This principle ensures that we isolate the effect of context alone.

The core research problem we address using causal theory is: *How can we reduce context-induced bias in emotion recognition by adjusting the causal influence of biased context features prior to fusion of face and context features.* The objective of AG-CIM module is to perform fine-grained, instance-specific refinement of context representations in isolation (prior to fusion), while preserving informative contextual semantics, ensuring that only meaningful context contributes to the final prediction.

4 Proposed Method

To tackle the issue of context bias in context aware emotion recognition, we proposed Attention Guided Context Debiasing Network (AGCD-Net), a model designed for robust emotion recognition in dynamic environment. As context and facial cues, both are important for FER, AGCD-Net extracts facial and contextual cues using an **Attention-Based Dual Encoding Network** while actively correcting for biased context through an **Attention Guided - Causal Intervention Module (AG-CIM)** before fusing them for final classification.

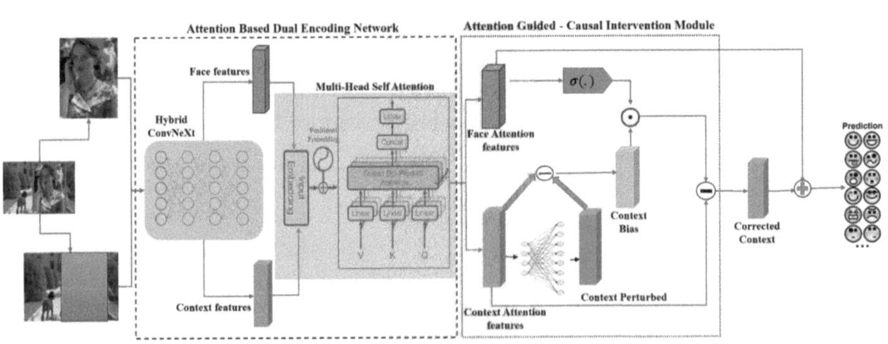

Fig. 1. Proposed Attention Guided Context Debiasing Model for Context-Aware Emotion Recognition

Specifically, let an image be represented as I with a discrete emotion label y from a set of K emotions $\{y_1, \ldots, y_K\}$. Our goal is to predict the emotion label

y by processing separate facial and contextual features, modeled as ϕ_f and ϕ_c, respectively. To achieve this, we propose AGCD-Net with three components: i) *Attention-Based Dual Encoding Network* for independently encoding and focusing on face and context, ii) *Attention Guided - Causal Intervention Module* to correct contextual biases and eliminate influences from irrelevant context features, and iii) *Final Classification Module* to produce the predicted emotion based on the debiased and fused features, as illustrated in Fig. 1.

4.1 Attention-Based Dual Encoding Network

The Attention-Based Dual Encoding Network is meticulously designed to process the face and context streams independently, thereby enabling the model to effectively distinguish between emotional cues derived from facial expressions and those arising from the surrounding context.

Hybrid ConvNeXt: Encoder CNN for Feature Extraction. To extract robust features from facial and contextual images, we introduce Hybrid ConvNeXt, an enhanced version of the ConvNeXt architecture tailored for computer vision classification, with specific design elements aimed at improving context-aware emotion recognition. The enhancements focus on improving feature extraction and robustness to spatial variations in both facial and contextual features, as illustrated in Fig. 2. The Core Components of Hybrid ConvNeXt are:

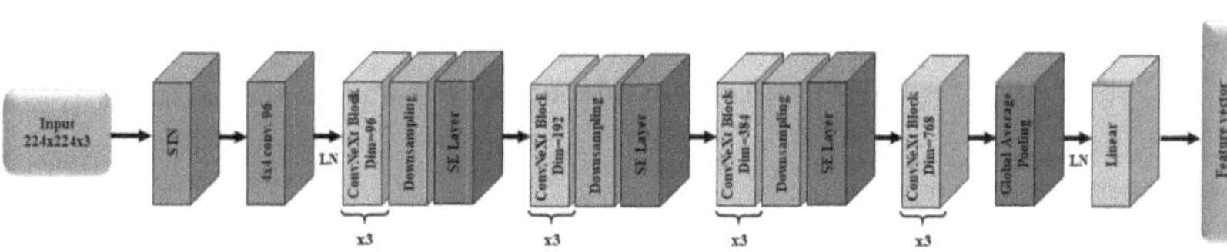

Fig. 2. Proposed Hybrid ConvNeXt Encoding Network illustrating the key components such as ConvNeXt backbone, STN, and SE blocks

- ConvNeXt Backbone: The foundation of Hybrid ConvNeXt is the hierarchical multi-stage design of ConvNeXt [18], which employs patchified convolutions for downsampling, depthwise convolutions for efficient feature extraction, and advanced normalization techniques like LayerNorm. This structure enables the extraction of robust, multiscale spatial features. As illustrated in Fig. 3(a), ConvNeXt leverages larger kernel sizes and replaces the ReLU activation function with GELU for smoother gradient flow, contributing to the improved performance of Hybrid ConvNeXt.
- Spatial Transformer Network (STN): An STN [10] is integrated at the input to enhance the model's ability to handle spatial variations (scale, rotation, and translation) by learning spatial transformations directly from the data.

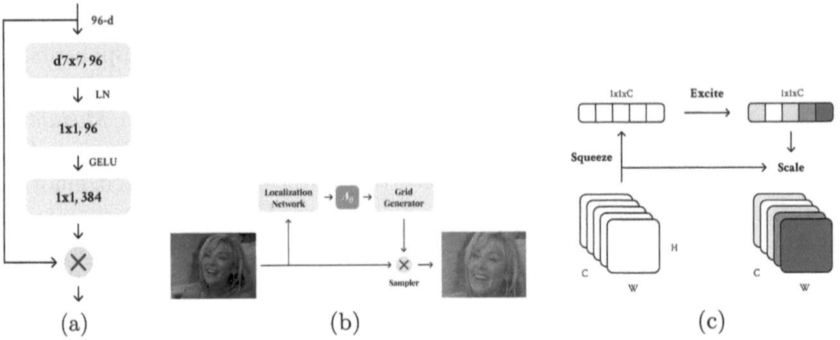

Fig. 3. Architectural components used in the Hybrid ConvNeXt: (a) ConvNeXt block, (b) Spatial Transformer module, and (c) Squeeze-and-Excitation block.

STN consists of three key components, as illustrated in Fig. 3(b) : a **localization network** that predicts the transformation parameters, a **grid generator** that produces sampling coordinates, and a **sampler** that applies the transformation to the input image. The transformation parameters can be represented using a 3×2 affine transformation matrix A_θ:

$$A_\theta = \begin{bmatrix} s_x \cos(\theta) & -s_y \sin(\theta) & t_x \\ s_x \sin(\theta) & s_y \cos(\theta) & t_y \end{bmatrix} \tag{1}$$

where t_x, t_y are the translation parameters, s_x, s_y are the scale factors, and θ is the rotation angle. The grid generator uses A_θ to produce a set of sampling grid coordinates, and the sampler applies these to the input image via bilinear interpolation, ensuring precise alignment of facial regions before feature extraction.

- Squeeze-and-Excitation (SE) Blocks: SE blocks [9] are applied after the ConvNeXt feature extraction stages to recalibrate channel-wise feature responses adaptively. The SE mechanism comprises two main operations: squeeze and excitation, as illustrated in Fig. 3(c) .

  In the squeeze operation, global average pooling is applied to each channel to capture the global distribution:

$$z_c = \frac{1}{H \times W} \sum_{i=1}^{H} \sum_{j=1}^{W} X_{i,j,c} \tag{2}$$

  where $X_{i,j,c}$ represents the value of the c-th channel at spatial position (i, j). The excitation operation models the interdependencies among channels and recalibrates them using two fully connected layers with non-linear activations:

$$s = \sigma(W_2 \cdot \text{ReLU}(W_1 \cdot z)) \tag{3}$$

  where W_1 and W_2 are weight matrices, z is the descriptor vector from the squeeze operation, and σ denotes the sigmoid function. The resulting scaling factors are applied to recalibrate the feature maps.

Feature Extraction Process. For our context-aware emotion recognition system, Hybrid ConvNeXt serves as the feature encoder for both facial and contextual images. This design allows the network to leverage spatially and contextually aware feature embeddings from both the face and the background of each image. We ensure that the extracted features from face (ϕ_f) and context (ϕ_c) remain generalizable and ready for downstream processing in the next modules: the Multi-Head Self-Attention (MHSA) and the Attention Guided - Causal Intervention Module.

Multi-Head Self Attention. Multi-Head Self-Attention (MHSA) enhances the model's ability to encode diverse information by allowing it to focus on multiple parts of the face and context features simultaneously. This is achieved by linearly transforming the features from face (ϕ_f) and context (ϕ_c) into multiple query, key, and value vectors, and combining the outputs to produce attention-refined face ($\phi_{f,\text{att}}$) and context ($\phi_{c,\text{att}}$) features.

4.2 Attention Guided Causal Intervention Module (AG-CIM)

The Attention Guided Context Debiasing Network (AGCD-Net) model incorporates causal theory (Sect. 3) to address spurious correlations in context features that often misleads prediction model. The Attention Guided - Causal Intervention Module (AG-CIM) is designed to correct these biases by simulating counterfactual scenarios and adjusting context features, using face features to guide the corrections.

Counterfactual Generation. The first step of AG-CIM involves generating counterfactual representations of the context to simulate how the context features appear if the spurious correlation between context and emotion class were minimized. The attended context features $\phi_{c,\text{att}}$ are perturbed through learnable transformation to generate the counterfactual representation, as shown in Fig. 1. Mathematically it is represented as:

$$\phi_{c,\text{pert}} = \mathbf{W}_p \cdot \phi_{c,\text{att}} \tag{4}$$

where $\mathbf{W}_p$ is a learnable weight matrix trained via backpropagation to generate counterfactual context features by introducing controlled perturbations that suppress spurious or irrelevant patterns.

Context Bias Computation. Next, the bias in context features is quantified as the difference between original and perturbed (counterfactual) features:

$$\Delta\phi_c = \phi_{c,\text{att}} - \phi_{c,\text{pert}} \tag{5}$$

This difference represents the spurious parts of the context that should be removed to prevent bias. If the difference $\Delta\phi_c$ is significant, it suggests that the context itself plays a major role in biasing the output (emotion prediction). By computing the difference between the original and perturbed context features, AGCD-Net identifies spurious correlations in the context and applies a controlled correction before making predictions.

Bias Correction Using Face Attention. To apply correction and decide how much weight to assign to context bias, we apply another linear transformation $\mathbf{W}_c$ and then adjust it with a sigmoid function. If the bias (Eq. 5) is large, the sigmoid function will allow more correction; if it's small, it dials down the correction. Additionally, a learnable parameter α is used to dynamically adjust the influence of face features to control their relative importance in the correction process as follows:

$$\phi_{c,\text{corr}} = \phi_{c,\text{att}} - \mathbf{W}_c \Delta\phi_c \cdot \sigma(\alpha \times \phi_{f,\text{att}}) \tag{6}$$

where $\mathbf{W}_c$ is a learnable parameter matrix. The α is learnable parameter initialized to 1, and the sigmoid function learns and adjusts the strength of the correction based on the face features informativeness. This approach minimizes the impact of irrelevant context, reducing bias.

4.3 Fusion and Classification

The face features and unbiased context features are fused via element-wise addition and passed through fully connected layer and softmax function to estimate the probability distribution over emotion classes:

$$\hat{y} = \text{softmax}(\mathbf{W}_f \cdot (\phi_{f,\text{att}} + \phi_{c,\text{corr}})) \tag{7}$$

where $\mathbf{W}_f$ maps the fused feature space to the emotion label space.
To optimize the model's performance, we used Cross-Entropy Loss ($\mathcal{L}_{\text{CE}}$) for classification. Additionally, we introduce the ***Attention Loss*** **($\mathcal{L}_{\text{att}}$)** to regularize the attention weights. The attention loss encourages balanced attention between the face and context features, defined as:

$$\mathcal{L}_{\text{att}} = \frac{1}{N} \sum_{i=1}^{N} \left(|h_{\text{face}}| + |h_{\text{context}}| \right) \tag{8}$$

where h_{face} and h_{context} represent the attention weights. This regularization term is added to the Cross-Entropy Loss, resulting in the final loss function:

$$\mathcal{L}_{\text{final}} = \mathcal{L}_{\text{CE}} + \mathcal{L}_{\text{att}} \tag{9}$$

5 Experiments and Results

5.1 Dataset and Evaluation Metrics

We used the CAER-S dataset with 70k images from 79 indoor and outdoor TV shows, covering diverse scenes and labeled across seven emotion classes. It was split into 70% training, 10% validation, and 20% testing. Standard classification accuracy was used, following the official PyTorch implementation[1].

5.2 Implementation Details

AGCD-Net was implemented in PyTorch using face and context features extracted by the proposed Hybrid ConvNeXt trained on CAER-S. Training employed AdamW optimizer (lr $= 1 \times 10^{-5}$, weight decay $= 1 \times 10^{-4}$), a CosineAnnealingWarmRestarts scheduler ($T_0 = 128$, $T_{mult} = 2$), and a batch size of 128. Data augmentation was applied, and the loss combined cross-entropy with label smoothing ($\epsilon = 0.2$) and a self-attention loss to enforce embedding consistency.

5.3 Results and Discussion

Table 1 presents a clear comparison of baseline, SOTA, and ablation models (CIM modules & AGCD-Net) on CAER-S. The results demonstrate: i) the **superior performance of our proposed AGCD model** (ablation - last row), achieving 90.65% accuracy, and ii) the **effectiveness of our proposed Hybrid ConvNeXt model** *without the context debiasing (AG-CIM) module* (SOTA - last row), achieving 88.84%, compared to existing state-of-the-art CAER models. Additionally, combining CAER-Net [16] with our AG-CIM (Ablation - 2nd last row), achieved a notable accuracy of 76.03%, representing improvements of 1.63% and 0.22% compared to CCIM [26] and CLEF [27], respectively. This shows the effectiveness of Hybrid ConvNeXt and face-guided, early-stage context debiasing over rigid and post-hoc prototypes like CCIM [26] and CLEF [27].

The normalized confusion matrix, shown in Fig. 4, evaluates the model's performance across emotion categories. The diagonal values represent high classification accuracy, with Fear (97%) and Disgust (95%) achieving the best results. However, the model struggles to predict the Neutral (81%) and Happy (87%) classes due to significant misclassification between the two. This issue arises from the high correlation (facial and contextual similarities) between these emotions in the dataset, leading to overlapping feature spaces. The highest off-diagonal values in the Neutral column correspond to Happy and vice versa, underscoring the challenge. Addressing this issue, possibly through more refined data collection for these classes in addition to the data augmentation techniques already implemented in our model, could significantly enhance prediction accuracy.

5.4 Ablation Study

Table 2 evaluates the impact of key components on model performance by systematically removing or including the Face MHS-Attention, Context MHS-Attention, and the AG-CIM. **Model A**, including all components, achieves the

[1] https://github.com/ndkhanh360/CAER.

Table 1. Performance comparison of CAER models on the CAER-S test set, including pre-trained ImageNet models on CAER-S, fine-tuned ImageNet models on CAER-S, SOTA, and ablation results with Causal Intervention Modules.

Category	Methods	Year	Accuracy (%)
Pre-trained	ImageNet-AlexNet [14]	2012	47.36
	ImageNet-VGGNet [22]	2015	49.89
	ImageNet-ResNet [8]	2016	57.33
Fine-tuned	Fine-tuned AlexNet [14]	2012	61.73
	Fine-tuned VGGNet [22]	2015	64.85
	Fine-tuned ResNet [8]	2016	68.46
SOTA	CAER-Net-S [16]	2019	73.52
	EMOT-Net [13]	2019	74.51
	GNN-CNN [28]	2019	77.21
	GLAMOR-Net [15]	2022	77.90
	CAHFW-Net [30]	2023	83.76
	EfficientFace [29]	2021	85.87
	Hybrid ConvNeXt w/o AG-CIM (Ours)	2025	**88.84**
Ablation	CAER-Net-S + CCIM [26]	2024	74.81
	CAER-Net-S + CLEF [27]	2024	75.86
	CAER-Net-S + AG-CIM (Our)	2025	**76.03**
	AGCD (ours)	2025	**90.65**

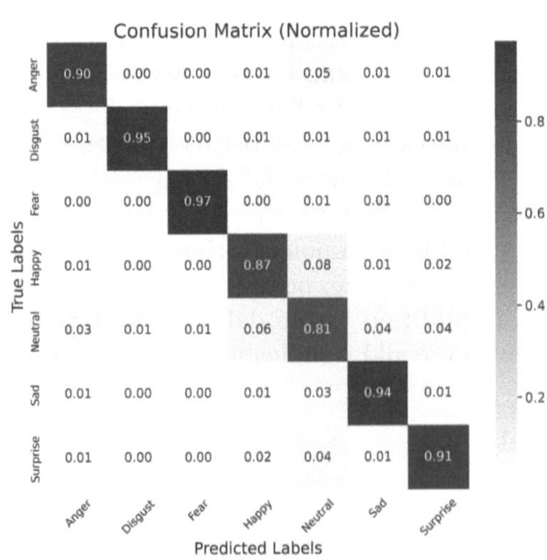

Fig. 4. Normalized Confusion Matrix of AGCD-Net on CAER-S Dataset

Table 2. Ablation study on the impact of AGCD-Net components. ✓ = active, × = inactive.

Model	Face MHS-Attention	Context MHS-Attention	Causal Intervention Module (AG-CIM)	Overall Accuracy (in %)
A	✓	✓	✓	90.65
B	×	×	✓	90.40
C	✓	✓	×	90.27
D	×	✓	✓	90.47
E	×	×	×	88.84

highest accuracy of 90.65%. **Model B**, without MHS-Attention but with AG-CIM, slightly decreases to 90.40%, showing the robustness of AG-CIM. **Model C**, with both MHS-Attentions but without AG-CIM, reaches 90.27%, indicating AG-CIM's beneficial boost. **Model D**, missing Face MHS-Attention but retaining Context MHS-Attention and AG-CIM, performs at 90.47%, highlighting i) the effectiveness of context attention and causal intervention, and ii) the strength of Hybrid ConvNeXt. **Model E**, using only Hybrid ConvNeXt without MHS-Attention or AG-CIM, records 88.84%, still surpassing state-of-the-art techniques (Table 1), emphasizing the importance of these components.

6 Conclusion and Future Directions

We introduced AGCD-Net, an Attention Guided Context Debiasing model designed for emotion recognition in uncontrolled and dynamic environments. Leveraging the Hybrid ConvNeXt model with a dual attention mechanism, our framework extracts high-level context-aware features that outperform existing state-of-the-art methods. By integrating the proposed Attention-Guided Causal Intervention Module (AG-CIM), AGCD-Net effectively reduces context-induced bias and enhances classification accuracy. Extensive experiments on the CAER-S dataset demonstrate that AGCD-Net achieves state-of-the-art performance, significantly improving emotion recognition in challenging scenarios. These results underscore the effectiveness of causal debiasing and attention-driven correction in addressing spurious correlations. As future work, we plan to validate AGCD-Net on additional benchmarks, explore lightweight variants for edge deployment, and adapt it to healthcare scenarios involving emotionally vulnerable or minimally communicative individuals—such as patients with cognitive impairments.

References

1. Ahmed, F., Bari, A.H., Gavrilova, M.L.: Emotion recognition from body movement. IEEE Access **8**, 11761–11781 (2019)

2. Bohi, A., Boudouri, Y.E., Sfeir, I.: A novel deep learning approach for facial emotion recognition: application to detecting emotional responses in elderly individuals with alzheimer's disease. Neural Comput. Appl. 1–19 (2024)
3. Drimalla, H., et al.: Towards the automatic detection of social biomarkers in autism spectrum disorder: introducing the simulated interaction task (sit). NPJ Digit. Med. **3**(1), 25 (2020)
4. El Boudouri, Y., Bohi, A.: Emonext: an adapted convnext for facial emotion recognition. In: 2023 IEEE 25th International Workshop on Multimedia Signal Processing (MMSP), pp. 1–6. IEEE (2023)
5. Fabian Benitez-Quiroz, C., Srinivasan, R., Martinez, A.M.: Emotionet: an accurate, real-time algorithm for the automatic annotation of a million facial expressions in the wild. In: Proceedings of the IEEE Conference on Computer Vision and Pattern Recognition, pp. 5562–5570 (2016)
6. Greco, A., Roberto, A., Saggese, A., Vento, M., Vigilante, V.: Emotion analysis from faces for social robotics. In: 2019 IEEE International Conference on Systems, Man and Cybernetics (SMC), pp. 358–364. IEEE (2019)
7. Gursesli, M.C., Lombardi, S., Duradoni, M., Bocchi, L., Guazzini, A., Lanata, A.: Facial emotion recognition (fer) through custom lightweight cnn model: performance evaluation in public datasets. IEEE Access (2024)
8. He, K., Zhang, X., Ren, S., Sun, J.: Deep residual learning for image recognition. In: Proceedings of the IEEE Conference on Computer Vision and Pattern Recognition, pp. 770–778 (2016)
9. Hu, J., Shen, L., Sun, G.: Squeeze-and-excitation networks. In: Proceedings of the IEEE Conference on Computer Vision and Pattern Recognition, pp. 7132–7141 (2018)
10. Jaderberg, M., Simonyan, K., Zisserman, A., et al.: Spatial transformer networks. In: Advances in Neural Information Processing Systems, vol. 28 (2015)
11. Jiang, X., et al.: Dfew: a large-scale database for recognizing dynamic facial expressions in the wild. In: Proceedings of the 28th ACM International Conference on Multimedia, pp. 2881–2889 (2020)
12. Kossaifi, J., Toisoul, A., Bulat, A., Panagakis, Y., Hospedales, T.M., Pantic, M.: Factorized higher-order cnns with an application to spatio-temporal emotion estimation. In: Proceedings of the IEEE/CVF Conference on Computer Vision and Pattern Recognition, pp. 6060–6069 (2020)
13. Kosti, R., Alvarez, J.M., Recasens, A., Lapedriza, A.: Context based emotion recognition using emotic dataset. IEEE Trans. Pattern Anal. Mach. Intell. **42**(11), 2755–2766 (2019)
14. Krizhevsky, A., Sutskever, I., Hinton, G.E.: Imagenet classification with deep convolutional neural networks. In: Advances in Neural Information Processing Systems, vol. 25 (2012)
15. Le, N., Nguyen, K., Nguyen, A., Le, B.: Global-local attention for emotion recognition. Neural Comput. Appl. **34**(24), 21625–21639 (2022)
16. Lee, J., Kim, S., Kim, S., Park, J., Sohn, K.: Context-aware emotion recognition networks. In: Proceedings of the IEEE/CVF International Conference on Computer Vision, pp. 10143–10152 (2019)
17. Li, X., Peng, X., Ding, C.: Sequential interactive biased network for context-aware emotion recognition. In: 2021 IEEE International Joint Conference on Biometrics (IJCB), pp. 1–6. IEEE (2021)
18. Liu, Z., Mao, H., Wu, C.Y., Feichtenhofer, C., Darrell, T., Xie, S.: A convnet for the 2020s. In: Proceedings of the IEEE/CVF Conference on Computer Vision and Pattern Recognition, pp. 11976–11986 (2022)

19. Pearl, J.: [bayesian analysis in expert systems]: comment: graphical models, causality and intervention. Stat. Sci. **8**(3), 266–269 (1993)
20. Schuller, B., Vlasenko, B., Eyben, F., Rigoll, G., Wendemuth, A.: Acoustic emotion recognition: a benchmark comparison of performances. In: 2009 IEEE Workshop on Automatic Speech Recognition & Understanding, pp. 552–557. IEEE (2009)
21. Shan, C., Gong, S., McOwan, P.W.: Facial expression recognition based on local binary patterns: a comprehensive study. Image Vis. Comput. **27**(6), 803–816 (2009)
22. Simonyan, K., Zisserman, A.: Very deep convolutional networks for large-scale image recognition. arXiv preprint arXiv:1409.1556 (2014)
23. Wang, Z., Li, Y., Wang, S., Ji, Q.: Capturing global semantic relationships for facial action unit recognition. In: Proceedings of the IEEE International Conference on Computer Vision, pp. 3304–3311 (2013)
24. Yahyaoui, M.A., Oujabour, M., Ben Letaifa, L., Bohi, A.: Multi-face emotion detection for effective human-robot interaction. In: Proceedings of the 17th International Conference on Agents and Artificial Intelligence, vol. 1, ICAART, pp. 91–99. INSTICC, SciTePress (2025). https://doi.org/10.5220/0013170300003890
25. Yan, J., et al.: Multimodal emotion recognition based on facial expressions, speech, and body gestures. Electronics (2079-9292) **13**(18) (2024)
26. Yang, D., Yang, K., Kuang, H., Chen, Z., Wang, Y., Zhang, L.: Towards context-aware emotion recognition debiasing from a causal demystification perspective via de-confounded training. IEEE Trans. Pattern Anal. Mach. Intell. (2024)
27. Yang, D., Yang, K., Li, M., Wang, S., Wang, S., Zhang, L.: Robust emotion recognition in context debiasing. In: Proceedings of the IEEE/CVF Conference on Computer Vision and Pattern Recognition, pp. 12447–12457 (2024)
28. Zhang, M., Liang, Y., Ma, H.: Context-aware affective graph reasoning for emotion recognition. In: 2019 IEEE International Conference on Multimedia and Expo (ICME), pp. 151–156. IEEE (2019)
29. Zhao, Z., Liu, Q., Zhou, F.: Robust lightweight facial expression recognition network with label distribution training. In: Proceedings of the AAAI Conference on Artificial Intelligence, vol. 35, pp. 3510–3519 (2021)
30. Zhou, S., Wu, X., Jiang, F., Huang, Q., Huang, C.: Emotion recognition from large-scale video clips with cross-attention and hybrid feature weighting neural networks. Int. J. Environ. Res. Public Health **20**(2), 1400 (2023)

Prompt-Based Consistent Video Colorization

Silvia Dani[1], Tiberio Uricchio[2], and Lorenzo Seidenari[1](✉)

[1] University of Firenze, Florence, Italy
{silvia.dani,lorenzo.seidenari}@unifi.it
[2] University of Pisa, Pisa, Italy
tiberio.uricchio@unipi.it

Abstract. Existing video colorization methods struggle with temporal flickering or demand extensive manual input. We propose a novel approach automating high-fidelity video colorization using rich semantic guidance derived from language and segmentation. We employ a language-conditioned diffusion model to colorize grayscale frames. Guidance is provided via automatically generated object masks and textual prompts; our primary automatic method uses a generic prompt, achieving state-of-the-art results without specific color input. Temporal stability is achieved by warping color information from previous frames using optical flow (RAFT); a correction step detects and fixes inconsistencies introduced by warping. Evaluations on standard benchmarks (DAVIS30, VIDEVO20) show our method achieves state-of-the-art performance in colorization accuracy (PSNR) and visual realism (Colorfulness, CDC), demonstrating the efficacy of automated prompt-based guidance for consistent video colorization.

Keywords: Video colorization · Temporal consistency · Automatic colorization

1 Introduction

Video colorization represents an important area of research with applications spanning historical footage restoration, media production, and artistic content creation. While significant advances have been made in this domain, existing techniques face several key challenges that limit their practical applications. Traditional colorization approaches rely heavily on user-provided color hints for grayscale frames, which can be both time-consuming to generate and often result in inconsistent colorization across frames. Furthermore, maintaining temporal consistency across colorized frames remains a significant hurdle, with independently colorized frames frequently resulting in abrupt changes and visually disruptive flickering effects.

In this paper, we explore a novel approach that leverages neural-language based methods for video colorization. Our key insight is that color information

E. Rodolà et al. (Eds.): ICIAP 2025, LNCS 16167, pp. 546–558, 2026.
https://doi.org/10.1007/978-3-032-10185-3_43

can be effectively guided through a combination of grayscale frames, automatically generated object masks, and textual prompts. Our primary automatic approach leverages a generic prompt, eliminating the need for specific color input, while we also explore the use of detailed descriptions for enhanced control.

Existing video colorization methods can be broadly categorized into reference-based approaches, which propagate colors from a reference frame throughout the video, and automatic colorization methods, which colorize each frame independently. Reference-based methods like FAVC [9] and TCVC [10] are effective when suitable reference frames are available but struggle with scene changes and with long-term consistency. Automatic methods such as those by Zhang et al. [21] and Iizuka et al. [4], coming mainly from the image colorization domain, provide frame-by-frame colorization but often produce temporally inconsistent results.

Our method addresses several key limitations of existing techniques:

- We eliminate the need for manual color annotation by leveraging language models for guidance, achieving state-of-the-art automatic colorization even with generic prompts.
- We achieve temporal consistency through a dual approach: using an optical flow-based method to propagate color information between frames while implementing a correction mechanism to address areas with significant color deviation.
- Extensive experiments on the DAVIS30 [12] and VIDEVO20 [8] datasets demonstrate that our approach achieves superior performance compared to existing colorization methods, both in terms of colorization quality and temporal consistency.

Our method is particularly beneficial for colorizing common footage where consistent and semantically accurate colors are desired without requiring extensive manual intervention. The textual prompt-based approach enables intuitive control over the colorization process, allowing for easy specification of desired color schemes through natural language.

The remainder of this paper is organized as follows: Sect. 2 reviews related work in video and image colorization. Section 3 details our proposed method, including the colorization pipeline, textual prompt generation, object mask creation, and temporal consistency maintenance. Section 4 presents our experimental results, including comparisons with state-of-the-art methods and ablation studies. Finally, Sect. 5 concludes the paper with our final remarks.

2 Related Works

In this section, we review relevant literature in both video and image colorization, with a particular focus on approaches that inform our proposed method.

2.1 Video Colorization

Video colorization methods are broadly reference-based or automatic. Reference-based methods [16,17] propagate color from keyframes but struggle with long-term consistency and scene changes. FAVC [9] constructs KNN graphs on ground-truth color videos with temporal losses for consistency. TCVC [10] employs bidirectional feature propagation to enhance consistency, though challenges with complex motion persist. Welsh et al. [17] exploit luminance features using a consistent colorized target image, identifying corresponding pixels in luminance space. Vondrick et al. [16] developed self-supervised tracking via video colorization, transferring colors from reference frames. Their pointing mechanism reduces inconsistencies but occasionally produces unclear color boundaries. Liu et al. [10]'s TCVC combines a colorization network with temporal coherence mechanisms, though still faces challenges with significant motion and occlusion. Automatic methods [20] typically extend image colorization techniques but often lack robust temporal handling.

2.2 Image Colorization

While our focus is on video colorization, our approach builds upon advancements in image colorization. Early learning-based approaches treat colorization as a prediction problem. Iizuka et al. [4] merged local patch information with global priors for complex scene colorization. Zhang et al. [21] formulated colorization as classification, predicting color distributions to capture the multi-modal nature of the problem. Su et al. [13] introduced an instance-aware model (InsColor) combining object detection and colorization, demonstrating that segmentation information improves quality, particularly for distinct objects. This insight informs our use of SAM [6] for creating object masks. Language-based image colorization has emerged as a powerful paradigm for user control. Manjunatha et al. [11] pioneered using textual descriptions to guide colorization. Recent works include L-CoDe by Weng et al. [19] using color-object decoupled conditions, Chang et al. [1] transformer-based L-CoDer, and Chang et al. [2] diffusion priors for language-based colorization. Transformer-based approaches include Kumar et al. [7] Colorization Transformer leveraging attention for long-range dependencies, Weng et al. [18] CT^2 using color tokens, and Ji et al. [5] ColorFormer employing hybrid-attention and color memory.

2.3 Temporal Consistency in Video Processing

Ensuring temporal consistency is a critical challenge in video colorization. Optical flow has been widely used to enforce consistency by warping frames or features. The RAFT architecture by Teed and Deng [15] provides accurate flow estimation for various video processing tasks. Methods like TCVC [10] propagate deep features between frames, proving more robust than direct color propagation, especially with occlusions and large motions. Lei and Chen [9] introduced self-regularization terms that penalize discrepancies between adjacent frames, reducing flickering without explicit correspondence matching.

Our work builds upon these advances by combining language-guided colorization with instance-aware segmentation and effective temporal consistency mechanisms. We automatically generate textual descriptions at key points, segment objects using SAM, and maintain temporal consistency through a combination of optical flow and targeted correction of problematic regions.

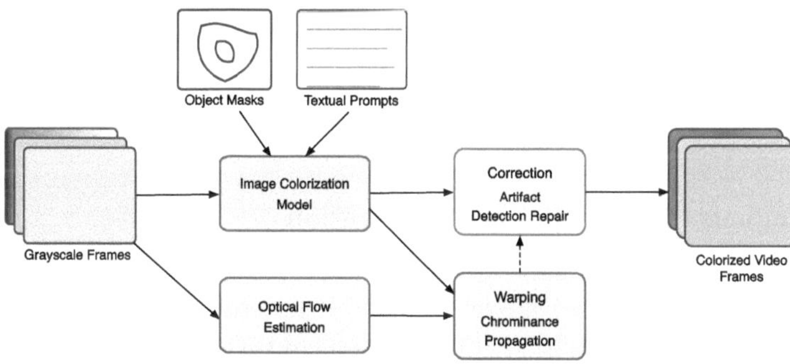

Fig. 1. Overview of the proposed framework. Our method takes grayscale video frames as input, along with automatically generated object masks from SAM and a textual prompt (generic for automatic mode, detailed for guided mode).

3 Method

Our proposed method achieves temporally consistent video colorization guided by textual prompts and object segmentation masks. The core idea is to colorize grayscale video frames using a language-conditioned image colorization model, while enforcing temporal coherence through optical flow-based color propagation and a targeted correction mechanism. The process operates primarily in the CIE LAB color space, leveraging the separation of luminance (L) and chrominance (A, B channels). An overview of the pipeline for processing frame $t+1$ given frame t is illustrated in Fig. 1.

3.1 Overview

Given an input grayscale video sequence $\{L_t\}$, where L_t is the luminance channel of the frame at time t, our goal is to generate a corresponding colorized sequence $\{I_{final}(t)\}$. For each frame L_t, we automatically generate a set of object masks M_t. We then utilize a textual prompt P_t to guide the colorization. For our main automatic approach evaluated against the state-of-the-art, we use a generic prompt (e.g., "a colorful image") to ensure no external color information biases the comparison. We also explore the use of detailed, content-specific prompts to demonstrate the framework's controllability.

These inputs (L_t, M_t, P_t) are fed into a language-guided image colorization network (L-CAD [2]) to produce an initial colorized frame $I_{L-CAD}(t)$. To ensure temporal consistency, we compute the optical flow $F_{t \to t+1}$ between L_t and L_{t+1} using RAFT [15]. We then warp the chrominance channels (A_t, B_t) of the previous final colorized frame $I_{final}(t)$ using $F_{t \to t+1}$ to obtain warped chrominance channels (A'_{t+1}, B'_{t+1}). These are combined with the input luminance L_{t+1} to form a preliminary warped frame $I_{warp}(t+1)$. A correction step identifies regions where warping failed or introduced significant artifacts by comparing $I_{warp}(t+1)$ with $I_{final}(t)$. These regions are filled using the colors from the L-CAD output for the current frame, $I_{L-CAD}(t+1)$, resulting in the final colorized frame $I_{final}(t+1)$.

3.2 Language-Guided Frame Colorization

We employ the L-CAD model [2] as our core per-frame colorization engine. L-CAD is a diffusion-based model capable of generating diverse and controllable colorization based on textual descriptions of varying detail levels. For each grayscale input frame L_t, L-CAD takes the following as input: i) the grayscale frame L_t; ii) a textual prompt P_t; iii) a set of automatically generated object segmentation masks M_t. The model leverages these inputs to predict the corresponding chrominance channels $(A_{L-CAD}(t), B_{L-CAD}(t))$. When provided with a *generic prompt* like "a colorful image", L-CAD infers plausible colors primarily based on the grayscale structure (L_t), the spatial guidance from masks (M_t), and its internal knowledge about typical object colors learned during pre-training. When provided with detailed prompts containing color information, L-CAD uses this explicit guidance to achieve more specific colorization. The result is the colorized frame $I_{L-CAD}(t)$ in the LAB space.

Effective guidance for L-CAD involves both spatial and semantic information.

For spatial information, we utilize the Segment Anything Model (SAM) [6] to automatically generate instance segmentation masks for each grayscale frame L_t. SAM produces highly accurate masks for various objects without requiring class labels. These masks M_t are provided as input to L-CAD to help delineate object boundaries and assign colors accurately within detected regions. While SAM can generate many masks, we currently utilize all generated masks in our experiments, which was found to be optimal.

Regarding semantic information, we use the fixed, generic prompt P_t = "a colorful image" for all frames. This ensures a fair comparison by providing no specific semantic or color guidance beyond the request for colorization itself. To demonstrate the potential for user control and evaluate the impact of dynamic guidance, we also utilize detailed prompts. These prompts describe scene content and specific colors (examples in Fig. 2a, b). Since generating such colorful prompts automatically from grayscale input is impossible unless you have already a model with color priors, these were generated using the GPT-3.5-turbo API applied to the original ground truth color frames, simulating an oracle or ideal user providing accurate descriptions. For the dynamic prompt ablation study, a new detailed prompt is generated under two conditions: i) at the beginning of a new scene (automatically detected on the grayscale video), and ii) at regular intervals of $\Delta t = 1$ second. Between generation points, the most recent detailed prompt P_t is used for subsequent frames ($P_{t+1} = P_t$).

(a)

(b)

Fig. 2. Textual prompt generated using ground truth color information to simulate ideal guidance from image a: *Cow brown and white in the center. Grass green in the background. Soil dark brown in the foreground. Fence dark brown at the bottom.* Textual prompt from image b: *Horse is brown and is on the left. Rider is wearing white and is on the horse. Fence is white and is in the center. Ground is beige and covers the bottom. Trees are green and are in the background. Sky is blue and is at the top. Clouds are white and are in the sky. Plants are green and are in the foreground.*

3.3 Temporal Consistency

Maintaining color stability across frames is crucial for avoiding flickering artifacts. We achieve this through a combination of optical flow-based propagation and targeted correction. We use the RAFT model [15] to estimate the dense optical flow field $F_{t \to t+1}$ between consecutive grayscale frames L_t and L_{t+1}. The chrominance channels (A_t, B_t) from the final colorized frame of the previous step, $I_{final}(t)$, are warped using the backward optical flow $F_{t \to t+1}$ to predict the chrominance for the current frame $t + 1$.

$$(A'_{t+1}(x, y), B'_{t+1}(x, y)) = (A_t(p), B_t(p)) \quad (1)$$

where $p = (x + F_{t \to t+1,x}(x, y), y + F_{t \to t+1,y}(x, y))$. Here, (x, y) are coordinates in frame $t + 1$, and p represents the corresponding coordinates in frame t according to the flow field. The warped chrominance (A'_{t+1}, B'_{t+1}) is combined with the input luminance L_{t+1} to form the warped frame $I_{warp}(t+1)$. This step effectively transfers stable colors from the previous frame based on motion.

An example of a disocclusion artifact introduced by warping is shown in Fig. 3.

Fig. 3. Example of color shifting from initial frame a to the 11th frame b.

If frame $t+1$ is detected as the start of a new scene, the warping step (Eq. 1) is skipped. In this case, the output from the L-CAD model for the current frame is used directly as the preliminary result: $I_{prelim}(t+1) = I_{L-CAD}(t+1)$. This prevents propagating colors across unrelated scenes.

Correction of Warping Artifacts. Optical flow estimation and warping are not always perfect, leading to artifacts, color shifting, or uncolored (grayscale) regions, especially due to occlusions or complex motion. We identify these problematic areas by comparing the warped frame $I_{warp}(t+1)$ with the previous final frame $I_{final}(t)$. A binary mask $M_{corr}(t+1)$ identifies pixels where the distortion is significant, typically measured by a low pixel-wise Peak Signal-to-Noise Ratio (PSNR) below a threshold τ:

$$M_{corr}(t+1)(x,y) = \begin{cases} 1 & \text{if } 20 \cdot \log_{10}\left(\frac{\max[I_{final}(t)]}{\|I_{warp}^{x,y}(t+1) - I_{final}^{x,y}(t)\|}\right) < \tau \\ 0 & \text{otherwise} \end{cases} \tag{2}$$

Pixels marked by $M_{corr}(t+1) = 1$ indicate regions requiring correction.

The final colorized frame $I_{final}(t+1)$ is composed by taking the warped colors $I_{warp}(t+1)$ in regions deemed reliable ($M_{corr} = 0$) and filling the corrupted regions ($M_{corr} = 1$) with the colors generated by the L-CAD model for the current frame, $I_{L-CAD}(t+1)$:

$$I_{final}(t+1) = (1 - M_{corr}(t+1)) \odot I_{warp}(t+1) + M_{corr}(t+1) \odot I_{L-CAD}(t+1) \tag{3}$$

where $\odot$ denotes element-wise multiplication. This selective correction ensures that we rely on propagated colors for temporal stability where possible, but fall back to the current frame's L-CAD prediction in problematic areas, maintaining both consistency and accuracy.

3.4 Implementation Details

Our framework integrates several pre-existing models. We use the L-CAD model [2] with its publicly available weights as the core image colorization engine.

Object masks are obtained using the default Segment Anything Model (SAM) [6]. For the main automatic method, the generic prompt "a colorful image" is used. For the guided method, detailed prompts were generated using the GPT-3.5-turbo API from ground truth color frames, with new prompts generated at scene changes and every $\Delta t = 1$ second. For optical flow, we employ the RAFT-Large model [15] with the weights provided by the PyTorch Vision library [14]. The threshold τ for detecting corrupted warping areas via PSNR (Eq. 2) is set to 25 dB. All experiments were conducted using PyTorch on NVIDIA RTX 4090 GPUs.

4 Experiments

In this section, we evaluate the performance of our proposed prompt-based video colorization method. We first present quantitative and qualitative comparisons against state-of-the-art methods and then provide ablation studies to validate the contributions of key components of our framework.

4.1 Datasets

We follow the experimental setup of [9,10] using the test sets of DAVIS30 [12] and Videvo20 [8] datasets. Originally created for video object segmentation, the DAVIS dataset features diverse scenes with significant object motion and appearance changes. We use the standard test split containing 30 videos. Videvo20 was introduced for evaluating temporal consistency in video processing tasks. We use the test split defined in [8], which consists of 20 videos, often characterized by smoother camera motion compared to DAVIS. As our method is training-free, we only utilize the test splits of these datasets for evaluation.

Following existing approaches, we compare our results against the ground truth color videos using Peak Signal-to-Noise Ratio (PSNR), Colorfulness [3] and Color Distribution Consistency (CDC) [9]. Colorfulness is a no-reference metric quantifying the perceived richness and intensity of colors in an image. We also report the ratio of the colorfulness score of our results to that of the ground truth ($\frac{\text{Colorfulness}}{\text{True Colorfulness}}$). A ratio closer to 1 indicates a similar level of color vibrancy as the original video. CDC is also a no-reference metric designed to evaluate temporal consistency by measuring the similarity of color distributions between adjacent frames. We report the ratio relative to the ground truth ($\frac{\text{CDC}}{\text{True CDC}}$). A ratio closer to 1 signifies better temporal stability, closer to the consistency level of the original video.

4.2 Comparison with the State of the Art

We compare our proposed method against several representative image and video colorization techniques. Table 1 presents the quantitative comparison on the DAVIS30 and VIDEVO20 test sets. We evaluate two versions of our approach: first, our main automatic approach using only the generic prompt "a colorful image" (*proposed method*). This represents a fair comparison against other automatic methods that do not use external color guidance. Second, a version that uses detailed, colorful prompts generated from ground truth (simulating ideal user guidance) to showcase the framework's potential when specific instructions are provided.

Table 1. Comparison of different methods on DAVIS30 and VIDEVO20. * Best value is closer to 1.

Method	DAVIS30			VIDEVO20		
	$\frac{\text{CDC}}{\text{True CDC}}$ *	PSNR↑	$\frac{\text{Colorfulness}}{\text{True Colorfulness}}$ *	$\frac{\text{CDC}}{\text{True CDC}}$ *	PSNR↑	$\frac{\text{Colorfulness}}{\text{True Colorfulness}}$ *
Iizuka et al. [4]	1.83	23.85	0.62	2.96	23.69	0.77
CIC [21]	1.97	23.19	0.94	2.21	22.51	1.08
IDC [22]	1.60	25.42	0.67	1.58	25.35	0.70
InsColor [13]	3.05	24.51	0.61	4.92	24.80	0.75
FAVC [9]	1.35	24.38	0.57	1.15	24.81	0.60
TCVC+ (b:IDC) [10]	**1.19**	25.50	0.67	**1.01**	25.43	0.70
Proposed + prompt	1.25	**29.63**	**0.94**	1.08	**30.26**	**1.05**
Proposed method	1.36	**28.63**	**0.93**	1.18	**29.54**	**1.02**

Our method ("Proposed method") achieves significantly higher PSNR scores than all competing methods on both datasets, indicating superior reconstruction fidelity to the ground truth colors, even without specific guidance. When provided with detailed guidance ("Proposed + prompt"), the PSNR improves further, highlighting the benefit of semantic control.

In terms of Colorfulness, both versions of our method produce results much closer to the ground truth vibrancy (ratio near 1.0) compared to most prior works (which often yield ratios <0.7), except for CIC which can oversaturate. Regarding temporal consistency (CDC), our main automatic approach ('Proposed method') achieves scores (1.36 on DAVIS, 1.18 on VIDEVO) reasonably close to the state-of-the-art TCVC+ (1.19 on DAVIS, 1.01 on VIDEVO), demonstrating that our flow-based propagation and correction strategy effectively maintains temporal smoothness even with generic guidance. Providing detailed prompts ('Proposed + prompt') further improves CDC (1.25 on DAVIS, 1.08 on VIDEVO), bringing it even closer to methods specifically optimized for consistency via bidirectional feature propagation.

5 Ablation Studies

Impact of Dynamic Prompts. To assess the benefit of updating guidance over time when detailed information is available, we compare our "Proposed + prompt" method (using detailed prompts generated from ground truth, updated at scene changes and every second) against a variant that uses only the detailed prompt generated for the first frame throughout the entire video. Table 2 shows the results on the DAVIS30 pairs dataset (a synthetic dataset created by concatenating random pairs from DAVIS30 to ensure each test sample contains at least one scene change).

Table 2. Comparison between the proposed method and its version with one fixed prompt. * Best value is closer to 1.

Method	DAVIS30 pairs		
	PSNR ↑	$\frac{\text{Colorfulness}}{\text{True Colorfulness}}$ *	$\frac{\text{CDC}}{\text{True CDC}}$ *
Without changing prompt	28.23	**0.98**	1.22
Changing prompt	**29.42**	0.97	**1.15**

Using dynamic detailed prompts significantly improves PSNR (29.42 vs. 28.23) and slightly improves temporal consistency (CDC ratio 1.15 vs. 1.22). While the fixed-prompt version shows a slightly better Colorfulness ratio, qualitative results, Fig. 4 highlight the importance of adapting guidance. A fixed detailed prompt can lead desaturated or incorrect colors as the scene content diverges from the initial description (Fig. 4b), or conversely, overly saturated and unnatural colors if the initial prompt locks onto specific colors that are no longer appropriate (Fig. 4c). Dynamically updated detailed prompts allow the colorization to adapt to evolving content, leading to more accurate and plausible results overall when such guidance is available.

(a) Changing prompt (b) w/o changing prompt - content diverged (c) w/o changing prompt - inappropriate colors

Fig. 4. Example of frame colorization with and without changing prompt every Δt and change of scene in DAVIS30 pairs.

Impact of Warping Correction. We evaluate the necessity of our warping correction mechanism by comparing the full method against a version where the warped frame $I_{warp}(t+1)$ is used directly as the final output, without correcting corrupted areas identified by Eq. 2. Table 3 shows the results. Without correction, PSNR and CDC scores are slightly better on average. However, the Colorfulness ratio drops dramatically (0.65 vs. 0.94 on DAVIS, 0.66 vs. 1.05

on Videvo), indicating significantly desaturated results. Qualitative examples in Fig. 5 clearly illustrate why correction is crucial. Without it, areas with occlusions, disocclusions, or complex motion where optical flow fails remain uncolored (appearing gray) or exhibit severe color shifting artifacts (Fig. 5b). The correction mechanism (Eq. 3) effectively fills these regions using the current frame's L-CAD prediction, restoring plausible colors and significantly improving the visual quality and vibrancy (Fig. 5a), even if the filled colors slightly deviate from the ground truth, minimally impacting average PSNR/CDC. This demonstrates that the correction step is essential for perceptual quality and avoiding distracting artifacts.

Table 3. Comparison between the proposed method and its version without correcting gray areas. * Best value is closer to 1.

Method	DAVIS30			VIDEVO20		
	$\frac{\text{CDC}}{\text{True CDC}}$ *	PSNR↑	$\frac{\text{Colorfulness}}{\text{True Colorfulness}}$ *	$\frac{\text{CDC}}{\text{True CDC}}$ *	PSNR↑	$\frac{\text{Colorfulness}}{\text{True Colorfulness}}$ *
Without warping correction	**1.25**	**29.64**	0.65	**0.98**	**30.62**	0.66
With warping correction	1.25	29.63	**0.94**	1.08	30.26	**1.05**

(a) With warping correction (b) Without warping correction

Fig. 5. Example of frame colorization with and without warping correction.

6 Conclusions

This paper introduced a novel framework for automatic video colorization that leverages language-based guidance and robust temporal consistency enforcement. By combining a language-conditioned diffusion model with automatically generated object masks and textual prompts, our method achieves detailed and plausible colorization. Temporal coherence is maintained through optical flow-based color propagation combined with a crucial correction step that addresses warping artifacts, ensuring smooth and visually pleasing results. We report state-of-the-art performance on standard benchmarks using only a generic prompt, significantly improving reconstruction fidelity and color vibrancy over previous automatic methods while maintaining competitive temporal consistency. Furthermore, we show that incorporating detailed, color-specific prompts (simulating user guidance) further enhances performance and allows for semantic control over the colorization process.

Acknowledgments. Work partially supported by the Italian Ministry of Education and Research (MUR) in the framework of the FoReLab project (Departments of Excellence).

References

1. Chang, Z., Weng, S., Zhang, P., Li, Y., Li, S., Shi, B.: L-coder: language-based colorization with color-object decoupling transformer (2022). arXiv preprint arXiv:2211.12120
2. Zheng, C., Shuchen, W., Peixuan, Z., Yu, L., Si, L., Boxin, S.: L-cad: language-based colorization with any-level descriptions using diffusion priors. In: Advances in Neural Information Processing Systems (NeurIPS) (2023)
3. Hasler, D., Süsstrunk, S.: Measuring colourfulness in natural images. In: IS&T/SPIE Electronic Imaging (2003)
4. Iizuka, S., Simo-Serra, E., Ishikawa, H.: Let there be color! joint end-to-end learning of global and local image priors for automatic image colorization with simultaneous classification. ACM Trans. Graph. (TOG) **35**(4), 110:1–110:11 (2016)
5. Ji, X., Jiang, B., Luo, D., Tao, G., Chu, W., Xie, Z., Wang, C.: Colorformer: image colorization via color memory assisted hybrid-attention transformer. Pattern Recogn. **132**, 108923 (2022)
6. Kirillov, A., Mintun, E., Ravi, N., Mao, H., Rolland, C., Gustafson, L., Xiao, T., Whitehead, S., Berg, A.C., Lo, W.-Y., et al.: Segment anything. In: 2023 IEEE/CVF International Conference on Computer Vision (ICCV), pp. 3992–4003. IEEE (2023)
7. Kumar, M., Weissenborn, D., Kalchbrenner, N.: Colorization transformer. In: International Conference on Learning Representations (ICLR) (2021)
8. Lai, W.-S., Huang, J.-B., Wang, O., Shechtman, E., Yumer, E., Yang, M.-H.: Learning blind video temporal consistency. In: Proceedings of the European Conference on Computer Vision (ECCV), pp. 170–185 (2018)
9. Lei, C., Che, Q.: Fully automatic video colorization with self-regularization and diversity. In: IEEE/CVF Conference on Computer Vision and Pattern Recognition (CVPR) (2019)
10. Liu, Y., Zhao, H., Chan, K.C.K., Wang, X., Loy, C.C., Qiao, Y., Don, C.:: Temporally consistent video colorization with deep feature propagation and self-regularization learning. In: IEEE Transactions on Pattern Analysis and Machine Intelligence (TPAMI) (2021)
11. Manjunatha, V., Iyyer, M., Boyd-Graber, J., Davis, L.: Learning to color from language. In: Conference of the North American Chapter of the Association for Computational Linguistics: Human Language Technologies (NAACL-HLT), pp. 764–769 (2018)

12. Perazzi, F., Pont-Tuset, J., McWilliams, B., Van Gool, L., Gross, M., Sorkine-Hornung, A.: A benchmark dataset and evaluation methodology for video object segmentation. In: IEEE Conference on Computer Vision and Pattern Recognition (CVPR) (2016)
13. Su, J.-W., Chu, H.-K., Huan, J.-B.: Instance-aware image colorization. In: IEEE/CVF Conference on Computer Vision and Pattern Recognition (CVPR), pp. 7968–7977 (2020)
14. PyTorch Team. Raft: Optical flow implementation in Pytorch, 2020. PyTorch Vision Library
15. Teed, Z., Deng, J.: Raft: recurrent all-pairs field transforms for optical flow. In: European Conference on Computer Vision (ECCV) (2020)
16. Vondrick, C., Shrivastava, A., Fathi, A., Guadarrama, S., Murphy, K.: Tracking emerges by colorizing videos. In: Proceedings of the European Conference on Computer Vision (ECCV), pp. 391–408 (2018)
17. Welsh, T., Ashikhmin, M., Mueller, K.: Transferring color to greyscale images. In: Proceedings of the 29th Annual Conference on Computer Graphics and Interactive Techniques, pp. 277–280 (2002)
18. Weng, S., Sun, J., Li, Y., Li, S., Shi, B.: Ct2: colorization transformer via color tokens. In: European Conference on Computer Vision (ECCV) (2022)
19. Weng, S., Wu, H., Chang, Z., Tang, J., Li, S., Shi, B.: L-code: language-based colorization using color-object decoupled conditions. In: IEEE/CVF Conference on Computer Vision and Pattern Recognition (CVPR), pp. 13061–13070 (2022)
20. Zhang, B., He, M., Liao, J., Sander, P.V., Yuan, L., Bermak, A., Chen, D.: Deep exemplar-based video colorization. In: Proceedings of the IEEE/CVF Conference on Computer Vision and Pattern Recognition, pp. 8052–8061 (2019)
21. Zhang, R., Isola, P., Efros, A.A.: Colorful image colorization. In: European Conference on Computer Vision (ECCV), pp. 649–666 (2016)
22. Zhang, R., Zhu, J.-Y., Isola, P., Geng, X., Lin, A.S., Yu, T., Efros, A.A.: Real-time user-guided image colorization with learned deep priors. ACM Trans. Graph. (TOG) **36**(4) (2017)

An Investigation on Incremental Learning from Unbalanced Streamed Data

Guido Borghi[1(✉)], Gabriele Graffieti[2], and Roberto Vezzani[1]

[1] University of Modena and Reggio Emilia, Modena, Italy
{guido.borghi,roberto.vezzani}@unimore.it
[2] University of Bologna, Cesena, Italy
gabriele.graffieti@unibo.it

Abstract. In real-world contexts, data can be characterized by a streaming nature, unbalanced distribution, data drift over a long time frame, and strong correlation of samples in short time ranges. Moreover, a clear separation between the traditional training and deployment phases is usually lacking. This data organization and fruition represents a challenging scenario for incremental learning agents, *i.e.* agents that have the ability to improve their knowledge through past experiences incrementally. In this paper, we investigate the combined impact of temporal similarity and unbalanced data distribution on the classification performance of various algorithms from different research fields, including Continual and Online Learning. Starting from three datasets, we create and release different benchmarks that replicate the investigated setting to analyze the performance of each tested method.

1 Introduction

The ability to incrementally learn from continuous data sequences, in which temporally close samples can be very similar and, at the same time, the same object or scene can abruptly vary in long-time acquisitions, is a desirable skill for real-world AI agents [11,33]. These agents are systems that can continuously learn from their experience, operating in real-world contexts [4,10,28]. This is the case, for instance, of a domestic robot that acquires the environment and classifies surrounding objects that may change their visual appearance due to different light sources. Another example is a vision-based system for Advanced Driver-Assistance Systems that classifies vehicles to support the driving activity: the data may vary depending on the road type, the weather, or the time of day.

To maintain or even improve a reliable classification capability, these learning agents should be able to quickly adapt their knowledge to the surrounding world, which continuously evolves and visually changes. In this context, the learning process greatly differs from the classic Machine Learning (ML) setting [35], in which the agent's life is usually divided into two sequential, fixed, and separated phases: training and deployment. In a real-world context, these two phases are not realistic since all training data is not available at the beginning of the learning phase, and there is no single deployment procedure.

E. Rodolà et al. (Eds.): ICIAP 2025, LNCS 16167, pp. 559–570, 2026.
https://doi.org/10.1007/978-3-032-10185-3_44

On the contrary, the agent must be able to learn from new observations, continually updating its knowledge while maintaining the knowledge acquired in the past, without a definite boundary between training and deployment.

Indeed, from a formal point of view, the learning paradigm described above shares some elements with the Continual [28] and Online [29] learning paradigms. In particular, it is based on the following peculiar elements:

1. Input data is available as a continuous temporal stream. Formally, in a certain time instant t, only a single data point x_t is available as input. Then, given a time range $[t, t+k]$, multiple sequential data samples are available for input $\{x_t, x_t+1, \ldots, x_t+k\}$. This streaming nature of training data makes it harder to have a fixed training set that fully represents the current and future distribution of data. Therefore, the learning phase must be continuous and distributed over time [14].
2. The distribution of the input data is influenced by the realistic fruition and interaction with the environment by the learning agent. Therefore, data can be characterized by a high visual similarity in a short time range (*e.g.* the domestic robot is handling a particular object), which can abruptly change in long-time ranges (*e.g.* the light source has changed) [10]. The natural data acquisition also influences the class distribution: likely, objects are not uniformly available in input data; then, input data can be unbalanced. These important aspects are further discussed and analyzed in Sect. 2.1.

Finally, it is desirable that these learning agents, especially if implemented in embedded systems, *i.e.* systems with some hardware limitations [38], such as computational power, memory size, power consumption, and similar, should also have the following characteristics [13,14]:

- **Fast update**: new knowledge, *i.e.* new available input data $\{x_t, x_t + 1, \ldots, x_t + k\}$, should be used immediately. For this reason, it is advisable to accumulate only a relatively small amount of input data samples before triggering the learning process of the agent. We observe that the definition of the proper amount of data is tricky, since the computational load, the training time, and the required memory are directly related.
- **Single-epoch learning**: a learning agent sees every data point x_t only once. The learning procedure should be fast, then not based on multi-epoch learning iterations, a common setting in the ML scenario [35].
- **Limited storage memory**: a learning agent must limit the memory usage, in terms of the number of input samples stored in the replay memory [30], even though studies [10,28] have shown that the replay memory efficiently contrasts the catastrophic forgetting [27].

Therefore, in this paper, we aim to investigate the impact of streamed unbalanced data on several methods available in the literature which have been created to handle a continuous learning procedure [28] and rely on streaming input data [17,29]. We perform experimental evaluations using three existing datasets, namely Core50 [24], OpenLoris [32] and CLAD [36] that we adapted to produce unbalanced input streams and temporal similarity or dissimilarity in specific time ranges.

A similar investigation to our work is reported in [15] in which the focus is on embedded devices, with limited memory and computational capacity. Experiments are based on efficient models, such as EfficientNet [34] and MobileNet [31], but also common networks like ResNet-18 [16], and three different datasets: OpenLoris [32], Places-365 [25] and Places-Long-Tail [23]. However, it is worth noting that the combined impact of temporal similarity and unbalanced data distribution (on the whole datasets, at the same time) has not been evaluated.

2 Proposed Benchmarks

We introduce two concepts that we are going to use in the rest of the paper:

- **Experience**: an amount of data that the learning agent can accumulate in its temporary buffer before triggering the model update, *i.e.* the starting of the learning procedure. In all considered benchmarks, data belonging to different experiences are not shared, and, in particular, data in past experiences are not available, with the exception of those data stored in the replay memory.
- **Macro-experience**: a set of single experiences acquired in similar conditions, therefore data from the same macro-experience presents similar characteristics, while data from different macro-experiences may greatly differ from each other. In our benchmarks, it is important to note that the learning agent is agnostic with respect to the boundaries across different macro-experiences. In other words, macro-experiences are a virtual division of input data, introduced only for testing purposes (*i.e.* testing the model in specific moments during the training process to evaluate the performance).

2.1 Data Organization

Temporal similarity and unbalanced data distribution strongly affect the composition and the organization of the input streaming data as follows:

- **Temporal Similarity**: due to the streaming nature of data, input samples can be highly temporally correlated in short time ranges. From a formal point of view:

$$|F(x_t) - F(x_{t+k})| \propto k \tag{1}$$

  where $F()$ is a feature extractor (for instance, a network trained on the Imagenet [8] dataset), x_t is a particular data sample available in the range $[t, \ldots, t+k, \ldots t+n]$, and then k is a specific instant in the time range. In other words, short-range consecutive input samples may be characterized by very similar visual appearance, while data drifts can happen in long-time ranges. Therefore, a sequence of learning experiences, i.e. a training procedure conducted on a certain amount of input data, can be aggregated in a single macro-experience as defined above. In case of data drift, a switch from one macro-experience to another one happens: in this case, the conditions can completely change, but the class knowledge needs to be properly adjusted without forgetting [19] since old conditions are likely to return in the future.

- **Unbalanced Data Distribution**: the input sequence is likely to contain non-iid, or unbalanced, data, *i.e.* for each class, a different amount of data is available in a specific time range. It is important to observe that unbalanced data can characterize not only a single experience, but also a macro-experience or the whole dataset. It is not guaranteed that the learning agent receives in input a balanced amount of data belonging to each class, and this is a challenging element for a proper and effective learning procedure [3,18].

2.2 Data Sources

To create benchmarks, we were forced to select only datasets with temporal sequences of data, *i.e.* images are not randomly sampled or acquired in different contexts, as in the majority of datasets commonly used in Continual Learning (*e.g.* Cifar-10/100 [20]), but are organized in temporally ordered video frames.

In the following, we detail the dataset selection in addition to the procedure that we apply to adapt the datasets:

OpenLoris [32]: this robotic vision dataset consists of several videos collected with an RGB-D camera mounted on different mobile robots moving in real environments. This dataset presents realistic challenges that a robot usually faces, such as illumination changes, the presence of occlusions, variations in object size and camera-object distance, and cluttering. These challenges are quantized at different levels. 40 different classes are available, we maintain the original training and testing splits.

We create an unbalanced version of this dataset relying on the *Zipf*'s law [21]. Indeed, as suggested in recent literature [5], Zipf's law is a good approximation of how we naturally interact with objects in the environment. In more detail, we interact for the majority of the time with a few common objects, while we interact sporadically with all the others. We apply Zipf's law on the whole dataset, obtaining a long-tail data distribution. From this general distribution of data across the whole dataset, we produce 9 different macro-experiences. The data distribution of each macro-experience is different from each other, and some classes may not be present in some macro-experiences.

Core50 UB [24]: this dataset collects videos of hand held items acquired in 11 sessions (8 indoor and 3 outdoor) with different backgrounds. A total of 50 objects that belong to 10 different categories are included. To increase the level of complexity, in our experiments we consider categories as classes: in this way, in the same class, there are objects even very different from a visual point of view (*e.g.* different models and types of light bulbs belong to the same category). In addition, objects are partially occluded by the grasping hand. We split the dataset as reported in the original paper, *i.e.* we put 8 sessions for training and 3 sessions for testing. In this setting, each session corresponds to a different macro-experience.

We create a new version of this dataset following the procedure outlined in [10]: in each macro-experience, not all classes are always available; in addition, only a certain proportion of the original samples for each class is available. The input stream consists of video sequences that can be interleaved at the frame level with other sequences: then, the temporal coherence is quite coarse, since in a given time range, only frames belonging to the same object are temporally connected. We observe that in this case, differently from the OpenLoris dataset, the same data distribution is used for each macro-experience.

CLAD-C [36]: is a classification benchmark based on SODA10M [12], a dataset acquired in the automotive setting. Indeed, there are 6 classes (pedestrian, bicycle, car, truck, tram, and tricycle) acquired across 32 Asiatic cities. Objects of interest are provided in manually annotated bounding boxes, for a total of 22 object-centered. In SODA10M, different macro-experiences correspond to scenes acquired in different hours (*e.g.* day/night), weather conditions (*e.g.* sun, rain), and different road types (*e.g.* highway, country roads). The main challenges of this dataset are related to occlusions (images are acquired through a camera located in the frontal part of the car, in a low position), and the variability of the distance between objects and the camera which can negatively affect the spatial resolution of objects. In addition, some classes (*e.g.* "car") represent the large majority of the data distribution, while other classes (*e.g.* "tricycle") is largely underrepresented. The input stream consists of the sequence of the bounding boxes localized in images: since a single frame can have one or more bounding boxes, the resulting streaming has a huge variability in terms of the number of classes available in a certain time range. We observe that this stream setting is quite different from the previous ones, and thus represents an interesting test for the investigated algorithms.

For OpenLoris and Core50 datasets we also create a balanced version, *i.e.* we maintain the same amount of data of the unbalanced versions (lower than the original ones) but we balance the number and the composition of each class. In this manner, it is possible to train a method with the same amount of training data, varying only the data distribution.

2.3 Experimental Setup

We organize the evaluation in two different steps, in order to investigate the impact of the temporal similarity and unbalanced data distribution (see Sect. 2.1). Specifically, the steps are detailed as follows:

1. **Temporal Similarity vs Shuffling**: in this experiment, we aim to assess the effects of temporal similarity on input streaming data. In particular, we compare methods running the learning phase on the balanced versions of Core50 and OpenLoris datasets with and without shuffling data before training. In

that way, we aim to highlight and isolate the impact of temporal similarity on the final performance of classifiers.

2. **Balanced vs Unbalanced**: once the impact of temporal similarity has been investigated, this second experiment is designed to assess the impact of the unbalanced distribution of input data already affected by temporal similarity. In this way, the input stream includes both temporal similarity and unbalanced data distribution. As described, it is important to note that for a fair comparison, the balanced versions of each dataset contain the same amount of data as the unbalanced ones.

3 Experiments

3.1 Metric

To assess the performance of the tested methods, we rely on a specific metric, *i.e.* the Average Mean Class Accuracy (AMCA) [10] defined as:

$$AMCA = \frac{1}{N_c \cdot N_{me}} \sum_{c=1}^{N_c} \sum_{e=1}^{N_{me}} \frac{corr_e^c}{tot_e^c} \quad (2)$$

where the number of classes and macro-experience is N_c and N_{me}, respectively. $corr_e^c$ and tot_e^c are the number of correctly classified test images of class c on macro-experience e, and the total number of test images of class c in macro-experience e. AMCA is useful to investigate the capability of a learning agent to incrementally learn, *i.e.* to improve the classification ability over time, across different macro-experiences.

3.2 Investigated Methods

Many of the investigated methods are based on a combination of Deep Learning architectures and approaches that belong to different research fields, *i.e.* Streaming [17], Online [29] and Continual Learning [28].

For each investigated method, in the experimental evaluation, we use the implementation of the authors if available and the training parameters reported in the original paper.

ExStream [13] method essentially consists of a strategy to cluster buffers of prototypes of each class. Then, the last fully connected layer of a CNN is trained using all the samples stored in the buffers, following a batch-based learning procedure similar to the common Machine Learning paradigm.

In [14], the authors revisit the streaming linear discriminant analysis to handle streaming input data, with a method here referred to as **SLDA**. In particular, the authors propose to compute for each class a mean vector and a covariance matrix that is updated or used fixed after the base initialization. The covariance matrix is then used to compute the final prediction. Features are extracted with a pre-trained and frozen feature extractor [16]. We observe that the covariance matrix is sensitive to the class order, but it is less sensitive to catastrophic forgetting since its running means are independent for each class. Moreover, this method is not based on the common backpropagation-based training procedure, avoiding issues related to the batch size, their composition, and the choice of hyperparameters (optimizer, learning rate, and so on).

Differently from the previous works, in [7] authors investigate different strategies to populate the replay memory in the Continual Learning learning paradigm, without any assumptions about the input data stream (in particular, no i.i.d. data assumptions), to train a conventional neural network [16]. In particular, the authors propose to select samples that are representative of all previously seen data for the replay buffer without the knowledge about task boundaries that correspond, in our benchmarks, to the macro-experiences. Authors investigate four different memory population strategies: i) Class-Balancing Reservoir Sampling (**CBRS**): this strategy aims to preserve the distribution of each class of the training dataset, altering the input stream distribution so that unbalanced data distribution is mitigated. Specifically, as long as the memory is not filled, all the input instances are stored, otherwise, instances are stored relying on the status of the specific incoming class (largest, full, and filled); ii) Gradient-based Sample Selection [2] (**GSS**): this method greedily maximizes the variance of the samples contained in the replay memory, analyzing the gradient directions. A similar analysis is conducted between the input instance and some randomly sampled stored instances. We observe that the main drawback of this approach is the high computational load since several feed-forward operations are required for each new input and related comparisons. iii) **Reservoir** [37]: the memory population strategy is divided into two steps: in the first, similar to the CBRS approach, all input instances are stored until the memory is filled; in the second one, the current input instance is stored with a probability of $\frac{m}{n}$, with m as memory size and n number of instances encountered so far. iv) **Random**: the current input instance is stored with 50% of probability in a randomly picked store location.

In our investigation, we also test the performance of the well-known Learning without Forgetting [22] (**LwF**) method, a Continual Learning algorithm expressively designed to contrast the catastrophic forgetting. In particular, LwF proposes the use of the knowledge distillation [9], consisting of two models, one of the current experience and one of the past one: the old model is frozen and used to distil knowledge into the current model.

Finally, we include in our analysis also the work proposed in [10], developed for the SSLAD competition [1] (and then referred here as **SSLAD**), developed for the Streaming Learning, that uses the CopyWeights with Re-init [26] (CWR) algorithm to contrast the catastrophic forgetting [27] and a double-memory to tackle the unbalanced data distribution in input, exploiting a reservoir mechanism in the first level of the memory. Unfortunately, the proposed method relies on the knowledge of the macro-experience boundaries, contrasting the assumptions described above.

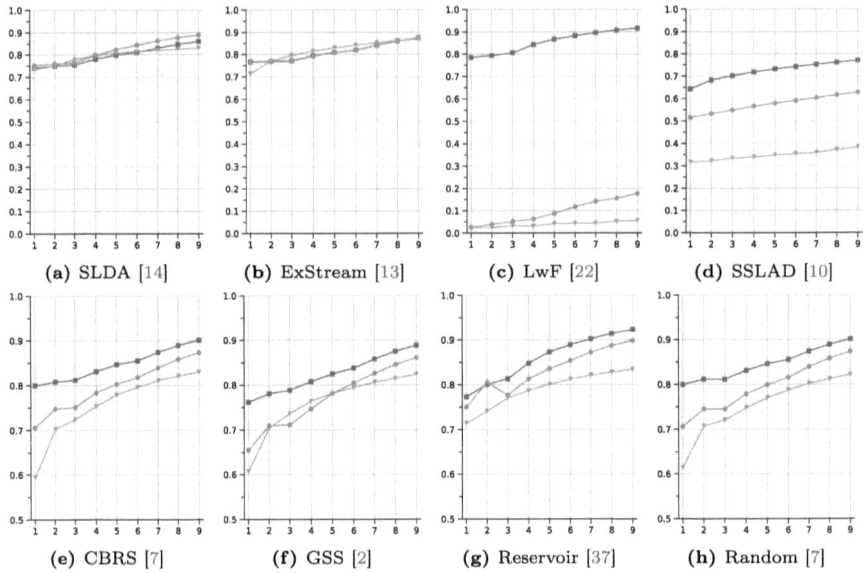

Fig. 1. Results in terms of AMCA obtained on the OpenLoris [32] dataset. The blue line denotes performance with balanced and shuffled data, the orange one the balanced data with temporal similarity (no shuffling operation introduced), while the green line shows the performance in the presence of both unbalanced data distribution and temporal similarity (see Sect. 2.1). (Color figure online)

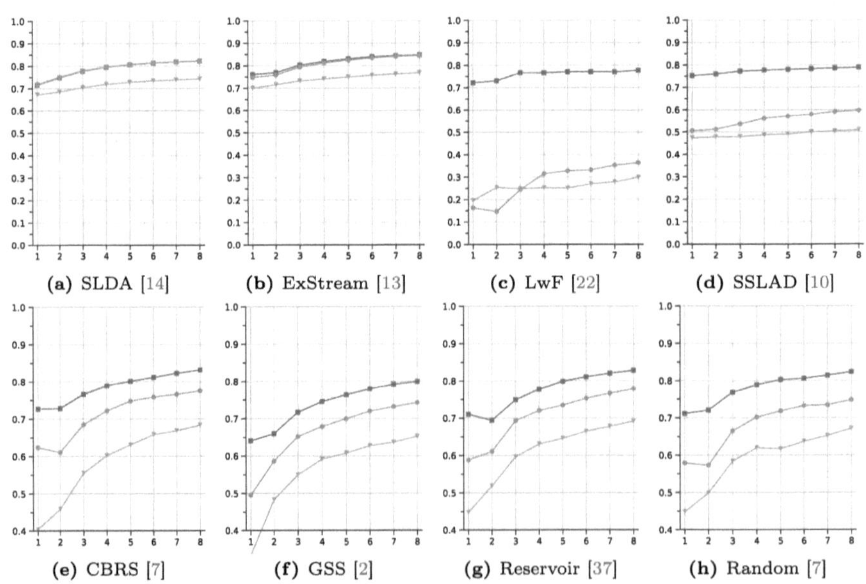

Fig. 2. Results obtained on the Core50 [24] dataset. Further details about the plot lines are reported in Fig. 1. (Color figure online)

3.3 Experimental Results

In all experiments, the experience size, *i.e.* the amount of data that can be accumulated before triggering the learning process (see Sect. 2), is set to 10, as also used in [7,13]. When possible, we force learning agents to iterate on training data for only one epoch. All methods receive input images with a 128×128 spatial resolution. Finally, it is important to note that macro-experience boundaries are not provided during the training, but they are used only to compute AMCA.

Results obtained on the OpenLoris and Core50 datasets are reported in Fig. 1 and Fig. 2, respectively.

In both, the blue lines show the performance of each method with the balanced version of the two datasets as input. In this case, the whole set of training data is shuffled so the stream of input data is not temporally continuous, *i.e.* is not a video stream but a stream of images without any temporal coherence. This setting is close to the common Machine Learning scenario, in which usually (but not mandatory) input data is balanced across available classes and the shuffling operation is performed before each epoch [6]. As expected, in this scenario a large part of the investigated methods achieve the best accuracy across other evaluations: the models continuously learn across different macro-experiences achieving a high final AMCA (~90% with OpenLoris and ~80% with Core50), with an improvement with respect to the first macro-experience of about 15% with both datasets.

The orange line represents the performance of the evaluated strategies receiving as input the balanced version of the datasets, but without any shuffling operation: this aims to highlight the impact of the time similarity. From a general point of view, all the models obtain worse results w.r.t. the previous experiment, with a worsening of about ~10% on OpenLoris and ~20% on Core50. SLDA and ExStream represent an exception since we observe that their performance is not affected by temporal similarity. In particular, SLDA is based on a pre-trained and blocked feature extractor, and only a covariance matrix is updated, thus limiting the overfitting phenomena that can occur during the usual training of deep learning architectures, in which all the weights of the model are updated. Besides, we observe that in ExStream the model is fine-tuned using the instances stored in several replay buffers (one for each class), limited in size, and shuffled before each iteration. Unfortunately, ExStream tends to extend the training time as buffers are filled since the model can do several iterations (epochs) on buffered data. Moreover, we note that LwF strongly suffers temporal similarity in input data, confirming that a Continual Learning model without any mechanism (such as but not only a replay memory) to deal with unbalanced streamed data, performs poorly in this specific scenario, with a worsening of about −70% on OpenLoris and about −40% on Core50.

Finally, the green line shows the behaviour of the investigated methods in the presence of unbalanced data distributions and temporal similarity. On Core50 dataset, we observe that the performance of all the methods tested degraded significantly, including SLDA and ExStream, while on OpenLoris the performances remain more similar, probably due to the fact that in the OpenLoris dataset the test frames are similar to the training ones (they are sampled from the same video sequence), and then the classification task is easier. This consideration is also supported by the greater performance generally achieved on the OpenLoris dataset. In ExStream, the use of different buffers is capable of partially mitigating the problem of unbalanced data, especially when buffers are filled, but contrasts with the single-epoch learning paradigm (see Sect. 1). It is also worth noting that even CBRS, which is specifically designed to contrast data imbalance, performs way poorer in this setting than in the previous ones, confirming the difficulties posed by the investigated setting.

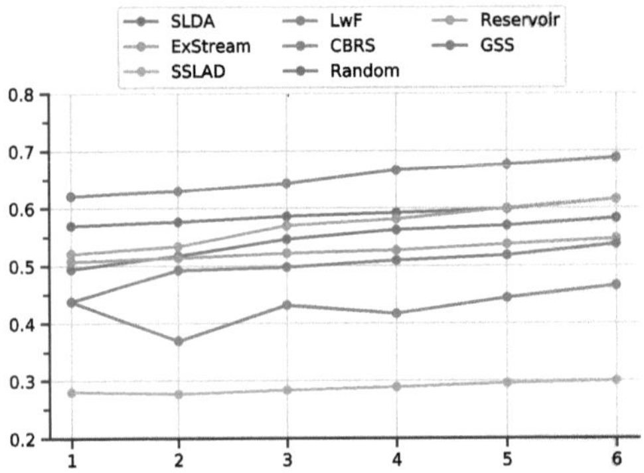

Fig. 3. Results obtained on the CLAD-C [36] benchmark.

Finally, we run the methods on the CLAD-C benchmark, collecting the results reported in Fig. 3. Probably CLAD-C represents the most challenging test set since it combines difficulties for the classification task related to a real-world acquisition (*e.g.* low quality images, extreme object variability, persistent occlusions) and unbalanced data distribution. Indeed, as shown, tested methods are able to learn incrementally with only a limited margin. Secondly, CBRS method overcomes the SLDA approach, probably due to the presence of an unblocked neural network and replay memory: on one hand, this suggests the importance to have past samples and a not-freeze neural network to improve the classification accuracy, while it is also important to note that, on the other hand, a blocked feature extractor prevents overfitting phenomena.

4 Conclusion

In this paper, we have proposed an investigation on the impact of unbalanced streamed data on training real-world learning agents that interact with the surrounding world and aim to continuously learn from the experience. Experimental results obtained comparing several methods on three different datasets for the classification task confirm that unbalanced data with temporal similarity (or dissimilarity) represent an interesting research field and a challenging benchmark that still needs to be further investigated in future works. To this aim, we have formally defined the data organization and the benchmarks that can be exploited in future research. Finally, we observe the need to introduce a proper dataset specifically acquired following the above-mentioned considerations on unbalanced streamed data, to improve the realism and the efficacy of the testing set and to address different vision-based tasks.

References

1. ICCV SSLAD competition. https://sslad2021.github.io/index.html. Accessed 1 Dec. 2022
2. Aljundi, R., Lin, M., Goujaud, B., Bengio, Y.: Gradient based sample selection for online continual learning. Adv. Neural Inform. Process. Syst. **32** (2019)
3. Batista, G.E., Prati, R.C., Monard, M.C.: A study of the behavior of several methods for balancing machine learning training data. ACM SIGKDD Explor. Newsl **6**(1), 20–29 (2004)
4. Borghi, G., Graffieti, G., Franco, A., Maltoni, D.: Incremental training of face morphing detectors. In: 2022 26th International Conference on Pattern Recognition (ICPR), pp. 914–921. IEEE (2022)
5. Chan, S.C.Y., Lampinen, A.K., Richemond, P.H., Hill, F.: Zipfian environments for reinforcement learning (2022). https://doi.org/10.48550/ARXIV.2203.08222, https://arxiv.org/abs/2203.08222
6. Chicco, D.: Ten quick tips for machine learning in computational biology. BioData Min. **10**(1), 1–17 (2017)
7. Chrysakis, A., Moens, M.F.: Online continual learning from imbalanced data. In: International Conference on Machine Learning. PMLR (2020)
8. Deng, J., Dong, W., Socher, R., Li, L.J., Li, K., Fei-Fei, L.: Imagenet: a large-scale hierarchical image database. In: 2009 IEEE Conference on Computer Vision and Pattern Recognition, pp. 248–255. IEEE (2009)
9. Gou, J., Yu, B., Maybank, S.J., Tao, D.: Knowledge distillation: a survey. Int. J. Comput. Vision **129**(6), 1789–1819 (2021)
10. Graffieti, G., Borghi, G., Maltoni, D.: Continual learning in real-life applications. IEEE Robot. Autom. Lett. **7**(3), 6195–6202 (2022)
11. Gunasekara, N., Pfahringer, B., Gomes, H.M., Bifet, A.: Survey on online streaming continual learning. In: Proceedings of the Thirty-Second International Joint Conference on Artificial Intelligence, pp. 6628–6637 (2023)
12. Han, J., Liang, X., Xu, H., Chen, K., Hong, L., Ye, C., Zhang, W., Li, Z., Liang, X., Xu, C.: Soda10m: towards large-scale object detection benchmark for autonomous driving. arXiv e-prints pp. arXiv–2106 (2021)
13. Hayes, T.L., Cahill, N.D., Kanan, C.: Memory efficient experience replay for streaming learning. In: 2019 International Conference on Robotics and Automation (ICRA). IEEE (2019)
14. Hayes, T.L., Kanan, C.: Lifelong machine learning with deep streaming linear discriminant analysis. In: Proceedings of the IEEE/CVF Conference on Computer Vision and Pattern Recognition Workshops (2020)
15. Hayes, T.L., Kanan, C.: Online continual learning for embedded devices (2022). arXiv preprint arXiv:2203.10681
16. He, K., Zhang, X., Ren, S., Sun, J.: Deep residual learning for image recognition. In: Proceedings of the IEEE Conference on Computer Vision and Pattern Recognition, pp. 770–778 (2016)
17. Hoens, T.R., Polikar, R., Chawla, N.V.: Learning from streaming data with concept drift and imbalance: an overview. Progr. Artif. Intell. **1**(1), 89–101 (2012)
18. Japkowicz, N., et al.: Learning from imbalanced data sets: a comparison of various strategies. In: AAAI Workshop on Learning from Imbalanced Data Sets, vol. 68, pp. 10–15. AAAI Press Menlo Park, CA (2000)
19. Kirkpatrick, J., Pascanu, R., Rabinowitz, N., Veness, J., Desjardins, G., Rusu, A.A., Milan, K., Quan, J., Ramalho, T., Grabska-Barwinska, A., et al.: Overcoming

catastrophic forgetting in neural networks. Proc. Natl. Acad. Sci. **114**(13), 3521–3526 (2017)
20. Krizhevsky, A., Hinton, G., et al.: Learning multiple layers of features from tiny images (2009)
21. Li, W.: Zipf's law everywhere. Glottometrics **5**(2002), 14–21 (2002)
22. Li, Z., Hoiem, D.: Learning without forgetting. IEEE Trans. Pattern Anal. Mach. Intell. **40**(12), 2935–2947 (2017)
23. Liu, Z., Miao, Z., Zhan, X., Wang, J., Gong, B., Yu, S.X.: Large-scale long-tailed recognition in an open world. In: Proceedings of the IEEE/CVF Conference on Computer Vision and Pattern Recognition, pp. 2537–2546 (2019)
24. Lomonaco, V., Maltoni, D.: Core50: a new dataset and benchmark for continuous object recognition. In: Conference on Robot Learning
25. López-Cifuentes, A., Escudero-Viñolo, M., Bescós, J., García-Martín, Á.: Semantic-aware scene recognition. Pattern Recogn. **102**, 107256 (2020)
26. Maltoni, D., Lomonaco, V.: Continuous learning in single-incremental-task scenarios. Neural Netw. **116**, 56–73 (2019)
27. McCloskey, M., Cohen, N.J.: Catastrophic interference in connectionist networks: The sequential learning problem. In: Psychology of Learning and Motivation, vol. 24, pp. 109–165. Elsevier (1989)
28. Parisi, G.I., Kemker, R., Part, J.L., Kanan, C., Wermter, S.: Continual lifelong learning with neural networks: a review. Neural Netw. **113** (2019)
29. Perry, E.H., Pilati, M.L.: Online learning. New Dir. Teach. Learn. **128** (2011)
30. Rolnick, D., Ahuja, A., Schwarz, J., Lillicrap, T., Wayne, G.: Experience replay for continual learning. In: Advances in Neural Information Processing Systems
31. Sandler, M., Howard, A., Zhu, M., Zhmoginov, A., Chen, L.C.: Mobilenetv2: inverted residuals and linear bottlenecks. In: Proceedings of the IEEE Conference on Computer Vision and Pattern Recognition (2018)
32. She, Q., Feng, F., Hao, X., Yang, Q., Lan, C., Lomonaco, V., Shi, X., Wang, Z., Guo, Y., Zhang, Y., Qiao, F., Chan, R.H.M.: OpenLORIS-Object: a robotic vision dataset and benchmark for lifelong deep learning. In: 2020 International Conference on Robotics and Automation (ICRA), pp. 4767–4773 (2020)
33. Souza, V.M.A., dos Reis, D.M., Maletzke, A.G., Batista, G.E.A.P.A.: Challenges in benchmarking stream learning algorithms with real-world data. Data Min. Knowl. Disc. **34**(6), 1805–1858 (2020). https://doi.org/10.1007/s10618-020-00698-5
34. Tan, M., Le, Q.: Efficientnet: rethinking model scaling for convolutional neural networks. In: International Conference on Machine Learning. PMLR (2019)
35. Verbraeken, J., Wolting, M., Katzy, J., Kloppenburg, J., Verbelen, T., Rellermeyer, J.S.: A survey on distributed machine learning. ACM. Comput. Surv. (CSUR) **53**(2), 1–33 (2020)
36. Verwimp, E., et al.: Clad: a realistic continual learning benchmark for autonomous driving (2022). arXiv preprint arXiv:2210.03482
37. Vitter, J.S.: Random sampling with a reservoir. ACM Trans. Math. Softw. (TOMS) **11**(1), 37–57 (1985)
38. Vorabbi, L., Maltoni, D., Borghi, G., Santi, S., et al.: Enabling on-device continual learning with binary neural networks and latent replay. In: Proceedings of the 19th International Joint Conference on Computer Vision, Imaging and Computer Graphics Theory and Applications—VISAPP. vol. 2, pp. 25–36. Science and Technology Publications, LDA (2024)

Improving Generative Data Augmentation with Prior-Knowledge for Dysplasia Grading of Colorectal Polyps

Roberto Craparotta[1], Desislav Ivanov[1], Carlo Alberto Barbano[1(✉)], Enzo Tartaglione[1,3], Alessandro Gambella[2], Luca Cavallo[2], Paola Cassoni[2], Luca Bertero[2], and Marco Grangetto[1]

[1] Computer Science Department, University of Turin, 10149 Turin, Italy
carlo.barbano@unito.it
[2] Medical Sciences Department, University of Turin, 10126 Turin, Italy
[3] LTCI, Télécom Paris, Institut Polytechnique de Paris, Palaiseau, France

Abstract. Generative models (GANs or Diffusion Models) have achieved state-of-the-art performance in image synthesis in various domains, including the medical field. Many works have shown that it is possible to fully train deep models (e.g. classifiers) solely based on synthesized data and that augmenting real data with synthesized samples can lead to superior discriminative power. However, large datasets are needed to train realistic generative models. On the other hand, this can be an issue in certain domains, such as medical imaging, where datasets are typically smaller and more costly to annotate. Small sample size poses a fundamental problem in the applicability of generative models for data augmentation, as a generative model will not necessarily approximate the real data distribution better than a simple classifier in a low-data regime. To tackle this issue, we introduce prior information in the generative process to compensate for the lack of data, unlocking generative augmentation in low-data settings. In this work, we focus on computational pathology, specifically on the sensitive topic of the classification of colorectal polyp dysplasia. To guide the generative process, we take advantage of medical knowledge on tissue morphology taken from the World Health Organization (WHO) guidelines for the classification of dysplasia. By incorporating our proposed generative pipeline into a contrastive learning framework, we achieve state-of-the-art results in the detection of high-grade dysplasia on the UnitoPatho dataset.

Keywords: Generative models · Contrastive learning · Computational pathology

D. Ivanov—Work done while at University of Turin. Now at Google.

Supplementary Information The online version contains supplementary material available at https://doi.org/10.1007/978-3-032-10185-3_45.

E. Rodolà et al. (Eds.): ICIAP 2025, LNCS 16167, pp. 571–583, 2026.
https://doi.org/10.1007/978-3-032-10185-3_45

1 Introduction

The advances of deep learning in classification and computer vision have pushed the frontier of automatic medical diagnosis and computer-aided diagnosis. Recently, the application of deep learning to medical diagnosis has become a topic of great interest, including the fields of radiology [25], cardiology [36], and computational pathology [31]. In particular, deep learning in computational pathology has shown promising results, with the development of algorithms that can be trained on large amounts of data to, for example, automatically detect and segment tumor tissue in digitized pathology images [8]. These computer-aided tumor detection and segmentation tools can potentially support pathologists in their diagnostic tasks. The adoption of such models can have an enormous impact on the clinical decisions made by pathologists, so it is crucial to ensure that the models are accurate and robust enough. One way to achieve high precision and accuracy without overfitting the models is to use large amounts of data to correctly capture the variety of cases that can be seen in the clinical world. However, collecting large databases of real-world data for a niche diagnosis field can be challenging. First, the data collection is costly because it requires special equipment, trained medical personnel, and regulatory approval. Second, data should be collected from multiple sources and institutions to avoid bias [26]. Finally, data sharing is also challenging because of privacy and regulatory concerns.

In this context, generative models have shown great promise because they can generate realistic synthetic images that can be used to train models and perform data augmentation. Using synthetic images can have several benefits, including the generation of large datasets that can be used to train robust models and the ability to generate images that can augment the training dataset to improve the performance of existing models. In particular, Generative Adversarial Networks (GANs) [9] have shown noticeable results in the generation of realistic images in a wide range of applications, including the synthesis of realistic photographs of faces, natural scenes, and medical images [16]. In addition, GANs can be used to generate new images from control points [7], to interpolate between existing images [42], and to generate sequences of images [41]. Furthermore, the idea of using GANs for medical imaging has been researched in several different studies and tasks [19,37], ranging from segmentation [21,35], to synthesis [38] and semi-supervised classification [14].

We focus our research on UniToPatho [3], a dataset containing Hematoxylin and Eosin stained (H&E) colorectal histopathological scans. One of the main purposes of the dataset is to enable the development of automatic prediction of dysplasia grade of colorectal polyps to support pathologists for more accurate diagnosis, which is a crucial topic in pathology [11,12,23]. The core obstacle of the dataset is represented by the lack of high dysplasia grade samples, which produces a strong imbalance in the data distribution. For these reasons, we focus on building new generative models capable of synthesizing new samples that can be used to augment the UnitoPatho dataset, as illustrated in Fig. 1. The purpose of such augmentations is to allow for an improved accuracy of automatic diagnosis. Our contribution is two-fold.

1. We propose a *conditional GAN model* that takes nuclei segmentation maps as input and *produces realistic-looking histological tissue.* The generation of images in GANs is usually guided by normally distributed random values. However, the synthesis can be conditioned by different factors, which, in our case, are segmentation masks of nuclei. By conditioning the generator on these masks, we can alter the morphological structure of the tissue by rearranging the nuclei in the masks (e.g. increase cellular density), which is simpler to perform on binary segmentation masks than on real tissue images.
2. We *integrate our generative model as an augmentation module in a contrastive learning framework*, reaching state-of-the-art results on dysplasia grading on UnitoPatho.

While in this work we focus on GAN architecture and we target a specific computational pathology task, our proposed approach is general and can be applied to different architectures and domains.

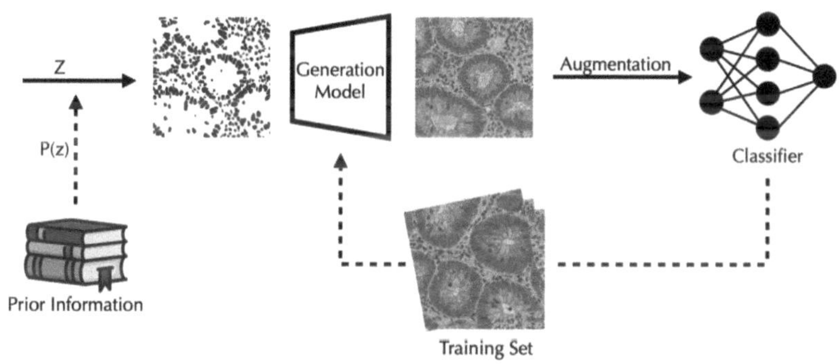

Fig. 1. Overview of our framework. We guide the generative process with prior information extracted from medical guidelines (WHO); the generative model is employed in a contrastive learning setup to synthesize novel positive samples with the ultimate goal of improving the model's generalization.

2 Related Works

The application of deep learning to digital pathology has shown significant promise, particularly in tasks involving whole slide image (WSI) analysis. Despite this progress, relatively few works have addressed the dual challenge of classifying both histological type and dysplasia grade in colorectal polyps. Korbar et al. [22] introduced a crop-based approach using residual networks for polyp classification, demonstrating the potential of convolutional neural networks (CNNs) for WSI analysis. Subsequent work has explored the interpretability of the model through attention mechanisms [30], and Bychkov et al. [4] extended this by incorporating spatial context through recurrent models to predict the results of colorectal cancer. In a more recent and comprehensive study, Perlo et al. [27] proposed a deep learning pipeline for dysplasia grading of colorectal adenomas using WSIs. Their hierarchical classification strategy, distinguishing high-grade from low-grade dysplasia, achieves performance levels comparable to expert pathologists. This work highlights the importance of resolution-aware, texture-focused modeling in histopathological classification tasks and provides a strong baseline for future research involving WSI-based grading systems. In [3], Barbano et al. presented the UniToPatho dataset, a collection of Hematoxylin and Eosin stained (H&E) colorectal histopathological scans, and developed a model for automatic tissue classification. These studies underscore the challenges posed by small dataset sizes and the imbalance of high dysplasia grade samples, highlighting the opportunities for advanced generative models to augment the dataset and improve diagnostic precision.

GAN-based data augmentation in digital pathology. Digital pathology, with its high-resolution WSIs, presents a rich but computationally intensive environment for deep learning models. GANs have been increasingly applied to synthesize realistic histopathological images in [28]. However, training GANs on histopathological data introduces several challenges: the high resolution of WSIs requires large computational resources, and more importantly, the number of labeled samples, particularly those with rare or high-grade abnormalities, is often limited. These limitations frequently lead to overfitting, unstable training, or mode collapse. Moreover, the generated images may lack biological plausibility unless domain-specific constraints are incorporated. StyleGAN [17] has been adapted for several medical applications, such as retinal image synthesis [39] and mammography [40]. Despite these successes, few works have explored its use in histopathology or its potential for integrating domain-specific priors. Recent efforts have started conditioning GANs on auxiliary inputs, such as segmentation maps or style codes, to improve relevance and realism [6]. However, the majority of these approaches still rely on large-scale datasets and do not address the data scarcity and granularity issues typical of rare pathological conditions.

3 Method

An overview of the method is provided in Fig. 1. Here, we present the different components of our framework, namely: generative model, prior-guided nuclei augmentation, and contrastive learning.

3.1 Segmentation-Conditioned Generative Model

The first key component of our pipeline is a generative model that can translate nuclei segmentation masks into realistic tissue. Our GAN architecture is based on StyleGAN-2 [18] To condition the generator with a segmentation mask, we introduce a convolutional encoder based on ResNet [13] that, similarly to U-Net [29], concatenates the output of each residual block to the corresponding block in the decoder network. Additionally, before concatenation, we use an additional instance normalization layer [33]. A high-level design of the generator network is summarized in Fig. 2. We use a non-saturating adversarial loss, path-length regularization, and R_1 gradient penalty [24]. Furthermore, we add a reconstruction term based on L1 loss between the generated output and the ground truth patch corresponding to the provided segmentation mask. The generator and discriminator objectives ($\mathcal{L}_g$ and $\mathcal{L}_d$ respectively) are:

$$\mathcal{L}_g = -\mathbb{E}_z[\log(d(g(z)))] + \gamma_{pl} R^{PL} + \lambda \mathbb{E}_{z,s,x}[\|x - g(z,s)\|_1] \quad (1)$$

$$\mathcal{L}_d = -\mathbb{E}_x[\log(d(x))] - \mathbb{E}_z[\log(1 - d(g(z)))] + \gamma_{R_1} \mathbb{E}_x[\|\nabla_x d(x)\|^2] \quad (2)$$

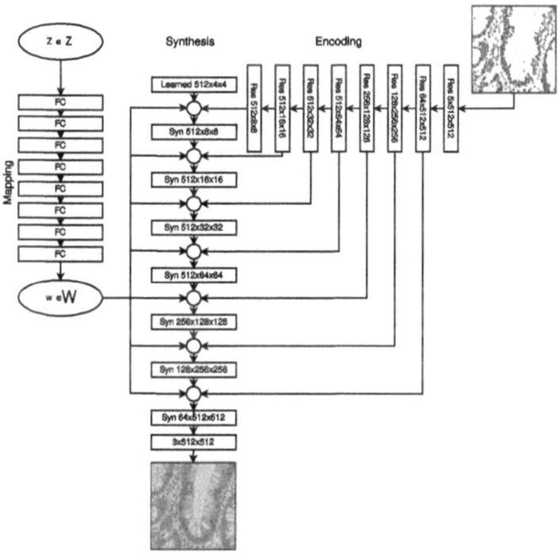

Fig. 2. Architecture of the extended generator. The Mapping and Synthesis networks follow the StyleGAN2 [18] architecture. The Encoder network uses skip connections similar to the U-Net [29] architecture and is used to condition the generative process with an input segmentation mask.

where g is the generator, d is the discriminator, x is the ground truth image, and s is the input segmentation mask. This adversarial loss is an alternative to the original min-max loss proposed in [9] that avoids the saturation of the Generator network. The regularization term R^{PL}, weighted by $\gamma_{pl} \geq 0$, in (1) is the path-length regularization, as defined in [18], that encourages the Generator to avoid drastic changes in the neighborhood of sampled latent vectors mapped into the $\mathcal{W}$ space; whereas the reconstruction term is weighted by $\lambda \geq 0$. The regularization term weighted by $\gamma_{R_1} \geq 0$ in (2) is the R_1 gradient penalty, which forces the discriminator to avoid producing gradients without suffering from the adversarial loss when the generator reaches the true real data distribution. The structure of the two losses is the one proposed in [18]. Additionally, we also pass the segmentation mask as input to the discriminator network to better learn correct relationships between segmentation masks and real images.

3.2 Nuclei Augmentation

After training the generative model to produce realistic-looking tissue, as described in Sect. 3.1, the second key component of our pipeline is leveraging prior medical information to design biologically plausible data augmentation for the segmentation masks in order to synthesize novel high-grade samples. Towards this end, we developed augmentation techniques aimed at reproducing morphological variations consistent with the cellular alterations typically observed in high-grade dysplasia. These transformations were defined based on features that expert pathologists consider key indicators in the assessment of dysplasia grade:

> "*High-grade Dysplasia (HGD) is characterized by marked complex glandular crowding and irregularity of glands, cribriform architecture, and intraluminal necrosis as architectural features. These can be observed at low magnification. For the diagnosis of HGD, these must be accompanied by cytological features of HGD, including substantial loss of cell polarity, markedly enlarged nuclei with prominent nucleoli, and a dispersed chromatin pattern, often with atypical mitotic figures.*" (The WHO guidelines [32])

Among the highlighted features, in this work, we focus on increased cellular density and irregularity of glands.

Cellular Density. This augmentation is designed to simulate increased cellular density, which is a common feature of HGD tissue. This is achieved by duplicating and repositioning existing cells, creating a visual effect of cellular crowding. Individual cells are first located with the connected components algorithms provided by scikit-image [34] and are then copy-pasted within a certain range from their initial position.

Irregularity of Glands. The second augmentation focuses on the irregularity of glands, reproducing the misalignment and irregular spatial arrangement of cells relative to their original architecture. This latter aspect is particularly useful for representing structural instability. Irregular gland shapes are a key indicator of HGD tissue. Towards this end, each cell is first located as an independent connected component, and its position is shifted by a small random amount in the direction spanning from the starting position and the centroid of the corresponding gland. Thus, we may have some cells closer to the centroid of the glands and others farther away. To compute the glands' centroids, we employ a gland segmentation model pre-trained on the GlaS dataset [1].

In summary, the proposed approach leverages biologically inspired transformations to generate synthetic variants of HGD patches with the goal of facilitating the automatic recognition of patterns that clinicians use in practice to distinguish between different grades of dysplasia. This method not only increases dataset diversity but also helps the model focus on morphologically relevant features for diagnostic purposes. Figure 4 shows some examples of the proposed augmentations.

3.3 Supervised Contrastive Learning

The final part of our pipeline is represented by the Contrastive Learning (CL) training of the patch encoder. CL is more robust than traditional end-to-end approaches, such as cross-entropy, against noise in the data or the labels, resulting in better generalizing models [2,10,20]. CL has been popularized by self-supervised methods such as SimCLR [5], and supervised methods such as SupCon [20]. CL aims at pulling representations of similar samples (called *positive*) close together in the representation space while repelling the representations of dissimilar samples (called *negative*). In self-supervised CL [5], positive samples are obtained as augmented views of a starting *anchor* sample (e.g. by applying color jittering, cropping, etc.), while in supervised CL, labels are also used to define positive pairs. We use SupCon [20] as the objective function to leverage the available labels of the UnitoPatho dataset, and we leverage our generative model to synthesize a novel positive sample starting from the anchor's segmentation mask, which we augment with the steps presented in Sect. 3.2. After training the encoder with SupCon, a linear classifier is trained on top of the frozen backbone to discriminate between low-grade and high-grade dysplasia using standard cross-entropy.

4 Results

Here we report our main results. Additional results, ablations, and experimental details can be found in the supplementary material.

4.1 Generation of Histological Tissue

As shown in Fig. 3, our model successfully produces high-quality and high-resolution reconstructions[1]. We experiment with different weights for the reconstruction loss and, contrary to other works [15], this loss term does not provide any advantage in the synthesis. In fact, visual quality improves for lower weights of the term. In addition, the lowest FID50k result that we achieve is 3.46, without any reconstruction loss. We argue that this behavior is caused by the fact that our synthesis is also guided by random latent vectors defining the style that are independent from the segmentation masks. This leads the reconstruction loss to try associating the same image with multiple style vectors, which creates instability in the training of the Generator. Furthermore, the removal of the reconstruction loss demonstrates that the Discriminator network is capable of autonomously learning the relationship between segmentations and images solely from the adversarial loss. We also compare our proposed model with the state-of-the-art image-to-image model pix2pix. The results, in terms of FID50k, are shown in Table 1.

Table 1. Comparison of the FID metric between our model and pix2pix [15].

Model	512px		2048px	
	Train	Test	Train	Test
Pix2Pix [15]	37.6	36.2	82.1	88.1
Ours	**3.46**	**4.32**	**5.33**	**6.12**

[1] Additional examples of generation can be found in the supplementary files.

4.2 Biologically Plausible Data Augmentation

Examples of the proposed nuclei augmentations can be found in Fig. 4. As the results show, thanks to our conditional generative approach, we can synthesize realistic samples with altered morphological structures. To further preserve the biological plausibility of the generated synthetic images, a segmentation model was integrated to accurately isolate glandular structures within histological images. Gland segmentation is a crucial step, as these structures represent the primary region of interest for pathologists when assessing dysplasia grade. The adopted model is based on a U-Net architecture and pre-trained on the GlaS 2015 dataset [1]. The binary masks generated by this model are used to delineate the regions where augmentation operations are applied. This approach restricts augmentations to glandular areas only, avoiding alterations in diagnostically irrelevant regions such as the stroma. The integration of gland segmentation helps maintain a high level of biological fidelity in the synthetic images, allowing the model to focus on morphologically relevant features. The segmentation model is executed during the pre-processing phase, producing masks that are then used in the image generation pipeline without affecting training efficiency.

4.3 Dysplasia Prediction Results

Lastly, we integrate our conditional generative model with the proposed augmentation in a contrastive framework to train a robust feature extractor based on ResNet-18. The extracted features are used to train a linear classifier to distinguish between low-grade and high-grade dysplasia. The final results in terms of dysplasia prediction can be found in Table 2. We compare with the approach proposed in [3], consisting of a ResNet-18 trained at 800 µm × 800 µm patches (1812 × 1812 pixels), resulting in a balanced accuracy (BA) of ∼80%. On the other hand, switching to a contrastive learning framework leads to an improvement in accuracy to 82.10%. This confirms that CL typically learns better representations than standard cross-entropy approaches. To study the impact of our

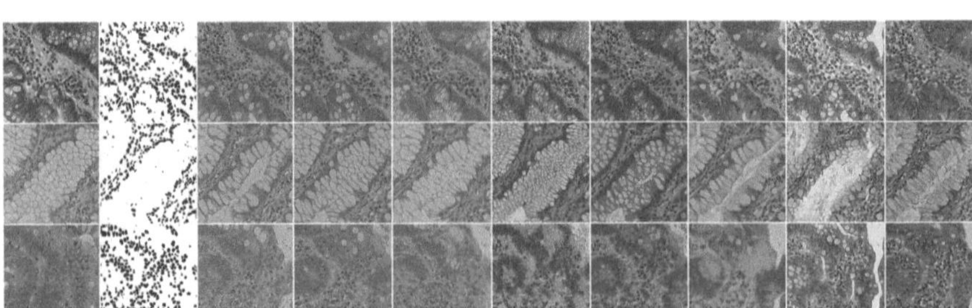

Fig. 3. Images generated by our model. The first column contains the real images. The second column contains the segmentation masks. The rest of the columns are generated by the GAN. Each of the generations' columns represents a randomly sampled latent vector, which enforces the same style across different input masks.

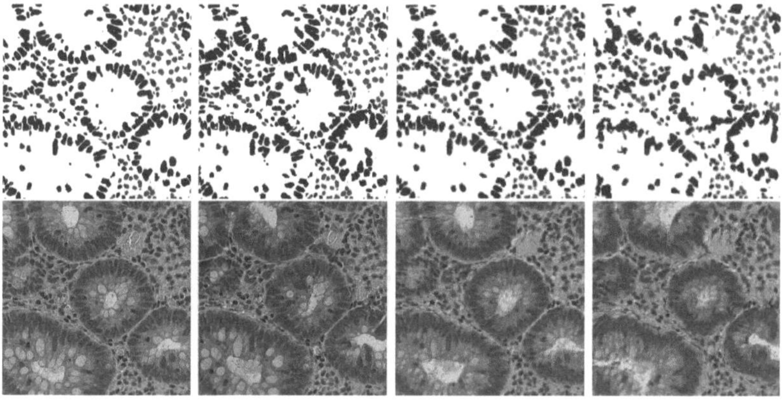

(a) Cellular Density (b) Glands Irregularity

Fig. 4. Example of our biology-inspired data augmentation. We alter the input segmentation mask to the generative model, to obtain a realistic reconstruction of (a) increased cellular density (c) irregularity in glands.

Table 2. Results in dysplasia grading. With our approach, we achieve the best results overall, surpassing previous methods on the UnitoPatho dataset.

Method	Architecture	Balanced Acc. (%)	Recall LG	Recall HG
Cross-Entropy [3]	ResNet-18	80.6%±0.01	85.9±0.1	73.86±0.2
SupCon	ResNet-18	82.10±0.03	87.75±1.05	76.46±0.54
SupCon + Generative (baseline)	ResNet-18	82.29±0.05	86.84±2.04	**77.74**±0.78
SupCon + Prior-guided Generative (**ours**)	ResNet-18	**83.01**±0.02	**89.15**±0.89	76.86±0.45

conditional generative models, we first integrate it without employing our proposed augmentation scheme. The resulting accuracy does change significantly from the previous results, showing that our generator can plausibly reconstruct tissue. Lastly, we introduce the augmentation discussed in Sect. 3.2, and we observe an increase in accuracy to 83.01%. The results show that our framework can synthesize novel HG samples based on the prior medical knowledge captured by our proposed data augmentation.

5 Conclusions

We have proposed a generative augmentation approach that can integrate prior medical knowledge in the form of data augmentation to make up for the lack of data for the desired target class. In this work, we focus on dysplasia grading of colorectal polyps, and we leverage pathologists' and WHO's guidelines to craft specific data augmentation aimed at increasing cellular density and gland irregularity in the tissue patches. The augmentation is performed on cellular segmentation masks that are fed to our state-of-the-art tissue generator, turning them into realistic-looking high-grade patches. Coupled with a contrastive learning framework, this approach achieves the best results in dysplasia grading on the UnitoPatho dataset. Although in this work we focused on GANs and on the specific task of dysplasia grading, our proposed approach is general and can be adapted to different architectures (e.g., diffusion models) and to different downstream tasks.

References

1. Gland segmentation in colon histology images: the glas challenge contest. Med. Image Anal. **35**, 489–502 (2017)
2. Barbano, C.A., Dufumier, B., Tartaglione, E., Grangetto, M., Gori, P.: Unbiased supervised contrastive learning. In: The Eleventh International Conference on Learning Representations (2023). https://openreview.net/forum?id=Ph5cJSfD2XN
3. Barbano, C.A., Perlo, D., Tartaglione, E., Fiandrotti, A., Bertero, L., Cassoni, P., Grangetto, M.: Unitopatho, a labeled histopathological dataset for colorectal polyps classification and adenoma dysplasia grading. In: 2021 IEEE International Conference on Image Processing (ICIP), pp. 76–80 (2021). https://doi.org/10.1109/ICIP42928.2021.9506198
4. Bychkov, D., Linder, N., Turkki, R., Nordling, S., Kovanen, P.E., Verrill, C., Walliander, M., Lundin, M., Haglund, C., Lundin, J.: Deep learning based tissue analysis predicts outcome in colorectal cancer. Sci. Rep. Nat. **8** (2018)
5. Chen, T., Kornblith, S., Norouzi, M., Hinton, G.: A simple framework for contrastive learning of visual representations. In: International Conference on Machine Learning, pp. 1597–1607. PmLR (2020)
6. Chen, X., et al.: Synthetic data generation in medical imaging using conditional generative adversarial networks. Artif. Intell. Med. **128**, 102283 (2022)
7. Chen, X., Duan, Y., Houthooft, R., Schulman, J., Sutskever, I., Abbeel, P.: Infogan: interpretable representation learning by information maximizing generative adversarial nets (2016). https://doi.org/10.48550/ARXIV.1606.03657, https://arxiv.org/abs/1606.03657
8. Gamper, J., Koohbanani, N.A., Benet, K., Khuram, A., Rajpoot, N.: Pannuke: an open pan-cancer histology dataset for nuclei instance segmentation and classification. In: European Congress on Digital Pathology, pp. 11–19. Springer (2019)
9. Goodfellow, I.J., Pouget-Abadie, J., Mirza, M., Xu, B., Warde-Farley, D., Ozair, S., Courville, A., Bengio, Y.: Generative adversarial networks (2014). https://doi.org/10.48550/ARXIV.1406.2661, https://arxiv.org/abs/1406.2661
10. Graf, F., Hofer, C., Niethammer, M., Kwitt, R.: Dissecting supervised contrastive learning. In: International Conference on Machine Learning, pp. 3821–3830. PMLR (2021)

11. Gupta, S., Lieberman, D., Anderson, J.C., Burke, C.A., Dominitz, J.A., Kaltenbach, T., Robertson, D.J., Shaukat, A., Syngal, S., Rex, D.K.: Recommendations for follow-up after colonoscopy and polypectomy: a consensus update by the US multi-society task force on colorectal cancer. Gastroenterology **158**(4), 1131 1153.e5 (2020)
12. Hassan, C., Antonelli, G., Dumonceau, J.M., Regula, J., Bretthauer, M., Chaussade, S., Dekker, E., Ferlitsch, M., Gimeno-Garcia, A., Jover, R., Kalager, M., Pellisé, M., Pox, C., Ricciardiello, L., Rutter, M., Helsingen, L.M., Bleijenberg, A., Senore, C., van Hooft, J.E., Dinis-Ribeiro, M., Quintero, E.: Post-polypectomy colonoscopy surveillance: European society of gastrointestinal endoscopy (ESGE) guideline—update 2020. Endoscopy **52**(8), 687–700 (2020)
13. He, K., Zhang, X., Ren, S., Sun, J.: Deep residual learning for image recognition (2015). https://doi.org/10.48550/ARXIV.1512.03385, https://arxiv.org/abs/1512.03385
14. Hu, B., Tang, Y., Chang, E.I.C., Fan, Y., Lai, M., Xu, Y.: Unsupervised learning for cell-level visual representation in histopathology images with generative adversarial networks. IEEE J. Biomed. Health Inform. **23**(3), 1316–1328 (2019). https://doi.org/10.1109/jbhi.2018.2852639
15. Isola, P., Zhu, J.Y., Zhou, T., Efros, A.A.: Image-to-image translation with conditional adversarial networks (2016). https://doi.org/10.48550/ARXIV.1611.07004, https://arxiv.org/abs/1611.07004
16. Karras, T., Aila, T., Laine, S., Lehtinen, J.: Progressive growing of GANs for improved quality, stability, and variation (2017). https://doi.org/10.48550/ARXIV.1710.10196, https://arxiv.org/abs/1710.10196
17. Karras, T., Laine, S., Aila, T.: A style-based generator architecture for generative adversarial networks. In: Proceedings of the IEEE/CVF Conference on Computer Vision and Pattern Recognition, pp. 4401–4410 (2019)
18. Karras, T., Laine, S., Aittala, M., Hellsten, J., Lehtinen, J., Aila, T.: Analyzing and improving the image quality of StyleGAN (2019). https://doi.org/10.48550/ARXIV.1912.04958, https://arxiv.org/abs/1912.04958
19. Kazeminia, S., Baur, C., Kuijper, A., van Ginneken, B., Navab, N., Albarqouni, S., Mukhopadhyay, A.: GANs for medical image analysis (2018). https://doi.org/10.48550/ARXIV.1809.06222, https://arxiv.org/abs/1809.06222
20. Khosla, P., Teterwak, P., Wang, C., Sarna, A., Tian, Y., Isola, P., Maschinot, A., Liu, C., Krishnan, D.: Supervised contrastive learning. Adv. Neural. Inf. Process. Syst. **33**, 18661–18673 (2020)
21. Kohl, S., Bonekamp, D., Schlemmer, H.P., Yaqubi, K., Hohenfellner, M., Hadaschik, B., Radtke, J.P., Maier-Hein, K.: Adversarial networks for the detection of aggressive prostate cancer (2017). https://doi.org/10.48550/ARXIV.1702.08014, https://arxiv.org/abs/1702.08014
22. Korbar, B., Olofson, A.M., Miraflor, A.P., Nicka, C.M., Suriawinata, M.A., Torresani, L., Suriawinata, A.A., Hassanpour, S.: Deep learning for classification of colorectal polyps on whole-slide images. J. Pathol. Inform. **8** (2017)
23. Matsuda, T., Chiu, H.M., Sano, Y., Fujii, T., Ono, A., Saito, Y.: Surveillance colonoscopy after endoscopic treatment for colorectal neoplasia: from the standpoint of the Asia-Pacific region. Dig. Endosc. **28**(3), 342–347 (2016)
24. Mescheder, L., Geiger, A., Nowozin, S.: Which training methods for GANs do actually converge? (2018). https://doi.org/10.48550/ARXIV.1801.04406, https://arxiv.org/abs/1801.04406
25. Najjar, R.: Redefining radiology: a review of artificial intelligence integration in medical imaging. Diagnostics **13**(17), 2760 (2023)

26. Norori, N., Hu, Q., Aellen, F.M., Faraci, F.D., Tzovara, A.: Addressing bias in big data and AI for health care: a call for open science. Patterns (N Y) **2**(10), 100347 (2021)
27. Perlo, D., Tartaglione, E., Bertero, L., Cassoni, P., Grangetto, M.: Dysplasia grading of colorectal polyps through convolutional neural network analysis of whole slide images. In: Medical Imaging with Deep Learning (MICAD) (2021). https://dx.doi.org/10.21227/9fsv-tm25, mICAD 2021
28. Quiros, A.C., Murray-Smith, R., Yuan, K.: Pathologygan: learning deep representations of cancer tissue. In: Arbel, T., Ben Ayed, I., de Bruijne, M., Descoteaux, M., Lombaert, H., Pal, C. (eds.) Proceedings of the Third Conference on Medical Imaging with Deep Learning. Proceedings of Machine Learning Research, vol. 121, pp. 669–695. PMLR (2020). https://proceedings.mlr.press/v121/quiros20a.html
29. Ronneberger, O., Fischer, P., Brox, T.: U-net: convolutional networks for biomedical image segmentation (2015). https://doi.org/10.48550/ARXIV.1505.04597, https://arxiv.org/abs/1505.04597
30. Selvaraju, R.R., Cogswell, M., Das, A., Vedantam, R., Parikh, D., Batra, D.: Grad-CAM: visual explanations from deep networks via gradient-based localization. In: ICCV, pp. 618–626. IEEE Computer Society (2017)
31. Song, A.H., Jaume, G., Williamson, D.F., Lu, M.Y., Vaidya, A., Miller, T.R., Mahmood, F.: Artificial intelligence for digital and computational pathology. Nat. Rev. Bioeng. **1**(12), 930–949 (2023)
32. WHO Classification of Tumours. Digestive System Tumours: WHO Classification of Tumours, vol. 1. World Health Organization (2019). https://books.google.it/books?id=l5l_xgEACAAJ
33. Ulyanov, D., Vedaldi, A., Lempitsky, V.: Instance normalization: the missing ingredient for fast stylization (2016). https://doi.org/10.48550/ARXIV.1607.08022, https://arxiv.org/abs/1607.08022
34. van der Walt, S., Schönberger, J.L., Nunez-Iglesias, J., Boulogne, F., Warner, J.D., Yager, N., Gouillart, E., Yu, T.: The scikit-image contributors: scikit-image: image processing in Python. PeerJ **2**, e453 (6 2014). https://doi.org/10.7717/peerj.453, https://doi.org/10.7717/peerj.453
35. Wang, D., Gu, C., Wu, K., Guan, X.: Adversarial neural networks for basal membrane segmentation of microinvasive cervix carcinoma in histopathology images. In: 2017 International Conference on Machine Learning and Cybernetics (ICMLC), vol. 2, pp. 385–389 (2017). https://doi.org/10.1109/ICMLC.2017.8108952
36. Wehbe, R.M., Katsaggelos, A.K., Hammond, K.J., Hong, H., Ahmad, F.S., Ouyang, D., Shah, S.J., McCarthy, P.M., Thomas, J.D.: Deep learning for cardiovascular imaging: a review. JAMA Cardiol. **8**(11), 1089–1098 (2023)
37. Wolterink, J.M., Kamnitsas, K., Ledig, C., Išgum, I.: Generative adversarial networks and adversarial methods in biomedical image analysis (2018). https://doi.org/10.48550/ARXIV.1810.10352, https://arxiv.org/abs/1810.10352
38. Yi, X., Walia, E., Babyn, P.: Generative adversarial network in medical imaging: a review. Med. Image Anal. **58**, 101552 (2019). https://doi.org/10.1016/j.media.2019.101552, https://www.sciencedirect.com/science/article/pii/S1361841518308430 2
39. Yu, B., Zhao, C., et al.: Medical image synthesis for data augmentation and anonymization using generative adversarial networks. Artif. Intell. Med. **124**, 102147 (2022)
40. Zhang, Y., et al.: Synthesis of high-resolution mammograms with progressive generative adversarial networks. Med. Image Anal. **68**, 101907 (2021)

41. Zhu, J.Y., Park, T., Isola, P., Efros, A.A.: Unpaired image-to-image translation using cycle-consistent adversarial networks (2017). https://doi.org/10.48550/ARXIV.1703.10593, https://arxiv.org/abs/1703.10593
42. Zhu, J.Y., Zhang, R., Pathak, D., Darrell, T., Efros, A.A., Wang, O., Shechtman, E.: Toward multimodal image-to-image translation (2017). https://doi.org/10.48550/ARXIV.1711.11586, https://arxiv.org/abs/1711.11586

Automatic Benchmarking of Large Multimodal Models via Iterative Experiment Programming

Alessandro Conti[1(✉)], Massimiliano Mancini[1], Enrico Fini[2], Yiming Wang[3], Paolo Rota[1], and Elisa Ricci[1,3]

[1] University of Trento, Trento, Italy
alessandro.conti-1@unitn.it
[2] Trento, Italy
[3] Fondazione Bruno Kessler, Trento, Italy

Abstract. Assessing the capabilities of large multimodal models (LMMs) often requires the creation of ad-hoc evaluations. Currently, building new benchmarks requires tremendous amounts of manual work for each specific analysis. This makes the evaluation process tedious and costly. In this paper, we present APEx, Automatic Programming of Experiments, the first framework for automatic benchmarking of LMMs. Given a research question expressed in natural language, APEx leverages a Large Language Model (LLM) and a library of pre-specified tools to generate a set of experiments for the model at hand, and progressively compile a scientific report. The report drives the testing procedure: based on the current status of the investigation, APEx chooses which experiments to perform and whether the results are sufficient to draw conclusions. Finally, the LLM refines the report, presenting the results to the user in natural language. Thanks to its modularity, our framework is flexible and extensible as new tools become available. Empirically, APEx reproduces the findings of existing studies while allowing for arbitrary analyses and hypothesis testing.

Keywords: Automatic Evaluation · Large Multimodal Models · Large Language Models

1 Introduction

As machine learning models grow more powerful, so does the interest in probing their capabilities and limitations. Large Multimodal Models (LMMs) [17,19] have been extensively analyzed [13,42,46]. Recent studies [5,48] typically follow a standard pipeline: (i) selecting models, (ii) building task-specific benchmarks (often requiring data collection/annotation), and (iii) analyzing their results.

Consider a simple query: "Do LMMs understand colors?" Answering it requires testing models on color recognition tasks. But how many colors should

E. Fini—Independent researcher

E. Rodolà et al. (Eds.): ICIAP 2025, LNCS 16167, pp. 584–596, 2026.
https://doi.org/10.1007/978-3-032-10185-3_46

be tested? How complex should the palette be? Is further testing needed if a model performs well on initial examples? Such evaluations demand extensive manual effort—designing benchmarks, running tests, analyzing outcomes, and judging whether conclusions are sufficiently supported. These tasks are tedious, domain-dependent, and often subjective, motivating the need for an automated framework capable of designing and executing a broad range of experiments.

We address this challenge by applying visual tool programming via Large Language Models (LLMs) [41]. Our framework, AUTOMATIC PROGRAMMING OF EXPERIMENTS (APEx), answers user-specified inquiries by generating a structured scientific report that is iteratively refined through experiments. An LLM: (i) creates a benchmark (via retrieval, generation, or augmentation), (ii) selects and runs models, (iii) analyzes results, and (iv) updates the report. This process repeats until the LLM deems the report sufficient to answer the original query.

We evaluate APEx across diverse questions, showing it flexibly builds benchmarks, runs experiments, and draws insightful conclusions. APEx reproduces key findings from prior hand-crafted benchmarks [42], handles queries at multiple granularities, and supports modular extensions with new tools.

Contributions.

- We introduce APEx, the first automated framework for benchmarking LMMs based on user-specified queries;
- APEx automates benchmark design, experiment execution, and result analysis, with a modular architecture for future expansion;
- We demonstrate that APEx produces valid and insightful reports, reproducing prior findings;
- APEx enables discovery of new strengths and limitations of LMMs by handling diverse, open-ended inquiries.

2 Related Work

Visual Programming. Composing atomic operations into complex functions is a common approach in VQA [2]. Modular neural networks [1,37] decompose questions (*e.g.*, "How many dogs are in the image?") into sub-tasks (*e.g.*, detection, counting), chained to solve the full query. This paradigm evolved with LLMs [25], especially in NLP, where frameworks augment LLMs with tool usage [26,28,33] by prompting them with APIs and examples for in-context learning [3]. In vision, VisProg [12] and ViperGPT [38] adopt the same idea, using LLMs to generate programs executed by pre-trained vision models—acting as parser and subtask decomposer [1]. Subsequent work either enhanced pipelines (*e.g.*, expanding tools [35], verification [21], self-training [16]) or targeted specific tasks (*e.g.*, behavior understanding [44], 3D grounding [45], text-to-image [8]). Shaham et al. [34] also used tools to interpret deep network neurons. APEx builds on LLM-based visual programming [12,38], but shifts the focus from solving tasks to generating programs for systematically testing LMMs.

Analyzing Vision Language Models. Several studies analyzed LMM capabilities and limitations. Ma et al. [23,40], Thrush et al. , showed struggles with fine-grained textual differences; [46] highlighted bag-of-words behavior when relationships are ignored during training. Others revealed biases: attribute associations [39], shape vs. texture [11], and frequency-driven generalization [10, 24,43]. Task-specific evaluations include time/location [47], egocentric vision [7], retrieval [6], spatial reasoning [15], and general multimodal understanding [4,36]. Robustness has also been tested across distribution shifts [29], corruptions [5], adversarial settings [48], and missing modalities [22]. Notably, most works require handcrafted benchmarks, which are sometimes later revised [13]. In contrast, we provide a general-purpose tool for automatically generating benchmarks and probing LMM behavior. As APEx will be made open-source, researchers can both use and extend it by contributing new modules.

3 APEx: Automatic Programming of Experiments

APEx is a framework that autonomously conducts experiments on LMMs to answer a user-defined query. The goal is to return an answer a in natural language, supported by experimental evidence.

APEx is modular (Fig. 1) and comprises several components: the orchestrator (an LLM that designs and interprets experiments), the generator (handling image retrieval, generation, and transformation), the library (a set of pre-trained LMMs), and the engine (the execution backend). Its architecture is extensible—supporting any future LMMs, LLMs, tools, or generators. Given a query (*e.g.*, "Is BLIP-2 robust to left rotations?"), APEx selects tools, runs experiments, and analyzes results iteratively until a conclusion is reached. Results are stored in a structured *report*, updated throughout the experimentation loop.

3.1 Initialization

Upon receiving a query, APEx is initialized with a specific library of models, generator tools, and the LLM-powered orchestrator.

The system instantiates an empty JSON-formatted report containing the query and selected models. This serves as a read-write memory buffer for the orchestrator to drive decisions and track results. The report is initialized via: $\text{report} = \text{orchestrator}([\text{prompt}_S, \text{prompt}_I, \text{query}])$, where prompt_S defines the task and available tools/models, and prompt_I initializes the process. Unlike in-context learning [3], APEx uses tool documentation, improving scalability [28].

The initialized report contains the query and the models to test. Queries may target all models (*e.g.*, "Can models...?") or specific ones (*e.g.*, "Can BLIP-2...?").

3.2 Iterative Experimentation

At the core of APEx is an iterative experimentation loop. Each cycle, the orchestrator receives the report and user query, then decides whether further

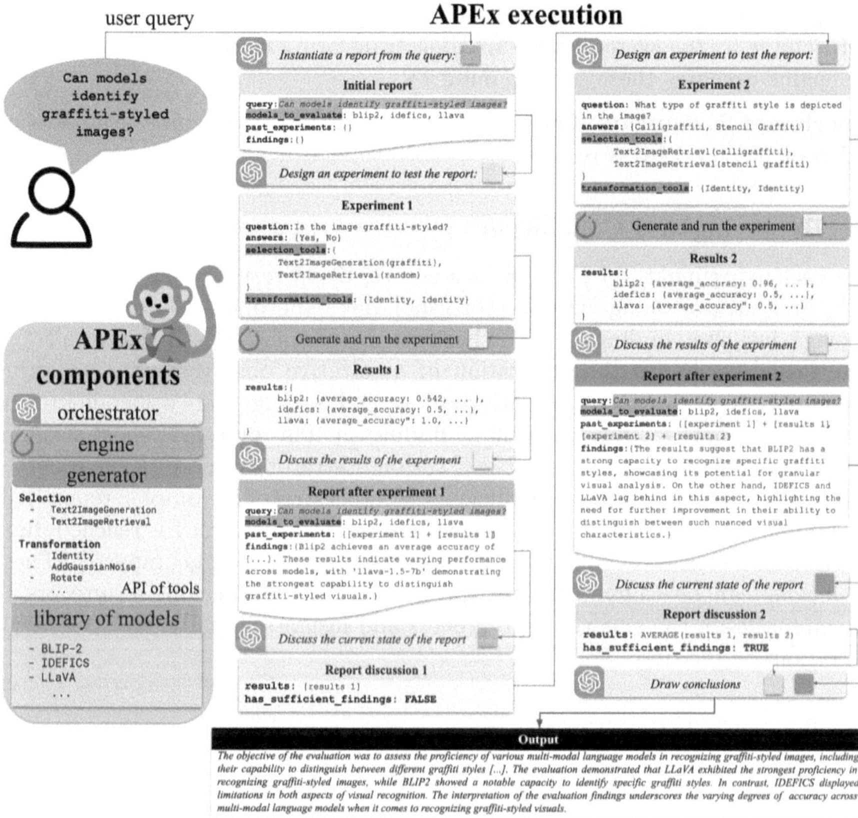

Fig. 1. APEx. APEx has four components: an orchestrator for reasoning, an engine for function execution, a benchmark generator with image selection and manipulation tools, and a library of LMMs. Given a user query, the orchestrator instantiates a report and generates the first experiment for the engine to execute. The orchestrator then collects, discusses, and adds the results to the report. The process is repeated until it is deemed sufficient to answer the query. Finally, the orchestrator summarizes the report and answers the user. (Color figure online)

experiments are needed. To define a new experiment, we use: experiment = orchestrator ([prompt$_S$, prompt$_E$, report]). The experiment contains a *goal* and a *benchmark generation* process.

Goal Definition. The goal defines a question (*e.g.*, "Is the image blue?") and multiple choices (*e.g.*, {"yes", "no"}), in a VQA format. Performance is measured via answer ranking [20]. We include "Unknown" as a valid abstention option.

Benchmark Generation. The benchmark $\mathcal{D} = \{(x_i, \text{question}, \text{choice}_i)\}_{i=1}^{N}$, is created using tools tools$_E$ from the experiment. Tools can collect images (*e.g.*, from ImageNet [32]), generate them (*e.g.*, with Stable Diffusion [31]), or apply transformations (*e.g.*, rotation). Transformations are structured as Python calls interpretable by the system, *e.g.*, "src.tools.transform.OverlayColor([255, 0, 0])". The engine executes these calls to build $\mathcal{D}$.

Experiment Execution. Once the dataset is ready, the engine evaluates all models on $\mathcal{D}$ and returns results_E.

3.3 Reporting and Conclusion

After each experiment, APEx collects a new result set results_E. These are passed to the orchestrator along with the corresponding experiment to extract natural language insights: $\text{findings}_E = \text{orchestrator}([\text{prompt}_F, \text{experiment}, \text{results}_E])$. Here, prompt_F guides the generation of qualitative observations, supporting reasoning beyond raw metrics. Both results_E and findings_E are appended to the report with: $\text{report} \leftarrow [\text{report}, \text{results}_E, \text{findings}_E]$. This updated report accumulates all evidence gathered so far. Next, the orchestrator evaluates if the report is sufficient to answer the query: $\text{sufficiency} = \text{orchestrator}([\text{prompt}_B, \text{report}])$. If not, the loop continues with a new experiment. Once sufficiency is reached, final conclusions are drawn: $\text{conclusions} = \text{orchestrator}([\text{prompt}_C, \text{report}])$. These conclusions can be tailored to user preferences, *i.e.*, from concise answers to full summaries of the experimental process and insights. Finally, APEx returns conclusions to the user.

3.4 Implementation Details

Orchestrator. The core component of APEx is the orchestrator, implemented as an LLM. We used GPT-3.5 [25] for its efficiency and widespread use. To instruct the model to produce structured outputs, we use `function calling`[1], meaning the LLM outputs adhere to the structured JSON format rather than being unrestricted text, ensuring easy mapping of the model outputs to functions/tools available in APEx. In cases where the orchestrator's output is invalid, *e.g.*, it lacks JSON fields, or it requests non-available tools, APEx informs the orchestrator of the exception message (within the prompt), using this information as prior to *self-heal* the previous output. In the case of complex healing (*e.g.*, multiple sequential failures), a new experiment is designed from scratch. To avoid infinite loops, we set the maximum number of self-healing retries to 3 before regenerating the experiment from scratch. Moreover, in all of our evaluations, we limit the maximum number of experiments the orchestrator can design to 5.

Benchmark Generation. As stated in Sect. 3.2 the data benchmark generation step in APEx comprises both selection and transformation tools. For selection, we use Stable Diffusion XL Turbo [27] to generate images due to its efficiency, and the ImageNet dataset [32] for retrieval due to its large number of classes. We additionally expand the dataset with the metaclasses from ImageNet-X [14], *i.e.*, clustering the 1000 classes into 17 groups (*e.g.*, plants, fruits). With both tools, it is

[1] https://platform.openai.com/docs/guides/function-calling.

possible to define an image type (*e.g.*, photo, oil painting) and a class name (*e.g.*, Siamese cat, sofa). When the retrieval query is impossible, *e.g.*, the requested class is not available in the dataset, APEx automatically resorts to generation. We also use the unique keyword "random" to denote sampling any semantic class in the dataset. This is used whenever the experiment does not require any specific class, and all of them can be considered for the benchmark. Regarding the transformation tools, *i.e.*, tools to manipulate the input samples for visual variations, they can be roughly split into geometric, pixel, semantic, and style transforms, and the complete list of transforms currently implemented are AddGaussianNoise, AddJPEGCompression, ApplyCutMix, ApplyMixUp, ChangeBrightness, ChangeContrast, CropRandomShuffleAndRecompose, DefocusBlurImage, EditImageStyle, EditImageWeather, FlipImage, Identity, OverlayColor, PasteGeneratedObjectAtRandomPosition, PasteGeometricShapeAtRandomPosition, PasteTextAtRandomPosition, RotateImage, and ZoomAtRandomPosition. We devise the transformation tools to produce data types defined in [42]. All data types are covered, but not all are explicitly defined with a tool (*e.g.*, we do not design specific tools for style transfer to cartoons, but provide a generic tool for style editing).

LMMs Evaluation. While in principle APEx supports a large variety of LMMs, in our implementation we demonstrate its capabilities using three widely used auto-regressively trained VLMs (referred as LMMs in [42]): BLIP-2 [19], IDEFICS [17], and LLaVA [18]. All the models are downloaded from HuggingFace[2] and are extended to support answer ranking. We focus on small models, using `blip2-opt-2.7b`, `idefics-9b-instruct`, and `llava-1.5-7b`. We quantize all the models to 8 bits and use a single NVIDIA RTX A4000 for the experiments. About the quantitative metrics in results_E, we evaluate the methods with average accuracy and class-wise accuracy. When the models abstain from answering (*i.e.*, answering with "Unknown" to the question), we compute the abstention rate. The quantitative metrics are reported in a key-value format organized by evaluated models (see Fig. 1, Results box).

4 Experiments

APEx produces free-form text summarizing experimental findings. A qualitative example is shown in Fig. 1. Here, we focus on quantitative evaluations.

First, we assess whether APEx can replicate findings from a manually-designed benchmark [42] on data type identification, *i.e.*, recognizing image alterations (*e.g.*, "Can models identify vertical flip/Gaussian noise/cartoon-styled images?"). Second, we demonstrate APEx's flexibility across diverse queries, including transformation group analysis, semantic group classification (*e.g.*, dog breeds, plants, fruits), and recognition under style alterations (*e.g.*, "Can models recognize domestic animals with style-type transformations?").

[2] https://huggingface.co.

4.1 Reproducing Prior Findings: Data Type Identification

We compare APEx's conclusions with those from [42], which introduced two datasets (natural and synthetic) with single-animal images altered using 27 transformations across four categories: geometric, pixel, style, and semantic. The benchmark evaluated both contrastive (*e.g.*, CLIP [30]) and auto-regressive models (*e.g.*, LLaVA) by asking them to identify the applied transformation. Auto-regressive models were assessed via answer ranking [9,20]. Their main finding is that these models capture semantic content but struggle with visual data-type understanding.

To replicate this, we prompt APEx with queries like: "Can models identify {data-type} in images?", letting it autonomously define, run, and analyze the experiments. We rephrased ambiguous prompts when necessary (*e.g.*, "multi-different" became "multiple objects of different classes").

Unlike [42], which evaluated all data types jointly, APEx tests each independently. Since our accuracy scores are generally higher, we report normalized values in Fig. 2, ordering data types as in the original benchmark.

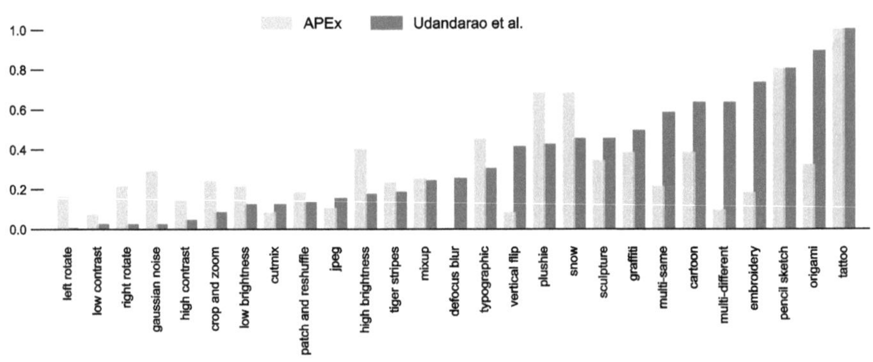

Fig. 2. Data types identification. Summary of the normalized results (vertical axis) achieved by the models across the 27 data type identification tasks, in comparison to those obtained in [42]. The values equal the min-max normalized performance across the set of experiments designed by APEx.

Despite differences in exact scores, APEx recovers similar trends—especially when grouped by category (geometric, pixel, style, semantic), as shown in Table 1. For instance, both methods find LMMs perform better on style/semantic transformations (*e.g.*, tattoos, embroidery) than on low-level changes (*e.g.*, low contrast, jpeg). This supports the key insight: LMMs are stronger in style/semantic recognition than in pixel/geometric understanding.

4.2 Flexibility Analyses

Coarse *Query* Handling. We test APEx's ability to handle high-level queries using transformation groups from [42]: geometric, pixel, semantic, and style.

Queries follow the template: "Can models identify {transformation-group} transformations?" This setting requires APEx to reason hierarchically, *i.e.*, selecting relevant tools (*e.g.*, geometric transformations) and designing diverse experiments, from binary detection to fine-grained classification within the same group. Results in Table 1 confirm previous trends: both APEx and [42] find geometric transformations harder (0.50 *vs.*0.10) and style manipulations easier (0.78 *vs.*0.46). APEx also tends to explore multiple variations within a group (*e.g.*, compression, noise, grayscale for pixel-level), and when queries are vague (*e.g.*, "style-type"), it runs broader comparisons, *e.g.*, dog oil paintings *vs.*bird pastels, or art deco *vs.* baroque structures.

Table 1. Data type group ranking Ranking of data type group recognition performance from the best (top) to worst (bottom). APEx achieves the same ranking as [42] both when testing each data type independently and aggregating metrics by their group (Avg. types), and when directly querying for the group understanding (Groups).

Rank	APEx		[42]
	Avg. types	Groups	
1	Style	Style	Style
2	Semantic	Semantic	Semantic
3	Pixel	Pixel	Pixel
4	Geometric	Geometric	Geometric

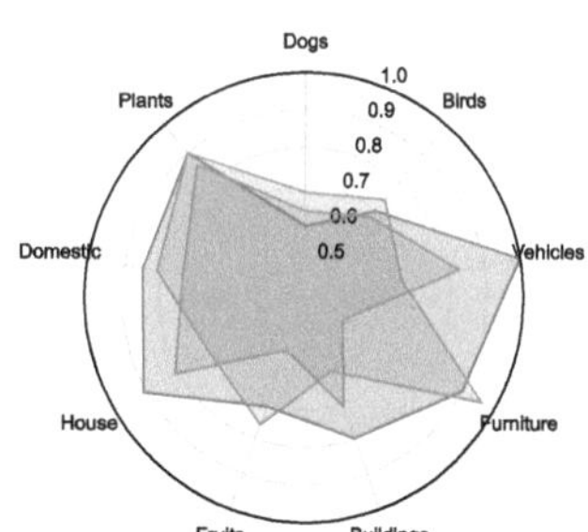

Fig. 3. Data classes recognition Accuracy averaged over experiments of APEx for BLIP-2, IDEFICS, and LLaVA on the nine recognition tasks. (Color figure online)

Data Class Recognition. To assess semantic understanding, we prompt APEx with queries over nine fine-grained class groups (from ImageNet-X [14]): dog breeds, birds, vehicles, buildings, furniture, fruits, plants, household objects, and domestic animals.[3] Example query: "Can models recognize dog breeds?"

[3] We exclude overly generic or redundant groups (*e.g.*, "other", "commodity").

Fig. 3 shows average class-wise accuracy per model. BLIP-2 generally leads, particularly on vehicles, buildings, and household items. LLaVA excels on birds and dogs, while BLIP-2 dominates broader categories (*e.g.*, domestic animals). Notable examples include: (i) only BLIP-2 distinguishes cathedrals from skyscrapers (0.88 accuracy); all others confuse them; (ii) LLaVA fails on fruit color recognition (random accuracy), while BLIP-2 achieves perfect performance; (iii) fine-grained distinctions like husky *vs.*malamute or Persian *vs.*Siamese are correctly addressed by APEx's generated experiments.

Combined queries: class under transformation. Finally, we combine semantic and transformation dimensions with queries like: "Can models recognize {class} under {data-type}?" across the 9 class groups and 4 transformation groups. Results in Table 2 reflect earlier trends: performance improves from geometric to style transformations, with model strengths varying by task. BLIP-2 excels on objects (*e.g.*, vehicles), LLaVA on animals, and IDEFICS occasionally outperforms both (*e.g.*, fruits under geometric/semantic shifts). These insights highlight how APEx facilitates broad and deep analysis with minimal manual effort, adapting to both query granularity and domain complexity.

Table 2. Data classes robustness to data types. Summary of the average accuracy achieved by BLIP-2 , IDEFICS , and LLaVA across the nine recognition tasks when adding different data type transformations. Accuracy is averaged across the set of experiments designed by APEx.

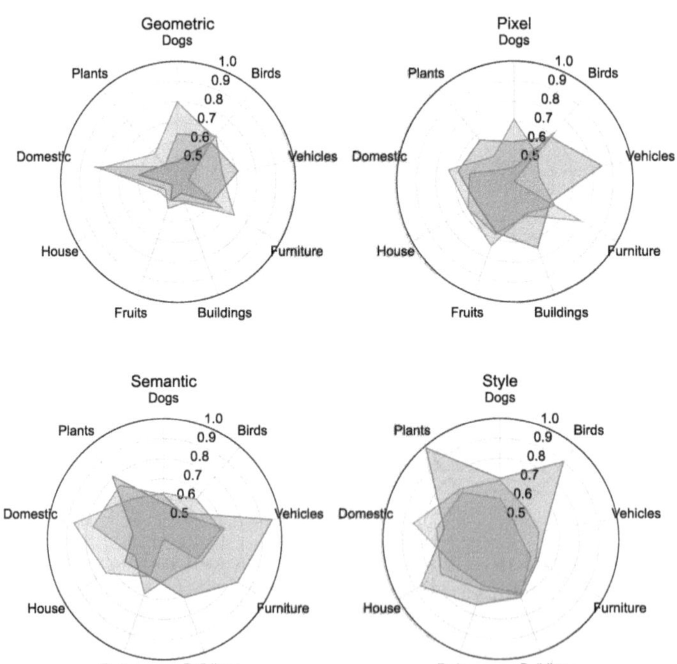

5 Conclusion

APEx introduces a new paradigm for benchmarking Large Multimodal Models (LMMs), automating the full evaluation pipeline, *i.e.*, from experimental design to execution and analysis. By substantially reducing the manual effort typically required for benchmarking, APEx lowers the risk of subjective biases and increases reproducibility. Its modular design also ensures extensibility, allowing seamless integration of additional tools and functionalities as the ecosystem evolves.

Through extensive experiments, we demonstrate that APEx can reproduce findings from prior manually curated studies, validating its reliability in generating meaningful insights. Beyond its direct application to LMM evaluation, APEx contributes to the broader AI community by offering an automated, general-purpose, and scalable benchmarking framework—supporting more systematic and comprehensive model analysis.

Limitations. Despite its strong performance, APEx inherits key limitations of visual programming frameworks [12,34,38], particularly the reliance on the reasoning capabilities of the underlying language model. We observe failure cases where the LLM misunderstands the report or query (*e.g.*, repeating questions), misinterprets the task (*e.g.*, selecting irrelevant transformations), or generates ambiguous or incorrect answers. Some failure modes are more subtle, such as "meta" questions where APEx inadvertently shifts focus to analyzing the LMM itself instead of testing it. While these issues present challenges, many can be mitigated through self-healing control mechanisms (see Sect. 3.4) or resolved by adopting more advanced LLMs (*e.g.*, GPT-4o), which can be easily integrated thanks to the system's modularity.

Acknowledgments. This work was sponsored by the MUR PNRR project FAIR Future AI Research (PE00000013), funded by NextGeneration, and the Italian Ministerial grants PRIN 2022: "BFAIR: Bias-Free Artificial Intelligence methods for automated visual Recognition" (CUP E53D23008010006), and the EU project ELIAS (No. 01120237).

Disclosure of Interests. The authors have no competing interests to declare that are relevant to the content of this article.

References

1. Andreas, J., Rohrbach, M., Darrell, T., Klein, D.: Neural module networks. In: CVPR (2016)
2. Antol, S., Agrawal, A., Lu, J., Mitchell, M., Batra, D., Zitnick, C.L., Parikh, D.: VQA: visual question answering. In: ICCV (2015)
3. Brown, T., Mann, B., Ryder, N., Subbiah, M., Kaplan, J.D., Dhariwal, P., Neelakantan, A., Shyam, P., Sastry, G., Askell, A., et al.: Language models are few-shot learners. In: NeurIPS (2020)

4. Chen, L., Zhang, Y., Ren, S., Zhao, H., Cai, Z., Wang, Y., Liu, T., Chang, B.: Towards end-to-end embodied decision making with multi-modal large language model: explorations with GPT4-vision and beyond. In: NeurIPS-WS (2023)
5. Chen, S., Gu, J., Han, Z., Ma, Y., Torr, P., Tresp, V.: Benchmarking robustness of adaptation methods on pre-trained vision-language models. In: NeurIPS (2024)
6. Chen, Y., Hu, H., Luan, Y., Sun, H., Changpinyo, S., Ritter, A., Chang, M.W.: Can pre-trained vision and language models answer visual information-seeking questions? In: EMNLP (2023)
7. Cheng, S., Guo, Z., Wu, J., Fang, K., Li, P., Liu, H., Liu, Y.: Can vision-language models think from a first-person perspective? (2023) arXiv preprint arXiv:2311.15596
8. Cho, J., Zala, A., Bansal, M.: Visual programming for step-by-step text-to-image generation and evaluation. In: NeurIPS (2024)
9. Dai, W., Li, J., Li, D., Tiong, A.M.H., Zhao, J., Wang, W., Li, B., Fung, P.N., Hoi, S.: Instructblip: towards general-purpose vision-language models with instruction tuning. In: NeurIPS (2024)
10. Fang, A., Ilharco, G., Wortsman, M., Wan, Y., Shankar, V., Dave, A., Schmidt, L.: Data determines distributional robustness in contrastive language image pre-training (clip). In: ICML. PMLR (2022)
11. Gavrikov, P., Lukasik, J., Jung, S., Geirhos, R., Lamm, B., Mirza, M.J., Keuper, M., Keuper, J.: Are vision language models texture or shape biased and can we steer them? (2024). arXiv preprint arXiv:2403.09193
12. Gupta, T., Kembhavi, A.: Visual programming: compositional visual reasoning without training. In: CVPR (2023)
13. Hsieh, C.Y., Zhang, J., Ma, Z., Kembhavi, A., Krishna, R.: Sugarcrepe: fixing hackable benchmarks for vision-language compositionality. In: NeurIPS (2024)
14. Idrissi, B.Y., Bouchacourt, D., Balestriero, R., Evtimov, I., Hazirbas, C., Ballas, N., Vincent, P., Drozdzal, M., Lopez-Paz, D., Ibrahim, M.: Imagenet-X: understanding model mistakes with factor of variation annotations (2022). arXiv preprint arXiv:2211.01866
15. Kamath, A., Hessel, J., Chang, K.W.: What's "up" with vision-language models? Investigating their struggle with spatial reasoning. In: The 2023 Conference on Empirical Methods in Natural Language Processing (2023)
16. Khan, Z., BG, V.K., Schulter, S., Fu, Y., Chandraker, M.: Self-training large language models for improved visual program synthesis with visual reinforcement. In: CVPR (2024)
17. Laurençon, H., Saulnier, L., Tronchon, L., Bekman, S., Singh, A., Lozhkov, A., Wang, T., Karamcheti, S., Rush, A., Kiela, D., et al.: Obelics: an open web-scale filtered dataset of interleaved image-text documents. In: NeurIPS (2023)
18. Li, C., Wong, C., Zhang, S., Usuyama, N., Liu, H., Yang, J., Naumann, T., Poon, H., Gao, J.: LLaVA-Med: training a large language-and-vision assistant for biomedicine in one day. In: NeurIPS (2024)
19. Li, J., Li, D., Savarese, S., Hoi, S.: Blip-2: bootstrapping language-image pre-training with frozen image encoders and large language models. In: ICML (2023)
20. Lin, S., Hilton, J., Evans, O.: TruthfulQA: measuring how models mimic human falsehoods. In: ACL (2022)
21. Lu, P., Peng, B., Cheng, H., Galley, M., Chang, K.W., Wu, Y.N., Zhu, S.C., Gao, J.: Chameleon: plug-and-play compositional reasoning with large language models. In: NeurIPS (2024)
22. Ma, M., Ren, J., Zhao, L., Testuggine, D., Peng, X.: Are multimodal transformers robust to missing modality? In: CVPR (2022)

23. Ma, Z., Hong, J., Gul, M.O., Gandhi, M., Gao, I., Krishna, R.: Crepe: can vision-language foundation models reason compositionally? In: CVPR (2023)
24. Mayilvahanan, P., Wiedemer, T., Rusak, E., Bethge, M., Brendel, W.: Does clip's generalization performance mainly stem from high train-test similarity? In: NeurIPS-WS (2023)
25. OpenAI: Introducing chatGPT (2022). https://openai.com/blog/chatgpt
26. Parisi, A., Zhao, Y., Fiedel, N.: TALM: tool augmented language models (2022). arXiv preprint arXiv:2205.12255
27. Podell, D., English, Z., Lacey, K., Blattmann, A., Dockhorn, T., Müller, J., Penna, J., Rombach, R.: SDXL: improving latent diffusion models for high-resolution image synthesis (2023). arXiv preprint arXiv:2307.01952
28. Qin, Y., Liang, S., Ye, Y., Zhu, K., Yan, L., Lu, Y., Lin, Y., Cong, X., Tang, X., Qian, B., et al.: ToolLLM: facilitating large language models to master 16000+ real-world APIs. In: ICLR (2024)
29. Qiu, J., Zhu, Y., Shi, X., Wenzel, F., Tang, Z., Zhao, D., Li, B., Li, M.: Benchmarking robustness of multimodal image-text models under distribution shift. J. Data-Centric Mach. Learn. Res. (2023)
30. Radford, A., Kim, J.W., Hallacy, C., Ramesh, A., Goh, G., Agarwal, S., Sastry, G., Askell, A., Mishkin, P., Clark, J., et al.: Learning transferable visual models from natural language supervision. In: ICML (2021)
31. Rombach, R., Blattmann, A., Lorenz, D., Esser, P., Ommer, B.: High-resolution image synthesis with latent diffusion models. In: CVPR (2022)
32. Russakovsky, O., Deng, J., Su, H., Krause, J., Satheesh, S., Ma, S., Huang, Z., Karpathy, A., Khosla, A., Bernstein, M., et al.: Imagenet large scale visual recognition challenge. IIJCV **115**, 211–252 (2015)
33. Schick, T., Dwivedi-Yu, J., Dessì, R., Raileanu, R., Lomeli, M., Hambro, E., Zettlemoyer, L., Cancedda, N., Scialom, T.: Toolformer: language models can teach themselves to use tools. In: NeurIPS (2024)
34. Shaham, T.R., Schwettmann, S., Wang, F., Rajaram, A., Hernandez, E., Andreas, J., Torralba, A.: A multimodal automated interpretability agent (2024). arXiv preprint arXiv:2404.14394
35. Shen, Y., Song, K., Tan, X., Li, D., Lu, W., Zhuang, Y.: HuggingGPT: solving AI tasks with chatGPT and its friends in hugging face. In: NeurIPS (2024)
36. Shi, Z., Wang, Z., Fan, H., Yin, Z., Sheng, L., Qiao, Y., Shao, J.: Chef: a comprehensive evaluation framework for standardized assessment of multimodal large language models (2023). arXiv preprint arXiv:2311.02692
37. Subramanian, S., Narasimhan, M., Khangaonkar, K., Yang, K., Nagrani, A., Schmid, C., Zeng, A., Darrell, T., Klein, D.: Modular visual question answering via code generation. In: ACL (2023)
38. Surís, D., Menon, S., Vondrick, C.: ViperGPT: visual inference via python execution for reasoning. In: ICCV (2023)
39. Tang, Y., Yamada, Y., Zhang, Y., Yildirim, I.: When are lemons purple? The concept association bias of vision-language models. In: EMNLP (2023)
40. Thrush, T., Jiang, R., Bartolo, M., Singh, A., Williams, A., Kiela, D., Ross, C.: Winoground: probing vision and language models for visio-linguistic compositionality. In: CVPR (2022)
41. Touvron, H., Lavril, T., Izacard, G., Martinet, X., Lachaux, M.A., Lacroix, T., Rozière, B., Goyal, N., Hambro, E., Azhar, F., Rodriguez, A., Joulin, A., Grave, E., Lample, G.: Llama: open and efficient foundation language models (2023). arXiv preprint arXiv:2302.13971

42. Udandarao, V., Burg, M.F., Albanie, S., Bethge, M.: Visual data-type understanding does not emerge from scaling vision-language models. In: ICLR (2024)
43. Udandarao, V., Prabhu, A., Ghosh, A., Sharma, Y., Torr, P.H., Bibi, A., Albanie, S., Bethge, M.: No "zero-shot" without exponential data: pretraining concept frequency determines multimodal model performance (2024). arXiv preprint arXiv:2404.04125
44. Ye, S., Lauer, J., Zhou, M., Mathis, A., Mathis, M.: AmadeusGPT: a natural language interface for interactive animal behavioral analysis. In: NeurIPS (2024)
45. Yuan, Z., Ren, J., Feng, C.M., Zhao, H., Cui, S., Li, Z.: Visual programming for zero-shot open-vocabulary 3D visual grounding. In: CVPR (2024)
46. Yuksekgonul, M., Bianchi, F., Kalluri, P., Jurafsky, D., Zou, J.: When and why vision-language models behave like bags-of-words, and what to do about it? In: ICLR (2022)
47. Zhang, G., Zhang, Y., Zhang, K., Tresp, V.: Can vision-language models be a good guesser? Exploring VLMS for times and location reasoning. In: WACV (2024)
48. Zhao, Y., Pang, T., Du, C., Yang, X., Li, C., Cheung, N.M.M., Lin, M.: On evaluating adversarial robustness of large vision-language models. In: NeurIPS (2024)

High-Fidelity Infrared Scenario Simulation with Conditional CycleGAN

Margherita Barbuti[1(✉)], Fabio Dell'Acqua[2], Roberto Conte[1], and Simone Finelli[1]

[1] Leonardo S.p.A., Nerviano 20014, Italy
{margherita.barbuti,roberto.conte,simone.finelli}@leonardo.com
[2] Department of Industrial and Information Engineering, University of Pavia, Pavia 27100, Italy
fabio.dellacqua@unipv.it

Abstract. Infrared (IR) scenario generators are essential for development and validation of IR-based imaging and surveillance systems using synthetic signals in place of costly and scarce real-world acquisitions. Traditional physics-based simulation models struggle to reproduce realistic terrain textures and atmospheric artifacts such as clouds. To address this limitation, we propose a Conditional CycleGAN model for enhancing simulated IR images, allowing for the guided generation of specific scene features. Our approach translates low-fidelity simulated data into high-fidelity IR images while enabling user control over scene attributes. Experimental evaluations demonstrate that our method produces visually and statistically accurate textures, improving the realism of synthetic IR data.

Keywords: Infrared simulation · Conditional CycleGAN · Texture synthesis · Remote sensing · Deep learning

1 Introduction

The generation of synthetic infrared (IR) signals and images is widely used to validate systems based on IR sensors. While current physical models reliably simulate radiation sources and propagation, they struggle to reproduce complex textures and environmental artifacts, limiting realism in synthetic IR imagery. *Texture synthesis* is essential for overcoming this limitation, enhancing the realism of synthetic landscapes in virtual environments and terrain modeling.

Traditional texturing methods often require manual design and high computational resources, making them impractical for large-scale IR datasets. The advent of generative networks, particularly Generative Adversarial Networks (GANs), has transformed landscape texturing by automating the creation of high-quality, realistic textures. GAN-based models, such as those developed by Efros and Leung [2] and Efros and Freeman [3], replicate the intricate statistical

E. Rodolà et al. (Eds.): ICIAP 2025, LNCS 16167, pp. 597–607, 2026.
https://doi.org/10.1007/978-3-032-10185-3_47

properties of natural textures more effectively than earlier procedural methods. More recent advancements, including image-to-image translation models like Conditional GANs [5] and CycleGAN [15], enable the generation of diverse and structurally coherent landscapes from simple sketches or segmentation maps. These models have significantly improved synthetic landscape realism, benefiting applications in remote sensing, video games, and algorithmic simulation [10].

In this work, we enhance the realism of synthetic IR imaging by introducing a Conditional CycleGAN, an extension of CycleGAN that enables guided image-to-image translation. Unlike traditional CycleGAN, our model incorporates labeled information, allowing explicit control over key environmental elements such as clouds. Realistic landscape generation in IR imaging extends beyond visual realism, requiring statistical consistency in texture synthesis. Standard GAN-based image synthesis often overlooks the spatial and statistical coherence needed for IR scene simulation. Ensuring that synthetic IR textures preserve correct spatial distributions and statistical correlations is essential for producing reliable datasets. To address this, textural metrics are employed, such as the Gray-Level Co-Occurrence Matrix (GLCM) [6], commonly used in computer vision, to quantitatively assess the realism of generated textures.

To our knowledge, this is the first application of Conditional CycleGANs for the controlled generation of clouds in IR imagery. By enabling more flexible and realistic synthetic data generation, our approach has the potential to simplify dataset creation for training and evaluating IR-based systems, such as detection and tracking, ultimately improving model robustness and generalization in this domain.

2 Background

2.1 Generative Adversarial Networks

Generative Adversarial Networks (GANs) [4] consist of a generator and a discriminator competing in a minimax process to synthesize realistic data samples, including images. Conditional GANs (cGANs) [12] extend this framework by incorporating auxiliary information, such as class labels or additional conditioning variables, to guide the generation process. This modification enhances control over the generated outputs, enabling targeted feature manipulation and improved relevance to specific tasks.

2.2 CycleGAN

The CycleGAN, proposed by Zhu et al. [15], builds upon the GAN framework to facilitate image-to-image translation between two unpaired domains. It introduces a cycle consistency constraint, ensuring that an image translated from one domain to another can be mapped back to its original form, preserving essential content. While CycleGAN effectively learns domain mappings without paired training data, it lacks explicit control over specific attributes of the generated images, a limitation addressed by cGANs through guided conditioning mechanisms.

2.3 Conditional CycleGAN

A conditional variant to the CycleGAN model has been previously explored in other works by [11] and [13], for guided generation of facial features and expressions. We take inspiration from these works in how the conditional information is provided to the model, merging the contributions of both papers. The aim of this work is to implement and train this conditional variant of the CycleGAN model, which we call Conditional CycleGAN (C-CycleGAN), for specialized, attribute-guided image generation.

3 Methodology

3.1 Dataset

The dataset used in our experiments includes real infrared (IR) images collected from an airborne sensor, alongside synthetic images produced using a physics-based IR scenario simulator. The real images contain diverse landscapes with complex textures and environmental variations, capturing essential elements like vegetation, urban and industrial structures, and cloud formations. Two examples of the same scene captured in the real world and in the scenario simulator are shown in Fig. 1.

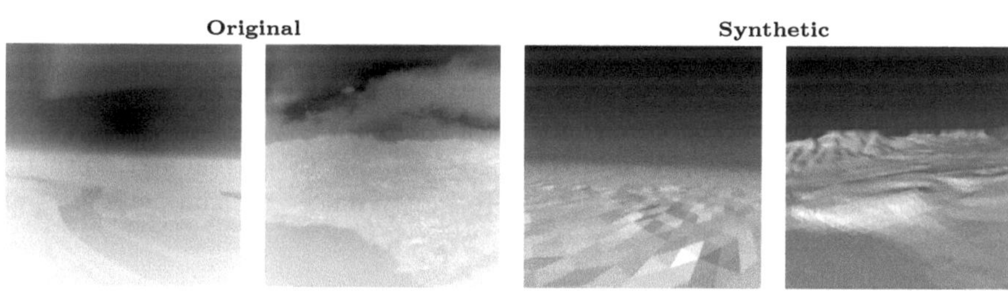

Fig. 1. Example of real and synthetic IR images of the same scene. Real images include complex environmental structures such as clouds, while synthetic ones lack these features.

The dataset is limited to a single accessible recording, duc to industrial security reasons; the original video is sampled at 1 Hz, resulting in 2606 frames comprising both train and test sets. The diversity of the data is constrained: landscape variations depend on the flight path (e.g., mountainous terrain only appears in a small portion of the video), and weather conditions are mostly stable, leading to consistent lighting and irradiance.

These limitations affect the model's ability to generalize, as it is inherently bounded by the variability in the training data. Still, these images provide a valuable reference for training deep learning models to replicate real-world infrared scenarios.

Conversely, synthetic images, although useful for testing, lack the detailed terrain features and realistic cloud structures necessary for effective training and evaluation of IR detection systems. The synthetic sequences are generated to closely match the available video recording in terms of acquisition geometry and received radiance. This is achieved by replicating the sensor's geographic location, configuration, and current environmental conditions. The optical behavior of the sensor is also modeled and these parameters are tuned to approximate the pixel value distribution of real images as closely as possible. The output of the simulator is 16-bit images at the same resolution as the original ones (see Sect. 3.3, where the received irradiance at the sensor aperture is linearly converted to digital numbers using gain and offset parameters.

IR Simulator Description. The scenario simulator used to produce synthetic images models the behavior of objects in the emissive infrared spectrum, assuming them to be gray bodies. Scenarios are built on Digital Elevation Models, and the terrain emission is modeled based on its heat capacity, temperature and solar exposure, depending its geometry and the sun position. The mesh grid is composed of triangles, each responding independently to these physical parameters. The simulator cannot produce visible artifacts such as clouds or urban areas, resulting in relatively plain landscapes, shaped only by terrain geometry. The radiance emitted by each surface (and the background) is attenuated by a transmissivity factor, computed using a custom implementation based on MODTRAN's [1] radiative transfer principles at different atmospheric layers. This allows to simulate atmospheric effects under various conditions; in this case, the Mid-Latitude Summer model was adopted in clear sky.

By using C-CycleGAN, we aim to bridge the gap between real and synthetic data by enhancing the synthetic dataset to better reflect the characteristics of real-world IR imagery.

3.2 Model Architecture

The C-CycleGAN framework is based on the architecture of the original CycleGAN [15], and consists of two generator-discriminator networks: one that transforms simulated images into realistic IR images and another that converts them back to the simulated domain to ensure consistency.

This bidirectional mapping allows for the preservation of critical scene details while introducing the necessary atmospheric and terrain-related enhancements. The generator and the discriminator networks are adopted from the CycleGAN [15]; the former consists of two stride-2 convolutions, several residual blocks [7], and two $\frac{1}{2}$-strided convolutions, while the latter is a 70×70 PatchGAN [9,15]. To control specific scene elements, such as cloud presence, we integrate conditional labels (e.g., "cloudy" vs. "clear sky") into the generator's input. This conditioning mechanism ensures that the model can selectively adjust specific scene attributes while maintaining structural coherence. During implementation,

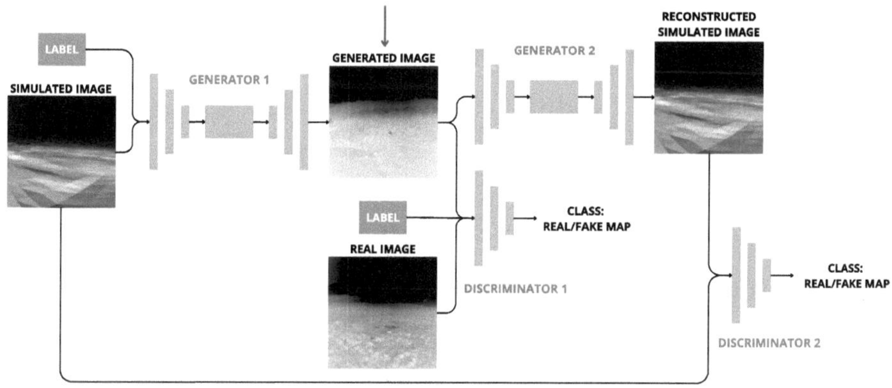

Fig. 2. Simplified architecture of the C-CycleGAN. The red arrow indicates the experiment output image. (Color figure online)

the conditional label is incorporated into $Generator_1$ [2] at three stages during training:

- At the beginning: given a $1 \times 256 \times 256$ grayscale image, the input of the model is a $2 \times 256 \times 256$ matrix, where the first channel is the image and the second channel is the homogeneous label matrix.
- Before the ResNet blocks (after downsampling): at this stage the image has a size of $256 \times 64 \times 64$, and the input is $257 \times 64 \times 64$ with the concatenated label.
- After the ResNet blocks (before upsampling): the ResNet blocks maintain the input size. By adding one extra channel for the label, the input of the third part of the model has dimension $258 \times 64 \times 64$.

The diagram in Fig. 2 provides a high-level overview of the model's architecture.

3.3 Training Details

The loss function used follows the standard CycleGAN formulation [15], modified to make $Generator_1$ and $Discriminator_1$ explicitly conditioned on an attribute label a. Due to synthetic images being always cloud-free, conditioning $Generator_2$ and $Discriminator_2$ was deemed unnecessary. Our results confirm that this design choice does not hinder performance.

The objective function is defined as:

$$\begin{aligned} L^{cond}_{cyclegan} =& L^{cond}_{GAN}(\mathcal{G}_1, \mathcal{D}_1, a) + L_{GAN}(\mathcal{G}_2, \mathcal{D}_2) \\ &+ \lambda_1 L^{cond}_{cycle}(\mathcal{G}_1, \mathcal{G}_2, a) + \lambda_2 L^{cond}_{id}(\mathcal{G}_1, \mathcal{G}_2, a) \end{aligned} \quad (1)$$

The composing loss terms are described below. Hyperparameters λ_1 and λ_2 control the relative importance of the cycle-consistency and identity losses. In this notation, $x \in \mathcal{X}$ denotes real IR images, $z \in \mathcal{Z}$ denotes simulated (synthetic) IR images, and $a \in \mathcal{A}$ is the conditioning attribute.

- **Adversarial Loss (Conditioned)** $L_{\text{GAN}}^{\text{cond}}$ encourages generated images to be indistinguishable from real ones:

$$L_{\text{GAN}}^{\text{cond}}(\mathcal{G}_1, \mathcal{D}_1, a) = \mathbb{E}_{x,a}\left[\log \mathcal{D}_1(x, a)\right] + \mathbb{E}_{z,a}\left[\log(1 - \mathcal{D}_1(\mathcal{G}_1(z, a), a))\right] \tag{2}$$

- **Adversarial Loss (Unconditioned)** L_{GAN} for the reverse mapping:

$$L_{\text{GAN}}(\mathcal{G}_2, \mathcal{D}_2) = \mathbb{E}_{z}\left[\log \mathcal{D}_2(z)\right] + \mathbb{E}_{x}\left[\log(1 - \mathcal{D}_2(\mathcal{G}_2(x)))\right] \tag{3}$$

- **Cycle-Consistency Loss** $L_{\text{cycle}}^{\text{cond}}$ ensures that image content is preserved across domain translations:

$$L_{\text{cycle}}^{\text{cond}}(\mathcal{G}_1, \mathcal{G}_2, a) = \mathbb{E}_{z,a}\left[\|\mathcal{G}_2(\mathcal{G}_1(z, a)) - z\|_1\right] + \mathbb{E}_{x,a}\left[\|\mathcal{G}_1(\mathcal{G}_2(x), a) - x\|_1\right] \tag{4}$$

- **Identity Loss** $L_{\text{id}}^{\text{cond}}$ discourages unnecessary modifications to images already in the target domain:

$$L_{\text{id}}^{\text{cond}}(\mathcal{G}_1, \mathcal{G}_2, a) = \mathbb{E}_{x,a}\left[\|\mathcal{G}_1(x, a) - x\|_1\right] + \mathbb{E}_{z}\left[\|\mathcal{G}_2(z) - z\|_1\right] \tag{5}$$

The model was trained on a dataset split into 85% training and 15% testing images. The training data was manually labeled with binary values: 0 for clear sky and 1 for cloudy sky images, taking advantage of the dataset's sequential nature for time efficiency. The test set was labeled based on the desired output (e.g., 0 for clear sky when the target is clear sky). Optimization was performed using the Adam optimizer with a learning rate of 0.0002, and training lasted 200 epochs. The input training images were rescaled, followed by a 256 × 256 random crop and a 50% chance of flipping. Test images were rescaled and center-cropped to 256 × 256. Both input and output images were in 16-bit format and were passed to the model in batches of 2.

For comparison purposes, two state of the art non-conditional image-to-image translation models were also trained on our dataset, namely the Pix2Pix model [9] and the CycleGAN [15].

3.4 Test and Evaluation

The test phase was conducted on 391 images, for two rounds. In the first round, all images were conditioned to generate a clear sky, in the second round, to generate clouds.

We assessed overall visual similarity, terrain consistency, and texture reproduction (clouds, terrain, and sky). Statistically, the texture of synthetic scene components is compared to that of real ones, such as clouds or terrain patches. This is done using histograms, as well as contrast and energy metrics derived from the Gray Level Co-Occurrence Matrix (GLCM), computed on the patches under analysis. The quality of the generated clear sky is compared to real and synthetic ones with vertical profile plots.

To compute clouds and clear sky statistics a non-fixed number of samples was selected, to capture the overall behavior of such attributes. Clouds were extracted from the sky regions using automatic segmentation based on binary clustering. Instead, due to their variability, single terrain patches were chosen based on visual similarity between real and generated samples, and compared with textural metrics. To improve the robustness of terrain evaluation, future work will focus on developing a semi-automatic method that enables comparison across a larger subset of the test data.

Perceptual and distribution-based metrics, such as LPIPS [14] and FID [8], which are commonly used in the evaluation of generative models, were not included in this study due to time constraints. Their integration is planned for future developments, to complement the existing visual and texture-based assessment.

4 Results

4.1 Visual Analysis

Figure 3 illustrate the highly realistic results achieved by C-CycleGAN in the translation of the two images, one per row, using label "clear sky" (left) and "cloudy sky" (right). Our approach successfully preserves critical terrain structures while introducing user-defined atmospheric variations. Generated clouds are visually similar to real ones, and variable in shape and size. The terrain texture in the generated instances is not affected by the conditional label, and it can deviate from the original terrain, highlighting the ability of the model to generate unseen patterns. The model's ability to adaptively modify atmospheric features while maintaining structural consistency makes it particularly advantageous for practical applications in IR-based detection systems.

4.2 Statistical Evaluation

Statistical metrics were evaluated separately for clouds, terrain and clear sky, with comparison between ground truth and synthetic instances.

Clouds. Figure 4 shows the distribution of mean Normalized Contrast and Energy in generated clouds, derived from the GLCM as described in [6], which describe contrast and pattern repetitivity within the image analyzed. The metrics were computed for a set of 40 samples for the ground truth clouds and the synthetic ones generated by C-CycleGAN, Pix2Pix, and CycleGAN. The contrast plot indicates that the C-CycleGAN, with respect to the other two models,

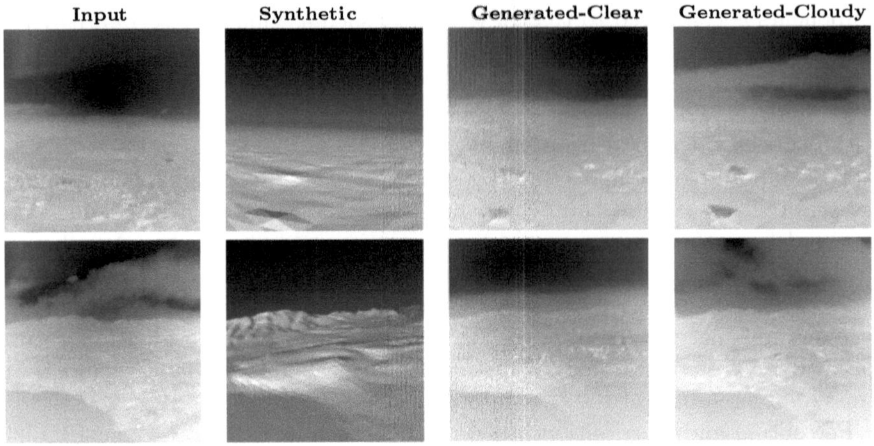

Fig. 3. Examples of generated IR images using the C-CycleGAN. For each test case (row), the left column shows the original and synthetic input; the right columns show the outputs conditioned to generate "clear sky" and "cloudy sky," respectively.

may favor more uniform cloud structure with less intensity variation. The distribution of mean energy for the C-CycleGAN shows high-fidelity to that of real data, albeit with less variability; the C-CycleGAN performs significantly better than the other two models in reproducing the spatial patterns found in real data.

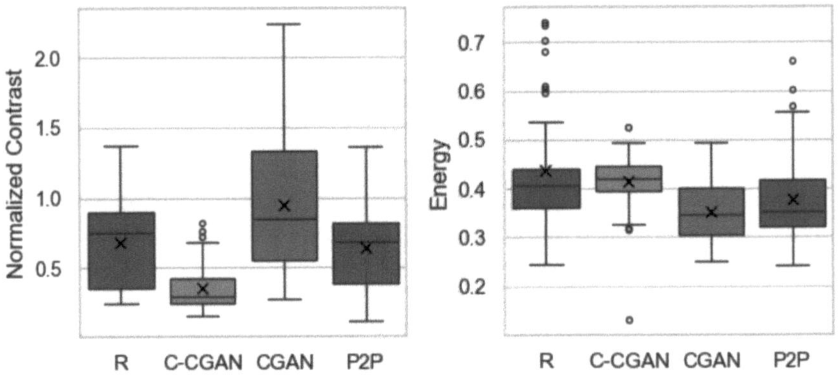

Fig. 4. Boxplot comparison of Normalized Contrast and Energy for real clouds and those generated by each model. C-CycleGAN more closely matches the real distribution, especially in spatial pattern fidelity.

Terrain. Terrain statistics were evaluated on similar patches extracted from a real image and a generated one for each generator model. The selection is made based on visual similarity to a real terrain patch in terms of colors and geometry, suggesting that the models are attempting to replicate the same textural distribution. The analyzed cropped area belongs to an urban landscape, characterized by bright buildings and dark spots; Fig. 5 shows the real patch and that extracted from a C-CycleGAN image.

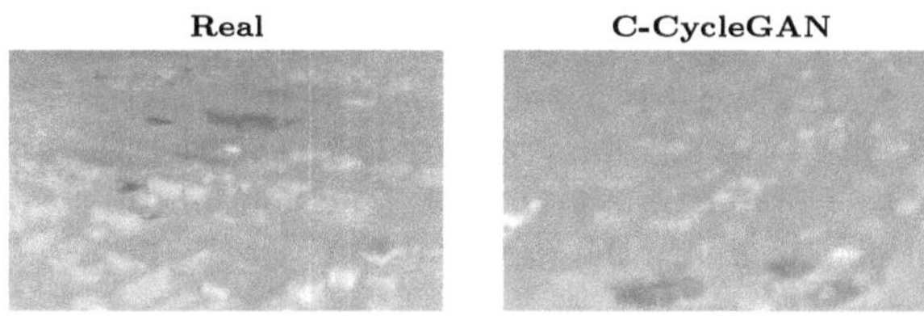

Fig. 5. Visual comparison of terrain patches from a real IR image and a C-CycleGAN-generated image. The patches show similar structure and energy values, confirming spatial consistency.

Table 1 reports pixel intensity ranges and standard deviation from the histograms built on the selected patches, as well as the values of normalized contrast and energy from the GLCM. Results show that the C-CycleGAN pixel distribution covers a narrower range of values to towards brighter values of the real data distribution. The lower standard deviation of the C-CycleGAN is reflected in a lower contrast value with respect to the the real data. Energy values are consistent across all analysed patches, signifying the presence of similar spatial patterns.

Table 1. Quantitative metrics computed on real and generated terrain patches. C-CycleGAN demonstrates a balanced compromise between brightness range, contrast, and energy, closely aligning with real data characteristics.

Dataset	Value range	St. Dev.	Norm. Contrast	Energy
Real	[49905, 51873]	224	0.12	0.73
Pix2Pix	[49572, 51904]	223	0.14	0.73
CycleGAN	[49974, 51948]	175	0.11	0.74
C-CycleGAN	[50196, 51968]	158	0.10	0.75

5 Conclusion

In this paper, we proposed a Conditional CycleGAN (C-CycleGAN) model for high-fidelity IR scenario simulation, enabling controlled generation of realistic textures and cloud formations. The approach improves the realism of synthetic

IR data while preserving structural consistency, making it suitable for training object detection algorithms. Our model achieves comparable visual and statistical fidelity to a traditional CycleGAN on the image test set, though it would benefit from a more varied dataset to improve generalization. The results highlight the effectiveness of the conditioning mechanism and suggest potential for further research.

As future work, we plan to assess the usefulness of the generated images in downstream tasks. Specifically, we will explore their impact on infrared object detection, by testing standard detectors on the enhanced synthetic data and testing on real scenes. We also aim to extend the model for temporal synthesis, allowing us to test on object tracking and motion analysis tasks. These experiments will help evaluate the practical value of the proposed approach for IR-based systems.

References

1. Berk, A., et al.: MODTRAN5: a reformulated atmospheric band model with auxiliary species and practical multiple scattering options. In: Larar, A.M., Suzuki, M., Tong, Q. (eds.) Multispectral and Hyperspectral Remote Sensing Instruments and Applications II. vol. 5655, pp. 88 – 95. International Society for Optics and Photonics, SPIE (2005). https://doi.org/10.1117/12.578758
2. Efros, A., Leung, T.: Texture synthesis by non-parametric sampling. In: Proceedings of the Seventh IEEE International Conference on Computer Vision. vol. 2, pp. 1033–1038 (1999). https://doi.org/10.1109/ICCV.1999.790383
3. Efros, A.A., Freeman, W.T.: Image quilting for texture synthesis and transfer. Association for Computing Machinery, New York 1 edn. (2023). https://doi.org/10.1145/3596711.3596771
4. Goodfellow, I.J., et al.: Generative adversarial networks (2014). https://doi.org/10.48550/arXiv.1406.2661
5. Guérin, E., Digne, J., Galin, E., Peytavie, A., Wolf, C., Benes, B., Martinez, B.: Interactive example-based terrain authoring with conditional generative adversarial networks. ACM Trans. Graph. **36**(6) (2017). https://doi.org/10.1145/3130800.3130804
6. Haralick, R.M., Shanmugam, K., Dinstein, I.: Textural features for image classification. IEEE Trans. Syst. Man Cybern. **SMC-3**(6), 610–621 (1973). https://doi.org/10.1109/TSMC.1973.4309314
7. He, K., Zhang, X., Ren, S., Sun, J.: Deep residual learning for image recognition. CoRR abs/1512.03385 (2015). https://doi.org/10.48550/arXiv.1512.03385
8. Heusel, M., Ramsauer, H., Unterthiner, T., Nessler, B., Hochreiter, S.: Gans trained by a two time-scale update rule converge to a local nash equilibrium. In: Guyon, I., Luxburg, U.V., Bengio, S., Wallach, H., Fergus, R., Vishwanathan, S., Garnett, R. (eds.) Advances in Neural Information Processing Systems. vol. 30. Curran Associates, Inc. (2017). https://doi.org/10.48550/arXiv.1706.08500
9. Isola, P., Zhu, J.Y., Zhou, T., Efros, A.A.: Image-to-image translation with conditional adversarial networks (2016). https://doi.org/10.48550/arXiv.1611.07004
10. Lee, G., Yim, J., Kim, C., Kim, M.: Stylandgan: a stylegan based landscape image synthesis using depth-map (2022). https://doi.org/10.48550/arXiv.2205.06611

11. Lu, Y., Tai, Y.W., Tang, C.K.: Attribute-guided face generation using conditional cyclegan (2018). https://doi.org/10.48550/arXiv.1705.09966
12. Mirza, M., Osindero, S.: Conditional generative adversarial nets (2014). https://doi.org/10.48550/arXiv.1411.1784
13. Tesei, G.: Generating realistic facial expressions through conditional cycle-consistent generative adversarial networks (ccyclegan) (2019)
14. Zhang, R., Isola, P., Efros, A.A., Shechtman, E., Wang, O.: The unreasonable effectiveness of deep features as a perceptual metric. arXiv preprint arXiv:1801.03924 abs/1801.03924 (2018). https://doi.org/10.48550/arXiv.1801.03924
15. Zhu, J.Y., Park, T., Isola, P., Efros, A.A., Research, B.A.: Unpaired image-to-image translation using cycle-consistent adversarial networks (2017). https://doi.org/10.48550/arXiv.1703.10593

Multi-Frame Alignment for Video Super-Resolution Using Attention

Marco Di Rienzo[1], Matteo Bruni[2], Leonardo Galteri[3], Federico Becattini[4](✉), and Marco Bertini[1]

[1] University of Florence, Florence, Italy
{marco.dirienzo,marco.bertini}@unifi.it
[2] Small Pixels, Florence, Italy
matteo.bruni@small-pixels.com
[3] Pegaso Telematic University, Naples, Italy
leonardo.galteri@unipegaso.it
[4] University of Siena, Siena, Italy
federico.becattini@unisi.it

Abstract. This paper presents a novel attention-based video super-resolution (VSR) method that avoids costly optical flow estimation while effectively exploiting temporal correlations between frames. We propose an aligner module that utilizes cross-attention to blend relevant patches from adjacent frames, gathering information from multiple frames simultaneously. This method improves upon traditional flow-based approaches by working at a block level and enabling the blending of several pixels, yielding better alignment for larger motions. The proposed VSR technique can upscale videos up to 4x while simultaneously removing compression artifacts, enhancing both resolution and quality. Experimental results demonstrate the effectiveness of this approach compared to classic flow-based methods, particularly in handling compressed videos where compression artifacts can severely impact optical flow estimation.

Keywords: Video super-resolution · Attention · Artifact Removal

1 Introduction

Video super-resolution (VSR) techniques aim to improve the quality of a sequence of low-resolution frames by generating a high-resolution sequence. Single-Image Super-Resolution (SISR) techniques can be trivially applied to an entire video, obtaining a frame from each individual low-resolution input frame; however, in this way the temporal correlations that exist between frames are not exploited to further improve the produced output. Thus, the exploitation of temporal dependencies turns out to be essential for the proper functioning of VSR methods. Several VSR methods [3,6,14,23,25,36] rely on the alignment of temporal features within the pixel space using optical flow, which is used to estimate the motion between frames and warp them accordingly. Their standard

E. Rodolà et al. (Eds.): ICIAP 2025, LNCS 16167, pp. 608–620, 2026.
https://doi.org/10.1007/978-3-032-10185-3_48

pipeline involves feature alignment as the first step, followed by the actual super resolution task, where high-resolution details are reconstructed from the aligned frames. However, optical flow techniques are usually computationally expensive and thus are not viable for applications where real-time processing represents a hard constraint. Furthermore, optical flow limits alignment to adjacent frames: even though it is possible to align temporally distant frames, with large time offsets the method is prone to errors. In this case, temporal inaccuracies can significantly impact the motion estimation and the overall perceived quality of the super-resolved video, making it challenging to maintain high fidelity across video frames. Another serious complicating factor emerges when trying to increase the resolution of compressed videos [31], as the presence of compression artifacts and the removal of important information at the pixel level often become detrimental to optical flow estimation. In particular, the estimated motion can be severely distorted due to different noise patterns, resulting in poor performance in the alignment phase.

In this work, we adhere to the two-step paradigm involving an alignment module followed by a super-resolution convolutional model. However, instead of relying on costly flow-estimation steps, we improve the alignment module with an attention-based mechanism, capable of blending relevant patches from adjacent frames. In this way, we can easily gather information from multiple adjacent frames. The proposed solution is general and can replace any flow-based alignment strategy found in super-resolution pipelines. In addition, since we train the whole pipeline end-to-end, we can jointly tackle super-resolution and video artifact removal, which proves to help obtain better alignments and overall better visual quality in the reconstructed frames.

Our main contributions are the following:

- We propose an attention-based video upscaling method, that retrieves similar patches to gather information and improve the visual quality.
- Our solution avoids costly flow-alignment by building aligning patches via cross-attention. This removes the overhead of estimating a flow field and allows us to align information from multiple frames. In principle, our approach can be plugged-in in any existing method that relies on flow-alignment.
- The proposed method can upscale videos up to 4x and simultaneously removes compression artifacts, enhancing both the resolution and the quality of the videos. Experimental results demonstrate the effectiveness of our approach against classic flow-based approaches.

2 Related Work

A great amount of image super-resolution techniques have been developed, including SRCNN [10], FSRCNN [11], VDSR [15], ESPCN [28], and SRGAN [18]. These models leverage convolutional layers to extract features from the input low-resolution (LR) images. By incorporating deconvolutional layers in the final stages of the network, the image is upscaled directly within the model. Significant improvements have been achieved with recent transformer-based methods. Notable methods include IPT [7], SwinIR [21], UFormer [34], ART [37], CAT [9] and HAT [8]. VSR techniques instead enhance video quality by leveraging temporal information across multiple frames. Kappeler et al. [14] introduced VSRnet, a pioneering VSR technique using CNNs. This model aligns and processes consecutive frames to improve resolution. GAN-based techniques as in [32], employ adversarial training to produce high-resolution videos with enhanced perceptual quality. RNN-based methods such as [30], utilize temporal dependencies in video sequences to better capture motion and maintain consistency across frames. The typical VSR pipeline includes alignment, feature extraction and fusion, and reconstruction modules. Alignment is crucial to ensure that frames are accurately aligned before the super-resolution step. However, optical flow methods like SpyNet [26] can struggle with large motions. Attention-based alignment methods, such as the one proposed in this work, offer a more flexible approach by blending multiple reference blocks. Feature extraction and fusion from multiple frames allow the model to gather comprehensive information, improving the quality of the upscaled video, and the final reconstruction module refines the output, enhancing details and reducing artifacts. Recent innovations like BasicVSR [5] employ a bidirectional recurrent network to integrate backward and forward propagation, significantly improving temporal coherence and detail reconstruction, while EGVSR [4] introduces an end-to-end trainable system that relies on optical flow and warping for frame alignment, enhancing computational efficiency. Other recent strategies have shifted to use the flow to guide deformable convolutions [6,20,29] and deformable attention schemes [22]. To further enhance robustness against unaccurate optical flow, the approach proposed in [27] introduces patch alignment, which averages the motion vectors within predefined grid areas.

3 Method

We aim to predict the high-resolution frame SR_t using an input combination of the low-resolution frame I_t and a sequence of neighboring low-resolution frames $(\ldots, I_{t-2}, I_{t-1}, I_{t+1}, I_{t+2}, \ldots)$. The network is parameterized by weights θ where each θ^l denotes the weights for a certain layer l. For a training set of input low-resolution frames I_i, and the corresponding target high-resolution frames T_i, with $i \in \{1, ..., N\}$, where i represents a specific training sample index, the model is trained for the parameters $\hat{\theta}$ that minimize the error between the target and the prediction. The model learns the function h_θ where $SR_i = h_\theta(I_i)$, and approximates SR_i as close as possible to the target reference frame T_i.

Video super-resolution pipelines often involve a first step of alignment where the neighboring frames are aligned towards the current frame. In this work, we propose a novel alignment method solely based on cross-attention, avoiding typical optical flow estimation and warping operations.

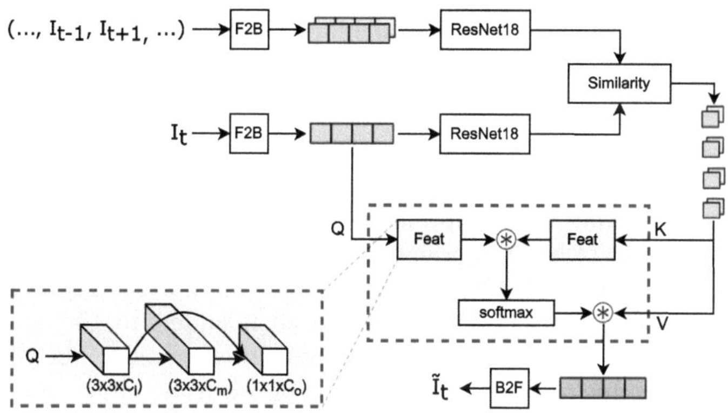

Fig. 1. Aligner module. Blocks inside the shaded box compute the cross-attention. The features for keys and queries are extracted by the convolutional layers shown in the yellow box, C indicates hyper-parameters for the channels dimension. (Color figure online)

Aligner Module. The proposed alignment method is based on cross-attention and its scheme is shown in Fig. 1. The first step of our method is to divide the reference frames $(\ldots, I_{t-2}, I_{t-1}, I_{t+1}, I_{t+2}, \ldots)$ and the target frame (I_t) into non-overlapping blocks (F2B operation in the scheme) and stacking them along the batch dimension. Then, for each block of the target frame, we find the most similar blocks in each of the reference frames: to achieve this, features of the blocks are extracted by passing them through a pre-trained ResNet18, and then a similarity between target and reference blocks is computed to only keep the k most similar blocks. This step allows computing the cross attention on a limited number of blocks, significantly reducing the computational burden.

Once we have the blocks from the target frame and the top-k blocks for each reference frame, we compare them via cross-attention. We use the reference blocks as V (value), while we use cascaded residual blocks, shown in the yellow box in Fig. 1, to extract features from the two sets of target and reference blocks, which we use as Q (query) and K (key), respectively. Q and K are matrix multiplied to create a dot-product attention matrix $A = \text{softmax}\left(\frac{QK^T}{\sqrt{d_k}}\right) V$. Note that the matrix A maps images from the K domain to the Q domain via an adaptive linear combination; that is, the matrix represents a linear combination map that allows pixel values from the reference blocks to be blended together to reconstruct pixels of the target block. This linear combination map is used to align references to the target. The new set of aligned blocks is then reorganized back to compose an image (B2F operation in the scheme), i.e., producing the aligned frame $\tilde{I}_t$. This method is advantageous over the typical optical flow and warping because optical flow computes the planar displacement between frames at a pixel level, while our method works at a block level and permits the blending of several pixels in the reference blocks to reconstruct a single pixel in the target block. This yields a better alignment on larger motions. By keeping the top-k

blocks, our method can also find similar blocks in distant areas of the reference frame, and use that in the subsequent cross-attention. Moreover, more reference frames can be considered and aligned at once.

Upsample Module. Once the aligned frame has been obtained, we concatenate it with the low-resolution frame along the channel dimension and send the resulting tensor as input to the upsample module. The upsample module draws inspiration from that of ESPCN [28]: it is composed of several consecutive residual blocks and a pixel-shuffle [28] layer at the end that performs the upsampling.

The output of the upsample module is then added to the original low-resolution target frame I_t upsampled via bicubic interpolation. The effectiveness of this operation is due to the high correlation between the input and the output frame, allowing the network to focus on predicting high frequencies instead of the whole image. The result is the estimated high-resolution frame SR_t. The overall scheme of the proposed architecture is shown in Fig. 2.

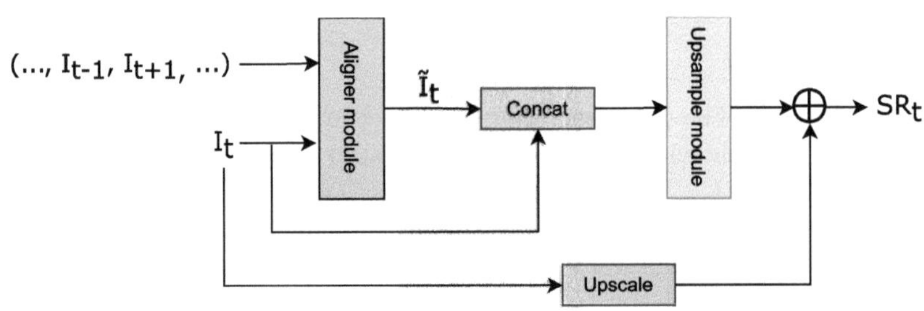

Fig. 2. Full scheme of our architecture.

Objective Function. During the training process, we aim to recover high-resolution signals from the low-resolution input images. Our objective function comprises several loss terms, which we outline in detail.

Pixel Loss. This term is a signal-based loss and is the most commonly used in SISR and VSR; it computes the L1 or Charbonnier distance [2] between each generated pixel in the predicted HR frame and the corresponding pixel in the ground truth HR frame. We use the Charbonnier-loss for better stability: $\mathcal{L}_{pixel} = \frac{1}{N}\sum_{i=1}^{N}\sqrt{(SR_i - T_i)^2 + \epsilon^2}$ with N the batch size, ϵ 10^{-3}, SR the predicted HR frame and T the HR frame.

Perceptual Loss. This term measures the perceptual similarity between the predicted HR frame and the ground truth HR frame. We use the LPIPS loss [35] with AlexNet [17] as backbone: $\mathcal{L}_{perceptual} = \frac{1}{N}\sum_{i=1}^{N} d_{\text{LPIPS}}(SR_i, T_i)$.

Alignment Loss. This term is used to push the aligner module toward a better reconstructed low-resolution frame and is similar to the loss used in optical flow-based methods, usually called the warping loss. It is a pixel loss with Charbonnier-distance between the aligned LR frame and the target LR frame:

$\mathcal{L}_{align} = \frac{1}{N}\sum_{i=1}^{N}\sqrt{(\tilde{I}_i - I_i)^2 + \epsilon^2}$ where N is the batch size, ϵ is 10^{-3}, $\tilde{I}$ is the aligned LR frame and I is the target LR frame. The total loss is a weighted sum of the aforementioned loss terms: $\mathcal{L}_{total} = \alpha\mathcal{L}_{pixel} + \beta\mathcal{L}_{perceptual} + \gamma\mathcal{L}_{align}$ where α, β and γ are the weights for each term.

(a) Motion image (b) SpyNet+warp [26] (c) Attention-based 2f

Fig. 3. Different alignments. (a) Average between reference and target; (b) alignment of SpyNet+Warp [26]; (c) attention-based alignment.

4 Experiments

In this section, we present the results of different experiments. All experiments are done on a single NVIDIA GeForce RTX 2080 Ti GPU with 10GB of memory.

Dataset and Metrics. We conduct all experiments on the *BVI-DVC* dataset [24]. The dataset consists of 200 sequences of frames, each truncated at the 64th frame, regardless of the frame rate, which ranges from 25 to 120. These sequences encompass various content types, including natural scenes, man-made objects, and city scenes. We created downscaled versions at 1080p and 270p resolutions from the 2160p originals, which will be our ground-truth and training data, respectively. From the 1080p downscaled sequences, we removed 4 sequences as test set[1]. The remaining sequences serve as the ground truth for our training data. To generate the compressed input data, each 270p sequence is compressed using the H.264 codec with a Constant Rate Factor (CRF) of 22. We chose a fixed CRF to mitigate the mode-collapse issue discussed in [12]. However, it is important to note that a fixed CRF does not ensure consistent quality across all frames; frames with motion may have lower quality compared to static frames. Given our objective of applying super-resolution to compressed videos (which involves both artifact reduction and super-resolution), it is crucial to train the models on the compressed videos rather than just downscaled versions of high-quality videos. Training the model solely for super-resolution could result in failure to detect features to super-resolve or even enlarge compression artifacts,

[1] *AmericanFootballS3Harmonics*, *BasketballS3YonseiUniversity*, *CityStreetS7IRIS*, *SquareS2IRIS*.

thereby diminishing the overall quality. In this study, we aim to produce results centered on both reconstruction accuracy and perceptual naturalness. Therefore, we employ metrics such as PSNR [13], SSIM [33], LPIPS [35]. These are full-reference metrics, meaning that they require both the target and the reconstructed output for quality assessment. Furthermore, we consider VMAF [19] which is a widely used full reference video metric, and ARNIQA [1], which is a no-reference metric, therefore not requiring the target image for the assessment. All metric evaluations are performed on the 3 RGB channels. Image quality metrics are individually calculated for each frame and averaged throughout the test video. On the other hand, the VMAF value is calculated by encoding the generated frames back into a video and evaluated against the target video directly.

Alignment Performance. In this experiment, we compare the alignment capabilities of the motion estimation/motion compensation with optical flow and warp, with our attention-based aligner module described in Sect. 3 with various temporal window sizes. We first warp the reference (e.g., the previous frame) to the target frame (i.e., the current frame) with optical flow estimated by SpyNet [26]. Then we perform the same alignment with our attention-based method. All experiments are performed with a block size of 16 with a stride of 16, and $k = 1$. An example pair and alignment results of both methods can be seen in Fig. 3.

The effectiveness of the optical flow-based techniques diminishes when confronted with significant motion. This decline primarily stems from the fact that this method constrains the changes between frames to planar geometric motion with a one-to-one correspondence of pixels. From the perspective of the attention mechanism, these constraints are equivalent to restricting the attention matrix to a permutation matrix where each row has only one entry as 1. However, in our approach, we can loosen this constraint by ensuring that the sum of each row equals 1 (achieved through softmax), instead of only one element. This relaxation permits the involvement of multiple pixels and blending, facilitating the alignment of the reference image to the target image which yields enhanced performance as shown in Table 1. For our attention-based alignment, we report the results taking into account a variable number N of reference images, where $\lceil N/2 \rceil$ images are taken from the previous frames and $\lfloor N/2 \rfloor$ images are taken from the subsequent frames. Interestingly, increasing the number of reference frames provides more relevant information to the model, yet even the single-reference aligner proves to be much more effective than the flow-based aligner.

Table 1. Alignment varying the amount of reference frames.

Method	#Ref	PSNR	SSIM	LPIPS
SpyNet+warp	1	23.21	0.873	0.096
Attention-based	1	29.47	0.966	0.042
Attention-based	2	36.95	0.973	0.034
Attention-based	4	**37.17**	**0.976**	**0.026**

Upscaling Videos by 4x. We upsample the input sequences of size (480 × 270 × 3) to a size of (1920 × 1080 × 3), for an upscale factor of 4×. Both lossless and compressed videos with a CRF of 22 are considered. Under our training setup, because we have access to both the input and the target data, the model is trained in a fully supervised fashion. Hence, the network is trained to super-resolve the input frame, for which the ground truth frame is available.

We randomly crop patches of input size (64 × 64 × 3) from the input frames for efficient training. The patches are randomly augmented with mirror flips and rotations. The entire pipeline is trained end-to-end using the loss described in Sect. 3. Adam optimizer [16] with default settings is used to train the model for 100 epochs with a batch size of 8. The learning rate is initialized to $5e-4$. The model is compared against a classic bicubic upsampling, i.e., the upsample module described in Sect. 3 without the aligner module (which we name SRNet - Super Resolution Network), and an optical flow based network similar to EGVSR [4], trained with similar hyperparameters and the same upsample module of our network for a fair comparison.

Table 2. Evaluation of 4× upsampling on videos without compression. Each metric is calculated on RGB channels.

Model	#Params	PSNR	SSIM	LPIPS	VMAF	ARNIQA
Bicubic	–	29.98	0.829	0.316	68.32	0.254
SRNet	1.3M	31.02	0.837	0.170	84.03	0.393
Flow+SRNet	2.7M	32.52	0.866	0.095	87.05	0.411
XA+SRNet	1.4M	**32.84**	**0.865**	**0.086**	**88.26**	**0.424**

Uncompressed Videos. We first train and test our model on uncompressed videos.The average metric values measured for the test videos are shown in Table 2. It can be seen that the SRNet model alone is better than the bicubic approach on PSNR and large improvements can be observed on perceptual metrics. VSR models outperform the SRNet method on every metric, underlining the importance of utilizing the additional information contained in the temporal dimension of videos. Our method also gains an advantage over the optical flow method. Examples of generated SR frames using the different models and the bicubic method are shown in Fig. 4 together with the input and target frames. The bicubic method only generates averaged pixel values and has issues generating the edges and the finer details, leading to the degradation of high-frequency information and thus a blurry image. In contrast, the frames generated by the models are clearer and have sharper edges compared to the bicubic method. Additionally, VSR models improve over the SR model by utilizing additional information from other frames, leading to an even clearer image and improved consistency between frames, according to the VMAF metric. Our method can recover more details than the flow-based model.

Fig. 4. 4× upsample on test clip *basket* and *square* not compressed.

Compressed Videos. In this experiment, we train our network on compressed video with H.264 codec and CRF = 22. These lossy/compressed images lack rich detail and often contain visual artifacts and aliasing. The average metric values measured for the test videos are shown in Table 3. It can be seen that on lossy images, the bicubic method does worse by a significant margin even on PSNR, due to the lack of ability to reduce compression artifacts and create sharp details. Similarly to the experiment on uncompressed videos, the VSR models outperform the SR model, especially on perceptual metrics. Samples of generated SR frames are shown in Fig. 5. Again, the bicubic method reconstructs a blurry image, but this time it also amplifies compression artifacts like ringing. On the other hand, the models can reconstruct a sharper image, and also reduce the compression artifacts. As before, VSR models improve over the SR model by utilizing additional information from other frames. In this case, the effect of leveraging temporal information is bigger due to the additional loss of information due to compression.

Table 3. Evaluation of 4× upsampling on videos compressed with CRF = 22. Each metric is calculated on RGB channels.

Model	#Params	PSNR	SSIM	LPIPS	VMAF	ARNIQA
Bicubic	-	28.64	0.787	0.395	49.16	0.235
SRNet	1.3M	29.26	0.799	0.229	61.17	0.336
Flow+SRNet	2.7M	29.67	0.795	0.176	64.69	0.366
XA+SRNet	1.4M	**29.99**	**0.802**	**0.155**	**65.11**	**0.372**

Fig. 5. 4× upsample on test clip *basket* and *square* compressed with CRF 22.

(a) no LL. (b) 1 LL frames. (c) 2 LL frames. (d) All LL. (e) GT

Fig. 6. 4× upsample with varying number of lossless (LL) frames. Improvements can be seen when lossless frames are available, especially between the upsample with no lossless frames and 1 lossless frame.

Table 4. 4× upsampling on videos compressed with CRF 22, varying the lossless frames in the aligner window.

Frames	PSNR	SSIM	LPIPS	VMAF	ARNIQA
1	30.83	0.838	0.126	68.88	0.374
2	31.12	0.830	0.110	71.03	0.378
3	31.15	0.831	0.105	71.93	0.380
4	31.16	0.831	0.104	72.13	0.381
5	31.16	0.832	0.102	**72.21**	0.381
6	**31.16**	**0.832**	**0.102**	72.20	**0.381**

Compressed Videos with Lossless Frames Available. In this experiment, we simulate a scenario in which a compressed video has to be upscaled, but also some lossless frames are available. A possible application scenario is a streaming setting where a subset of the frames are streamed at a high bitrate and the rest are streamed after being compressed. We use these additional lossless frames by inserting them in the window of the aligner module.

The resulting metrics on the test videos are shown in Table 4. It can be seen that between the model with all compressed frames and the model with just 1 lossless frame, there is a considerable increase of 0.84dB in PSNR and 3.77 points of VMAF. The increase is still sensible in PSNR and VMAF with an additional frame, to decrease to a negligible amount with additional lossless frames.

Table 5. Performance evaluation of 4× upsampling when the temporal window size is changed.

Frames	PSNR	SSIM	LPIPS	VMAF	ARNIQA
2	31.88	0.855	0.118	85.69	0.409
3	32.71	0.863	0.090	87.92	0.420
4	32.80	0.864	0.088	88.06	0.422
5	32.83	0.865	0.087	88.22	0.424
6	32.84	0.865	0.087	88.25	0.424
7	**32.84**	**0.865**	**0.086**	**88.26**	**0.424**

A visual comparison between upsampled compressed frames with a varying number of lossless frames in the temporal window is shown in Fig. 6. The difference is especially visible between Fig. 6 awhere no lossless frames are available and Fig. 6 bwhere 1 lossless frame is available. We can see that some color is added back to the image, e.g. the red stripes on the pole, the shapes are clearer and some background noise is removed. In Fig. 6 cwe can see the result with 2 lossless frames, where the image is even clearer and the shapes more defined. For comparison, the result using only lossless frames is shown in Fig. 6 d. Note that we use a window of 3 past frames and 3 future frames. The main difference lies in the shapes in the image. Compared with the ground truth in Fig. 6 e, the image using only lossless frames has consistent shapes, while the image using only two lossless frames still maintains the deformed shapes of the compressed source. This stems from the fact that the compressed image is still used as the query in our alignment module, making the lossless frames help in improving the general quality of the image, but leaving the deformed shapes as they are.

Ablation Studies. We first vary the temporal length of the input sequence from 2 (fusing 1 previous and reference frames) to 7 (fusing 3 previous, reference, and 3 future frames) to super-resolve one frame. The metrics on test videos are reported in Table 5. If we compare these results with those in Table 2, we see that with only 2 frames (previous and reference) the metrics are higher by a considerable margin than the upsample module alone (single frame), denoted as SRNet in the table, with an improvement of 0.86dB in PSNR and about 1.6 points of VMAF, but lower than the Flow+SRNet due to the additional information. With 3 frames (1 previous, reference, 1 future) the metrics surpass the flow method by about 0.2dB of PSNR and 0.87 points of VMAF, underlining the importance of considering the future frame to predict the reference one.

Then, we compare the results varying the number of top-k blocks in the aligner module. The idea is that also considering blocks that are not the exact same object but have similar content, can help the aligner. Results are shown in Table 6. For the considered test videos, only minimal gains are achieved when changing from $k = 1$ to $k = 2$, and no improvement in visual quality can be noticed.

Table 6. Evaluation of 4× upsampling when $k = 1$ and $k = 2$.

k	PSNR	SSIM	LPIPS	VMAF	ARNIQA
1	29.99	0.802	0.155	65.11	0.372
2	**30.08**	**0.803**	**0.148**	**65.28**	**0.374**

5 Conclusions

In this paper, we introduced an attention-based method for patch alignment in a video super-resolution pipeline. The advantage of our proposed aligner is that it avoids costly optical flow computations, commonly used in state of the art methods, and can be easily extended to take into account multiple frames. We empirically demonstrated the effectiveness of the proposed approach for video super-resolution up to 4x, while also addressing the issue of compression artifact removal. The proposed method is model agnostic and can be plugged into any super-resolution pipeline that leverages an alignment module, potentially improving any flow-based method.

References

1. Agnolucci, L., Galteri, L., Bertini, M., Del Bimbo, A.: Arniqa: learning distortion manifold for image quality assessment. In: Proc. of WACV (2024)
2. Barron, J.T.: A general and adaptive robust loss function (2019)
3. Caballero, J., et al.: Real-time video super-resolution with spatio-temporal networks and motion compensation. In: Proc. of CVPR, pp. 4778–4787 (2017)
4. Cao, Y., Wang, C., Song, C., Tang, Y., Li, H.: Real-time super-resolution system of 4k-video based on deep learning (2021)
5. Chan, K.C.K., Wang, X., Yu, K., Dong, C., Loy, C.C.: Basicvsr: the search for essential components in video super-resolution and beyond (2021)
6. Chan, K.C., Zhou, S., Xu, X., Loy, C.C.: Basicvsr++: improving video super-resolution with enhanced propagation and alignment. In: Proc. of CVPR (2022)
7. Chen, H., et al.: Pre-trained image processing transformer. In: Proc. of CVPR (2021)
8. Chen, X., Wang, X., Zhou, J., Qiao, Y., Dong, C.: Activating more pixels in image super-resolution transformer. In: Proc. of CVPR, pp. 22367–22377 (2023)
9. Chen, Z., Zhang, Y., Gu, J., Kong, L., Yuan, X., et al.: Cross aggregation transformer for image restoration. NeurIPS **35**, 25478–25490 (2022)
10. Dong, C., Loy, C.C., He, K., Tang, X.: Image super-resolution using deep convolutional networks (2015)
11. Dong, C., Loy, C.C., Tang, X.: Accelerating the super-resolution convolutional neural network (2016)
12. Galteri, L., Seidenari, L., Bertini, M., Bimbo, A.: Deep universal generative adversarial compression artifact removal. IEEE Trans. Multimedia **21** (2019)
13. Horé, A., Ziou, D.: Image quality metrics: PSNR vs. SSIM. In: 2010 20th International Conference on Pattern Recognition, pp. 2366–2369 (2010). https://api.semanticscholar.org/CorpusID:9506273
14. Kappeler, A., Yoo, S., Dai, Q., Katsaggelos, A.K.: Video super-resolution with convolutional neural networks. IEEE Trans. Comput. Imaging **2**, 109–122 (2016). https://api.semanticscholar.org/CorpusID:9356783
15. Kim, J., Lee, J.K., Lee, K.M.: Accurate image super-resolution using very deep convolutional networks (2016)
16. Kingma, D.P., Ba, J.: Adam: a method for stochastic optimization (2017)
17. Krizhevsky, A., Sutskever, I., Hinton, G.E.: Imagenet classification with deep convolutional neural networks. Commun. ACM **60**, 84 – 90 (2012). https://api.semanticscholar.org/CorpusID:195908774

18. Ledig, C., et al.: Photo-realistic single image super-resolution using a generative adversarial network (2017)
19. Li, Z., Aaron, A., Katsavounidis, I., Moorthy, A., Manohara, M.: Toward a practical perceptual video quality metric (2016). https://netflixtechblog.com
20. Liang, J., et al.: VRT: a video restoration transformer. IEEE Trans. Image Process. (2024)
21. Liang, J., Cao, J., Sun, G., Zhang, K., Van Gool, L., Timofte, R.: Swinir: image restoration using swin transformer. In: Proc. of ICCV, pp. 1833–1844 (2021)
22. Liang, J., et al.: Recurrent video restoration transformer with guided deformable attention (2022)
23. Liu, H., et al.: Video super resolution based on deep learning: a comprehensive survey (2022)
24. Ma, D., Zhang, F., Bull, D.R.: BVI-DVC: a training database for deep video compression. IEEE Trans. Multimedia **24**, 3847–3858 (2022)
25. Qing, T., Ying, X., Sha, Z., Wu, J.: Video super-resolution with pyramid flow-guided deformable alignment network. In: Proc. of ICEEMT. IEEE (2023)
26. Ranjan, A., Black, M.J.: Optical flow estimation using a spatial pyramid network (2016)
27. Shi, S., Gu, J., Xie, L., Wang, X., Yang, Y., Dong, C.: Rethinking alignment in video super-resolution transformers. NeurIPS **35**, 36081–36093 (2022)
28. Shi, W., et al.: Real-time single image and video super-resolution using an efficient sub-pixel convolutional neural network. Proc. CVPR (2016)
29. Teed, Z., Deng, J.: RAFT: recurrent all-pairs field transforms for optical flow. In: Vedaldi, A., Bischof, H., Brox, T., Frahm, J.-M. (eds.) ECCV 2020. LNCS, vol. 12347, pp. 402–419. Springer, Cham (2020). https://doi.org/10.1007/978-3-030-58536-5_24
30. Tian, Y., Zhang, Y., Fu, Y., Xu, C.: TDAN: temporally deformable alignment network for video super-resolution (2018)
31. Wang, J., Shao, Z., Huang, X., Lu, T., Zhang, R., Li, Y.: From artifact removal to super-resolution. IEEE Trans. Geosci. Remote Sens. **60**, 1–15 (2022)
32. Wang, X., Chan, K.C.K., Yu, K., Dong, C., Loy, C.C.: Edvr: video restoration with enhanced deformable convolutional networks (2019)
33. Wang, Z., Bovik, A.C., Sheikh, H.R., Simoncelli, E.P.: Image quality assessment: from error visibility to structural similarity. IEEE Trans. Image Process. **13**, 600–612 (2004). https://api.semanticscholar.org/CorpusID:207761262
34. Zamir, S.W., Arora, A., Khan, S., Hayat, M., Khan, F.S., Yang, M.H.: Restormer: efficient transformer for high-resolution image restoration. In: Proc. of CVPR, pp. 5728–5739 (2022)
35. Zhang, R., Isola, P., Efros, A.A., Shechtman, E., Wang, O.: The unreasonable effectiveness of deep features as a perceptual metric (2018)
36. Zhu, Q., Chen, F., Zhu, S., Liu, Y., Zhou, X., Xiong, R., Zeng, B.: Dvsrnet: deep video super-resolution based on progressive deformable alignment and temporal-sparse enhancement. IEEE Trans. Neural Netw. Learn. Syst. (2024)
37. Zhu, Q., Li, P., Li, Q.: Attention retractable frequency fusion transformer for image super resolution. In: Proc. of CVPR, pp. 1756–1763 (2023)

D^2R^2: Deep-Learning Based Real-Time Dense Object Reconstruction

Marco Lombardi[1], Mattia Savardi[2(✉)], and Alberto Signoroni[2]

[1] Department of Information Engineering, University of Brescia, Brescia, Italy
Marco.Lombardi@unibs.it
[2] Department of Medical and Surgical specialtics, Radiological Sciences and Public Health, University of Brescia, Brescia, Italy
{Mattia.Savardi,Alberto.Signoroni}@unibs.it

Abstract. The geometric reconstruction of real objects using short-range 3D measurements typically carried out with portable RGB-D scanners, presents several critical aspects that have not yet been fully resolved. One of the most challenging aspects is the recovery from tracking loss with respect to the forming model, especially in a real-time acquisition and reconstruction context. This paper presents a fully 3D pipeline for high-quality 3D object reconstruction. The pipeline integrates a traditional cumulative approach with robust tracking recovery solutions based on geometric deep learning and dynamic pose optimization. The solution is completely geometric, based on learned 3D feature extraction and matching. It significantly outperforms state-of-the-art reconstruction based on 2D features extracted from image data.

Keywords: 3D Object Reconstruction · Global Pose Optimization · Deep Learning-based view alignment · Real-time · Dense point clouds

1 Introduction

3D object reconstruction is crucial in fields such as manufacturing, reverse engineering, 3D printing prototyping, cultural heritage conservation, virtual reality, and robotics [9,21]. High-quality 3D models are typically generated from 2.5D partial views captured using handheld RGB-D scanners [28]. These devices combine RGB cameras and depth sensors to capture both visual and spatial information, providing real-time feedback to help users navigate around objects and ensure comprehensive surface data acquisition.

While 3D reconstruction of indoor environments has been extensively studied, single-object reconstruction demands higher geometric detail. This requires dense point clouds, accurate depth data (e.g., from infrared structured-light projection), and short-range scanning to exclude background interference. Such setups often introduce rapid changes in the scanning angle relative to the object surface, increasing the risk of tracking interruptions during reconstruction. Achieving geometric accuracy, real-time performance, and robustness to

E. Rodolà et al. (Eds.): ICIAP 2025, LNCS 16167, pp. 621–632, 2026.
https://doi.org/10.1007/978-3-032-10185-3_49

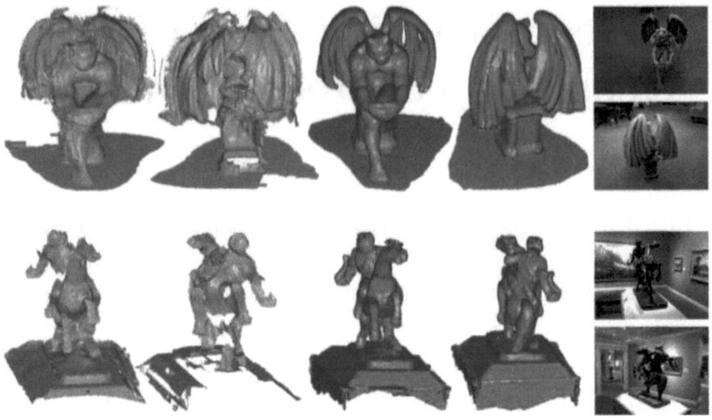

Fig. 1. 3D reconstruction of two objects from the Redwood dataset [4]. From left to right: BundleFusion (front and back, blue shading); our method (front and back, green shading); a few images from the acquired scenes.(Color figure online)

tracking loss remains a significant challenge that traditional and neural methods only partially address [13].

This work proposes a novel neural-driven pipeline for real-time 3D object reconstruction from high-frame-rate, high-density RGB-D streams. Our approach prioritizes geometric accuracy, usability, and resilience against tracking loss from accidental or voluntary scan interruptions. Unlike methods designed for indoor environments [24,28], our pipeline specifically targets the unique demands of short-range object scanning.

Building on the deep learning-based tracking recovery module proposed by [15], we introduce two key innovations: a globally consistent pose optimization and a dynamic model update mechanism. By integrating these elements with traditional methods, our pipeline achieves fast, robust, and high-quality 3D reconstruction tailored for handheld RGB-D scanners.

Our contribution can be resumed as follows:

- We propose a novel 3D reconstruction pipeline for handheld RGB-D scanning that combines deep learning with traditional methods, integrating globally consistent pose optimization and dynamic model updating.
- We introduce two new criteria to identify when fine registration fails, triggering recovery mechanisms while maintaining a frame rate near 20fps.
- We conduct a comprehensive evaluation on multiple challenging datasets, demonstrating superior accuracy, robustness, and computational efficiency over state-of-the-art approaches.
- We validate the real-world applicability of our tracking pipeline by integrating it into a complete 3D reconstruction system, enabling high-quality object modeling from handheld RGB-D scans, with applications in 3D modeling, augmented reality, and robotics.

2 Related Work

Real-time 3D reconstruction from high-frame-rate RGB and RGB-D streams is a well-studied challenge due to its demanding requirements [8,12,24]. Unlike multiview RGB-based systems [3,10,23], RGB-D methods leverage geometric data to improve multiview alignment, especially when using handheld scanners Fig. 1.

Traditional RGB-D SLAM solutions [8] have been enhanced by neural methods [22,27], with a focus on indoor environment reconstruction. However, these methods, often designed for low-cost scanners prioritizing image texture over geometric detail [14,17,19], are less suitable for single-object reconstruction, where geometric accuracy and high point cloud density are paramount.

Relying solely on visual texture features for single object reconstruction is often unreliable due to issues like lack of background, lighting and perspective changes, and uniform surfaces. As a result, indoor reconstruction methods tend to perform poorly in this setting, requiring specialized solutions. Deep learning offers potential here, particularly to enhance traditional real-time pipelines. With sufficient geometric data, volumetric approaches based on recent data-driven 3D feature extraction methods are worth exploring. [1,18].

While indoor environment datasets are abundant [7,25], datasets for accurate single object reconstruction are scarce. To our knowledge, the only datasets suitable for benchmarking dense and accurate surface reconstruction, consistent with the goal of this work, are [16] and [4]. These datasets enable the development and thorough testing of approaches like ours, which focus on tracking recovery by exploiting geometric features while considering real-time reconstruction constraints.

Our approach balances geometric quality and speed, differing from static, high-precision scanning. High-precision scanners use repeated 3D feature matching and optimization [5,11,26], but do not require real-time operation or continuous pose estimation.

In real-time applications, there are interesting traditional baselines to leverage, such as BundleFusion (BF) [8]. Advances in the neural direction for robustness against loss and tracking recovery in the case of object reconstruction have been proposed in [15] with the introduction of 3D neural feature exploitation, although this approach still has unresolved aspects, such as global alignment optimisation and loop closure with dynamic model updating.

3 Method

We present a real-time 3D object reconstruction pipeline designed for dense data from handheld optical devices. Our system leverages RGB-D streams obtained via stereo triangulation, following principles introduced in KinectFusion [12]. Specifically, we estimate a Truncated Signed Distance Function (TSDF) within a regular voxel grid to incrementally form an implicit object surface representation [6]. Pose estimation is performed using an optimized Iterative Closest Point (ICP) algorithm [2].

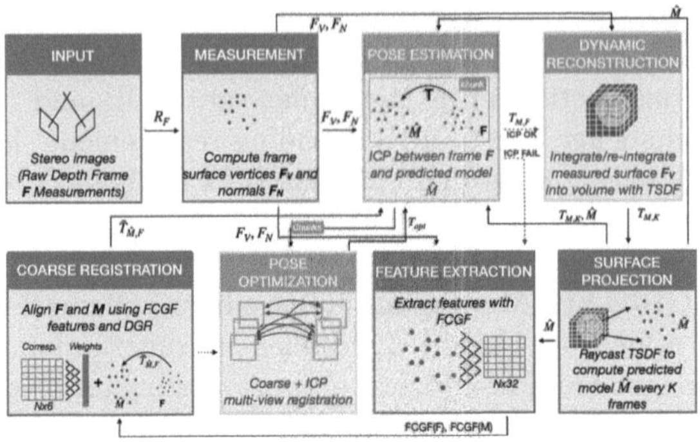

Fig. 2. D^2R^2 pipeline. The blocks with a green border are those relating to the main contribution of this work. (Color figure online)

A core challenge in real-time systems is tracking loss, whether accidental or intentional, which breaks ICP assumptions (as ICP relies on good initial pose estimates and significant overlap between consecutive frames) and necessitates on-the-fly recovery. Our pipeline, named D^2R^2, builds on Lombardi et al. [15], which introduced a data-driven safeguard module for tracking recovery. We extend this work with three key innovations: Refined Tracking Recovery Activation; Novel Pose Optimization, and Dynamic Reconstruction with an adaptive model that updates for continuity during interruptions.

The new components, highlighted in green in Fig. 2, collectively enhance system robustness and performance under challenging conditions.

3.1 Camera Tracking

The results in [15] demonstrate that deep learning-driven 3D feature extraction and alignment solutions can be effective to perform on-the-fly coarse registration for track recovery, even in complex scenarios. The handheld nature of the scanner, with live feedback during 3D reconstruction, allows users to seamlessly exploit this algorithmic intervention when issues arise. At first, to better assess when to activate the recovery system, we introduce two new criteria for frame rejection, namely the `fitness ratio` and the `camera movement`. These criteria serve as thresholds to effectively detect critical misalignments, thereby increasing the robustness of the reconstruction. ICP fitness represents the percentage of source points successfully aligned to the reference, meaning they are within the defined maximum Euclidean distance threshold. With the `fitness ratio` rejection criterion we want to detect situations when the scene changes significantly in just one frame (which is suspicious since we assume we are scanning at a high rate) or when ICP misalignment occurs (more likely). We formulate it as:

$$\begin{cases} \text{Valid} \quad \text{if} \quad \frac{Fitness_{curr}}{Fitness_{prev}} < \tau_{fit\_ratio}, \\ \text{Invalid} \quad \text{elsewhere} \end{cases} \tag{1}$$

We set the threshold τ_{fit_ratio} to 70% to be sensitive only to extreme situations (e.g. a fitness drop from 90%, which is roughly the value we get in good situations, to a lower one equal to 60% which can depict a just acceptable value).

The second criterion has a similar aim but works in the metric space of the camera movement. It requires to evaluate two thresholds: the maximum frame-to-frame rotation and translation the camera is meant to have. The `camera movement` rejection criterion is defined as:

$$\begin{cases} \text{Valid} \quad \text{if} \quad \Delta_{rot} < \tau_{rot} \quad \text{and} \quad \Delta_{transl} < \tau_{transl}, \\ \text{Invalid} \quad \text{elsewhere} \end{cases} \tag{2}$$

where Δ_{rot} and Δ_{transl} are the experienced rotation and translation between two frames. The optimal value for these thresholds depends on the type of scanner since we expect to be allowed to move it faster or slower according to its maximum frame rate. The second criterion, `camera movement`, complements the `fitness ratio` by directly assessing the physical motion of the scanner. While a low fitness ratio indicates a potential misalignment based on the data, the camera movement criterion helps detect issues arising from excessively rapid or erratic scanner motion, even if the ICP still manages to find a temporary (but potentially incorrect) alignment. For instance, if the user suddenly jerks the scanner, the frame-to-frame movement (Δ_{rot} and Δ_{transl}) will likely exceed the defined thresholds (τ_{rot} and τ_{transl}), indicating an unreliable frame even if the fitness ratio hasn't dropped drastically yet. This is particularly important for maintaining the temporal consistency of the reconstruction. For a scanner with a frame rate of 30 Hz, we might set τ_{rot} to 5°C and τ_{transl} to 2 cm, allowing for smooth and controlled movements while flagging rapid motions.

3.2 Pose Optimization

Similar to BF [8], our approach involves dividing the frames into chunks. However, unlike BF, which identifies two moments of pose optimisation, one intra- and one inter-chunk, we omit the intra-chunk step. In our setting, experiments revealed that using intra-chunk optimization is not advantageous since the produced poses closely match the ones already provided by fine registration using ICP, with the failure detection module already effectively providing robustness against false alignments, avoiding ill-conditioned chunks. In contrast, BF employs a different strategy based on sparse-then-dense registration, but it significantly down-samples input data for efficiency, resulting in less accurate dense alignments leading to the need for an intra-chunk multiview refinement.

We utilize chunks as fundamental units for constructing our 3D model. In contrast to BundleFusion, which relies on keyframes to represent chunks, we introduce an auxiliary, smaller voxel volume that operates in parallel with the main reconstruction volume. As new frames arrive within a chunk, they are integrated into this supplementary volume. Upon the completion of a chunk (defined by a fixed number of frames or a significant tracking event), we extract the complete surface from this small volume, generating a dense 3D point cloud that represents the entire chunk. This approach offers a significant advantage over using sparse keyframes as it captures the full geometric information of the chunk, leading to a more accurate representation for subsequent pose optimization. While integrating into an additional volume and performing surface extraction might seem computationally intensive due to repeated integration, we mitigate this by clearing the Y volume after each chunk is processed (keeping only the resulting point cloud) and by leveraging the parallel processing capabilities of the GPU for fast voxel integration, ensuring that the time consumption remains below 2 ms per integration.

Eventually, for multi-view alignment, we exploit the Choi *et al.*'s solution [3], also implemented in Open3D, which builds a pose graph where nodes represent individual views associated with pose matrices T_i. Edges connect nodes using transformation matrices (e.g., $T_{i,j}$), aligning adjacent views. The unknown set of pose matrices T_i are estimated via graph optimization. To define graph edges, we perform pairwise registrations across all collected chunks and use the safeguard module to align from scratch distant views.

The computational complexity of the proposed solution becomes apparent due to the substantial burden arising from multiple pairwise registrations. Computed sequentially, these coarse registrations would greatly exceed our real-time constraint. To address this challenge, we adopt a *divide-et-impera* approach. Initially, we set the maximum number of chunks to 50 and distribute them into 5 macro containers, performing $C(10) = \frac{10!}{2 \cdot (n-2)!} = 45$ pairwise registrations in each container (where n is the number of chunks in each macro container). Subsequently, we aggregate the point clouds from these containers, repeating the multi-view alignment at the macro level. This results in a total of $45 \times 5 + C(5) = 235$ registrations, significantly reducing the computational load. Furthermore, leveraging the inherent parallelism of our networks, we process batches of 4 point clouds and 2 sets of correspondences simultaneously for feature extraction and registration, further diminishing the computational impact. This entire process is executed in parallel threads, yielding a refined model with the 50 chunks in a few seconds, enabling continuous scanning and ensuring real-time compliance.

Upon completion of the reconstruction process, a set of point clouds awaits alignment. Here, we harness the power of graph optimization. Notably, the datasets under consideration do not feature any acquisition set exceeding 1500 frames, resulting in a maximum of three macro chunks after the process.

3.3 Dynamic Reconstruction

As illustrated in Fig. 2, the output of the pose estimation feeds a dynamic reconstruction module. Although the currently estimated poses are not definitive, global evaluation may yield optimized versions. To refine the model post-pose optimization, frames that need adjustment are de-integrated and reintegrated. During frame integration $\mathcal{F}_i$, each voxel v in the volume is updated as follows:

$$\begin{cases} \mathbf{D}'(v) = \frac{\mathbf{D}(v)\mathbf{W}(v) + w_i(v)d_i(v)}{\mathbf{W}(v) + w_i(v)} \\ \mathbf{W}'(v) = \mathbf{W}(v) + w_i(v) \end{cases} \tag{3}$$

where $\boldsymbol{D}(v)$ is the signed distance of the voxel, $\boldsymbol{W}(v)$ is the voxel weight, $d_i(v)$ is a projective distance (along the z-axis) between a voxel and $\mathcal{F}_i$, and $w_i(v)$ is the integration weight for a sample of $\mathcal{F}_i$. Instead, to update a voxel after removing a frame $\mathcal{F}_i$ it is possible to revert the sign of summation and change it to a subtraction:

$$\begin{cases} \mathbf{D}'(v) = \frac{\mathbf{D}(v)\mathbf{W}(v) - w_i(v)d_i(v)}{\mathbf{W}(v) - w_i(v)} \\ \mathbf{W}'(v) = \mathbf{W}(v) - w_i(v) \end{cases} \tag{4}$$

So we only adjust frames in the model if the refined pose is significantly different from the current one to ensure that the number of operations is minimised and the user experience is not burdened with unnecessary latency.

4 Data Selection

While effective for large-scale scenes, BF and similar methods face limitations in short-range 3D object reconstruction. Their reliance on repeated sparse-to-dense alignment per frame adds complexity and struggles with objects lacking distinctive features or undergoing frequent viewpoint changes. Moreover, the extensive search for 2D correspondences can be a major point of failure for object reconstruction due to the lack of consistent, viewpoint-invariant features on individual objects compared to larger scenes with more structural context. These pipelines are optimized for large-scale scenes with rich textures and structural features, unlike short-range object reconstruction, which often lacks such characteristics. Short-range object reconstruction, therefore, necessitates methods that emphasize 3D geometric features over potentially unreliable 2D visual cues. The only public datasets we found adequate to the development and testing of our technique are the DenseMatch [16] and the Redwood [4] datasets. Despite similar in terms of the target content (short-range scanning of single objects and geometrically interesting artefacts), they are obtained with two different hand-held RGB-D sensors, so, taken together, they are also well representative of possible data variability.

We also included additional scenes from the DenseMatch dataset provided by the authors. The considered scenes from the DenseMatch dataset are illustrated

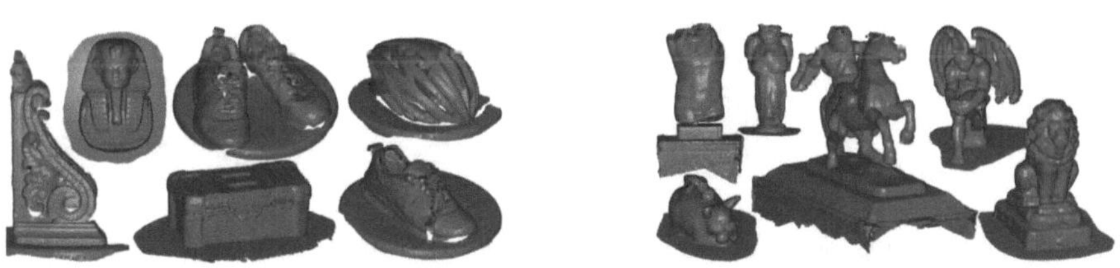

Fig. 3. DenseMatch test scenes (Left). From left to right: `column`, `statue`, `shoes`, `toolbox`, `helmet` and `shoes`. Redwood test scenes (Right). From left to right: `body`, `pig`, `angel`, `knight`, `gargoyle` and `lion`.

in Fig. 3 (Left). Each DenseMatch scene consists of multiple independent acquisition sessions (two to four), which we aggregated to test our pipeline's ability to handle discontinuous scanning and loop closure. While single acquisition sessions only cover the object partially, their aggregation covers (almost) the entire object. Our pipeline's pose optimization and dynamic reconstruction modules are designed to handle these multi-session aggregations, enabling effective loop closure despite repeated revisits from varying viewpoints.

The Redwood dataset [4] presents many more objects acquired with a lower-quality scanner with respect to the device used in DenseMatch. From it, we selected scenes in the *statues* and *toys* classes, depicted in Fig. 3(Right), with sufficient geometric quality for our application target.

5 Results

We evaluate our D^2R^2 method against BundleFusion (BF) as a baseline through quantitative and qualitative assessments, using a NVIDIA Titan V with 12GB of memory. On the Redwood dataset, visual analysis is employed due to the lack of ground truth. Additionally, we assess real-time feasibility through timing analysis and conduct an ablation study of the pose optimization module.

5.1 Comparison with BundleFusion

Visual assessment and comparisons are shown in Figs. 4 for the DenseMatch and Redwood datasets respectively, from which it is evident how D^2R^2 significantly outperforms BF in 3D object reconstruction across both test datasets.

BF's limitations are evident on the DenseMatch dataset. For instance, in the `shoe` and `statue` scenes, BF lost tracking early, producing misaligned, fragmented models. The `column` reconstruction showed mid-section misalignment,

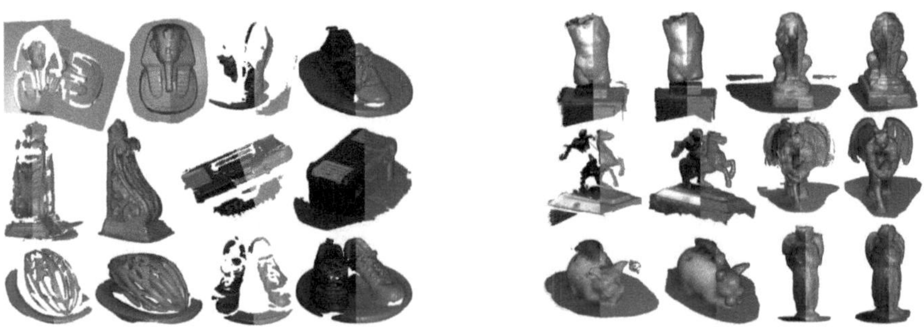

Fig. 4. (Left) Results on DenseMatch test scenes. For all reconstructed scenes, both textured and shaded meshes are shown (split views). (Right) Results on Redwood objects test scenes. For all reconstructed scenes, both textured and shaded meshes are shown (split views). In blue/green: results for BF/D^2R^2. (Color figure online)

Table 1. Quantitative alignment results: number of frames for each scene and the total frames integrated by each method. For DenseMatch point-to-point distance between the reconstructed model and the ground truth is also reported.

DenseMatch [16]						Redwood Objects [4]			
Scene	# Frames	Integrated		Dist [mm]		Scene	# Frames	Integrated	
		BF	D^2R^2	BF	D^2R^2			BF	D^2R^2
`column`	1292	472	1256	1.6	0.2	`angel`	1542	1533	1540
`helmet`	460	258	452	0.2	0.1	`body`	863	833	863
`shoe`	1059	102	1041	fail	0.1	`gargoyle`	1856	900	1855
`shoes`	1007	149	976	0.8	0.2	`knight`	3885	3283	3875
`statue`	709	86	709	fail	0.0	`lion`	1281	1265	1271
`toolbox`	1504	134	1477	0.7	0.4	`pig`	2905	2905	2900
Tot.	6026	1201 (19.9%)	5911 (98.0%)			**Tot.**	12332	10719 (86.9%)	12304 (99.7%)

while the two `shoes` yielded slightly better results due to the presence of a second object aiding disambiguation. However, tracking issues persisted, resulting in incomplete models. Although the `helmet` reconstruction was relatively more accurate, it still underperformed compared to D^2R^2.

On the Redwood dataset, D^2R^2 continued to surpass BF's performance, though BF performed slightly better on scenes like `body` and `lion` due to the larger field of view and longer scanning distance of the PrimeSense scanner. However, BF struggled with objects like the `gargoyle` and `knight`, losing tracking when moving around them. The absence of background context and challenging viewing angles hindered BF's 2D-based tracking, reinforcing D^2R^2's superior handling of real-time object reconstruction challenges.

In Table 1 details on the number of frames integrated using both methods are provided. For the DenseMatch dataset, our method achieved an impressive 98% integration rate, and 99.7% for Redwood dataset. In contrast, BF integrated only 19.9% with DenseMatch and 86.9% of frames with Redwood. Furthermore, our method consistently outperformed BF in terms of average point-to-point distances between real-time automatically reconstructed surfaces and the hand-driven alignment ground truth.

5.2 Timings

Table 2 presents the average reconstruction time performance of our system (Fig. 2) using data from the DenseMatch dataset. The first three operations are performed at each iteration, while the remaining ones are executed only in the presence of alignment errors and loss of tracking. With GPU support, our system achieves an average frame rate slightly below 20 fps. Combined with model accuracy this is evidence of the effectiveness of the new camera tracking criteria described in 3.1. Even when using CPU only, the average reconstruction rate remains near 7 fps (only occasionally dropping to 5 fps in challenging situations). This 7 fps average speed is borderline yet acceptable for a real-time application, albeit requiring users to move more carefully to avoid frequent tracking losses.

Table 2. D^2R^2 timings with DenseMatch data. In the third column, we highlight which operations are involved at each iteration. The total time additionally considers some miscellaneous which are not tracked specifically (for instance data transfer, chunks initialization, data sample and other minor computations).

Operation	Scope	Every Iteration?	CPU		GPU	
			T Avg [ms]	T Std [ms]	T Avg [ms]	T Std [ms]
Frame creation	Input	yes	42.1	5.6	24.4	3.2
Frame integration [6,20]	Integration	yes	69.6	2.4	2.1	0.8
Pose estimation (ICP pt2plane) [2]	Registration	yes	11.1	2.4	-	-
Surface projection	Model update	no	119.2	4.2	34.7	2.8
Feature extraction	Registration	no	183.2	5.2	35.7	4.6
Coarse reg. (DGR w/o opt) [5]	Registration	no	12.8	1.1	2.4	0.7
Total time for one iteration (fps in brackets $\{\cdot\}$)			150.1 {6.66}	8.5	55.2 {18.2}	6.3

In contrast, the standard approach with GPU is fully supported and reliable for real-time applications. For reference, we were able to run BF at nearly 22 fps when running our data. Additionally, we note that the feature-driven coarse registration DGR [5] is able to run faster than most of the operations.

5.3 Pose Optimization and Dynamic Reconstruction Contribution

We now investigate the contribution of the pose optimization and dynamic reconstruction modules characterizing our method. By removing these blocks, we virtually reproduce the working conditions of [15], with the difference we found that only partial models were correctly handled and reconstructed [15] while our solution allows complete and robust reconstruction also with data coming from multiple partial acquisitions aimed at the complete objecrt surface coverage. In Fig. 5, some meaningful cases are illustrated to demonstrate how the optimization modules played a crucial role in resolving minor and major issues, according to the scene. The most challenging case was the `helmet` scene, where a repetitive pattern on the helmet surface caused a partition of the data into two distinct sub-models. The optimization modules successfully corrected this issue by re-evaluating all chunks globally. Additionally, they refined the model by adjusting frames that had drifted during alignment. Overall, they significantly improve quality and correct critical errors that may occur during free-hand scanning. However, we observed that symmetric objects, as the toolbox scene, presented unique challenges due to the ambiguity in feature matching. In these cases, even with the optimization modules, misalignments can occur if the initial coarse registration is incorrect. Future work will explore strategies to mitigate these issues, such as incorporating symmetry detection and handling mechanisms.

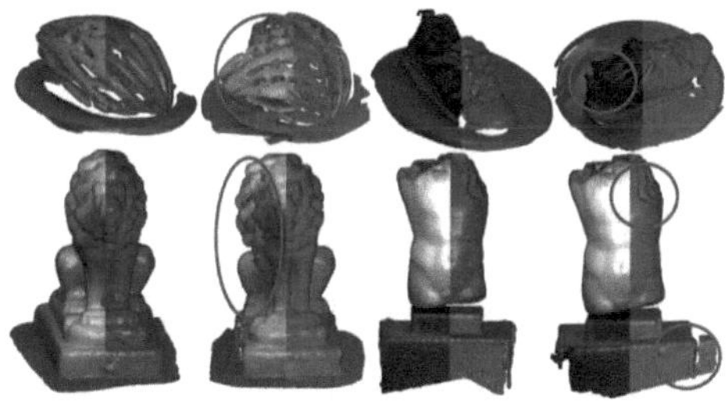

Fig. 5. Results of D^2R^2 on different scenes from DenseMatch (top) and Redwood (bottom). Each model is shown with both its texture and a shaded uniform colour. In green the results with the optimization modules active, in purple when they are turned off. Red circles highlight reconstruction errors. (Color figure online)

6 Conclusion

We introduced D^2R^2, a real-time 3D object reconstruction pipeline that integrates deep learning with traditional methods to enhance handheld RGB-D scanner performance. Key innovations include a safeguard module for tracking recovery, globally consistent pose optimization, and dynamic model updating.

Experimental results on the DenseMatch and Redwood datasets highlight D^2R^2's superior performance over state-of-the-art methods, achieving higher integration rates and reduced point-to-point errors. The pose optimization module notably improves reconstruction quality by addressing both minor and major alignment issues. Despite these advancements, challenges remain, particularly with symmetric or repetitive objects. Additionally, our current implementation does not explicitly address reflective or transparent surfaces, which can introduce noise and inaccuracies in the reconstruction.

In summary, D^2R^2 advances real-time 3D object reconstruction, leveraging deep learning to improve accuracy and usability, with potential applications spanning manufacturing to cultural heritage preservation.

References

1. Bai, X., Luo, Z., Zhou, L., Fu, H., Quan, L., Tai, C.L.: D3feat: joint learning of dense detection and description of 3D local features. In: CVPR (2020)
2. Besl, P.J., McKay, N.D.: A method for registration of 3-D shapes. IEEE Trans. Pattern Anal. Mach. Intell. **14**(2), 239–256 (1992)
3. Choi, S., Zhou, Q.Y., Koltun, V.: Robust reconstruction of indoor scenes. In: CVPR, pp. 5556–5565 (2015)
4. Choi, S., Zhou, Q.Y., Miller, S., Koltun, V.: A large dataset of object scans. arXiv:1602.02481 (2016)
5. Choy, C., Dong, W., Koltun, V.: Deep global registration. In: CVPR (2020)
6. Curless, B., Levoy, M.: A volumetric method for building complex models from range images. In: 23rd Annual Conference on Computer Graphics and Interactive Techniques, pp. 303–312. SIGGRAPH '96, ACM, New York (1996)

7. Dai, A., Chang, A.X., Savva, M., Halber, M., Funkhouser, T., Nießner, M.: Scannet: richly-annotated 3D reconstructions of indoor scenes. In: CVPR, p. 5828 (2017)
8. Dai, A., Nießner, M., Zollhöfer, M., Izadi, S., Theobalt, C.: Bundlefusion: real-time globally consistent 3D reconstruction using on-the-fly surface reintegration. ACM Trans. Graph. **36**(3) (2017)
9. Engelmann, F., Rematas, K., Leibe, B., Ferrari, V.: From points to multi-object 3D reconstruction. In: CVPR, pp. 4588–4597 (2021)
10. Gherardi, R., Farenzena, M., Fusiello, A.: Improving the efficiency of hierarchical structure-and-motion. In: CVPR, pp. 1594–1600 (2010)
11. Gojcic, Z., Zhou, C., Wegner, J.D., Guibas, L.J., Birdal, T.: Learning multiview 3D point cloud registration. In: CVPR, pp. 1756–1766 (2020)
12. Izadi, S., et al.: Kinectfusion: real-time dynamic 3D surface reconstruction and interaction. In: ACM SIGGRAPH 2011 Talks. SIGGRAPH '11, Association for Computing Machinery, New York (2011)
13. K., Z., Y., J., Y., Z., C., S.: A review of techniques for 3D reconstruction of indoor environments. Int. J. Geo-Inf. **9**(5) (2020)
14. Kerbl, B., Kopanas, G., Leimkühler, T., Drettakis, G.: 3D gaussian splatting for real-time radiance field rendering. ACM Trans. Graphics **42**(4) (2023)
15. Lombardi, M., Savardi, M., Signoroni, A.: Cross-domain assessment of deep learning-based alignment solutions for real-time 3D reconstruction. Comput. Graph. **99**, 54–69 (2021)
16. Lombardi, M., Savardi, M., Signoroni, A.: Densematch: a dataset for real-time 3D reconstruction. Data Brief **39**, 107476 (2021)
17. Mildenhall, B., Srinivasan, P.P., Tancik, M., Barron, J.T., Ramamoorthi, R., Ng, R.: NeRF: representing scenes as neural radiance fields for view synthesis. Commun. ACM **65**(1), 99–106 (2021)
18. Mohamed, A., Grandhe, R., Joseph, K., Khan, S., Khan, F.: D3former: debiased dual distilled transformer for incremental learning. In: CVPR, p. 2420 (2023)
19. Mur-Artal, R., Tardós, J.D.: ORB-SLAM2: an open-source SLAM system for monocular, stereo, and RGB-D cameras. IEEE Trans. Robot. **33**(5) (2017)
20. Nießner, M., Zollhöfer, M., Izadi, S., Stamminger, M.: Real-time 3D reconstruction at scale using voxel hashing. ACM Trans. Graph. **32**(6) (Nov 2013)
21. Runz, M., et al.: Frodo: from detections to 3D objects. In: CVPR, pp. 14720–14729 (2020)
22. Sandström, E., Li, Y., Van Gool, L., R. Oswald, M.: Point-SLAM: dense neural point cloud-based SLAM. In: 2023 IEEE/CVF International Conference on Computer Vision (ICCV) (2023)
23. Schmied, A., Fischer, T., Danelljan, M., Pollefeys, M., Yu, F.: R3D3: dense 3D reconstruction of dynamic scenes from multiple cameras. In: 2023 IEEE/CVF International Conference on Computer Vision, pp. 3216–3226 (2023)
24. Schops, T., Sattler, T., Pollefeys, M.: Bad SLAM: bundle adjusted direct RGB-D SLAM. In: CVPR, pp. 134–144 (2019)
25. Straub, J., et al.: The replica dataset: a digital replica of indoor spaces. arXiv preprint arXiv:1906.05797 (2019)
26. Zeng, A., Song, S., Nießner, M., Fisher, M., Xiao, J., Funkhouser, T.: 3DMatch: learning local geometric descriptors from RGB-D reconstructions. In: CVPR (2017)
27. Zhang, Y., Tosi, F., Mattoccia, S., Poggi, M.: GO-SLAM: global optimization for consistent 3D instant reconstruction. In: 2023 IEEE/CVF International Conference on Computer Vision, pp. 3727–3737 (2023)
28. Zollhöfer, M., et al.: State of the art on 3D reconstruction with RGB-D cameras. Comput. Graph. Forum **37**, 625–652 (2018)

Unsupervised Anomaly Localization In the Wild via Token Optimization and Test-Time Score Correction

Naoki Kawamura(✉), Gaku Minamoto, Tomohiro Nakai, Satoshi Ito, Osamu Yamaguchi, and Takahiro Takimoto

Toshiba Corporation, 1, Komukai-Toshiba-cho, Saiwai-ku, Kawasaki-shi 212-8582, Japan
naoki5.kawamura@toshiba.co.jp

Abstract. We propose a framework for reducing overdetection on in-the-wild data by unsupervised anomaly detection, where only anomaly-free images are provided for training. In recent years, the accuracy of anomaly detection has significantly improved through vision-language models (VLMs). However, related approaches suffer from severe overdetection for images captured in real-world scenarios such as car-mounted and drone-captured imagery because they often involve a wider angle of view and extremely complicated backgrounds. Therefore, in this paper, we tackle this challenge for such in-the-wild datasets by using token optimization (TO) and score correction (SC) using anomaly-free images. TO aims to obtain richer prompt description, which lead to better VLM performance while SC can directly reduce overdetection in the image domain. In the proposed framework, VLMs can effectively reduce overdetection through both TO and SC approaches. Experimental results demonstrate the superior performance of the proposed method, using LostandFound and ShanghaiTech-Campus datasets, which are publicly available in-the-wild datasets.

Keywords: Unsupervised anomaly detection · Vision-language model · Prompt learning

1 Introduction

Image anomaly detection and localization is one of the main computer vision tasks, with a huge variety of applications such as industrial inspection [1,3,4], surveillance systems [10,17], medical interpretation, and infrastructure maintenance [10,20,22]. Numerous tasks are handled by different supervision settings, including unsupervised [2,8,14,15,23], weakly supervised [7,19,26], and zero-shot settings [5,11,12,16]. In this paper, we focus on unsupervised anomaly detection in which only normal (i.e., anomaly-free) images are available as prior information.

E. Rodolà et al. (Eds.): ICIAP 2025, LNCS 16167, pp. 633–646, 2026.
https://doi.org/10.1007/978-3-032-10185-3_50

In the field of anomaly detection and many other computer vision topics, large pretrained vision-language models (VLMs) have attracted much attention in recent years. Derived from contrastive language-image pretraining (CLIP) [21], VLMs have come to be used for anomaly detection [5,15,26,29]. Typically, in VLM-based anomaly detection, text-based prompts are provided to help distinguish anomalous and normal regions within a test image. Those VLM-based methods have achieved remarkable performance through leveraging strong relationships between language and vision.

However, significant issues remain for practical application of existing VLM-based approaches. In most of them, datasets with limited practicality are focused on, which leads to performance limitations in practical usage. These datasets include simple images taken in controlled environments, such as having a single object with simple backgrounds, with few changes in angle of view or object location between images. This is undoubtedly that these fundamental settings are directly applicable for industrial inspection [1,3,4], or sub-image anomaly detection [8,9]. However, there would be demand for more practical application settings, particularly in infrastructure inspection and maintenance, such as road inspection using car cameras [20,22], patrols of factories using web cameras [18], and surveillance systems using drones [25,27]. Such datasets in practical environments, or in-the-wild datasets, are significantly more challenging, because there are wider angles of view, variations in camera locations, and more complicated backgrounds. Conventional approaches specifically formulated for the basic settings above would cause severe overdetection and poor detection performance in such scenarios. In order to ensure the performance of VLMs on in-the-wild datasets, it is important to prepare appropriate and sufficient domain-aware prompts [15,29]. Furthermore, even if VLMs have sufficient prompts, they tend to exhibit high score even for normal images, resulting in poor localization as shown in Fig. 1.

In this paper, we introduce two proposals into VLM-based methods for improving pixel-level anomaly detection performance on in-the-wild datasets. The proposed framework under the unsupervised setting is intended to bidirectionally improve VLM-based methods from the prompt description based on token optimization (TO) and test-time score correction (SC).

Fig. 1. Example of the proposed method. In the input image, anomaly object on the road is indicated by the orange circle. Poor prompts to a VLM can cause severe overdetection. Overdetection is reduced using both of TO and test-time SC using normal images.

Concerning TO, in conventional prompt learning methods [6,15,28,29], we think that there is a clear opportunity for improvement since they contain text-tokens or the frozen embeddings that are strongly dependent on the initial values. These frozen tokens are intended for domain-aware image subjects to detect and are usually given by the user. In the proposed method, the tokens describing the subjects are also fine-tuned using the normal images and their score map criteria to obtain a more appropriate prompt description.

For SC, we aim to significantly reduce overdetection based on the relationships between an input image and normal images. An effective technique used for a vision-only model [14] is combined with VLM-based methods. Such combination enables the proposed method to incorporate both advantages of a VLM with appropriate prompts and a vision-only statistic method. Our approach can be directly applied to any VLM-based anomaly detectors since the VLM itself is used as a pretrained feature extractor.

The contributions of this paper include the following:

- We propose a novel token optimization framework that trains the description of the conventionally frozen parts.
- We propose a simple unsupervised pixel-level anomaly detection framework that offers the advantages of both VLMs and vision-only models.
- We confirm that the proposed method can be applied to multiple kinds of VLM-based methods as long as normal images are available as prior information.
- Our empirical evidence indicates that the proposed framework outperforms SOTA VLM-based methods using publicly available in-the-wild datasets by successfully reducing overdetection.

The remainder of this paper is organized as follows. Section 2 introduces related work. Section 3 describes the proposed approach, and Sect. 4 presents our experimental results. Sect. 5 provides a summary of our conclusions.

2 Related Work

2.1 Anomaly Detection with VLM

In the past few years, VLMs have been attracting more and more attention. Main applications include zero-shot anomaly detection or few-normal-shot anomaly detection as well as several cases of weakly-supervised ones [12,26]. Here, we introduce zero-shot/few-shot VLM-based methods, where only prompts or additional few-normal-shot images are required as prior information, since our proposals could be applied to any of these methods with additional normal images. In some cases, the VLM-based methods outperform the dedicated vision-only methods [5,12,15,29].

Based on CLIP [21], which is the representative VLM for image classification, the contributions of VLM-based methods often focus on prompt engineering or prompt-based learning. WinCLIP and WinCLIP+ [12] were breakthroughs in improvement for anomaly detection. By applying prompt ensembling to manually create sufficient prompts, these showed the potential performance of VLMs for zero-shot and few-shot anomaly detection for industrial inspection and sliding-window processing. In APRIL-GAN [5], trainable linear layers are added to pretrained CLIP models to fine-tune for the anomaly detection task since the original CLIP is dedicated for natural image classification task. PromptAD [15] proposed a novel framework for prompt learning in unsupervised anomaly detection. The method outperformed other methods based on multi-class, or fully supervised, prompt learning under a few-shot setting.

In order to ensure the VLM performance, it is important to prepare appropriate and sufficient prompts depending on the model, task, and data domain [5,6,15,28,29].

2.2 Prompt Learning

When preparing prompts for VLMs, manual prompt ensembling may occasionally require hundreds or thousands of variations of prompts to obtain a stable solution [15]. Prompt learning methods can automatically generate prompts based on minimal priors as well as optimizing the prompt description. CoOp [28] is a representative method proposed for image classification, while PromptAD [15], as previously mentioned, represents a well-organized variant designed for unsupervised anomaly detection. AnomalyCLIP [29] is a zero-shot anomaly detector that introduces object-agnostic learnable prompts for general anomaly detection. The effectiveness was demonstrated using industrial data as well as medical images under the assumption that images have definite foregrounds and backgrounds. Whereas the above methods are based on metric learning, in which the prompts themselves are learned as parameters [15,29], there are also model-based-prompt generators [6].

In prompt learning methods, users are typically required to set small parts of the prompts, as tokens describing the image subject and the anomalies to detect [6,15,29].

3 Proposed Method

The proposed method is composed of two independent parts, namely, token optimization (TO) for prompts, and test-time score correction (SC) for VLMs (Fig. 2). Both parts are performed in an unsupervised setting, where only normal images are given in advance. In this section, we give preliminary information in Sect. 3.1, and then explain details of the respective two proposals in Sect. 3.2 and 3.3.

3.1 Preliminary

Anomaly Detection Using a VLM. Let an input image be denoted by $\boldsymbol{X}$, and a set of multiple text-prompts $\{\boldsymbol{t}_1^a, \boldsymbol{t}_2^a, \cdots, \boldsymbol{t}_{J_a}^a, \boldsymbol{t}_1^n, \boldsymbol{t}_2^n, \cdots, \boldsymbol{t}_{J_n}^n\}$. A text-prompt $\boldsymbol{t}_j^{\{a,n\}}$ is a tokenized sentence, and there are two types of prompts, one

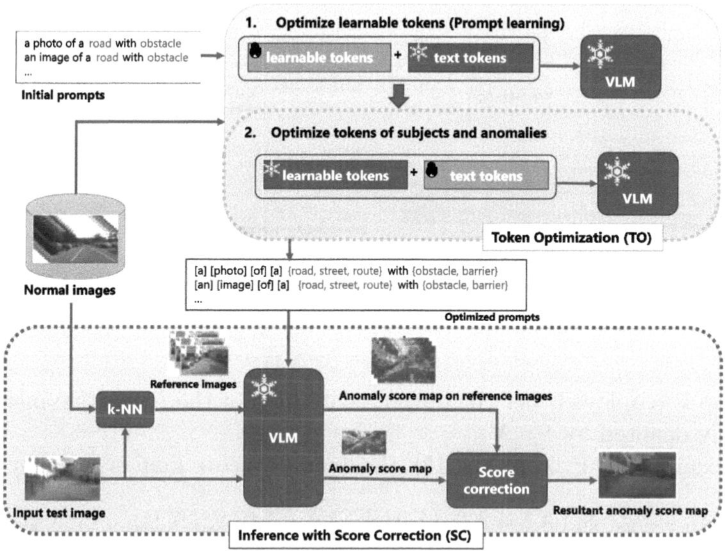

Fig. 2. Overview of the proposed method composed of two components: Prompt learning and TO indicated by purple dotted-line and inference with SC indicated by green dotted-line. Tokens denoted with $[\cdot]$ correspond to learnable tokens, and those in green and red are target tokens describing subjects and anomalies, respectively. Note that the prompt learning is skipped in cases involving manual prompt ensembling. (Color figure online)

for anomalies denoted by $\boldsymbol{t}_j^a$ and the other for normal descriptions denoted by $\boldsymbol{t}_j^n$. Anomaly prompts can be written, like "a photo of $\boldsymbol{q}_{\text{sbj}}$ with $\boldsymbol{q}_{\text{a}}$", where $\boldsymbol{q}_{\text{sbj}}$ is the text-token describing a subject of the image and $\boldsymbol{q}_{\text{a}}$ is that of the anomaly to detect. On the other hand, prompts for normal patterns can be described without $\boldsymbol{q}_{\text{a}}$, such as "a photo of $\boldsymbol{q}_{\text{sbj}}$".

Let an embedding of a prompt $\boldsymbol{t}_j^{\{a,n\}}$ using a pretrained text encoder be denoted by $g(\boldsymbol{t}_j^{\{a,n\}})$. Before inputting prompts to VLM, the embeddings are averaged such as $\bar{\boldsymbol{w}}^{\{a,n\}} = \sum_j g(t_j^{\{a,n\}})/J_{\{a,n\}}$. In anomaly detectors based on CLIP [21], the pixel-level anomaly score map of visual-language similarity can be formulated as

$$S_{\boldsymbol{p}}^t(\boldsymbol{X}) = \frac{\exp(\langle f(\boldsymbol{X},\boldsymbol{p}), \bar{\boldsymbol{w}}^a\rangle/\tau)}{\exp(\langle f(\boldsymbol{X},\boldsymbol{p}), \bar{\boldsymbol{w}}^a\rangle/\tau) + \exp(\langle f(\boldsymbol{X},\boldsymbol{p}), \bar{\boldsymbol{w}}^n\rangle/\tau)}, \tag{1}$$

where $\langle\cdot,\cdot\rangle$ is the cosine similarity operator, $f(\cdot)$ is a pretrained visual encoder, $\boldsymbol{p}$ is a coordinate index, and τ is the temperature hyperparameter. Note that the score maps of the respective CLIP-based methods [5,12,15] actually have their own variants or post-processing, and in most of them, a mixture of vision-language-based score map like Equation (1) and conventional vision-only-based

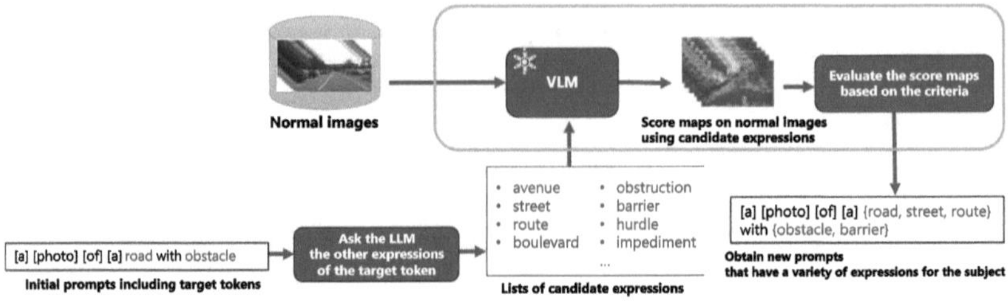

Fig. 3. Flowchart of TO. An example inside the process with gray box is shown in Fig. 4.

score map is employed. Let the output score map of the respective methods be commonly denoted by $S_{\boldsymbol{p}}(\boldsymbol{X})$.

For example, in PromptAD [15], the anomaly score map is given by

$$S_{\boldsymbol{p}}(\boldsymbol{X}) = 1.0/\{1.0/S_{\boldsymbol{p}}^{t}(\boldsymbol{X}) + 1.0/S_{\boldsymbol{p}}^{v}(\boldsymbol{X})\}, \quad (2)$$

where $S_{\boldsymbol{p}}^{v}(\boldsymbol{X})$ is computed as vision-guided anomaly score map based on the vision-only similarity between input and normal images using a visual encoder [8, 15,23].

Prompt Learning. Prompt learning is a task that automatically learns efficient prompts for VLMs. We introduce CoOp [28]-based methods including PromptAD [15], where the prompts themselves have learnable parameters. For example, the learnable prompt for the anomaly detection can be denoted by

$$\begin{aligned} \boldsymbol{s}_j^a &= [\boldsymbol{P}_1][\boldsymbol{P}_2]\cdots[\boldsymbol{P}_E][g(\boldsymbol{q}_{\text{sbj}})]\cdots[g(\boldsymbol{q}_{\text{a}})], \\ \boldsymbol{s}_j^n &= [\boldsymbol{P}_1][\boldsymbol{P}_2]\cdots[\boldsymbol{P}_E][g(\boldsymbol{q}_{\text{sbj}})], \end{aligned} \quad (3)$$

where $[\boldsymbol{P}_1][\boldsymbol{P}_2]\cdots[\boldsymbol{P}_E]$ are learnable tokens [15], and they correspond to normal templates in manual ensembling [5,12] such as"[a][photo][of]". $\boldsymbol{q}_{\text{sbj}}$ and $\boldsymbol{q}_{\text{a}}$ are initially given by users, and they are frozen during training in conventional methods. CoOp-based prompt learning methods train the parameters by fine-tuning the embedding vector of each text-token while keeping the initial description of the image subjects. Given a set of training images, metric learning is performed to search for the optimal description of Equation (3) [15].

3.2 Token Optimization (TO) for appropriate prompts

In existing prompt learning methods, the initial description of the subject and the anomaly is frozen, which we believe is room for improvement.

Similar to existing prompt learning [15], we have two kinds of inputs: one is a set of normal images, and the others are the text-token describing the image subject $\boldsymbol{q}_{\text{sbj}}$ and that describing the anomalies to detect $\boldsymbol{q}_{\text{a}}$. Note that in conventional methods [6,15,29] $\boldsymbol{q}_{\text{sbj}}$ and $\boldsymbol{q}_{\text{a}}$ are initially designed by users depending on the task and the dataset. The objective is to search for better descriptions for $\boldsymbol{q}_{\text{sbj}}$ and $\boldsymbol{q}_{\text{a}}$, and consequently achieve the improvement for $S_{\boldsymbol{p}}(\boldsymbol{X})$ through $S_{\boldsymbol{p}}^{t}(\boldsymbol{X})$.

Figure 3 shows an overview of the proposed TO method. Let $\mathcal{G}$ be a training dataset composed of normal images, and $\boldsymbol{b}$ be a set of prompts. $\boldsymbol{b}$ can be either text-prompts manually designed by users or learned prompts in vector format since $\boldsymbol{q}_{\text{sbj}}$ and $\boldsymbol{q}_{\text{a}}$ coincides with $g(\boldsymbol{q}_{\text{sbj}})$ and $g(\boldsymbol{q}_{\text{a}})$. In order to obtain better descriptions for $\boldsymbol{q}_{\text{sbj}}$ and $\boldsymbol{q}_{\text{a}}$, we leverage a large language model (LLM) and score maps of the normal images.

Listing Candidate Tokens. Given initial $\boldsymbol{q}_{\text{sbj}}$ and $\boldsymbol{q}_{\text{a}}$, the proposed method first lists the candidates for other expressions of $\boldsymbol{q}_{\text{sbj}}$ and $\boldsymbol{q}_{\text{a}}$. This process is performed by asking an LLM other expressions of $\boldsymbol{q}_{\text{sbj}}$ and $\boldsymbol{q}_{\text{a}}$ respectively. For example, when $\boldsymbol{q}_{\text{sbj}}$ ='road', ask an LLM the question, "What is the other expressions for'road'? List in bullet points." After the lists of other expressions of $\boldsymbol{q}_{\text{sbj}}$ and $\boldsymbol{q}_{\text{a}}$ are obtained, the candidates for the optimal prompts can be made using other expressions. Let $\boldsymbol{c}_{\boldsymbol{q}_{\text{sbj}}}^{n}$ be the n-th other expression of $\boldsymbol{q}_{\text{sbj}}$, and let $\boldsymbol{b}(\boldsymbol{q}_{\text{sbj}}, \boldsymbol{q}_{\text{a}})$ be a set of prompts using $\boldsymbol{q}_{\text{sbj}}$ and $\boldsymbol{q}_{\text{a}}$, where $\boldsymbol{q}_{\text{sbj}}$ and $\boldsymbol{q}_{\text{a}}$ are commonly used in the set of prompts. Consequently $\boldsymbol{b}(\boldsymbol{c}_{\boldsymbol{q}_{\text{sbj}}}^{n}, \boldsymbol{q}_{\text{a}})$ is the set of prompts using $\boldsymbol{c}_{\boldsymbol{q}_{\text{sbj}}}^{n}$ instead of $\boldsymbol{q}_{\text{sbj}}$.

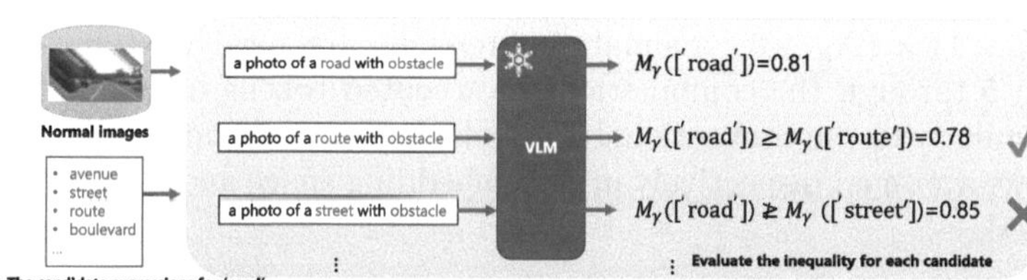

Fig. 4. Example of evaluating candidate tokens in TO. Each pair of the candidate prompt and normal images is input to a pretrained VLM to evaluate the inequality (5).

Evaluation of the Candidate Tokens to Accept. Next, score map $S_{\boldsymbol{p}}(\boldsymbol{X})$ is calculated using the pair of normal images $\boldsymbol{G}_m \in \mathcal{G}$ and the candidate prompts. Let $S_{\boldsymbol{p}}(\boldsymbol{G}_m; \boldsymbol{b}(\cdot))$ be the anomaly score map calculated using $\boldsymbol{G}_m$ and the prompt $\boldsymbol{b}(\cdot)$ out-placing the original prompt. Let a metric for the tokens based on the score map of normal images be denoted by

$$M_{\gamma}(\boldsymbol{q}_{\text{sbj}}, \boldsymbol{q}_{\text{a}}) = \sum_{m,p} \text{Th}(S_{\boldsymbol{p}}(\boldsymbol{G}_m; \boldsymbol{b}(\boldsymbol{q}_{\text{sbj}}, \boldsymbol{q}_{\text{a}})), \gamma), \tag{4}$$

where $\text{Th}(\cdot, \gamma)$ is the thresholding operator, which sets the score to zero if the score is lower than γ. The inequality (5) is then evaluated where the effectiveness over the normal images is compared between the initial prompt and the candidate prompt.

$$M_\gamma(\boldsymbol{q}_{\mathrm{sbj}}, \boldsymbol{q}_{\mathrm{a}}) \geq M_\gamma(\boldsymbol{c}^n_{\boldsymbol{q}_{\mathrm{sbj}}}, \boldsymbol{q}_{\mathrm{a}}). \tag{5}$$

The inequality (5) evaluates whether the candidate $\boldsymbol{c}^n_{\boldsymbol{q}_{\mathrm{sbj}}}$ describes the training image better than the initial $\boldsymbol{q}_{\mathrm{sbj}}$. The motivation is to obtain other expressions of the image subject in a broader sense by searching the candidate list. An example of the accepting process is shown in Fig. 4. The inequality is based on the degree of heat region in score maps of normal images. The lower the score maps react to the prompt involving anomaly description, the better the prompt is thought to describe normal images. That is, the candidate token is accepted if it better describes the image subject of training images. Similar to the inequality (5), the inequality for $\boldsymbol{c}^n_{\boldsymbol{q}_{\mathrm{a}}}$ and $\boldsymbol{q}_{\mathrm{a}}$ can be evaluated by

$$M_\gamma(\boldsymbol{q}_{\mathrm{sbj}}, \boldsymbol{q}_{\mathrm{a}}) \geq M_\gamma(\boldsymbol{q}_{\mathrm{sbj}}, \boldsymbol{c}^n_{\boldsymbol{q}_{\mathrm{a}}}). \tag{6}$$

The candidate tokens $\boldsymbol{c}^n_{\boldsymbol{q}_{\mathrm{sbj}}}$ and $\boldsymbol{c}^n_{\boldsymbol{q}_{\mathrm{a}}}$, which satisfy the inequalities (5) and (6), respectively, are finally accepted to construct the prompts. Let the number of accepted tokens from $\boldsymbol{c}^n_{\boldsymbol{q}_{\mathrm{sbj}}}$ be denoted by N_{sbj}, and that of $\boldsymbol{c}^n_{\boldsymbol{q}_{\mathrm{a}}}$ be denoted by N_a. The initial tokens $\boldsymbol{q}_{\mathrm{sbj}}, \boldsymbol{q}_{\mathrm{a}}$ are also adopted. Then, $N_{sbj} + 1$ normal expressions and $(N_{sbj} + 1) \times (N_a + 1)$ anomaly expressions are newly obtained for each prompt in $\boldsymbol{b}$ through the combinations of accepted tokens including the initial tokens. Similarly to the conventional methods [5,15,29], the anomaly and normal prompts are averaged respectively in the embedding space and used for the input to the VLM.

3.3 Score Correction (SC) using normal images for VLMs

In the anomaly detection in-the-wild by VLM, it is difficult to achieve sufficient performance using only the appropriate prompts, and severe overdetection can occur. In order to further reduce overdetection, we propose a combination of a vision-only statistic approach using normal images [14] with VLM-based methods.

Figure 2 (in green dotted-line) shows an overview of the proposed framework. Given a normal image dataset, it is assumed that the dataset contains several normal samples, each of which resembles the input image. The objective is then to correct the anomaly score map of an input image by using intra-class anomaly scores calculated among normal images resembling the input image. Each anomaly score map of an input image and the resembling normal images is calculated by the VLM [5,15,29], where the appropriate prompting is configured based on the TO method explained in Sect. 3.2.

Let $\boldsymbol{X}$ be an input image and $\mathcal{G}$ be a set of normal images. Each image with prompts is first input to a pretrained VLM to obtain the anomaly score map $\boldsymbol{S}$

as well as visual feature description $F(\cdot)$. Let the K-nearest samples to $\boldsymbol{X}$ in $\mathcal{G}$ be denoted by $\boldsymbol{R}_k$ ($k = 1, 2, \cdots, K$), which is obtained by a traditional k-NN search based on the Euclidean distance in the $F(\cdot)$-feature domain [14]. After each $\boldsymbol{R}_k$ is obtained, the anomaly score map for $\boldsymbol{R}_k$, which is denoted by $\boldsymbol{V}_k$, is calculated using the same VLM anomaly detector. Using $\boldsymbol{V}_k$, the anomaly score map $\boldsymbol{S}$ is corrected by Equation (7).

$$\hat{\boldsymbol{S}} = \sum_{k} [\boldsymbol{S} - \alpha \boldsymbol{V}_k]_+, \tag{7}$$

where α is a weight parameter that controls the balance between anomaly patterns and normal ones, and $[\cdot]_+$ is a tensor operator that sets negative elements to zero.

4 Experimental Results

4.1 Experimental Settings

Datasets. We examined the performance of the proposed method using the LostandFound dataset [20] and the ShanghaiTech Campus (STC) dataset [17].

The LostandFound dataset contains car-camera images captured at 13 streets. Each street view includes about 10 sequences containing a variety of fallen objects on the road. Each sequence is composed of 30 - 50 images, and there are 1,203 test images and 1,036 training images available.

The STC dataset contains videos of surveillance captured by static cameras at 12 scene locations. Each scene consists of several hundred test frames and tens of thousands of anomaly-free training frames. Following the protocol in [8,9], all the image frames were resized to 256 × 256 pixels, and frames were sub-sampled by a factor of five. There are 8,576 test images and 27,450 training images available in total, and the evaluation is conducted depending on respective scenes.

Implementation Details. For comparison with other methods, their respective official PyTorch implementations were used in all methods, both of which are based on OpenCLIP [21]. These source codes were customized to apply the proposed method. For APRIL-GAN [5], proper prompts for each dataset were first set manually, and then the proposed method was applied. For PromptAD [15], prompt learning was first applied and then our TO was performed. The number of learnable prompts of normal patterns (or that of templates of normal prompts in manual ensembling [5]) was set to six. On the other hand, the number of initial tokens of anomaly patterns was set to one, and there were nine variants of the token in total for each dataset. Microsoft Copilot [24] was employed as the LLM for TO in the proposed method. In our experiment with few-normal-shot VLMs, few-normal-shot images were given by reference images obtained by the proposed method, which are usually given as manually fixed images [5,15]. For the k-NN-based search used in the proposed method, we employed the Faiss library [13] to obtain the approximated k-NN results for computational acceleration. For the hyperparameters, $\gamma = 0.725$, $K = 5$, and $\alpha = 0.675$ were fixed.

The default setting of the OpenCLIP model [21] was used for the temperature τ.

Table 1. Comparison on the LostandFound [20] and STC [17] dataset for two evaluation metrics, pixel-average precision (pixel-AP [%](↑)), and false-positive rate at 95% true-positive rate (FPR [%](↓) @ TPR=95%). Each metric is the averaged results from nine pairs of initial tokens.

Methods		TO	SC	LostandFound		STC	
				pixel-AP(↑)	FPR(↓)	pixel-AP(↑)	FPR(↓)
Baseline	APRIL-GAN			22.9	43.5	23.4	32.1
Ours	APRIL-GAN+TO	✓		23.4	35.1	24.5	25.7
	APRIL-GAN+SC		✓	26.0	24.8	27.1	20.7
	APRIL-GAN++	✓	✓	**26.9**	**19.3**	**28.3**	**16.1**
Baseline	PromptAD			22.5	38.3	20.4	30.6
Ours	PromptAD+TO	✓		23.0	36.2	21.7	27.8
	PromptAD+SC		✓	25.5	29.1	24.0	25.4
	PromptAD++	✓	✓	**25.7**	**25.4**	**26.1**	**21.9**

Table 2. Comparison on the LostandFound [20] dataset in FPR [%] @ TPR=95%(↓). Scores in row order indicate the variations with respect to the initial tokens. S:{·} denotes the token describing the image subject, and A:{·} denotes that describing the anomaly to detect.

Methods	S:{road} A:{obstacle}	S:{road} A:{barrier}	S:{road} A:{block}	S:{street} A:{obstacle}	S:{street} A:{barrier}	S:{street} A:{block}	S:{route} A:{obstacle}	S:{route} A:{barrier}	S:{route} A:{block}
APRIL-GAN	42.4	41.9	44.6	42.0	43.7	46.4	42.8	43.2	45.2
APRIL-GAN+TO	35.1	34.8	35.7	34.3	34.8	36.2	34.3	34.6	36.4
APRIL-GAN++	18.3	17.7	19.8	17.2	18.8	18.6	20.5	21.1	21.3
PromptAD	36.4	39.4	39.1	36.1	37.8	36.9	38.5	40.3	40.5
PromptAD+TO	34.7	36.4	35.9	34.8	35.6	35.2	37.3	38.0	37.8
PromptAD++	23.9	24.2	24.0	24.8	26.7	26.1	25.8	27.1	26.7

4.2 Results

Table 1 shows the comparison results for the proposed TO/SC framework applied to VLM-based methods. The evaluation metrics are the pixel-level average precision and the false-positive rate at 95% true-positive rate. The former evaluates confidence of the prediction, which is the higher-the-better metric, and the latter evaluates overdetection at relatively high recall, which is the lower-the-better metric. Both metrics of a prompt-learning based VLM (PromptAD) and a manual prompt-ensembling VLM (APRIL-GAN) improved through the proposed TO method, where the FPR metric is improved by 4.9% on average. The performance was further improved by 10.5% when SC is also applied. Table 2 shows detailed examples with LostandFound when the initial tokens describing the subjects and the anomalies are changed. Table 3 shows the candidate tokens obtained by asking the LLM for the other expressions of each initial token, and Table 4 shows lists of accepted tokens in each case. It can be observed that the performance changes depending on the tokens initially given by users, and the performance was improved using the optimized tokens. An additional one to three text-tokens were accepted from ten respective candidates using the proposed method.

Figure 5 shows examples of the qualitative evaluation. The original APRIL-GAN suffers from severe overdetection. Through TO, the anomaly scores of GT regions were higher than those of other regions. With SC, overdetection was considerably reduced while the scores of GT regions were also reduced a little. The combination of TO and SC greatly reduced overdetection while keeping the higher scores of GT regions.

Table 3. List of candidate tokens obtained from the LLM (Microsoft-Copilot).

Initial tokens	Candidate tokens
S:{road}	S:{way, path, route, direction, course, nchorage, harbor, port, mooring, street}
S:{street}	S:{road, thoroughfare, way, avenue, drive, street, boulevard, route, expressway, roadway}
S:{route}	S:{way, path, course, direction, road, highway, thoroughfare, passage, avenue}
A:{obstacle}	A:{obstruction, barrier, hurdle, stumbling block, bar, block, impediment, hindrance, interference}
A:{barrier}	A:{fence, railing, barricade, hurdle, bar, wicket, wicket gate, lychgate, turnstile, port}
A:{block}	A:{chunk, hunk, brick, slab, lump, pad, notebook, tablet, sketchbook, scratch pad}

Table 4. Accepted token list of the proposed method.

Methods	S:{road} A:{obstacle}	S:{road} A:{barrier}	S:{road} A:{block}	S:{street} A:{obstacle}	S:{street} A:{barrier}	S:{street} A:{block}	S:{route} A:{obstacle}	S:{route} A:{barrier}	S:{route} A:{block}
APRIL-GAN+	S:{road, way, route} A:{ obstacle, barrier, hurdle}	S:{road, way, course} A:{barrier, fence, hurdle}	S:{road, path, course} A:{block, brick, hunk}	S:{street, way, road, drive} A:{obstacle, barrier}	S:{street, road, drive, route} A:{obstacle, fence}	S:{street, way, road, avenue} A:{block, brick, hunk}	S:{route, way} A:{obstacle, impediment}	S:{route, way, path} A:{barrier, barricade}	S:{route, avenue} A:{block, brick}
PromptAD+	S:{road, way, route} A:{obstacle, obstruction, barrier}	S:{road, way, route} A:{barrier, barricade}	S:{road, way, route} A:{block, brick, lump}	S:{street, road, drive} A:{obstacle, obstruction, barrier}	S:{street, road, drive} A:{barrier, barricade}	S:{street, road, drive} A:{block, brick, lump}	S:{route, street, road, way} A:{obstacle, obstruction}	S:{route, street, road, way} A:{barrier, barricade,}	S:{route, street, road, way} A:{block, brick}

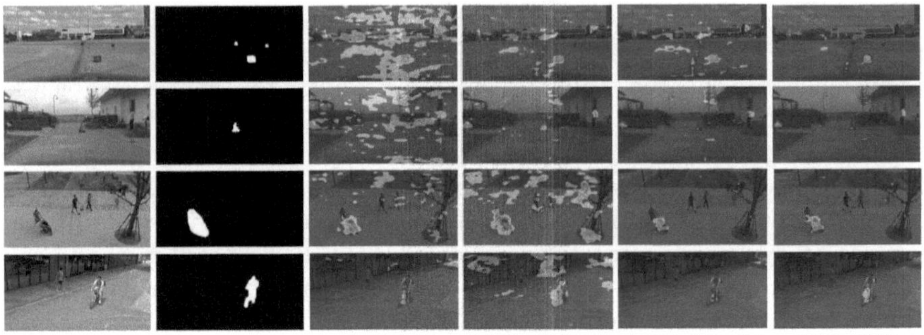

Fig. 5. Illustration of the anomaly score maps of a VLM-based method and the proposed method. Samples from LostandFound (top two rows) and STC (bottom two rows).

5 Conclusion

In this paper, we proposed a novel anomaly localization framework under the unsupervised setting. The proposed method has two independent components for improving existing VLM-based methods, namely, TO and a test-time SC framework using normal images. TO optimizes the tokens describing image subjects and anomalies in order to obtain more appropriate prompt description for both manual and learnable prompt cases. For the SC framework, the proposed method combines the advantages of a conventional vision-only statistic approach and modern VLM-based methods in order to prevent overdetection on in-the-wild data. The proposed method can be applied to any type of VLM-based anomaly detector. The experimental results showed that the proposed framework improves state-of-the-art VLMs and outperforms them by drastically reducing overdetection.

References

1. Bao, T., et al.: Miad: a maintenance inspection dataset for unsupervised anomaly detection. In: Proc. ICCV, pp. 993–1002 (2023)
2. Batzner, K., Heckler, L., König, R.: Efficientad: accurate visual anomaly detection at millisecond-level latencies. In: Proc. WACV, pp. 128–138 (2024)
3. Bergmann, P., Batzner, K., Fauser, M., Sattlegger, D., Steger, C.: The mvtec anomaly detection dataset: a comprehensive real-world dataset for unsupervised anomaly detection. IJCV **129**(4), 1038–1059 (2021)
4. Bergmann, P., Batzner, K., Fauser, M., Sattlegger, D., Steger, C.: Beyond dents and scratches: logical constraints in unsupervised anomaly detection and localization. IJCV **130**(4), 947–969 (2022)
5. Chen, X., Han, Y., Zhang, J.: A zero-/few-shot anomaly classification and segmentation method for CVPR 2023 VAND workshop challenge tracks 1&2: 1st place on zero-shot ad and 4th place on few-shot ad. arXiv preprint arXiv:2305.17382 (2023)

6. Cheng, J., et al.: Black-box prompt optimization: aligning large language models without model training. arXiv preprint arXiv:2311.04155 (2023)
7. Cho, M., Kim, M., Hwang, S., Park, C., Lee, K., Lee, S.: Look around for anomalies: weakly-supervised anomaly detection via context-motion relational learning. In: Proc. CVPR, pp. 12137–12146 (2023)
8. Cohen, N., Hoshen, Y.: Sub-image anomaly detection with deep pyramid correspondences. arXiv preprint arXiv:2005.02357 (2020)
9. Defard, T., Setkov, A., Loesch, A., Audigier, R.: Padim: a patch distribution modeling framework for anomaly detection and localization. arXiv preprint arXiv:2011.08785 (2020)
10. Duong, H.T., Le, V.T., Hoang, V.T.: Deep learning-based anomaly detection in video surveillance: a survey. Sensors **23**(11), 5024 (2023)
11. Gu, Z., Zhu, B., Zhu, G., Chen, Y., Tang, M., Wang, J.: Anomalygpt: detecting industrial anomalies using large vision-language models. In: Proc. AAAI Conference on Artificial Intelligence, vol. 38, no. 3, pp. 1932–1940 (2024)
12. Jeong, J., Zou, Y., Kim, T., Zhang, D., Ravichandran, A., Dabeer, O.: Winclip: zero-/few-shot anomaly classification and segmentation. In: Proc. CVPR, pp. 19606–19616 (2023)
13. Johnson, J., Douze, M., Jégou, H.: Billion-scale similarity search with GPUS. IEEE Trans. Big Data **7**(3), 535–547 (2019)
14. Kawamura, N.: Unsupervised anomaly localization using locally adaptive query-dependent scores. In: Proc. ICIAP, pp. 300–311. Springer (2022)
15. Li, X., et al.: Promptad: learning prompts with only normal samples for few-shot anomaly detection. In: Proc. CVPR, pp. 16838–16848 (2024)
16. Liu, S., Roy, S.: Zero/few-shot anomaly localisation using language prompts. In: Proc. of CVCI. vol. 13169, pp. 51–63. SPIE (2024)
17. Liu, W., W. Luo, D.L., Gao, S.: Future frame prediction for anomaly detection – a new baseline. In: Proc. CVPR (2018)
18. Luu, P., Weed, J., Rodriguez, S., Akhtar, S.: An AI-based web surveillance system using raspberry pi. J. Adv. Technol. Eng. Res. **5**(6), 231–242 (2019)
19. Pang, G., Shen, C., Jin, H., van den Hengel, A.: Deep weakly-supervised anomaly detection. In: Proc. 29th ACM SIGKDD Conference on Knowledge Discovery and Data Mining, pp. 1795–1807 (2023)
20. Pinggera, P., Ramos, S., Gehrig, S., Franke, U., Rother, C., Mester, R.: Lost and found: detecting small road hazards for self-driving vehicles. In: Proc. IROS, pp. 1099–1106. IEEE (2016)
21. Radford, A., et al.: Learning transferable visual models from natural language supervision. In: Proc. ICML, pp. 8748–8763. PMLR (2021)
22. Rai, S.N., Cermelli, F., Fontanel, D., Masone, C., Caputo, B.: Unmasking anomalies in road-scene segmentation. In: Proc. ICCV, pp. 4037–4046 (2023)
23. Roth, K., Pemula, L., Zepeda, J., Schölkopf, B., Brox, T., Gehler, P.: Towards total recall in industrial anomaly detection. In: Proc. CVPR, pp. 14318–14328 (2022)
24. Stratton, J.: An introduction to microsoft copilot. In: Copilot for Microsoft 365: Harness the Power of Generative AI in the Microsoft Apps You Use Every Day, pp. 19–35. Springer (2024)
25. Tran, T.M., Vu, T.N., Nguyen, T.V., Nguyen, K.: Uit-adrone: a novel drone dataset for traffic anomaly detection. IEEE J. Selected Top. Appl. Earth Observ. Remote Sens. **16**, 5590–5601 (2023)
26. Wu, P., et al.: Vadclip: adapting vision-language models for weakly supervised video anomaly detection. In: Proc. AAAI Conference on Artificial Intelligence, vol. 38, no. 6, pp. 6074–6082 (2024)

27. Zhai, X., Liu, K., Nash, W., Castineira, D.: Smart autopilot drone system for surface surveillance and anomaly detection via customizable deep neural network. In: Proc. IPTC, p. D021S053R001. IPTC (2020)
28. Zhou, K., Yang, J., Loy, C.C., Liu, Z.: Learning to prompt for vision-language models. IJCV **130**(9), 2337–2348 (2022)
29. Zhou, Q., Pang, G., Tian, Y., He, S., Chen, J.: Anomalyclip: object-agnostic prompt learning for zero-shot anomaly detection. arXiv preprint arXiv:2310.18961 (2023)

Enhancing Scene Segmentation Using Spatial Relationships: A Case Study on Byzantine Seals

Ege Şendoğan[1](✉), Victoria Eyharabide[2], and Isabelle Bloch[1]

[1] Sorbonne Université, CNRS, LIP6, Paris, France
ege.sendogan@lip6.fr

[2] Sorbonne Université, STIH, Paris, France

Abstract. The analysis of iconographic scenes on Byzantine seals provides invaluable insights into the empire's religious, social, and political trends. The segmentation of iconographic elements in Byzantine seals enables the extraction of detailed, quantifiable information about each design element, facilitating digital analysis of iconographic scenes. However, the task is non-trivial due to challenges such as seal deterioration, poor image acquisition, and the lack of data. In this work, we propose a loss function that incorporates structural constraints based on the expected spatial relationships between iconographic elements in the training of a neural network. The results of the proposed method are evaluated quantitatively and qualitatively. They show the benefit of integrating prior knowledge of spatial relationships into the network to improve performance in a challenging semantic segmentation task.

Keywords: Segmentation · Cultural heritage · Hybrid AI · Directional spatial relationships · Byzantine seals

1 Introduction

Byzantine lead seals are small, circular objects that people used not only to secure and authenticate documents but also to convey their identity. Their personalized designs may feature monograms and inscriptions that signify the owner's status, position, and family. Seals may also bear religious images, symbolizing the personal piety of their owners. Their analysis serves historians as a valuable resource for exploring the empire's religious, social, and political dynamics. Understanding the iconographic scenes on Byzantine seals necessitates a careful analysis of every element within the seal. However, this task is non-trivial and requires advanced expertise in sigillography and a profound knowledge of the Byzantine world. The automatic analysis of iconographic scenes on Byzantine seals, based on semantic segmentation, is thus important for assisting historians.

This work has been partially funded by the French National Research Agency (ANR), grant number ANR-21-CE38-0001, https://anr.fr/Projet-ANR-21-CE38-0001.

E. Rodolà et al. (Eds.): ICIAP 2025, LNCS 16167, pp. 647–659, 2026.
https://doi.org/10.1007/978-3-032-10185-3_51

The condition of Byzantine seals and the quality of their acquired images limit the performance of traditional segmentation methods. Lead seals are often deteriorated because of lead corrosion, wear, and other forms of deterioration, while poor lighting conditions are frequent among their images. These factors obscure key features, presenting high intra-class variance between visual appearances, making it difficult to identify and differentiate patterns reliably. Fully Convolutional Networks (FCNs) [9] have revolutionized image segmentation through their ability to capture local features such as edges and textures. However, segmenting the iconographic elements in Byzantine seals requires understanding the broader context. Additionally, FCNs are sensitive to class imbalance, which is common in iconographic scenes.

Several architecture-based solutions have been proposed to overcome the limitations of FCNs. U-Net [13] employs an architecture that consists of a down-sampling path that encodes multi-scale features and an up-sampling path that decodes these features. These paths are connected with skip connections, which help preserve features at multiple scales and accurately localize image features. SegNet [1] also employs an encoder-decoder architecture but uses max-pooling indices from the encoder to improve spatial context information without increasing the computational burden. DeepLab [5] consists of an FCN backbone that employs atrous convolution, which enables expanding the receptive field of convolutional filters to enhance the global context without increasing the number of parameters. PSPNet [16] is another network that employs the concept of atrous convolution. It also incorporates a pyramid pooling module that captures features at various scales. DeepLabv3+ [6] adopts the Atrous Spatial Pyramid Pooling (ASPP) module, which applies parallel atrous convolutional layers to the extracted feature maps to capture the multi-scale context and address the challenges of segmenting objects at various scales. The attention mechanism is another architectural innovation that helps to discern contextual dependencies beyond the scope of the convolutional kernels. PSANet [17] employs a self-attention module on top of an FCN backbone to learn contextually relevant positions and propagate information between them.

While architectural developments help improve the performance of CNNs in segmentation tasks, their ability to reason from contextual information, provided the receptive field is large enough, is limited by the data available to them [12]. Several studies have made use of external knowledge to enhance contextual reasoning. In [15], Wu et al. integrated the external linguistic knowledge derived from the dataset into the scene parsing through a graph convolutional network (GCN) approach. In [2], Benkirane et al. used spatial relationship priors to improve the performance of deep neural networks for panoptic segmentation. Specifically, they modeled spatial relationships by using region connection calculus (RCC) and integrated them into network training through a novel loss function. Riva also leveraged spatial relationship knowledge in [11], where two novel loss functions based on directional relationship satisfaction were proposed.

The limited number of Byzantine seals and the challenges of annotating them prevent the creation of a large annotated dataset. As a result, previous architectural advancements have struggled to improve CNN training for the semantic segmentation of the iconographic scenes on Byzantine seals. In this work, we introduce a loss function that integrates structural constraints, based on the expected spatial relationships between iconographic elements during network training. Byzantine seals feature structured religious scenes where design elements of seals that bear similar iconography follow consistent relative positioning. By guiding the network training with prior knowledge of the structure of iconographic scenes, we aim to help the network better understand the overall design, i.e., the general context of the seal, ultimately improving segmentation results. The main contributions of our paper are as follows:

- An automated method for segmenting and identifying iconographic elements in images of Byzantine seals.
- A new loss function to guide neural network training by leveraging prior knowledge of spatial relationships.
- Results on a series of seals, confirming the usefulness of spatial relationships to improve semantic segmentation.

The remainder of the paper is organized as follows. Section 2 introduces the proposed method. Section 3 explains our experiments. Section 4 shows the results we obtained, and Sect. 5 concludes the paper.

2 Proposed Method

To improve the segmentation and recognition of iconography elements of Byzantine seals, we propose a methodology that leverages prior knowledge about the spatial arrangement of these elements. At the core of this approach, we introduce a novel loss function based on spatial relationships, enabling the integration of the Byzantine seals' inherent structure (i.e. spatial organization) into a neural network for multi-object segmentation. This approach addresses challenges arising from the current state of Byzantine seals, as well as issues related to their image acquisition, such as poor lighting conditions. Since spatial relationships in images are prone to imprecision, we rely on fuzzy sets theory, which provides a robust framework for modeling this imprecision [4]. Spatial relations are computed between any two objects both in the reference segmentation and in the prediction, yielding two scores, which we define in a fuzzy set framework as degrees of satisfaction of relations. The loss function used in the neural network compares the two relation scores, and its minimization favors semantic segmentation satisfying the expected relations. Given a set R of source-target object pairs, where $|R| \geq 1$, we define a loss function that compares relation scores as follows:

$$\mathcal{L}_{\text{Rel}} = \frac{1}{|R|} \sum_{(S,T)\in R} |\mu(S,T) - \mu(\hat{S},\hat{T})|, \tag{1}$$

where $\mu(S,T)$ denotes the score of a relation between the source object S and the target object T in the reference segmentation, and $\mu(\hat{S},\hat{T})$ the score of the same relation computed for $\hat{S}$ and $\hat{T}$, predicted counterparts of S and T.

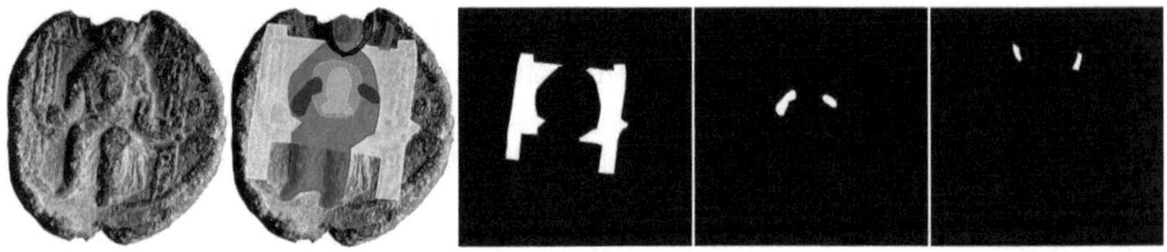

Fig. 1. A Byzantine seal (a) and its label map, including the Theotokos with a veil and a nimbus seated on a throne and holding the medallion of the Child with two hands (b). Segmentation maps of the throne, hand and nimbus classes. Throne and the Theotokos overlap, resulting in throne being represented as a multi-component structure that includes two disconnected and visible parts on either side of the Theotokos (c). Hands are inherently built of two components (d), and nimbus is split into two parts because of the missing part of the seal (e).

2.1 Encoding Spatial Relationships

This section explains how we model and encode the spatial relationships between seal objects using a fuzzy morphological approach. A morphological fuzzy pattern matching approach was proposed in [3] to compare the relative positions of two objects. This approach relies on fuzzy dilation using a fuzzy structuring element that characterizes the semantics of the application. In this paper, we use mostly directional relations, for which the structuring element is modeled as a fuzzy cone, as follows. The direction of interest for two objects S and T is defined by the angle $\theta_{(S,T)}$ between the horizontal direction and the line joining the centers of S and T. Then, for each point (i, j) of the structuring element, the angle between the line connecting the origin to this point and the desired direction is calculated, and the membership value at this point is a decreasing function of this angle [3,11]:

$$\kappa_{(S,T)}(i,j) = \max\left(1 - \frac{2\left|\arctan\left(\frac{j}{i+\epsilon}\right) - \theta_{(S,T)}\right|}{\eta}, 0\right). \qquad (2)$$

The parameter η controls the decay rate of the membership degrees, defining how strictly or loosely a spatial relationship between S and T is interpreted.

While this structuring element definition handles one-to-one relationships well, many real-world scenarios involve more complex spatial interactions. Some objects consist of multiple connected components (see Fig. 1c-e), meaning that a single source object may simultaneously relate to several parts of a target object. Two objects can also satisfy multiple relations, for instance, when one object surrounds another one, i.e., a target object may be in all directions with respect to the source (e.g., the relationship between the Child and the medallion, see Fig. 1). To model these complex spatial interactions (whether for union of components or union of relations), we define the structuring element as a square array with N cone-shaped distributions centered at $(0,0)$. For instance, N represents the number of directions necessary to capture surrounding relationships or the number of connected components of the target object T, where $T = \cup_{n=1}^{N} t_n, N \geq 1$. The structuring element $\kappa_{(S,t_n)}$ that represents the relationship between the source object S and each component t_n of T is computed by using Eq. (2). These individual structuring elements are then aggregated into a unified representation $\kappa'_{(S,T)}$, using the max as fuzzy union:

$$\kappa'_{(S,T)}(i,j) = \max_{n \in \{1 \ldots N\}} \kappa_{(S,t_n)}(i,j). \tag{3}$$

To ensure uniformity for this operation, each $\kappa_{(S,t_n)}$ is first padded with zeros to match the largest kernel size among all $\kappa_{(S,t_n)}$.

Once the structuring element $\kappa'_{(S,T)}$ is obtained, it is used to construct the fuzzy landscape $\varphi^{*(S,T)}$ around S, where the value at each point represents the degree of satisfaction of the spatial relation of interest. In [3], $\varphi^{*(S,T)}$ is modeled as a morphological dilation of S with $\kappa'_{(S,T)}$. However, morphological operations are non-differentiable, which poses challenges in gradient-based optimization within deep learning frameworks. To address this, a common strategy is to use differentiable approximations of morphological operations, such as counter-harmonic mean (CHM) [10], or α-softmax and LogSumExp functions [8,14]. Another approach involves modifying the backpropagation process, enabling the direct use of morphological operators [7]. As shown in [11], a simple convolution can effectively approximate the dilation operation, providing a solution that is computationally efficient, hyperparameter-free, and well-suited for gradient-based network optimization. In this work, we adopt this convolution-based approach:

$$\varphi^{*(S,T)}(i,j) = (S * \kappa'_{(S,T)})(i,j), \tag{4}$$

where $*$ denotes the convolution operation.

Finally, we compute a relationship score to compare the target object T with the fuzzy landscape generated around the source object S, defined as the average of the intersection (defined as a product here) of T and $\varphi^{(S,T)}$:

$$\mu_{(S,T)} = \frac{\sum_{(i,j)\in\Omega} T(i,j)\varphi^{(S,T)}(i,j)}{\sum_{(i,j)\in\Omega} T(i,j)}, \tag{5}$$

where Ω is the coordinate space, and $T(i,j)$ is the membership value of the target object T at pixel (i,j). The term $\varphi^{(S,T)}$ represents normalized fuzzy landscape $\varphi^{*(S,T)}$, scaled to the range [0,1]. This score reflects how well T fits within the fuzzy landscape that highlights regions of interaction between S and its surroundings in a specific direction, with a score of 0 indicating no alignment and a score of 1 indicating perfect spatial satisfaction.

2.2 Training with Spatial Relationship Priors

Enhancing neural network training with external knowledge of spatial relationships can be achieved by using a dataset that includes relationship priors. A conventional semantic segmentation dataset, which comprises pairs of seal images and their corresponding label maps (x^k, y^k), can be enhanced by adding the encoding of each relationship (S, T) that belongs to R, the set of the relationships which are present in the kth image. The encoding of a relationship (S, T) consists of $\kappa'_{(S,T)}$ and $\mu_{(S,T)}$. In this expanded dataset, called the informed dataset, each sample now consists of a seal image (x^k), its label map (y^k), a set of structuring elements representing spatial relationships (κ'^k), and the corresponding relationship scores (μ^k).

A traditional convolutional neural network for multi-object segmentation, which is parameterized by ψ, can be formulated as:

$$f_\psi : \mathbb{R}^{H\times W\times C} \to [0,1]^{H\times W\times Z}, \tag{6}$$

where $\mathbb{R}^{H\times W\times C}$ is the spatial domain of the input image with height H, width W and C channels, and Z is the number of objects. Such a network can be trained with a pixel-wise loss function, such as a cross-entropy loss function:

$$\mathcal{L}_{CE}(\hat{y}^k, y^k) = -\frac{1}{H\cdot W}\sum_{h=1}^{H}\sum_{w=1}^{W}\sum_{z=0}^{Z-1} y^k_{(h,w,z)} \log(\hat{y}^k_{(h,w,z)}), \tag{7}$$

where $\hat{y}$ represents the output of f_ψ, and $\sum_{z=0}^{Z-1} \hat{y}^k_{(h,w,z)} = 1, \quad \forall h, w$. To train the network with the new loss function, $\mathcal{L}_{Rel}$, we introduce a novel module that computes the spatial relationship score $\mu_{(\hat{S},\hat{T})}$ between the predicted counterparts of the ground truth objects (S, T). This module takes the output of f_ψ as an input. For each relationship (S, T) present in the image, it extracts the predicted segmentation maps $\hat{S}$, and $\hat{T}$. The fuzzy landscape $\varphi^{(\hat{S},\hat{T})}$ is then computed using $\hat{S}$ and structuring element for the ground truth relation (S, T), $\kappa'_{(S,T)}$, as defined by Eq. (4). Finally, the relation score $\mu_{(\hat{S},\hat{T})}$, allowing us to compute $\mathcal{L}_{Rel}$, obtained by using $\varphi^{(\hat{S},\hat{T})}$ and $\hat{T}$ as defined by Eq. (5).

This module can be incorporated into widely used CNN architectures with only minimal modifications to the existing network structure. For our experiments, we used the DeepLabV3+ architecture, which is widely used in semantic segmentation tasks. DeepLabV3+ was chosen for its strong ability to capture local and global context using atrous spatial pyramid pooling (ASPP). To optimize the model, we combined the proposed loss function with $\mathcal{L}_{CE}$:

$$\mathcal{L} = \mathcal{L}_{CE} + \alpha\mathcal{L}_{Rel}. \tag{8}$$

3 Experiments

We used three prestigious byzantine seal collections as our main data sources: the Zacos and Seyrig collections, housed in the French National Library (BnF), and Tatış collection, a private collection published by Jean-Claude Cheynet, a world-renowned sigillographer. We worked with the published seals from these collections. We owe access to these resources to the BnF and to Jean-Claude

Cheynet. From these collections, we have sampled 108 seal images that feature Theotokos (Virgin Mary) as the central Figure (see Fig. 1), since the Theotokos holds a significant place in Byzantine sigillography, and accounts for a considerable number of representations. These images have been annotated by experts in Byzantine sigillography, identifying eight distinct target classes. Given the intrinsic complexity of annotating Byzantine seals, the process involved reaching a consensus among multiple annotators, resulting in the identification of eight distinct target classes. These classes and their respective cardinalities are as follows: Theotokos (108), veil (107), nimbus (99), child (79), hands (88), medallion (44), nimbus crucified (28), and throne (9). The **Theotokos** is represented often with a **veil** and a **nimbus** above her head. Also, the Theotokos often holds the **Child** having a **crucified nimbus**. The **hands** of the Theotokos are depicted in a variety of gestures, including an orans position with hands raised, a prayerful posture with hands uplifted, a guiding gesture as she presents the Christ Child and other variations seen in different seal iconography. Two common depictions in Byzantine seals—the enthroned Theotokos, and Theotokos holding a Medallion of Christ Child—are also present in our dataset. Consequently, both the **throne** and **medallion** are among the objects identified by the annotators.

Spatial relationships between these iconographic elements exhibit consistent properties across different seals. For instance, when they exist, the throne is always positioned to the left and right of Theotokos, while the veil is consistently above her head. Similarly, the nimbus and the crucified nimbus are always above the Theotokos and the Child, respectively. The medallion always surrounds the Child. When they exist, we encode and add (Theotokos, throne), (Theotokos, veil), (Theotokos, nimbus), (Child, crucified nimbus), (Child, medallion) relations to the informed dataset to enhance the network training while leveraging the Theotokos and the Child as the reference (i.e., source) objects in each relationship, as all other objects are contextually linked to the Theotokos and the Child. Moreover, the relative position of the hands of the Theotokos and the Child varies depending on the iconographic style, yet it remains consistent within seals of the same style. Some seals depict the Theotokos holding the Child or the medallion of the Child, with her hands positioned near him. In these cases, we use the (Child, hands) relation. For seals with the orans or prayer posture, we apply the (Theotokos, hands) relation. Additionally, we encode the (Theotokos, Child) relation when the Theotokos is depicted holding the Child by her side.

We selectively chose these specific relationships to maximize the relative position information captured without introducing excessive computational complexity in the network training. The informed dataset is obtained as explained in Sect. 2.2. Three examples of $\kappa'_{(S,T)}$ (its symmetrical version because of the convolution) and $\mu_{(S,T)}$ are illustrated in the second and fourth columns of Fig. 2, respectively.

We conducted our experiments by using 5-fold cross-validation. We used different values for α during experiments to assess the effect of $\mathcal{L}_{Rel}$. We employed stochastic gradient descent optimization with a momentum of 0.9 and set the learning rate to 0.01. During training, we gradually decreased the learning rate to

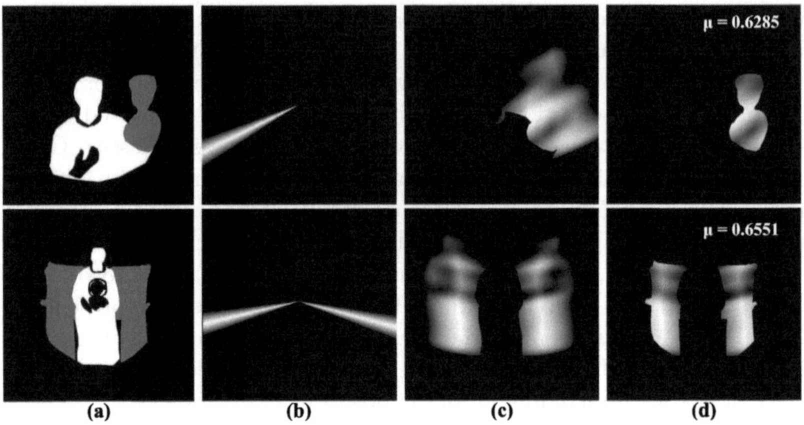

Fig. 2. First row: (a) label map containing the Theotokos and the Child, (b) structuring element $\kappa'_{(Theotokos,child)}$, (c) fuzzy landscape, (d) its intersection with the Child and relation score $\mu_{(Theotokos,child)}$. Second row: (a) label map containing the Theotokos and throne, (b) structuring element $\kappa'_{(Theotokos,throne)}$, (c) fuzzy landscape, (d) its intersection with the throne and the relation score $\mu_{(Theotokos,throne)}$.

0.0001. We set the batch size to 2 and trained the models for 25000 iterations. We performed data augmentation on the fly using random gamma correction and contrast-limited adaptive histogram equalization, as well as random flipping with a probability of 0.5. We also performed random resizing and random cropping. When the input image is flipped or resized, the same transformation is applied to the structuring elements. If a relationship disappears due to the crop operation, it is excluded from the computation of $\mathcal{L}_{Rel}$. Experiments lasted 3 to 8 h on NVIDIA A100 Tensor Core GPU with Cuda 12.0.

4 Results and Discussion

We evaluated the results qualitatively and quantitatively. Qualitative analysis was performed through visual inspection, which allows us to assess how meaningful, relevant, and consistent the results are for iconographic scene analysis. The quantitative assessment was carried out by using pixel-level measures, specifically, Dice coefficient (D), precision (P), and recall (R), which are widely used in semantic segmentation tasks. We present both overall performance results, mean Dice (mD), mean Precision (mP), and mean Recall (mR), averaged over all object classes (Table 1) and class-based results (Table 2). We compare our method to the reference model, which is trained using only the cross-entropy loss (i.e., $\alpha = 0$).

Table 1 presents the results for the mean ± standard deviation of mD, mP, mR metrics, computed over 5 folds. The highest mD score (75.7 ± 1.9) and mR (73.2 ± 2.5) were achieved using the proposed loss function with $\alpha = 0.33$. The best mP, 80.4 ± 1.9, is achieved with $\alpha = 1.25$, also using the proposed loss function. An improvement in each measure is also observed when $\alpha = 0.33$, compared to the reference model. The proposed loss function yields consistent results across different data splits in mD and mR metrics, as indicated by a reduced standard deviation for many values of α greater than 0.

Table 1. Mean Dice (mD), mean Precision (mP), and mean Recall (mR) scores. Values are given as mean ± standard deviation over 5 folds, for different values of α, weighting our novel loss function.

	0	0.2	0.25	0.33	0.75	1	1.25
mD	75 ± 2.4	74.4 ± 0.6	74.1 ± 1.7	**75.7 ± 1.9**	73.9 ± 1.1	74.6 ± 0.7	75.1 ± 2.4
mP	79.8 ± 0.9	78.7 ± 1.3	79.5 ± 2.5	**80.1 ± 1.3**	79.6 ± 1.1	78.5 ± 1.3	80.4 ± 1.9
mR	72.6 ± 3.6	72.2 ± 1.1	71.4 ± 3.6	**73.2 ± 2.5**	71 ± 1.2	72.3 ± 1.3	72.3 ± 3.9

Table 2 shows the class-based results for precision and recall. The highest precision for each object class was achieved when $\alpha > 0$ except for the nimbus class. A mid-range value (0.25 - 0.75) of α often provides a better precision than the reference model. The highest recall for each object class was achieved when $\alpha > 0$ except for the throne. Throne, medallion, and crucified nimbus are the least represented classes in the dataset. The reference model presents high recall standard deviation for these classes. The recall shows a notable standard deviation reduction when $\alpha > 0$. Recall standard deviation of the throne class drops from 19.1 at $\alpha = 0$ to much lower values (8.3 - 11.4) when α is between 0.25 and 0.75. Medallion and crucified nimbus recall standard deviations decrease from 7.9 and 9.3 at $\alpha = 0$ to 1.8 and 5.2, respectively, when $\alpha = 0.33$.

Minor pixel-level improvements, that do not significantly affect the pixel-based measures, still can carry substantial semantic importance in the analysis of iconographic scenes. The presence or absence of specific objects— such as a throne, medallion, Child, or hands— as well as their spatial arrangement and the number of components, are critical for determining the specific type of Theotokos depicted. These elements play a decisive role in interpreting the scene. Also, seal deterioration obscures the seal objects, making annotations subject to uncertainty. This introduces noise into the ground truth, making pixel-based measures less reliable as indicators of true performance. We thus considered another evaluation through visual inspection. Figure 3 shows qualitative results. In the first row of Fig. 3, the reference model incorrectly detected the elements that should not be in the scene (as shown in the canvas, a medallion and two

Table 2. Precision (P) and Recall (R) for each object class: throne (1), Theotokos (2), veil (3), nimbus (4), medallion (5), Child (6), crucified nimbus (7) and hands (8).

		0	0.2	0.25	0.33	0.75	1	1.25
1	P	76.5 ± 5	69.6 ± 4.1	76 ± 11	$\mathbf{76.6 \pm 8.8}$	70 ± 7.1	73.1 ± 8.9	75.4 ± 9.4
	R	$\mathbf{70.9 \pm 19.1}$	68.1 ± 12.7	61.8 ± 11.4	66.1 ± 9.2	65 ± 12.9	66.6 ± 11.2	67.4 ± 18
2	P	80.2 ± 2.6	79.5 ± 2.1	$\mathbf{82.3 \pm 2}$	81.2 ± 2.3	79.4 ± 2.8	80.1 ± 3.2	80.3 ± 2.6
	R	87.6 ± 1.6	$\mathbf{88.8 \pm 1}$	85.7 ± 2.8	87.8 ± 2.2	88.1 ± 1	88 ± 1.1	$\mathbf{88.8 \pm 1.1}$
3	P	86 ± 2.2	85.8 ± 1	85.6 ± 1.8	86.2 ± 1.4	85.9 ± 1.6	85.8 ± 1.1	$\mathbf{86.8 \pm 1.1}$
	R	80.3 ± 3	80.7 ± 1.5	80.3 ± 3.4	79.7 ± 1.2	80.3 ± 2.1	$\mathbf{80.8 \pm 0.9}$	79.4 ± 2.2
4	P	$\mathbf{77.1 \pm 1.9}$	75.7 ± 2	76.8 ± 2.8	76.8 ± 1	76.4 ± 0.6	76.7 ± 1.3	76.4 ± 1.6
	R	70.1 ± 1.8	$\mathbf{70.5 \pm 1.2}$	67.5 ± 2.8	70 ± 1.2	68.3 ± 1.8	67.3 ± 2.2	69.4 ± 1.9
5	P	71.6 ± 1.8	72.4 ± 3.6	71.3 ± 4.3	72.7 ± 5.7	71.2 ± 3.9	71.9 ± 3.3	$\mathbf{74 \pm 2.6}$
	R	72.4 ± 7.9	74 ± 4	72.4 ± 4.8	$\mathbf{75.7 \pm 1.8}$	73.1 ± 3.4	74.5 ± 1.3	72.5 ± 2.9
6	P	83 ± 3.4	83.2 ± 4.2	81.6 ± 4.6	82.4 ± 3.1	$\mathbf{83.8 \pm 3.2}$	81.1 ± 4.5	82.7 ± 4.3
	R	76.4 ± 3.6	76.9 ± 1.5	$\mathbf{79.8 \pm 5.3}$	79.2 ± 4.7	75.8 ± 2	79 ± 2.3	76.1 ± 2.9
7	P	67.8 ± 6.2	65.1 ± 4.2	66.8 ± 8.7	69.3 ± 8.6	$\mathbf{71.3 \pm 3.6}$	61.2 ± 5	69.2 ± 6.4
	R	32.6 ± 9.3	30.2 ± 5.6	32.5 ± 11	$\mathbf{34.9 \pm 5.2}$	30.7 ± 4.9	34 ± 9.8	34 ± 8
8	P	78.2 ± 2.4	79.2 ± 1.9	77.6 ± 4	77.3 ± 2.7	80.3 ± 1.8	78.8 ± 2.2	$\mathbf{81 \pm 2.6}$
	R	64.9 ± 5.6	62.8 ± 3.2	64.6 ± 3	$\mathbf{67.1 \pm 3.5}$	59.3 ± 5.4	62.3 ± 2.8	65 ± 3.1

Child elements are detected). Such false predictions lead to a misinterpretation of the Theotokos' type in the seal, which is worse than falsely predicting a class that is part of the scene. The proposed method not only improves the result by omitting the medallion and the first Child class, but also significantly reduces the presence of the second Child class. Additionally, our method improved the form of the Theotokos and her hands. In the second row of Fig. 3, the hands of the Theotokos holding the medallion, an element of critical importance in this iconographic depiction of the Theotokos, are better localized and detected when using the proposed loss function. Also, the proposed method improved the form of the medallion and the nimbus. The visual interpretation demonstrates that guiding neural network training with prior knowledge of spatial relationships improves the identification of key elements that define a particular type of iconographic scene while also helping to omit the presence of irrelevant classes, leading to a more accurate interpretation of the overall scene.

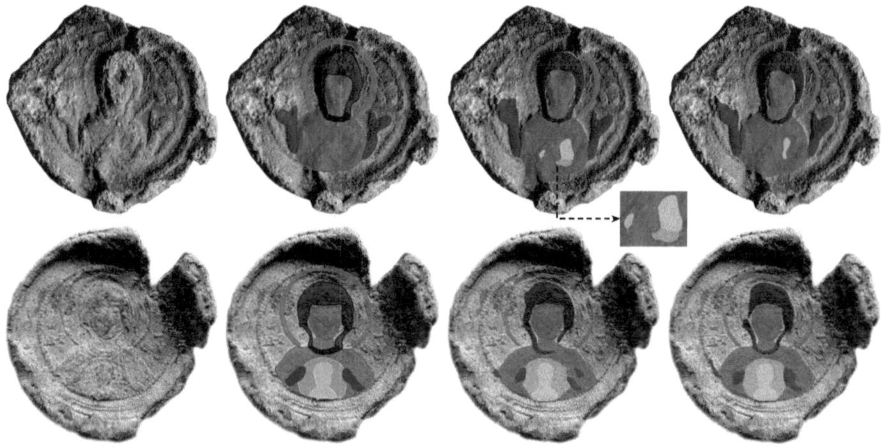

Fig. 3. Seal images (a) and their label maps (b) containing the Theotokos, veil, nimbus, and hands objects. The label map in the second row also contains the Child, and medallion objects. (c) Predictions of the reference model are shown. In the first row, a small canvas also added to better show the prediction. (d) Results of the proposed approach.

5 Conclusion

This work represents the first attempt to automatically analyze iconographic scenes on Byzantine seals through semantic segmentation of their constituent design elements. We employed DeepLabv3+ architecture with a ResNet-101 network backbone as the reference model. To improve the segmentation performance of the network, we introduced $\mathcal{L}_{Rel}$, a novel loss function that helps integrate structural constraints reflecting the expected spatial relationships between the design elements of the scene depicted on the seal. This approach involves encoding relationships in the reference segmentation and in its predicted counterpart. Then, the difference between them is measured. Spatial relationships are encoded using a morphological fuzzy pattern matching approach [3]. We combined $\mathcal{L}_{Rel}$ with cross-entropy in the loss function by assigning a weight α to $\mathcal{L}_{Rel}$, which ranges in [0, 1.25] in our experiments.

The results are promising and demonstrate the positive impact of integrating prior knowledge of spatial relationships into network training to enhance semantic segmentation in structured scenes such as those found on Byzantine seals. By guiding the network with structural constraints through the proposed $\mathcal{L}_{Rel}$ loss function, we improved the identification of key iconographic elements that define specific scene types. Additionally, this approach helped suppress irrelevant classes, leading to a more coherent and accurate semantic interpretation of the iconographic scenes.

Overall, our work highlights the potential of combining deep learning with external knowledge in the iconographic scenes for automatizing their analysis, an important task in studying Byzantine seals. Future work will focus on modeling other relations and designing evaluation measures that better reflect the significance of the segmentation results in the context of iconographic scenes of Byzantine seals. In addition, we plan to address the classification of Theotokos types based on the segmented design elements.

References

1. Badrinarayanan, V., Kendall, A., Cipolla, R.: SegNet: a deep convolutional encoder-decoder architecture for image segmentation. IEEE Trans. Patt. Analy. Mach. Intell. **39**(12), 2481–2495 (2017). https://doi.org/10.1109/TPAMI.2016.2644615
2. Benkirane, F.E., Crombez, N., Hilaire, V., Ruichek, Y.: Hybrid AI for panoptic segmentation: an informed deep learning approach with integration of prior spatial relationships knowledge. Comput. Vision Image Understand. **240**, 103909 (2024). https://doi.org/10.1016/j.cviu.2023.103909
3. Bloch, I.: Fuzzy relative position between objects in image processing: a morphological approach. IEEE Trans. Patt. Analy. Mach. Intell. **21**(7), 657–664 (1999). https://doi.org/10.1109/34.777378
4. Bloch, I.: Fuzzy sets for image processing and understanding. Fuzzy Sets Syst. **281**, 280–291 (2015). https://doi.org/10.1016/j.fss.2015.06.017
5. Chen, L.C., Papandreou, G., Kokkinos, I., Murphy, K., Yuille, A.L.: Semantic image segmentation with deep convolutional nets and fully connected CRFs. arXiv preprint arXiv:1412.7062 (2014)
6. Chen, L.C., Zhu, Y., Papandreou, G., Schroff, F., Adam, H.: Encoder-decoder with atrous separable convolution for semantic image segmentation. In: Eur. Conf. on Computer Vision (ECCV), pp. 801–818 (2018). https://doi.org/10.1007/978-3-030-01234-2_49
7. Franchi, G., Fehri, A., Yao, A.: Deep morphological networks. Pattern Recogn. **102**, 107246 (2020). https://doi.org/10.1016/j.patcog.2020.107246
8. Kirszenberg, A., Tochon, G., Puybareau, É., Angulo, J.: Going beyond p-convolutions to learn grayscale morphological operators. In: Int. Conf. on Discrete Geometry and Mathematical Morphology, pp. 470–482. Springer (2021). https://doi.org/10.1007/978-3-030-76657-3_34
9. Long, J., Shelhamer, E., Darrell, T.: Fully convolutional networks for semantic segmentation. In: IEEE Conf. on Computer Vision and Pattern Recognition (CVPR), pp. 3431–3440 (2015). https://doi.org/10.1109/CVPR.2015.7298965
10. Masci, J., Angulo, J., Schmidhuber, J.: A learning framework for morphological operators using counter–harmonic mean. In: Mathematical Morphology and Its Applications to Signal and Image Processing (ISMM), pp. 329–340 (2013). https://doi.org/10.1007/978-3-642-38294-9_28
11. Riva, M.: Spatial Relational Reasoning in Machine Learning: Deep Learning and Graph Clustering. Ph.D. thesis, Institut Polytechnique de Paris, Télécom Paris (2022)
12. Riva, M., Gori, P., Yger, F., Bloch, I.: Is the U-Net directional-relationship aware? In: IEEE Int. Conf. on Image Processing (ICIP), pp. 3391–3395 (2022). https://doi.org/10.1109/ICIP46576.2022.9897715

13. Ronneberger, O., Fischer, P., Brox, T.: U-Net: convolutional networks for biomedical image segmentation. In: Int. Conf. on Medical Image Computing and Computer-Assisted Intervention (MICCAI), pp. 234–241 (2015). https://doi.org/10.1007/978-3-319-24574-4_28
14. Shen, Y., Shih, F.Y., Zhong, X., Chang, I.C.: Deep morphological neural networks. Int. J. Patt. Recogn. Artif. Intell. **36**(12), 2252023 (2022). https://doi.org/10.1142/S0218001422520231
15. Wu, T., et al.: Ginet: Graph interaction network for scene parsing. In: Eur. Conf. on Computer Vision (ECCV), pp. 34–51. Springer (2020). https://doi.org/10.1007/978-3-030-58520-4_3
16. Zhao, H., Shi, J., Qi, X., Wang, X., Jia, J.: Pyramid scene parsing network. In: IEEE Conf. on Computer Vision and Pattern Recognition (CVPR), pp. 6230–6239 (2017). https://doi.org/10.1109/CVPR.2017.660
17. Zhao, H., et al.: PSANet: point-wise spatial attention network for scene parsing. In: Eur. Conf. on Computer Vision (ECCV) (2018). https://doi.org/10.1007/978-3-030-01240-3_17

Image Specific Protection Against Manipulation

Filippo Bartolucci(✉) and Giuseppe Lisanti

CVLab – Department of Computer Science and Engineering, University of Bologna, Bologna, Italy
{filippo.bartolucci3,giuseppe.lisanti}@unibo.it

Abstract. The rapid development of Generative Models (GMs) for image synthesis poses challenges to identifying manipulated images accurately. Traditionally, detection methods are trained to identify image manipulations after they occur. However, these approaches often struggle to generalize to GMs not seen during training. Proactive methods have been introduced to address this limitation. In this paradigm, images are embedded with a protection template as a form of proactive defense. Previous methods exploited a finite set of templates to proactively counter image manipulation, however, this raises concerns about potential vulnerabilities since a finite number of templates can provide a predictable exploit for malicious attackers. This work presents a template-based detection system that is designed to generate personalized templates for each image. Our approach enhances protection while also improving the accuracy of manipulation detection. Our solution achieves high detection accuracy on unseen GMs while also outperforming existing solutions.

Keywords: image manipulation · proactive detection · media forensics

1 Introduction

State-of-the-art generative models (GMs) [1,2] for image synthesis are revolutionizing the field of image creation, demonstrating impressive versatility across tasks such as image-to-image translation and generating highly realistic images. However, their remarkable capabilities also pose critical challenges: can we reliably determine if an image has been manipulated? Fake image generation can be classified into two categories: complete image generation, where entirely new images are created by inputting noise into a generative model, and partial image manipulation, which selectively modifies real images. Partial manipulations can significantly alter the semantics of genuine images, raising serious concerns about trust and security. As GMs become more accessible these risks increase, highlighting the need for robust and generalizable detection mechanisms.

Several methods have been proposed to detect images generated or manipulated by a GM [3–6]. These approaches, which are trained directly on real and manipulated images, are classified as *passive* since countermeasures are implemented only after the manipulation has taken place. However, when evaluated

E. Rodolà et al. (Eds.): ICIAP 2025, LNCS 16167, pp. 660–671, 2026.
https://doi.org/10.1007/978-3-032-10185-3_52

on GMs unseen during training, these solutions often struggle to generalize to novel manipulation techniques or newly developed GMs. This limitation arises because these methods typically rely on detecting specific artifacts that may not be present in GMs not seen at training time, requiring retraining and limiting adaptability. Recent advancements have introduced *proactive* approaches, where countermeasures are implemented before manipulation. One promising direction is template protection [7,8], where an undetectable signal, known as a template, is embedded within the image to serve as a marker of integrity. If the image undergoes manipulation, the signal is disrupted, providing an unambiguous indicator of alteration. Although effective, current template-based methods are limited by the use of a finite number of templates.

To address these challenges, we propose a novel approach that strengthens template protection by generating personalized templates for each image. Our approach follows an unsupervised learning scenario, using tailored loss functions to enforce key characteristics during training. The results demonstrate substantial improvements over prior work, achieving high detection accuracy and strong generalization capabilities on image manipulations performed by GMs unseen during training. Our main contributions can be summarized as follows:

- We introduce a novel proactive defense framework that embeds image-specific protection signals using a transformer-based encryption module;
- We demonstrate the generalization capability of our approach to manipulations produced by GMs not seen during training and to different data domains, not seen during training.
- Our method introduces minimal visual distortion while achieving high diversity across the image-specific protection signals.

2 Related Works

Early passive methods for manipulation detection trained binary classifiers to distinguish between real and generated images by identifying artifacts introduced by GMs [3,5]. The solution proposed by [4] extended this approach by leveraging face-context inconsistencies to detect discrepancies between facial features and their surrounding context. Ojha et al. [9] found that detection models often misclassify unseen GMs due to learning GM-specific artifacts, and addressed this by using a pre-trained, artifact-agnostic CLIP model. A key limitation of these passive detection approaches is their poor generalization to unseen manipulation techniques. Different GMs introduce distinct artifacts, leading to a significant drop in performance when these methods are applied to novel manipulations outside of their training distribution.

Proactive approaches shift the focus from artifact detection to actively countering manipulations. Adversarial disruption [10] disrupts GMs' outputs by introducing imperceptible perturbations, obtained using attacks such as FGSM and PGD, to the input images. While effective, it relies on gray-box and white-box

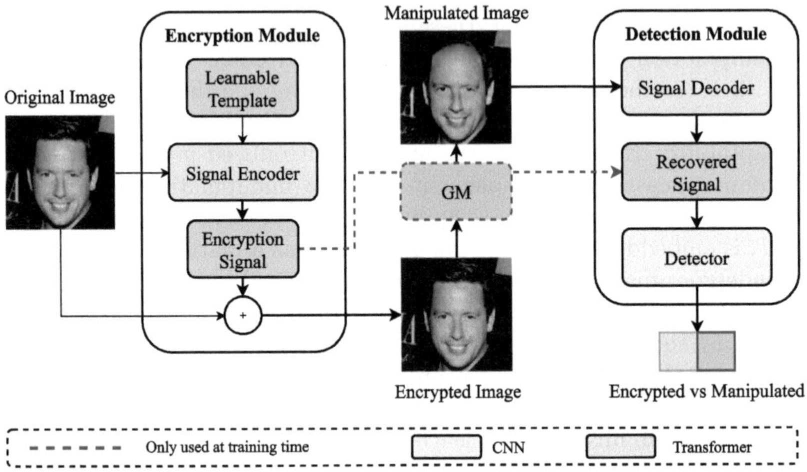

Fig. 1. Proposed architecture: an input image is processed by the encryption module, which generates an image-specific encryption signal. The encryption signal is added to the image, which is then manipulated using a generative model. The detection module extracts the encryption signal from the image and performs image manipulation detection.

settings, where the attacked GM is at least partially known, limiting its applicability in black-box scenarios. Another proactive method, FakeTagger [11], embeds hidden traceable tags within images to aid in manipulation tracking.

Template-based methods have also been explored for proactive defense. The framework introduced in [7] enhances manipulation detection by embedding templates in real images, which can be later recovered after GM manipulation. This process acts as an encryption mechanism, making it easier to differentiate between genuine and manipulated versions by simply verifying the integrity of the embedded template. MaLP [8] builds upon this work and adds manipulation localization capabilities. It uses optimized templates and Convolutional Neural Networks (CNN) to generate fakeness maps and localize manipulations. Although both [7] and [8] achieve interesting results, relying on fixed template sets represents a limitation.

3 Method

Our framework aims to detect image manipulations using a proactive approach. The proposed architecture is composed of two main components, an encryption module that secures real images and a detection module that performs binary classification to distinguish between encrypted and manipulated images. An overview of the proposed method is provided in Fig. 1.

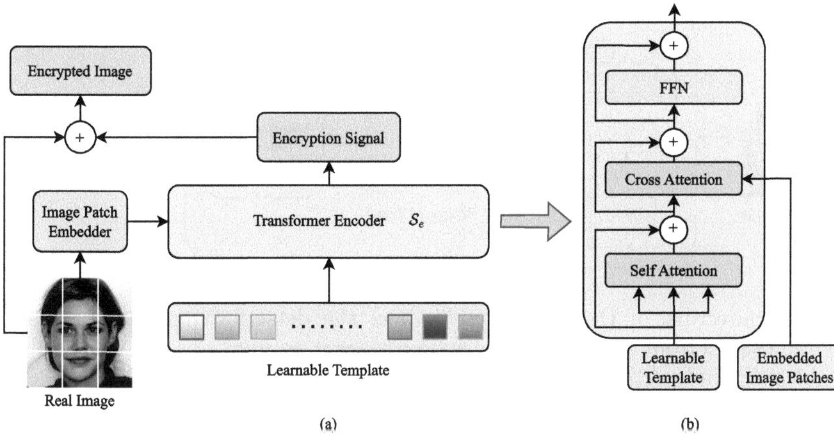

Fig. 2. a) Architecture of the Encryption Module: The image is divided into patches and processed by a Transformer encoder to generate the image-specific signal. b) Transformer Block of the Encryption Module: The transformer block consists of a self-attention layer, a cross-attention layer, and a feed-forward network. It takes a learnable template as input and is conditioned through cross-attention by the image for which the encryption signal must be generated.

3.1 Encryption Module

The encryption module safeguards an image by embedding an imperceptible signal in it. In particular, given a real image I, the encryption module $\mathcal{E}$ generates an image specific signal s_{IS} which is then embedded into the original image, producing an encrypted image $\mathcal{E}(I)$:

$$\mathcal{E}(I) = I + m\, s_{IS} \quad \text{with} \quad s_{IS} = \mathcal{S}_E(T, I) \tag{1}$$

where m is a fixed scaling parameter controlling the strength of the added signal while $\mathcal{S}_E$ is the Signal Encoder. The latter is based on a single block transformer encoder [12] and takes as input a learnable template, T, represented as a sequence of tokens, and the image to be encrypted, I, divided into patches. The learnable tokens are initialized randomly and have a dimensionality d while the image patches are mapped to the same dimension d, exploiting a patch embedder. An overview of the module can be seen in Fig. 2. The transformer block of the signal encoder consists of three key components: self-attention, cross-attention, and a feed-forward network.

Self-Attention is first computed on the learnable template, treating it as the queries, keys, and values, ensuring that it is refined independently of the image content. The output from this stage is then passed to the Cross-Attention, where the processed template tokens serve as the queries while the embedded image patches act as keys and values. This mechanism enables each template token to attend to different regions of the image based on content relevance, making the

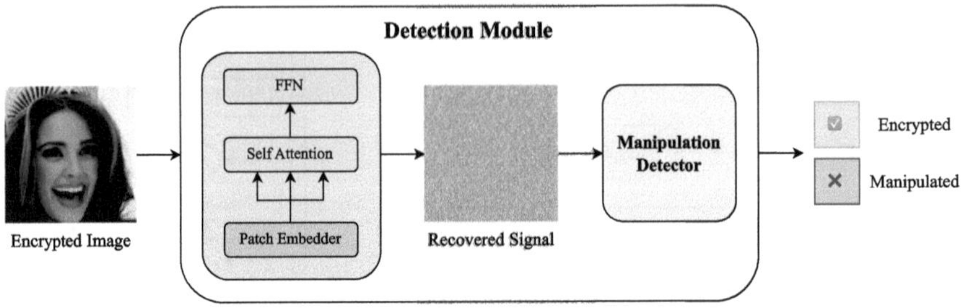

Fig. 3. Architecture of the Detection Module: the detection module extracts the encryption signal from the input image and provides it as input to a CNN-based detector, which determines whether the image has been manipulated or not.

template image-aware and enabling it to adapt its representation to the specific visual patterns in the input image.

Finally, the Feed-Forward Network (FFN) further processes the attended features, refining the output signal. This refined signal is then embedded into the image following Eq. 1 and the output is clipped in the range $[-1, 1]$.

3.2 Detection Module

The detection module is designed to recover the signal from an encrypted image and exploit it to detect if the encrypted image has been manipulated or not. As shown in Fig. 3, this module comprises two main components, a signal decoder, $\mathcal{S}_D$, and a manipulation detector, $\mathcal{D}$. The first component is a transformer block that extracts the encryption signal from an input image. Different from the signal encoder described in Sect. 3.1, the transformer block in the signal decoder relies solely on self-attention. The second component of the Detection Module is the manipulation detector, responsible for determining whether an image has been manipulated or not. This detector is based on a CNN architecture that receives as input the recovered signal and is trained with a binary cross-entropy loss to distinguish between images manipulated or not.

For an encrypted image, the goal of the detection module is to reconstruct a signal that closely resembles the original encryption signal, s_{IS}. Conversely, for a manipulated image, the extracted signal should differ significantly from the authentic one, facilitating manipulation detection. If an image has been altered, the model should be unable to recover the original signal, whereas, for encrypted images, it should preserve it.

3.3 Training Process and Loss Functions

During training the encryption module, $\mathcal{E}$, receives as input a real image, I, and embeds the template, s_{IS}, generated by the signal encoder, $\mathcal{S}_E$. Subsequently, STGAN is employed to obtain the manipulated version of the encrypted image, $\mathcal{G}(\mathcal{E}(I))$. Both $\mathcal{E}(I)$ and $\mathcal{G}(\mathcal{E}(I))$ are provided as input to the signal decoder that estimates the signal, s_{Real} for the encrypted image and the signal, s_{Fake} for the manipulated image:

$$s_{Real} = \mathcal{S}_D(\mathcal{E}(I)), \quad s_{Fake} = \mathcal{S}_D(\mathcal{G}(\mathcal{E}(I))). \tag{2}$$

The signals are then fed to the manipulation detector to predict tampering.

The whole training process is driven by several loss components. For the encryption module, the first loss component focuses on keeping the template as subtle as possible to avoid artifacts in the protected image. In particular, to this end, we minimize the magnitude of s_{IS}, such that:

$$\mathcal{L}_{mag} = \lambda_{mag} \times \|s_{IS}\|_2 \tag{3}$$

In addition, to enforce diversity in s_{IS} across different images and prevent mode collapse (i.e., the model learns a fixed signal), we also minimize the pairwise cosine similarity across a batch:

$$\mathcal{L}_d = \lambda_d \sum_{i=1}^{N} \sum_{j=i+1}^{N} \max(CS(s_{IS}[i], s_{IS}[j]), 0). \tag{4}$$

where N is the number of samples in a batch. Clamping negative values to zero prevents the model from learning only two opposite signals, forcing orthogonality to achieve greater signal diversity.

As regards the detection module, the first loss component focuses on keeping distinct real and fake signals, defined in Eq. 2, by minimizing their cosine similarity:

$$\mathcal{L}_{sig} = \lambda_{sig} \times CS(s_{Real}, s_{Fake}). \tag{5}$$

The smaller this loss, the more likely it is that s_{Real} and s_{Fake} are distinct from each other. In addition, the detection module is trained using a binary cross-entropy loss term on the prediction of the manipulation detector

$$\mathcal{L}_{det} = -\lambda_{det} \times \sum_{i=1}^{N} [y_i \log(\hat{y}_i) + (1 - y_i) \log(1 - \hat{y}_i)] \tag{6}$$

where y_i is the true label (0 for encrypted or 1 for manipulated), and $\hat{y}_i$ is the manipulation probability predicted by the detector.

The encryption and detection modules are trained jointly in an end to end manner, combining all the loss terms described above. The λ terms in Eqs. (3), (4), (5) and (6) control the influence of each loss term to the overall training process.

4 Experimental Results

Datasets. Following the protocol defined in [8], each of our models has been trained on the CelebA dataset, which is composed of 18700 images, using the train split released in their benchmark. For manipulating training images, we

Fig. 4. Examples of (a) real images, (b) images encrypted with our solution and (c) encrypted images manipulated using STGAN.

employed STGAN as GM to perform a fair comparison with the state-of-the-art [8]. STGAN is a pre-trained GM for facial attribute manipulation, i.e., it allows the manipulation of specific facial features. During training, we applied STGAN to modify all images using the "bald" attribute. Examples of real, encrypted, and manipulated images are shown in Fig. 4.

The evaluation was conducted on datasets not seen during training, as in [8]. In particular, the results reported in Table 1 were carried out on the CelebA-HQ dataset, while Table 2 includes results obtained on both CelebA-HQ [13] and Horse2Zebra [14]. Test splits for both datasets were provided by [8]. For manipulating test images, we used a set of different GMs, unseen at training time, namely, CycleGAN [14], StarGANv2 [15], BlendGAN [16], Stable Diffusion 1.5 (SD 1.5) [17], and Stable Diffusion XL (SDXL) Each test set consists of 200 real images and, following [8], we generate their corresponding manipulated versions using different GMs, resulting in a total of 400 images per GM—200 protected and 200 manipulated.

Implementation Details. The transformer blocks of the signal encoder and the signal decoder use 8 heads each with a dimension set to 64, therefore, the inner representation used by the transformer block is 512. The manipulation detector architecture, instead, is composed of eight convolutional layers and three fully connected layers. We empirically set $m = 0.01$ for Eq. (1); more details are provided in Sect. 4.3. All models have been optimized using AdamW with a batch size of 32 and a learning rate of 1×10^{-4}. All experiments have been conducted on an RTX 3090 with 24GB of VRAM.

Evaluation Metrics and Compared Methods. To evaluate the performance, we compute the binary detection accuracy for all tested methods, along with False Acceptance Rate (FAR) and False Rejection Rate (FRR). While to quantify image degradation caused by the introduction of the encryption signal s_{IS}, two metrics have been computed: mean squared error (MSE) and learned perceptual image patch similarity (LPIPS [18]) between the original and encrypted version of an image. MSE assesses pixel-wise differences, while LPIPS captures perceptual similarity, offering complementary perspectives on image degradation. We compare our method to the state-of-the-art MaLP [8] and three passive baselines: the approaches from [19] and [9], both trained on ProGAN, and a ViT-based passive detector [20] trained on STGAN manipulations.

Table 1. Performance on STGAN attribute manipulations. Higher accuracy indicates better overall performance, while lower FAR and FRR reflect better discrimination. Best results in **bold**, second best <u>underlined</u>.

Method	Bushy Eyebrows			Mustache			Eyeglasses			Young		
	Acc	FAR	FRR	Acc	FAR	FRR	Acc	FAR	FRR	Acc	FAR	FRR
Wang et al. [19]	0.585	0.015	0.815	0.573	0.015	0.840	0.568	0.015	0.850	0.583	0.015	0.820
Ojha et al. [9]	0.776	0.228	0.171	0.750	0.280	0.164	0.810	0.191	0.082	0.728	0.300	0.167
ViT [20]	0.970	<u>0.005</u>	0.055	0.980	<u>0.005</u>	0.035	0.970	<u>0.005</u>	0.055	0.975	<u>0.005</u>	0.045
MaLP [8]	<u>0.980</u>	0.015	<u>0.025</u>	<u>0.986</u>	<u>0.005</u>	<u>0.025</u>	<u>0.985</u>	<u>0.005</u>	<u>0.025</u>	<u>0.980</u>	0.015	<u>0.025</u>
Ours	**1.00**	**0.000**	**0.000**	**1.00**	**0.000**	**0.000**	**1.00**	**0.000**	**0.000**	**1.00**	**0.000**	**0.000**

4.1 Detection Performance Comparison Across Unseen Manipulations

We first evaluate the performance of all tested models using STGAN as the manipulator. Unlike during training, we apply unseen attribute manipulations to assess the generalization of these models to a known GM with unseen manipulations. As shown in Table 1, all models trained on STGAN generalize well, likely due to the distinctive artifacts it introduces. Our method achieves 100% accuracy with zero FAR and FRR across all attributes, demonstrating superior robustness to unseen manipulations.

4.2 Detection Performance Comparison Across Unseen GMs

To further stress the generalization capabilities, we tested all methods on images manipulated by GMs not seen during training, including StarGANv2 [15], BlendGAN [16], SD1.5 [17], and SDXL. In addition, following the benchmark from [8], we carried out an experiment considering a different domain, i.e., Horse2Zebra, and a different GM specifically trained to manipulate this dataset, i.e., CycleGAN [14].

Table 2 reports accuracy, FAR, and FRR for each method across both unseen GMs and unseen datasets. While MaLP achieves near-perfect performance on StarGANv2, its effectiveness declines when images are manipulated using BlendGAN or diffusion-based models (e.g., SD1.5 and SDXL). This may be because StarGANv2 typically alters the entire image, making manipulations more detectable, whereas diffusion-based models apply more subtle modifications, making detection more challenging. Similarly, passive methods from [19] and [9], both trained on ProGAN-generated images, achieve comparable performance on GAN-based GMs but show worst results on SD1.5 and SDXL,

Table 2. Performance on unseen GMs. Higher accuracy indicates better overall performance, while lower FAR and FRR reflect better discrimination between protected and manipulated images. Best results in **bold**. Second best underlined.

Method	CelebA-HQ												Horse2Zebra		
	StarGANv2			BlendGAN			SD 1.5			SDXL			CycleGAN		
	Acc	FAR	FRR	Acc	FAR	FRR	Acc	FAR	FRR	Acc	FAR	FRR	Acc	FAR	FRR
Wang et al. [19]	0.863	**0.005**	0.270	0.963	**0.000**	0.075	0.500	**0.000**	1.000	0.500	**0.000**	1.000	0.807	0.025	0.530
Ojha et al. [9]	0.960	0.015	0.043	0.923	0.044	0.025	0.805	0.161	0.147	0.510	0.284	0.360	0.935	0.062	0.031
ViT [20]	0.768	0.410	0.055	0.475	0.995	0.055	0.545	0.885	0.025	0.565	0.845	0.025	0.648	0.675	0.190
MaLP [8]	**0.998**	**0.005**	**0.000**	0.700	**0.000**	0.605	0.667	0.500	0.375	0.678	0.500	0.375	0.675	0.345	0.310
Ours	0.910	0.180	**0.000**	**0.998**	0.005	**0.000**	**0.975**	0.050	**0.000**	**0.975**	0.050	**0.000**	**0.995**	**0.000**	**0.010**

Fig. 5. Comparison of (a) real images, (b) images encrypted with our solution and (c) images encrypted using MaLP [8].

revealing poor adaptability to models outside their training distribution. In contrast, our method exhibits strong generalization across all tested GMs, achieving high accuracy and consistently low FAR and FRR. Notably, it outperforms both proactive and passive methods on BlendGAN, SD1.5, SDXL, and CycleGAN. The low FAR confirms that our model rarely misclassifies protected images, while the low FRR indicates strong detection of manipulated content.

Compared to the results reported in Sect. 4.1, it becomes evident that both the ViT baseline and MaLP rely heavily on detecting STGAN-specific artifacts. As a result, they generalize poorly to unseen GMs. In contrast, our approach focuses on changes in the encrypted signal rather than model-specific artifacts, enabling robust and reliable detection across a wide range of manipulation types.

4.3 Image Degradation After Encryption Signal Embedding

A strong signal may degrade image quality excessively, making images unusable. Therefore, we assess the extent of image degradation caused by the encryption process described in Sect. 3.1.

We computed the average MSE and LPIPS between the original images and their protected versions. Our method minimizes image degradation at both

Fig. 6. Examples of images specific signals (*top*) for different images (*bottom*). The signals have been normalized for visualization purposes.

the pixel and perceptual levels, achieving significantly lower values than MaLP. Specifically, our solution attains an average MSE of 0.0001, compared to MaLP's higher MSE of 0.003. Similarly, in terms of perceptual similarity, our approach achieves an average LPIPS of 0.0005, substantially lower than MaLPâĂŹs 0.19, demonstrating superior visual fidelity preservation. This is achieved by optimizing the signal magnitude using the loss in Eq. (3) and the parameter m.

From Fig. 5, it is possible to compare real images to their encrypted counterparts using both our method and the state-of-the-art. The embedded encryption signal does not introduce noticeable differences in color or structure. However, a slight increase in noise is visible, particularly in brighter regions, demonstrating the subtle yet effective integration of the protection signal. On the contrary, the template added by MaLP [8] is more noticeable, Fig. 5(c).

4.4 Evaluating Signal Diversity

A key feature of our framework is its ability to generate customized signals for individual images. To assess signal diversity, we compute the average pairwise cosine similarity (CS) between encryption signals across different images. CS has been chosen for its invariance to magnitude, allowing us to focus solely on structural differences between signals. Using a batch of 200 unseen images, we generated the corresponding signals and computed the average CS across all pairs, obtaining a value of 2.93×10^{-3}. This indicates that the signals are nearly orthogonal to each other, ensuring that different images receive distinct protection. This result is achieved through the personalization introduced by the cross-attention mechanism in the encryption module (Sect. 3.1), combined during training with the loss defined in Equation (4). Figure 6 illustrates examples of images alongside their corresponding signals.

4.5 Ablation

We conducted an ablation study to evaluate the contribution of each loss component to the performance of our model.

The $\mathcal{L}_{mag}$ loss in Eq. 3 controls the strength of the protection signal. Without this loss term, the perturbation becomes unbounded, consequently compromising visual fidelity. This is further corroborated by the fact that images protected with a model trained without this loss exhibit higher MSE and LPIPS, 0.0004 and 0.0030, respectively, compared to 0.0001 and 0.0005 reported in Sect. 4.3.

The $\mathcal{L}_{div}$ in Eq. 4 encourages the encoder to generate distinct perturbations for each image. In its absence, the generation collapses to a single, universal perturbation applied across all images. This behavior is confirmed by a CS of 1.0 between the perturbations of different images, indicating they are identical.

Finally, the L_{sig} term is used to enforce the decoder to correctly recover the protection signal. Without this loss, the model may bypass the signal during detection. In fact, omitting this loss component during training results in an average accuracy of 93.06 across all unseen GMs, compared to a higher accuracy of 97.6 when the loss is used to train the model.

5 Conclusions

This work proposes a novel approach for proactive detection of manipulations by GMs, focusing on image-specific protection. In particular, our solution introduces an encryption module that customizes a learnable template for each input image using the cross-attention mechanism. The resulting encrypted image can be safely shared while embedding a structured protection signal. A corresponding detection module is then employed to extract the encryption signal and determine whether an image has been tampered with or not. Unlike prior approaches that rely on a fixed set of templates, our solution generates content-aware encryption signals, providing stronger protection against malicious attackers.

Experimental results demonstrate that our approach, compared to the state-of-the-art in proactive defense [8], preserves image quality while also exhibiting superior robustness to manipulations of different attributes or from different GMs, unseen during training. These findings highlight the potential of our approach as an effective and adaptable defense mechanism in the evolving landscape of generative image manipulations.

Acknowledgment. Funded by the European Union – Next Generation EU within the framework of the National Recovery and Resilience Plan NRRP – Mission 4 "Education and Research" – Component 2 - Investment 1.1 "National Research Program and Projects of Significant National Interest Fund (PRIN)" (Call D.D. MUR n. 104/2022) – PRIN2022 – Project reference: Adversarial Venture, the Mixed Blessing of Adversarial Attacks 20227YET9B_002 J53D23007030006.

References

1. Goodfellow, I.J., et al.: Generative adversarial networks. arXiv preprint arXiv:1406.2661 [stat.ML] (2014)
2. Rombach, R., Blattmann, A., Lorenz, D., Esser, P., Ommer, B.: High-resolution image synthesis with latent diffusion models. In: Proceedings of the IEEE/CVF Conference on Computer Vision and Pattern Recognition, pp. 10684–10695 (2022)
3. Li, Y., Lyu, S.: Exposing deepfake videos by detecting face warping artifacts. arXiv preprint arXiv:1811.00656 (2018)
4. Nirkin, Y., Wolf, L., Keller, Y., Hassner, T.: Deepfake detection based on the discrepancy between the face and its context. arXiv preprint arXiv:2008.12262 (2020)
5. Rossler, A., Cozzolino, D., Verdoliva, L., Riess, C., Thies, J., Nießner, M.: Faceforensics++: learning to detect manipulated facial images. In: Proceedings of the IEEE/CVF International Conference on Computer Vision, pp. 1–11 (2019)
6. Zhou, P., Han, X., Morariu, V.I., Davis, L.S.: Learning rich features for image manipulation detection. In: Proceedings of the IEEE Conference on Computer Vision and Pattern Recognition, pp. 1053–1061 (2018)
7. Asnani, V., Yin, X., Hassner, T., Liu, S., Liu, X.: Proactive image manipulation detection. In: Proceedings of the IEEE/CVF Conference on Computer Vision and Pattern Recognition, pp. 15386–15395 (2022)
8. Asnani, V., Yin, X., Hassner, T., Liu, X.: Manipulation localization using a proactive scheme, Malp (2023)
9. Ojha, U., Li, Y., Rajan, A.S., Lee, Y.J.: Towards universal fake image detectors that generalize across generative models. In: CVPR (2023)
10. Ruiz, N., Bargal, S.A., Stan, S.: Adversarial attacks against conditional image translation networks and facial manipulation systems, Disrupting deepfakes (2020)
11. Wang, R., Juefei-Xu, F., Luo, M., Liu, Y., Wang, L.: Robust safeguards against deepfake dissemination via provenance tracking, Faketagger (2021)
12. Vaswani, A., et al.: Attention is all you need, Gomez (2017)
13. Karras, T., Aila, T., Laine, S., Lehtinen, J.: Progressive growing of GANs for improved quality, stability, and variation. arXiv preprint arXiv:1710.10196 (2017)
14. Zhu, J. Y., Park, T., Isola, P., Efros, A.A.: Unpaired image-to-image translation using cycle-consistent adversarial networks. In: Computer Vision (ICCV), 2017 IEEE International Conference on (2017)
15. Choi, Y., Uh, Y., Yoo, J., Ha, J.W.: Stargan v2: diverse image synthesis for multiple domains. In: Proceedings of the IEEE/CVF Conference on Computer Vision and Pattern Recognition, pp. 8188–8197 (2020)
16. Liu, M., Li, Q., Qin, Z., Zhang, G., Wan, P., Zheng, W.: Blendgan: implicitly GAN blending for arbitrary stylized face generation. Adv. Neural. Inf. Process. Syst. **34**, 29710–29722 (2021)
17. Rombach, R., Blattmann, A., Lorenz, D., Esser, P., Ommer, B.: High-resolution image synthesis with latent diffusion models. In: Proceedings of the IEEE/CVF Conference on Computer Vision and Pattern Recognition (CVPR), pp. 10684–10695 (2022)
18. Zhang, R., Isola, P., Efros, A.A., Shechtman, E., Wang, O.: The unreasonable effectiveness of deep features as a perceptual metric (2018)
19. Wang, S. Y., Wang, O., Zhang, R., Owens, A., Efros, A.A.: CNN-generated images are surprisingly easy to spot... for now (2020)
20. Dosovitskiy, A., et al.: An image is worth 16x16 words: transformers for image recognition at scale. In: ICLR (2021)

Author Index

E. Rodolà et al. (Eds.): ICIAP 2025, LNCS 16167, pp. 673–675, 2026.
https://doi.org/10.1007/978-3-032-10185-3

MIX
Papier aus verantwortungsvollen Quellen
Paper from responsible sources
FSC® C105338

If you have any concerns about our products,
you can contact us on
ProductSafety@springernature.com

In case Publisher is established outside the EU,
the EU authorized representative is:
Springer Nature Customer Service Center GmbH
Europaplatz 3, 69115 Heidelberg, Germany

Printed by Libri Plureos GmbH
in Hamburg, Germany